Management of Microbial Resources in the Environment

Abdul Malik • Elisabeth Grohmann
Madalena Alves
Editors

Management of Microbial Resources in the Environment

Springer

Editors
Abdul Malik
Department of Agricultural Microbiology
Faculty of Agricultural Sciences
Aligarh Muslim University
Aligarh, India

Visiting Professor/Scientist
Division of Infectious Diseases
University Medical Center Freiburg
Freiburg, Germany

Madalena Alves
Center of Biological Engineering
University of Minho
Braga, Portugal

Elisabeth Grohmann
Division of Infectious Diseases
University Medical Center Freiburg
Freiburg, Germany

ISBN 978-94-007-5930-5 ISBN 978-94-007-5931-2 (eBook)
DOI 10.1007/978-94-007-5931-2
Springer Dordrecht Heidelberg New York London

Library of Congress Control Number: 2013933022

Printed on acid-free paper

Springer is part of Springer Science+Business Media (www.springer.com)

Preface

Microorganisms are microscopic forms of life that include bacteria, protozoa, algae, fungi and viruses, and comprise the greatest number of individual organisms on the earth. At best, however, science has identified and is aware of only a minute fraction of them. Somewhere between 0.1 and 1.0% of all bacterial species have been described and the vast majority of the remaining bacterial species is unknown. Uncultured organisms comprise the vast majority of the microbial world. Although culturing has been indispensable for increasing our understanding of specific organisms, problems with using culturing for community analysis arise from the fact that an artificial homogenous medium typically allows growth of only a small fraction of the organisms. Culturing fails to reproduce the ecological niches and symbiotic relationships encountered in complex natural environments that are required to support the full spectrum of microbial diversity. Microorganisms are valuable genetic resources, which contribute to the global economic and social development. Microbial Resources include the exploration, collection, characterization, evaluation and conservation of microbes for sustainable utilization in the development of the global/national economy, i.e. Agriculture, Ecosystem, Environment, Industry and Medicine. Microbes play a critical role in natural biogeochemical cycles and make up about 60% of the Earth's biomass. Many research institutes/universities all over the world are carrying out microbiological and biotechnological research, which results in generating lots of genomic resources like cDNA libraries, gene constructs, promoter regions, transgenes etc. These are valuable resources for gene discovery and transgenic product development.

In this book we provide up-to-date information about the management of microbial resources in the environment. This book covers the ecology of microorganisms in natural and engineered environments; genomic, metagenomic and molecular advances in understanding of microbial interactions; microbial diversity and phylogeny; ecological studies of human, animal and plant microbiology and disease; microbial processes and interactions in the environment; and technological developments. This book is not intended to serve as an encyclopedic review of the subject. However, the various chapters incorporate both theoretical and practical aspects and may serve as baseline information for future research through which significant development is possible.

The book has twenty chapters, with each focused on a specific topic to cover diverse perspective topics. An introductory chapter on management of microbial resources in the environment is included. Topics include: recent development in the methods of studying microbial diversity, microbial resource centers towards harnessing microbial diversity for human welfare, polyphasic identification and preservation of fungal diversity, fungal biodiversity: a potential tool in plant disease management, bioinformatics approaches in studying microbial diversity, recent advances in metagenomic studies of soil microbial communities, mobile genetic elements (MGEs) carrying catabolic genes, conjugative plasmids in anthropogenic soils, potential eco-friendly soil microorganisms, current aspects of metal resistant bacteria in bioremediation, anaerobic digestion of the organic fraction of municipal solid waste, microbial insecticides: food security and human health, microbes (PGPR) in sustainable agriculture, antibiotic resistance gene pool and bacterial adaptation to xenobiotics in the environment, synthetic lethal genetic interaction networks and their utility for anticancer therapy, adaptation of *Candida albicans* for growth within the host, role of marine anaerobic bacteria and archaea in bioenergy production, bacteriocins: natural weapons for control of food pathogens, and anaerobic degradation of lindane and other HCH isomers.

With great pleasure, we extend our sincere thanks to all our well-qualified and internationally renowned contributors from different countries for providing the important, authoritative and cutting edge scientific information/technology to make this book a reality. All chapters are well illustrated with appropriately placed tables and figures and enriched with up to date information. We are also thankful to the reviewers who carefully and timely reviewed the manuscripts. Dr. Abdul Malik is also thankful to the Department of Biotechnology, Govt. of India, New Delhi for DBT CREST award/Fellowship during the preparation of the book.

We are extremely thankful to Springer, Dordrecht, the Netherlands for completing the review process expeditiously to grant acceptance for publication. We appreciate the great efforts of the book publishing team, especially of Dr. Alexandrine Cheronet, Senior Publishing Editor Environmental Sciences in responding to all queries very promptly.

We express sincere thanks to our family members for all the support they provided, and regret the neglect and loss they suffered during the preparation of this book.

Aligarh, India — Abdul Malik
Freiburg, Germany — Elisabeth Grohmann
Braga, Portugal — Madalena Alves

Contents

Contributors

A.A. Abreu IBB – Institute for Biotechnology and Bioengineering, Centre of Biological Engineering, University of Minho, Braga, Portugal

Hikmate Abriouel Área de Microbiología, Departamento de Ciencias de la Salud, Facultad de Ciencias Experimentales, Edif. B3, Universidad de Jaén, Jaén, Spain

Nadeem Ahmad Department of Plant Protection, Faculty of Agricultural Sciences, Aligarh Muslim University, Aligarh, India

Salman Ahmad Department of Plant Protection, Faculty of Agricultural Sciences, Aligarh Muslim University, Aligarh, India

Tufail Ahmad Department of Plant Protection, Faculty of Agricultural Sciences, Aligarh Muslim University, Aligarh, India

Shakil Ahmed Molecular and Structural Biology Division, CSIR – Central Drug Research Institute, Lucknow, India

Muna Albanna Water and Environmental Engineering Department, School of Natural Resources Engineering, German Jordanian University, Amman, Jordan

M.M. Alves IBB – Institute for Biotechnology and Bioengineering, Centre of Biological Engineering, University of Minho, Braga, Portugal

E.E. Andronov All-Russia Research Institute for Agricultural Microbiology, Saint-Petersburg, Russia

M. Shafiq Ansari Department of Plant Protection, Faculty of Agricultural Sciences, Aligarh Muslim University, Aligarh, India

Mohd Ikram Ansari Department of Agricultural Microbiology, Faculty of Agricultural Sciences, Aligarh Muslim University, Aligarh, India

Basma Rajeh Mohammad Abu Arra Biology Programme, School of Distance Education, Universiti Sains Malaysia, Minden, Penang, Malaysia

Shabbir Ashraf Department of Plant Protection, Faculty of Agriculture Sciences, Aligarh Muslim University, Aligarh, Uttar Pradesh, India

Shafquat Azim Department of Biochemistry, Faculty of Life Sciences, Aligarh Muslim University, Aligarh, Uttar Pradesh, India

Satinder Kaur Brar INRS-ETE, Université du Québec, Québec, Canada

A.J. Cavaleiro IBB – Institute for Biotechnology and Bioengineering, Centre of Biological Engineering, University of Minho, Braga, Portugal

Ramesh Chand Department of Mycology & Plant Pathology, Institute of Agricultural Sciences, Banaras Hindu University (BHU), Varanasi, Uttar Pradesh, India

Vijay Bahadur Chauhan Department of Mycology & Plant Pathology, Institute of Agricultural Sciences, Banaras Hindu University (BHU), Varanasi, Uttar Pradesh, India

Gurpreet Singh Dhillon INRS-ETE, Université du Québec, Québec, QC, Canada

Djamel Drider Laboratoire des Procédés Biologiques, Génie Enzymatique et Microbien (ProBioGEM), UPRES-EA 1026, Polytech'Lille/IUTA, Université Lille Nord de France, Villeneuve d'Ascq Cedex, France

Miguel Ángel Ferandez-Fuentes Área de Microbiología, Departamento de Ciencias de la Salud, Facultad de Ciencias Experimentales, Edif. B3, Universidad de Jaén, Jaén, Spain

Laboratoire des Procédés Biologiques, Génie Enzymatique et Microbien (ProBioGEM), UPRES-EA 1026, Polytech'Lille/IUTA, Université Lille Nord de France, Villeneuve d'Ascq Cedex, France

Ismail Fliss STELA Dairy Research Center, Nutraceuticals and Functional Foods Institute, Université Laval, Québec, QC, Canada

Antonio Galvez Área de Microbiología, Departamento de Ciencias de la Salud, Facultad de Ciencias Experimentales, Edif. B3, Universidad de Jaén, Jaén, Spain

Sumathi Ganeshan Biology Programme, School of Distance Education, Universiti Sains Malaysia, Minden, Penang, Malaysia

Elisabeth Grohmann Division of Infectious Diseases, University Medical Center Freiburg, Freiburg, Germany

Fazil Hasan Department of Plant Protection, Faculty of Agricultural Sciences, Aligarh Muslim University, Aligarh, India

Mohammad Aamir Hussain Department of Biochemistry, Faculty of Life Sciences, Aligarh Muslim University, Aligarh, Uttar Pradesh, India

Hassan Mubarak Ishqi Department of Biochemistry, Faculty of Life Sciences, Aligarh Muslim University, Aligarh, Uttar Pradesh, India

Surinder Kaur Department of Mycology & Plant Pathology, Institute of Agricultural Sciences, Banaras Hindu University (BHU), Varanasi, Uttar Pradesh, India

INRS-ETE, Université du Québec, Québec, Canada

Saman Khan Molecular and Structural Biology Division, CSIR – Central Drug Research Institute, Lucknow, India

Alette A.M. Langenhoff Sub-department of Environmental Technology, Wageningen University, Wageningen, The Netherlands

Nelson Lima IBB – Biological Engineering Centre, Applied Mycology Group, University of Minho, Braga, Portugal

Showkat Ahmad Lone Biotechnology and Climate Change, National Research Centre on Plant Biotechnology, New Delhi, India

Department of Agricultural Microbiology, Faculty of Agricultural Sciences, Aligarh Muslim University, Aligarh, Uttar Pradesh, India

Abdul Malik Department of Agricultural Microbiology, Faculty of Agricultural Sciences, Aligarh Muslim University, Aligarh, Uttar Pradesh, India

Farhana Masood Department of Agricultural Microbiology, Faculty of Agricultural Sciences, Aligarh Muslim University, Aligarh, Uttar Pradesh, India

Farrakh Mehboob Laboratory of Microbiology, Wageningen University, Wageningen, The Netherlands

Ecotoxicology Research Programme, National Institute of Bio-remediation, National Agriculture Research Centre, Islamabad, Pakistan

Hideaki Nojiri Biotechnology Research Center, The University of Tokyo, Bunkyo-ku, Tokyo, Japan

Agricultural Bioinformatics Research Unit, The University of Tokyo, Bunkyo-ku, Tokyo, Japan

Nabil Ben Omar Área de Microbiología, Departamento de Ciencias de la Salud, Facultad de Ciencias Experimentales, Edif. B3, Universidad de Jaén, Jaén, Spain

Jasdeep Chatrath Padaria Biotechnology and Climate Change, National Research Centre on Plant Biotechnology, New Delhi, India

Leonel Pereira IBB – Biological Engineering Centre, Applied Mycology Group, University of Minho, Braga, Portugal

M.A. Pereira IBB – Institute for Biotechnology and Bioengineering, Centre of Biological Engineering, Universidade do Minho, Braga, Portugal

E.V. Pershina All-Russia Research Institute for Agricultural Microbiology, Saint-Petersburg, Russia

A.G. Pinaev All-Russia Research Institute for Agricultural Microbiology, Saint-Petersburg, Russia

N.A. Provorov All-Russia Research Institute for Agricultural Microbiology, Saint-Petersburg, Russia

Sayeed Ur Rehman Department of Biochemistry, Faculty of Life Sciences, Aligarh Muslim University, Aligarh, Uttar Pradesh, India

Cledir Santos IBB – Biological Engineering Centre, Applied Mycology Group, University of Minho, Braga, Portugal

Tarique Sarwar Department of Biochemistry, Faculty of Life Sciences, Aligarh Muslim University, Aligarh, Uttar Pradesh, India

Gosse Schraa Laboratory of Microbiology, Wageningen University, Wageningen, The Netherlands

Masaki Shintani Department of Materials Science and Chemical Engineering, Faculty of Engineering, Shizuoka University, Shizuoka, Japan

Japan Collection of Microorganisms, RIKEN BioResource Centre, Tsukuba, Ibaraki 305-0074, Japan

Marta F. Simões IBB – Biological Engineering Centre, Applied Mycology Group, University of Minho, Braga, Portugal

D.P. Singh Department of Environmental Science, Babasaheb Bhimrao Ambedkar (Central) University, Lucknow, Uttar Pradesh, India

Jay Shankar Singh Department of Environmental Microbiology, Babasaheb Bhimrao Ambedkar (Central) University, Lucknow, Uttar Pradesh, India

Amit Kumar Sonkar Molecular and Structural Biology Division, CSIR – Central Drug Research Institute, Lucknow, India

Diana Sousa IBB – Institute for Biotechnology and Bioengineering, Centre of Biological Engineering, University of Minho, Braga, Portugal

Alfons J.M. Stams Laboratory of Microbiology, Wageningen University, Wageningen, The Netherlands

Centre of Biological Engineering, University of Minho, Braga, Portugal

Mohammad Tabish Department of Biochemistry, Faculty of Life Sciences, Aligarh Muslim University, Aligarh, Uttar Pradesh, India

Mausum Verma Institut de recherche et de développement en agroenvironnement inc. (IRDA), Québec, QC, Canada

Zuraini Zakaria Biology Programme, School of Distance Education, Universiti Sains Malaysia, Minden, Penang, Malaysia

Mohammad Zuhaib Department of Plant Protection, Faculty of Agriculture Sciences, Aligarh Muslim University, Aligarh, Uttar Pradesh, India

Chapter 1
Management of Microbial Resources in the Environment: A Broad Perspective

Abdul Malik, Farhana Masood, and Elisabeth Grohmann

Abstract The earth contains a huge number of largely uncharacterized Bacteria and Archaea. Microbiologists are struggling to summarize their genetic diversity and classify them, which has resulted in heated debates on methods for defining species, mechanisms that lead to speciation and whether microbial species even exist. New molecular microbiological techniques allow for environmental screening to determine the presence of nucleic acids in environmental samples. These molecular genetic techniques allow screening for organisms that could be maintained in culture along with those that cannot be identified by standard non molecular means as they cannot be cultured. Although not allowing the description of specific organisms, this technique permits determining numbers and lineages of microorganisms in environmental samples, notably phylogenetic relationships and genetic similarity to sequences in established databases. Recent progress has revealed that the capture of genetic resources from complex microbial communities allows the discovery of a richness of new enzymatic diversity that had not previously been imagined. This new diversity, constitutes a large potential of new and improved applications in industry, medicine, agriculture, bioenergy etc., and promises to facilitate in a significant manner, our transition to a sustainable society, by contributing to the transition to renewable sources of energy, chemicals and materials, the reduction of pollutant burdens. Hiding within the as-yet-undiscovered microorganisms are cures

A. Malik (✉) • F. Masood
Department of Agricultural Microbiology, Faculty of Agricultural Sciences,
Aligarh Muslim University, Aligarh, Uttar Pradesh 202002, India
e-mail: ab_malik30@yahoo.com; farhanamasud4@gmail.com

E. Grohmann
Division of Infectious Diseases, University Medical Center Freiburg,
Hugstetter Strasse 55, Freiburg 79106, Germany
e-mail: elisabeth.grohmann@uniklinik-freiburg.de

A. Malik et al. (eds.), *Management of Microbial Resources in the Environment*,
DOI 10.1007/978-94-007-5931-2_1, © Springer Science+Business Media Dordrecht 2013

for diseases, means to clean polluted environments, new food sources, and better ways to manufacture products daily used in modern society. This chapter focuses on the structural and functional diversity of the microbes around the globe, which are the primary and richest source of natural genetic resources that can be utilized for the improvement of agriculture, food production, and human health as well as for the welfare of the environment and ecosystems.

Keywords Microbial resources • Environment • Genetic diversity • Molecular tools • Metagenomics

1 Introduction

Microorganisms are a highly diverse group of organisms and constitute about 60% of the Earth's biomass (Singh et al. 2009). In aquatic environments, such as the oceans, the number of microbial cells has been estimated to be approximately 1.2×10^{29}, while in terrestrial environments, soil sustains as many as $4\text{–}5 \times 10^{30}$ microbial cells (Singh et al. 2009). Owing to such enormous numbers, microorganisms are essential components of the Earth's biota and represent a large unexplored reservoir of genetic diversity. Understanding this unexplored genetic diversity is a high-priority issue in microbial ecology from perspectives such as global climate change and the greenhouse effect. In fact, all organisms in the biosphere either directly or indirectly depend on microbial activities. In soil ecosystems, microorganisms are pivotal in suppressing soil-borne plant diseases, promoting plant growth, and in promoting changes in vegetation (Garbeva et al. 2004). An understanding of microbial dynamics and their interactions with biotic and abiotic factors is indispensable in bioremediation techniques, energy generation processes, and in pharmaceuticals, food, chemical and mining industries.

A plethora of biochemical and molecular methods have been applied to reveal the microbial community composition over time and space in response to environmental changes. These new approaches allow linkage between ecological processes in the environment with specific microbial populations and help us to answer important questions in microbial ecology such as what factors and resources govern the enormous genetic and metabolic diversity in an environment. In this context, it is important to review to what extent microbial ecology, particularly the management of the various kinds of microbial resources, can offer great new potentials to address these super challenges.

2 Microbial Resources

Microbial culture collections currently contain more than a million different strains (Verstraete et al. 2007) and thus are a testimony of the efforts made for the conservation of biodiversity and the desire to make these potentials available to the public.

To what extent these collections can and need to be expanded is debatable, since it is generally accepted that microorganisms tend to act not alone but in association with others, it is obvious therefore, that at present considerable effort should be devoted to the preservation and collection of novel microbial associations in natural samples and enrichment cultures. However, preservation of the habitats in which they thrive is also needed. Up to now, attention has mainly been focused on various unique sites such as hot springs or pristine places (e.g., the Arctic/Antarctic region). The latter, for instance, has given rise in the last decade to an enormous expansion in the knowledge of novel polar microbial taxa (van Trappen et al. 2005) which in turn has led to industrial applications such as cold-adapted enzymes (Siddiqui and Cavicchioli 2006), anti-freeze products (Gilbert et al. 2004) and strains capable to bioremediate in cold soils (Margesin et al. 2003). We should explore new frontier habitats such as the deep sea, the deep underground and the deep intestine. Indeed, such environments harbor a wealth of putatively useful processes and products.

Most importantly, not only these "natural" habitats are of value, but also a number of sites altered by industrial actions, often unwanted, are now to be earmarked as "resources" of microbial diversity. Examples are sites with acid mine drainage, which recently showed potentials for the production of anti-cancer drugs (Yamada et al. 2004) and aquifers polluted with chloro-organics which have yielded very interesting halo-respiring micro-organisms (de Wildeman and Verstraete 2003; Smidt and de Vos 2004). It becomes obvious that not only the maintenance of microbial culture collections can be justified, but just as well the preservation of special sites, as sources of ongoing microbial evolution, selection and development of special microbial interactions, processes and products.

3 Storage of Microbial Resources

For the effective conservation and utilization of microbial resources which have been found, all developed countries have established microbial culture collections, some of which have history up to 100 years. They take sound management mechanisms, act relatively independently, and have strong research capabilities. Not only microbial resources are available in the culture collection, technical services can also be provided. Microbial resource centers (MRCs) harbour collections of culturable organisms (e.g. algae, bacteria, fungi, including yeasts, protozoa and viruses), their replicable parts (e.g. genomes, plasmids, viruses, cDNAs), viable but not yet culturable organisms, cells and tissues, databases containing molecular, physiological and structural information relevant to these collections and allied bioinformatics.

3.1 Role of Culture Collections

The major culture collections throughout the world have as their primary commission, the preservation and distribution of germplasm that has been demonstrated to have significance to the microbiology community. The importance of a particular strain may be as a reference for medical or taxonomic research, as an assay organism for testing or screening, or as an essential component of a patent application for a product or process in which it is involved. Alternatively, the strain may be placed in a collection with reference to the publication in which it was cited as part of the investigation. This latter form of deposition is essential on account of the inherent transience of researchers and their research programs, making it possible for later investigators to repeat or advance published research that would be impossible in the absence of the strains involved. As mentioned above, the many national reference and service collections have succeeded in preserving, for later generations of microbiologists, many of the private and specialized collections of microorganisms that may represent an entire career of one microbiologist. In other cases, however, the acquisition of a collection of strains may well result from a change in the direction of the research program in a scientist's laboratory.

The major culture collections of the world also serve as centers for excellence in research in systematics and taxonomy. In large part the identification and characterization of strains is an integral function of collections, and the availability of a large collection of strains is essential for this type of research. Culture collections that have contract identification services are also continually searching for faster and more reliable methods to characterize unknown strains for their clients. In many cases, the strains maintained in any collection will directly reflect the taxonomic interests of the curators, in terms of the depth and breadth of particular taxonomic groups.

3.2 *Bioinformatics and Culture Collection*

The meticulous and thorough characterization of microorganisms, the storage and analysis of the generated information and its intelligent interrogation will provide us with microbial solutions to critical challenges of the twenty first century. It is imperative that scientists, researchers in biosystematics and taxonomy employ modern tools of informatics and data processing to make best use of our microbial resources. The elements of taxonomy such as species description, development of identification keys, scientific nomenclature, treatment of morphological, nutritional and physiological traits, are increasingly being computerized to meet the challenge.

The process of storage of genetic information with digital techniques for archiving, interpreting and quantifying of data in artificial systems is an important feature of bioinformatics (Wuyts et al. 2001, 2002). Microbial taxonomists and curators must take full advantage of the available technology that has been so ably adopted in other biological fields. Sequence data available on the web for many years for public access and utilization (http://www.ncbi.nih.gov/Genbank/GenbankSearch.html); EMBL (http://www.ebi.ac.uk/embl/) are two key resources.

However, Bridge et al. (2003) suggested that up to 20% of publicly available, taxonomically important DNA sequences for three randomly chosen groups of fungi were probably incorrectly named, chimaeric, of poor quality or incomplete for reliable comparison. MRCs have a role to play, in providing information based on authentic and stable strains with validated sequence data. MRCs need to harness the new bioinformatics technologies and begin with networking processes to establish a global network. The groundwork can be laid by stimulating collaborative molecular taxonomic research and novel database development.

4 Microbial Resource Management (MRM)

In order to deal with MRM in environmental practice, a number of approaches are currently in use. We can at present link high throughput/quick scan molecular analyses with performance data. We can for instance line up data obtained by DGGE, T-RFLP, micro-arrays, etc., to biomolecular databases resp. to diagnoses/prediction/prevention/control (Denef et al. 2004; van der Gast et al. 2006). Yet, although the "omic" methods are capable to provide an "avalanche" of information, the latter is difficult to interpret. In case one is interested in straightforward information on process performance and end product output, the process engineer will continue to find the conventional physical/chemical parameters to be the best source of information. However, in case the in-depth questions about the coherence of the microbial community are of interest, then the molecular analyses can provide information, which is not so much of direct tactical importance but rather essential in terms of strategic considerations on the behavior of the "health" of the microbial community.

Verstraete et al. (2007) developed the Microbial Resource Management (MRM) concept. This is a practical mindset that has been developed as a concept to solve practical problems through the use of microorganisms. For this reason, the inception of MRM was shortly followed by a complementary set of tools to deal with it (Marzorati et al. 2008) which come in the form of a three stage analysis independent of the technique and its settings. These tools help to provide an ecological and predictive value of the analysis which incorporates the structure and diversity of the microbial community being examined.

The most popular molecular fingerprinting techniques including temperature gradient gel electrophoresis (TGGE) and denaturing-gradient gel electrophoresis (DGGE), terminal-restriction fragment length polymorphism (t-RFLP), 16S rRNA gene clone libraries, and length heterogeneity-polymerase chain reaction (LH-PCR) are commonly used to study the structure and composition of the microbial communities. Interpreting and comparing this type of data became easier due to the initiation of MRM and the three-stage tool set developed by Marzorati et al. (2008). The parameters within the tool set include range weighted richness (Rr), dynamics (Dy), and functional organization (Fo). When using these tools in combination, they can provide us with an ecological interpretation of the raw data describing the structure of the community. This has been demonstrated successfully in various environments over the last few years including the human gut (Grootaert et al. 2009; van den Abbeele et al. 2010); wastewater treatment systems (Wittebolle et al. 2008, 2009a, b; Vlaeminck et al. 2009); prebiotics and human gut microbial diversity (Marzorati et al. 2010b; Possemiers et al. 2010); microbial community related to celiac health issues (Schippa et al. 2010); drinking water (Lautenschlager et al. 2010); anaerobic digestion (Carballa et al. 2011; Pycke et al. 2011); aquaculture (Qi et al. 2009; De Schryver et al. 2010, 2011; Crab et al. 2009; Prol-Garcia et al. 2010); these above mentioned studies, which commonly used fingerprinting techniques such as DGGE, fatty acid methyl ester (FAME), clone libraries, and t-RFLP, have helped us to elucidate unknown characteristics of natural prokaryotic ecosystems in these various areas utilizing MRM. The first parameter, Rr, was originally introduced to establish a technique-specific range of values which indicate the richness and genetic diversity (based on the polymorphism of the 16S rRNA gene sequence) of species within an indigenous bacterial community (Marzorati et al. 2008). Rr was based on the DGGE gel patterns derived from the GC content and positioning of the sequences from complex microbial communities. A high Rr value indicates an environment with a high carrying capacity, an environment that can host several species with a wide GC variability. It has been successfully used with rRNA intergenic spacer analysis (RISA; Rojas-Oropeza et al. 2010), TGGE (Schippa et al. 2010), and clone libraries (Marzorati et al. 2010a).

Dy, the second parameter in the tool set, was used to determine the rate of change within the same community over a fixed time interval. It refers to the number of species, on average, that are detected to be of significance in a given environment at a certain time point, thus providing a large picture of the dynamics within a community. Dy can be used as a standalone parameter as seen previously in a study looking at the changing community during bioaugmentation of activated sludge (Bathe et al. 2009).

The third complementary parameter is the functional organization (Fo; Marzorati et al. 2008). This parameter initially was designed to determine the resulting action of which microorganisms were suited to the ongoing environmental-microbiological interactions. This should inevitably give them a selective advantage over the other bacteria, thus increasing their dominance among the other species in the microbial community being examined. Similarly, Fo was successfully used to demonstrate changes in evenness in various areas of research including wastewater and MFCs. Fo was renamed as community organization (Co), as a parameter that describes the microbial community in terms of degrees of evenness.

4.1 Advances in Microbial Resource Management

The recent development of new technologies providing high-throughput, low-cost sequencing methods has provided us with alternatives including Lynx Therapeutics' Massively Parallel Signature Sequencing (MPSS; Reinartz et al. 2002), 454 pyrosequencing (Edwards et al. 2006), Illumina (Solexa) sequencing (Whiteford et al. 2009), and ABI SOLiD (Sequencing by Oligonucleotide Ligation and Detection) sequencing (Valouev et al. 2008). These technologies are more sensitive than the traditional DGGE and other fingerprinting methods and provide us with a broader taxonomic coverage of the unknown and often unculturable microbial communities. Therefore, the basic parameters of the MRM tool set have to be reengineered and adapted to provide a universal platform with which to compare and contrast this new immense amount of data. Rr, based on community fingerprinting methods, was often limited in comparing richness in complex communities due to their low detection limit (Bent et al. 2007). These techniques often considerably underestimate the actual richness of communities and hence the actual diversity, which is why we must now be more vigilant in the face of new sequencing technologies. In fact, high throughput sequencing provides a much deeper insight of the actual diversity of the analyzed microbial community. As a result, the analysis provides not only the sequences of the dominant microorganisms but also of all those less abundant microorganisms normally not detected with the above mentioned fingerprinting techniques due to the low detection limit.

To know about the specific community dynamics a new technique, stable isotope probing (SIP) combined with DGGE is used. SIP is based on an ultracentrifugation step that fractionates a microbial sample previously incubated with, the compound of interest which was isotopically labelled (Neufeld et al. 2007). The result is a separation of the microbial sample according to its weight, which is dependent on its isotope content. SIP is an excellent technique to identify key microbial players in mixed communities and to look at a specific process in a controlled environment. With the key players identified, their physiology and ecology can be investigated, therefore providing important tools for MRM. Through SIP, we can determine which microorganisms are responsible for the process of interest in addition to interactions within the community providing us with information on the functional organization of the community.

5 Utilization of Microbial Resources

In modern times, as biotechnology advances constantly, microorganisms have been applied to all aspects of industrial and agricultural production. Microorganisms have generated enormous social and economic benefits. It mainly refers to green chemistry and engineering, environmental bioremediation, renewable energy, natural medicine, food production and processing. Developing agricultural microbial resources is of vital importance. In recent years, research and development on new agricultural production technology have made great progress. It is mainly represented by microbial feed, microbial fertilizers, microbial pesticides, and microbial food.

As people seize natural resources crazily and over depend on fossil energy, issues such as severe energy depletion, resources shortage and environmental pollution have come up. Industrial production and discharge by traditional chemical method are also one of the major reasons for environmental pollution. A sustainable society should be less dependent on unsustainable resources and pollution caused by fossil resources should be reduced. It's vital to make full use of the abundant natural resources, to replace backward, polluted chemical industry with innovative and advanced bioeconomy.

5.1 Environmental Management

Microbial resources will become an important force in solving the environmental problems. Microbes play an important role in water environment, such as purification and pollution. Stabilization lagoons and bio-membranes are two classic approaches dealing with polluted water. When the polluted water is pumped into the reaction pool, microbes in it transform organics into inorganics by degradation, nitrification and also photosynthesis. Usually several pools are used together, including facultative anaerobic, anaerobic and aerobic ones. For bio-membranes, various kinds of microbes will attach to the membrane and form an ecosystem, which results in a very high speed of the degradation of organic matter and also in a very high quality of the water obtained.

Waste is an unavoidable by product of human activities. Rapid population growth, urbanization and industrial growth have led to increase the quantity and complexity of generated waste and severe waste management problems in most cities of third world countries. The large quantity of waste generated necessitates a system of collection, transportation and disposal. Landfill and composting methods could be used in dealing with solid waste. Landfilling involves the controlled disposal of wastes on or in the earth's mantle. Landfills are used to dispose of solid waste that cannot be recycled and is of no further use, the residual matter remaining after solid wastes have been pre-sorted at a materials recovery facility and the residual matter remaining after the recovery of conversion products or energy. It is by far the most common method of ultimate disposal for waste residuals. Many countries use uninhabited land, quarries, mines and pits as landfill sites. Biological reprocessing methods like composting and anaerobic digestion are natural ways to decompose solid organic waste. Composting is nature's way of recycling organic wastes. Composting is a method of decomposing waste for disposal by microorganisms (mainly bacteria and fungi) to produce a humus-like substance that can be used as a fertilizer. This process converts waste which is organic in nature to inorganic materials that can be returned to the soil as fertilizer i.e. biological stabilization of organic material in such a manner that most of the nutrient and humus that are so necessary for plant growth are returned to the soil.

5.2 Energy Development

Microorganisms will play an irreplaceable role in the process of searching for new energy, compounding new energy and energy re-synthesis. Modification and utilization of the existing microorganisms and the exploration of new microorganisms resources for renewable bio-energy manufacturing are the new novel perspectives. By the year 2120, 3.6% of electric power and 6–7% of the total energy will come from renewable resources (Lakó et al. 2008).

The waste generated in the process of industrial and agricultural production, such as crop straw, weeds, manure, sludge, organic industrial waste, garbage, etc., can be used as raw materials, and transformed into combustible gas or liquid bio-fuels by the microorganisms. Fatty acid esters (fatty acid methyl ester or ethyl ester) are major components of biodiesel. The current bio-diesel limited to the sources of animal fats or vegetable oil has certain limitations in its development. New technologies that transform the biomass directly into fatty acid ester through microbial fermentation need to be developed, at the same time, microbial resources, such as microalgae which can synthesize oil naturally, yeast, etc., should be explored.

The main component of biogas is methane, which is the product of organic matter decomposed by microorganisms under strict anaerobic conditions. Methane fermentation can release about 90% of the chemical energy in organic matter, which can be transformed into mechanical energy, electrical energy and heat energy. At present, gas as a fuel source has been transported around the world through pipelines for domestic and industrial use or converted to methanol as a supplementary fuel for internal combustion engines.

5.3 *Bio-chemical Refining*

The biorefining concept is an analogue of today's petroleum refineries producing multiple fuels and products from petroleum. By combining chemistry, biotechnology, engineering and systems approach, biorefinery could produce food, fertilizers, industrial chemicals, fuels, and power from biomass (Kamm and Kamm 2004). With the rapid consumption of fossil resources and the increasingly serious issue of the environment security, the refining of bio-based chemicals has become more and more popular. Many chemical companies are increasing their investments to produce 'green' chemicals through the utilization of microbial genes and enzyme resources as well as the biotransformation method in place of bio-chemical conversion. Bio-based chemicals have many types, including biological ethylene, optically pure D-or L-lactic acid, 1, 3-propanediol, 1, 4-butanediol, 3-hydroxy-propionic acid, acrylic acid, n-butanol, butyric acid, succinic acid, adipic acid, etc. They are not only important chemicals, but also important chemicals after being transformed. In addition, microbes can utilize biomass to synthesize lots of compounds that possess potential application value. As a result, further exploiting microbial resources and accelerating the biorefinery of the above chemicals will play an important role in development of bio-economy. Besides, exploration of the genes and enzymes from microorganisms and replacement of chemical transformation by bio-transformation are also important future directions.

5.4 *Industrial Enzymes and Biocatalysts*

The variety of microorganisms able to degrade natural and synthetic organic compounds can be used for applications in environmental biotechnology as well as in industrial synthetic chemistry. In particular, the latter approach to use enzymes for biotransformation is of growing interest. Biotransformations are chemical reactions that are catalyzed by microorganisms in terms of growing or resting cells or that are catalyzed by isolated enzymes. Because of the high stereo- or regioselectivity combined with high product purity and high enantiomeric excesses, biotransformations can be technically superior to traditional chemical synthesis. The use of biotransformations for industrial synthetic chemistry is an interdisciplinary, and therefore very exciting, field that needs the close cooperation of microbiologists, molecular biologists, chemists, and engineers. Besides classical methods, new technologies including the screening for non-culturable microorganisms and high throughput screening techniques are speeding up the discovery of new biocatalysts. The key of biotransformation is to develop highly efficient, highly selective biological catalysts, which can be realized by combining basic genomic technology with high-throughput screening technology and genome database mining. Meanwhile, further improvement and optimization of the directed evolution technology is another important force.

5.5 *Utilization of Microbial Resources in Extreme Environments*

The microbial resources in extreme environments which have a unique type of genes, special physiological mechanisms and metabolic products, are a kind of treasure-house of new resources. Enzymes from extreme environments have a very strong catalytic effect on environmental friendly products, and can be applied to a variety of special reaction systems. Psychrophilic enzymes can reduce the energy consumption in industry, while thermophilic enzymes are the important source of thermo stable enzymes and leaching bacteria. Its application in food, energy, environment, metabolic projects, mineral exploration, etc. can provide opportunities for the development of new chemicals, drugs and biological products.

6 Recent Development in Microbial Resource Utilization

6.1 *Isolation and Microorganism Culture Technology and Metagenomics Technology*

Microorganisms play an important role in recycling of materials and life continuance, their diversity is used to monitor and predict environmental changes. Lots of unknown microorganisms have never been cultured. In recent years, scientists have developed several new methods to isolate and culture microorganisms. The key point of acquiring new functional microorganisms is to create high throughput, rapid and efficient technologies towards isolation and culture of microorganisms.

Meanwhile, genomics and modern molecular biology technologies are getting more mature. These modern technologies gradually infiltrate into the entire field of life sciences. They also represent new research methods for microbiological research. A recently developed metagenomic approach employs cloning of the total microbial genome, the so-called 'metagenome', directly isolated from natural environments in culturable bacteria such as *Escherichia coli* (Beja et al. 2000; Handelsman et al. 1998; Rondon et al. 1999, 2000) and discovering novel microbial resources (Handelsman 2004). The basis of metagenomic approach originated from the molecular ecological studies of microbial diversity, indicating that majority of micro-organisms in nature was not cultivated by standard cultivation techniques. In addition, the combination of phylogenetic marker screening of metagenomic libraries and genomics could reveal the physiology of as-yet-uncultured microorganisms, only identified by culture-independent studies (Quaiser et al. 2002; Liles et al. 2003).

A series of engineering systems and technical platforms have been built for genetic engineering, protein engineering, metabolic engineering, synthetic biology, and bioprocess studies. In 1990s, with the development of gene function studies by applying functional genomics, molecular biology, molecular pathology and cell biology, the number of targets for drug screening has been growing in an unprecedented rate. High-throughput screening (HTS) is one of the newest techniques used in drug design and may be applied in biological and chemical sciences. This method, due to utilization of robots, detectors and software that regulate the whole process, enables a series of analyses of chemical compounds to be conducted in a short time and the affinity of biological structures which is often related to toxicity to be defined. The HTS method is more frequently utilized in conjunction with analytical techniques such as NMR or coupled methods e.g., LC-MS/MS.

6.2 Combinatorial Biochemistry Technology

Combinatorial biochemistry and combinatorial biosynthesis make use of microorganisms to synthesize a wide range of compounds at the gene level. Combinatorial biosynthesis has been successfully employed for e.g. carotenoid- and antibiotic polyketide-producing micro-organisms. It involves interchanging secondary metabolism genes between micro-organisms to create unnatural gene combinations or hybrid genes. Novel metabolites can be made due to the effect of new enzyme-metabolic pathway combinations or to the formation of proteins with new enzymatic properties. Combinatorial biosynthesis technology can help us to discover new compounds. Combinatorial biosynthesis combined with genetic engineering and high-throughput screening technology will make it possible to carry out drug development in vivo by microorganisms via modern biology and chemistry. By this approach, we will find better strategies to synthesize new drugs and get more compounds with new structure.

6.3 Directed Evolution Technology

The development of industrial biotechnology focusing on biocatalysis needs to exploit new efficient biocatalysts. These biocatalysts should meet the demands of new catalytic activity and high productivity. They can adapt to unsuitable environments and satisfy industrial development. To this end, we can utilize microbial resources to develop new enzymes. And we can artificially modify enzymes' biological activity in accordance with special needs.

In recent years, with the development of directed evolution technology, there is no need to solve the protein's three dimensional structure and enzymatic mechanism in advance. The natural evolution mechanism of artificial simulation (random mutation, recombination, natural selection) modified enzyme genes *in vitro*. It can directly select for enzyme mutants whose function might meet special requirements. By this way, we can get such enzymes in a few days or weeks, much quicker than millions of years nature needs to achieve it. It is an important method of finding novel bioactive molecules and biotransformation pathways. The newborn directed evolution technology has greatly expanded the range of protein engineering research and application. It has opened up a new way for enzyme's structure and function research. Meanwhile, it gradually demonstrates its vitality in the fields of industry, agriculture, medicine, etc.

6.4 Biological Information Technology

Modern biological technologies result in an enormous accumulation of biological data, including microbial strain resources and related genetic resources. They enable us to restudy biological problems on basis of the whole genome, which will lead to major scientific discoveries. The huge amount of data generated from genomics research (transcriptomics, proteomics, and metabolomics) is what traditional methods cannot do, so bioinformatics came into being. Future development direction is to utilize ongoing biological information technologies and develop new software algorithms.

7 Conclusions

Microorganisms are an almost unlimited source of metabolic capabilities ready to be exploited for multiple purposes. A combination of several techniques should be applied to interrogate the diversity, function, and ecology of microorganisms. Culture-based and culture-independent molecular techniques are neither contradictory nor excluding and should be considered as complementary. An interdisciplinary systems approach embracing several "omics" technologies to reveal the interactions between genes, proteins, and environmental factors will be needed to provide new insights into environmental microbiology. Environmental metagenomic libraries have proved to be great resources for new microbial enzymes and antibiotics with potential applications in biotechnology, medicine, and industry respectively. Massive construction of metagenomic libraries and development of high throughput screening technologies will be necessary to obtain valuable microbial resources. Development of multi-"omics" approaches will be a high-priority area of research in the coming years. Microbial resource management is the basis of a number of new developments in domains such as environmental safety and health, renewable energy production, closing environmental cycles and providing new materials.

References

Bathe S, Schwarzenbeck N, Hausner M (2009) Bioaugmentation of activated sludge towards 3-chloroaniline removal with a mixed bacterial population carrying a degradative plasmid. Bioresour Technol 100:2902–2909

Beja O, Suzuki MT, Koonin EV, Aravind L, Hadd A, Nguyen LP, Villacorta R, Amjadi M, Garrigues C, Javanovich SB, Feldman RA, Delong EF (2000) Construction and analysis of bacterial artificial chromosome libraries from a marine microbial assemblage. Environ Microbiol 2:516–529

Bent SJ, Pierson JD, Forney LJ, Danovaro R, Luna GM, Dell'anno A, Pietrangeli B (2007) Measuring species richness based on microbial community fingerprints: the emperor has no clothes. Appl Environ Microbiol 73:2399–2401

Bridge PD, Roberts PJ, Spooner BM, Panchal G (2003) On the unreliability of published DNA sequences. New Phytol 160:43–48

Carballa M, Smits M, Etcherbehere C et al (2011) Correlations between molecular and operational parameters in continuous lab scale anaerobic reactors. Appl Microbiol Biotechnol 89:303–314

Crab R, Kochva M, Verstraete W, Avnimelech Y (2009) Bio-flocs technology application in over-wintering of tilapia. Aquacult Eng 40:105–112

De Schryver P, Sinha AK, Kunwar PS, Baruah K, Verstraete W, Boon N, De Boeck G, Bossier P (2010) Poly-beta-hydroxybutyrate (PHB) increases growth performance and intestinal bacterial range weighted richness in juvenile European sea bass, *Dicentrarchus labrax*. Appl Microbiol Biotechnol 86:1535–1541

De Schryver P, Dierckens K, Thi QQ, Amalia R, Marzorati M, Bossier P, Boon N, Verstraete W (2011) Convergent dynamics of the juvenile European sea bass gut microbiota induced by poly-beta-hydroxybutyrate. Environ Microbiol 13:1042–1051

de Wildeman S, Verstraete W (2003) The quest for microbial reductive dechlorination of C_2 to C_4 chloroalkanes is warranted. Appl Microbiol Biotechnol 61:94–102

Denef V, Park J, Tsoi T, Rouillard J, Zhang H, Wibbenmeyer JA, Verstraete W, Gulari E, Hashsham SA, Tiedje JM (2004) Biphenyl and benzoate metabolism in a genomic context: outlining genome-wide metabolic networks in *Burkholderia xenovorans* LB400. Appl Environ Microbiol 70:4961–4970

Edwards RA, Rodriguez-Brito B, Wegley L, Haynes M, Breitbart M, Peterson DM, Saar MO, Alexander S, Alexander EC Jr, Rohwer F (2006) Using pyrosequencing to shed light on deep mine microbial ecology. BMC Genomics 7:57

Garbeva P, van Veen JA, van Elsas JD (2004) Microbial diversity in soil: selection microbial populations by plant and soil type and implications for disease suppressiveness. Annu Rev Phytopathol 42:243–270

Gilbert JA, Hill PJ, Dodd CER, Laybourn-Parry J (2004) Demonstration of antifreeze protein activity in Antarctic lake bacteria. Microbiology 150:171–180

Grootaert C, van den Abbeele P, Marzorati M, Broekaert WF, Courtin CM, Delcour JA, Verstraete W, van de Wiele T (2009) Comparison of prebiotic effects of arabinoxylan oligosaccharides and inulin in a simulator of the human intestinal microbial ecosystem. FEMS Microbiol Ecol 69:231–242

Handelsman J (2004) Metagenomics: application of genomics to uncultured microorganisms. Microbiol Mol Biol Rev 68:669–685

Handelsman J, Rondon MR, Brady SP, Clady J, Goodman RM (1998) Molecular biological access to the chemistry of unknown soil microbes: a new frontier for natural products. Chem Biol 5:245–249

Lakó J, Hancsók J, Yuzhakova T, Marton G, Utasi A, Rédey Á (2008) Biomass – a source of chemicals and energy for sustainable development. Environ Eng Manag J 7:499–509

Lautenschlager K, Boon N, Wang Y, Egli T, Hammes F (2010) Overnight stagnation of drinking water in household taps induces microbial growth and changes in community composition. Water Res 44:4868–4877

Liles MR, Manske BF, Bintrim SB, Handelsman J, Goodman RM (2003) A census of rRNA genes and linked genomic sequences within a soil metagenomic library. Appl Environ Microbiol 69:2684–2691

Kamm B, Kamm M (2004) Principles of biorefineries. Appl Microbiol Biotechnol 64:137–145

Margesin R, Gander S, Zacke G, Gounot AM, Schinner F (2003) Hydrocarbon degradation and enzyme activities of cold adapted bacteria and yeast. Extremophiles 7:451–458

Marzorati M, Wittebolle L, Boon N, Daffonchio D, Verstraete W (2008) How to get more out of molecular fingerprints: practical tools for microbial ecology. Environ Microbiol 10:1571–1581

Marzorati M, Balloi A, de Ferra F, Corallo L, Carpani G, Wittebolle L, Verstraete W, Daffonchio D (2010a) Bacterial diversity and reductive dehalogenase redundancy in a 1,2-dichloroethanedegrading bacterial consortium enriched from a contaminated aquifer. Microb Cell Fact 19:9–12

Marzorati M, Verhelst A, Luta G, Sinnott R, Verstraete W, Van de Wiele T, Possemiers S (2010b) In vitro modulation of the human gastrointestinal microbial community by plant-derived polysaccharide-rich dietary supplements. Int J Food Microbiol 139:168–176

Neufeld JD, Vohra J, Dumont MG, Lueders T, Manefield M, Friedrich MW, Murrell JC (2007) DNA stable-isotope probing. Nat Protoc 2:860–866

Possemiers S, Marzorati M, Verstraete W, Van de Wiele T (2010) Bacteria and chocolate: a successful combination for probiotic delivery. Int J Food Microbiol 141:97–103

Prol-Garcia MJ, Planas M, Pintado J (2010) Different colonization and residence time of *Listonella anguillarum* and *Vibrio splendidus* in the rotifer *Brachionus plicatilis* determined by real-time PCR and DGGE. Aquaculture 302:26–35

Pycke B, Etcherbehere C, Van de Caveye P, Negroni A, Verstraete W, Boon N (2011) A time course analysis of four full-scale anaerobic digesters in relation to the dynamics of change of their microbial communities. Water Sci Technol 63:769–775

Qi ZZ, Dierckens K, Defoirdt T, Sorgeloos P, Boon N, Bao ZM, Bossier P (2009) Analysis of the evolution of microbial communities associated with different cultures of rotifer strains belonging to different cryptic species of the *Brachionus plicatilis* species complex. Aquaculture 292:23–29

Quaiser A, Ochsenreiter T, Klenk H-P, Kletzin A, Treusch AH, Meurer G, Eck J, Sensen CW, Schleper G (2002) First insight into the genome of an uncultivated crenarchaeote from soil. Environ Microbiol 4:603–611

Reinartz J, Bruyns E, Lin JZ, Burcham T, Brenner S, Bowen B, Kramer M, Woychik R (2002) Massively parallel signature sequencing (MPSS) as a tool for in-depth quantitative gene expression profiling in all organisms. Brief Funct Genomic Proteomic 1:95–104

Rojas-Oropeza M, Dendooven L, Garza-Avendano L, Souza V, Philippot L, Cabirol N (2010) Effects of biosolids application on nitrogen dynamics and microbial structure in a saline-sodic soil of the former Lake Texcoco (Mexico). Bioresour Technol 101:2491–2498

Rondon MR, Goodman RM, Handelsman J (1999) The earth's bounty: assessing and accessing soil microbial diversity. Trends Biotechnol 17:403–409

Rondon MR, August PR, Bettermann AD, Brady SF, Grossman TH, Liles MR, Loiacono KA, Lynch BA, MacNeil IA, Minor C, Tiong CL, Gilman M, Osburne MS, Clardy J, Handelsman J, Goodman RM (2000) Cloning the soil metagenome: a strategy for accessing the genetic and functional diversity of uncultured microorganisms. Appl Environ Microbiol 66:2541–2547

Schippa S, Iebba V, Barbato M, Di Nardo G, Totino V, Checchi MP, Longhi C, Maiella G, Cucchiara S, Conte MP (2010) A distinctive 'microbial signature' in celiac pediatric patients. BMC Microbiol 17:175

Siddiqui KS, Cavicchioli R (2006) Cold-adapted enzymes. Annu Rev Biochem 75:403–433

Singh BK, Campbell CD, Sorenson SJ, Zhou J (2009) Soil genomics. Nat Rev Microbiol 7:756. doi:10.1038/nrmicro2119-c1

Smidt H, de Vos WM (2004) Anaerobic microbial dehalogenation. Annu Rev Microbiol 58:43–73

Valouev A, Ichikawa J, Tonthat T, Stuart J, Ranade S, Peckham H, Zeng K, Malek JA, Costa G, McKernan K, Sidow A, Fire A, Johnson SM (2008) A high-resolution, nucleosome position map of *C. elegans* reveals a lack of universal sequence-dictated positioning. Genome Res 18:1051–1063

Van den Abbeele P, Grootaert C, Marzorati M, Possemiers S, Verstraete W, Gerard P, Rabot S, Bruneau A, El Aidy S, Derrien M, Zoetendal E, Kleerebezem M, Smidt H, Van de Wiele T (2010) Microbial community development in a dynamic gut model is reproducible, colon region specific, and selective for *Bacteroidetes* and *Clostridium* Cluster IX. Appl Environ Microbiol 76:5237–5246

van der Gast CJ, Jefferson B, Reid E (2006) Bacterial diversity is determined by volume in membrane bioreactors. Environ Microbiol 8:1048–1055

van Trappen S, Vandecandelaere I, Mergaert J, Swings J (2005) *Flavobacterium fryxellicola* sp. *nov*. and *Flavobacterium psychrolimnae* sp. *nov*, novel psychrophilic bacteria isolated from microbial mats in Antarctic lakes. Int J Syst Evol Microbiol 55:769–772

Verstraete W, Wittelbolle L, Heylen K, Vanparys B, de Vos P, van de Wiele T, Boon N (2007) Microbial resource management: the road to go for environmental biotechnology. Eng Life Sci 7:117–126

Vlaeminck SE, Terada A, Smets BF, Van der Linden D, Boon N, Verstraete W, Carballa M (2009) Nitrogen removal from digested black water by one-stage partial nitritation and anammox. Environ Sci Technol 43:5035–5041

Whiteford N, Skelly T, Curtis C, Ritchie ME, Lohr A, Zaranek AW, Abnizova I, Brown C (2009) Swift: primary data analysis for the Illumina Solexa sequencing platform. Bioinformatics 25:2194–2199

Wittebolle L, Vervaeren H, Verstraete W, Boon N (2008) Quantifying community dynamics of nitrifiers in functionally stable reactors. Appl Environ Microbiol 74:286–293

Wittebolle L, Verstraete W, Boon N (2009b) The inoculum effect on the ammonia-oxidizing bacterial communities in parallel sequential batch reactors. Water Res 43:4149–4158

Wuyts J, De Rijk P, Van de Peer Y, Winkelmans T, De Wachter R (2001) The European large subunit ribosomal RNA database. Nucleic Acids Res 29:175–177

Wuyts J, Van de Peen Y, Winkelmans T, De Wachter R (2002) The European database on small subunit ribosomal RNA. Nucleic Acids Res 30:183–185

Yamada T, Hiraoka Y, Das Gupta TK, Chakrabarty AM (2004) *Rusticyanin*, a bacterial electron transfer protein causes G (1) arrest in J774 and apoptosis in human cancer cells. Cell Cycle 3:1182–1187

Chapter 2
Recent Development in the Methods of Studying Microbial Diversity

Mohd Ikram Ansari and Abdul Malik

Abstract Soils harbor an exceptional diversity of prokaryotes that are largely undescribed beyond the level of ribotype and are a potentially vast resource for natural product discovery. It is estimated that in 1 g of soil there are 4,000 different bacterial "genomic units" based on DNA–DNA reassociation. Approximately 1% of the soil bacterial population can be cultured by standard laboratory practices. Soil structure and physicochemical properties limit the applicability and value of methods involving direct observation, and ecological studies have focused on communities and populations, rather than single cells or microcolonies. Ecological studies of soil microorganisms require reliable techniques for assessment of microbial community composition, abundance, growth, and activity. The methods for composition and diversity analysis of soil microbes has advantaged in the past years. Traditionally, taxonomic classification of bacteria has been performed based on metabolic, morphologic, and physiological traits. This approach emulates the methodological approach of botanists and zoologists; however, it requires the isolation and cultivation of individual bacterial species. Assessments of bacterial communities from a number of environments have found that the fraction of cells that may be cultured is not representative of the abundance or diversity of the microbial community present in the environment; it is often observed that direct microscopic counts exceed viable cell counts by several orders of magnitude. The widespread use of molecular techniques in studying microbial communities has greatly enhanced our understanding of microbial diversity and function in the natural environment and contributed to an

M.I. Ansari (✉)
Department of Agricultural Microbiology, Aligarh Muslim University, Aligarh, Uttar Pradesh 202002, India
e-mail: ikram_ansari21@yahoo.com

A. Malik
Department of Agricultural Microbiology, Faculty of Agricultural Sciences, Aligarh Muslim University, Aligarh, Uttar Pradesh 202002, India
e-mail: ab_malik30@yahoo.com

A. Malik et al. (eds.), *Management of Microbial Resources in the Environment*,
DOI 10.1007/978-94-007-5931-2_2, © Springer Science+Business Media Dordrecht 2013

explosion of novel commercially viable enzymes. The analysis of amplified and sequenced 16S rRNA genes has become the most important single approach for microbial diversity studies. This chapter describes the methods used for studying microbial diversity and the recent development in these methods. The new sequencing technologies used for sequencing thousands of reads in a single run and a cost-effective method to run many samples.

Keywords Microbial diversity • Cultural methods • Molecular methods • Single cell analysis • 16S rRNA • Sequencing

1 Introduction

Microbial diversity is a general term used to include genetic diversity, that is, the amount and distribution of genetic information, within microbial species; diversity of bacterial and fungal species in microbial communities; and ecological diversity, that is, variation in community structure, complexity of interactions, number of trophic levels, and number of guilds. Microbial diversity is measured by various techniques such as traditional plate counting and direct counts as well as the newer molecular-based procedures and fatty acid analysis (Hill et al. 2000; Kirk et al. 2004; Faulwetter et al. 2009).

Microbial characteristics of soils are being evaluated increasingly as sensitive indicators of soil health because of the clear relationships between microbial diversity, soil and plant quality, and ecosystem sustainability (Doran et al. 1994). While the understanding of microbial properties such as biomass, activity, and diversity are important to scientists in furthering knowledge of the factors contributing to soil health, results of such analyses may also be useful to extension personnel and farmers in devising practical measures of soil quality (Hill et al. 2000). The establishment of comparative small subunit rRNA sequence analysis as initially introduced by Fox et al. (1977) certainly represents a major milestone in the concomitant 'evolution' of systematic and methods for characterization. These molecules almost perfectly meet the basic requirements of a general phylogenetic marker i.e. ubiquitous distribution, functional constancy, low frequency of lateral gene transfer and recombination combined with a comprehensive database of primary structures (Cole et al. 2007; Ludwig 2007).

Over the past two decades, the approach to analyzing soil microbial communities has changed dramatically. Many new methods and approaches are now available, allowing soil microbiologists to gain access to more of the microorganisms residing in soil and allowing for better assessments of microbial diversity (Nannipieri et al. 2003; Fierer and Lennon 2011).

A number of methods are currently available for studies on soil microbial communities. The use of molecular techniques for investigating microbial diversity in soil communities continues to provide new understanding of the distribution and diversity of organisms in soil habitats. The use of SSU rRNA or rDNA sequences, combined with fluorescent oligonucleotide probes provides a powerful approach for studying soil microorganisms that may not be amenable to current culturing techniques (Borneman and Becke 2007).

Despite the utility of culture-independent techniques such as SSU rRNA or rDNA analyses, there remains a general need to cultivate microorganisms from soil habitats to better understand their role in soil processes. Future studies of soil microbial communities must necessarily rely on a combination of both culture-dependent and culture-independent methods and approaches. Only then we will be able to develop a more complete picture of the contribution of specific microbial communities to the overall quality and health of agricultural soils. In this chapter, we briefly discuss some of the most important approaches for studying soil microbial communities. Our goal is to place the newer methods in perspective with the traditional culture-based approaches for assessing microbial diversity.

2 Culture-Dependent Methods of Community Analysis

About 99% of soil microorganisms are unculturable. Advances in molecular biology methods have assisted our understanding of microbial functions, their interactions with other organisms and their environment. These studies have primarily been conducted in the laboratory, with less research on microorganisms in their natural habitat. It was recently estimated that less than 1% of the 10^9 bacterial cells in a gram of soil were culturable in laboratory media (Davis et al. 2005). One can appreciate the lack of knowledge on the unculturable 99% of the microbial population, in terms of their contribution to nutrient and energy flow, soil respiration, gene transfer, degradation of pollutants, diseases and quorum sensing, all of whose mechanisms have yet to be better understood. Researchers have developed new methods that allow new knowledge to be forthcoming (Saleh-Lakha et al. 2005). Culture-based approaches, while extremely useful for understanding the physiological potential of isolated organisms, do not necessarily provide a comprehensive information on the composition of microbial communities. The results obtained by culture-dependent techniques covered only those few organisms that could be cultivated. Due to this well documented disparity between cultivable and *in situ* diversity, it is often difficult to assess the significance of cultured members in microbial communities. Several studies have employed culture-independent techniques to show that cultivated microorganisms from diverse environments often may represent very minor components of the microbial community as a whole (Ward et al. 1990; Amann et al. 1995; Larsen et al. 2008).

2.1 Dilution Plating and Culturing Methods

Traditionally, the analysis of soil microbial communities has relied on culturing techniques using a variety of culture media designed to maximize the recovery of different microbial species. This is particularly the case for soil health studies. There are numerous examples where these techniques have revealed a diversity of microorganisms associated with various soil quality parameters such as disease suppression and organic matter decomposition (Alvarez et al. 1995; van Hu and Bruggen 1997; Maloney et al. 1997; Garbeva et al. 2004). Although there have been recent attempts to devise suites of culture media to maximize the recovery of diverse microbial groups from soils (Balestra and Misaghi 1997; Mitsui et al. 1997), it has been estimated that less than 0.1% of the microorganisms found in typical agricultural soils are culturable using current culture media formulations (Atlas and Bartha 1998). This is based on comparisons between direct microscopic counts of microbes in soil samples and recoverable colony forming units (Hill et al. 2000; Juan et al. 2008).

Collado et al. (2007) found that high-throughput bacterial cultivation has improved the recovery of slow-growing and previously uncultured bacteria. The most robust high-throughput methods are based on techniques of 'dilution to extinction' or 'extinction culturing'. The low-density partitioning of Colony Forming Units (CFUs) in tubes or micro-wells exploits the fact that the number of culturable species typically increases as inoculum density decreases. Bacterial high-throughput culturing methods were adapted to fungi to generate large numbers of fungal extinction cultures. The efficiency of extinction culturing was assessed by comparing it with particle filtration and automated plate-streaking. Compared with plating methods using continuous surfaces, extinction culturing distributes fungal propagules over partitioned surfaces. Inter colony interactions are reduced, permitting longer incubation times, and colony initiation and recovery improved. Efforts to evaluate and recover colonies from fungal isolation plates were substantially reduced (Collado et al. 2007).

2.2 Community-Level Physiological Profiles

Traditionally, methods to analyze soil microorganisms have been based on cultivation and isolation (van Elsas et al. 1998); a wide variety of culture media has therefore been designed to maximize the recovery of diverse microbial groups. A Biolog-based method for directly analyzing the potential activity of soil microbial communities, denoted community-level physiological profiling (CLPP) (Garland 1996a; Garbeva et al. 2004) has also been introduced. Microbial community analyses based on CLPPs have been corroborated by other microbial community measures, including plate counts (Soderber et al. 2002), fatty acid methyl ester and phospholipid fatty acid analysis (Widmer et al. 2001; Soderber et al. 2002), API 20NE enzyme and C tests, and an array of molecular assays (Ovreas and Torsvik 1998; Di Giovanni et al. 1999a; Widmer et al. 2001). In addition, previous research has demonstrated that CLPPs are highly reproducible (Haack et al. 1995; Di Giovanni et al. 1999b; Classen et al. 2003).

One of the more widely used culture-dependent methods for analyzing soil microbial communities has been that of community-level physiological profiles

(Garland and Mills 1991; Konopka et al. 1998). This technique takes advantage of the traditional methods of bacterial taxonomy in which bacterial species are identified based on their utilization of different carbon sources. Community-level physiological profiles have been facilitated by the use of a commercial taxonomic system, known as the BIOLOG system, which is currently available and has been used extensively for the analysis of soil microbial communities (Lehman et al. 1995; Garland 1996b; Hill et al. 2000; Liu et al. 2006).

This BIOLOG system is based on the utilization of a suite of 95 different carbon sources (Garland and Mills 1991). Utilization of each substrate is detected by the reduction of a tetrazolium dye, which results in a color change that can be quantified spectrophotometrically. The pattern of substrates that are oxidized can be compared among different soil samples from a series of times or locations as an indication of differences in the physiological functions of microbial communities. Most commonly, multivariate statistical techniques are necessary to analyze the substrate utilization profile data (Hackett and Griffiths 1997; Hitzl et al. 1997).

The CLPP method provides an exciting opportunity to overcome the drawbacks of alternative time consuming culture-based analyses or biochemical tests (Schutter and Dick 2001; Preston-Mafham et al. 2002). The CLPP approach is frequently employed to determine the effect of various environmental factors on the biological status of particular soil sites by following catabolic traits (Haack et al. 1995). On the other hand as (Garland 1996a, 1997) noticed, the metabolic growth response, which involves cooperative as well as competitive effects in BIOLOG Eco Plates wells, might be a major drawback of the CLPP method (Frac et al. 2012).

The metabolic diversity of microbial communities is fundamental for the multiple soil functions mediated by microorganisms. Community level physiological profiles based on sole C source oxidation have been used as a fast and reproducible tool to study soil microbial functional diversity because the utilisation of available carbon is the key factor governing microbial growth in soil. This study reveals that significantly different CLPP patterns can be generated on the basis of only 3–4 genera, as reflected by PCR-DGGE analysis. Also for this reason, CLPPs based on incubations of soil suspensions should just be used as a screening method and always be accompanied by other techniques for community analysis. The CLPP methods can discriminate between different soil microbial communities, CLPPs may provide little insight about the function of the community *in situ*. The CLPPs have the greatest utility when they are combined with other microbial methods that do not rely on the culturing of the soil microflora (Classen et al. 2003; Ros et al. 2008). Despite the fact that culture-dependent techniques are not ideal for studies of the composition of natural microbial communities when used alone, they provide one of the most useful means of understanding the growth habit, development, and potential function of microorganisms from soil habitats. A combination of culture-based and culture-independent approaches is likely to reveal more complete information regarding the composition of soil microbial communities (Liesack et al. 1997).

3 Culture-Independent Methods of Community Analysis

Because of the inherent limitations of culture-based methods, soil microbial ecologists are turning increasingly to culture-independent methods of community analysis. Using culture-independent methods, the composition of communities can be inferred based on the extraction, quantification, and identification of molecules from soil that are specific to certain microorganisms or microbial groups; or advanced fluorescence microscopic techniques. Useful molecules for such studies include phospholipid fatty acids and nucleic acids (Morgan and Winstanley 1997) whereas the microscopic techniques involve either the hybridization of fluorescent-labeled nucleic acid probes with total RNA extracted from soils or hybridizations with cells *in situ* (Collins et al. 2006). Sequence analysis of the 16S rRNA gene is the dominant method of determining identity and phylogenetic relatedness of microorganisms (Curtis et al. 2006), although other genes, such as *rpoB*, may provide greater resolution in phylogenetic associations at the species and subspecies levels (Case et al. 2007; Little et al. 2008).

3.1 Phospholipid Fatty Acid Analysis

Phospholipid fatty acid (PLFA) analysis has been used as a culture-independent method of assessing the structure of soil microbial communities and determining gross changes that accompany soil disturbances such as cropping practices (Zelles et al. 1995), pollution (Frostegard et al. 1993), fumigation (Macalady et al. 1998), and changes in soil quality (Petersen et al. 1998). Phospholipid fatty acids are potentially useful signature molecules due to their presence in all living cells. In microorganisms, phospholipids are found exclusively in cell membranes and not in other parts of the cell as storage products (Mathew et al. 2012). Our knowledge of such signature molecules comes from the use of fatty acid analysis for bacterial taxonomy, in which specific fatty acid methyl esters (FAMEs) have been used as an accepted taxonomic discriminator for species identification. Furthermore, phospholipid fatty acids are easily extracted from microbial cells in soil (Zelles and Bai 1993) allowing access to a greater proportion of the microbial community resident in soil than would otherwise be accessed during culture-dependent methods of analysis. The presence and abundance of these signature fatty acids in soil reveals the presence and abundance of particular organisms or groups of organisms in which those signatures can be found.

Phospholipid fatty acids have been widely used to characterize environmental microbial communities, generating community profiles that can distinguish phylogenetic or functional microbial groups within the community (Guckert et al. 1985). PLFAs have the potential to serve as biomarkers for changes in community composition following perturbation due to the rapid decomposition of PLFAs following cell death. However, because neither specific PLFA molecules nor categories of PLFAs have been consistently assigned to particular ecological categories, it can be difficult to precisely ascribe PLFA biomarkers to community responses (Zelles 1997), thus motivating biochemical research to resolve this (Zelles 1997). Examples of these correlations with varied ecological categories include: cy19:0 is indicative of a condition, microbial stress (Li et al. 2007); branched fatty acids are associated with a physiological trait, Gram positive microorganisms (Haubert et al. 2006), a physiological requirement, anaerobes (Nakamura et al. 2003), and a functional capability, metal reducers (Pfiffner et al. 2006; Schryver et al. 2006); and under specific growth conditions, branched odd-chain fatty acids may indicate narrower taxonomic groups of microorganisms, *Desulfococcus* or *Desulfosarcina* (Webster et al. 2006). Using a PLFA profile to describe the microbial community is also a daunting task, given that many PLFA biomarkers are poorly validated or may be valid only under particular conditions (Robertson et al. 2011).

The conventional analysis of phospholipid fatty acids involves lipid extraction and consecutive chromatographic separation of phospholipids from other lipid fractions, which is time-consuming and costly. In recent years, different investigators have tried to overcome these limitations by using other biological markers or by modifying the analytical procedures used to assess n-3 fatty acid status (Klingler and Koletzko 2012).

Despite the usefulness of this method, there are some important limitations (Haack et al. 1994). First, appropriate signature molecules are not known for all organisms in a soil sample and, in a number of cases, a specific fatty acid present in a soil sample cannot be linked with a specific microorganism or group of microorganisms. In general, the method cannot be used to characterize microorganisms to species level. Second, since the method relies heavily on signature fatty acids to determine gross community structure, any variation in these signatures would give rise to false community estimates created by artifacts in the methods. Third, bacteria and fungi produce widely different amounts of PLFA and the types of fatty acids vary with growth conditions and environmental stresses. Although signature PLFAs can be correlated with the presence of some groups of organisms, they may not necessarily be unique to only those groups under all conditions. Consequently, this could give rise to false community signatures (Frostegard et al. 2010).

3.2 Nucleic Acid Techniques

Of all the cell component molecules tested to date, nucleic acids have been the most useful in providing a new understanding of the structure of microbial communities. For example, in studies of soil microbial diversity, Torsvik and colleagues (Torsvik et al. 1996; Ovreas and Torsvik 1998) compared the re-association kinetics of DNA isolated from soil with that of pure cultures of microorganisms. They reasoned that the greater the sequence diversity of the DNA (and hence the microbial diversity), the greater the DNA reannealing time. The greatest advantage of the analysis of SSU rDNA is that microorganisms from natural habitats can be studied and characterized without culturing. Various studies have shown that rDNA from over 90% of the microorganisms that can be observed microscopically in situ can be extracted and analyzed (Porteous et al. 1997) as compared with less than 0.1% of the microorganisms observed in soil that can be recovered on culture media. Numerous studies have applied these techniques to the study of soil microbial communities (Borneman and Triplett 1997; Grosskopf et al. 1998).

All DNA extraction techniques are based on methods developed over the past 20 years. Once the microbial community rDNA is amplified from soil samples using PCR, individual amplicons must be separated prior to sequence analysis. Methods used most commonly for the separation of individual amplicons have been standard cloning procedures using a variety of *Escherichia coli* vectors. Recently, as a complement to cloning procedures, the use of denaturing gradient and temperature gradient gel electrophoresis (DGGE/TGGE) for separating individual amplicons has been described (Heuer et al. 1997; Muyzer and Smalla 1998). This technique allows one to separate mixtures of PCR products that are of the same length but differ only in sequence. The separation power of this technique rests with the melting behavior of the double stranded DNA molecule. As DNA molecules are electrophoresed in an increasing gradient of denaturant or in an increasing temperature gradient, they remain double-stranded until they reach the denaturant concentration or temperature that melts the double-stranded molecule. As the DNA melts, it branches, thus reducing the mobility in the gel. Since the melting behavior is largely dictated by the nucleotide sequence, the separation will resolve individual bands, each corresponding to a unique sequence. A number of different technologies are available for nucleic acid based systematics and identification. These methods can be assigned to two major groups: direct approaches determining or targeting sequence stretches and indirect procedures providing differentiating information without exact knowledge of the respective primary structure regions or the sequence (Ludwig 2007). A selection of methods by which nucleic acid inherited information is visualised without primary structure sequencing is described in the following section. Two major groups can be defined. The first group comprises methods for measurement of nucleic acid composition and similarity, the second concerns pattern techniques.

3.2.1 Nucleic Acid Composition and Similarity

Guanine Plus Cytosine (G+C) Content

Difference in the guanine plus cytosine (G+C) content of DNA can be used to study soil microbial diversity. This procedure is based on the knowledge that microorganisms differ in their G+C content and that taxonomically related groups only differ between 3 and 5% (Tiedje et al. 1999; McCutcheon et al. 2009). This method provides a coarse level of resolution as different taxonomic groups may share the same mol percentage range of G+C. Determination and comparison of genomic DNA G+C content were among the first approaches for characterisation of genomes. The molar G+C composition of the respective genomic DNA is still an intrinsic parameter of the minimum description of taxonomic units. Determination of the transition temperature of double stranded to single stranded DNA or the buoyant density of

DNA in a density gradient is one of the most commonly applied methods (Tamaoka 1994). The melting curves provide microbial community profiles indicative of the overall genetic diversity. Even if this analysis is considered to be low resolution, it can be used to indicate overall changes in microbial community structure, especially when the diversity is low. An advantage of this approach is that no PCR is used with all DNA extractions, quantification, or detecting rare members in the microbial populations. Thus, some of the less dominant microorganisms in the community that PCR might not detect without fractionation can be detected and analyzed. However, it requires large quantities of DNA (Tiedje et al. 1999; Medlin and Kooistra 2010).

The G+C approach does not provide any phylogenetic information nor allows to assign an organism to a certain taxon, it rather shows discriminating capacity. Different G+C contents indicate different organisms, whereas identical values *per se* do not necessarily characterise closely related taxa. The base composition of DNA expressed in mol% G+C widely varies among the prokaryotes ranging from 24 to 76 mol% (Tamaoka 1994; Ludwig 2007).

Quantitative Nucleic Acid Hybridization

DNA reassociation methods roughly allow estimating genomic similarity of pairs of strains or closely related species. The underlying principle quantifies the reassociation of complementary regions or stretches of single stranded DNA of heterologous origin. A variety of alternative techniques and formats has been developed measuring the amount of heterologous hybrids (end point measurement) (Grimont et al. 1980), the kinetics of heterologous hybridization (Huss et al. 1983), or the stability of heterologous hybrids (De Ley et al. 1973). Initially rather laborious techniques requiring large amounts of purified DNA have been miniaturised. Nowadays micro plate formats are mainly in use for endpoint (Ziemke et al. 1998; Christensen et al. 2000) and stability measurement (Mehlen et al. 2004). In both cases target DNA is immobilised in the cavities and labelled driver DNA is added for hybridisation. After removing not bound driver DNA the retained label is measured or the melting behaviour of the hybrids is followed applying temperature or denaturing buffer gradients for end point or stability measurement, respectively. The differences in binding capacity or melting temperature (Δ Tm) of homologous and heterologous settings allow estimating overall similarity. The amount of purified DNA needed for the measurements can be reduced by linker mediated PCR amplification of driver and target DNA (Mehlen et al. 2004). Large amounts of highly purified DNA are needed applying optical methods for comparative monitoring of the kinetics of hybrid formation (Huss et al. 1983). Therefore, the latter methods currently are not in common use. Nucleic acid-based techniques have a higher sensitivity, therefore requiring a higher level of quality control to prevent contamination, increasing the importance of effective sample preparation as a critical step for successful detection. Consideration of contamination, inhibitors in the specimen sample, and DNA degradation due to unfavorable conditions must be accounted for in the sensor design to help reduce the incidence of fal e positive or false negative results (Liu et al. 2009).

Estimating DNA similarity by quantitative hybridization, one has to be aware that only the minor part of the genome contributes to the measured values. The real sequence complementarity of the involved genomes has to be at least 80–85% to allow formation of heterologous hybrids (Lengler et al. 2009). There are some general drawbacks of quantitative DNA–DNA hybridization methods. Differences in genome size and DNA concentration heavily influence the obtained results except in the case of hybrid stability determination. Furthermore, the results depend on the experimental parameters and are not cumulative. Different experiments cannot directly be compared. The respective references (type strains) have to be included in every individual experiment. Reciprocal arrangements may produce similarity values differing by up to 10%. In addition, plasmid and chromosomal as well as core genome and foreign DNA (acquired by horizontal gene transfer) are not discriminated. Despite all these drawbacks, quantitative DNA hybridization is the only generally applicable method to estimate relationships at lower taxonomic levels where conserved phylogenetic markers fail. The current prokaryotic species concept is still based on the 70% (measured) similarity or 5° Δ Tm criterion (Stackebrandt et al. 2002). Quantitative DNA–DNA hybridization data are still essential components of species description. Consequently, DNA reassociation techniques were also frequently used for strain assignment and species description of acetic acid bacteria (Yukphan et al. 2004; Tanasupawat et al. 2004; Dellaglio et al. 2005; Ishida et al. 2005; Muthukumarasamy et al. 2005; Lisdiyanti et al. 2006; Silva et al. 2006).

3.2.2 Pattern Techniques

A number of approaches generating and visualising DNA fragments are available for DNA based differentiation. In general, two groups of procedures can be distinguished: site specific fragmentation and primer directed PCR (polymerase chain reaction) amplification of purified DNA.

Restriction Fragment Length Polymorphism (RFLP)

RFLP is a culture-independent technique for assessing microbial community diversity using agarose gel electrophoresis. To obtain useful results, one must ensure digestion completeness and the reproducibility of the RFLP banding pattern. In general, PCR-amplified rDNA is digested with specific restriction enzyme. Different lengths are detected using agarose or non-denaturing polyacrylamide gel electrophoresis (PAGE) in the case of community analysis (Tiedje et al. 1999). Digestion of purified genomic DNA using appropriate (mixtures of) restriction endonucleases followed by pulsed field or conventional gel electrophoretic separation of the generated fragments results in usually rather complex patterns after visualisation by DNA staining. Comparative analysis of these patterns by staining followed by software assisted interpretation of the resulting profiles can be used for differentiation of even closely related strains. The power of the method rather concerns differentiation than identification of closely related strains. Different patterns represent different organisms, however, identical patterns do not necessarily indicate the same strain. Nucleic acid fragments sharing the same size might substantially differ with respect to their primary structures (Gonzalez et al. 2005).

Ribotyping

The complexity of RFLP patterns can be substantially reduced by localising certain genes performing Southern hybridization of the fragments. If hybridization probes targeting rDNA are used, the procedure is known as ribotyping (Regnault et al. 1997). Given the ubiquitous occurrence of highly conserved rDNA targets, a set of a few general probes allows to study diverse organisms. The great importance of the approach in the past nowadays has been overcome by modern rapid rDNA sequencing techniques.

Random Amplified Polymorphic DNA (RAPD)

Randomly amplified polymorphic DNA polymerase chain reaction (RAPD-PCR) is a simple and rapid method for identification of useful genetic markers and determination of organismal genetic diversity at various taxonomic levels (Liu et al. 2006). With this technique, PCR is performed with random primers; the amplified products are analyzed with electrophoresis, the gels are stained with ethidum bromide; and the gel images are analyzed with imaging systems. Specific DNA fragment patterns can be generated by PCR using single primer pairs or mixtures of primers targeting multiple sites on the genome. The primary structure of such primers can be designed randomly or target specific. Binding to multiple targets of partial sequence complementarity on the genome is achieved by primer annealing at relaxed hybridization conditions during initial PCR cycles. The resulting random fragments provide the templates in subsequent PCR cycles following primer annealing at stringent hybridisation conditions. Then RAPD bands are scored as binary presence or absence characters, to assemble a matrix of RAPD phenotypes. The percentage of polymorphic bands is utilized to measure genetic diversity. Compared with other molecular markers, advantages of this method are that it is simpler, fast, and genomic abundance. Nevertheless, the short primers result in low repetition. High standardisation of the experimental parameters and equipment is needed to ensure reproducibility of the fragment patterns (Gonzalez et al. 2005; Ludwig 2007).

Amplified Fragment-Length Polymorphism (AFLP)

The AFLP (amplified fragment-length polymorphism) procedure (Vos et al. 1995) combines DNA restriction and PCR amplification. AFLPs are PCR-based markers for the rapid screening of genetic diversity. The restriction fragments are ligated to linkers providing targets for PCR primers. These primers contain additional selective bases at their 3 ends reducing the number of perfectly matched targets in the mixture of linker tailed restriction fragments. Thus, only a subset of the genomic restriction fragments is amplified by PCR. The complexity of the resulting patterns is substantially reduced in comparison with RFLP and RAPD approaches. The key feature of AFLP-PCR is its capacity for simultaneous screening of many different DNA regions distributed randomly throughout the genome. In essence, AFLP methods allow PCR amplification to detect polymorphisms of genomic restriction fragments. AFLP markers have proven useful for assessing genetic differences among individuals, populations, and independently evolving lineages, such as species (Vos et al. 1995). The main disadvantage of AFLP-PCR is the difficulty in identifying homologous markers (alleles), rendering this method less useful for studies that require precise assignment of allelic states, such as heterozygosity analysis. However, because of the rapidity and ease with which reliable, high-resolution marks can be generated, AFLPs are emerging as a powerful addition to the molecular toolkit of ecologists and evolutionary biologists (Mueller and Wolfenbarger 1999).

Automated Ribosomal Intergenic Spacer Analysis (ARISA) and Ribosomal Intergenic Spacer Analysis (RISA)

Automated ribosomal intergenic spacer analysis (ARISA), a commonly used DNA-based community fingerprinting method (Ovreas 2000), is a high-resolution, highly reproducible technique for detecting differences among complex fungal communities (Ranjard et al. 2001). RISA is based on the length polymorphism of the ribosomal intergenic spacer region between the 16S and 23srRNA genes (Borneman and Triplett 1997). The non-coding ribosomal internal spacer region is variable in both size and nucleotide sequence even within closely related strains and the method has been successfully used to characterize, classify, and type strains, as well as to fingerprint simple communities and mixed populations (Ranjard et al. 2000). Ribosomal intergenic spacer analysis (RISA) exploits variability in the length of the internal transcribed spacer regions of rRNA genes to sort samples rapidly into operational taxonomic units (OTUs). Members of different species may share the same ITS fragment size (Ranjard et al. 2001). Although ARISA assays a different taxonomic resolution than species level, it is a consistent measure of community composition. Consequently, differences between two OTU assemblages directly reflect changes in species composition (Green et al. 2004).

Single-Strand Conformational Polymorphism (SSCP)

SSCP, based on separation of PCR-amplified rRNA and rDNA molecules, has been used successfully to analyze the structure and dynamics of microbial communities

(Schwieger and Tebbe 1998). The method is based on the differential intra-molecular folding of single-stranded DNA that is itself dependent upon DNA sequence variations. Thus, DNA secondary structure alters the electrophoresis mobility of the single-stranded PCR amplifications enabling them to be resolved. SSCP has been used to differentiate between pure cultures of soil microorganisms and to distinguish community fingerprints of uncultivated rhizospheric microbial communities from different plants (Schwieger and Tebbe 1998). SSCP analysis should, in principle, be easier to carry out than DGGE or TGGE, as no primers with GC-clamp or specific apparatus for gradient gels are required.

A limitation of the method, in addition to potential PCR bias, however, is that a single bacterial species may yield several bands due to the presence of several operons or more than one conformation of the single-stranded PCR amplifications. Another approach to identify community members is to apply specific enrichments to enhance the growth of the microorganism of interest. This strategy is particularly useful in studies of functional groups or guilds (Lynch et al. 2004).

Amplified DNA Restriction Analysis (ADRA)

The combined application of site specific PCR amplification and restriction fragment polymorphism analysis usually provides profiles of low complexity. In principle, the approach can be used for comparative analysis of any region of genomic DNA carrying primer targets common to the desired target organisms and appropriate for PCR amplification. Whereas the former two techniques do only provide anonymous patterns, the assignment of the resulting fragments to given homologous genome regions is provided by the site specificity of the PCR amplification. Amplified rDNA restriction analysis (ARDRA) represents the most commonly used special format of the methodology. The conserved character of rDNA facilitates the design of general amplification primers, however, limits pattern variety. Consequently, the discriminatory power is limited. In many cases, ADRA targeting the intergenic spacers of rRNA genes (ITS) allows differentiation of closely related strains. This approach currently is popular for strain differentiation of acetic acid bacteria (Tanasupawat et al. 2004; Yukphan et al. 2004, 2006; Kretova and Grones 2005; Malimas et al. 2006). Nevertheless, ARDRA patterns are often overvalued and misinterpreted with respect to their relevance for phylogeny inference. Although the presence or absence of a restriction site may represent a valuable diagnostic feature, a potential phylogenetic relevance cannot be postulated without assignment of the respective organism to a given phylogenetic group. The latter, however, requires either knowledge of at least neighboured sequence stretches or other appropriate data for phylogenetic assignment. Thus, ARDRA patterns *per se* only provide differentiating information. Anyway, the former importance of the technique has been overcome by the modern methods of rapid sequencing rDNA and ITS DNA providing much more information for identification and phylogeny inference.

Denaturing Gradient Gel Electrophoresis (DGGE) and Temperature Gradient Gel Electrophoresis (TGGE)

DGGE and TGGE are two similar methods for studying microbial diversity. Theoretically, DGGE can separate DNA with one base-pair difference (Miller et al. 1999). TGGE uses the same principle as DGGE except the gradient is temperature rather than chemical denaturants. These techniques were originally developed to detect point mutations in DNA sequences. Advantages of DGGE and TGGE include being reliable, reproducible, rapid, and somewhat inexpensive; providing concurrent analysis of multiple samples; and having the ability to follow changes in microbial populations (Muyzer 1999).

DGGE is used to detect polymorphisms of DNA fragments not correlated with fragment size. Small restriction or PCR generated fragments are separated by gel electrophoresis while passing a low to high denaturant or temperature gradient. Fragments differing in sequence composition may exhibit different melting behaviour when progressing into higher denaturing conditions. This can be monitored by mobility shifts. Although this technique is of greater interest in microbial ecology – as a rapid method for estimating complexity – it has been also successfully applied for the identification of pure cultures. In both cases most frequently an appropriate part of usually PCR amplified rDNA is subjected to denaturing electrophoresis (Lopez et al. 2003; DeVero et al. 2006; Ludwig 2007).

3.3 Phylogenetic Analysis

The success of any of the preceding methods for community characterization relies on a suitable phylogenetic analysis because many of the organisms that are likely to be described from soil communities have not been studied previously. A number of phylogenetic methods have been utilized in studies of microbial ecology (Woese 1987). While rDNA and rRNA are commonly used as characters in phylogenetic analysis, the list of characters is extensive and can range from molecular to morphological traits (Olsen and Woese 1993). For microorganisms, molecular data often provide the greatest wealth of information because microorganisms such as bacteria simply do not have the large morphological diversity to make morphological characteristics useful in establishing phylogenies. Aside from the derivation of taxonomies, phylogenetic analyses are important in identifying similarities between organisms, leading to the ability to understand the physiology and ecology of as yet non-culturable species. Unfortunately for taxonomists, phylogenetic analyses have at least one major drawback. The fact that an analysis based on a single type of molecule results in a close relationship between taxa. However the analysis on the same taxa with another equally suitable molecule does not necessarily support the result (Olsen and Woese 1993). When based on a limited set of taxonomic criteria, it is difficult to say with certainty whether or not those criteria can resolve an unknown microorganism from other known microorganisms. Therefore, microbial phylogenies should be interpreted with caution when used in soil microbial community analyses.

3.4 Nucleic Acid Hybridization and Fluorescent In Situ Hybridization (FISH)

Nucleic acid hybridization using specific probes is an important qualitative tool in molecular bacterial ecology (Theron and Cloete 2000). These hybridization techniques can be performed on extracted DNA and RNA, or *in situ* hybridization can be conducted at the cellular level. The FISH method has also been used successfully to study the spatial distribution of bacteria in biofilms (Thurnheer et al. 2004). However, with respect to sensitivity, some limitations to the standard FISH method that prevents detection of cells with low ribosome content have been noted. Low physiological activity was often correlated with low ribosome content per cell, therefore slow-growing or starving cells may not be detected (Amann et al. 1995). To overcome this limitation, FISH has adopted a tyramime signal amplification technique, which allows the analysis of slow-growing microorganisms (Pernthaler et al. 2002). Also, a disadvantage of nucleic acid hybridization or FISH in particular is the lack of sensitivity unless sequences are present in high copy number (Pernthaler et al. 2002).

Fluorescent in situ hybridization (FISH) has been used primarily with prokaryotic communities and allows the direct identification and quantification of specific and/or general taxonomic groups of microorganisms within their natural microhabitat (Amann et al. 1995; Kenzaka et al. 1998). In FISH, whole cells are fixed, their 16S or 23S rRNA is hybridized with fluorescently-labeled taxon-specific oligonucleotide probes, and then the labeled cells are viewed by scanning confocal laser microscopy (SCLM). Because whole cells are hybridized, artifacts arising from biases in DNA extraction, PCR amplification, and cloning are avoided (Ludwig et al. 1997; Felske et al. 1998). FISH has two advantages over immunofluorescence techniques. First, FISH can detect microorganisms across all phylogenetic levels, whereas immunofluorescence techniques are limited to the species and sub-species levels. Second, FISH is more sensitive than immunofluorescence because non-specific binding to soil particles does not typically occur (Amann et al. 1995). FISH probes can be generated without prior isolation of the microorganism, whereas pure cultures are needed in immunofluorescence studies for generating specific antibodies (Hahn et al. 1992). Scanning confocal laser microscopy (SCLM) surpasses epifluorescence microscopy in sensitivity and has the ability to view the distribution of several taxonomic groups simultaneously as a three-dimensional image (Assmus et al. 1995; Kirchhof et al. 1997). Use of distinctive fluorescent dyes and corresponding filter sets allows the observer to differentiate fluorescing microbes from autofluorescent soil particles and plant debris (Assmus et al. 1995; MacNaughton et al. 1996). FISH provides a more accurate quantification of cells as compared to the rough estimates obtained from dot blot assays (Amann et al. 1995) in which microbial DNA is blotted onto a membrane than fluorescent oligonucleotide probe is visualized.

The sensitivity of FISH has been greatly improved to afford the detection of single cells within complex environments such as rhizosphere and bulk soils

(MacNaughton et al. 1996; Zarda et al. 1997; Felske et al. 1998). Strongly fluorescing dyes can be used or multiple probes can be designed to target different regions of the same 16S or 23S rRNA molecule, thus increasing the strength of the signal (Amann et al. 1995; Ludwig et al. 1997). Probes for kingdoms (Eubacteria, Archaea, Eucarya), families, genera, species, or sub-species can be differentially labeled and used in combination to view the occurrence and distribution of several taxonomic groups simultaneously within a single soil sample (Amann et al. 1995; Zarda et al. 1997).

To be detected, soil microbes must be metabolically active and possess cell walls sufficiently permeable to allow penetration of the probe (Christensen and Poulsen 1994; Amann et al. 1995). Penetration of cells with such probes is a problem in nutrient-poor soils and in soils where microorganisms are dormant or quiescent (Hahn et al. 1992; Fischer et al. 1995) because cells are generally smaller and cell walls relatively thicker under these conditions. However, progress is being made to overcome these problems with groups such as actinobacteria and *Bacillus* spores (Fischer et al. 1995). To address the problem of low metabolic activity in soil, some researchers have added nutrients to stimulate microbial activity (Hahn et al. 1992).

However, so as not to bias the community profile, the amendments should equally stimulate all members of the community. FISH can be used to visualize soil microorganisms that have not yet been cultured, and is useful in studying the ecological distribution of microorganisms throughout diverse habitats (Ludwig et al. 1997; Zarda et al. 1997; van Wullings et al. 1998). When using FISH to examine all members within a given taxon, one must keep in mind that the probe being used is only as good as the representative members that were used to generate it (Amann et al. 1995). Other, non-cultured organisms may not be detected with this probe or cross-hybridization to related organisms may occur (Hahn et al. 1992; MacNaughton et al. 1996; Felske et al. 1998).

FISH can be combined with cultivation techniques, immunofluorescence, nucleotide probes targeting structural genes or mRNAs, reporter genes, microsensors, or flow cytometry to gain information regarding the structure and function of microorganisms within a complex microbial community (Amann and Kuhl 1998). FISH is a powerful tool that can be used not only for studying individuals within a population, but also has potential uses for studying population dynamics, tracking microorganisms released into the environment (e.g. for biological control or bioremediation), epidemiology, and microbial ecology of economically important plant pathogens in agricultural soils (Hahn et al. 1992; Kirchhof et al. 1997; van Wullings et al. 1998).

4 Single-Cell Analysis Methods

Isolating and analyzing individual microbial cells from the total population have enabled us to explore the taxonomy, physiology, and activity of microbes at the single-cell level (Brehm-Stecher and Johnson 2004; Zengler 2009). Single cell isolation techniques are the methods to physically separate individual cells from each other and/or from matrix materials (e.g., soil particles) (Zengler 2009).

Single-cell isolation techniques can be used for three major purposes in microbiology and biotechnology: (1) to cultivate previously uncultured microbes; (2) to assess and monitor cell physiology and function; and (3) to screen for novel microbiological products such as enzymes and antibiotics. Modern molecular approaches, which are based primarily on ribosomal RNA gene sequences, have revealed the enormous diversity of the microbial world present in soil, water, and other natural and artificial environments (Keller and Zengler 2004; Christen 2008). However, a majority of microbes are still difficult to cultivate. Several approaches have been applied to culture these previously uncultivated microbes, including media refinements, specific enrichment of target microorganisms, cultivation under simulated natural conditions or low nutrient conditions, *in situ* cultivation, use of syntrophic interactions, and use of single-cell isolation techniques (Alain and Querellou 2009; Kikuchi 2009; Yamada and Sekiguchi 2009; Zengler 2009; Ishii et al. 2010). Single-cell isolation techniques produce an environment without resource competition, thereby allowing microbes, especially those that grow very slowly, to multiply without the interference of fast-growing organisms.

4.1 Dilution-to-Extinction Method

Probably the simplest method to obtain single cells from heterogeneous populations is to serially dilute a sample solution until only single cells remain (Button et al. 1993). Growth is measured after incubating the diluted cultures. This approach is similar to the most probable number technique in which populations of microbes are measured by serially diluting a culture until they become extinct (Haas 1989). By using dilution-to-extinction culturing, single cells of the most abundant microbes, rather than those of the most nutrient-tolerant and fast-growing organisms, can be obtained (Button et al. 1993). In addition, the use of microtiter plates allows relatively high-throughput screening of dilution-to-extinction cultures (Connon and Giovannoni 2002).

4.2 Micromanipulation

A more technical approach to obtain single cells is to use a micromanipulator. There are two types of micromanipulation: mechanical and optical micromanipulation (Brehm-Stecher and Johnson 2004). In mechanical micromanipulation, single cells are individually captured from a heterogeneous population using microcapillary tubes and transferred to a culture medium or other reaction solution (Fröhlich and König 2000). The isolated single cells can be subsequently used for cultivation (Fröhlich and König 2000; Ishoey et al. 2008; Ashida et al. 2010) or culture-independent analysis

(Kvist et al. 2007). Various microorganisms, including potentially novel strains, have been isolated from termite guts (Fröhlich and König 2000), hot spring samples (Ishoey et al. 2008), and wetland rice soils (Ashida et al. 2010) by using mechanical micromanipulators. In addition, endosymbiotic protozoans have also been isolated from the termite gut (Hongoh et al. 2008; Sato et al. 2009; Ishii et al. 2010).

4.3 Flow Cytometry and Cell Sorting

Flow cytometry (FCM) is an approach to quantitatively analyze multiple characteristics of millions of single cells and other particulate matter from a heterogeneous population (Brehm-Stecher and Johnson 2004). The cell sample is mixed with a carrier fluid called sheath fluid, which is forced through an orifice to generate a stream. By controlling the pressure of the system and orifice size, a laminar flow regime is established. This hydrodynamic focusing allows only single cells to pass through an illumination zone where forward and side light scatter, and several fluorescence parameters can be measured (Link et al. 2007; Czechowska et al. 2008). An important extension of FCM for microbiological and biotechnological studies is cell sorting (Link et al. 2007). The technique of sorting fluorescent-labeled cells, which is most frequently used in combination with FCM, is called fluorescence-activated cell sorting (FACS) (Czechowska et al. 2008). Droplet deflection is used for high-throughput cell sorting. In this technique, drops, each containing only a single cell, are formed by intense vibration of the stream. A droplet containing a cell of interest is charged and deflected into a collection tube or microtiter plate. Cells are sorted as a population when collected in a tube; whereas single cells can be obtained when cells are collected in a microtiter plate. The FCM and FACS systems can perform high-throughput single-cell analysis and sorting with a rate of $>10^4$ cells/s. Isolated or sorted cells can be used for subsequent culture-based (Kalyuzhnaya et al. 2008; Wang et al. 2009) or culture-independent analyses (Kalyuzhnaya et al. 2008; Fujii and Hiraishi 2009). Wang et al. (2009) sorted low nucleic acid contents (LNA) bacteria (as a population) by FACS with SYBR green-stained freshwater samples. They applied the dilution-to-extinction culturing method to the sorted populations and obtained a pure culture of LNA bacteria closely related to the *Polynucleobacter* cluster (Wang et al. 2009; Ishii et al. 2010).

4.4 Microfluidics

Several microfluidic devices have been developed to isolate single cells and to sort cells of interest. Microfabricated FACS (μFACS) offers sorting of various particles including *E. coli* cells (Fu et al. 1999). In μFACS, sorting is performed by changing the flow direction after detection of cells of interest (Fu et al. 1999). Hu et al. (2005) developed a microfluidic cell sorter using dielectrophoresis to separate target cells from the background population. In a dielectrophoresis-activated cell sorter (DACS), target cells are labeled with antibody that is subsequently bound to polystyrene beads. The bead-labeled cells are repelled from the electrodes and collected as they flow. This system offers high-throughput screening of cells with a rate of $>10^4$ cells/s (Hu et al. 2005; Bessette et al. 2007). High-throughput and sensitive detection make it possible to detect a rare event (Hu et al. 2005). A similar device was also developed and applied to separate airborne microbes and dust particles (Moon et al. 2009). When magnetic beads, instead of polystyrene beads, are bound to cells, magnetic fields can be used to deflect and sort labeled cells (magnetic-activated cell sorting; MACS) (Adams et al. 2008). A label-free, microfluidic device was recently used to manipulate and sort cells (Kose et al. 2009). This device utilizes a carrier fluid containing magnetic nanoparticles (ferrofluid). When an external magnetic field is produced by the electric current, magnetic nanoparticles are attracted toward the electrode. As a result, non-magnetic particles (e.g., cells) are pushed away from the electrode and collected in a specific location. The critical current frequency causing movement of nonmagnetic particles depends on the particle size, thereby allowing size-dependent cell sorting (Kose et al. 2009).

In addition to cell sorting, microfluidic devices also offer powerful single-cell-based analysis. Ottesen et al. (2006) performed multigene PCR on single cells separated and partitioned by a microfluidic digital PCR system. Sequence analysis of the PCR amplicons retrieved from each chamber allowed the linking of phylogeny to the function of the previously uncultivated microorganisms. For example, using this approach, it was discovered that the previously uncultivated *Treponema*, which resides inside the guts of wood-feeding termites, harbored key metabolic enzymes for CO_2-reductive homoacetogenesis (Ottesen et al. 2006).

Microfluidic devices that allow single-cell isolation and incubation have also been reported. Yamaguchi et al. (2009) developed a microfluidic device to capture, incubate, and release single cells by controlling the flow rate of the device. Liu et al. (2009) developed a unique and potentially powerful device that allows single-cell isolation, incubation, and separation of the clonal population for further analyses using a plug-based microfluidic approach (Boedicker et al. 2008). In this device, single cells from a mixed population are stochastically isolated into plugs. The plugs are incubated to grow microcolonies of the isolated single cells, and the clonal populations are split into subpopulations for further analyses, such as cultivation, cryopreservation, physiological assays, and culture-independent analyses (Liu et al. 2009).

4.5 Compartmentalization of Single Cells

Single cells can be partitioned into small compartments to physically separate them from each other. Compartmentalization can be achieved by creating cell-like structures (Link et al. 2007; Bershtein and Tawfik 2008; Bergquist et al. 2009). The compartments were initially created to screen the desired gene and protein mutants.

In this technique, single genes, components required for in vitro transcription and translation, and enzyme substrates were encapsulated in a water-in-oil emulsion. By testing these compartments, enzymes that catalyze the desired reaction can be detected. This technique was further improved to encapsulate single *E. coli* transformants in a water-in-oil-in-water emulsion so that FCM and FACS can be applied to screen $>10^7$ mutants (Aharoni et al. 2005). Compartmentalization is extremely powerful when the fluorescent product of an enzymatic reaction is diffusible (Bershtein and Tawfik 2008; Link et al. 2007). This approach has the potential to be applied to fields beyond directed evolution; for example, to screen for potentially novel enzymes from metagenome libraries and to isolate unique microbes with a target function (Ishii et al. 2010).

Cell compartmentalization (or encapsulation) was also used to cultivate yet-uncultured microbes (Zengler et al. 2002; Keller and Zengler 2004). Zengler et al. (2002) used gel micro droplets to encapsulate single cells obtained from environmental samples and incubated them under simulated natural conditions to form microcolonies. Microcapsules containing the microcolonies were separated from free-living cells and empty microcapsules, and individually placed in 96-well microtiter plates by FACS for further cultivation. By this approach, novel and diverse bacteria were isolated (Zengler et al. 2002). Detection of unique microbial signatures in these microcapsules may facilitate screening for potentially novel microbes (Bergquist et al. 2009).

5 Combined Use with Other Techniques

The single-cell isolation techniques described above become more useful when combined with other techniques, such as viable staining, direct viable count, antibody staining, *in situ* and *in vivo* hybridization, and those using autofluorescence protein reporters, depending on the purpose of the study. These techniques provide information on the taxonomy, function, activity, and viability of microbes at the single-cell level.

5.1 Viable-Cell Staining

Several fluorescent stains have been used to detect physiological activities in cells (e.g., membrane integrity, enzyme activity, and intracellular pH), and thereby assess cell viability (Joux and Lebaron 2000). Several researchers have combined these fluorescent stains with single-cell isolation and cell-sorting techniques. This combination is especially useful when isolated single cells are subjected to cultivation, since cultivation is only possible if the isolated cells are alive and able to multiply (Huber et al. 2000). Ferrari and Gillings (2009) used the LIVE/DEAD BacLight kit to stain viable microcolonies formed in the soil-substrate membrane system, which was previously shown to promote the growth of yet-uncultivated soil bacteria (Ferrari et al. 2005). The viable microcolonies were then individually isolated using a micromanipulator and transferred to a culture medium (Ferrari and Gillings 2009). This approach allowed them to obtain diverse and novel isolates. Similarly, Ashida et al. (2010) individually isolated viable cells stained with CFDA-AM using a micromanipulator.

Flow cytometry and FACS have also been used to separate viable cells from the total population. For example, Kalyuzhnaya et al. (2008) used RSG to detect actively respiring microbial populations in the presence or absence of C1 compounds (e.g., methane). These actively respiring populations were separated from the total population by FACS and then used as inocula for the enriched growth of methylotrophs. Similarly, Fujii and Hiraishi (2009) sorted CTC-positive cells from composting samples. DGGE of PCR-amplified 16S rRNA gene fragments showed that the FACS-sorted population has a different community structure from the total population. These studies suggest the usefulness of the combination of viable-cell staining and single-cell isolation and cell-sorting techniques.

5.2 Direct Viable Count

Kogure et al. (1979) reported a method to count viable and growing cells directly under a microscope (direct viable count; DVC). In the DVC method, viable cells are elongated by the addition of a growth-promoting substrate (e.g., yeast extract) and cell-division inhibitors (e.g., nalidixic acid, cephalexin). After staining, elongated viable cells can be clearly distinguished from non-elongated cells (Kogure et al. 1979, 1984; Joux and Lebaron 1997).

Although the DVC method has been widely used for the quantification of viable bacteria, isolation of these viable bacteria was not achieved until recently. Ashida et al. (2010) used the mechanical micromanipulator to isolate single cells that were elongated under denitrification inducing conditions. The elongated cells were individually captured and transferred to a medium for growth. Since single cells with a specific function (e.g., denitrification) can be selectively isolated, Ashida et al. (2010) termed this method as functional single-cell (FSC) isolation.

5.3 Antibody Staining

Antibodies can recognize specific groups of microbes by binding to antigens, such as capsular, flagellar, or cell wall antigens (Brehm-Stecher and Johnson 2004). Antibodies can be tagged with fluorophore (fluorescent antibody or immunofluorescence) so that antibody-bound cells can be recognized by their fluorescence. Antibodies can be also tagged with magnetic beads (immunomagnetic beads) or polystyrene beads, allowing magnetic (Pernthaler et al. 2008) or dielectrophoretic (Hu et al. 2005) separation of the antibody-bound cells. Although the application of antibody-based detection of microbes is limited because of its cross-reactivity, it can be used for live cells allowing subsequent culture-based analyses. However, to develop antibodies specific to a group of microorganisms, we need to have a population (dead or alive) of these microbes. This requirement may limit the application of antibody-based analysis to study previously uncultivated microbes.

Fluorescent antibodies have been widely used to specifically detect pathogenic bacteria, including infectious adenoviruses (Li et al. 2010), *Escherichia coli* O157 serotype (Pyle et al. 1999; Shelton and Karns 2001), and *Salmonella* (McClelland and Pinder 1994). When combined with flow cytometry, fluorescent antibodies provide rapid detection and quantification of microbes of interest (Veal et al. 2000; Li et al. 2010). Antibody-bound cells could also be sorted by FACS followed by a culture-based analysis. Antibody-based cell recognition is also often used in microfluidic cell sorting. Cells with fluorophore-tagged, magnetic bead-tagged, or polystyrene bead-tagged antibodies were efficiently separated from background populations by μFACS (Fu et al. 1999), MACS (Adams et al. 2008), or DACS (Hu et al. 2005), respectively.

5.4 In Situ Hybridization and Other Culture-Independent Analysis

In situ hybridization, especially fluorescence in situ hybridization (FISH), has been widely used in combination with single-cell isolation techniques, since FISH can identify the taxonomy and potential function of microbes at the single-cell level (Amann and Fuchs 2008). In the FISH technique, fluorescently labeled nucleic acid probes are hybridized to the target sequences within whole, permeabilized cells (Brehm-Stecher and Johnson 2004). Ribosomal RNA is most frequently used as a target for FISH probes because taxonomic identification is possible based on rRNA sequences. In addition, rRNA is abundant in single cells (10^3–10^5 copies); therefore, strong fluorescent signals can be achieved by rRNA-based FISH analysis. However, recent methodological improvements, including catalyzed reporter deposition-FISH (CARD-FISH; Schönhuber et al. 1997), development of individual genes-FISH (RING-FISH; Zwirglmaier et al. 2004), and use of near-infrared dye (Coleman et al. 2007), enable us to target functional genes with low abundance (Pratscher et al. 2009; Kawakami et al. 2010).

The amount of DNA in sorted or isolated single cells is often too small to perform subsequent culture-independent analyses. Therefore, multiple displacement amplification (MDA) is frequently employed, which can amplify genomic DNA from single cells using Phi29 DNA polymerase (Lasken 2007; Marcy et al. 2007a). Stepanauskas and Sieracki (2007) sorted marine bacterial cells individually into a 96-well plate by FACS, and amplified their genome by MDA. PCR analysis was performed using the MDA product as a template. Since these MDA products originate from single cells, this approach could match the phylogeny and metabolism of the uncultivated organism. The combination of FISH, single-cell isolation, and MDA also allows us to sequence the whole genome of uncultivated microorganisms. Hongoh et al. (2008) successfully sequenced the whole genome of an uncultivated symbiont of the wood-feeding termite. Similarly, flow cytometry (Podar et al. 2007; Rodrigue et al. 2009) and microfluidic devices (Marcy et al. 2007b) have been successfully used to isolate single cells, including uncultivated TM7 microbes, which are subsequently used for MDA and genome sequencing. This field of microbiology is now called "single-cell genomics" (Walker and Parkhill 2008).

Although the FISH process itself kills microbial cells, it can be applied to a subpopulation originating from single cells, while keeping other subpopulations for subsequent culture-based analyses. This approach was used to verify the identity of cells obtained by dilution-to-extinction culturing (Rappé et al. 2002) and microfluidics (Liu et al. 2009). Since many oligotrophic bacteria do not produce colonies on agar plates or turbidity in liquid media (Simu and Hagström 2004), FISH-based, single-cell identification is very useful (Ishii et al. 2010).

5.5 *In Vivo Hybridization*

Silverman and Kool (2005) performed hybridization using quenched auto ligation (QUAL) probes targeting alive, unfixed bacteria. They used a small amount of detergent (0.05% SDS) to soften the bacterial cell wall and introduced the QUAL probes into living cells, thereby allowing "*in vivo* hybridization" (Czechowska et al. 2008). The fluorescence signal of the QUAL probes is exposed only after hybridization and ligation; therefore, washing of unbound probe is not necessary, unlike conventional FISH analysis. The QUAL probes used in the abovementioned study were specifically hybridized with 16S rRNA sequences of *E. coli*, *Salmonella*, and *Pseudomonas*, and the hybridized cells were successfully differentiated using FCM (Silverman and Kool 2005). However, there is a controversy whether the hybridized cells are alive or not, since Amann and Fuchs (2008) reported that the SDS treatment killed the majority of *E. coli* cell suspensions. More studies are necessary to establish *in vivo* hybridization in microbial cells. In vivo hybridization is also possible with other quenched probes, such as molecular beacons (Santangelo et al. 2004), thiazole orange "light-up" probes (Privat et al. 2001), and reduction triggered fluorescence probes (Abe et al. 2008; Franzini and Kool 2009). In addition, the gene expression level in live cells can be examined using *in vivo* hybridization targeting mRNA (Tyagi 2009). However, most of these studies targeted eukaryotic cells and used microinjection to introduce probes.

5.6 *Autofluorescent Proteins*

Autofluorescent proteins (AFPs), such as green fluorescent protein (GFP), have been broadly used to visualize protein expression, localization, and functionality *in vivo* (Link et al. 2007; Southward and Surette 2002). Although applications of AFPs are generally limited to microbes that are transformable, they are extremely useful for studying microbial pathogenesis, biotechnology, and systems biology. The combination of AFPs and FACS can be used to screen genes of pathogenic microbes (e.g., *Salmonella enterica* serovar Typhimurium) that express preferentially in host mammalian cells (Valdivia and Falkow 1997). Their strategy was to use *Salmonella* cells transformed with genomic fragments fused upstream of GFP obtained from a plasmid library. *Salmonella*-infected mammalian cells that showed strong fluorescence because of GFP expression were sorted by FACS. Bacteria were isolated from the sorted mammalian cells and grown *ex vivo* (i.e., outside their host), and then those showing low fluorescence were sorted by FACS. The resulting sorted bacterial population would harbor plasmids that are preferentially expressed when they infect mammalian host cells (Valdivia and Falkow 1997). Similarly, AFPs and FACS have also been used to screen substrate-induced genes in metagenome libraries (Uchiyama et al. 2005). Another interesting application of AFPs and FACS is to study horizontal gene transfer (Sørensen et al. 2003; Musovic et al. 2006). Musovic et al. (2006) sorted indigenous rhizosphere bacteria that acquired GFP-encoding plasmid through horizontal gene transfer from a *Pseudomonas putida* donor. Diverse bacterial populations were sorted, indicating that the plasmid can be transferred to various indigenous bacteria (Musovic et al. 2006).

6 Conclusion

Molecular techniques can enhance our understanding of phytoplankton biodiversity in an environment as vast as the world's oceans and in organisms so tiny that they can only be reliably counted using flow cytometry. We can never set a universal standard judging each method because every molecular biologic technique finds its own (or shared) advantages. Various single-cell isolation techniques have been developed and applied in combination with other techniques for microbiological and biotechnological studies. Each technique has its advantages and disadvantages; therefore, the choice of technique depends on the purpose of the study. These technologies continue to be improved for high-throughput, precision, and low-cost analysis. Phylogenetic diversity can be recovered without dependence on more traditional, often biased, preservation or culturing methods. Molecular techniques can reconstruct the phylogenetic history of a group and can document the spatial and temporal structuring of genetic diversity, *i.e.*, biodiversity below the species level. A variety of molecular tools may need to be invoked in order to find the resolution needed to separates species, populations or individuals. The incorporation of all facets of the biology of the phytoplankton is essential to formulate a multidisciplinary definition of a species and to reconstruct its phylogenetic history.

The modern high throughput sequencing and microarray techniques most probably will allow rapid full genome sequencing of new isolates in the near future.

The perspectives to gain this information from single cells certainly will open new horizons in the study of complex microbial communities. Bioinformatics tools have to be developed for data mining in the enormous amount of data to be expected. Only powerful tools which can be trained (machine learning) will enable the researchers to extract and interpret that parts of information which describe and identify an organism.

References

Abe H, Wang J, Furukawa K, Oki K, Uda M, Tsuneda S, Ito Y (2008) A reduction-triggered fluorescence probe for sensing nucleic acids. Bioconjug Chem 19:1219–1226

Adams JD, Kim U, Soh HT (2008) Multitarget magnetic activated cell sorter. Proc Natl Acad Sci 105:18165–18170

Aharoni A, Amitai G, Bernath K, Magdassi S, Tawfik DS (2005) High-throughput screening of enzyme libraries: thiolactonases evolved by fluorescence-activated sorting of single cells in emulsion compartments. Chem Biol 12:1281–1289

Alain K, Querellou J (2009) Cultivation the uncultured: limits advances and future challenges. Extremophiles 13:583–594

Alvarez MAD, Gagne S, Antoun L (1995) Effect of compost on rhizosphere microflora of the tomato and on the incidence of plant growth-promoting rhizobacteria. Appl Environ Microbiol 61:194–199

Amann R, Fuchs BM (2008) Single-cell identification in microbial communities by improved fluorescence in situ hybridization techniques. Nat Rev Microbiol 6:339–348

Amann R, Kuhl M (1998) In situ methods for assessment of microorganisms and their activities. Curr Opin Microbiol 1:352–358

Amann RI, Ludwig W, Schleifer KH (1995) Phylogenetic identification and *in situ* detection of individual microbial cells without cultivation. Appl Environ Microbiol 59:143–169

Ashida N, Ishii S, Hayano S, Tago K, Tsuji T, Yoshimura Y, Otsuka S, Senoo K (2010) Isolation of functional single cells from environments using a micromanipulator: application to study denitrifying bacteria. Appl Microbiol Biotechnol 85:1211–1217

Assmus B, Hutzler P, Kirchhof G, Amann R, Lawrence JR, Hartmann A (1995) In situ localization of *Azospirillum brasilense* in the rhizosphere of wheat with fluorescently labeled rRNA-targeted oligonucleotide probes and scanning confocal laser microscopy. Appl Environ Microbiol 61:1013–1019

Atlas RM, Bartha R (1998) Microbial ecology: fundamentals and applications. Benjamin/Cummings, Redwood City, 694 pp

Balestra GM, Misaghi IJ (1997) Increasing the efficiency of the plate counting method for estimating bacterial diversity. J Microbiol Methods 30:111–117

Bergquist PL, Hardiman EM, Ferrari BC, Winsley T (2009) Applications of flow cytometry in environmental microbiology and biotechnology. Extremophiles 13:389–401

Bershtein S, Tawfik DS (2008) Advances in laboratory evolution of enzymes. Curr Opin Chem Biol 12:151–158

Bessette PH, Hu X, Soh HT, Daugherty PS (2007) Microfluidic library screening for mapping antibody epitopes. Anal Chem 79:2174–2178

Boedicker JQ, Li L, Kline TR, Ismagilov RF (2008) Detecting bacteria and determining their susceptibility to antibiotics by stochastic confinement in nanoliter droplets using plug-based microfluidics. Lab Chip 8:1265–1272

Borneman J, Becke JO (2007) Identifying microorganisms involved in specific pathogen suppression in soil. Annu Rev Phytopathol 45:153–172

Borneman J, Triplett EW (1997) Molecular microbial diversity in soils from eastern Amazonia: evidence for unusual microorganisms and microbial population shifts associated with deforestation. Appl Environ Microbiol 63:2647–2653

Brehm-Stecher BF, Johnson EA (2004) Single-cell microbiology: tools technologies and applications. Microbiol Mol Biol Rev 68:538–559

Button DK, Schut F, Quang P, Martin R, Robertson BR (1993) Viability and isolation of marine bacteria by dilution culture: theory procedures and initial results. Appl Environ Microbiol 59:881–891

Case RJ, Boucher Y, Dahllof I, Holmstrom C, Doolittle WF, Kjelleberg S (2007) Use of 16S rRNA and *rpoB* genes as molecular markers for microbial ecology studies. Appl Environ Microbiol 73:278–288

Christen R (2008) Global sequencing: a review of current molecular data and new methods available to assess microbial diversity. Microbes Environ 23:253–268

Christensen H, Poulsen LK (1994) Detection of *Pseudomonas* in soil by rRNA targeted in situ hybridization. Soil Biol Biochem 26:1093–1096

Christensen H, Angen Ø, Olsen JE, Bisgaard M (2000) DNA–DNA hybridization determined in micro-wells utilizing covalent attachment of DNA. Int J Syst Evol Microbiol 50:1095–1102

Classen AT, Boyle SI, Haskins KE, Overby ST, Hart SC (2003) Community-level physiological profiles of bacteria and fungi: plate type and incubation temperature influences on contrasting soils. FEMS Microbiol Ecol 44:319–328

Cole JR, Chai B, Farris RJ, Wang Q, Kulam-Syed-Mohideen AS, McGarrell DM, Bandela AM, Cardenas E, Garrity GM, Tiedje JM (2007) The ribosomal database project (RDP-II): introducing myRDP space and quality controlled public data. Nucleic Acids Res 35:D169–D172 (Database issue)

Coleman JR, Culley DE, Chrisler WB, Brockman FJ (2007) mRNA-targeted fluorescent in situ hybridization (FISH) of gram-negative bacteria without template amplification or tyramide signal amplification. J Microbiol Methods 71:246–255

Collado J, Platas G, Paulus B, Bills GF (2007) High-throughput culturing of fungi from plant litter by a dilution-to-extinction technique. FEMS Microbiol Ecol 60(3):521–533

Collins G, Kavanagh S, McHugh S, Connaughton S, Kearney A, Rice O, Carrigg C, Scully C, Bhreathnach N, Mahony T, Madden P, Enright AM, Flaherty VO (2006) Accessing the black box of microbial diversity and ecophysiology: recent advances through polyphasic experiments. J Environ Sci Health Part A 41:897–922

Connon SA, Giovannoni SJ (2002) High-throughput methods for culturing microorganisms in very-low-nutrient media yield diverse new marine isolates. Appl Environ Microbiol 68:3878–3885

Curtis TP, Head IM, Lunn M, Woodcock S, Schloss PD, Sloan WT (2006) What is the extent of prokaryotic diversity? Philos Trans R Soc Lond Ser B 361:2023–2037

Czechowska K, Johnson DR, van der Meer JR (2008) Use of flow cytometric methods for single-cell analysis in environmental microbiology. Curr Opin Microbiol 11:205–212

Davis KER, Joseph SJ, Janssen PH (2005) Effects of growth medium inoculum size and incubation time on culturability and isolation of soil bacteria. Appl Environ Microbiol 71:826–834

De Ley J, De Tijtgart R, Smedt J, Michiels M (1973) Thermal stability of DANN–DANN hybrids within the genus *Agrobacterium*. J Gen Microbiol 78:241–252

Dellaglio F, Cleenwerck I, Felis GE, Engelbeen K, Janssens D, Marzotto M (2005) Description of *Gluconacetobacter swingsii* sp nov and *Gluconacetobacter rhaeticus* sp nov isolated from Italian apple fruit. Int J Syst Evol Microbiol 55:2365–2370

DeVero L, Gala E, Gullo M, Solieri L, Landi S, Giudici P (2006) Application of denaturing gradient gel electrophoresis (DGGE) analysis to evaluate acetic acid bacteria in traditional balsamic vinegar. Food Microbiol 23:809–813

Di Giovanni GD, Watrud LS, Seidler RJ, Widmer F (1999a) Comparison of parental and transgenic alfalfa rhizosphere bacterial communities using Biolog GN metabolic fingerprinting and enterobacterial repetitive intergenic consensus sequence-PCR (ERIC-PCR). Microb Ecol 37:129–139

Di Giovanni GD, Watrud LS, Seidler RJ, Widmer F (1999b) Fingerprinting of mixed bacterial strains and Biolog gramnegative (GN) substrate communities by enterobacterial repetitive intergenic consensus sequence-PCR (ERIC-PCR). Curr Microbiol 38:217–223

Doran JW, Coleman DC, Bezdicek DF, Stewart BA (1994) Defining soil quality for a sustainable environment. Am Soc Agro SSSA Special Publication No. 35. Soil Science Society of America, Madison

Faulwetter JL, Gagnon V, Sundberg C, Chazaren F, Burra MD, Brisson J, Campera AK, Stein OR (2009) Microbial processes influencing performance of treatment wetlands: a review. Ecol Eng 35:987–1004

Felske A, Akkermans ADL, De Vos W (1998) In situ detection of an uncultured predominant *Bacillus* in Dutch grassland soils. Appl Environ Microbiol 64:4588–4590

Ferrari BC, Gillings MR (2009) Cultivation of fastidious bacteria by viability staining and micromanipulation in a soil substrate membrane system. Appl Environ Microbiol 75:3352–3354

Ferrari BC, Binnerup SJ, Gillings M (2005) Microcolony cultivation on a soil substrate membrane system selects for previously uncultured soil bacteria. Appl Environ Microbiol 71:8714–8720

Fierer N, Lennon JT (2011) The generation and maintenance of diversity in microbial communities. Am J Bot 98(3):439–448

Fischer K, Hahn D, Honerlage W, Schonholzer F, Zeyer J (1995) In situ detection of spores and vegetative cells of *Bacillus megaterium* in soil by whole cell hybridization. Syst Appl Microbiol 18:265–273

Fox GE, Pechman KR, Woese CR (1977) Comparative cataloging of 16S ribosomal ribonucleic acid: molecular approach to prokaryotic systematic. Int J Syst Bacteriol 27:44–57

Frac M, Oszust K, Lipiec J (2012) Community level physiological profiles (CLPP) characterization and microbial activity of soil amended with dairy sewage sludge. Sensors 12:3253–3268

Franzini RM, Kool ET (2009) Efficient nucleic acid detection by templated reductive quencher release. J Am Chem Soc 131:16021–16023

Fröhlich J, König H (2000) New techniques for isolation of single prokaryotic cells. FEMS Microbiol Rev 24:567–572

Frostegard A, Tunlid A, Baath E (1993) Phospholipid fatty acid composition biomass and activity of microbial communities from two soil types experimentally exposed to different heavy metals. Appl Environ Microbiol 59:3605–3617

Frostegard A, Tunlid A et al (2010) Use and misuse of PLFA measurements in soils. Soil Biol Biochem. doi:101016/jsoilbio201011021

Fu AY, Spence C, Scherer A, Arnold FH, Quake SR (1999) A microfabricated fluorescence-activated cell sorter. Nat Biotechnol 17:1109–1111

Fujii Y, Hiraishi A (2009) Combined use of cyanoditolyl tetrazolium staining and flow cytometry for detection of metabolically active bacteria in a fed-batch composting process. Microbes Environ 24:57–63

Garbeva P, van Veen JA, van Elsas JD (2004) Microbial diversity in soil: selection of microbial populations by plant and soil type and implications for disease suppressiveness. Annu Rev Phytopathol 42:243–270

Garland JL (1996a) Analytical approaches to the characterization of samples of microbial communities using patterns of potential C utilization. Soil Biol Biochem 28(213)

Garland JL (1996b) Patterns of potential C source utilization by rhizosphere communities. Soil Biol Biochem 28:223–230

Garland J (1997) Analysis and interpretation of community-level physiological profiles in microbial ecology. FEMS Microbiol Ecol 24:289–300

Garland JL, Mills AL (1991) Classification and characterization of heterotrophic microbial communities on the basis of patterns of community-level sole-carbon-source utilization. Appl Environ Microbiol 57:2351–2359

Gonzalez A, Hierro N, PobletMMas A, Guillamon JM (2005) Application of molecular methods to demonstrate species and strain evolution of acetic acid bacteria population during wine production. Int J Food Microbiol 102:295–304

Green JL, Holmes AJ, Westoby M, Oliver L, Briscoe D, Dangerfield M, Gillings M, Beattie AJ (2004) Spatial scaling of microbial eukaryote diversity. Nature 432:747–750
Grimont PAD, Popoff MY, Grimont F, Coynault C, Lemelin M (1980) Reproducibility and correlation study of three deoxyribonucleic acid hybridization procedures. Curr Microbiol 4:325–330
Grosskopf R, Stubner S, Liesack W (1998) Novel euryarchaeotal lineages detected on rice roots and in the anoxic bulk soil of flooded rice microcosms. Appl Environ Microbiol 64:4983–4989
Guckert JB, Antworth CP, Nichols PD, White DC (1985) Phospholipid ester linked fatty acid profiles as reproducible assays for changes in prokaryotic community structure of estuarine sediments. FEMS Microb Ecol 31:147–158
Haack SK, Garchow H, Odelson DA, Forney LJ, Klug MJ (1994) Accuracy reproducibility and interpretation of fatty acid methyl ester profiles of model bacterial communities. Appl Environ Microbiol 60:2483–2493
Haack SK, Garchow H, Klug M, Forney LJ (1995) Analysis of factors affecting the accuracy reproducibility and interpretation of microbial community carbon source utilization patterns. Appl Environ Ecol 61:1458–1468
Haas CN (1989) Estimation of microbial densities from dilution count experiments. Appl Environ Microbiol 44:1934–1942
Hackett CA, Griffiths BS (1997) Statistical analysis of the time course of Biolog substrate utilization. J Microbiol Methods 30:63–69
Hahn D, Amann RI, Ludwig W, Akkermans ADL, Schleifer KH (1992) Detection of microorganisms in soil after in situ hybridization with rRNA-targeted fluorescently labeled oligonucleotides. J Gen Microbiol 138:879–887
Haubert D, Haggblom MM, Langel R, Scheu S, Ruess L (2006) Trophic shift of stable isotopes and fatty acids in Collembola on bacterial diets. Soil Biol Biochem 38:2004–2007
Heuer H, Krsek M, Baker P, Smalla K, Wellington EMH (1997) Analysis of actinomycete communities by specific amplification of genes encoding 16S rRNA and gel-electrophoretic separation in denaturing gradients. Appl Environ Microbiol 63:3233–3241
Hill GT, Mitkowskia NA, Aldrich-Wolfe L, Emelea LR, Jurkonie DD, Ficke A, Maldonado-Ramireza S, Lyncha ST, Nelson EB (2000) Methods for assessing the composition and diversity of soil microbial communities. Appl Soil Ecol 15:25–36
Hitzl W, Henrich M, Kessel M, Insam H (1997) Applications of multivariate analysis of variance and related techniques in soil studies with substrate utilization tests. J Microbiol Methods 30:81–89
Hongoh Y, Sharma VK, Prakash T, Noda S, Taylor TD, Kudo T, Sakaki Y, Toyoda A, Hattori M, Ohkuma M (2008) Complete genome of the uncultured Termite Group 1 bacteria in a single host protest cell. Proc Natl Acad Sci USA 105:5555–5560
Hu S, van Bruggen AHC (1997) Microbial dynamics associated with multiphasic decomposition of ^{14}C-labeled cellulose in soil. Microb Ecol 33:134–143
Hu X, Bessette PH, Qian J, Meinhart CD, Daugherty PS, Soh HT (2005) Marker-specific sorting of rare cells using dielectrophoresis. Proc Natl Acad Sci USA 102:15757–15761
Huber R, Huber H, Stettera KO (2000) Towards the ecology of hyperthermophiles: biotopes new isolation strategies and novel metabolic properties. FEMS Microbiol Rev 24:615–623
Huss VAR, Festl H, Schleifer KH (1983) Studies on the spectrophotometric determination of DNA hybridization from renaturation rates. Syst Appl Microbiol 4:184–192
Ishida T, Yokota A, Umezawa Y, Toda T, Yamada K (2005) Identification and characterization of lactococcal and Acetobacter strains isolated from traditional Caucasusian fermented milk. J Nutr Sci Vitaminol 51:187–193
Ishii S, Tago K, Senoo K (2010) Single-cell analysis and isolation for microbiology and biotechnology: methods and applications. Appl Microbiol Biotechnol 86:1281–1292
Ishoey T, Woyke T, Stepanauskas R, Novotny M, Lasken RS (2008) Genomic sequencing of single microbial cells from environmental samples. Curr Opin Microbiol 11:198–204
Joux F, Lebaron P (1997) Ecological implications of an improved direct viable count method for aquatic bacteria. Appl Environ Microbiol 63:3643–3647

Joux F, Lebaron P (2000) Use of fluorescent probes to assess physiological functions of bacteria at single-cell level. Microbes Infect 2:1523–1535

Juan Z, Weihua G, Renqing W, Xuemei H, Qiang W (2008) Microbial community diversity in the profile of an agricultural soil in northern China. J Environ Sci 20:981–988

Kalyuzhnaya MG, Lidstrom ME, Chistoserdova L (2008) Real-time detection of actively metabolizing microbes by redox sensing as applied to methylotroph populations in Lake Washington. ISME J 2:696–706

Kawakami S, Kubota K, Imachi H, Yamaguchi T, Harada H, Ohashi A (2010) Detection of single copy genes by two-pass tyramide signal amplification fluorescence in situ hybridization (Two-Pass TSA-FISH) with single oligonucleotide probes. Microbes Environ 25:15–21

Keller M, Zengler K (2004) Tapping into microbial diversity. Nat Rev Microbiol 2:141–150

Kenzaka T, Yamaguchi N, Tani K, Nasu M (1998) rRNA-Targeted fluorescent in situ hybridization analysis of bacterial community structure in river water. Microbiology 144:2085–2093

Kikuchi Y (2009) Endosymbiotic bacteria in insects: their diversity and culturability. Microbes Environ 24:195–204

Kirchhof G, Schloter M, Assmus B, Hartmann A (1997) Molecular microbial ecology approaches applied to diazotrophs associated with non-legumes. Soil Biol Biochem 29:853–862

Kirk JL, Beaudette LA, Hart M, Moutoglis P, Klironomos JN, Lee H, Trevors JT (2004) Methods of studying soil microbial diversity. J Microbiol Methods 58:169–188

Klingler M, Koletzko B (2012) Novel methodologies for assessing omega-3 fatty acid status – a systematic review. Br J Nutr 107:S53–S63

Kogure K, Shimidu U, Taga N (1979) A tentative direct microscopic method for counting living marine bacteria. Can J Microbiol 25:415–420

Kogure K, Shimidu U, Taga N (1984) An improved direct viable count method for aquatic bacteria. Arch Hydrobiol 102:117–122

Konopka A, Oliver L, Turco RF Jr (1998) The use of carbon substrate utilization patterns in environmental and ecological microbiology. Microb Ecol 35:103–115

Kose AR, Fischer B, Mao L, Koser H (2009) Label-free cellular manipulation and sorting via biocompatible ferrofluids. Proc Natl Acad Sci USA 106:21478–21483

Kretova M, Grones J (2005) Molecular analysis of 16S–23S spacer regions of *Acetobacter* species. Folia Microbiol 50:288–292

Kvist T, Ahring BK, Lasken RS, Westerman P (2007) Specific single-cell isolation and genomic amplification of uncultured microorganisms. Appl Microbiol Biotechnol 74:926–935

Larsen MKS, Thomsen TR, Moser C, Høiby N, Nielsen PH (2008) Use of cultivation-dependent and -independent techniques to assess contamination of central venous catheters: a pilot study. BMC Clin Pathol 8:10

Lasken RS (2007) Single-cell genomic sequencing using multiple displacement amplification. Curr Opin Microbiol 10:510–516

Lehman RM, Colwell FS, Ringelberg DB, White DC (1995) Combined microbial community-level analyses for quality assurance of terrestrial subsurface cores. J Microbiol Methods 22:263–281

Lengler JW, Drews G, Schlegel HG (eds) (2009) Prokaryotic diversity and systematics. In: Biology of the Prokaryotes. Blackwell Science Ltd, Oxford. doi: 10.1002/9781444313314.ch29

Li YL, Peacock AD, White DC, Geyer R, Zhang CLL (2007) Spatial patterns of bacterial signature biomarkers in marine sediments of the Gulf of Mexico. Chem Geol 238:168–179

Li D, He M, Jiang SC (2010) Detection of infectious adenoviruses in environmental waters by fluorescence-activated cell sorting assay. Appl Environ Microbiol 76:1442–1448

Liesack W, Janssen PH, Rainey FA, Ward-Rainey NL, Stackenbrandt E (1997) Microbial diversity in soil: the need for a combined approach using molecular and cultivation techniques. In: van Elsas JD, Trevors JT, Wellington EMH (eds) Modern soil microbiology. Marcel Dekker, New York, pp 375–439

Link AJ, Jeong KJ, Georgiou G (2007) Beyond toothpicks: new methods for isolating mutant bacteria. Nat Rev Microbiol 5:680–688

Lisdiyanti P, Navarro RR, Uchimura T, Komagata K (2006) Reclassification of *Gluconacetobacter hansenii* strains and proposals of *Gluconacetobacter saccharivorans* sp nov and *Gluconacetobacter nataicola* sp nov. Int J Syst Evol Microbiol 56:2101–2111

Little AEF, Robinson CJ, Peterson SB, Raffa KF, Handelsman J (2008) Rules of engagement: interspecies interactions that regulate microbial communities. Annu Rev Microbiol 62: 375–401

Liu BR, Jia GM, Chen J, Wang G (2006) A review of methods for studying microbial diversity in soils. Pedosphere 16(1):18–24

Liu W, Kim HJ, Lucchetta EM, Du W, Ismagilov RF (2009) Isolation incubation and parallel functional testing and identification by FISH of rare microbial single-copy cells from multi-species mixtures using the combination of chemistrode and stochastic confinement. Lab Chip 9:2153–2162

Lopez I, Ruiz-Larrea F, Cocolin L, Orr E, Phister T, Marshall M, VanderGheynst J, Mills DA (2003) Design and evaluation of PCR primers for analysis of bacterial populations in wine by denaturing gradient gel electrophoresis. Appl Environ Microbiol 69:6801–6807

Ludwig W (2007) Nucleic acid techniques in bacterial systematics and identification. Int J Food Microbiol 120:225–236

Ludwig W, Bauer SH, Bauer M, Held I, Kirchhof G, Schulze R, Huber I, Spring S, Hartmann A, Schleifer KH (1997) Detection and in situ identification of representatives of a widely distributed new bacterial phylum. FEMS Microbiol Lett 153:181–190

Lynch JM, Benedetti A, Insam HM, Nuti P, Smalla K, Torsvik V, Nannipieri P (2004) Microbial diversity in soil: ecological theories the contribution of molecular techniques and the impact of transgenic plants and transgenic microorganisms. Biol Fertil Soils 40:363–385

Macalady JL, Fuller ME, Scow KM (1998) Effects of metan sodium fumigation on soil microbial activity and community structure. J Environ Qual 27:53–63

MacNaughton SJ, Booth T, Embley TM, O'Donnell AG (1996) Physical stabilization and confocal microscopy of bacteria on roots using 16S rRNA targeted fluorescent-labeled oligonucleotide probes. J Microbiol Methods 26:279–285

Malimas T, Yukphan P, Takahashi M, Potacharoen W, Tanasupawat S, Nakagawa Y, Tanticharoen M, Yamada Y (2006) Heterogeneity of strains assigned to Gluconobacter frateurii Mason and Claus 1989 based on restriction analysis of 16S–23S rDNA internal transcribed spacer regions. Biosci Biotechnol Biochem 70:684–690

Maloney PE, van Bruggen AHC, Hu S (1997) Bacterial community structure in relation to the carbon environments in lettuce and tomato rhizospheres and in bulk soil. Microb Ecol 34:109–117

Marcy Y, Ishoey T, Lasken RS, Stockwell TB, Walenz BP, Halpern AL, Beeson KY, Goldberg SMD, Quake SR (2007a) Nanoliter reactors improve multiple displacement amplification of genomes from single cells. PLoS Genet 3:e155

Marcy Y, Ouverney C, Bik EM, Losekann T, Ivanova N, Martin HG, Szeto E, Platt D, Hugenholtz P, Relman DA, Quake SR (2007b) Dissecting biological "dark matter" with single-cell genetic analysis of rare and uncultivated TM7 microbes from the human mouth. Proc Natl Acad Sci USA 104:11889–11894

Mathew RP, Feng Y, Githinji L, Ankumah R, Balkcom KS (2012) Impact of no-tillage and conventional tillage systems on soil microbial communities. Appl Environ Soil Sci. doi:101155/2012/548620

McClelland RG, Pinder AC (1994) Detection of *Salmonella typhimurium* in dairy products with flow cytometry and monoclonal antibodies. Appl Environ Microbiol 60:4255–4262

McCutcheon JP, McDonald BR, Moran NA (2009) Origin of an alternative genetic code in the extremely small and GC–rich genome of a bacterial symbiont. PLoS Genet 5(7):e1000565. doi:101371/journalpgen1000565

Medlin LK, Kooistra WHCF (2010) Methods to estimate the diversity in the marine photosynthetic protist community with illustrations from case studies: a review. Diversity 2:973–1014

Mehlen A, Goeldner M, Ried S, Stindl S, Ludwig W, Schleifer KH (2004) Development of a fast DNA–DNA hybridization method based on melting profiles in microplates. Syst Appl Microbiol 27:689–695

Miller KM, Ming TJ, Schulze AD, Withler RE (1999) Denaturing gradient gel electrophoresis (DGGE): a rapid and sensitive technique to screen nucleotide sequence variation in populations. Biotechniques 27:1016–1030

Mitsui H, Gorlach K, Lee H-J Hattori R, Hattori T (1997) Incubation time and media requirements of culturable bacteria from different phylogenetic groups. J Microbiol Methods 30:103–110

Moon HS, Nam YW, Park JC, Jung HI (2009) Dielectrophoretic separation of airborne microbes and dust particles using a microfluidic channel for real-time bioaerosol monitoring. Environ Sci Technol 43:5857–5863

Morgan JAW, Winstanley C (1997) Microbial biomarkers. In: van Elsas JD, Trevors JT, Wellington EMH (eds) Modern soil microbiology. Marcel Dekker, New York, pp 331–352

Mueller V, Wolfenbarger V (1999) AFLP genotyping and fingerprinting. Trends Ecol Evol 14:389–394

Musovic S, Oregaard G, Kroer N, Sørensen SJ (2006) Cultivation independent examination of horizontal transfer and host range of an incP-1 plasmid among gram-positive and gram-negative bacteria indigenous to the barley rhizosphere. Appl Environ Microbiol 72:6687–6692

Muthukumarasamy R, Cleenwerck I, Revathi G, Vadivelu M, Janssens D, Hoste B, Gum KU, Park KD, Son CY, Sa T, Caballero-Mellado JC (2005) Natural association of *Gluconacetobacter diazotrophicus* and diazotrophic *Acetobacter peroxydans* with wetland rice. Syst Appl Microbiol 28:277–286

Muyzer G (1999) DGGE/TGGE a method for identifying genes from natural ecosystems. Curr Opin Microbiol 2:317–322

Muyzer G, Smalla K (1998) Application of denaturing gradient gel electrophoresis (DGGE) and temperature gradient gel electrophoresis (TGGE) in microbial ecology. Antonie Van Leeuwenhoek 73:127–141

Nakamura A, Tun CC, Asakawa S, Kimura M (2003) Microbial community responsible for the decomposition of rice straw in a paddy field: estimation by phospholipid fatty acid analysis. Biol Fertil Soils 38:288–295

Nannipieri P, Ascher J, Ceccherini MT, Landi L, Pietramellara G, Renella G (2003) Microbial diversity and soil functions. Eur J Soil Sci 54:655–670

Olsen GJ, Woese CR (1993) Ribosomal RNA: a key to phylogeny. FASEB J 7:113–123

Ottesen EA, Hong JW, Quake SR, Leadbetter JR (2006) Microfluidic digital PCR enables multigene analysis of individual environmental bacteria. Science 314:1464–1467

Ovreas L (2000) Population and community level approaches for analyzing microbial diversity in natural environments. Ecol Lett 3:236–251

Ovreas L, Torsvik V (1998) Microbial diversity and community structure in two different agricultural soil communities. Microb Ecol 36:303–315

Pernthaler A, Pernthaler J, Amann R (2002) Fluorescence in situ hybridization and catalyzed reporter deposition for the identification of marine bacteria. Appl Environ Microbiol 68:3094–3101

Pernthaler A, Dekas AE, Brown CT, Goffredi SK, Embaye T, Orphan VJ (2008) Diverse syntrophic partnerships from deep-sea methane vents revealed by direct cell capture and metagenomics. Proc Natl Acad Sci USA 105:7052–7057

Petersen SO, Debosz K, Schjonning P, Christensen BT, Elmholt S (1998) Phospholipid fatty acid profiles and C availability in wet-stable macro-aggregates from conventionally and organically farmed soils. Geoderma 78:181–196

Pfiffner SM, Cantu JM, Smithgall A, Peacock AD, White DC, Moser DP, Onstott TC, van Heerden E (2006) Deep subsurface microbial biomass and community structure in Witwatersrand Basin mines. Geomicrobiol J 23:431–442

Podar M, Abulencia CB, Walcher M, Hutchison D, Zengler K, Garcia JA, Holland T, Cotton D, Hauser L, Keller M (2007) Targeted access to the genomes of low-abundance organisms in complex microbial communities. Appl Environ Microbiol 73:3205–3214

Porteous LA, Seidler RJ, Watrud LS (1997) An improved method for purifying DNA from soil for polymerase chain reaction amplification and molecular ecology applications. Mol Ecol 6:787–791

Pratscher J, Stichternoth C, Fichtl K, Schleifer KH, Braker G (2009) Application of recognition of individual genes-fluorescence in situ hybridization (RING-FISH) to detect nitrite reductase genes (nirK) of denitrifiers in pure cultures and environmental samples. Appl Environ Microbiol 75:802–810

Preston-Mafham J, Boddy L, Randerson P (2002) Analysis of microbial community functional diversity using sole-carbon-source utilization profiles – a critique. FEMS Microbiol Ecol 42(1)

Privat E, Melvin T, Asseline U, Vigny P (2001) Oligonucleotide conjugated thiazole orange probes as "light-up" probes for messenger ribonucleic acid molecules in living cells. Photochem Photobiol 74:532–541

Pyle BH, Broadaway SC, McFeters GA (1999) Sensitive detection of *Escherichia coli* O157:H7 in food and water by immunomagnetic separation and solid-phase laser cytometry. Appl Environ Microbiol 65:1966–1972

Ranjard L, Brothier E, Nazaret S (2000) Sequencing bands of RISA fingerprints for the characterization and the microscale distribution of soil bacterial populations responding to mercury spiking. Appl Environ Microbiol 66:5334–5339

Ranjard L, Poly F, Lata J-C Mougel C, Thioulouse J, Nazaret S (2001) Characterization of bacterial and fungal soil communities by automated ribosomal intergenic spacer analysis fingerprints: biological and methodological variability. Appl Environ Microbiol 67:4479–4487

Rappé MS, Connon SA, Vergin KL, Giovannoni SJ (2002) Cultivation of the ubiquitous SAR11 marine bacterioplankton clade. Nature 418:630–633

Regnault B, Grimont F, Grimont PAD (1997) Universal ribotyping method using a chemically labelled oligonucleotide probe mixture. Res Microbiol 148:649–659

Robertson BJW, Bunn AL, Bailey VL (2011) Phospholipid fatty acid biomarkers in a freshwater periphyton community exposed to uranium: discovery by non-linear statistical learning. J Environ Radioact 102(1):64–71

Rodrigue S, Malmstron RR, Berlin AM, Birren BW, Henn MR, Chisholm SW (2009) Whole genome amplification and de novo assembly of single bacterial cells. PLoS One 4:e6864

Ros M, Goberna M, Pascual JA, Klammer S, Insam H (2008) 16S rDNA analysis reveals low microbial diversity in community level physiological profile assays. J Microbiol Methods 72(3):221–226

Saleh-Lakha S, Miller M, Campbell RG, Schneider K, Elahimanesh P, Harta MM, Trevors JT (2005) Microbial gene expression in soil: methods applications and challenges. J Microbiol Methods 63:1–19

Santangelo PJ, Nix B, Tsourkas A, Bao G (2004) Dual FRET molecular beacons form RNA detection in living cells. Nucleic Acids Res 32:e57

Sato T, Hongoh Y, Noda S, Hattori S, Ui S, Ohkuma M (2009) *Candidatus desulfovibrio trichonymphae* a novel intracellular symbiont of the flagellate *Trichonympha agilis* in termite gut. Environ Microbiol 11:1007–1015

Schönhuber W, Fuchs B, Juretschko S, Amann R (1997) Improved sensitivity of whole-cell hybridization by the combination of horseradish peroxidase-labeled oligonucleotides and tyramide signal amplification. Appl Environ Microbiol 63:3268–3273

Schryver JC, Brandt CC, Pfiffner SM, Palumbo AV, Peacock AD, White DC, McKinley JP, Long PE (2006) Application of nonlinear analysis methods for identifying relationships between microbial community structure and groundwater geochemistry. Microb Ecol 51:177–188

Schutter M, Dick R (2001) Shifts in substrate utilization potential and structure of soil microbial communities in response to carbon substrates. Soil Biol Biochem 33:1481–1491

Schwieger F, Tebbe CC (1998) A new approach to utilize PCR single-strand-conformation polymorphism for 16s rRNA gene-based microbial community analysis. Appl Environ Microbiol 64:4870–4876

Shelton DR, Karns JS (2001) Quantitative detection of *Escherichia coli* O157 in surface waters by using immunomagnetic electrochemiluminescence. Appl Environ Microbiol 67:2908–2915

Silva LR, Cleenwerck I, Rivas R, Swings J, Trujillo ME, Willems A, Velazquez E (2006) *Acetobacter oeni* sp nov isolated from spoiled red wine. Int J Syst Evol Microbiol 56:21–24

Silverman AP, Kool ET (2005) Quenched autoligation probes allow discrimination of live bacterial species by single nucleotide differences in rRNA. Nucleic Acids Res 33:4978–4986

Simu K, Hagström A (2004) Oligotrophic bacterioplankton with a novel single-cell life strategy. Appl Environ Microbiol 70:2445–2451

Soderber KH, Olsson PA, Bafiafith E (2002) Structure and activity of the bacterial community in the rhizosphere of different plant species and the effect of arbuscular mycorrhizal colonization. FEMS Microbiol Ecol 40:223–231

Sørensen SJ, Sørensen AH, Hansen LH, Oregaard G, Veal D (2003) Direct detection and quantification of horizontal gene transfer by using flow cytometry and gfp as a reporter gene. Curr Microbiol 47:0129–0133

Southward CM, Surette MG (2002) The dynamic microbe: green fluorescent protein brings bacteria to light. Mol Microbiol 45:1191–1196

Stackebrandt E, Frederiksen W, Garrity GM, Grimont PAD, Kämpfer P, Maiden MCJ, Nesme X, Rossello-Mora R, Swings J, Trüper HG, Vauterin L, Ward AC, Whitman WB (2002) Report of the adhoc committee for the re-evaluation of the species definition in bacteriology. Int J Syst Evol Microbiol 52:1043–1047

Stepanauskas R, Sieracki ME (2007) Matching phylogeny and metabolism in the uncultured marine bacteria one cell at a time. Proc Natl Acad Sci USA 104:9052–9057

Tamaoka J (1994) Determination of DNA base composition. In: Goodfellow M, O'Donnell AG (eds) Chemical methods in prokaryotic systematics. Wiley, Chichester, pp 463–469

Tanasupawat S, Thawai C, Yukphan P, Moonmangmee D, Itoh T, Adachi O, Yamada Y (2004) *Gluconobacter thailandicus* sp nov an acetic acid bacterium in the alpha-Proteobacteria. J Gen Appl Microbiol 50:159–167

Theron J, Cloete TE (2000) Molecular techniques for determining microbial diversity and community structure in natural environments. Crit Rev Microbiol 26:37–57

Thurnheer T, Gmur R, Guggenheim B (2004) Multiplex FISH analysis *of* a six-species bacterial biofilm. J Microbiol Methods 56:37–47

Tiedje JM, Asuming-Brempong S, Nusslein K, Marsh TL, Flynn SJ (1999) Opening the black box *of* soil microbial diversity. Appl Soil Ecol 13:109–122

Torsvik V, Sorheim R, Goksoyr J (1996) Total bacterial diversity in soil and sediment communities — a review. J Ind Microbiol 17:170–178

Tyagi S (2009) Imaging intracellular RNA distribution and dynamics in living cells. Nat Methods 6:331–338

Uchiyama T, Abe T, Ikemura T, Watanabe K (2005) Substrate-induced gene-expression screening of environmental metagenome libraries for isolation of catabolic genes. Nat Biotechnol 23:88–93

Valdivia RH, Falkow S (1997) Fluorescence-based isolation of bacterial genes expressed within host cells. Science 277:2007–2011

van Elsas JD, Duarte GE, Rosado AS, Smalla K (1998) Microbiological and molecular methods for monitoring microbial inoculants and their effects in the environment. J Microbiol Methods 32:133–154

Veal DA, Deere D, Ferrari B, Piper J, Attfield PV (2000) Fluorescence staining and flow cytometry for monitoring microbial cells. J Immunol Methods 243:191–210

Vos P, Hogers R, Bleeker M, Reijans M, van de Lee T, Hornes M, Friters A, Pot J, Paleman J, Kuiper M, Zabeau M (1995) AFLP: a new technique for DNA fingerprinting. Nucleic Acids Res 23:4407–4414

Walker A, Parkhill J (2008) Single-cell genomics. Nat Rev Microbiol 6:176–177

Wang Y, Hammes F, Boon N, Chami M, Egli T (2009) Isolation and characterization of low nucleic acid (LNA)-content bacteria. ISME J 3:889–902

Ward DM, Weller R, Bateson MM (1990) 16S rRNA sequences reveal numerous uncultured microorganisms in a natural community. Nature 345:63–65

Webster G, Watt LC, Rinna J, Fry JC, Evershed RP, Parkes RJ, Weightman AJ (2006) A comparison of stable-isotope probing of DNA and phospholipid fatty acids to study prokaryotic functional diversity in sulfate-reducing marine sediment enrichment slurries. Environ Microbiol 8:1575–1589

Widmer F, FlieMbach A, Laczko E, Schulze-Aurich J, Zeyer J (2001) Assessing soil biological characteristics: a comparison of bulk soil community DNA- PFLA- and Biolog-analyses. Soil Biol Biochem 33:1029–1036

Woese CR (1987) Bacterial evolution. Microbiol Rev 51:221–271

Wullings BA, van Beuningen AR, Janse JD, Akkermanns ADL (1998) Detection of *Ralstonia solanacearum* which causes brown rot of potato by fluorescent in situ hybridization with 23S rRNA-targeted probes. Appl Environ Microbiol 64:4546–4554

Yamada T, Sekiguchi Y (2009) Cultivation of uncultured Chloroflexi subphyla: significance and ecophysiology of formerly uncultured Chloroflexi 'Subphylum I' with natural and biotechnological relevance. Microbes Environ 24:205–216

Yamaguchi Y, Arakawa T, Takeda N, Edagawa Y, Shoji S (2009) Development of a poly-dimethylsiloxane microfluidic device for single cell isolation and incubation. Sens Actuators B Chem 136:555–561

Yukphan P, Potacharoen W, Nakagawa Y, Tanticharoen M, Yamada Y (2004) Identification of strains assigned to the genus *Gluconobacter Asai* 1935 based on the sequence and the restriction analyses of the 16S–23S rDNA internal transcribed spacer regions. J Gen Appl Microbiol 50:9–15

Yukphan P, Malimas T, Potacharoen W, Tanasupawat S, Tanticharoen M, Yamada Y (2006) Identification of strains assigned to the genus Asaia Yamada et al. 2000 based on restriction analysis of 16S–23S rDNA internal transcribed spacer regions. J Gen Appl Microbiol 52:55–62

Zarda B, Hahn D, Chatzinotas A, Schonhuber W, Neef J, Amann RI, Zeyer J (1997) Analysis of bacterial community structure in bulk soil by in situ hybridization. Arch Microbiol 168:185–192

Zelles L (1997) Phospholipid fatty acid profiles in selected members of soil microbial communities. Chemosphere 35:275–294

Zelles L, Bai QY (1993) Fractionation of fatty acids derived from soil lipids by solid phase extraction and their quantitative analysis by GC-MS. Soil Biol Biochem 25:495–507

Zelles L, Rackwitz R, Bai QY, Beck T, Beese F (1995) Discrimination of microbial diversity by fatty acid profiles of phospholipids and lipopolysaccharides in differently cultivated soils. Plant Soil 170:115–122

Zengler K (2009) Central role of the cell in microbial ecology. Microbiol Mol Biol Rev 73:712–729

Zengler K, Toledo G, Rappé M, Elkins J, Mathur EJ, Short JM, Keller M (2002) Cultivating the uncultured. Proc Natl Acad Sci USA 99:15681–15686

Ziemke F, Hofle MG, Laculat J, Rossello-Mora R (1998) Reclassification of *Shewanella putrefaciens* Owen's genomic group II as *Shewanella baltica* sp nov. Int J Syst Bacteriol 48:179–186

Zwirglmaier K, Ludwig W, Schleifer KH (2004) Recognition of individual genes in a single bacterial cell by fluorescence in situ hybridization – RING-FISH. Mol Microbiol 51:89–96

Chapter 3
Microbial Resource Centers Towards Harnessing Microbial Diversity for Human Welfare

Showkat Ahmad Lone, Abdul Malik, and Jasdeep Chatrath Padaria

Abstract Microbes are known to play an important role in numerous metabolic processes like nutrient cycling, environmental detoxification, production of antibiotics, vitamins, industrial enzymes etc. Therefore it is important to efficiently harness and utilize the biologically important properties of microbes and their products to tackle the ever growing challenges of food security, healthcare and environmental pollution. A complete knowledge of these microbes with respect to the role played by them in ecosystem function is essential to fully exploit them for the benefit of mankind. Unfortunately only 4–5% of the microbes have been explored so far whereas the rest ≈95% is are still un-culturable. A number of uncertainties still exist with respect to the microbial diversity as knowledge regarding the number of species of microorganisms that exist, their distribution, stability in the environment and the important roles played by them are lacking to a greater extent. Microbial diversity is an unseen global resource that deserves to be conserved and utilized judiciously. Microbial resource centers play an important role in this regard as they act as living libraries holding microorganisms. The primary function of these centers is to collect,

S.A. Lone
Biotechnology and Climate Change, National Research Centre on Plant Biotechnology, New Delhi 110012, India

Department of Agricultural Microbiology, Faculty of Agricultural Sciences, Aligarh Muslim University, Aligarh, Uttar Pradesh 202002, India
e-mail: showkat.microbio@gmail.com

A. Malik
Department of Agricultural Microbiology, Faculty of Agricultural Sciences, Aligarh Muslim University, Aligarh, Uttar Pradesh 202002, India
e-mail: ab_malik30@yahoo.com

J.C. Padaria (✉)
Biotechnology and Climate Change, National Research Centre on Plant Biotechnology, New Delhi 110012, India
e-mail: jasdeep_kaur64@yahoo.co.in

A. Malik et al. (eds.), *Management of Microbial Resources in the Environment*,
DOI 10.1007/978-94-007-5931-2_3,

maintain and distribute microbial strains and/or their products to researchers and industrialists all over the world. The role of microbial Culture Collections with respect to conservation and propagation of microbial resources and the difficulties and uncertainties of conservation faced are discussed.

Keywords Microbial resource centers (MRCs) • World Data Centre for Microorganisms (WDCM) • World Federation for Culture Collections (WFCC)

3.1 Introduction

Microorganisms–Bacteria, Viruses, Viroids, Filamentous Fungi, Yeast, Microalgae and Protozoa comprise the group containing the highest number of organisms on the Earth (Colwell 1997). They are ubiquitous in distribution, occurring in a wide range of environments such as hydrothermal vents (deep sea smokers) (Kato et al. 2009), hot springs (Kumar et al. 2004), acid mine drainages and rivers (Amils et al. 2007), in gypsum halite crusts and in NaCl crystals (Mancinelli et al. 2004), in polar regions (Deming 2002) and even in nuclear reactors (Rothschild and Mancinelli 2001). Microorganisms are indispensable as they perform numerous functions to support life on this planet. The increasing world population has put enormous pressure on our natural resources which are depleting at an alarming rate. Microbial Biotechnology and the management of natural processes have a vital role to play in protecting these depleting resources. Microorganisms also play a key role in addressing big challenges in health care (Production of new antibiotics and vaccines), nutritional security (Single cell proteins and production of dairy products), and climate change (Production of effective Biofertilizers and Biopesticides). But to harness above benefits, less than 1% of the estimated number of microbial species are described and available to man. As new species are discovered, the expertise is difficult to locate to ensure their correct identification as a result this human resource is diminishing day by day. It is thus crucial that the microbial diversity of the world is not lost and that it is identified, characterized and exploited in a sustainable way for the benefit of humankind. Culture Collections have an important role in safeguarding of microbial diversity for future use, thereby providing the biological resource to underpin research and development. Culture Collection Centers, or Microbial Resource Centers or Microbial Gene Banks are centers where microorganisms of scientific or industrial research are maintained in viable form. In addition to collection and maintenance of microorganisms, microbial resource centers also distribute authentic microbial strains (algae, bacteria, fungi, yeast, protozoa and viruses) and/or their products (Genomes, Plamids, cDNAs) to the researchers and industrialists all over the world. Microbial resource centers are regarded as living libraries holding microbes and represent dynamic institutions of learning, research, scientific culture and information (Arora et al. 2005). There is a need to ensure that these centers take advantage of the latest technologies available and the constraints faced by them in terms of financial support be overcome so that they can deliver long lasting solutions to the aforementioned problems in addition to the enhancement of knowledge. The increasing demand for microbial resource and the size of this relatively untapped and hidden resource offers justification for an increase in number, scope and quality of Microbial Resource Centers. For the sake of convenience the term "Culture Collection Centers" instead of "Microbial Resource Centers" will be used here after.

3.2 History of Culture Collections

Collection of microorganisms have come a long way, from beginnings in the 1890s with Kral's collection in Prague and the collection at the Institute Pasteur, Paris, to the resource centers of 2010 driven by excellence. The discovery of the pure culture technique by Robert Koch lead to the idea of Culture Collections Centers, and the first Culture Collection to provide services was established by Prof. Frantisek Král in 1890 at the German University of Prague (Czech Republic) (Kocur 1990). Kral (1846–1911) worked for about 30 years for the glass manufacturing company Venceslaw Batka; afterwards, he worked as a technician at the Institute of Hygiene of the German University of Prague. His experience with manufacturing laboratory glass products was the reason he was subsequently chosen as director of the bacterial collection by Prof. Soyka. Then because of his experience in isolating, cultivating and maintaining microorganisms, he was appointed Associate Professor of Bacteriology. In 1900, Kral published the first catalogue of microorganisms from a Culture Collection (Uruburu 2003). After his death in 1911, the charge was taken over by Professor Ernst Pribham who transferred it to the University of Vienna where he issued several catalogues listing the holdings of the collection. Part of this collection was brought to Loyola University in Chicago by Prof. Pribham in the 1930s. After Prof. Pribhams death many of the collection's cultures were subsequently transferred to the American Type Culture Collection (ATCC). Unfortunately, the Vienna portion of the Pribham's Collection was largely lost during World War II (Brenner et al. 2005). Following Kral's collection, many Culture Collection centers were established. Currently, the oldest known collections are the Mycothèque de l'Universitée Catholique de Louvain (MUCL) established in 1894, in Louvain, Belgium, and the Collection of the Centraalbureau voor Schimmelcultures (CBS) founded in 1906, in Utrecht, the Netherlands (Uruburu 2003).

3.2.1 Establishment of the World Federation of Culture Collections (WFCC)

The foundation stone for coordination among Culture Collections was laid down by Prof. P. Hauduroy in 1946 with the establishment of the centralized information facility at the University of Lausanne, Switzerland (Uruburu 2003). The facility maintained information about the strains deposited in different collections and also published an information bulletin regarding the same. The Lausanne Center later became associated with the International Association of Microbiological Societies (IAMS, now named the International Union of Microbiological Societies, IUMS) and in cooperation with it; an International Federation of Type Cultures was formed with the aim of repairing the damage caused to Culture Collections during World War II. In 1962, a Culture Collections Conference was held in Ottawa (Canada), and the IAMS was recommended that a section on Culture Collections be set up, which came into being in 1963. Prof. Skerman (Australia) was the first chairman of the section; other members of the steering committee were Krasilnikow (Russia), Asai (Japan), Van Beverwijk (Netherlands), Martin (Canada), Donovick (USA) and Steel (UK). After reorganization of IAMS in the year 1970, this section became the World Federation for Culture Collections (WFCC), which has been active since then. The World Federation for Culture Collections (WFCC) is a COMCOF (Committees, Commissions and Federations) of the International Union of Microbiological Societies (IUMS) and a scientific member of the International Union of Biological Sciences (IUBS). Its key objective is the promotion and development of collections of cultures of microorganisms and cultured cells. Retention and support of existing collections, as well as assistance and advice to help new collections become established remain key activities. The World Data Center for Microorganisms (WDCM) was set up as a data center of WFCC; the WDCM is a vehicle for networking microbial resource centers of various types of microbes. It also serves as an information resource for the customers of the microbial resource centers. WFCC consists of several committees and organizes the International Congress of Culture Collections (ICCC) every 4 years.

3.3 Status of Culture Collections

There are 619 Culture Collection centers in 71 countries registered in the World Data Center for Microorganisms (WDCM) (Table 3.1). The WDCM holds information of 2,015,030 microorganisms among which 958,973 are bacteria, 534,464 are fungi, 32,916 are viruses and 8,703 are cell lines. The numbers of representative and type strains of microorganisms held in Culture Collections of the world are shown in Table 3.2. These collections are supported by different sources of funding (Table 3.3). Two hundreden sixty one collections out of 619 produce catalogues of holdings and update it at regular intervals. The number of cultures maintained in each collection varies from country to country depending upon the type of microorganisms they preserve, and the number of cultures they receive. As per WDCM the top 20 countries holding the largest number of cultures are (Table 3.4). The WDCM data reflect the world-wide imbalance in collection holdings. While most countries such as classified as mega diverse countries host 60–70% of the planets biodiversity, while greatest diversity lies outside these regions. As an example Europe has a lower presence of biodiversity compared to other continents, this lower biodiversity is

Table 3.1 Culture Collection centers registered with WDCM

Country	Culture collections	Cultures
Europe		
Armenia	1	17,805
Austria	2	6,070
Belarus	1	1,175
Belgium	6	58,252
Bulgaria	4	13,354
Czech Republic	13	9,251
Denmark	3	86,951
Estonia	4	13,300
Finland	2	7,713
France	37	67,257
Germany	13	51,582
Greece (Hellenic Rep.)	6	5,292
Hungary	8	11,390
Ireland	1	380
Italy	10	14,670
Kazakhstan	2	199
Latvia	1	692
Netherlands	6	79,775
Norway	2	2,602
Poland	9	8,464
Portugal	5	7,035
Romania	2	760
Russian Federation	17	48,920
Slovakia	3	4,916
Slovenia	2	4,160
Spain	4	9,027
Sweden	3	52,700
Switzerland	3	3,598
Turkey	9	4,364
U.K.	18	82,132
Ukraine	7	4,583
Uzbekistan	3	1,443
Yugoslavia	2	897
Total	209	680,709
Africa		
Egypt	1	1,808
Morocco	1	913
Nigeria	2	223
Senegal	1	210
South Africa	3	10,860
Uganda	1	550
Zimbabwe	2	702
Total	11	15,266

(continued)

Table 3.1 (continued)

Country	Culture collections	Cultures
America		
Alaska	2	NA
Argentina	12	7,073
Brazil	60	145,992
Canada	19	77,841
Chile	1	NA
Colombia	2	4,347
Cuba	11	7,791
Mexico	16	7,824
U.S.A.	21	210,276
Venezuela	3	258
Total	147	461,402
Asia		
Bangladesh	2	NA
China	24	88,373
Hong Kong	1	60
India	25	215,359
Indonesia	17	10,505
Iran	8	6,517
Israel	4	776
Japan	25	232,717
Korea (Rep. of)	21	122,096
Malaysia	6	2,513
Mongolia	1	1,500
Pakistan	6	2,653
Philippines	6	3,392
Singapore	3	1,389
Sri Lanka	4	136
Taiwan	2	29,692
Thailand	59	43,106
Vietnam	1	6,449
Total	215	767,233
Oceania		
Australia	34	72,040
New Zealand	7	16,947
Papua New Guinea	1	270
Total	42	89,257

Source: WDCM (2012), *NA* data not available

attributed to its small size, distance from tropics and mountain ranges resulting in less species migration (Steck and Pautasso 2008). Despite less biodiversity European countries hold 33% of the Culture Collections and 41% of the microorganisms. On the other hand Colombia is one of the 17 countries of mega diversity, but it has only two Culture Collections holding a total of 4,092 cultures. Similarly, China has a huge biodiversity because of a vast territory of complex climates and very diverse geography. China is also one of the largest countries with respect to agricultural

Table 3.2 Representative and type strains of microorganisms collected in the world

Strain	No. of species/sub-species
Algae	3,060
Archaea	460
Bacteria	16,495
c DNA	15
Cell lines_animal	401
Cell lines_plants	0
Fungi	25,592
Hybridomas_animal	0
Hybridomas_plants	0
Lichens	0
Plasmids	648
Protozoa	60
Vectors	1,783
Viruses_animal	66
Viruses_bacteria	976
Viruses_plants	84
Yeasts	1,200

Source: WDCM (2012)

Table 3.3 Type of financial sources of the Culture Collections registered with WDCM

Type of support	No. of collections
Governmental	245
University	232
Semi-governmental	58
Private	35
Industry	17

Source: WDCM (2012)

productivity harboring more than 30,000 flowering plants, 30 species of grain, 200 types of vegetables and 300 types of fruit trees. With respect to microbial diversity China harbors 30,000 fungi and 16,000 bacteria. However according to Hawksworth's (Hawksworth 2001) formula for 30,000 flowering plants there must be 180,000 fungi (Smith 2003). But there are only 24 collections registered at WDCM with 88,373 strains of bacteria and fungi. India is one among the 12 countries of megadiversity endowed with enormous variability with respect to microbes (bacteria, cyanobacteria, fungi and viruses) (Arora et al. 2005). But lack of adequate support and expertise hampers the discovery of novel microbes and their products. There are 25 Culture Collections in India registered with WDCM holding 215,359 microbes. Despite a large number of new microbial genera and species being discovered and their biotechnological potential being emphasized, the data on Indian microbial resources remain mostly with the investigators and in papers published by them (Arora et al. 2005). There is no cohesive information available about how many of these potential microbes have been preserved for future use; as a result the status of

Table 3.4 Top 20 strain holders registered with WDCM

Rank	Country	Total hold
1	Japan	232,717
2	India	215,359
3	U.S.A.	210,276
4	Brazil	145,992
5	Korea (Rep. of)	122,096
6	China	88,373
7	Denmark	86,951
8	U.K.	82,132
9	Netherlands	79,775
10	Canada	77,841
11	Australia	72,040
12	France	67,257
13	Belgium	58,252
14	Sweden	52,700
15	Germany	51,582
16	Russian Federation	48,920
17	Thailand	43,106
18	Taiwan	29,692
19	Armenia	17,805
20	New Zealand	16,947

Source: WDCM (2012)

the culture preservation remains unclear. There is an urgent need for awareness among researchers regarding importance of culture preservation in India; the same can be achieved by conducting workshops and trainings regarding the importance of strain conservation by Culture Collections of the country in a coordinated manner. Meanwhile a database containing collective information of the existing microbial diversity should be developed and made public. This will help researchers to avoid repetition with respect to microbial isolation and to focus on existing but unexplored taxa and their exploitation.

3.4 Roles of Culture Collections

Since the beginning of microbiology as science, a huge number of microorganisms was isolated from a wide variety of natural sources, and used for scientific research and the industrial fermentation. However, a large number of important microorganisms had been lost in the past due to the change of interest of researchers and/or difficulties in maintaining the cultures in their original form. Though potential properties of microorganisms have been developed over a period of time by studying the microbial cultures maintained in Culture Collections, adequate and reliable sources of properly preserved cultures are needed to conduct application oriented research. In order to maintain a large number of microorganisms being isolated through various

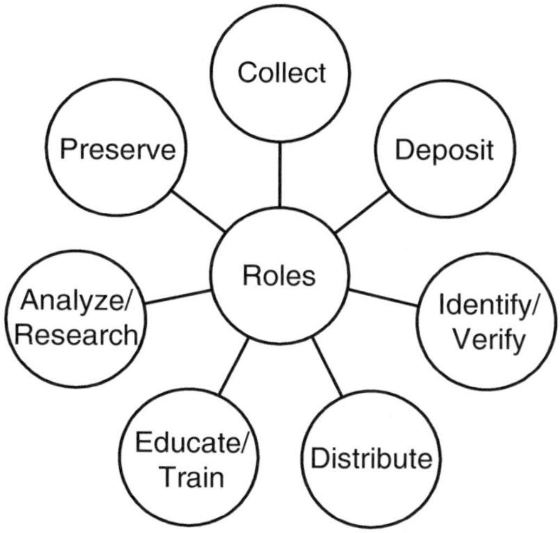

Fig. 3.1 Roles of culture collections (*Source*: Sigler 2004)

microbial diversity programs and to cope up with the improvement of existing strains, well-organized collections are needed as depositories and for the promotion of research. Collections serve as repositories for valuable isolates of historical, geographical, taxonomic, agricultural, medical, veterinary, or industrial significance (Sigler 2004). Roles of Culture Collections are to collect, maintain and distribute authentic cultures and information about them to the researchers all over the world (Fig. 3.1).

The range of services provided by individual collections may vary considerably; the services offered by some major Culture Collections worldwide are summarized as:

3.4.1 Accession and Deposition of Cultures

Acquisition of new strains and/or their genetic elements forms an essential part of Culture Collections. Each collection has developed certain criteria for accepting new strains depending upon interests, resources and assessment of probable future needs. Specific groups of microorganisms are dealt by specialized curators, who conduct research in their particular field of interest, maintain contact with active workers in the field and are responsible for enhancement of their collections with new acquisitions. Some general categories of microorganisms considered for acquisition are as follows

(A) Published strains of newly named taxa
(B) Type, neotype and selected reference strains, specific and unique biotypes
(C) Strains with special properties and application (bioassay, quality control, resistance test, degradation etc.)
(D) Strains of special significance to agriculture, biotechnology, medicine, education etc.

(E) Strains mentioned in patent applications
(F) Genetically manipulated strains, plasmid carriers, special mutants etc.

According to the *International Code of Bacteriological Nomenclature*, for a valid publication of a new species its type strain must be designated and be deposited in one or more of the established Culture Collections. Similarly, for adequate documentation of each newly isolated strain, the editors of scientific journals recommend proper deposition of new isolates before publication. Most of the culture collections require completion of a requisition form for each deposit in which specific information such as the source of isolation, history, taxonomic status, published data, special properties, reason(s) for deposit, growth medium, incubation temperature and generation time, pH and optimum procedure for long term preservation are to be given.

3.4.2 Preservation

In nature microorganisms occur in mixed populations of diverse or closely related species. Different strains or species exhibit different properties, thus require special preservation methods to ensure optimal viability, storage, and purity. In order to minimize the probability of strains being lost, each strain should, whenever practical, be maintained by at least two different procedures. At least one of these should be by freeze-drying (lyophilisation) or storage at ultra low temperature in liquid nitrogen or mechanical freezers maintaining temperatures of −140° C or lower (cryopreservation); these are the best methods for minimizing the risks of genetic change. In some cases, for example cell lines, where only freezing is available, duplicates should be stored in separate refrigerators with different electrical supplies.

3.4.3 Supply of Culture

Since the Culture Collections are authentic repositories for proper maintenance of valuable microbes and their products, it is expected that they supply viable and authentic strains for research; teaching or applied purposes and they do so with highest level of accuracy. Normally, a nominal fee is charged for each culture supplied but as a general rule, Culture Collections supply cultures to other collections on an exchange basis. Patent strains are usually restricted and are supplied to only authorized persons under certain conditions (Crespr 1985). Cultures purchased from a collection are guaranteed for viability, purity and to certain extent, according to the properties cited in publications or collections' catalogue. However, the characteristics of certain strains having plasmid(s), mutants and phage hosts may change during storage and may differ from the properties cited in literature. For this reason, Culture Collections often emphasize that recipients of strains should report discrepancies if any faced by them.

Certain regulations must be observed while importing cultures from other countries: license or import permits may be required for this purpose. Universal Postal Union lays down common regulations about packing and shipping of cultures. The details on export and import restrictions of microorganisms are often published in the *News Letter of WFCC*.

3.4.4 Research, Consultancy and Training

In addition to service functions, Culture Collections carry out research on taxonomy and identification of microorganisms. Several collections offer training facilities in their field of specialization and provide advisory and consultation services on such matters to the collections' staff from medical, environmental, industry, or government laboratories, who have responsibilities for isolating and identifying microorganisms, diagnosing disease, quality control, fermentation and collection management. Training courses on management of Culture Collections in general, modern preservation techniques and in applied microbiological research are organized and conducted by experienced staff of major Culture Collections of national and international level. Under the auspices of WFCC such training courses are mostly linked with the *International Conference on Culture Collections* (ICCC).

3.4.5 Strain Data and Information

As per WFCC guidelines it is mandatory for a Culture Collection to maintain the records of their holdings. Records of each strain held should at least contain following categories of information:

- Place
- Substrate or host
- Date of isolation
- Name of person isolating the strain
- Depositor (or other source of the strain, such as from another Collection)
- Name of the person identifying the strain
- Preservation procedures used
- Optimal growth media and temperatures
- Data on biochemical or other characteristics
- Regulatory conditions applying (relating for example to quarantine, containment levels and patent status)

WDCM provides for an efficient coding of the strains by defining a collection acronym and WFCC number which allows each Culture Collection to give a Globally Unique Identifier (GUID) to each strain of its holding, combining their acronym with their own internal numbering. The pioneering work of WDCM enables an appropriate recording and management of the documentation related to the strains. Collections should use this system to be part of the WDCM network and to be connected to the international scientific community.

Printed or on line catalogues of the strains containing above mentioned information should be produced and updated at regular intervals. Cultures with restricted distribution should be clearly marked and the ones that are not available for distribution should not be mentioned in catalogues or publicly accessible databases. Whenever resources permit, the records should be computerized also.

3.5 Constraints Faced by Culture Collection Centers

The effective curation and management of a Culture Collection is a demanding task. Since growth patterns and preservation methods vary among microorganisms, special attention has to be paid to each and every individual microbe in order to ensure optimal viability and ultimate purity during storage. Commitment to the maintenance of the collection and its services in the long-term should therefore be included in the strategic plans or objectives of the parent organization as appropriate. Following are a few major constraints faced by microbial Culture Collections with respect to providing up to date facilities for strain acquisition, preservation and distribution of strains.

3.5.1 Funding Levels

Culture Collection centers are critical components of the scientific infrastructure. The consequence of inadequate funding of Culture Collections on the scientific community would result in inadequately characterized and documented biological materials, and inadequately specialized expertise in preservation, maintenance and characterization of biological materials. This could ultimately lead to piling up of unrecognized errors. Moreover, the scientific community depends upon research done by individual workers that includes cultures also, unnecessary duplications would occur resulting in squandering of public investment into biological research. Finally, inadequate support of such an important and dynamic scientific infrastructure could lead to the loss of important microbes of commercial importance. Moreover lack of research training opportunities as a result of limited funds could lead to disintegration of research in Culture Collection (isolation and rapid identification of strains and microbial systematics) with other fields of research (genomics, ecology and molecular evolution). Keeping in mind the potential threats posed to the microbial resources it is of immense importance the federal agencies put in more efforts in terms of providing sufficient funds to the resource centers so that microbes of importance can be stored and propagated in the most efficient way.

3.5.2 *Orphaned Collections*

With the change in research directions or priorities and/or loss of the key position associated, Culture Collections are often rendered vulnerable to being orphaned, i.e. abandoned by the institution holding it. Once abandoned it becomes hard to mount an effective campaign to save them, which ultimately results in the loss of a natural bounty of microorganisms and the decades of toil that have been put in the collection, characterization and establishment of effective strains goes astray.

3.5.3 *Staff*

Since most of the living materials/microorganisms are sensitive they must be dealt in a timely manner or else there is a chance of losing them. Since routine accessions, preservation, maintenance and viability checking are time consuming and often involve expertise, an adequate number of well trained permanent technical staff must be recruited to perform the job. Untrained or poorly trained individuals are often an impediment to short-staffed collections because if left unsupervised, they may introduce contamination or replacement errors that may not be discovered until it is too late to avoid permanent loss. Such errors undermine the credibility of the collection as scientists are reluctant to acquire misidentified material. As noted above, professional staff, especially taxonomists, are crucial to the long-term future and viability of collections. They are required to secure funding, provide advisory or consulting services, keep accessions taxonomically up-to-date, and ensure continued development of the collection through acquisition of new material.

3.5.4 *Space and Equipment*

In order to accommodate a large number of collections large space and equipment is required. Most Collection Centers in general and the ones at universities in particular are short of sufficient space and equipment, the latter are required for basic work like, autoclaving, microscopy, refrigeration and cryopreservation of the obtained cultures.

3.5.5 *Information and Website Management*

In order to allow users easy and rapid access to information on microbial resources, database searching, access to protocols, preserved stocks, shipping forms and catalogues should be made easier and user friendly. Moreover the information regarding acquisition of new strains should be updated periodically. Though major collections have searchable internet databases that are easily accessible by users worldwide, from the user's prospective there is a disadvantage that Web-based data from different collections are not harmonized, and it may be difficult for users to accumulate the data from different sources. Keeping the above difficulty in mind, it is of prime importance that the information in a particular database of a Culture Collection synchronizes with that of the other Collections i.e. the web layout, formats and procedures for availing stains must be unified across all collection centers.

3.5.6 *Culture Exchange*

Distribution of cultures is a fundamental role of any Culture Collection, but this important feature is negatively impacted by regulations governing safe handling, containment, classification of agents according to risk, packaging and shipping of biological materials and infectious agents, and prohibitions governing agents deemed as possible biological weapons. Although need for some regulations with respect to safe handling and transport of microorganisms is understood, at the same time the regulations must be harmonized with respect to stringency. As regulations become more stringent, the suppliers (Collections) as well as the users (Scientists) become over burdened with the increasing administrative procedures and cost involved (see Padhye et al. 1998). Priorities for collections at both the international and national level are to (1) develop consensus in risk (hazard) classification of organisms, (2) change regulations governing transport of Risk level 2 organisms, and (3) provide a more streamlined system for permits (Sigler 2004).

3.6 Related Links

Biological Resource Centers (BRCs): are both service providers and repositories of living cells, genomes of organisms, and information relating to heredity and the functions of biological systems. BRCs contain collections of culturable organisms (e.g. microorganisms, plants, animal and human cells), replicable parts of these (e.g. genomes, plasmids, viruses, and cDNAs), viable but not yet culturable organisms, cells and tissues, as well as databases containing molecular, physiological and structural information relevant to these collections and related bioinformatics (http://www.oecd.org/document/51/0,3343,fr_2649_37437_33791027_1_1_1_37437,00.html).

Global Biological Resource Centre Network (GBRCN): a network designed to accommodate the future needs of biotechnology and biomedicine (http://www.gbrcn.org/).

Convention on Biological Diversity (CBD): an international treaty (1992) to sustain the diversity of life on Earth (http://www.cbd.int/convention/).

World Data Centre for Microorganisms (WDCM): an electronic gateway to databases on microbes and cell lines and resources on biodiversity, molecular biology and genomes (http://www.wfcc.info/datacenter.html).

European Consortium of Microbial Resources Centers (EMbaRC): an EU funded project that aims to improve, coordinate and validate microbial resource centre delivery to researchers from both public and private sectors. The EMbaRC project is a mixture of networking, access, training and research (http://www.embarc.eu/).

Organization for Economic Cooperation and Development (OECD): an international organization helping governments tackle the economic, social and governance challenges of a global economy. OECD Best Practice Guidelines for BRCs are given in (www.oecd.org).

World Federation for Culture Collections (WFCC): a federation within the International Union of Microbiological Societies (IUMS) concerned with the collection, authentication, maintenance and distribution of cultures of microorganisms and cultured cells (http://www.wfcc.info/datacenter.html).

Common Access to Biological Resources and Information (CABRI): a previously funded EU project aiming at providing biological resources and quality guidelines to users (http://www.cabri.org/).

European Culture Collections' Organization (ECCO): a consortium of European collections to promote collaboration and exchange of ideas and information about all aspects of Culture Collection activity (http://www.eccosite.org/).

European Biological Resource Centers Network (EBRCN): a previously funded EU project dealing with issues raised by the OECD Initiative on BRCs (http://www.cabri.org/FAQ/faq.html).

International Code of Nomenclature of Prokaryotes: governs the scientific names for prokaryotes and the rules for naming taxa of bacteria (http://www.ncbi.nlm.nih.gov/bookshelf/br.fcgi?book=icnb&part=A185).

Knowledge-based bioeconomy: a concept that transforms life-sciences knowledge into new, sustainable, eco-efficient and competitive products (http://cordis.europa.eu/fp7/kbbe/about-kbbe_en.html).

3.7 Conclusion and Future Prospects

In summary, Culture Collections differ with respect to diversification and focus of phylogenetic and metabolic diversity of the taxa maintained. Financial support and history of collections are the two main criteria that determine size of individual holdings and the expertise of curators.

The World Data Centre for Microorganisms (WDCM) currently lists 619 collections, some public, others academic or commercial, in 71 countries, but this apparently impressive number does not reflect the ongoing struggle of the vast majority of collections, for at least medium term funding. Only a few collections established in the early era of microbiology survived the politically and scientifically turbulent twentieth century. There is no data available on the number of microbial collections, their founding and their shut- down dates over the 120 year history of microbial holdings. No one can estimate the number of once properly maintained, irrecoverable valuable strains that were discarded due to the abandonment of facilities. These figures reflect a major loss and mankind has suffered, as many of these strains would have been exploited for various purposes related to agricultural productivity, crop protection, health care and bioremediation aspects. It is thus essential that specialized databases on microbial genetic diversity be established and integrated in a more coordinated manner in order to prevent loss of valuable microbial strains.

With focus on biodiversity and bio-prospecting, Culture Collections of microorganisms play an important role in an ongoing exploration of the microbial world. In such a situation Culture Collections are encouraged to implement the best techniques available (both biochemical and molecular) for prospecting of novel microorganisms with novel beneficial attributes, which can later be tailor made to find solutions to problems humanity is facing in terms of nutritional security and environmental resilience. In order to strengthen the status of microbial collections, following key issues are to be addressed (1) collections that aim to receive the status of being a Biological Resource Center (BRC) need improvement and expansion in research, training, bioinformatics and data management, (2) any expansion to the areas mentioned therefore will need more expert curators and technical staff, (3) strategies that encourage authors to deposit a larger number of strains to the public domain are to be devised and executed, (4) the possibility that many collections cover the same range of microorganisms may be excluded by expanding collections in a specific area, availing diverse funding schemes and by varying collections' focus. Moreover, the Culture Collection centers must be in touch with all microbiologists, collection users, editors and research program funders to ensure that all important strains with key attributes arising from different research programs are preserved for future use.

References

Amils R, Gonzalez-Toril E, Fernandez-Remolar D, Gomez F, Aguilera A, Rodrıguez N, Malki M, Garcıa-Moyano A, Fairen AG, de la Fuente V, Sanz JL (2007) Extreme environments as mars terrestrial analogs: the Rio Tinto case. Planet Space Sci 55:370–381

Arora DK, Saikia R, Dwievdi R, Smith D (2005) Current status, strategy and future prospects of microbial resource collections. Curr Sci 89(3):488–495

Brenner DJ, Krieg NR, Staley JT, Garrity GM (2005) Bergey's manual of systematic bacteriology, 2nd edn. Springer, New York, pp 111–112

Cnespr RS (1985) Microbiological inventions and the patent law-the international dimension. In: Russel GE (ed) Biotechnology and genetic engineering reviews, vol 3. Intercept, New Castle upon Tyne, 37 pp

Colwell RR (1997) Microbial diversity: the importance of exploration and conservation. J Ind Microbiol Biotechnol 18:302–307

Deming JW (2002) Psychrophiles and polar regions. Curr Opin Microbiol 5:301–309

Hawksworth DL (2001) The magnitude of fungal diversity: the 1.5 million species estimate revisited. Mycol Res 105:1422–1432

Convention on Biological diversity (2012) Home pages of History of Convention. http:// www.cbd.int/history/. Accessed 24 June 2012

Kato S, Hara K, Kasai H, Teramura T, Sunamura M, Ishibashi J, Kakegawa T, Yamanaka T, Kimura H, Maruma K, Urabe T, Yamagishi A (2009) Spatial distribution, diversity and composition of bacterial communities in subseafloor fluids at a deep-sea hydrothermal field of the Suiyo Seamount. Deep-Sea Res 56:1844–1855

Kocur M (1990) History of the Kral collection. In: Sly L, Kirsop B (eds) 100 years of culture collections. Institute for Fermentation, Osaka

Kumar B, Trivedi P, Mishra AK, Pandey A, Palni LMS (2004) Microbial diversity of soil from two hot springs in Uttaranchal Himalaya. Microbiol Res 159:141–146

Mancinelli RL, Fahlen TF, Landheim R, Klovstad MR (2004) Brines and evaporites: analogs for Martian life. Adv Space Res 33:1244–1246

Padhye AA, Bennett JE, McGinnis MR, Sigler L, Flis A, Salkin IF (1998) Biosafety considerations in handling medically important fungi. Med Mycol 36:258–265

Rothschild LJ, Mancinelli RL (2001) Life in extreme environments. Nature 409:1092–1101

Sigler L (2004) Culture collections in Canada: perspectives and problems. Can J Plant Pathol 26:39–47

Smith D (2003) Culture collections over the world. Int Microbiol 6:95–100

Steck CE, Pautasso M (2008) Human population, grasshopper and plant species richness in European countries. Acta Oecol 34:303–310

Uruburu F (2003) History and services of culture collections. Int Microbiol 6:101–103

World Data Center for Microorganisms (WDCM) (2012) Home pages of culture collections in the world. http://wdcm.nig.ac.jp/hpcc.html. Accessed 23 June 2012

Chapter 4
Fungal Biodiversity: A Potential Tool in Plant Disease Management

Shabbir Ashraf and Mohammad Zuhaib

Abstract Fungal flora is considerably rich and diverse. Biodiversity of fungal flora in a region is represented by a number of taxonomic groups in different habitats. The rich diversity of fungus in the form of fungal biocontrol agents are now being potentially explored as an important tool in the management of plant diseases. Excessive use of pesticides results in the form of pest resistance, disturbance in ecosystem due to destruction of natural enemies and environmental pollution which leads to health problems. Thus fungal biocontrol agents have high potential to replace the use of synthetic chemicals. Some of the most widely used biocontrol agents in the world belong to the fungal genus *Trichoderma*. In particular isolates of *Trichoderma harzianum*, *T. virens*, *T. hamatum*, are used against diseases in a wide variety of economically important crops. They have been used with success against soilborne, seedborne, storage rots and diseases in the phyllosphere. *T. harzianum* and *Gliocladium virens* have been successfully used against *Botrytis cineria* in different crops. Many other fungi have been shown to antagonize and inhibit numerous fungal pathogens of aerial plant parts. *Chaetomium*, *Tuberculina maxima*, *Verticillium lecanii*, *Ampelomyces quisqualis*, *Tilletiopsis* and *Gonatobotrys simplex* are some of the most effectively used biocontrol agents against *Athelia bombacina*, *Venturia inequalis*, *Cronartium ribicola*, *Puccinia*, *Erysiphe ovata*, *Sphaerotheca fuliginea* and *Alternaria alternata* respectively. This review provides a broad perspective on the range of diversity of fungal biocontrol agents available for commercial exploitation, mechanism, commercial formulations in use and bottle necks in biocontrol of plant diseases. The fungal bioagents are expected to have great potential in addressing some of the key pest problems in the near future.

Keywords Fungal biodiversity • Biocontrol agent • Plant disease management • Hyperparasitism

---

S. Ashraf (✉) • M. Zuhaib
Department of Plant Protection, Faculty of Agricultural Sciences,
Aligarh Muslim University, Aligarh, Uttar Pradesh 202002, India
e-mail: shabbiragri@yahoo.co.uk

A. Malik et al. (eds.), *Management of Microbial Resources in the Environment*,
DOI 10.1007/978-94-007-5931-2_4, © Springer Science+Business Media Dordrecht 2013

1 Introduction

Presently "fungi" as a mega-diverse group, span three kingdoms, most belonging to the Fungi (Eumycota), while others are classified in the *Protozoa* and *Chromista* (*Straminipila*) (Cavalier-Smith 1998, James et al. 2006b). In the history of mycology, it was predicted that the fungi are the most diverse, in terms of species richness and render mycology to be larger than the rest of botany (Fries 1825). While Pascoe (1990) gave particular reference of Australia and suggested that there were at least ten times as many fungi as vascular plants. Data from 25 studies in different parts of Asia, Europe and North America were analyzed statistically by Schmitt and Mueller and the results showed that fungal species richness was much higher than that of the plants, and were consistent with the high estimates of species numbers made by Hawksworth (1991) which was about 1.5 M species and the figure has been widely cited and accepted. Based on information in the US National Fungus Collection database, Rossman (1994) studied thoroughly on fungal diversity by estimating the number of particular species found worldwide (Table 4.1). Among the fungal groups discussed, Basidiomycetes, Gasteromycetes, Ascomycetes, Myxomycetes and Chytridiomycetes comprise of important fungi responsible for plant diseases which are the major constraints in the profitable cultivation of crops. Potentially immortal fungi spread their tentacles everywhere. In 1845, when potato late blight fungus caused havoc in Ireland and soon after downey mildew fungus, *Plasmopara viticola* threatened the wine industry in France. Apple scab, which threatened apple cultivation, Panama disease of banana, Wilt disease of pigeonpea, chickpea, castor and guava, rust and smut of cereals are some of the serious fungal diseases.

For the management of fungal diseases, synthetic fungicides are usually applied as effective, dependable and economical control measures. However, the indiscriminate use of chemical fungicides has resulted in several problems, such as toxic residues in food, water and soil and disruption of the ecosystem, leading to the fear that their regular use may pollute the environment further. Hardly 0.1% of the agrochemicals used in crop protection reach the target pest, leaving the remaining 99.9% to enter the environment to cause hazard to non-target organisms, including humans (Pimentel and Levitan 1986). At present about 150 different fungicidal compounds, formulated and sold in several fold larger number of different proprietary products, are used in the agriculture world. The value of fungicides of the major crop groups and the importance of the dynamic nature of the fungicide market was reported by Phillips McDougall (2006) and who also gave the total value of fungicide sales which is approximately 7,491 million US dollars. According to WHO estimates, approximately 0.75 million people get ill every year with pesticide poisoning. Moreover resistance of pathogens to fungicides has rendered many fungicides ineffective, giving rise to new physiological races (*forma specialis*) of pathogen. Biological control is potentially a sustainable solution to manage plant disease in developing as well as developed countries, due to its long term effect and without any side effects. Host resistance augmented with biological control measures, especially with mycoparasites can be most useful and ecofriendly for

Table 4.1 Major groups of fungi and estimated world species numbers as compiled by Rossman (1994)

Group	Species world wide
Well-known	
Aphyllophorales s. lat.	20,000
Macrolichens	20,000
Moderately well-known	
Agaricales s. lat.	80,000
Dematiaceous and aquatic hyphomycetes	80,000
Uredinales	50,000
Hypocreales and *Xylariales*	50,000
Ustilaginales	15,000
Gasteromycetes	10,000
Erysiphales	10,000
Jelly fungi *s. lat.*	5,000
Pezizales	3,000
Myxomycetes	1,500
Endomycetales (true yeasts)	1,000
Poorly known	
Non-dematiaceous hyphomycetes	200,000
Coelomycetes	200,000
Other perithecioid ascomycetes	100,000
Helotiales	70,000
Insect-specific fungi	50,000
Crustose lichens	20,000
Mucorales	20,000
Oomycetes	20,000
Chytridiomycetes	20,000
Endogonales and *Glomales*	1,000
Total	10,28,500

controlling important fungal diseases. The biological control agents act against plant pathogens through different modes of action. Antagonistic interaction may include antibiosis, competition and hyperparasitism (Cook and Baker 1983). There is competition when two or more microorganisms compete for the same resources may be space, nutrition and oxygen. Antibiosis is release of antibiotics or toxic metabolites which have direct inhibitory effect on another (pathogen). Hyperparasitism or predation results from biotrophic or necrotrophic interactions that lead to parasitism of plant pathogen by the biological control agent. The most important well studied antagonists against several plant pathogens are fungi like *Trichoderma spp.* (particularly *T. hamatum*, *T. harzianum*, *T. viride*), *Ampelomyces quisqualis*, *Aspergillus*, (particularly *A. niger*, *A. terreus*), *Chaetomium globosum*, *Coniothyrium minitans*, *Fusarium spp.*, *Gliocladium virens*, *Penicillium citrinum*, *Peniophora gigantea, and Sporodesmium spp.*

It was no doubt to say that fungal biodiversity is a potential tool in plant disease management. This review will focus on the recent developments in the field of

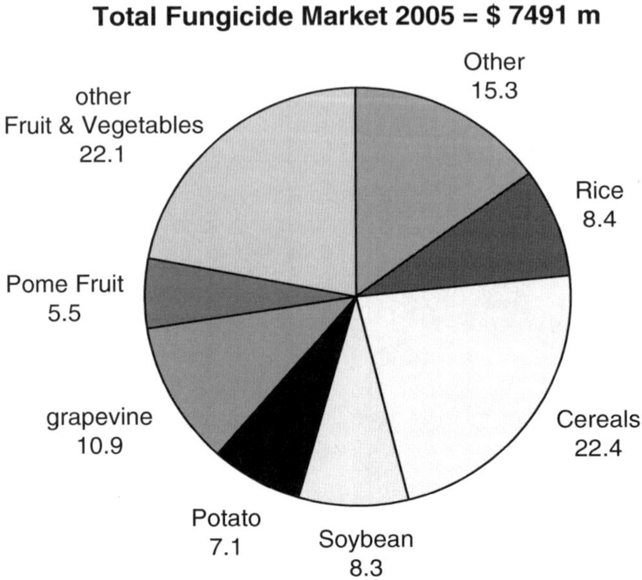

Fig. 4.1 Global fungicides market share for the major crop groups for 2005. Pome fruits: A fleshy fruit, such as apple, pear, or quince, having several seed chambers and an outer fleshy part largely derived from the hypanthium (*Source*: Phillips McDougall 2006)

biological control of plant diseases through fungal biocontrol agents, which will emphasize on the mechanism, commercialization and bottle necks of biological control of plant diseases (Fig. 4.1).

2 Biological Control of Plant Diseases

The use of microorganisms as biological control agents to control plant disease is a potentially powerful alternative method (Kulkarni et al. 2007). Because of their rich diversity, complexity of interactions and numerous metabolic pathways, microbes are an amazing resource for biological activity (Tejesvi et al. 2007; Mitchell et al. 2008; Raghukumar 2008). Biological control of pathogens is the total or partial destruction of the pathogen population by other organisms which occurs routinely in nature. There are many examples of management of several diseases in which the pathogen cannot develop in certain areas because the soil contains microorganisms antagonistic to these pathogens. It was also observed that the plant attacked by the pathogen has also been naturally inoculated with antagonistic microorganisms before or after the pathogen attack. Sometimes the antagonistic microorganisms may consist of avirulent strains of the same pathogen that destroy or inhibit the development of the pathogen, as happens in hypo virulence and cross protection.

In some cases even higher plants reduce the amount of inoculum either by trapping available pathogens (trap plants) or by releasing into the soil a substance toxic to the pathogens. Agriculturists have increased their efforts to take advantage of such natural biological control against several plant diseases. Biological antagonism, although subject to numerous ecological limitations is expected to become an important part of the control measures employed against many more diseases. Numerous kinds of antagonistic microorganisms have been found to increase in suppressive soils most commonly. However, pathogen and disease suppression have been shown to be caused by fungi, such as *Trichoderma* and *Sporidesmium sp*. Suppressive soil added to conducive soil can reduce the severity of disease by introducing microorganisms antagonistic to the pathogen.

Biological control practices for direct protection of plants from pathogen involve the deployment of antagonistic microorganisms at the infection court before and after infection takes place. Although hundreds, possibly thousands of microorganisms have been shown to interfere with the growth of plant pathogens in the laboratory, greenhouse, or fields and to provide some protection from the diseases caused by them, so far only six microorganisms have been registered and are commercially available for use. The six microorganisms include three fungi: *Gliocladium virens*, for control of seedling disease of ornamental and bedding plants; *Trichoderma harzianum* for control of several soil borne plant pathogenic fungi and *Trichoderma harzianum/T. polysporum* for the control of wood decays (Adam 1990; Swati and Adholeya 2008). Over the past 30 years, microorganisms have been described, characterized, and tested for their use as biocontrol agents against diseases caused by plant pathogens. Biocontrol agents and especially antagonistic fungi have been used to control plant diseases with 90% of applications being formulated using different strains of *Trichoderma*, e.g. *T. harzianum*, *T. virens*, *T. viride* (Benítez et al. 2004). Many species of *Chaetomium* e.g. *Chaetomium globosum*, *C. cochlioides*, *C. cupreum* can also be antagonistic against various soil microorganisms (Soytong et al. 2005; Kanokmedhakul et al. 2002, 2006). Some of the important genera of fungi used as biocontrol agents are *Trichoderma*, *Gliocladium*, *Aspergillus*, *Penicillium*, *Neurospora*, *Chaetomium*, *Dactylella*, *Arthrobotrys*, *Ampelomyces quisqualis*, *Glomus*, etc.

2.1 Classification of Biocontrol Agents

Mycoparasitism is the parasitism of one fungus by another fungus (Hawksworth et al. 1983). Mycoparasitism appears to be a complex process, involving recognition of plant pathogens by chemotropism, coiling around the pathogen and appresorial formation, followed by production of cell wall-degrading enzymes and peptaibols, mediated by heterotrimeric G-proteins and mitogen-activated protein (MAP) kinases (Druzhinina et al. 2011). Mycoparasitic fungi are widespread today and recognized as potential biological control agents of economically important plant pathogenic fungi (Baker and Cook 1974) they have been observed both under laboratory and field conditions (Jeffries and Young 1994). Barnett and Binder (1973) divided mycoparasites into two main categories depending on their mode of nutrition.

2.1.1 Necrotrophic or Destructive Mycoparasites

In this relationship, mycoparasites kill their host as a result of their parasitic activity, whereas biotrophic mycoparasites obtain their nutrients directly from the living mycelium of the hosts. In necrotrophic relationships the antagonistic action of the mycoparasite is strongly aggressive and the mycoparasites dominate the association. Hyphae of the parasite contact and grow in contact association with those of the host, sometimes coiling around them, and frequently penetrating. Secretion of hyphal wall degrading enzymes or exotoxins may cause the death of the cytoplasm of the host prior to hyphal contact, or alternatively cytoplasmic death may not occur until contact has been established. Necrotrophic parasites tend to have a broad range of host fungi, and are relatively unspecialized in their mechanism of parasitism, for example, they often release toxins and lytic enzymes into the environment, are overtly destructive and usually lack specialized infection structures. In this way their behavior parallels that of the necrotrophic fungi which parasitize plants.

2.1.2 Biotrophic Mycoparasites

In this relationship, the living host supports the growth of the parasites for an extended period of time, may not appear diseased, while its growth rate, sporulation and metabolism may appear overtly to be little affected, at least in the early stages of the relationship. The parasitic relationship is physiologically balanced and the parasite appears to be highly adapted to this mode of life. Biotrophic mycoparasites tend to have more restricted host ranges than necrotrophs, and often form specialized infection structures or a host parasite interface. Exotoxin production has not been demonstrated in any biotrophic mycoparasitic interaction.

2.2 Biocontrol Agent Antagonistic to Foliar Pathogen

Many fungi have been shown to antagonize and inhibit numerous foliar fungal pathogens of aerial plant parts, viz., *Chaetomium sp.* and *Athelia bombaciana* suppress *Venturia enequalis* ascospores and conidia production in fallen and growing leaves, respectively. *Tuberculina maxima* parasitizes the white pine blister rust fungus *Cronartium rubicola*; *Verticillium lecanii* and *Darluca filum* parasitize several rusts; *Ampelomyces quisqualis* parasitizes several powdery mildews, *Tilletiopsis* parasitizes cucumber powdery mildew fungus *Sphaerotheca fuliginea* and *Cladosporium herbarum* is reported antagonistic to *Alternaria sp.* (*A. alternata* and

Table 4.2 A list of fungi tested as potential biocontrol agents against powderymildews

Antagonist	Powdery mildew	Host plant	References
Acremonium byssoides	*Oidium heveae*	*Hevea brasiliensis*	Hawksworth (1981)
Alternaria alternatum	*Sphaerotheca fuliginea*	*Cucurbitaceae*	Hijwegen (1989)
A. strictum	*S. fuliginea*	*Cucumis sativus*	Hijwegen (1988, 1989)
Ampelomyces spp.	Many species	Many species	Jarvis et al. (1977); Hofstein et al. (1996); Philipp et al. (1990)
Aspergillus fumigatus	*E. cichoracearum*	*Cucurbita maxima*	Srivastava and Bisht (1986)
Cladosporium oxysporum	*Phyllactinia corylea*	*Morus alba*	Raghavendra Rao and Pavgi (1978)
C. spongiosum	*Phyllactinia dalbergiae*	*Dalbergia sissoo*	Mathur and Mukerji (1981)
Cephalosporium sp.	*Leveillula taurica*	*Capsicum annuum*	Diop-Bruckler and Molot (1987)
Pseudozyma spp.	*Erysiphe polygoni*	*Trifolium pratense*	Traquair et al. (1988)
Tilletiopsis albescens	*Sphaerotheca fuliginea*	*C. sativus*	Knudsen and Skou (1993); Hijwegen (1988, 1989)
T. minor	*E. martii*	*L. polyphyllus*	Hijwegen and Buchenauer (1984)
	S. fuliginea	*C. sativus*	Hijwegen (1992)
Verticillium lecanii	*Oidium tingtanium*	*Citrus spp.*	Raghavendra Rao and Pavgi (1978)
	Sphaerotheca fuliginea	*C. sativus*	Hijwegen (1988, 1989); Askary et al. (1998)
	Uncinula necator	*Vitis vinifera*	Heintz and Blaich (1990)

Source: Kiss (2003)

A. porii) in field condition while strains of *T. harzianum* (Imtiaz and Lee 2008) and *Myrothecium sp.* (Chawda and Rajasab 1992; Kumar 2007) were found effective *in-vitro* respectively. Powdery mildew fungi (Erysiphaceae) are one of the most conspicuous groups of plant pathogens, comprising more than 500 species that attack more than 1,500 plant genera. The pathogen is widely distributed and causes heavy losses while it was estimated that fungicide treatments against powdery mildews can cost as much as US$6,000 per hectare per year in rose production in Canada (Paulitz and Belanger 2001), $70–150 per hectare per year in apple orchards in Virginia, USA (Yoder 2000) and in Europe, the largest area of fungicide use is for the control of powdery mildews (Hewitt 1998; Legler et al. 2011). *Ampelomyces* and *Pseudozyma species*, *Verticillium lecanii*, *Tilletiopsis spp.* and *Acremonium alternatum* have been thoroughly studied fungal antagonists of powdery mildews (Belanger and Labbe 2002). Some of the fungi listed in Table 4.2 are well-known natural antagonists of powdery mildews. *Ampelomyces spp.* (Kiss 1998; Kiss et al. 2004) and *Tilletiopsis spp.* (Urquhart et al. 1994) were repeatedly isolated from plants infected with powdery mildew worldwide. In a study of evaluation of effective *Trichoderma* species against *Alternaria porii*, causing onion blotch, the percent inhibition of mycelial growth and conidial growth is highest for *Trichoderma virens* followed by *T. harzianum* and *T. pseudokoningii* (Imtiaz and Lee 2008).

2.3 Biocontrol Agents Antagonistic to Soil-Borne Fungal Pathogens

The principal fungi used as biological control agents against soilborne diseases include the two mentioned above as being used commercially, namely, *Trichoderma harzianum* and *Gliocladium virens*. They are effective against damping-off, root rot and wilt diseases of ornamental plants, vegetables and cereals caused by *Pythium*, *Phytophthora*, *Sclerotium*, *Fusarium* and some other fungi. In addition *Sporidesmium sclerotiorum*, *Conithyrium minitants*, *Talaromyces flavus* and others have been tested for control of some diseases caused by *Sclerotinia*, *Verticillium* and *Rhizoctonia*. Some species of *Pythium*, such as *Pythium nunn* and *P. oligandrum*, protect potted ornamental plants and vegetables from plant pathogenic species of *Pythium* and *Talaromyces flavus*. Some binucleate non-pathogenic strains of *Rhizoctonia* have been found to protect plants from the pathogenic *Rhizoctonia solani*. In another study two isolates of *T. koningii* reduced seedling death caused by *R. solani* and *Pythium ultimum var. sporangiiferum*. Neither isolate of *T. koningii* suppressed damping-off caused by either pathogen (*Pythium* and *R.solani*) as consistently as the binucleate strains of *Rhizoctonia* (Harris 1999). Experimental control of several other soil borne diseases has been obtained with many other fungi. There are many reports in literature of soil borne fungi which are susceptible to antagonism or mycoparasitism by other soil fungi. A well known example is the genus *Rhizoctonia*, the non sporulating anamorph of the mycelium of numerous isolates from soil and infected host plants. The pathogen infects the root tip region of young roots and can continue to infect newly produced root tips as the plant grows. The plants not killed by the pathogen may successfully outgrow the disease as *Rhizoctonia* is essentially a parasite of young tissues which have not developed resistance to the attack. The pathogenic *R. solani* is susceptible to attack by mycoparasitic fungi. Some of the parasites of *R. solani* are summarized in the Table 4.3. In another study, hyphae coiling of *R. solani* with the strains of *T. harzianum* was studied and only one of them (Th-9) was chosen for visualization via SEM. After 3 days of contact between the mycoparasites and the pathogen complete colonization of *R. solani* with several strains of *Trichoderma* was observed. The parasitic hyphae reached the host hyphae and grew on the surface always with coiling and later they penetrated the cell wall directly without formation of appressorium-like structures. The invaded hyphae of *R. solani* looked disintegrated. Furthermore, total disappearance of the host hyphae after 7 days of interaction was observed. Lytic enzymes seem to be capable of degrading the cell walls of *R. solani*, as discussed above (strain Th-9), which produces high amounts of β-glucosidase (Melo et al. 1997),

Table 4.3 Some parasites of *Rhizoctonia solani*

Mycoparasite	Reference
Arthrobotrytis oligospora	Persson (1991)
Gliocladium catenulatum	Jager et al. (1979)
Gliocladium spp.	Molan (2009)
G. roseum	Jager et al. (1979)
G.virens	
Pythium oligandrum	Boosalis (1956)
Trichoderma hamatum	Chet et al. (1981); Chet and Baker (1981); Elad et al. (1983b)
Trichoderma harzianum	Melo and Faull (2000)
T. koningii	Melo and Faull (2000)
Verticillium biguttatum	Boogert van den (1989)
V. chlamydosporium	Turhan (1990)
V. lamellicola	Kuter (1984)
V. lecani	Kuter (1984)
V. nigrescens	Kuter (1984)

is an example of cell wall degrading enzymes involved in antagonistic mechanism. A variety of extracellular lytic enzymes play an important role in the antagonistic mechanism by the parasite. High chitinase and β-(1,3)-glucanase activities have been reported to be produced by *T. harzianum* (Sivan and Chet 1989; Harman et al. 1981; Abubaker 2010) and there is a relationship between the production of these enzymes and the ability to control plant diseases (Ridout et al. 1988; Adams 2004). *Fusarial* wilts of several crops, such as celery, cucumber, and sweet potato, caused by the respective *formae specialis* of *Fusarium oxysporum* have been reported to be successfully controlled by inoculating transplants or cutting with nonpathogenic strains of the same fungus. Some of these strains have been isolated from the vascular tissues of host plants that remained healthy while nearby plants had been killed by the wilt inducing strains of the fungus. It is believed that the nonpathogenic strains not only compete with the pathogenic ones in the rhizosphere and for infection sites, but they also enhance the resistance of the host towards the pathogenic strains (Agrios 2005).

3 Mechanism Involved in Antagonism

Several mechanisms, operating alone or in concert are known to be involved in antagonistic interactions in the rhizosphere and as well as in phyllosphere. Nutrient competition, antibiosis and mycoparasitism are the major mechanisms. Additional mechanism such as induced resistance, interference with pathogen related enzymes, and a number of still unknown mechanisms, may complete the microbial arsenal (Elad 1996). Knowledge of mechanisms involved in biocontrol is important for estimating and predicting its reliability and selection of better strains. Besides other criteria, the choice of an antagonist with its characteristic mechanisms depends on the stage of the life cycle of the pathogen the antagonist is aimed at. Allowable interaction times and niche characteristics determine the suitability of certain modes of action during different development stages of the pathogen. The mechanisms of biocontrol mainly include antibiosis, competition, mycoparasitism, cell wall degrading enzymes, and induced resistance (Harman et al. 1998; Heidi and Abo-Elnaga 2012). These mechanisms are probably never mutually exclusive; these terms are meant to organize the examples into general groups to facilitate comparisons. A summarized data is given in Table 4.4.

3.1 Antibiosis

Antibiosis plays an important role in plant disease suppression by most of the fungi. The process has been defined as the interactions that involve a low-molecular weight compound or an antibiotic produced by a microorganism that has a direct effect on another microorganism (Weller 1988). Several antibiotic producing fungi have been used in biocontrol studies. Examples include Epicoccum *nigrum* which produces antibiotic compounds effective against *Botrytis cineria* (Hannusch and Boland 1996a) and *Sclerotinia sclerotiorum* (Hannusch and Boland 1996b) while; *Monilinia laxa* (Madrigal et al. 1994) and *Chetomium globosum* are effective antagonists of *Venturia inaequalis* (Boudreau and Andrews 1987). *Trichoderma virens*, which controls damping-off of cotton caused by *Pythium ultimum*, produces gliovirin. Mutant analysis has been used to demonstrate that the antibiotic gliovirin plays a role in biocontrol (Fravel 1988; Prasad et al. 2008). The importance of gliotoxin produced by *Trichoderma virens* in the suppression of *Pythium* damping-off of cotton seedlings has been confirmed by mutational analysis (DiPietro et al. 1993). Chaetomin is produced by *Chaetomium globosum*, peptaibols are produced by *Trichoderma harzianum*, and pyrones are produced by *Trichoderma* spp. (Schirmbock et al. 1994; Mukherjee et al. 2011). The antagonism of a strain of *Trichoderma harzianum* also seems to be based on antibiotics causing disorganization of the cytoplasm within 12 h and cause subsequent cell death of *B. cineria* (Belanger et al. 1995). The role of antibiotic production by antagonistic fungi has been less studied than antagonistic bacteria. One reason may be that these substances have merely been identified; currently scientists are now showing interest in exploiting the antibiotic synthesized during fungal interaction through molecular tools.

3.2 Competition

The nutrient sources in the soil and rhizosphere are frequently not sufficient for microorganisms. For a successful colonization of phyllosphere and rhizosphere a microbe must effectively compete for the available nutrients (Loper and Buyer 1991;

Table 4.4 Mechanisms of specific biocontrol agents for controlling plant pathogen

Biocontrol agents	Strain	Possible process/ metabolite	Target pathogen	Crop	Evidence for involvement
		Antibiosis			
Chaetomium globosum	Cg-13	Chaetomin	*P.ultimum*	Sugarbeet	*In-vitro* demonstration
Trichoderma (Gliocladium)		Gliovirin	*P.ultimum*	Cotton	Geneticanalysis *In-vivo*
Trichoderma virens	G-20	Gliotoxin	*P.ultimum*	Cotton	Geneticanalysis *In-vivo*
Trichoderma harzianum, Trichoderma koningii		Akylpyrones	Various fungi	Various crops	Antibiotics isolated *In-vitro*
Trichoderma harzianum	ATCC-36042	Peptaibol antibiotics	*Botrytiscinerea*, otherfungi	Grape vine	Antibiotics isolated *in-vitro*
		Competition			
Trichoderm harzianum		Nutrientsand space	Various fungi	Grapevine	Inferred from *In-vivo* activity
Phlebia (Peniophora) gigantea		Infection sites	*Heterobasidion (Fomesannosum), (Fomes) annosum*	Pine conifers	Infield demonstration
		Cell-wall degrading enzymes			
Serratia marcescens		Chitinolytic enzyme	Various fungi	Soybean	Geneticanalysis heteroexpression
Trichoderma harzianum	ATCC-36042	Chitinolytic enzymes, Glucanases	Various fungi	Pea, soybean	*In-vitro* demonstration
		Mycoparasitism			
Coniothyrium minitans		Mycoparasitism	*Sclerotinia*	Sunflower	Infield
Pythium spp		Mycoparasitism	*Pythium* spp.	Various crops	*In-vitro* demonstration
Trichoderma spp.		Mycoparasitism	*Various and numerous fungi*	Various crops	*In -vitro* demonstration
		Induced resistance			
Binucleate rhizoctoctnia	BNR-AG-K	Induced resistance	*Rhizoctonia solani AG-4*	Soybean	*In vitro* demonstration

Pal and Gardener 2006). On plant surfaces, host-supplied nutrients include exudates, leachates, or senesced tissue. There is a general belief that competition between pathogens and non-pathogens for nutrient resources is an important issue in biocontrol (Elad and Baker 1985; Kaur et al. 2010). In early studies (Fokkema 1971) on the role of pollen on infection of rye leaves by *Cochliobolus sativus*, *Saptoria nodorum* and *Puccinia recondita*, spore germination and superficial growth of mycelium of two necrotrophic pathogens was highly stimulated by the presence of pollens, resulting in more penetration sites and up to a tenfold increase of necrotic leaf area where as the infection by the biotrophic rust fungus was not enhanced. There was a positive correlation between the superficial mycelium density of *C. stivus* 2–3 days after inoculation and the necrotic leaf area. *Annosus* root disease caused by the fungus *Heterobasidion annosum* has been recognized as a worldwide problem in conifer forests for decades. *H. annosum* causes root and butt rot of conifers, infects freshly cut pine stumps and then spreads into the roots of the standing trees, which it kills. If the stump surface is inoculated with oidia of the fungus *Peniophora gigantea* immediately after the tree is felled *Peniophora* occupies the cut surface and spreads through the stump into the lateral roots and successfully competes with and replaces the pathogenic *Heterobasidion* in the stump, thus protecting nearby trees. It is also believed that competition for nutrients is more critical for soil borne pathogens, including *Fusarium* and *Pythium* species that infect through mycelial contact than foliar pathogens that germinate directly on plant surfaces and infect through appressoria and infection pegs (Keel et al. 1989). Sivan and Chet (1989) demonstrated that competition for nutrients is the major mechanism used by *T. harzianum* to control *F. oxysporum f.sp. melonis. T. harzianum* is able to control *B. cinerea* on grapes by colonizing blossom tissue and excluding the pathogen from its infection site (Gullino 1992; Vinalea et al. 2008). Moreover, *Trichoderma* has a strong capacity to mobilize and take up soil nutrients, thus making it more efficient and competitive than many other soil microbes (Benítez et al. 2004). Results of a study by Anderson et al. (1988) revealed that production of a particular plant glycoprotein called agglutinin was correlated with potential of pathogen to colonize the root system. The pathogen mutants deficient in this ability exhibited reduced capacity to colonize the rhizosphere and a corresponding increase in soilborne diseases.

3.3 Mycoparasitism

Mycoparasitism appears to be a complex process, involving recognition of plant pathogens by chemotropism, coiling around the pathogen and appresorial formation, followed by production of cell wall-degrading enzymes and peptaibols, mediated by heterotrimeric G-proteins and mitogen activated protein (MAP) kinases (Druzhinina et al. 2011). Parasitism is aimed at pathogenic mycelium already established and to reduce sporulation of the pathogen and hence to limit its dissemination. Enzymes degrading fungal cell walls such as chitinases and β-glucanases are commonly produced by hyperparasites while secretion of lytic enzymes including β-1,3-glucanases, proteinases, chitinases and lipases, by *Trichoderma* spp. was also reported by Harman (2001). Involvement of β-1,6-glucanases and β-1,4-glucanases may also play an important role in mycoparasitism (Thrane et al. 1997). Parasitism depends on close contact between antagonist and host, on the secretion of enzymes, and on the active growth of the hyperparasite into the host. These processes need time so it is unlikely that infection structures of pathogens can be parasitized and killed rapidly enough to prevent penetration of the host plant. When studying the chronological events during the interaction between *T. harzianum* and *Botrytis cinerea* at the ultrastructural level, Belanger et al. (1995) found evidence for antibiosis early during interaction within the first 12 h, but clear chitinolytic activity of the antagonist could not be demonstrated before the 10th day of interaction. *Trichoderma* recognizes signals from the host fungus, triggering coiling and host penetration. Different strains can follow different patterns of induction, but the fungi apparently always produce low levels of an extracellular exochitinase. The possible role of agglutinins in the recognition process determining fungal specificity has been examined recently. Barak et al. (1985) proposed that lectins of plant pathogenic fungi might play a role in recognition. Inbar and Chet (1992) proved the role of lectins in recognition during mycoparasitism using a biometric system. Ordentlinch et al. (1990) reported that there was no correlation between *in-vivo* and separated *in-vitro* dual culture or enzyme assays. Mycelial development of necrotrophic pathogens also can be controlled by mycoparasites. The mycoparasite *Coniothyrium minitans* penetrates mycelium of *S. sclerotiorum* that subsequently collapses and becomes necrotic (Whipps and Gerlagh 1992; Jones and Stewart 2011). The production of β-1,3 glucanases and chitinases enables the mycoparasite to utilize the host cells, and mycelium of the mycoparasite proliferates around dead hyphae of the host fungus. Above all, C. *minitans* is a potent parasite of sclerotia of *S. sclerotiorum*, as it reduces the survival of sclerotia, in field trials, by about 90% (Gerlagh et al. 1995). The antagonist *Limonomyces roseipellis* has chitinolytic activity and mycoparasitism may be one of the mechanisms involved in its antagonism against *Pyrenophora tritici-repentis* in the debris of wheat crops (Pfender 1988).

4 Commercialization of Bioagents

As mentioned earlier there are approximately 61 biocontrol products (i.e. including fungal and bacterial) being produced on a commercial basis all over the world. A successful biocontrol requires considerable understanding of cropping system, disease epidemiology, the biology, ecology, and population dynamics of biocontrol organisms; and the interactions among these variables (Lo et al. 1998). Understanding the mechanisms or activities for antagonist pathogen interactions will be one of the important steps because it may provide a reasonable basis for selection and construction of more effective biocontrol agents (Lo et al. 1997; Heydari and Pessarakli 2010). Over the past few years, the novel application s of molecular techniques has broadened our insight into the basis of biological control of plant diseases.

Table 4.5 List of fungal bioagents with their trade and manufacturers name

Commercial products	Bioagents used	Name of the manufacturer
AQ10 biofungicide	*Ampelomyces quisqualis* isolate M-10	Ecogen, Inc. Israel
Anti-Fungus	*Trichoderma* spp.	Grondortsmettingen De Cuester, Belgium
Biofungus	*Trichoderma* spp.	Grondortsmettingen De Cuester n. V.Belgium
Bas-derma	*T. viride*	Basarass Biocontrol Res. Lab., India
Binab T	*Trichoderma harzianum* (ATCC 20476) and *Trichoderma polysporum* (ATCC 20475)	Bio-Innovation AB, UK
Bioderma	*Trichoderma viride/T. harzianum*	Biotech International Ltd., India
Biofox C	*Fusarium oxysporum* (Non-pathogenic)	S. I. A. P. A., Italy
Prestop, Prirnastop	*Gliocladium catenulatum*	Kemira Agro. Oy, Finland
Root shield, Plant shield, T-22 Planter box	*Trichoderma harzianum* Rifai strain KRL-AG (T-22)	Bioworks Inc., USA
Root Pro, Root Prota to	*Trichoderma harzianum*	Efal Agr, Israel
Soilgard	*Gliocladium virens* strain GL-21	Thermo Trilogy, USA
Supresivit	*Trichoderma harzianum*	Borregaard and Reitzel, Czech Republic
T-22 G, T-22 HB	*Trichoderma harzianum* strain KRL-AG2	THT Inc., USA
Trichodex, Trichopel	*Trichoderma harzianum*	Makhteshim Chemical Works Ltd., USA
Trichopel, Trichoject, Trichodowels, Trichoseal	*Trichoderma harzianum* and *Trichoderma viride*	Agrimm Technologies Ltd., New Zealand
Trichopel	*Trichoderma harzianum* and *Trichoderma viride*	Agrimm Technologies Ltd., New Zealand
Trichoderma 2000	*Trichoderma* sp.	Myocontrol Ltd., Israel
Tri-control	*Trichoderma* spp.	Jeypee Biotechs, India
Trieco	*Trichoderma viride*	Ecosense Labs Pvt. Ltd., Mumbai, India
TY	*Trichoderma* sp.	Mycocontrol, Israel

New molecular approaches have been available for assessment of interaction between the antagonist and pathogen, ecological traits of antagonists in rhizosphere and improving the efficacy of fungal biocontrol agents (Harman et al. 1998; Perveen and Bokhari 2012). There has been a significant increase in the number of biological control agents registered or on the market worldwide in the last few years (Whipps 1998). Table 4.5 enlists the fungal bioagents with their trade and manufacturers name.

Despite all the success stories of effective biological control of plant diseases the rate of commercialization of biocontrol agents has been slow and therefore their application in the field is restricted to the 33 commercial biocontrol products listed by Fravel et al. (1999). Over half of them have applications in nursery or greenhouses and many were developed specifically for soil-borne plant pathogens e.g. *Pythium*, *Fusarium* and *Rhizoctonia*. The world's total greenhouse area is 307,000 ha including both plastic and glass (Gullinos et al. 1999) whereas the total land area under cultivation in 1998 was 1.51 billion hectare (FAOSTAT 2000). The use of biocontrol is more prevalent in green houses and protected structures than in open field in developed countries. This is another area of concern, which must be looked into before moving towards commercialization. Not many products are available which are broad-spectrum so that, they can be used on a large scale. Still, today important fungal plant diseases lacking commercialized biocontrol management and large number of diseases are still not covered by the commercialized biocontrol formulations. Thus, there is need for moving ahead with selecting the effective strains of biocontrol fungi and developing them as broad-spectrum formulations for use against a large number of soil-borne diseases in varying soil conditions. Another major obstacle is absence of infrastructure for scale up and commercialization of biocontrol products in most of the countries (Cook 1993). A system similar to that developed for release of new cultivars by breeders must be developed for commercialization and dissemination of biocontrol agents to the farmers.

4.1 Selection of Strain

Selection of a potential strain is the ideal condition required for the development of a biocontrol agent and its subsequent commercialization. The relative ease of finding the right antagonist strain under laboratory and field conditions is an added advantage but it does ensure that the agent would work consistently and effectively in field. The search for an effective strain involves years of screening and testing. The development of T-22 strain of *Trichoderma harzianum* by Harman and his associates took more than a decade and it was only in the last decade of last century that the sale of its product, Root Shield, started to pick up (Harman 2000; Shoresh and Harman 2008). Genetic engineering of effective strains or those strains, which do not have the desired traits, must also be tried. The demonstration by Huang et al. (1997) of the activity against *Rhizoctonia* root rot in addition to the ability of a phloroglucinol producing strain to produce phenazine-l-carboxylate shows that it may be possible to produce strains with ability to control more than one disease and possibly more than one crop. This may broaden the market potential needed to justify costs of development and registration of commercialized biological control products. There are some of the biological control agents like Plant Growth Promoting Rhizobacteria (PGPR) and some strains of *Trichoderma* and *Gliocladium*, which besides controlling various soil borne diseases also enhance the yield of host crop (Nautiyal 2000; Kotamraju 2010). It would be a boon to the industries associated with development of biocontrol products, if a strain that possesses both these properties and can be produced successfully on a large scale.

4.2 *Shelf Life and Storage*

For the establishment of a product containing microbial antagonists in the market it is essential that it has long shelf life and can be stored at room temperature for at least one crop season. Mathre et al. (1999) suggested that a shelf life of 18-months is preferable for a commercial microbial product with a requirement of storage at room temperature. The last condition is essential, as the end users can not afford to store commercial biocontrol agents at any temperature other than room temperature. Besides, it will render the biocontrol product economically unviable due to high cost of maintenance.

5 Conclusion

It has been already mentioned in the literature that presently, the "fungi" are a mega-diverse group, estimates for the number of fungi in the world range up to 13.5 M species (McNeely et al. 1990; Hawksworth 1991, 2001). The role of fungal bio-agents in the management of devastating fungal plant diseases is of great importance. It is the gift of nature that fungal antagonist of most plant pathogens are commonly found in the vicinity or associated with the pathogens. It was the famous scientist Weindling (1932) who made the world to know about mycoparasitism as a remedy for plant diseases, now it is fruitful to say that fungal diversity is an important tool in plant disease management. Still a large work has to be done in the field of biological control of plant diseases. Although there is a large chunk of biocontrol products commercialized all over the world for managing soil borne fungal diseases of plants but the commercial bio products have the disadvantage of controlling only one or two pathogens and have been tried on relatively few crops before being released in the market. Since the commercialized products available in the world market for disease management are either location specific or disease specific or both so extensive performance trials should be conducted before accepting one strain for advancement towards commercialization. The release of new wheat cultivars is done after conducting experiments for 100 site years (number of sites × number of years) so as to produce enough data extensively but only after being sure about commercialization of microbial agents for use in agriculture. *Trichoderma* spp. play a major role as biocontrol agents, owing to their capabilities of ameliorating crop-yields by a role of biopesticide, among fungal bipesticides they comprise of 65% share. In order to enhance marketability of these fungi as BCAs, feasible commercial production processes are of utmost importance. Their cheaper formulations and alternative substrates for their mass production, ease of delivery and application, compatibility with other bioagents are such bottle necks in biocontrol so strategies should be made to rule out these problems.

References

Abubaker K (2010) Cell wall degrading enzymes and interaction between *Trichoderma aggressivum* and *Agaricus bisporus*. Dissertation, University St. Catharines, St. Catharines

Adams PB (1990) The potential of mycoparasites for biological control of plant diseases. Annu Rev Phytopathol 28:59–72

Adams DJ (2004) Fungal cell wall chitinases and glucanases. Microbiology 150:2029–2035

Agrios GN (2005) Plant pathology, 5th edn. Elsevier Academic Press, Amsterdam

Anderson AJ, Tari PH, Tepper CS (1988) Genetic studies on the role of an agglutinin in root colonization by *Pseudomonas putida*. Appl Environ Microbiol 54:375–380

Askary H, Carriere Y, Bélanger RR, Brodeur J (1998) Pathogenicity of the fungus *Verticillium lecanii* to aphids and powdery mildew. Biocontrol Sci Technol 8:23–32

Baker KF, Cook RJ (eds) (1974) Biological control of plant pathogens. W. H. Freeman, San Francisco, 433 pp

Barak R, Elad Y, Mirelman D, Chet I (1985) Lectins: a possible basis for specific recognition in the interaction of Trichoderma and Sclerotium rolfsii. Phytopathology 75:458–462

Belanger RR, Labbe C (2002) Control of powdery mildews without chemicals: prophylactic and biological alternatives for horticultural crops. In: Belanger RR, Bushnell WR, Dik AJ, Carver TLW (eds) The powdery mildews: a comprehensive treatise. American Phytopathological Society, St. Paul, pp 256–267

Belanger RR, Dufour N, Caron J, Benhamou N (1995) Chronological events associated with the antagonistic properties of *Trichoderma harzianum* against *Botrytis cinerea*: indirect evidence for sequential role of antibiosis and parasitism. Biocontrol Sci Technol 5:41

Benítez T, Rincón MA, Limón MC, Codón CA (2004) Biocontrol mechanisms of *Trichoderma* strains. Int Microbiol 7:249–260

Bernett HL, Binder FL (1973) The fungal host parasite relationship. Annu Rev Phytopathol 11:273–292

Boogert van den PHJF (1989) Nutritional requirement of the Mycoparasitism fungus. *Verticillium biguttatum*. Neth J Plant Pathol 95:149–156

Boosalis MG (1956) Effect of soil temperature and green manure amendmend of unsterilized soil on parasitism of *Rhizoctonia solani* by *Penicillium vermiculatum* and *Trichoderma* spp. Phytopathology 46:473–478

Boudreau MA, Andrews JH (1987) Factors influencing antagonism of *Chaetomium globosum* to *Venturia inaequalis:* a case study in failed biocontrol. Phytopathology 77:1470

Cavalier-Smith T (1998) A revised six-kingdom system of life. Biol Rev 73:203–266

Chawda HT, Rajasab AH (1992) Effect of culture filtrate of Myrothecium verrucaria on the conidial germination of some pathogens of onion. Onion Newsl Tropics 4:62–65

Chet I, Baker R (1981) Isolation and biocontrol potential of *Trichoderma hamatum* from soil naturally suppressive to *Rhizoctonia solani*. Phytopathology 71:286–290

Chet I, Baker R, Harman GE (1981) *Trichoderma hamatum*: its hyphal interaction with *Rhizoctonia solani* and *Pythium* sp. Microb Ecol 7:29–38

Cook RJ, Baker KF (1983) The nature and practice of biological control of plant pathogens. The American Phytopathological Society, St. Paul

Cook RJ (1993) Making greater use of introduced microorganism for biological control of plant pathogens. Ann Rev Phytopathol 81:53–80

Diop-Bruckler M, Molot PM (1987) Intéret de quelques hyperparasites dans la lutte contre *Leveillula taurica*, *Sphaerotheca fuliginea*. Bull OEPP 17:593–600

DiPietro A, Lorito M, Hayes CK, Broadway RM, Harman GE (1993) Endochitinase from *Gliocladium virens*: isolation, characterization and synergistic antifungal activity in combination with gliotoxin. Phytopathology 83:308–313

Druzhinina IS, Seidl-Seiboth V, Herrera-Estrella A, Horwitz BA, Kenerle CM, Monte E, Mukherje PK, Zeilinger S, Grigorie IV, Kubicek CP (2011) *Trichoderma*: the genomics of opportunistic success. Nat Rev Microbiol 9(10):749–759

Elad Y, Chet I, Boyle P, Henis Y (1983) Parasitism of *Trichoderma* spp on *Rhizoctonia solani* and *Sclerotium rolfsii* scanning electron microscopy and fluorescence microscopy. Phytopathology 73:85–88

Elad Y, Baker R (1985) Influence of trace amounts of cations and siderophore-producing pseudomonads on chlamydospore germination of *Fusarium oxysporum*. Ecol Epidemiol 75:1047–1052

Elad Y (1996) Mechanisms involved in the biological control of Botrytis cinerea incited diseases. Eur J Plant Pathol 102:719–732

FAOSTA T (2000). Food and Agriculture Organization of United Nations. http://apps.fao.org/.

Fokkema NJ (1971) The effect of pollen in the phyllosphere of rye on colonization by Saprophytic fungi and on infection by Helminthosporium sativum and other leaf Pathogens. Neth J Plant Pathol 77(Suppl. 1):1–60

Fravel DR (1988) Role of antibiosis in the biocontrol of plant diseases. Annu Rev Phytopathol 26:75–91

Fravel DR, Rhodes DJ, Larkin RP (1999) Production and commercialization of biocontrol products. In: Albajes R, Gullino ML, van Lenteren JC, Elad Y (eds) Integrated pest and disease management in greenhouse crops. Kluwer Academic Publishers, Dordrecht, 545, pp 365–376

Fries EM (1825) Systema Orbis Vegetabilis, vol 1. Typographia Academica, Lund

Gerlagh M, Goossen-van der geijn HM, Verdam B, Fokkema NJ (1995) Biological control of white mould (*Scleotinia sclerotiorum*) in various crops by application of *Coniothyrium minitans*. In: Whipps JM, Gerlagh T (eds) Biological control of Sclerotium forming fungi, IOBC/WPRS Bulletin 18(3):13–17

Gullino ML (1992) Control of botrytis rot of grapes and vegetables with *Trichoderma* spp. In: Tjamos EC, Papavizas GC, Cook RJ (eds) Biological control of plant diseases, progress and challenges for the future. Plenum Press, New York, pp 125–132

Gullinos ML, Albajes R, van Lenteren JE (1999) Setting the stage: characteristics of protected cultivation and tools for sustainable crop production. In: Albajes R, Lodovica Gullino M, van Lenteren JE, Elad Y (eds) Integrated pest management in greenhouse crops. Kluwer Academic Publishers, Dorcrecht, pp 1–15, 545

Hannusch DJ, Boland L (1996a) Humidity on biological control of white mould of bean (*Sclerotinia sclerotiarum*). Phytopathology 86:156

Hannusch DJ, Boland GL (1996b) Humidity and biological control agents on grey mold bean. Eur J Plant Pathol 102:133

Harman GE (2001) In: Tzeng DDS, Huang JW (eds) Proceedings of international symposium on biological control of plant diseases for the new century – mode of action and application technology. National Chung Hsing University, Taichung City, pp 71–84

Harman GE (2000) Myths and dogmas of biological control: change in perceptions desired from research on *Trichoderma harzianum* T-22. Plant Dis 84(4):377–393

Harman GE, Björkman T (1998) Potential and existing uses of Trichoderma and Gliocladium for plant disease control and plant growth enhancement. In: Kubicek CK, Harman GE (eds) Trichoderma and Gliocladium, vol 2. Taylor and Francis, London, pp 229–265

Harris AR (1999) Biocontrol of *Rhizoctonia solani* and *Pythium ultimumon Capsicum* by *Trichoderma koningii* in potting medium. Microbiol Res 154:333–337

Hawksworth DL (1981) A survey of the fungicolous conidial fungi. In: Cole GT, Kendrick B (eds) Biology of conidial fungi, vol I. Academic, New York, pp 171–235

Hawksworth DL (1991) The fungal dimension of biodiversity: magnitude, significance, and conservation. Mycol Res 95:641–655
Hawksworth DL (2001) The magnitude of fungal diversity: the 1.5million species estimate revisited. Mycol Res 105:1422–1432
Hawksworth DL, Sutton BC, Ainsworth GC (1983) Ainsworth & Bisby's dictionary of the fungi, 7th edn. Commonwealth Mycological Institute, Kew
Heidi IG Abo-Elnaga (2012) Biological control of damping off and root rot of wheat and sugar beet with *Trichoderma harzianum*. Plant Pathol J 11:25–31
Heintz C, Blaich R (1990) *Verticillium lecanii* als hyperparasit des rebmehltaus (uncinula necator). Vitis 29:229–232
Hewitt HG (1998) Fungicides in crop protection. CABI Publishing, Wallingford
Heydari A, Pessarakli M (2010) A review on biological control of fungal plant pathogens using microbial antagonists. J Biol Sci 10:273–290
Hijwegen T (1989) Effect of culture filtrates of seventeen fungicolous fungi on sporulation of cucumber powdery mildew. Neth J Plant Pathol 95(Suppl 1):95–98
Hijwegen T (1992) Biological control of cucumber powdery mildew with Tilletiopsis minor under greenhouse conditions. Neth J Plant Pathol 98:221–225
Hijwegen T, Buchenauer H (1984) Isolation and identification of hyperparasitic fungi associated with Erysiphaceae. Neth J Plant Pathol 90:79–84
Hijwegen T (1988) Effect of seventeen fungicolous fungi on sporulation of cucumber powdery mildew. Neth J Plant Pathol 94:185–190
Hofstein R, Daoust RA, Aeschlimann JP (1996) Constraints to the development of biofungicides: the example of 'AQ-10', a new product for controlling powdery mildews. Entomophaga 41:455–460
Huang JW, Chen J, Berti WR, Cunningham S (1997) Phytoremediation of lead contaminated soils: role of synthetic chelates in lead phytoextraction. Environ Sci Technol 31:800–805
Imtiaj A, Tae L (2008) Antagonistic effect of three Trichoderma species on the Alternaria porri Pathogen of onion Blotch. World J Agric Sci 4:13–17
Inbar J, Chet I (1992) Biomimics of fungal cell – cell recognition by use of lectin-coated nylon fibers. J Bacteriol 174:1055–1059
Jager G, Ten Hoopen A, Velvis H (1979) Hyperparasites of *Rhizoctonia solani* in Dutch potatoe feilds. Neth J Plant Pathol 85:253–268
James TY, Kauff F, Schoch CL et al (2006) Reconstructing the early evolution of the fungi using a six gene phylogeny. Nature 443:818–822
Jarvis WR, Slingsby K (1977) The control of powdery mildew of greenhouse cucumber by water sprays and *Ampelomyces quisqualis*. Plant Dis Rep 61:728–730
Jeffries P, Young TWK (1994) Interfungal parasitic relationship. CAB International, Oxon, pp 9–12
Jones EE, Stewart A (2011) Coniothyrium minitans survival in soil and ability to infect sclerotia of *Sclerotinia sclerotiorum*. N Z Plant Prot 64:168–174
Kanokmedhakul S, Kanokmedhakul K, Phonkerd N, Soytong K, Kongsaree P, Suksamrarn A (2002) Antimycobacterial anthraquinonechromanone compound and diketopiperazine alkaloid from the fungus *Chaetomium globosum* KMITL-N0802. Planta Med 68:834–836
Kanokmedhakul S, Kanokmedhakul K, Nasomjai P, Loungsysouphanh S, Soytong K, Isobe M, Kongsaeree K, Prabpai S, Suksamran A (2006) Antifungal azaphilones from the fungus; *Chaetomium cupreum*. J Nat Prod 69:891–895
Kaur R, Kaur J, Singh RS (2010) Nonpathogenic *Fusarium* as a biological control agent. Plant Pathol J 9:79–91
Keel C, Voisard C, Berling CH, Kahir G, Defago G (1989) Iron sufficiency is a prerequisit for suppression of tobacco black root rot by *Pseudomonas fluorescnes* strain CHA0 under gnotobiotic contiditions. Phytopathology 79:584–589
Kiss L (1998) Natural occurrence of *Ampelomyces mycoparasites* in mycelia of powdery mildew fungi. New Phytol 140:709–714
Kiss L (2003) A review of fungal antagonists of powdery mildews and their potential as biocontrol agents. Pest Manag Sci 59:475–483

Kiss L, Russell JC, Sznentivanyi O, Xu X, Jeffries P (2004) Biology and biocontrol potential of *Ampelomyces mycoparasites*, natural antagonists of powdery mildew fungi. Biocontrol Sci Technol 17:635–651

Knudsen IMB, Skou JP (1993) The affectivity of *Tilletiopsis albescens* in biocontrol of powdery mildew. Ann Appl Biol 123:173–185

Kotamraju VKK (2010) Management of sheath blight and enhancement of growth and yield of rice with plant growth-promoting rhizobacteria. Dissertation, Auburn University Alabama

Kulkarni M, Chaudhari R, Chaudhari A (2007) Novel tensio-active microbial compounds for biocontrol applications. In: Ciancio A, Mukerji KG (eds) General concepts in integrated pest and disease management. Springer, Dordrecht, pp 295–304

Kumar P (2007) Biological management of *Alternaria* blight of onion. Dissertation, University of Agricultural Science, Dharwad

Kuter GA (1984) Hyphal interaction of *Rhizoctonia solani* and some *Verticillium* spp. Mycologia 76:936–940

Legler et al. (2011) New perspectives for the use of *Ampelomyces*-based biofungicides for effective control of powdery mildew on grapevine. Paper presented at the 4th international conference on non chemical crop protection methods, Lille, 8–10 Mar 2011

Lo CT, Nelson EB, Harman GE (1997) Biological control of Pythium, Rhizoctonia and Sclerotinia infected diseases of turfgrass with *Trichoderma harzianum*. Phytopathology 84:1372–1379

Lo CT, Nelson EB, Hayes CK, Harman GE (1998) Ecological studies of transformed *Trichoderma harzianum* strain 1295–22 in the rhizosphere and on the phylloplane of creeping bentgrass. Phytopathology 88:129–136

Loper JE, Buyer JS (1991) Siderophores in microbial interactions of plant surfaces. Mol Plant Microbe Interact 4:5–13

Madrigal C, Pascual S, Melgarejo P (1994) Biological control of peach twig blight (*Monilinia laxa*) with *Epicoccum nigrum*. Plant Pathol 43:554

Mathre DE, Cook RJ, Callan NW (1999) From discovery to use: traversing the world of commercializing biocontrol agents for plant disease control. Plant Dis 83:972–983

Mathur M, Mukerji KG (1981) Antagonistic behaviour of *Cladosporium spongiosum* against *Phyllactinia dalbergiae* on *Dalbergia sissoo*. Angew Bot 55:75–77

McDougall P. (2006) Agriservice Report. Pathhead, Midlothian, Scotland

McNeely JA, Miller KR, Mittermeier R, Werner TB (1990) Conserving the world's biological diversity. International Union for Conservation of Nature and Natural Resources, Gland

Melo IS, Faull JL (2000) Parasitism of *Rhizoctonia solani* by strains of *Trichoderm*

Melo IS, Faull JL, Graeme-Cook KA (1997) Relationship between in vitro cellulose production of UV-induced mutants of *Trichoderma harzianum* and their bean rhizosphere competence. Mycol Res 101:1389–1392

Mitchell AM, Strobel GA, Hess WM, Vargas PN, Ezra D (2008) *Muscodor crispans*, a novel endophyte from *Ananas ananassoides* in the Bolivian Amazon. Fungal Divers 31:37–43

Molan YY (2009) Detection of presumptive mycoparasites in soil placed on host-colonized agar plates in Riyadh region, Saudi Arabia. Asian J Plant Pathol 3:22–26

Mukherjee PK et al (2011) Secondary metabolism in *Trichoderma* – a genomic perspective. Microbiology 158:35–45. doi:10.1099/mic. 0.053629-0

Nautiyal CS (2000) Plant beneficial rhizosphere competent bacteria. Proc Natl Acad Sci India 70:107–123

Ordentlinch A, Nachmias A, Chet I (1990) Integrated control of Verticillium dahlia in potatoe by Trichoderma harzianum and captan. Crop Prot 9:363–366

Pal KK, Mc Spadden Gardener B (2006) Biological control of plant pathogens. Plant Health Instructor. doi:10.1094/PHI-A-2006-1117-02

Pascoe IG (1990) History of systematic mycology in Australia. In: Short PS (ed) History of systematic botany in Australia. Australian Systematic Botany Society, South Yarra, pp 259–264

Paulitz TC, Belanger RR (2001) Biological control in greenhouse systems. Annu Rev Phytopathol 39:103–133

Persson Y (1991) Mycoparasitism by the nematode trapping fungus *Arthrobotrys oligospora*. Ph.D. thesis, University of Lund, Lund

Perveen K, Bokhari NA (2012) Antagonistic activity of *Trichoderma harzianum* and *Trichoderma viride* isolated from soil of date palm field against *Fusarium oxysporum* African. J Microbiol Res 6(13):3348–3353

Pfender WF, Wootke SL (1988) Microbial communities of Pyrenophora-infested wheat straw as examined by multivariate analysis. Microb Ecol 15:95–113

Philipp WD, Beuther E, Hermann D, Klinkert F, Oberwalder C, Schmidtke M, Straub B (1990) Zur Formulierung des Mehltauhyperparasiten *Ampelomyces quisqualis* Ces. Z Pflkrankh Pflschutz 97:120–132

Pimentel D, Levitan L (1986) Pesticides: amounts applied amounts reaching pests. Biosciences 36:86–91

Prasad V et al (2008) Microorganism mediated management of plant pathogens. In: Maheshwari DK, Dubey RC (eds) Potential microorganism for sustainable agriculture. IK International Publishing House, New Delhi

Raghavendra Rao NN, Pavgi MS (1978) Two mycoparasites on powdery mildews. Sydowia 30:145–147

Raghukumar C (2008) Marine fungal biotechnology: an ecological perspective. Fungal Divers 31:19–35

Ridout CJ, Coley-Smith JR, Lynch JM (1988) Fractionation of extracellular enzymes from a mycoparasitic strain of *Trichoderma harzianum*. Enzyme Microb Technol 10:180–187

Rossman AY (1994) A strategy for an all-taxa inventory of fungal diversity. In: Peng CI, Chen CH (eds) Biodiversity and terrestrial ecosystems. Institute of Botany, Academia Sinica, Taipei, pp 169–194

Schirmbock M, Lorito M, Wang YL, Hayes CK, Arisan AI, Scala F, Harman GE, Kubicek C (1994) Parallel formation and synergism of hydrolytic enzymes and peptaibol antibiotics, molecular mechanisms involved in the antagonistic action of *Trichoderma harzianum* against phytopathogenicfungi. Appl Environ Microbiol 60:4364–4370

Shoresh M, Harman GE (2008) The molecular basis of shoot responses of maize seedlings to *Trichoderma harzianum* T22 inoculation of the root: a proteomic approach. Plant Physiol 147(4):2147–2163

Sivan A, Chet I (1989) The possible role of competition between *Trichoderma harzianum* and *Fusarium oxysporum* on rhizosphere colonization. Phytopathology 79:198–203

Soytong K, Srinon W, Rattanacherdchai K, Kanokmedhakul S, Kanokmedhakul K (2005) Application of antagonistic fungi to control antracnose disease of grape. Int J Agric Technol 1:33–41

Srivastava SL, Bisht S (1986) Inhibitory effect of some phylloplane fungi on powdery mildew disease development. Indian Phytopathol 39:83–86

Swati S, Adholeya A (2008) Biological control alternative paradigm for commercialization. In: Maheshwari DK, Dubey RC (eds) Potential microorganism for sustainable agriculture. IK International Publishing House, India

Tejesvi MV, Kini KR, Prakash HS, Subbiah V, Shetty HS (2007) Genetic diversity and antifungal activity of species of *Pestalotiopsis* isolated as endophytes from medicinal plants. Fungal Divers 24:37–54

Thrane C, Tronsmo A, Jensen DF (1997) Endo-1, 3-b-glucanase and cellulose from *Trichoderma harzianum*: purification and partial characterization, induction by and biological activity against plant pathogenic Pythium spp. Eur J Plant Pathol 103:331–344

Traquair JA, Shaw LA, Jarvis WR (1988) New species of *Stephanoascus* with *Sporothrix anamorphs*. Can J Bot 66:926–933

Turhan G (1990) Further hyperparasitism of rhizoctonia solani as promising candidate for biological control. Zeitshrift Pflanzenkrankheiten Pflanzenshutz 97:208–215

Urquhart EJ, Menzies JG, Punja ZK (1994) Growth and biological control activity of *Tilletiopsis* species against powdery mildew (*Sphaerotheca fuliginea*) on greenhouse cucumber. Phytopathology 84:341–351

Vinalea F, Sivasithamparamb K, Emilio L, Ghisalbertic MR, Sheridan L, Lorito WM (2008) Review article trichoderma–plant–pathogen interactions. Soil Biol Biochem 40:1–10

Weindling R (1932) Studies on a lethal principle defective in the parasitic action of *Trichoderma lignorum* on *Rizoctonia solani* and other soil fungi. Phytopathology 26:1068–1070

Weller DM (1988) Biological control of soil borne plant pathogens in the rhizosphere with bacteria. Annu Rev Phytopathol 26:379–407

Whipps JM, Gerlagh M (1992) Mycol Res 96:897–907

Whipps JM (1998) Microbial interactions as basis for biological control of fungi diseases. In: 7th international congress of plant pathology, invited papers abstract, vol 1:2.10.1S. BSPP, Edinburgh, Scotland, 1998

Yoder KS (2000) Effect of powdery mildew on apple yield and economic benefits of its management in Virginia. Plant Dis 84:1171–1176

Chapter 5
Polyphasic Identification and Preservation of Fungal Diversity: Concepts and Applications

Marta F. Simões, Leonel Pereira, Cledir Santos, and Nelson Lima

Abstract Fungi are a diverse group of unique eukaryotic organisms currently accepted as the Eumycota kingdom. The (under) estimated number of fungal species is 1.5×10^6 of which only a small number have been identified (*ca.* 8–10%). They are ubiquitous in nature with an extraordinary ability to decompose plant wastes while also causing much spoilage of food and other relevant commodities. Certain species are used directly as food and others in the manufacture of foodstuffs, antibiotics, enzymes, organic acids and alcohol. Still others can infect humans, animals and crops. Information about each microorganism (e.g. morphological and molecular descriptions, including modern spectral data – MALDI-TOF MS, physiological and biochemical features, ecological roles, and societal risks or benefits) is the key element in fungal identification. In order to attain a sound fungal identification a polyphasic approach is required. It is achieved through the integration of all biological traits data. Fungal service culture collections have well established management systems and preservation techniques that are of elemental importance and guarantee the proper identification and characterisation of environmental fungal isolates. They also assure the continuity of taxonomic and comparative studies and fungal availability for biotechnological exploitation. To foster bio-economy and sustain the biotechnological developments new demands for quality control of fungal holdings preserved in culture collections are in course. The quality control system is associated with new guidelines for the culture collections to operate at global level and to adapt the traditional fungal repositories into the new OECD concept of Biological Resource Centres (BRCs).

Keywords Fungal diversity • Polyphasic identification • Biological Resource Centres (BRCs) • Preservation • Culture Collections

---

M.F. Simões • L. Pereira • C. Santos • N. Lima (✉)
IBB – Biological Engineering Centre, Applied Mycology Group,
University of Minho, Campus of Gualtar, 4710-057 Braga, Portugal
e-mail: mfilipa@deb.uminho.pt; leoneljpp@deb.uminho.pt;
cledir.santos@deb.uminho.pt; nelson@ie.uminho.pt

A. Malik et al. (eds.), *Management of Microbial Resources in the Environment*,
DOI 10.1007/978-94-007-5931-2_5, © Springer Science+Business Media Dordrecht 2013

1 Introduction: The Eumycota Kingdom

1.1 *Phyla and Diversity*

Fungi are eukaryotic organisms with a cell wall rich in glucans and chitin that grow by absorption of nutrients after extracellular digestion of organic materials (organo-chemo-heterotrophism). They range in size from species with massive underground structures to microscopic single-celled yeasts. In addition, a large group is characterised by hyphal growth which absorbs nutrients and supports spores and are referred to as filamentous fungi.

Fungi are a ubiquitous and very diverse organisms referred as the Eumycota kingdom or, sometimes, as the "fifth kingdom". The concept of the Fungi as one of six kingdoms of life was introduced by Jahn and Jahn (1949) and a five kingdom system was advanced later by Whittaker (1959). However, from the formal point of view, neither of these works included a Latin diagnosis as required by the International Code of Botanical Nomenclature and the name "kingdom of Fungi" was therefore invalid until Moore (1980) has published it. This kingdom comprises seven currently recognised phyla: namely, Basidiomycota (mushrooms, rusts, smuts, etc.) and Ascomycota (sac fungi, yeasts, etc.) which belong both to the subkingdom Dikarya, and the basal fungi Glomeromycota (arbuscular mycorrhizal fungi and relatives), Microsporidia (microscopic parasitic group), Blastocladiomycota (zoosporic fungi found in soil and fresh water habitats) and its "sister" Neocallimastigomycota (microscopic anaerobic fungi), and the most ancient phyla, Chytridiomycota (microscopic and zoosporic fungi) (Hibbett et al. 2007). At this stage, this kingdom accepts one subkingdom, 10 subphyla, 35 classes, 12 subclasses, and 129 orders. However, fungal classification is very dynamic and the recent discovery reported by Jones and co-authors (2011) of the new proposed phyla Cryptomycota (fungi isolated from aquatic environments which include the genus *Rozella* considered one of the earliest diverging lineages of fungi) shows that even today we have only a scarce knowledge about fungi. The fungal taxonomy evolves continuously with successive redefinitions of the fungal tree of life. The constant changes on fungal taxonomic schemes as well as the new proposed concept of "one fungus one name", to avoid different names for the anamorphic and teleomorphic fungal states (Hawksworth 2011a, b), show that the fungal identification and classification remain difficult and for specialised experts only.

1.2 *Facts and Numbers*

In order to answer the question "How many fungal species exist in the Earth?" several series of scientific papers have been reported concerning the great gap between known and estimated species richness (Hawksworth and Rossman 1997; Hawksworth 2001; Schmit and Mueller 2007; May 2010; Bass and Richards 2011; Blackwell 2011; Mora et al. 2011).

Table 5.1 Most popular world fungal species number estimations

Publication	Estimated species number
Hawksworth (1991)	1,500,000
Hammond (1992)	1,000,000
Rossman (1994)	1,000,000
Hammond (1995)	1,500,000
Cannon (1997)	9,900,000
Fröhlich and Hyde (1999)	1,500,000
May (2000)	500,000
Hawksworth (2001)	2,270,000
O'Brein et al. (2005)	3,500,000–5,100,000
Schmit and Mueller (2007)	712,000
Mora et al. (2011)	611,000

Fungi are cryptic, understudied and hyper diverse organisms, and due to these peculiar features the information currently available for most of the species is very limited and incomplete and estimations of the number of species differ considerably (Branco 2011). Until now, there have been several attempts to estimate the total species richness of fungi (Table 5.1) that were based on different assumptions.

The most commonly accepted estimation is the 1.5 million species hypothesised by Hawksworth (1991) and revised 10 years later by himself (Hawksworth 2001). This assumption was based on the observed ratio between flowering plant diversity and fungal diversity growing in particularly well-studied European countries (Finland, Switzerland and United Kingdom), where fungi have been sufficiently well studied to enable a reasonable estimate of true diversity.

The ratio obtained by DL Hawksworth was 5:1–7:1 of fungal species to flowering plant species in these particular locations. Since the estimative of plant species diversity was 300,000; the extrapolated fungal species number was 1,500,000 for the ratio 5:1. Although this number is very large when compared with more conservative estimations such as of May (2000) and Schmit and Mueller (2007), DL Hawksworth admitted that the value achieved was also underestimated by the following reasons: (1) the ratio of fungal species to plant was based on the lower world region of flowering plant species richness when compared with other regions like North America and Central America; (2) it does not consider interactions with other organisms, such as insects; (3) ratios are based on data from geographical areas with few mycological information; and (4) no extrapolation is made for higher ratios like in tropical and polar regions (Hawksworth 1991, 2001).

An extensive survey of the fungi associated with six individual palms in the genus *Licuala* in Australia and Brunei Darussalam (Borneo) performed by Fröhlich and Hyde (1999) suggested an estimate of 1.5 million species of fungi. However, these authors have considered this estimation very conservative. On the other hand, analyses by Cannon (1997) based on similar studies in two small Caribbean areas of Puerto Rico and the Dominican Republic, using data from the Ascomycete family *Phyllachoraceae* pointed out that the Eumycota kingdom would have 9.9 million of species (Cannon 1997). Moreover, O'Brein and co-authors (2005), using the assumption internal transcribed spacer (ITS)-based, estimated soil fungal richness collected from pine and mixed hardwood, and from the vascular plant richness of the pine and the mixed hardwood, they reached a fungal species richness ranging from 3.5 to 5.1 million. However, they assumed that the richness was underestimated because of the conservative 97% ITS similarity clustering, used to designate Operational Taxonomic Units (OTUs), and the richness estimators continue to increase with increasing sampling.

Mueller and Schmit (2007) combined published and unpublished data on the diversity and geographic distribution of the Eumycota kingdom to estimate a minimum ratio between the number of fungi and plants from different regions of the world. In order to avoid the risk of overestimation they deliberately used conservative assumptions at all stages of their analyses. In this study authors reached a final value of 712,000 species (Schmit and Mueller 2007). Mora and co-authors (2011) using a predictive algorithm based on a correlation between the number of higher taxa and taxonomic rank across the tree of life, came up with a number of 611,000 species of fungi in the world. Because the higher taxa have not been fully identified until now, this dependence leads to an underestimation of the number of fungal species.

Most recent data from the Dictionary of Fungi reported about 100,000 different fungal species, this means that we know, at best, 16% of the fungus species on Earth, or probably, in true reality, only around 1%. In addition, most of these species belong to the Basidiomycetes and Ascomycetes groups which emphasize how much remains to be discovered in the Eumycota kingdom. New high level taxa have been detected and described, for example the phylum Cryptomycota as previously mentioned. The acceptance of Cryptomycota alone could radically increase the size of the Eumycota kingdom (Blackwell 2011).

1.3 Fungal Diversity and Culture Collections

This huge diversity (but at the same time small) can be exploited to provide solutions for many of the challenges that threaten the world today, such as climate change and global warming, loss of biodiversity, invasive alien species, population growth and over-exploitation, pollution, etc., all of them underpin biotechnology, sustainable development and sustainable agriculture. On the other hand, the exploitation of this resource is only possible if the diversity can be analysed, identified, characterised and authenticated by expert taxonomists, and other scientists (e.g., fungi conservationists, food, environmental and medical mycologists, mycotechnologists, etc.), within e.g., culture collections (Scoble 2010).

Culture collections are distributed worldwide and are responsible for holding the fungal diversity primary data. The size and organisation of this kind of infrastructures vary broadly, ranging from collections with small sets of strains to large, structured, well-documented and maintained collections. Examples of Culture

Table 5.2 Culture Collections in the world that preserve fungal holdings

Acronym	Name and country	Website URL
ATCC	American Type Culture Collection, USA	http://www.atcc.org
BCCM/MUCL	(Agro)Industrial Fungi & Yeasts Collection, Belgium	http://bccm.belspo.be/about/mucl.php
CBS	Centraalbureau voor Schimmelcultures, The Netherlands	http://www.cbs.knaw.nl
FGSC	Fungal Genetics Stock Center, USA	http://www.fgsc.net
IMI	CAB International, UK	http://cabi.bio-aware.com
MUM	Micoteca da Universidade do Minho, Portugal	http://www.micoteca.deb.uminho.pt
NCPF	National Collection of Pathogenic Fungi, UK	http://www.hpacultures.org.uk/collections/ncpf.jsp
NRRL	National Center for Agricultural Utilization Research, USDA, USA	http://www.ncaur.usda.gov
URM	University Recife Mycology, Brazil	http://www.ufpe.br/micoteca

Collections that preserve fungi are listed in Table 5.2. For a more comprehensive and detailed information, the World Data Centre of Microorganisms (WDCM, www.wdcm.org) which is the data centre of the World Federation of Culture Collections (WFCC, www.wfcc.info) gives several possibilities to find culture collections and information about fungal holdings around the world. From more than 600 Culture Collections distributed in 70 countries, one fourth fungal strains out of about two million microorganisms are available to be supplied to the different user communities.

The identification, characterisation, correct preservation, stable storage and supply of viable and authentic microorganisms in Culture Collections are a major contribution to the knowledge-based bio-economy. In addition, Culture Collections are becoming even more important with the necessity of authentication the fungal strains to avoid the frequent fungal misidentifications.

The authentication of fungal strains will be a step forward to decrease the fungal misidentifications that is pervasive within the scientific community. This problem does not only concern the scientific community but also the general society.

For decades Culture Collections have provided their public service information through their catalogues. Nowadays, computerisation of catalogues is becoming increasingly imperative. The most straightforward way to make the access to strain data more open and easier are web databases that are vital to distribute information and in turn create knowledge (Canhos et al. 2004; Baird 2010). During the last decade numerous biodiversity information databases have been launched for data collecting. The most important web databases for fungal diversity are: MycoBank (www.mycobank.org), CABRI HyperCatalogue (www.cabri.org), Index Fungorum (www.indexfungorum.org), StrainInfo (www.straininfo.net) and GBIF (www.gbif.org).

2 Fungal Identification

2.1 *Polyphasic Approach*

The identification of species is an important goal in taxonomic microbiology. Information about each microorganism (e.g., morphological description, physiological and biochemical properties, ecological roles, and societal risks or benefits) is a key element in this process. Identifications can be a long and seemingly never-ending process with frequent revisions of the taxonomic schemes. These changes make identification even more complicated for non-specialised researchers as every taxonomic group has specialised literature, terminology and features. This difficulty occur even a basic level that fungal identifications can only be undertaken by a narrow group of scientists especially skilled in the "art". The concept of species is clearly abstract and delimitations are very difficult, and often not consensual. Taking this into account, microbial taxonomy (more evident in fungal taxonomy) and their associate data can often be best applied at the moment where the data are used for a specific purpose: A pragmatic definition is "data fit for use". It is gradually becoming clearer that microbial identification and authentication require a multiple steps approach to generate accurate and useful data (Keys et al. 2004). In reality, this means that is necessary to combine more traditional phenotypic and physiological approaches with modern molecular biology (Rodrigues et al. 2009, 2011). Restriction fragment length polymorphism (RFPLs), random amplification of polymorphic DNA (RAPDs), amplified fragment length polymorphism-PCR (AFLPs-PCR), and DNA fingerprinting have all been used to distinguish microbial taxa that are difficult to characterise by traditional morphological means. It is assumed that the genotype of the species is only an indirect indication of phenotype and ecological adaptation. In other words, microbial species can be defined as the smallest aggregation of population with a common lineage that share unique diagnosable phenotypic features.

Recently, consistent identification and characterisation of species of filamentous fungi has been developed by application of the so-called polyphasic approach. The polyphasic approach consists of the use of different techniques based on the systematisation of the scientific knowledge. Different methodologies such as micro- and macro-morphology, biochemical and molecular biology analyses are applied. Microbial spectral analysis based on mass spectrometry (particularly matrix assisted laser desorption/ionisation time-of-flight mass spectrometry//MALDI-TOF MS) has been developed and used as an important step in the polyphasic identification of filamentous fungi from different fields (Santos and Lima 2010; Santos et al. 2010; Rodrigues et al. 2011; Dias et al. 2011). Information obtained by each technique presented above is compared with each other in a step-by-step based approach. The first

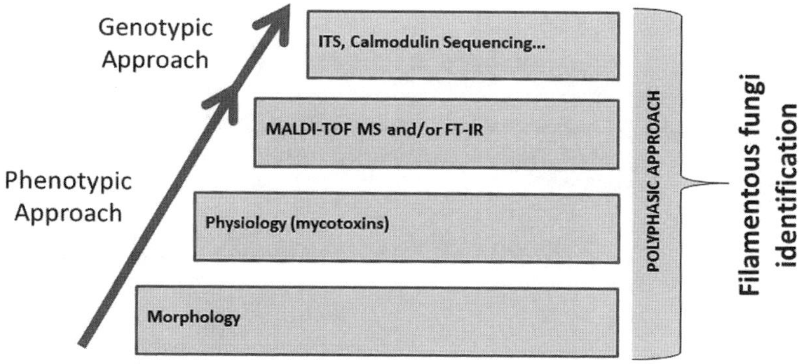

Fig. 5.1 Modern polyphasic methodology of filamentous fungi identification according to Santos and Lima (2010). This scheme recognises molecular biology as the gold standard technique in the step-by-step identification

approach commonly used in this process is morphology followed by biochemistry, spectral analysis and molecular biology. Due to the higher costs and longer answer time, when compared with MALDI-TOF MS, molecular biology in our polyphasic approach is normally used as the last methodology (Fig. 5.1). According to Santos and Lima (2010) molecular biology has been recognised as the gold standard technique in the modern polyphasic methodology of filamentous fungi identification.

2.2 Morphology

Internationally based culture collections are obliged to guarantee the authenticity of the fungi that they hold (Santos and Lima 2001). Identifications are time-consuming, and decisions regarding what represents a species tend to be subjective. The standard method for identifying and classifying filamentous fungi remains morphology because, in general, filamentous fungi have more distinctive morphology traits than, for example, single-cell bacteria and yeast (Santos et al. 2010). Moreover, identification of filamentous fungi by morphology is generally regarded to be very difficult, as features used for delimitation of species within a genus often show minute differences that can only be reliably evaluated by experienced mycologists. Even so, fungal identifications based on morphology only are prone of misidentifications. Furthermore, taxonomy of some filamentous fungi genera is subjected to frequent revision. Table 5.3 shows an example of taxonomy revision evolution for *Aspergillus* sub-genus, section and species within the last 25 years.

In general, however, filamentous fungi as a whole remain difficult to identify because of the biological complexity: classification undergoes almost continual change, particularly since the introduction of nucleic acid-based methodology (Hibbett et al. 2007) where 'cryptic species' are revealed. To illustrate the points

Table 5.3 Taxonomical revision evolution for *Aspergillus* sub-genus, section and species within the last 25 years

Taxonomy after Gams et al. (1985)		Taxonomy after Peterson (2000)		
Sub-genus	Section	Sub-genus	Section	Species
Aspergillus	*Aspergillus*	*Aspergillus*	*Aspergillus*	
	Restricti		*Restricti*	
			Cervini	
			Terrei	
			Flavipedis	
			Wentii	
			Flavi	
			Nigri	
			Circundati	
			Candidi	
			Cremei	
Fumigati	*Fumigati*	*Fumigati*	*Fumigati*	More than 250 species
	Cevini		*Clavati*	
Ornati	*Ornati*	–	–	
Clavati	*Clavati*	–	–	
Nidulantes	*Nidulantes*	*Nidulantes*	*Nidulantes*	
	Versicolores		*Ornati*	
	Usti		*Sparci*	
	Terrei			
	Flavipedes			
Circundati	*Wentii*	–	–	
	Flavi			
	Nigri			
	Circundati			
	Candidi			
	Cremei			
	Sparci			

made, some important penicillia, *Cordyceps* and *Ganoderma* remain problematic despite decades of study (Santos et al. 2010).

Daily routine identification of filamentous fungi genera with high relevance in the environmental field, such as *Aspergillus*, *Fusarium*, *Penicillium*, *Trichoderma*, among others, has been traditionally based on morphological characterisation. Conidial wall ornamentation is regarded as the primary morphological diagnostic character for identification of filamentous fungi species. For some of the known filamentous fungi, genera and/or species differentiation can be achieved through different growth conditions (Rodrigues et al. 2009). Once morphology and colony colour vary among media, use of standard culture media is crucial for the fungal morphology-based identification (Santos et al. 1998). Pitt (1988) used a loopful of spores suspended in 0.2% agar for *Penicillium* species identification. In this case, a suspension of spores is used for three-point inoculations on plates containing three different growth media that are incubated at three different temperatures. Cultures

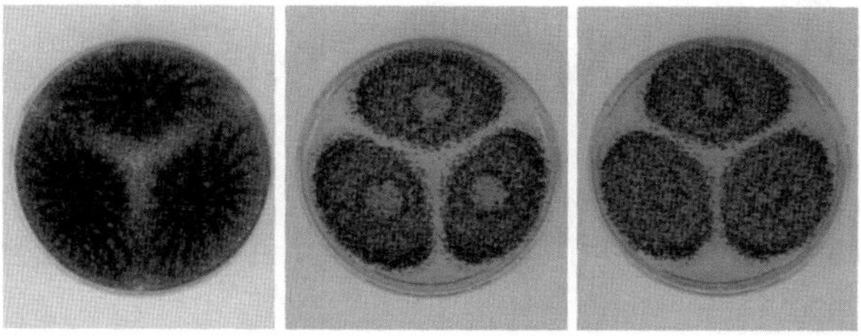

Fig. 5.2 Example of *Aspergillus ibericus* MUM 03.49 growth in CYA (*left*), PDA (*middle*) and MEA (*right*) media at 25°C for 7 days

are grown for (a) 7 days on Czapek Yeast Autolysate agar (CYA) at 5, 25 and 37°C; (b) Malt Extract agar (MEA) and (c) Glycerol Nitrate agar (G25N) both at 25°C. On the other hand, media commonly suggested for *Aspergillus* identification based on three-point inoculations are CYA, CYA with 20% saccharose (CY20S), Potato Dextrose Agar (PDA) and MEA (Fig. 5.2). Medium G25N can be useful for induction of sporulation in case of fungi isolated on "dry" substrates (xerophilic fungi). In both cases cultures are grown for 7 days in the dark at 25°C. Other important medium for *Aspergillus* species identification is *Aspergillus flavus* and *parasiticus* agar (AFPA; Pitt et al. 1983). The latter is used to confirm the identification of a filamentous fungi group based on colony reverse colour (e.g. *A. flavus*, *A. parasiticus* and *A. nomius* are distinguished by bright orange yellow reverse colour in contrast with the cream reverse colour of *A. tamarii*).

Conidia ornamentation of *Aspergillus* and *Penicillium* species are features that are independent of growth conditions (Santos et al. 1998) and can be used for species diagnosis. In addition, these fungi can produce different pigments depending on the growth conditions. Such pigments sometimes help in the fungal identification (Klich 2002; Samson et al. 2004).

2.3 *Biochemical Features and Secondary Metabolites*

The incorporation of biochemical features (e.g., enzymes and secondary metabolites such as mycotoxins) into fungal taxonomy has helped to solve morphological limitations (Petisco et al. 2008). Rapid and reliable physiological tests are available only for a limited number of taxa. Different classes of enzymes have been used as important biochemical traits for the characterisation of fungi. Generally fungi can live in limited conditions because they are able to produce a set of enzymes such as lignin peroxidases, manganese peroxidases, laccases, cellulases, pectinases, xylanases, esterases and lipases capable to convert wood, plastic, paints and jet fuel, among other materials, into useful nutrients.

White rot fungi, for instance, are able to produce several enzymes that have been related to their ability to degrade natural polymers, such as lignin and cellulose, and different synthetic chemicals, usually compounds recalcitrant to biodegradation (Martins et al. 2003). These enzymes are one of the fungal features used for the development of the fungal biochemical profile.

Another important biomarkers used in fungal identification are mycotoxins. Mycotoxins are toxic secondary metabolites produced by fungi, which have adverse health effects on animals and humans. Currently, hundreds of fungal toxins are known although only a small number are taken seriously as mycotoxins and an estimate of ten would be reasonable (Nielsen and Smedsgaard 2003; Paterson and Lima 2010). Most of them are produced by the genera *Aspergillus*, *Fusarium*, *Penicillium* and *Claviceps*. The most serious threats are from mycotoxins having significant oral toxicity and occurring naturally in crops, processed foods, and feed ingredients (Abramson et al. 2009).

Some mycotoxins produced by fungal species belonging to the genus *Aspergillus* section *Flavi*, for instance, are widely used in fungal species identification. Aflatoxins (AFs) B and G (AFBs and AFGs), cyclopiazonic acid (CPA), aspergillic acid and kojic acid (Hong et al. 2005; Samson et al. 2007; Varga et al. 2007a, b; Rodrigues et al. 2011) are the most common mycotoxins produced by this group of fungi. Generally, each fungal species is usually characterised by its specific mycotoxigenic profile (Frisvad 1989; Larsen et al. 2005). However, cases have been reported where different isolates belonging to the same species do not produce the expected secondary metabolites, and this is especially common regarding AFs production (Vaamonde et al. 2003; Giorni et al. 2007; Pildain et al. 2008; Rodrigues et al. 2009).

Another important mycotoxin used as biomarker in the biochemical identification of filamentous fungi species is ochratoxin A (OTA). Some of species belonging to *Aspergillus* and *Penicillium* are able to produce this secondary metabolite. Nowadays, there are about 20 species accepted as OTA producers, which are distributed in three phylogenetically related but distinct groups of aspergilli of the subgenus *Aspergillus* (e.g., *A. carbonarius*, *A. niger*, *A. ochraceus*, etc.) and in only two species of the subgenus *Penicillium*. At the moment, *P. verrucosum* and *P. nordicum* are the only OTA producing species accepted in the genus *Penicillium* (Cabañes et al. 2010).

The three major classes of mycotoxins produced by *Fusarium* species that have been proven to cause animal disease outbreaks are: trichothecenes, fumonisins, and zearalenones. In addition, mycotoxins belonging to the class of trichothecenes have been most strongly associated with chronic and fatal toxicoses of humans and animals, including Alimentary Toxic Aleukia in Russia and Central Asia, Akakabi-byo (red mold disease) in Japan, and swine feed refusal in the central United States (Desjardins and Proctor 2007).

2.4 New Spectral Analysis by MALDI-TOF MS

MALDI-TOF MS emerged in the late 1980s as a sound technique to investigate high molecular mass organic compounds. It runs through a soft ionisation of molecules resulting in minimum fragmentation (Tanaka et al. 1988). MALDI-TOF MS involves subjecting a sample covered with an UV-absorbing matrix that functions as an energy mediator, to a pulsed nitrogen laser. Matrices are chemical compounds generally containing aromatic moieties that transfer the absorbed photoenergy from the irradiation source to the surrounding sample molecules, resulting in minimum fragmentation (Hillenkamp and Peter-Katalinić 2007; Santos et al. 2010). Several matrices are commercially available and the choice depends on the laser wavelength used in each MALDI-TOF MS instrument (Table 5.4). Choosing the appropriate matrix for the identification of filamentous fungi is crucial. Matrices are used as liquid solution and its final composition is constituted by the organic chemical compound that is the matrix properly dissolved in an organic solvent, generally ethanol and/or acetonitrile, and water. In order to help analyte ionisation a strong acid such as trifluoroacetic acid (TFA) is normally employed during matrix liquid solution preparation. The use of the appropriate matrix leads to an optimal signal/noise ratio with narrowest analyte peaks and little signal suppression. Further details for identification of filamentous fungi by MALDI-TOF MS can be obtained elsewhere (Rodrigues et al. 2011).

Currently, the two most commonly used matrices for filamentous fungi identification are 2,5-dihydroxybenzoic acid (DHB) and α-cyano-4-hydroxycinnamic acid (CHCA). Both DHB and CHCA matrices are appropriate for the analysis of molecules with a mass range between 2 and 20 kDa. Constantly expressed and highly abundant proteins, such as ribosomal proteins that appear in this specific mass range can be used as biomarkers. Based on this knowledge, different studies have demonstrated the high potential of this technique for species and strain identification of filamentous fungi (Santos and Lima 2010; Santos et al. 2010; Dias et al. 2011; Rodrigues et al. 2011).

The seminal paper by Cain et al. (1994), where authors presented bacterial identification by MALDI-TOF MS based on a methodology with previous sample preparation with minimal purification of cell contents, was the important milestone for the use of MALDI-TOF MS technique in modern fungal taxonomy. Two years later Holland et al. (1996) described by the first time a method for the rapid identification of whole bacteria. In this work bacteria were sampled from colonies on an agar plate, mixed with the matrix, air-dried, and introduced in batches into the MALDI-TOF apparatus for analysis. This work represents the first reported example of successful bacterial chemotaxonomy by MALDI-TOF MS analysis of whole cells.

MALDI-TOF MS for the identification and classification of microorganisms needs dedicated software and database [e.g., BioTyper™ (Bruker Daltonics Inc., Bremen, Germany), SARAMIS™ (AnagnosTec GmbH, Potsdam-Golm, Germany) or VITEK® MS (bioMérieux, Craponne, France)] to enable comparisons of the unknown proteins with reference molecular masses. Ribosomal proteins are used normally as reference molecular masses as they are the most abundant ones in the cells, as described above.

External MALDI-TOF MS calibration has been performed by use of well characterised proteins from *Escherichia coli* strain K-12 (Ryzhov and Fenselau 2001). From tens of ribosomal proteins of intact *E. coli* K-12 cells 12 well defined proteins are used as MALDI-TOF MS standard (4,365.4, 5,096.8, 5,381.4, 6,241.4, 6,255.4, 6,316.2, 6,411.6, 6,856.1, 7,158.8, 7,274.5, 7,872.1, 9,742 and 12,227.3 Da). The low costs, simple growth conditions, and reliability found in the *E. coli* biomarkers make them the first choice to be use in microbial identification.

Table 5.4 Most common MALDI-TOF MS matrices commercially available and best laser wavelength where each matrix operates

Name	Abbreviation	Wavelength	MW	Chemical structure
α-cyano-4-hydroxy-cinnamic acid	CHCA	UV 337, 353 nm	189.1675	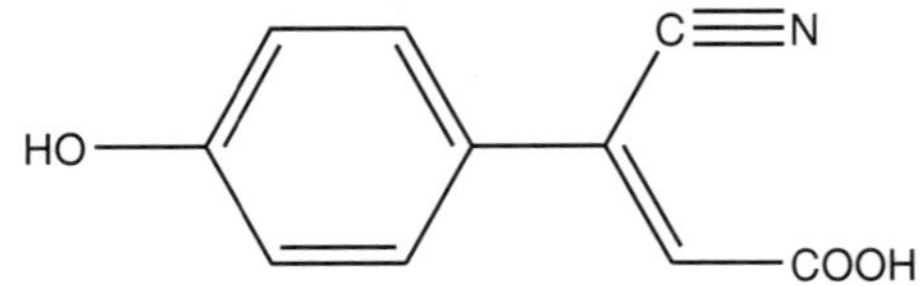
2,5-di-hydroxy-benzoic acid	DHB	UV 337, 353 nm	154.1201	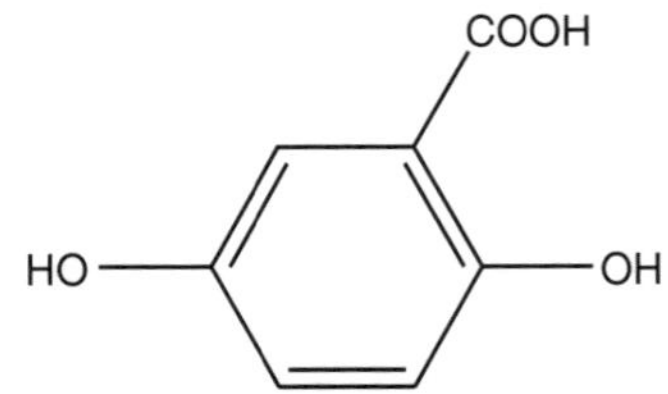
Sinapinic acid	SA	UV 337, 353 nm	224.2100	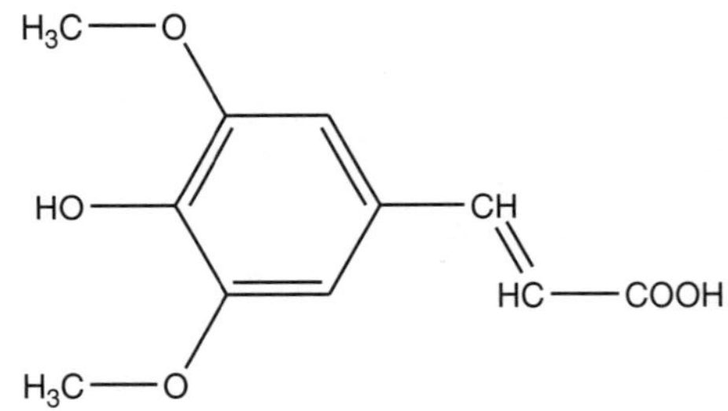

Nicotinic acid	NA	UV 266 nm	123.1094	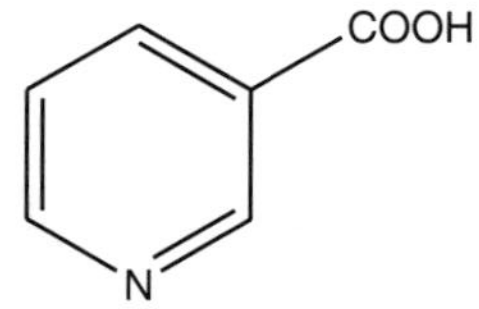
Succinic acid	SCA	IR 2.94, 2.79 μm	118.0266	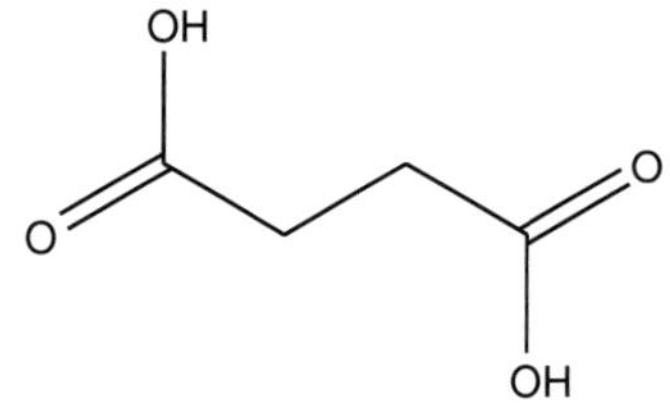

CHCA and DHB are the most used matrices for identification of fungi by MALDI-TOF MS

2.5 Molecular Biology

Advances in molecular tools, in particularly the advent of polymerase chain reaction (PCR) have formed the basis for a boost in studies concerning fungal diversity and allowed a much more straightforward and sensible approach to the study of cryptic organisms, such as fungi. With molecular data information a better understanding of the fungal relationships and the process of a more natural classification are possible (Begerow et al. 2010; Branco 2011). For example, cryptic species that are common in the Eumycota kingdom are only liable to be differentiated and correctly identified through the tools of molecular biology. An example was reported by Perrone and co-authors (2011) that recognised the cryptic species *Aspergillus awamorii* by phylogenetic analysis of DNA sequence data.

With the help of bioinformatics tools, new genetic databases are being created and used as sources for molecular barcodes. DNA barcoding is a technique for a simple characterisation of organisms at species level using a short DNA sequence from a standard and agreed-upon position in the genome. The technique is characterised for being accurate, rapid, cost-effective, culture-independent, universally accessible and usable by non-experts (Frezal and Leblois 2008). A fraction of the mitochondrial cytochrome c oxidase 1 (COX1) gene has become the first official DNA barcode for animals (Hebert et al. 2004; www.barcoding.si.edu) and for plants (Hollingsworth et al. 2009). However, since some fungal species possess distinct genotypes at the mtDNA COI locus, no official consensus DNA fragment was defined for fungi (Samson and Varga 2000; Seifert et al. 2007). Therefore, it becomes important to pinpoint genes which characterise fungal organisms at different taxonomic levels in a straightforward and standardised way.

The most frequently studied markers are one or more of the nuclear rRNA genes, for example, small subunit (SSU) rRNA gene (Wubet et al. 2006), the internal transcribed spacer (ITS) rDNA region including the 5.8S rRNA gene (Hempel et al. 2007; Sykorova et al. 2007), and a part of the large subunit (LSU) rRNA gene (Gollotte et al. 2004; Pivato et al. 2007; Rosendahl et al. 2009), as represented in the Fig. 5.3.

The most widely accepted and used DNA fragment for identifying fungi at the species level is the ITS region. In addition to the standard ITS1 + ITS4 primers used by most research groups, several taxon-specific primers have been described that allow selective amplification of fungal sequences (Table 5.5).

ITS can be used as universal fungal barcode locus for several reasons. Genomes include numerous ribosomal DNA encoding genes distributed in tandem arrays along the same or different chromosomes and they are assumed to be extremely similar (Rooney and Ward 2005). In addition, the ITS fragment is easily amplified by PCR, even from low-quality samples that make it a fast and easy tool to describe fungal diversity (Nilsson et al. 2010). However, there are also disadvantages in the choice and use of the ITS region since the capacity to discriminate at the species level differs considerably among fungal groups that in some cases the identification at species level could be ambiguous (Soares et al. 2012).

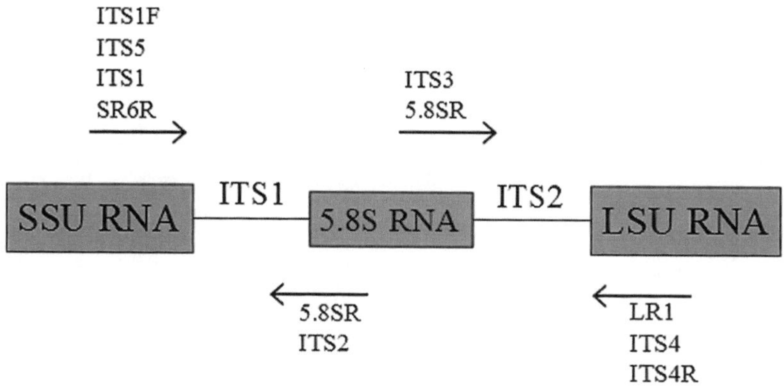

Fig. 5.3 Schematic representation of the internal transcribed spacer (ITS) in the ribosomal operon, the nuclear ribosomal repetitive unit used to describe fungi to the species level and the ITS primers. *SSU* small subunit rRNA gene, *LSU* large subunit rRNA gene

Table 5.5 Primers for ITS region

Primer name/Reference	Sequence (5′→3′)
ITS1 (White et al. 1990)	TCCGTAGGTGAACCTGCGG
ITS2 (White et al. 1990)	GCTGCGTTCTTCATCGATGC
ITS3 (White et al. 1990)	GCATCGATGAAGAACGCAGC
ITS4 (White et al. 1990)	TCCTCCGCTTATTGATATGC
ITS5 White et al. (1990)	GGAAGTAAAAGTCGTAACAAGG
ITS1-F (Gardes and Bruns 1993)	CTTGGTCATTTAGAGGAAGTAA
ITS4-B (Gardes and Bruns 1993)	CAGGAGACTTGTACACGGTCCAG
5.8S (Vilgalys and Hester 1990)	CGCTGCGTTCTTCATCG
5.8SR (Vilgalys and Hester 1990)	TCGATGAAGAACGCAGCG
SR6R (Vilgalys and Hester 1990)	AAGWAAAAGTCGTAACAAGG

3 Fungal Preservation

3.1 Preservation Techniques of Ex Situ Environmental Fungal Strains

Nowadays, it is a must to preserve our biological resources. Maintaining and preserving are essential elements of systematic and biodiversity studies. Environmental fungal strains can be used in research, as reference strains in diagnostic laboratories, for teaching purposes, in bio-industries (e.g., enzyme production, fermentation, bioconversion) and environmental biotechnology (e.g., bioremediation). Since fungi are such a diverse group, several different methods of cultivation and preservation are required to ensure the viability and morphological, physiological and genetic integrity of the cultures over time. The cost effectiveness and robustness of each preservation method are also important aspects to be taken into consideration.

"Off-site" or *ex-situ* preservation of fungal strains means conserving the fungi outside their natural environment and this can be done by several methods. The primary methods of culture preservation are continuous growth, drying and freezing. Continuous growth methods, in which cultures are grown on agar-solid media, are typically used for short-term storage. Such cultures are stored at room temperature between 5 and 25°C, or they may be frozen to increase the interval between subcultures. The methods are simple and inexpensive because specialised equipment is not required.

Drying is the most useful method of preservation for cultures that produce spores or other resting structures. Silica-gel, glass beads and soil are substrates commonly used in drying. The preservation of fungi on silica-gel has been successfully performed for periods over 10 years (Smith and Onions 1983). Drying methods are technically simple and do not require expensive equipment. Freezing methods, including cryopreservation, are versatile and widely applicable. Most fungi can be preserved, with or without cryoprotectants, in liquid nitrogen or in standard home freezers. With freeze-drying the fungal cultures are frozen and subsequently dried under vacuum. The method is highly successful with cultures that produce conidial forms. Freeze-drying and freezing are excellent methods for long-term preservation. However, both methods require specialised and expensive equipment. The choice of preservation methods depends on the species of concern, the resources available and the desired time of preservation. Some low cost methods of preservation are: storage in distilled water and silica-gel though they are not long-term preservation methods, the maximum duration for these methods is described as being 10 years or less (Nakasone et al. 2004).

It is important that biological resources are preserved in a physiological and genetically stable state; this is why, whenever possible, strains should be preserved with one of the permanent methods, like freeze-drying or cryopreservation, these being essential for strains with economically important characteristics and for type strains (Nagai et al. 2005). There are several points to ponder when choosing a preservation method, these being:

- viability maintenance, with minimal loss of viability during proceedings and protocols of preservation;
- prevention of changes in population, keeping characteristics close to the original in the viable population;
- prevention of genetic changes, with maintenance of genetic integrity of each strain;
- assurance of purity, with preservation without any contaminants;
- costs, evaluate the relation between capital costs with the labour costs associated with the preservation method;
- number of cultures, since a large number of cultures affects costs, storage space and manipulation time;
- value of the cultures, since important cultures must be preserved through more than one technique in order to minimise their possible loss and to guarantee a safe deposit of strains;

- frequency of usage, compromise between the regular or not so often use of the strains;
- supply and transport of cultures, must be done according to national and international norms and rules, under the present legislation.

To help in the decision making of preservation and storage method one can use decision-based keys, but it is important to consider that no preservation method should be assumed to guarantee total physiological and genetic stability of an isolate; this is why many scientists wish to preserve many replicates of their fungus and use more than one technique to reduce the chance of strain deterioration (Ryan et al. 2000).

Preservation techniques can be differentiated based on several parameters like the storage period, being short- or long-term, as described in Table 5.6.

The best way to preserve fungi and assure their long-term stability is to find conditions where storage guarantees that after retrieval there has not been any growth or reproduction. Furthermore, all the structural and functional characteristics shall be maintained. However, many parameters can influence or alter this goal. As discussed above, different studies and investigations aim to minimise the number of generations from the initial isolate and reduce to a minimal the cellular activity. Research in this area concerns the optimisation of existing protocols to achieve optimal survival and stability (e.g., delicate and recalcitrant strains), and the development of new methods to fulfil the demands of quality and stability for general research and Biological Resource Centres (BRCs) (Ryan et al. 2001; Santos et al. 2002a, b; Ryan et al. 2012).

3.2 Management Systems for Microbial Diversity Ex situ Preservation

From the social, industrial, economic and scientific point of view the biodiversity is very important for the world. There is a demand to assure and guarantee the conservation of biodiversity. Its sustainable uses and the equitable share of benefits, arising from the use of biodiversity through ethical sourcing practices between all the different BRCs and stakeholders. Efforts are being made to help diminish or stop the loss of biodiversity. One of these efforts is the CBD – Convention on Biological Diversity adopted in 1992 at Rio de Janeiro, Brazil (http://www.cbd.int/history). This is an international convention for the (1) conservation of biodiversity (see articles 8 and 9 for *in situ* and *ex situ* conservation, respectively); (2) sustainable use of the components of biodiversity (see article 10) and, (3) the equitable sharing of the benefits derived from the use of genetic resources (see article 15). The CBD has a universal participation of 193 parties (http://www.cbd.int/convention/text). Additionally, several protocols were created: the Cartagena Protocol was created in order to protect the biological diversity from the potential risks posed by living modified organisms which means any living organism that possesses a novel combination of genetic material obtained through the use of modern biotechnology (Mackenzie et al. 2003;

Table 5.6 Description of the main features of the most common methods of fungal preservation in the context of culture collections

Method		References
Serial transfer or subculturing	Short description: Growth in appropriate media and conditions; transfer an inoculum from the vigorous, healthy and actively growing fungus culture to test tubes or Petri dishes containing an agar medium of choice, in determined time periods; keeping the samples at room temperature or at 4°C in-between the transfer that usually occurs every 2 months. Advantages: Simple procedure, inexpensive and widely used. Disadvantages: Short-term preservation technique. Time consuming and labour intensive. Cultures require constant specialist supervision.	Smith and Onions (1983, 1994); Smith et al. (2001); Nakasone et al. (2004)
Preservation with mineral oil	Short description: Fungal cultures are grown on agar slants in tubes and afterwards covered with sterilised mineral oil or liquid paraffin. Advantages: Simple procedure. Low-cost and low-maintenance method. Appropriate for mycelial or nonsporulating cultures that are not adequate to be preserved by freezing or freeze-drying. Disadvantages: Subculturing the colony several times might be necessary in order to get a vigorous oil-free culture. Growth rates slow down as storage time increases. Requires enough storage space.	Smith and Onions (1994); Sharma and Smith (1999); Nakasone et al. (2004)
Preservation in water	Short description: First described by Castellani (1967). Cut disks from growing colony edges are transferred to sterile cotton-plugged or screw-cap test tubes filled with several millilitres of water; or cryovials are filled with several discs and topped with sterile distilled water. The tubes are stored at room temperature or at 4°C. Advantages: Inexpensive and low-maintenance method. It has been described that water suppresses morphological changes in most fungi. Avoids mite contamination. Disadvantages: Viability decreases with the increasing of storage time. Requires enough storage space.	Castellani (1967); Capriles et al. (1989); Smith and Onions (1994); Nakasone et al. (2004)
Silica gel	Short description: Screw-cap tubes are partially filled with silica without indicator dye, which has been sterilised by dry heat. Spores are suspended in a cooled solution of skimmed milk (5%). Silica gel is chilled and placed in an ice-water bath. The spore suspension is added to the silica gel to wet the gel. Tubes are stored with the caps loose at 25 or 30°C for several days. Viability is checked and if the cultures are viable, caps are tightened and the tubes are stored. Advantages: Storage at low temperatures can increase the survival period twofold to threefold over storage at room temperature. Therefore, it can be used as a medium term storage method, especially for sporulating fungi. It is an inexpensive, rapid and simple technique to use. Disadvantages: Some fungi such as *Pythium*, *Phytophthora* and some Oomycota species, do not survive to this process. It is limited to some fungi with delicate spores. Requires enough storage space.	Perkins (1962); Grivell and Jackson (1969); Smith and Onions (1994); Sharma and Smith (1999); Nakasone et al. (2004)

Deep-Freezing	Short description: Cultures are grown on agar slants in bottles or test tubes with screw-caps and can be placed directly in the freezer. However, in a different process, vigorous cultures can be grown in tubes and flooded with a solution containing 10% glycerol in water. In order to produce a suspension with spores and mycelium, cultures are scraped and then applied into cryovials with sterile glass beads inside. Finally, they are frozen at −80°C, with a cooling rate of −1°C per minute. Advantages: The retrieval procedure is very simple. It is a long-term preservation technique where most fungal cultures frozen at −20 to −80°C remain viable and successfully preserved for up to 5 years. Disadvantages: Repeated freezing and thawing will significantly reduce the culture viability and several factors can affect the retrieval viability and effectiveness.	Smith and Onions (1994); Baker and Jeffries (2006)
Cryopreservation in liquid nitrogen	Short description: For fungal cultures that do not sporulate or produce mycelia growing deep into the agar, sterilised screw-cap vials are filled with one quarter to half volume with a 10% glycerol solution. Plugs are cut from vigorously growing cultures and placed in a cryovial which is placed directly into the vapour phase of a liquid-nitrogen tank. Advantages: Storage of fungi using plastic straws instead of vials or tubes has been reported. Procedures used to harvest materials differ depending on whether the fungi sporulate, have mycelia that penetrate below the surface of the agar, or grow only in liquid culture. It is an effective, timesaving and reduced labour requirements. Method with increased assurance of long-term availability. Disadvantages: It is an expensive method due to constant maintenance by refilling of liquid nitrogen and high costs of the apparatus. It needs constant surveillance. Also there are space requirements for the refrigeration units.	Nakasone et al. (2004); Smith and Onions (1994)
Freeze-drying	Short description: The procedure involves a two-stage process: (1) ampoules containing the cultures as suspensions are cooled to −60°C; air pressure in the ampoules is reduced and water is removed by ablation as the temperature is allowed to rise; (2) the ampoules are placed on a vacuum manifold for secondary drying, evacuated, and sealed under vacuum. Finished freeze-dried ampoules are stored. Advantages: Long-term storage at room temperature. Low-cost form of permanent preservation. It is the method of choice for many spore-forming fungi that produce large numbers of small (10-μm or less in diameter) spores. Disadvantages: Not appropriate for all fungi. The technique is used primarily with species that form numerous, relatively small propagules.	Fry and Greaves (1951); Schipper and Bekker-Holtman (1976); Sharma and Smith (1999); Nakasone et al. (2004); Uzunova-Doneva and Donev (2005); Miyamoto-Shinohara et al. (2006)

http://bch.cbd.int/protocol); the Nagoya Protocol on Access and Benefit Sharing is a new international agreement under the CBD, adopted in 2010, that intends to facilitate the implementation of resolved principles (Convention on Biological Diversity 2011). These principles include the fact that research and development of biodiversity-based products can only take place with the approval of countries and communities involved as contracting parties, which must also share in the benefits. For these reasons, public and general participation and awareness of biodiversity is a must that is considered in all the actions taken and to be made in the future (Smith 2003).

Culture collections aim to collect, maintain and distribute microbial strains among microbiologists and are considered to be a means to preserve microbial diversity *ex situ*. Microbial collections originated when Koch's school introduced pure culture techniques in bacteriology, and the first culture collection to provide services was established by Professor Král, in 1890, at the German University of Prague, who published the first catalogue of strains from a culture collection in 1900. After the first collection of Král, many others were developed, being Mycothèque de l'Uiversitée Catholique de Louvain (BCCM/MUCL, see Table 5.2) in Belgium and the Collection of the Centraalbureau voor Schimmelcultures (CBS) in the Netherlands, two of the oldest working culture collections established. Since then, many others have been established, some preserving high diversity of microorganisms, others specialised in particular groups of microorganisms. In 1925, the ATCC – American Type Culture Collection was created in Washington and is now located in Manassas, Virginia. With time, international cooperation between culture collections were started (Uruburu 2003).

Nowadays, there is an increase in the global efforts in the identification, conservation, data generation and quality management of microorganisms.

Areas of the world rich in biodiversity (e.g., tropical regions), have few facilities for the *ex situ* preservation that is essential for compliance with the requirements of the CBD and that enable them to take advantage of the potential of microorganisms (Smith 2003).

Organisations such as the WFCC and the ECCO – European Culture Collections' Organization (www.eccosite.org), bring together a critical mass of culture collections and users to try and co-ordinate activities as well as to exchange information and technologies that facilitate progress. The WFCC was founded in 1968. It seeks to promote activities supporting the interests of culture collections and their users. In its newsletter the WFCC discloses the access to updated information on matters relevant to collections as well as committees reporting on patent depositions; postal, quarantine and safety regulations; the safeguarding of endangered collections; education, publicity, standards and biodiversity (Smith 2003, 2012). The ECCO was established in 1981. Its mission is to bring the managers of the major public service collections in Europe together to discuss common policy, exchange technologies and seek collaborative projects. It comprises above 60 members from 22 European countries that hold a total number over 350,000 strains representing filamentous fungi, yeasts, bacteria and also archaea, phages and plasmids, including plasmid bearing strains. ECCO also comprises members from the animal cells field, including human and hybridoma cell lines, as well as animal and plant viruses, plant cells, algae and protozoa. ECCO members have helped produce practical approaches to international rules and regulation (www.eccosite.org).

From ECCO, several actions have been developed and from the effort of all the members, other projects such as EMbaRC and MIRRI were established.

The EMbaRC – European Consortium of Microbial Resources Centres is a European Union funded project. It aims to facilitate the access of researchers from both public and private sectors, to improved, coordinated and validated microbial resources. This project is based on networking, access, training and research that count with the participation of all European Union members and associated countries (www.embarc.eu).

MIRRI – Microbial Resource Research Infrastructure is a European research infrastructure that intends to provide microorganisms services. It aims to facilitate access to high quality microorganisms, their derivatives (e.g., genomic DNA, cDNA, total mRNA, plasmids, etc.) and associated data for research, development and application. In order to deliver the resources and services more effectively and efficiently in biotechnology, MIRRI was developed with the main goal of connecting resource holders with researchers and policy makers. The infrastructure will work with over 70 microbial domain resource centres from all of the European Union members and associated countries. Collectively, they provide access to more than 350,000 strains of microorganisms. MIRRI is established on the ESFRI – European Strategy Forum for Research Infrastructures road map, to improve access to the microbial resources and services that are needed to accelerate research and discovery processes (www.mirri.org). In addition, MIRRI together with other activities supports a framework designated: GBRCN – Global Biological Resource Centre Network. The GBRCN is a project that followed the work done in the OECD – Organisation for Economic Cooperation and Development. In order to support research and biotechnology as a platform for a knowledge-based bio-economy, a strategy by OECD was developed to improve the access to high quality biological resources and information. This project implies the collaboration with existing networks such as ACM – Asian Consortium for Microorganisms, ECCO, WFCC, plus stakeholders and users to demonstrate the value of networking activities. Finally, it aims to develop common approaches and enhance coverage of available organisms and information to meet all the requirements of all users (www.gbrcn.org).

3.3 Culture Collections Versus Biological Resource Centres (BRCs)

Culture collections have been providing and maintaining organisms for over a century. They are the scaffold of scientific development in many areas of knowledge. In the beginnings, infrastructures and capabilities were very limited. Catalogues, when available, only existed in a printed version and most collections served only national users. Nowadays, science turns to new technologies and information that implies culture collections to adapt in order to provide products and services to their customers in a way that will facilitate their use.

The adoption of international scientific and performance criteria encouraged by several organisations adds value to strain holdings and enables networks to share strategies. These bases distinguish the BRCs from the culture collections. Today, culture collections must deal with the vast diversity of new genetic entities generated by scientific work. Due to the latest advances in fields such as genomics and proteomics, there is a continuous increase in the generation of large amounts of information and scientific data. These technological advances foster BRCs to have the responsibility of storing, maintaining, and disseminating biological material and generated information. All of these responsibilities require a different approach in both material and data management. Adoption of new methodologies to ensure stability of procedures and proceedings are also required. In addition, they have different customers and different financing sources. However, in general, all of the culture collections share the same goals and face the same challenges. Transition from culture collections to BRCs is the process of overcoming those challenges and reaching those goals. It shall be based on different steps such as: (1) assuring sustainability with creation and implementation of business plans supported by adequate business models; (2) compliance with legislation in order to operate on the national and international legal framework with special attention to the code of conduct for biosecurity; (3) implementation of quality management systems to be able to guarantee the minimal level required to operate in a global BRCs network; (4) guarantee of information and technology; (5) training and capacity building with main focus in creation of a new generation of fungal taxonomists able to incorporate different and modern technologies and bioinformatics; (6) taxonomic expertise; (7) application of new technologies and massive incorporation of biodiversity items (Smith and Ryan 2012).

It becomes obvious that BRCs are a key element for a sustainable international scientific infrastructure, which is necessary to underpin successful delivery of biotechnological benefits.

A common strategy is needed to address the incorporation of biodiversity. An example of the current gap in coverage is given by fungi. There are currently about 100,000 species of fungi named. However, there is still an estimated huge amount of fungal species yet to be discovered and described for the Science (Hawsworth 2001; Smith and Ryan 2012). New described fungal species require sound preservation technology. Acquired knowledge about these fungal species requires advances in preservation techniques. Moreover, establishment and maintenance of BRCs depend on the implementation of reliable preservation techniques and appropriate quality assurance to allow them to become effective and efficient.

Although it can be argued that sustainability is the prime challenge, there are many examples of how this can be achieved. Quality management has been addressed through OECD initiative and also EMbaRC and GBRCN activities. The network itself helps in accessing new technologies and available expertise through partnerships (Smith and Ryan 2012). Within this context, several guidelines were developed mainly by WFCC (2010) and OECD (2001, 2007), as well as by specific consortiums such as CABRI – Common Access to Biological Resources and Information (www.cabri.org). Special legislation for this context has been created.

In order to be applied to microbial laboratories, different standards such as the French norm NF S96-900, the ISO 9001, the ISO/IEC 17025 (IPAC 2010) and the ISO Guide 34, have been developed.

Acknowledgements The research leading to this work has received funding from the European Community's Seventh Framework Programme (FP7, 2007–2013), Research Infrastructures action, under the grant agreement No. FP7-228310 (EMbaRC project). M.F. Simões acknowledges FCT – Portugal for the scholarship SFRH/BD/64260/2009.

References

Abramson D, Lombaert G, Clear RM, Sholberg P, Trelka R, Rosin E (2009) Production of patulin and citrinin by *Penicillium expansum* from British Columbia (Canada) apples. Mycotoxin Res 25:85–88

Baird R (2010) Leveraging the fullest potential of scientific collections through digitization. Biodivers Inform 7:130–136

Baker M, Jeffries P (2006) Use of commercially available cryogenic vials for long-term preservation of dermatophyte fungi. J Clin Microbiol 44:617–618

Bass D, Richards TA (2011) Three reasons to re-evaluate fungal diversity 'on Earth and in the ocean'. Fungal Biol Rev 25:159–164

Begerow D, Nilsson H, Unterseher M, Maier W (2010) Current state and perspectives of fungal DNA barcoding and rapid identification procedures. Appl Microbiol Biotechnol 87:99–108

Blackwell M (2011) The fungi: 1, 2, 3 … 5.1 million species? Am J Bot 98:426–438

Branco S (2011) Fungal diversity – an overview. In: Venora G, Grillo O (eds) The dynamical processes of biodiversity – case studies of evolution and spatial distribution. InTech, Rijeka, pp 211–226

Cabañes FJ, Bragulat MR, Castellá G (2010) Ochratoxin a producing species in the genus *Penicillium*. Toxins 2:1111–1120

Cain TC, Lubman DM, Webber WJ Jr (1994) Differentiation of bacteria using protein profiles from MALDI-TOF-MS. Rapid Commun Mass Spectrom 8:1026–1030

Canhos VP, Souza S, Giovanni R, Canhos D (2004) Global biodiversity informatics: setting the scene for a "new world" of ecological modeling. Biodivers Inform 1:1–13

Cannon PF (1997) Diversity of the Phyllachoraceae with special reference to the tropics. In: Hyde KD (ed) Biodiversity of tropical microfungi. Hong Kong University Press, Hong Kong, pp 255–278

Capriles CH, Mata S, Middelveen M (1989) Preservation of fungi in water (Castellani): 20 years. Mycopathogia 106:73–79

Castellani A (1967) Maintenance and cultivation of common pathogenic fungi of man in sterile distilled water. Further researches. J Trop Med Hyg 70:181–184

Convention on Biological Diversity, United Nations (2011) Nagoya protocol on access to genetic resources and the fair and equitable sharing of benefits arising from their utilization to the convention on biological biodiversity – text and annex. Secretariat of the Convention on Biological Diversity, Montreal

Desjardins AE, Proctor RH (2007) Molecular biology of *Fusarium* mycotoxins. Int J Food Microbiol 119:47–50

Dias N, Santos C, Portela M, Lima N (2011) Toenail onychomycosis in a Portuguese geriatric population. Mycopathologia 172:55–61

Frezal L, Leblois R (2008) Four years of DNA barcoding: current advances and prospects. Infect Genet Evol 8:727–736

Frisvad JC (1989) The use of high-performance liquid chromatography and diode array detection in fungal chemotaxonomy based on profiles of secondary metabolites. Botanical J Linnean Soc 99:81–95

Fröhlich J, Hyde KD (1999) Biodiversity of palm fungi in the tropics: are global fungal diversity estimates realistic? Biodivers Conserv 8:977–1004

Fry RM, Greaves RIN (1951) The survival of bacteria during and after drying. J Hyg 49:220–246

Gams W, Christensen M, Onions AH, Pitt JI, Samson RA (1985) Infrageneric taxa of *Aspergillus*. In: Samson RA, Pitt JI (eds) Advances in *Penicillium* and *Aspergillus* systematics. Plenum Press, New York, pp 55–62

Gardes M, Bruns TD (1993) ITS primers with enhanced specificity for basidiomycetes – application to the identification of mycorrhizae and rusts. Mol Ecol 2:113–118

Giorni P, Magan N, Pietri A, Bertuzzi T, Battilani P (2007) Studies on *Aspergillus* section *Flavi* isolated from maize in northern Italy. Int J Food Microbiol 113:330–338

Gollotte A, Van Tuinen D, Atkinson D (2004) Diversity of arbuscular mycorrhizal fungi colonising roots of the grass species *Agrostis capillaris* and *Lolium perenne* in a field experiment. Mycorrhiza 14:111–117

Grivell AR, Jackson JF (1969) Microbial culture preservation with silica gel. J Gen Microbiol 58:423–425

Hammond PM (1992) Species inventory. In: Groombridge B (ed) Global biodiversity: status of the Earth's living resources. Chapman and Hall, London, pp 17–39

Hammond PM (1995) The current magnitude of biodiversity. In: Heywood VH (ed) Global biodiversity assessment. Cambridge University Press, Cambridge, pp 113–138

Hawksworth DL (1991) The fungal dimension of biodiversity: magnitude, significance, and conservation. Mycol Res 95:641–655

Hawksworth DL (2001) The magnitude of fungal diversity: the 1.5 million species estimate revisited. Mycol Res 105:1422–1432

Hawksworth DL (2011a) Naming *Aspergillus* species: progress towards one name for each species. Med Mycol 49:70–76

Hawksworth DL (2011b) A new dawn for the naming of fungi: impacts of decisions made in Melbourne in July 2011 on the future publication and regulation of fungal names. IMA Fungus 2:155–162

Hawksworth DL, Rossman AY (1997) Where are all the undescribed fungi? Phytopathology 87:888–891

Hebert PD, Stoeckle MY, Zemlak TS, Francis CM (2004) Identification of birds through DNA barcodes. PLoS Biol 2:1657–1663

Hempel S, Renker C, Buscot F (2007) Differences in the species composition of arbuscular mycorrhizal fungi in spore, root and soil communities in a grassland ecosystem. Environ Microbiol 9:1930–1938

Hibbett DS, Binder M, Bischoff JF, Blackwell M, Cannon PF et al (2007) A higher-level phylogenetic classification of the Fungi. Mycol Res 111:509–547

Hillenkamp F, Peter-Katalinic J (eds) (2007) MALDI MS: a practical guide to instrumentation, methods and applications. Wiley-VCH, Münster

Holland RD, Wilkes JG, Ralli F, Sutherland JB, Persons CC, Voorhees KJ, Lay JO Jr (1996) Rapid identification of intact whole bacteria based on spectral patterns using matrix-assisted laser desorption/ionization with time-of-flight mass spectrometry. Rapid Commun Mass Spectrom 10:1227–1232

Hollingsworth ML, Clark A, Forrest LL, Richardson J, Pennington RT, Long DG, Cowan R, Chase MW, Gaudeul M, Hollingsworth PM (2009) Selecting barcoding loci for plants: evaluation of seven candidate loci with species-level sampling in three divergent groups of land plants. Mol Ecol Resour 9:439–457

Hong S-B, Go S-J, Shin H-D, Frisvad JC, Samson RA (2005) Polyphasic taxonomy of *Aspergillus fumigatus* and related species. Mycologia 97:1316–1329

IPAC (2010) Guia para a aplicação da NP EN ISO/IEC 17025. http://www.ipac.pt/docs/publicdocs/regras/OGC001.pdf. Cited 01 Jun 2012

Jahn TL, Jahn FF (1949) How to know the protozoa. Wm C Brown, Dubuque

Jones MDM, Forn I, Gadelha C, Egan MJ, Bass D, Massana R, Richards TA (2011) Discovery of novel intermediate forms redefines the fungal tree of life. Nature 474:200–203

Keys CJ, Dare DJ, Sutton H, Wells G, Lunt M, McKenna T, McDowall M, Shah HN (2004) Infection. Genet Evol 4:221–242

Klich MA (2002) Identification of common *Aspergillus* species. CBS, Utrecht
Larsen TO, Smedsgaard J, Nielsen KF, Hansen ME, Frisvad JC (2005) Phenotypic taxonomy and metabolite profiling in microbial drug discovery. Nat Prod Rep 22:672–693
Mackenzie R, Burhenne-Guilmin F, La Viña AGM, Werksman JD, Ascencio A, Kinderlerer J, Kummer K, Tapper R (2003) An explanatory guide to the Cartagena protocol on biosafety. IUCN, Gland/Cambridge, p xvi + 295 pp
Martins MAM, Lima N, Silvestre AJD, Queiroz MJ (2003) Comparative studies of fungal degradation of single or mixed bioaccessible reactive azo dyes. Chemosphere 52:967–973
May RM (2000) The dimensions of life on Earth. In: Raven PH, Williams T (eds) Nature and human society: the quest for a sustainable world. National Academy, Washington, DC, pp 30–45
May RM (2010) Ecology. Tropical arthropod species, more or less? Science 329:41–42
Miyamoto-Shinohara Y, Sukenobe J, Imaizumi T, Nakahara T (2006) Survival curves for microbial species stored by freeze-drying. Cryobiology 52:27–32
Moore RT (1980) Taxonomic proposals for the classification of marine yeasts and other yeast-like fungi including the smuts. Botanica Marina 23:361–373
Mora C, Tittensor DP, Adl S et al (2011) How many species are there on Earth and in the ocean? PLoS Biol 9:1–8
Mueller G, Schmit JP (2007) Fungal biodiversity: what do we know? What can we predict? Biodivers Conserv 16:1–5
Nagai T, Tomioka K, Takeuchi K, Ilda M, Kawada M, Sato T (2005) Evaluation of preservation techniques of microorganism resources in the MAFF Genebank. Jpn Agric Res Q 39:19–27
Nakasone KK, Peterson SW, Jong SC (2004) Preservation and distribution of fungal cultures. In: Foster MS, Bills GF, Mueller GM (eds) Biodiversity of fungi: inventory and monitoring methods. Elsevier Academic Press, Amsterdam, pp 37–47
Nielsen KF, Smedsgaard J (2003) Fungal metabolite screening: database of 474 mycotoxins and fungal metabolites for dereplication by standardised liquid chromatography–UV–mass spectrometry methodology. J Chromatogr A 1002:111–136
Nilsson RH, Abarenkov K, Veldre V, Nylinder S, DE Wit P, Brosché S, Alfredsson JF, Ryberg M, Kristiansson E (2010) An open source chimera checker for the fungal ITS region. Mol Ecol Resour 10:1076–1081
O'Brien HE, Parrent JL, Jackson JA, Moncalvo J-M, Vilgalys R (2005) Fungal community analysis by large-scale sequencing of environmental samples. Appl Environ Microbiol 71:5544–5550
OECD (2001) Biological resource centres underpinning the future of life sciences and biotechnology. http://www.oecd.org/LongAbstract/0,3425,en2649347973168572611111,00.html. Cited 01 Jun 2012
OECD (2007) OEDC best practice guidelines for biological resource centres. Organisation for Economic Co-operation and Development. http://www.oecd.org/sti/biotechnologypolicies/38777417.pdf. Cited 01 Jun 2012
Paterson RRM, Lima N (2010) Toxicology of mycotoxins. In: Luch A (ed) Molecular, clinical and environmental toxicology, vol 2, Clinical toxicology. Birkhauser Verlag Ag, Basel, pp 31–63
Perkins DD (1962) Preservation of *Neurospora* stock cultures with anhydrous silica gel. Can J Microbiol 8:591–594
Perrone G, Stea G, Epifani F, Varga J, Frisvad JC, Samson RA (2011) *Aspergillus niger* contains the cryptic phylogenetic species *A. awamori*. Fungal Biol 115:1138–1150
Peterson SW (2000) Phylogenetic relationships in *Aspergillus* based on rDNA sequence analysis. In: Pitt JI, Samson RA (eds) Integration of modern taxonomic methods for *Penicillium* and *Aspergillus* classification. Harwood Academic Publishers, Amsterdam, pp 323–355
Petisco C, Downey G, Murray I, Zabalgogeazcoa I, García-Criado B, García-Ciudad A (2008) Direct classification of related species of fungal endophytes (*Epichloe* spp.) using visible and near-infrared spectroscopy and multivariate analysis. FEMS Microbiol Lett 284:135–141
Pildain MB, Frisvad JC, Vaamonde G, Cabral D, Varga J, Samson RA (2008) Two novel aflatoxin-producing *Aspergillus* species from Argentinean peanuts. Int J Syst Evol Microbiol 58:725–735
Pitt JI (1988) A laboratory guide to common *Penicillium* species. Commonwealth Scientific and Industrial Research Organization, Division of Food Processing, Australia
Pitt JI, Hocking AD, Glenn DR (1983) An improved medium for the detection of *Aspergillus flavus* and *A. parasiticus*. J Appl Microbiol 54:109–114

Pivato B, Mazurier S, Lemanceau P, Siblot S, Berta G, Mougel C, van Tuinen D (2007) Medicago species affect the community composition of arbuscular mycorrhizal fungi associated with roots. New Phytol 176:197–210

Rodrigues P, Venâncio A, Kozakiewicz Z, Lima N (2009) A polyphasic approach to the identification of aflatoxigenic and non-aflatoxigenic strains of *Aspergillus* section *Flavi* isolated from Portuguese almonds. Int J Food Microbiol 129:187–193

Rodrigues P, Santos C, Venâncio A, Lima N (2011) Species identification of *Aspergillus* section *Flavi* isolates from Portuguese almonds using phenotypic, including MALDI-TOF ICMS, and molecular approaches. J Appl Microbiol 111:877–892

Rooney AP, Ward TJ (2005) Evolution of a large ribosomal RNA multigene family in filamentous fungi: birth and death of a concerted evolution paradigm. Proc Natl Acad Sci 102:5084–5089

Rosendahl S, McGee P, Morton JB (2009) Lack of global population genetic differentiation in the arbuscular mycorrhizal fungus *Glomus mosseae* suggests a recent range expansion which may have coincided with the spread of agriculture. Mol Ecol 18:4316–4329

Rossman AY (1994) A strategy for an all-taxa inventory of fungal biodiversity. In: Peng CL, Chou CH (eds) Biodiversity and terrestrial ecosystems, Academia Sinica monograph series no. 14. Academia Sinica, Taipei, pp 169–194

Ryan MJ, Smith D, Jeffries P (2000) A decision-based key to determine the most appropriate protocol for the preservation of fungi. World J Microbiol Biotechnol 16:183–186

Ryan MJ, Jeffrie P, Bridge PD, Smith D (2001) Developing cryopreservation protocols to secure fungal gene function. Cryo Letters 22:115–124

Ryan MJ, Verkleij G, López-Ocaña L, Munaut F, Arahal DR, Simões MF et al (2012) Inter-laboratory evaluation, development and validation of fungal preservation regimes used in different European biological resources centres (BRCs). In: Paterson R, Simões MF, Pereira L, Santos C, Lima N (eds) Biological resource centres: closing the gap between science and society, Abstract book. Micoteca da Universidade do Minho, Braga, p 57

Ryzhov V, Fenselau C (2001) Characterization of the protein subset desorbed by MALDI from whole bacterial cell. Anal Chem 73:746–750

Samson RA, Varga J (2000) Molecular systematics of *Aspergillus* and its teleomorphs. In: Samson RA, Pitt JI (eds) *Aspergillus* molecular biology and genomics. Harwood Academic Publishers, Sendai, pp 9–49

Samson RA, Hoekstra ES, Frisvad JC (2004) Introduction to food- and airborne Fungi, 7th edn. CBS, Wageningen

Samson RA, Noonim P, Meijer M, Houbraken J, Frisvad JC, Varga J (2007) Diagnostic tools to identify black aspergilli. Stud Mycol 59:129–145

Santos IM, Lima N (2001) Criteria followed in the establishment of a filamentous fungi culture collection – Micoteca da Universidade do Minho (MUM). World J Microbiol Biotechnol 17:215–220

Santos C, Lima N (2010) A Identificação de fungos pela espectrometria de massa através da técnica de MALDI-TOF ICMS. In: Dianese JC, Santos LTP (eds) Proceedings of the 6th Brazilian congress of mycology, Brazilian Society of Mycology, CD-Rom, Brasilia, Brazil, 29th Nov to 2nd Dec 2010

Santos IM, Venâncio A, Lima N (1998) Fungos contaminantes na indústria alimentar. Micoteca da Universidade do Minho, Braga, pp 1–128

Santos IM, Abrunhosa L, Venâncio A, Lima N (2002a) The effect of culture preservation techniques on patulin and citrinin production by *Penicillium expansum* link. Lett Appl Microbiol 35:272–275

Santos IM, Lima N, Ryan MJ, Smith D (2002b) Metabolic response of filamentous fungi to preservation stress. In: Santos IM, Venâncio A, Lima N (eds) Ecologia dos fungos. Micoteca da Universidade do Minho, Braga, pp 55–66

Santos C, Paterson RMR, Venâncio A, Lima N (2010) Filamentous fungal characterisations by matrix-assisted laser desorption/ionisation time of flight mass spectrometry. J Appl Microbiol 108:375–385

Schipper MAA, Bekker-Holtman J (1976) Viability of lyophilized fungal cultures. Antonie Van Leeuwenhoek 42:325–328

Schmit JP, Mueller GM (2007) An estimate of the lower limit of global fungal diversity. Biodivers Conserv 16:99–111

Scoble MJ (2010) Natural history collections digitization: rationale and value. Biodivers Inform 7:77–80
Seifert KA, Samson RA, deWaard JR, Houbraken J, Lévesque CA, Moncalvo J-M, Louis-Seize G, Hebert PDN (2007) Prospects for fungus identification using CO1 DNA barcodes, with *Penicillium* as a test case. Proc Natl Acad Sci 104:3901–3906
Sharma B, Smith D (1999) Recovery of fungi after storage for over a quarter of a century. World J Microbiol Biotechnol 15:517–519
Smith D (2003) Culture collections over the world. Int Microbiol 6:95–100
Smith D (2012) Culture collections. In: Saraslani S, Gadd GM (eds) Advances in applied microbiology, vol 79. Academic, Burlington, pp 73–118
Smith D, Onions AHS (1983) A comparison of some preservation techniques for fungi. Trans Br Mycol Soc 81:535–540
Smith D, Onions AHS (1994) The preservation and maintenance of living fungi. IMI technical handbooks: No. 1, 2nd edn. International Mycological Institute, CAB International, Oxfordshire
Smith D, Ryan M (2012) Implementing best practices and validation of cryopreservation techniques for microorganisms. Scientific World Journal. doi:10.1100/2012/805659
Smith D, Ryan MJ, Day JG (eds) (2001) The UK National Culture Collection (UKNCC) biological resource: properties, maintenance and management. UKNCC, Surrey, United Kingdom
Soares C, Rodrigues P, Peterson SW, Lima N, Venâncio A (2012) Three new species of *Aspergillus* section *Flavi* isolated from almonds and maize in Portugal. Mycologia 104:682–697
Sykorova Z, Ineichen K, Wiemken A, Redecker D (2007) The cultivation bias: different communities of arbuscular mycorrhizal fungi detected in roots from the field, from bait plants transplanted to the field, and from a greenhouse trap experiment. Mycorrhiza 18:1–14
Tanaka K, Waki H, Ido Y, Akita S, Yoshida Y, Yoshida T (1988) Protein and polymer analyses up to m/z 100000 by laser ionization time-of-flight mass spectrometry. Rapid Commun Mass Spectrom 2:151–153
Uruburu F (2003) History and services of culture collections. Int Microbiol 6:101–103
Uzunova-Doneva T, Donev T (2005) Anabiosis and conservation of microorganisms. J Cult Collect 4:17–28
Vaamonde G, Patriarca A, Pinto VF, Comeria R, Degrossi C (2003) Variability of aflatoxin and cyclopiazonic acid production by *Aspergillus* section *Flavi* from different substrates in Argentina. Int J Food Microbiol 88:79–84
Varga J, Due M, Frisvad JC, Samson RA (2007a) Taxonomic revision of *Aspergillus* section *Clavati* based on molecular, morphological and physiological data. Stud Mycol 59:89–106
Varga J, Frisvad JC, Samson RA (2007b) Polyphasic taxonomy of *Aspergillus* section *Candidi* based on molecular, morphological and physiological data. Stud Mycol 59:75–88
Vilgalys R, Hester M (1990) Rapid genetic identification and mapping of enzymatically amplified ribosomal DNA from several *Cryptococcus* species. J Bacteriol 172:4238–4246
WFCC (2010) The WFCC guidelines for the establishment and operation of Culture Collections. http://www.wfcc.info/guidelines. Cited 01 Jun 2012
White TJ, Bruns T, Lee S, Taylor JW (1990) Amplification and direct sequencing of fungal ribosomal RNA genes for phylogenetics. In: Innis MA, Gelfand DH, Sninsky JJ, White TJ (eds) PCR protocols: a guide to methods and applications. Academic, New York, pp 315–322
Whittaker RH (1959) On the broad classification of organisms. Q Rev Biol 34:210–226
Wubet T, Weiss M, Kottke I, Teketay D, Oberwinkler F (2006) Phylogenetic analysis of nuclear small subunit rDNA sequences suggests that the endangered African Pencil Cedar, *Juniperus procera*, is associated with distinct members of Glomeraceae. Mycol Res 110:1059–1069

Chapter 6
Bioinformatics Approaches in Studying Microbial Diversity

Mohammad Tabish, Shafquat Azim, Mohammad Aamir Hussain, Sayeed Ur Rehman, Tarique Sarwar, and Hassan Mubarak Ishqi

Abstract Proper understanding of molecular sequences, identification and phylogenetics of microorganisms are very important in many branches of biological science. Generation of genomic DNA sequence data from different organisms including microbes requires the application of bioinformatics tools for their analysis. Bioinformatics tools including BLAST, multiple sequence alignment tools etc. are used to analyze nucleic acid and amino acid sequences for phylogenetic affiliation. The emerging fields of comparative genomics and phylogenomics require the substantial knowledge and understanding of computational methods to handle the large scale data involved. The introduction of comparative rRNA sequence analysis represents a major milestone in the history of microbiology. Also single gene based phylogenetic inference and alternative global markers including elongation and initiation factors, RNA polymerase subunits, DNA gyrases, heat shock and RecA proteins are of immense importance. The analysis of the sequence data involves four general steps including: (i) selection of a suitable molecule or molecules, (ii) acquisition of molecular sequences, (iii) multiple sequence alignment and (iv) phylogenetic evaluation. This chapter explains in detail how raw data of molecular sequences from any microbe may be used for the detection and identification of microorganisms using computer based bioinformatics tools.

Keyword Bioinformatics tool • BLAST • Sequence alignment • DNA Chips • Gene identification

M. Tabish (✉) • S. Azim • M.A. Hussain • S.U. Rehman • T. Sarwar • H.M. Ishqi
Department of Biochemistry, Faculty of Life Sciences, Aligarh Muslim University, Aligarh, Uttar Pradesh 202002, India
e-mail: tabish.biochem@gmail.com

A. Malik et al. (eds.), *Management of Microbial Resources in the Environment*,
DOI 10.1007/978-94-007-5931-2_6, © Springer Science+Business Media Dordrecht 2013

1 Introduction

There has been a flood of nucleic acid sequence information, bioinformatics tools and phylogenetic inference methods in public domain databases, literature and World Wide Web space. Last 20 years has seen the rapid development of prokaryotic genomics. Since the sequencing of *Haemophilus influenzae* in 1985 (Johnston 2010; Fleischmann et al. 1995), currently over 11,364 whole genome sequences organized in three major groups of organisms i.e. eukaryota, prokaryota (archaea and bacteria) and viruses are available in Genome database of NCBI including complete chromosomes, organelles and plasmids as well as draft genome assemblies. Out of 11,364 whole genome sequences, 7,473 genome projects running across the world belong only to microbes with 1,696 completed microbial genomes projects whereas assembly is being done for 2,247 organisms and 3,531 genome project are still unfinished (Benson et al. 2002). The developing technology of nucleic acid sequencing, together with the recognition that sequences of building blocks in informational macromolecules (nucleic acids, proteins) can be used as 'molecular clocks' that contain historical information, led to the development of the three-domain model in the late 1970s, primarily based on small subunit ribosomal RNA sequence comparisons. The information currently accumulating from complete genome sequences of an ever increasing number of prokaryotes are now leading to further modifications of our views on microbial phylogeny. Prokaryotic genomics has had a revolutionary impact on our view of the microbial world and also on the methodologies for microbiological studies.

2 Complexity of Microbial Genomes

Analysis of genomic sequences has revealed that microbial genomes are very diverse. This is due to the complicated nature of microbial evolution. Mutations play a key role in evolution of eukaryotic genomes whereas, the contents of prokaryotic genomes are also changed by gene losses, gene rearrangements, horizontal gene transfer, and so on (McHardy et al. 2007; Doolittle 1999; Woese 1987). This means that even strains from the same species can differ significantly. For example, two *Escherichia coli* strains O157:H7 and K-12 have more than 1,000 different genes (Perna et al. 2001). The dynamic nature of microbial genomes complicates several tasks in microbiological studies. One of these is the development of strategies to prevent and treat microbe related diseases. Since microbe related diseases are common threats to the public health, microbes especially bacteria have been studied for many years. One point of progress was the introduction of antibiotics to treat bacterial infections. However, the use of antibiotics has been challenged by the emergence of antibiotic resistance among bacteria.

Plasmids play an important role in conferring antibiotic resistance in microbes. It is believed that antibiotic resistance evolves via natural selection. However, antibiotic resistance can also be introduced to bacteria via horizontal gene transfer

(Boerlin and Reid-Smith 2008). Plasmids are extra-chromosomal genetic elements that constitute upto 10% of the total DNA found in many species of bacteria (Mølbak et al. 2003; Thomas 2000). Because plasmids are capable of cell-to-cell transfer between bacterial species, genes harboured by plasmids are widely shared, playing a critical role in the evolution of bacteria (Feinbaum 2001; Summers 1996). Establishing accurate relationships between plasmids will help us to understand an important factor in the dissemination of antibiotic resistance genes, and establishing accurate relationships between bacteria will help us to identify the factors that cause diseases, the risks of outbreaks, and methods for preventing disease transmission. Unfortunately, the complexity of microbial genomes is apparent when we try to compare the genetic contents of strains and to build a phylogeny tree from them (McHardy et al. 2007; Doolittle 1999).

3 Obtaining Data (Wet Lab Approach)

One characteristic of microbiological studies in the genomics era is that we can generate a huge amount of data efficiently. Numerous different genomics based experimental methods are available. These methods are usually called molecular methods since they are often based on genetic characteristics. Compared to traditional phenotype-based methods, molecular methods are cost effective, easy to implement, and generate highly discriminatory data (Foxman et al. 2005; Tenover et al. 1997). Of these methods, the most widely used method for nucleic acid amplification is the polymerase chain reaction assay i.e., PCR. This assay includes a specific primer pair to amplify a unique genomic target nucleotide sequence for analysis. Following PCR, a variety of post-amplification methods are used to evaluate the product such as direct sequence analysis, use of genus or species specific probes, and utilization of restriction enzymatic analysis of the product, e.g., restriction fragment length polymorphism analysis (RFLP). Pulse-field gel electrophoresis (PFGE) is also considered as the gold standard. Multiple locus variable-number tandem repeat analysis (MLVA) assays are also a potentially powerful alternative or complementary tool. Another most powerful technique is DNA microarrays which provide a powerful high-throughput genomic method that has been widely used in biological studies. To construct a DNA microarray, single-strand fragments of DNA (also called probes) representing the genes of an organism are attached to a surface of glass or plastic. Each fragment can bind to a complementary DNA or RNA strand. Typically, more than 30,000 spots can be put on one slide, and it is possible to create a microarray representing every gene in a genome. Thus, microarrays can provide genome wide information which allows a comprehensive genetic analysis of an organism or a sample. DNA microarrays have been used for genotyping, expression analysis, and studies of protein-DNA interactions (Bilitewski 2009). When used for assessing the genetic relationships of bacterial strains, microarrays may be prepared for whole genome composed of open reading frames (ORFs) of one complete genome sequence (Zhou 2003). However, this type of microarray is limited by the requirement of representing one complete reference sequence which may not contain genetic content specific to nonsequenced strains. One possible improvement is to include specific genes from multiple whole-genome sequences or to use mixed-genome microarrays (MGMs) which use randomly-selected gene fragments from many strains of bacteria as probes (Wan et al. 2007; Borucki et al. 2004; Call et al. 2003).

From the enormous data to knowledge of microbial genomic information makes it possible to study microorganisms systematically. Sequence-based identification requires the recognition of a molecular target that is large enough to allow discrimination of a wide variety of microbes. One such target area that has been recognized is the rDNA gene complex which is present in all microbial pathogens. In bacteria, this complex is composed of a 16S rRNA gene and a 23S rRNA gene separated by a genomic segment called the internal transcribed spacer (ITS). Within fungi there are three genes (18S, 5.8S, and 28S) with spacers located between the genes (ITS1 and ITS2). Located in the rDNA gene complex are highly variable sequences that provide unique signatures for the identification of species and also conserved regions that contain genomic codes for the structural restrains that are present within organism groups. It has been shown that the ITS regions contain the most variability and that these regions are useful under most circumstances for species recognition. The availability of these variable sequence regions (ITS) surrounded by conserved sequences (16S/23S and 18S/5.8S/28S) allows for the utilization of an amplification system using universal (or consensus) bacterial or fungal primers. Once amplification has occurred using the consensus primers, the sequence is determined and comparison analysis of the unknown sequence to known sequences contained within a large database (such as the National Center for Biological Information (NCBI), GenBank databases) can be done to determine similarity and subsequently may lead to species identification. However, how to manipulate the massive amount of available data, how to retrieve genomic information effectively, and how to process the large scale data efficiently are all challenging problems. Because of these problems, the field of bioinformatics has emerged and has become an integral part of microbial studies (Foster et al. 2012).

4 Bioinformatics

Bioinformatics has evolved into a full-fledged multidisciplinary subject that integrates developments in information and computer technology as applied to biotechnology and biological sciences. Bioinformatics uses computer software tools for database creation, data management, data warehousing, data mining and global communication networking. In this, knowledge of many branches are required like biology, mathematics, computer science, laws of physics & chemistry, and sound knowledge of information technology to analyze the data. Bioinformatics is not limited to the computing data, but in reality it can be used to solve many biological problems and find out how living things work. It is the comprehensive application of mathematics (e.g., probability and statistics), science including biochemistry, molecular biology and a core set of problem-solving methods e.g. computer algorithms to the understanding of living systems. Bioinformatics is the recording, annotation, storage, analysis, and searching/retrieval of nucleic acid sequence genes, RNAs, protein sequences and structural information. This includes databases of the sequences and structural information as well methods to access, search, visualize and retrieve the information.

Functional genomics, biomolecular structure, proteome analysis, cell metabolism, biodiversity, drug and vaccine designs are some of the areas in which bioinformatics is an integral component. Bioinformatics concern the creation and maintenance of databases of biological information whereby researchers can both access existing information and submit new entries. The most pressing tasks in bioinformatics involve the analysis of sequence information. Computational Biology is the name given to this process.

5 Bioinformatics and Its Scope

Bioinformatics has evolved into a full-fledged scientific discipline over the last decade. The definition of Bioinformatics is not restricted to computational molecular biology and computational structural biology. It now encompasses fields such as comparative genomics, structural genomics, transcriptomics, proteomics, cellunomics and metabolic pathway engineering. Developments in these fields have direct implications to healthcare, medicine, discovery of next generation drugs, development of agricultural products, renewable energy, environmental protection etc.

Bioinformatics integrates the advances in the areas of computer science, information science and information technology to solve complex problems in life sciences. The core data comprises of the genomes and proteomes of human to microbes, 3-D structures and functions of proteins, microarray data, metabolic pathways, cell lines, hybridoma and biodiversity etc. The sudden growth in the quantitative data in biology has rendered data capture, data warehousing and data mining as major issues for biotechnologists and biologist. Availability of enormous data has resulted in the realization of the inherent biocomplexity issues which call for innovative tools for synthesis of knowledge. Information technology, particularly the internet, is utilized to collect, distribute and access ever-increasing data which are later analyzed with mathematics and statistics-based tools. Bioinformatics has a key role to play in the cutting edge research and development areas such as functional genomics, proteomics, protein engineering, pharmacogenomics, discovery of new drugs and vaccines, molecular diagnostic kits, agro-biotechnology etc. This has attracted attention of several companies and entrepreneurs. As a result, a large number of bioinformatics based start-ups have been launched and the trend is likely to continue. A Bioinformatician must acquire/possess expertise in the essential multidisplinary fields that comprise the core of this new science. Quality research and education in bioinformatics are vital not only to meet the existing challenges but also to set and accomplish new goals in life sciences.

6 The Potential of Bioinformatics

The potential of bioinformatics in the identification of useful genes leading to the development of new gene products, drug discovery and drug development has led to a paradigm shift in biology and biotechnology. These fields are becoming more and more computationally intensive. The new paradigm, now emerging, is that all the genes will be known "in the sense of being resident in database available electronically", and the starting point of biological investigation will be theoretical and a scientist will begin with a theoretical conjecture and only then turning to experiment to follow or test the hypothesis. With a much deep understanding of the biological processes at the molecular level, the bioinformatics scientist have developed new techniques to analyze genes on an industrial scale resulting in a new area of science known as 'Genomics'. This is the science that deals with the study of whole genome, largely encompasses biology of genetics at molecular level i.e., the constitution of DNA and RNA, its analysis, translation of the chemical information carried over by these materials into biological data and digitizing that huge biological data through computational means.

The shift from gene biology has resulted in the development of strategies from lab techniques to computer programmes to analyze whole batch of genes at once. Genomics is revolutionizing drug development, gene therapy, and our entire approach to health care and human medicine. The genomic discoveries are getting translated into practical biomedical results through bioinformatics applications. Work on proteomics and genomics will continue using highly sophisticated software tools and data networks that can carry multimedia databases. Thus, the research will be in the development of multimedia databases in various areas of life sciences and biotechnology. There will be an urgent need for development of software tools for data mining, analysis and modelling, and downstream processing. It has now been universally recognized that bioinformatics is the key to the new grand data-intensive molecular biology that will lead us in this century.

7 Activities in Bioinformatics

We can split the activities in bioinformatics in two areas:

1. **The organization**: this includes the creation of databases of biological information and the maintenance of the databases. This is very important as we are sequencing tens of millions of bases a year and undertaking to sequence whole organism genomes. The growth of the sequence databases is an unbroken exponential.
2. **Analysis of the data**: this includes the following:
   - Development of methods to predict the structure and/or function of newly discovered proteins and structural RNA sequences.
   - Clustering protein sequences into families of related sequences and the development of protein models.

- Aligning similar proteins and generating phylogenetic trees to examine evolutionary relationships
- The development of new algorithms and statistics with which to assess relationships among members of large data sets.
- The development and implementation of tools that enable efficient access and management of different types of information and
- The analysis and interpretation of various types of data including nucleotide and amino acid sequences, protein domains, and protein structures in studying microbial diversity.

8 The Need for Bioinformatics

- Whole genome analysis and sequences
- Experimental analysis involving thousands of genes simultaneously
- DNA Chips and Array Analyses – expression arrays, Comparative analysis between species and strains
- Proteomics: 'Proteome' of an organism.
- Medical applications: Genetic Disease – Pharmaceutical and Biotech Industry
- Forensic applications
- Agricultural applications

9 Databases

Computational analysis and comparative microbial genomic studies are taking shape at a faster rate leading to the development of different types of function prediction concepts, most important of them being the gene context and gene content analysis. Gene content analysis is a comparison of gene repertoires across different genomes (Shah et al. 2005; Luscombe et al. 2001). The postgenomic problems like protein structural determination and issues of gene function identification become more promising (Gomez et al. 2008) with the rapidly increasing number of completely sequenced genomes. Predicting the structures of proteins encoded by genes of interest provides subtle clues regarding the functions of these proteins (Idekar et al. 2001).

Various databases have been established for storing genomic data, and the internet makes it possible for these data to be accessed and shared by the public. Since there are different types of genomic data, it is impossible to build one database containing all data. Currently there are two types of genomic databases. Primary databases contain sequences and structures (for example, NCBI GenBank) and related annotations, bibliographies, and cross-references to other databases and provide the basis for biological studies; secondary databases contain biological knowledge obtained by analyzing genomic sequences and structure data. The database of

Clusters of Orthologous Groups of proteins (COGs, http://www.ncbi.nlm.nih.gov/COG), for example, contains information for phylogenetic analysis (Tatusov et al. 1997, 2003). The Ribosomal Database Project (RDP) provides ribosome related data and annotated bacterial and archaeal small-subunit 16S rRNA sequences (Cole et al. 2005, 2009; Larsen et al. 1993). Knowledge from these databases can help to process biological data efficiently. For example, the Gene Ontology database has been used to process microarray datasets (Barrell et al. 2009; Harris et al. 2004). Nucleic acid sequence analysis has proven to be a valuable asset for organism identification in a number of applications. Some of the most interesting applications of this technology are for the identifications of variant strains of known species, the identification of un-cultivatable organisms in clinical samples and the recognition of new species.

10 Web-Based Resources for Microbial Genomics

MicrobesOnline: MicrobesOnline is a website for browsing and comparing prokaryotic genomes. MicrobesOnline is a product of the Virtual Institute for Microbial Stress and Survival, which is sponsored by the US Department of Energy Genomic Science Program.

Integrated Microbial Genomes (IMG): The Integrated Microbial Genomes (IMG) system serves as a community resource for comparative analysis and annotation of all publicly available genomes from three domains of life, in a uniquely integrated context.

CAMERA (Community Cyber infrastructure for Advanced Marine Microbial Ecology Research and Analysis): The aim of CAMERA is to serve the needs of the microbial ecology research community by creating a rich, distinctive data repository and a bioinformatics tools resource that will address many of the unique challenges of metagenomic analysis.

DOE JGI Microbial Genomics Database: From this site we can get details about JGI projects, or go directly to the individual microbial sites. All of the individual sites include direct access to download sequence file(s), BLAST, and view annotations.

GOLD™ Genomes OnLine Database: GOLD is a World Wide Web resource for comprehensive access to information regarding complete and ongoing genome projects, as well as metagenomes and metadata, around the world.

JCVI Comprehensive Microbial Resource (formerly The Institute for Genomic Research): The Comprehensive Microbial Resource (CMR) is a free website used to display information on all of the publicly available, complete prokaryotic genomes.

Sanger Centre Bacterial Genomes: The Sanger Institute bacterial sequencing effort is concentrated on pathogens and model organisms. The site provides a list of projects funded, underway or completed; all data from these projects are immediately and freely available.

Microbial Genomes from Genome Channel: Genome Channel is a computer-annotated listing of genomes maintained by the Computational Biology group at Oak Ridge National Laboratory.

Protein Data Bank: The RCSB PDB provides a variety of tools and resources for studying the structures of biological macromolecules and their relationships to sequence, function, and disease.

KEGG (Kyoto Encyclopaedia of Genes and Genomes): A grand challenge in the post-genomic era is a complete computer representation of the cell, the organism, and the biosphere, which will enable computational prediction of higher-level complexity of cellular processes and organism behaviours from genomic and molecular information. Towards this end a bioinformatics resource named KEGG has been developed as part of the research projects of the Kanehisa Laboratories in the Bioinformatics Centre of Kyoto University and the Human Genome Centre of the University of Tokyo.

11 Data Retrieval Methods and Online Resources for Microbial Diversity

In order to use the information available in databases, an efficient information retrieval method should be used to obtain all related information quickly. Such methods are different, depending on the type of data to be retrieved. FASTA and BLAST are the two most widely used methods for retrieving sequence data. FASTA was the first fast sequence searching algorithm used for comparing a query sequence against a database (Plewniak 2008; Pearson 1990). The FASTA algorithm performs a rapid and approximate search for matched sequence segments followed by application of the Smith-Waterman alignment algorithm (Plewniak 2008; Pearson 1991) to these segments. Depending upon the application there are several softwares available online for free to retrieve the microbial data. Some of them are briefly described below:

11.1 Pairwise Alignment

A number of computational methods have been developed and used in genomic studies. Of these methods, genetic sequence alignment is the foundation for many other methods and widely used in comparative genomics. A good alignment method should give biologically meaningful results and at the same time be computationally efficient. There are two types of alignment methods, local alignments and global alignments. The former methods try to identify similar segments between two sequences while the latter try to align the entire length of two sequences. Methods for aligning two sequences are called pairwise alignment methods. BLAST and

FASTA are two widely used pairwise alignment methods. BLAST (Basic Local Alignment Search Tool) is a rapid sequence database search tool which is more efficient than FASTA. The output of BLAST is a list of high-scoring segment pairs (HSPs) and an "E value" which is an estimate of the probability of finding an HSP with score S. The E value is often used as a standardized measure for estimating the statistical significance of sequence similarity.

These methods can be extended to multiple sequences; however, multiple sequence alignment (MSA) is more complicated. ClustalW (Larkin et al. 2007; Thompson et al. 2002) is a widely used MSA method which is efficient for aligning protein sequences and short nucleotide sequences. However, it may fail for distantly related sequences (Lin et al. 2011). PSI-BLAST (Lee et al. 2008; Schäffer et al. 2001; Altschul et al. 1997) is a very successful method for detecting weak similarities. Two recently developed algorithms, MLAGAN (Brudno et al. 2003) and MAVID (Dewey 2007; Bray and Pachter 2003, 2004), are designed for global alignment of both evolutionarily close and distant megabase length genomic sequences. However, a phylogenetic tree is assumed to be known for use with MLAGAN. MAVID is a progressive global alignment program that works by recursively aligning the 'alignments' at ancestral nodes of the guide phylogenetic tree. MAUVE is used for comparing long genome sequences efficiently and takes into account possible large-scale evolutionary events among sequences (Darling et al. 2004).

11.2 Phylogenetic Analysis

The goal of phylogenetic analysis is to reconstruct the evolutionary history of a set of organisms. In molecular epidemiology, it helps to elucidate mechanisms that lead to microbial outbreaks and epidemics. Phylogenetic analysis usually begins with multiple sequence alignment of the sequences of a set of organisms. After obtaining an MSA, a number of different phylogenetic methods can be used to compute phylogenetic trees. These methods can be broadly classified into maximum parsimony, distance, and maximum likelihood methods (Stark et al. 2010; Takahashi and Nei 2000). The difference between these methods is how they define which tree is best among all possible trees. Maximum parsimony tries to find an evolutionary tree or trees which require a minimum number of changes from the common ancestral sequences. For maximum likelihood methods, given the MSA, the probability of a specific tree occurring is computed, and the one or ones with the highest values are considered to be the evolutionary tree or trees. Distance-based methods construct a tree by hierarchical clustering methods using a distance matrix for all organisms that is computed using MSA. To use MSA for phylogenetic analysis, it is necessary to assume an underlying mutation model. Of the ones that have been proposed, the Jukes-Cantor (JC) model (Som 2006; Takahashi and Nei 2000) is the simplest one. In the JC model, each base in a DNA sequence has an equal mutation rate and all complementary pairs of the four nucleotides A, T, C and G have equal substitution rates. These assumptions are not realistic in practice, so many complex models have been proposed and tried. Successful phylogenetic analysis requires a suitable model. Phylogenetic analysis of microbial strains is problematic due to its dynamic nature (Wilmes et al. 2009). Different genes among strains may contain contradictory information about their evolution. Consensus trees have been suggested as a solution. An alternative is the introduction of networks that represent the evolutionary relationships between microbial strains.

11.3 AGeS: A Software System for Microbial Genome Sequence Annotation

AGeS is genome sequence annotation software which is a fully integrated with high performance software system to analyze DNA sequences and predict the protein-coding regions for completed and draft bacterial genomes. It predicts genomic features using a number of bioinformatics methods and provides visualization based on the familiar genome browser.

11.4 SILVA: A Comprehensive Online Resource for Quality Checked and Aligned Ribosomal RNA Sequence Data

Ribosomal RNA sequence data analyzing tool SILVA is available online at http://www.arb-silva.de/. Sequencing ribosomal RNA (rRNA) genes is currently the method of choice for phylogenetic reconstruction, nucleic acid based detection and quantification of microbial diversity (Pruesse et al. 2007). To cope with the flood of data, the SILVA system was implemented to provide a central, comprehensive web resource for up to date, quality controlled databases of aligned small and large subunit rRNA sequences from the bacteria and archaea domains. This programme is designed as a central comprehensive resource by integrating multiple taxonomic classifications and the latest validly described nomenclature as well as additional information, such as if a sequence was derived from a cultivated organism, a type strain, or belongs to a genome project.

11.5 16S and 23S Ribosomal RNA Mutation Database

Access to the expanded versions of the 16S and 23S Ribosomal RNA Mutation Databases has been improved to permit searches of the lists of alterations for all the data from (1) one specific organism, (2) one specific nucleotide position, (3) one specific phenotype. The URL for the searchable version of the Databases is: http://ribosome.fandm.edu.

11.6 5S Ribosomal RNA Database

5S Ribosomal RNA Database provides information on nucleotide sequences of 5S rRNAs and their genes. The sequences for particular organisms can be retrieved as single files using a taxonomic browser or in multiple sequence structural alignments. This programme is freely available at http://biobases.ibch.poznan.pl/5SData/.

11.7 Greengenes

This is an online full-length small-subunit (SSU) rRNA gene database called greengenes available at <http://greengenes.lbl.gov/> that keeps pace with public submissions of both archaeal and bacterial 16S rDNA sequences has been established (DeSantis et al. 2003). It addresses a number of limitations currently associated with SSU rRNA records in the public databases by providing automated chimera-screening, taxonomic placement of unclassified environmental sequences using multiple published taxonomies for each record, multiple standard alignments and uniform sequence-associated information curated from GenBank records. Greengenes also provides a suite of utensils for manipulation of sequences including an alignment tool and has been streamlined to interface with the widely used ARB program.

11.8 Ribosomal Database Project

The Ribosomal Database Project – II (RDP-II) (Maidak et al. 2001) available at <http://rdp.cme.msu.edu/> provides data, tools and services related to ribosomal RNA sequences to the research community. It offers aligned and annotated rRNA sequence data, analysis services, and phylogenetic inferences derived from these data. Currently available on the RDP-II website as a beta release, 9.0 provides over 50,000 annotated (eu) bacterial sequences aligned with a secondary-structure based alignment algorithm (Brown 2000). Data subsets are available for sequences of length 1,200 or greater and for sequences from type material. Annotation goals include up-to-date name, strain and culture deposit information, sequence length and quality information. In order to provide a phylogenetic context for the data, RDP-II makes available over 100 trees that span the phylogenetic breadth of life. Web based research tools are provided for comparing user submitted sequences to the RDP-II database (Sequence Match), aligning user sequences against the nearest RDP sequence (Sequence Aligner), examining probe and primer specificity (Probe Match), testing for chimeric sequences (Chimera Check), generating a distance matrix (Similarity Matrix), analyzing T-RFLP data (T-RFLP and TAP-TRFLP), a java-based phylogenetic tree browser (Sub Trees), a sequence search and selection tool (Hierarchy Browser) and a phylogenetic tree building and visualization tool (Phylip Interface). The latter tool has been enhanced to allow a choice of either the

Phylip neighbor-joining (Felsenstein 1993) or Weighbor weighted neighbor-joining (Bruno et al. 2000) programs for tree construction.

11.9 RISSC – Ribosomal Internal Spacer Sequence Collection

This is a database of ribosomal 16S-23S spacer sequences intended mainly for molecular biology studies in typing, phylogeny and population genetics. It compiles more than 2,500 entries of edited DNA sequence data from the 16S-23S ribosomal spacers present in most prokaryotes and organelles. Ribosomal spacers have proven to be extremely useful tools for typing and identifying closely related prokaryotes due to their high variability in size and/or sequence, much more so than the flanking 16S and 23S rRNA genes. These genes are commonly used to establish molecular relationships among microbes at a taxonomic level of species or higher (e.g. genus, domain). However their internal transcribed spacers (ITS) are much more useful to discriminate at the species or even strain level (Iwen et al. 2002). RISSC available at <http://ulises.umh.es/RISSC> provides the scientific community with a comprehensive set of ribosomal spacer sequences, fully edited and characterized with a key feature as is the presence/absence of tRNA genes within them, ready to be used and compared with their own ITS sequences.

11.10 probeBase

probeBase is a curated database of annotated rRNA-targeted oligonucleotide probes and supporting information (Loy et al. 2003, 2007). Rapid access to probe, microarray and reference data is achieved by powerful search tools and via different lists that are based on selected categories such as functional or taxonomic properties of the target organism(s), or the hybridization system in which the probes were applied. Additional information on probe coverage and specificity is available through direct submissions of probe sequences from probeBase to RDP-II and Greengenes, two major rRNA sequence databases.

ProbeBase available at <http://www.microbial-ecology.net/probebase> entries increased from 700 to more than 1,200 during the past 3 years. Several options for submission of single probes or entire probe sets, even prior to publication of newly developed probes, should further contribute to keeping probeBase an up-to-date and useful resource.

11.11 RRNDB

The Ribosomal RNA Operon Copy Number Database (RRNDB) available at <http://rrndb.cme.msu.edu/> contains annotated information on rRNA operon copy number among prokaryotes. Gene redundancy is uncommon in prokaryotic genomes, however rRNA genes can vary from one to as many as 15 copies. Despite the widespread use of 16S rRNA gene sequences for identification of prokaryotes, information on the number and sequence of individual rRNA genes on a genome is not readily accessible. Each entry in RRNDB contains detailed information linked directly to external websites including the Ribosomal Database Project, GenBank, PubMed, and several culture collections.

12 Identification of New Species or Variant Strains of Known Species

Bioinformatics has facilitated researchers to study microbial biodiversity because of its direct interventions in molecular identification, data storage and retrieval system that were the objects and the worrisome of systematic research. The bioinformatics driven approaches enabled people to work efficiently on microbial diversity, identification, characterization, molecular taxonomy and community analysis patterns of both culturable and unculturable organisms. Description of new species, genera and even molecular taxa emerged dramatically in the literature after 1990s and these efforts are largely driven by advances in sequencing technologies. The utilization of phenotypic identification methods classically requires a probability-based analysis to determine identity. In cases where identification probabilities are 98% with known species, the identification is generally considered acceptable. The lower the probability percentage however, the less accurate the identification becomes, frequently resulting in supplemental testing to resolve discrepancies among test results. It is not unusual for the laboratory to be unable to identify variant strains of known species using phenotypic methods. DNA sequencing now allows the laboratory a means to resolve those instances where phenotypic testing cannot differentiate among closely related organisms.

The recognition of a species that does not match known schemes for phenotypic identification may represent a previously unrecognized species (Relman 2002). Sequencing of areas within the rDNA complex may be useful to suggest a new species when there is a <98% of the sequence similarity with known species. The ability to separate a new species from an atypical strain of a known species is however, difficult. The first approach to recognition of a new species is to determine the phylogenetic position of the suspect new species compared to closely related known species. Phylogenetic trees using the 16S gene for bacteria and the 18S gene for fungi are commonly used for this type of analysis. The 16S rRNA approach is rooted in the concept of point mutation due to their slow mutation rate. Before microbial genomes were sequenced, using 16SrRNA database was considered and bacteria, archaea, and eukaryotes were identified.

A high degree of phenotypic consistency and rDNA sequence similarity as well as, a significant degree of DNA-DNA hybridization, is suggestive of a new species.

13 Bioinformatics Challenges

Many bioinformatics tools have been borrowed from the fields of artificial intelligence, data mining, and statistical methods. However, the characteristics of biological data may differ significantly from those of the original data for which the methods were developed. Though many computational methods have been introduced for genomic data analysis based on these methods, several challenges still exist. Though public databases such as GenBank are useful, the lack of quality sequences and the absence of sequence information on a large number of species as well as the availability of computational tools to reliably analyze the results are drawbacks to this technology. A typical DNA microarray might have thousands of features (probes) for, at most, one hundred samples. Feature reduction is typically required before these sorts of analyses can be performed (Al-Khaldi et al. 2012; Bier et al. 2008; Yauk and Berndt 2007). Another challenge is integrating data from different sources. These datasets might show a high degree of heterogeneity and might also vary in quality. They might be generated using different experimental platforms or based on different molecular methods. Using these data together efficiently requires developing suitable bioinformatics methods. Of these methods, the simplest one is to put several datasets together to build a larger dataset and then analyze this larger dataset. However, this method will not work if the formats of the original datasets differ. Furthermore, the best processing methods for different datasets are not the same. For example, Dice coefficents work well for some PFGE data but does not work well for some VNTR data. Thus, it might be an impossible task to choose an optimal method for a combined dataset. An alternate method is to process different datasets separately and then combine the results to obtain the final result. The difficulties with this kind of method, however, are determining the extent to which the different sources of data should contribute and explaining the combined results.

14 Conclusion

The development of computational methods based on the organized algorithms, interpretational skills and high storage capacities facilitated comparison of entire genomes and thus permit biologists to study more complex evolutionary trends like gene duplication, horizontal gene transfer and prediction of factors important in speciation (Nakashima et al. 2005). Bioinformatics researchers have compared extensively multiple genomes to correlate and classify the genomes into various families and to study evolution. It has been established by many researchers that overall evolution is a combination of point based mutation giving rise to restructuring of genomes based upon gene duplications, gene insertion, gene deletion, horizontal gene transfer etc. The ultimate aim of such studies lies in deciphering the evolutionary lineages among the group of organisms in a quest to determine the tree of life and the last universal common ancestor. The progress in bioinformatics and wet-lab techniques has to remain interdependent and focused complementing each other for their own progress and for the progress of biotechnology in future.

15 Some More Web Addresses for Bioinformatics Tools

Name of tool/database	Web address
ASD	http://www.ebi.ac.uk/asd
AUGUSTUS	http://augustus.gobics.de/bin/npsa_automat.pl?page=npsa_gor4.html
BLAST	http://www.ncbi.nlm.nih.gov/blast
CFSSP	http://www.biogem.org/tool/chou-fasman/
Clustal W	http://www.ebi.ac.uk/Tools/clustalw2/index.html
ComputpI/Mw	http://web.expasy.org/compute_pi/
CpG Island Searcher	http://www.uscnorris.com/cpgislands2/cpg.aspx
CpGPlot	http://www.ebi.ac.uk/Tools/emboss/cpgplot/index.html
DDBJ BLAST	http://blast.ddbj.nig.ac.jp
DNA tools	http://biology.semo.edu/cgi-bin/dnatools.pl
Entrez Gene	http://www.ncbi.nlm.nih.gov/sites/entrez
ESLPred2	http://www.imtech.res.in/raghava/eslpred2/
ExPaSy	http://expasy.org/tools/
FEX	http://www.softberry.ru/berry.phtml
FGENESH	http://www.softberry.ru/berry.phtml
GeneMark.hmm	http://www.itba.mi.cnr.it/webgene/
GOR	http://npsa-pbil.ibcp.fr/cgi-
HMMgene	http://www.cbs.dtu.dk/services/HMMgene/
MGI	http://www.informatics.jax.org/
MultiLoc2	http://abi.inf.uni-tuebingen.de/Services/MultiLoc2
Myristoylator	http://web.expasy.org/myristoylator/
NetAcet	http://www.cbs.dtu.dk/services/NetAcet/
NetOGlyc	http://www.cbs.dtu.dk/services/NetOGlyc
NetPhos	http://www.cbs.dtu.dk/services/NetPhos/
NetPhosK	http://www.cbs.dtu.dk/services/NetPhosK/
NetSurfP	http://www.cbs.dtu.dk/services/NetSurfP/
NMT	http://mendel.imp.ac.at/myristate/SUPLpredictor.htm
OligoCalc	http://www.basic.northwestern.edu/biotools/oligocalc.html
PSIPRED v3.0	http://bioinf.cs.ucl.ac.uk/psipred/
SherLoc2	http://abi.inf.uni-tuebingen.de/Services/SherLoc2
SIGSCAN	http://www-bimas.cit.nih.gov/molbio/signal/
SMS	http://www.bioinformatics.org/sms/
TermiNator	http://www.isv.cnrs-gif.fr/terminator3/index.html
TFBIND	http://tfbind.hgc.jp/
TFSEARCH	http://www.cbrc.jp/research/db/TFSEARCH.html

Glossary

Homology Searches: BLAST & FASTA Background Information: The three BLAST programs that one will commonly use are BLASTN, BLASTP and BLASTX. BLASTN will compare your DNA sequence with all the DNA sequences in the nonredundant database (nr). BLASTP will compare your protein sequence with all the protein sequences in nr. In BLASTX nucleotide sequence of interest will be translated in all six reading frames and the products compared with the nr protein database. A tutorial is also available at NCBI.

BLAST Homepage – (NCBI) Found at http://blast.ncbi.nlm.nih.gov/Blast.cgi. It is widely used for homology searches. BLAST stands for Basic Local Alignment Search Tool and it displays a number of organism and query specific blast.

Nucleotide BLAST (BLASTN) N.B. the default database is "human genomic and transcript" not "nucleotide collection (nt/nr)"

Protein BLAST (BLASTP) This program is also coupled with a motif search. If you suspect that your protein may only show weak sequence similarity to other proteins, I would suggest clicking on the PSI-BLAST (Position-Specific Iterated BLAST) feature. NCBI also provides a PSI-BLAST tutorial. PSI-BLAST searches to yield better delineation of true and false positives.

Translated BLAST (BLASTX) TBLASTX searches translated nucleotide databases using a translated nucleotide query; while TBLASTN searches translated nucleotide databases using a protein query. These are useful resources if you are interested in homologs in unfinished genomes. Undeter "Databases" select "genomic survey sequences", "High throughput genomic sequences" or "whole-genome shotgun reads"

Blast with Microbial Genomes (BLASTN, TBLASTN, TBLASTX etc.). It permits us to compare a nucleic acid or protein sequence against finished archaeal and bacterial genomes. Depending upon the time of day your results may appear almost immediately or the search may be delayed or not accepted at all. For PSI-BLAST, and other searches it is recommended to frequently enter information in the "Entrez Query" section e.g. *Escherichia coli* [organism] or Viruses [organism] to see "hits" specifically to E. coli or viruses/bacteriophages. It is advisable to always select "Show results in a new window"

EMB BLAST-(European Molecular Biology network). Very convenient since it permits one to specifically search databases such as prokaryote, bacteriophage, fungal, & 16S rRNA using BLASTN, and specific bacterial genomes or SwissProt using BLASTX or BLASTN.

ParAlign It employs a heuristic method for sequence alignment. In essence, ParAlign is about as sensitive as Smith-Waterman but runs at the speed of BLAST.

GTOP Sequence Homology Search (Laboratory for Gene-Product Informatics, National Institute of Genetics, Japan) – offers BLASTP search capability against individual archaea, bacteria, eukaryota, and viruses.

T4-like Phage NCBI MegaBLAST (Tulane Univ., New Orleans, U.S.A. & CNRS, Toulouse, France) This includes a growing list of T4-like completed phage sequences as well as those in the draft and contig stages of completion.

WU-BLAST (Washington University BLAST) The emphasis of this tool is to find regions of sequence similarity quickly, with minimum loss of sensitivity. This will yield functional and evolutionary clues about the structure and function of the novel sequence.

Batch BLAST (Greengene web server; University of Massachusetts, Lowell, U.S.A.) was developed by Michael V. Graves for DNA or protein BLAST sequence analysis against the NCBI databases. It allows one to submit a file that contains multiple sequences and then will organize the results by each individual sequence contained in the file.

HHPred Homology detection & structure prediction is a method for database searching and structure prediction that is as easy to use as BLAST but is much more sensitive in finding remote homologs. HHpred is the first server that is based on the pairwise comparison of profile hidden Markov models (HMMs). Whereas most conventional sequence search methods search sequence databases such as UniProt or the NR, HHpred searches alignment databases, like Pfam or SMART. This greatly simplifies the list of hits to a number of sequence families instead of a clutter of single sequences. HHpred accepts a single query sequence or a multiple alignment as input.

PSI-BLAST or PHI-BLAST search Position-Specific Iterative BLAST creates a profile after the initial search.

BLAST 2 BLAST two sequences against one another. This utilizes BLASTN, P, X as well as TBLASTN and TBLASTX.

Gene Context Tool It is an incredible tool for visualizing the genome context of a gene or group of genes.

TC-BLAST It scans the transport protein database (TC-DB) producing alignments and phylogenetic trees. The TC-DB details a comprehensive classification system for membrane transport proteins known as the Transport Commission (TC) system.

MEROPS BLAST This permits one to screen protein sequences against an extensive database of characterized peptidases.

SEARCHGTr It is web-based software for the analysis of glycosyltransferases involved in the biosynthesis of a variety of pharmaceutically important compounds like adriamycin, erythromycin, vancomycin etc. This software has been developed based on a comprehensive analysis of sequence/structural features of 102 GTrs of known specificity from 52 natural product biosynthetic gene clusters.

PipeAlign It offers an integrated approach to protein family analysis through a cascade of different sequence analysis programs (BALLAST, DbClustal multiple alignment program, Rascal alignment analysis) removal of any sequences that do not belong to the protein family are performed by the NorMD, and clustered into potential functional subfamilies using Secator or DPC.

MPsrch (EMBL-EBI) This sequence sequence comparison tool implements the true Smith and Waterman algorithm identifying hits in cases where Blast and Fasta fail and also reports fewer false-positives. This software provides information on Match %; % Query Match (% of the query sequence matched); Conservative changes; Mismatches; Indels; and Gaps.

GOAnno This web tool automatically annotates proteins according to the Gene Ontology using hierarchised multiple alignments. Positioning the query protein in its aligned functional subfamily represents a key step to obtain highly reliable predicted GO annotation based on the GOAnno algorithm.

COMPASS This is a profile-based method for the detection of remote sequence similarity and the prediction of protein structure. The server features three major developments: (i) improved statistical accuracy; (ii) increased speed from parallel implementation; and (iii) new functional features facilitating structure prediction. These features include visualization tools that allow the user to quickly and effectively analyze specific local structural region predictions suggested by COMPASS alignments.

MineBlast It performs BLASTP searches in UniProt to identify names and synonyms based on homologous proteins and subsequently queries PubMed, using combined search terms in order to find and present relevant literature.

Comparison of homology between two small genomes: SCAN2 (Softberry.com) It provides one with a colour-coded graphical alignment of genome length DNAs in Java. In the top panel regions of high sequence identity are presented in red. By highlighting the grey yellow, green, black boxes one can select specific regions for examination of the sequence alignment.

Advanced PipMaker It aligns two DNA sequences and returns a percent identity plot of that alignment, together with a traditional textual form of the alignment. We may need to download it for viewing and manipulating the output from pairwise alignment programs such as PipMaker representations of the alignments.

JDotter: A Java Dot Plot Viewer (Viral Bioinformatics Resource Center, University of Victoria, Canada) – a dot matrix plotter for Java. It produces similar diagrams to the above mentioned programs, but with better control on output.

multi-zPicture: multiple sequence alignment tool provides nice dotplot graphs and dynamic visualizations. If simple gene locations are provided in the form (e.g. >2,000–5,000 RNA_polymerase; indicates that the RNA polymerase gene is found on the plus strand between bases 2,000 and 5,000) this data will be added to the dynamic visualization. zPicture alignments can be automatically submitted to rVista to identify conserved transcription factor binding sites.

GeneOrder 3.0 This is ideal for comparing small GenBank genomes (up to 2 Mb). Each gene from the Query sequence is compared to all of the genes from the Reference sequence using BLASTP. There are two display formats: graphical and tabular.

CoreGenes This programme is designed to analyze two to five genomes simultaneously, generating a table of related genes – orthologs and putative orthologs. These entries are linked to their GenBank data. It has a limit of 0.35 Mb, while the newer version CoreGenes 2.0 extends the limit to approx. 2.0 Mb. If data is not present in GenBank, using this site will be very helpful.

CoreGenes 3.0 This is the latest member in the CoreGenes family of tools. It determines unique genes contained in a pair of proteomes. This has proved extremely useful in determining unique genes in comparisons between large Myoviridae.

References

Al-Khaldi SF, Mossoba MM, Allard MM, Lienau EK, Brown ED (2012) Bacterial identification and subtyping using DNA microarray and DNA sequencing. Methods Mol Biol 881:73–95

Altschul SF, Madden TL, Schäffer AA, Zhang J, Zhang Z, Miller W, Lipman DJ (1997) Gapped BLAST and PSI-BLAST: a new generation of protein database search programs. Nucleic Acids Res 25:3389–3402

Barrell D, Dimmer E, Huntley RP, Binns D, O'Donovan C, Apweiler R (2009) The GOA database in 2009–an integrated gene ontology annotation resource. Nucleic Acids Res 37:D396–D403

Benson DA, Karsch-Mizrachi I, Lipman DJ, Ostell J, Rapp BA, Wheeler DL (2002) GenBank. Nucleic Acids Res 28:15–18

Bier FF, von Nickisch-Rosenegk M, Ehrentreich-Förster E, Reiss E, Henkel J, Strehlow R, Andresen D (2008) DNA microarrays. Adv Biochem Eng Biotechnol 109:433–453

Bilitewski U (2009) DNA microarrays: an introduction to the technology. Methods Mol Biol 509:1–14

Boerlin P, Reid-Smith RJ (2008) Antimicrobial resistance: its emergence and transmission. Anim Health Res Rev 2:115–126

Borucki MK, Kim SH, Call DR, Smole SC, Pagotto F (2004) Selective discrimination of Listeria monocytogenes epidemic strains by a mixed-genome DNA microarray compared to discrimination by pulsed-field gel electrophoresis, ribotyping, and multilocus sequence typing. J Clin Microbiol 42:5270–5276

Bray N, Pachter L (2003) MAVID multiple alignment server. Nucleic Acids Res 31:3525–3526

Bray N, Pachter L (2004) MAVID: constrained ancestral alignment of multiple sequences. Genome Res 14:693–699

Brown MP (2000) Small subunit ribosomal RNA modeling using stochastic context free grammar. Proc Int Conf Intell Syst Mol Biol 8:57–66

Brudno M, Do CB, Cooper GM, Kim MF, Davydov E, Green ED, Sidow A, Batzoglou S, NISC Comparative Sequencing Program (2003) LAGAN and Multi-LAGAN: efficient tools for large-scale multiple alignment of genomic DNA. Genome Res 13:721–731

Bruno WJ, Socci ND, Halpern AL (2000) Weighted neighbor joining: a likelihood-based approach to distance-based phylogeny reconstruction. Mol Biol Evol 17:189–197

Call DR, Borucki MK, Besser TE (2003) Mixed-genome microarrays reveal multiple serotype and lineage-specific differences among strains of Listeria monocytogenes. J Clin Microbiol 41:632–639

Cole JR, Chai B, Farris RJ, Wang Q, Kulam SA, McGarrell DM, Garrity GM, Tiedje JM (2005) The ribosomal database project (RDP-II): sequences and tools for high-throughput rRNA analysis. Nucleic Acids Res 33(Database issue)

Cole JR, Wang Q, Cardenas E, Fish J, Chai B, Farris RJ, Kulam-Syed-Mohideen AS, McGarrell DM, Marsh T, Garrity GM, Tiedje JM (2009) The ribosomal database project: improved alignments and new tools for rRNA analysis. Nucleic Acids Res 37:D141–D145

Darling AC, Mau B, Blattner FR, Perna NT (2004) Mauve: multiple alignment of conserved genomic sequence with rearrangements. Genome Res 14:1394–1403

DeSantis TZ, Dubosarskiy I, Murray SR, Andersen GL (2003) Comprehensive aligned sequence construction for automated design of effective probes (CASCADE-P) using 16S rDNA. Bioinformatics 19:1461–1468

Dewey CN (2007) Aligning multiple whole genomes with Mercator and MAVID. Methods Mol Biol 395:221–236

Doolittle WF (1999) Phylogenetic classification and the universal tree. Science 284:2124–2128

Feinbaum R (2001) Introduction to plasmid biology. Curr Protoc Mol Biol Chapter 1:Unit 1.5

Felsenstein J (1993) PHYLIP (phylogeny inference package) version 3.5c. Department of Genetics, University of Washington, Seattle

Fleischmann RD, Adams MD, White O, Clayton RA, Kirkness EF, Kerlavage AR, Bult CJ, Tomb JF, Dougherty BA, Merrick JM et al (1995) Whole-genome random sequencing and assembly of *Haemophilus influenzae* Rd. Science 269:496–512

Foster JA, Moore JH, Gilbert JA, Bunge J (2012) Microbiome studies: analytical tools and techniques. Pac Symp Biocomput 17:200–202

Foxman B, Zhang L, Koopman JS, Manning SD, Marrs CF (2005) Choosing an appropriate bacterial typing technique for epidemiologic studies. Epidemiol Perspect Innov 25:2–10

Gomez SM, Choi K, Wu Y (2008) Prediction of protein-protein interaction networks. Curr Protoc Bioinform Chapter 8:Unit 8.2

Harris MA, Clark J, Ireland A, Lomax J, Ashburner M, Foulger R, Eilbeck K, Lewis S, Marshall B, Mungall C, Richter J, Rubin GM, Blake JA, Bult C, Dolan M, Drabkin H, Eppig JT, Hill DP, Ni L, Ringwald M, Balakrishnan R, Cherry JM, Christie KR, Costanzo MC, Dwight SS, Engel S, Fisk DG, Hirschman JE, Hong EL, Nash RS, Sethuraman A, Theesfeld CL, Botstein D, Dolinski K, Feierbach B, Berardini T, Mundodi S, Rhee SY, Apweiler R, Barrell D, Camon E, Dimmer E, Lee V, Chisholm R, Gaudet P, Kibbe W, Kishore R, Schwarz EM, Sternberg P, Gwinn M, Hannick L, Wortman J, Berriman M, Wood V, de la Cruz N, Tonellato P, Jaiswal P, Seigfried T, White R, Gene Ontology Consortium (2004) The gene ontology (GO) database and informatics resource. Nucleic Acids Res 32:D258–D261

Idekar T, Galitski T, Hood L (2001) A new approach to decoding life: systems biology. Annu Rev Genomics Hum Genet 2:343–372

Iwen PC, Hinrichs SH, Rupp ME (2002) Utilization of the internal transcribed spacer regions as molecular targets to detect and identify human fungal pathogens. Med Mycol 40:87–109

Johnston JW (2010). Laboratory growth and maintenance of *Haemophilus influenzae*. Curr Protoc Microbiol Chapter 6:Unit 6D.1

Larkin MA, Blackshields G, Brown NP, Chenna R, McGettigan PA, McWilliam H, Valentin F, Wallace IM, Wilm A, Lopez R, Thompson JD, Gibson TJ, Higgins DG (2007) Clustal W and Clustal X version 2.0. Bioinformatics 23:2947–2948

Larsen N, Olsen GJ, Maidak BL, McCaughey MJ, Overbeek R, Macke TJ, Marsh TL, Woese CR (1993) The ribosomal database project. Nucleic Acids Res 21:3021–3023

Lee MM, Chan MK, Bundschuh R (2008) Simple is beautiful: a straightforward approach to improve the delineation of true and false positives in PSI-BLAST searches. Bioinformatics 24:1339–1343

Lin HN, Notredame C, Chang JM, Sung TY, Hsu WL (2011) Improving the alignment quality of consistency based aligners with an evaluation function using synonymous protein words. PLoS One 6:e27872

Loy A, Horn M, Wagner M (2003) probeBase: an online resource for rRNA-targeted oligonucleotide probes. Nucleic Acids Res 31:514–516

Loy A, Maixner F, Wagner M, Horn M (2007) probeBase – an online resource for rRNA-targeted oligonucleotide probes: new features 2007. Nucleic Acids Res 35(Database issue):D800–D804

Luscombe NM, Greenbaum D, Gerstein M (2001) What is bioinformatics? A proposed definition and overview of the field. Methods Inform Med 40:346–358

Maidak BL, Cole JR, Lilburn TG, Parker CT, Saxman PR, Farris RJ, Garrity GM, Olsen GJ, Schmidt TM, Tiedje JM (2001) The RDP-II (Ribosomal Database Project). Nucleic Acids Res 29:173–174

McHardy AC, Martín HG, Tsirigos A, Hugenholtz P, Rigoutsos I (2007) Accurate phylogenetic classification of variable-length DNA fragments. Nat Methods 4:63–72

Mølbak L, Tett A, Ussery DW, Wall K, Turner S, Bailey M, Field D (2003) The plasmid genome database. Microbiology 149:3043–3045

Nakashima N, Mitani Y, Tamura T (2005) Actinomycetes as host cells for production of recombinant proteins. Microb Cell Fact 4:7

Pearson WR (1990) Rapid and sensitive sequence comparison with FASTP and FASTA. Methods Enzymol 183:63–98

Pearson WR (1991) Searching protein sequence libraries: comparison of the sensitivity and selectivity of the Smith-Waterman and FASTA algorithms. Genomics 11:635–650

Perna NT, Plunkett G 3rd, Burland V, Mau B, Glasner JD, Rose DJ, Mayhew GF, Evans PS, Gregor J, Kirkpatrick HA, Pósfai G, Hackett J, Klink S, Boutin A, Shao Y, Miller L, Grotbeck EJ, Davis NW, Lim A, Dimalanta ET, Potamousis KD, Apodaca J, Anantharaman TS, Lin J, Yen G, Schwartz DC, Welch RA, Blattner FR (2001) Genome sequence of enterohaemorrhagic *Escherichia coli* O157:H7. Nature 409:529–533

Plewniak F (2008) Database similarity searches. Methods Mol Biol 484:361–378

Pruesse E, Quast C, Knittel K, Fuchs BM, Ludwig WG, Peplies J, Glöckner FO (2007) SILVA: a comprehensive online resource for quality checked and aligned ribosomal RNA sequence data compatible with ARB. Nucleic Acids Res 35:7188–7196

Relman DA (2002) New technologies, human-microbe interactions, and the search for previously unrecognized pathogens. J Infect Dis 186(Suppl 2):254–258

Schäffer AA, Aravind L, Madden TL, Shavirin S, Spouge JL, Wolf YI, Koonin EV, Altschul SF (2001) Improving the accuracy of PSI-BLAST protein database searches with composition-based statistics and other refinements. Nucleic Acids Res 29:2994–3005

Shah SP, Huang Y, Xu T, Yuen MM, Ling J, Ouellette BF (2005) Atlas – a data warehouse for integrative bioinformatics. BMC Bioinformatics 21:6–34

Som A (2006) Theoretical foundation to estimate the relative efficiencies of the Jukes-Cantor + gamma model and the Jukes-Cantor model in obtaining the correct phylogenetic tree. Gene 385:103–110

Stark M, Berger SA, Stamatakis A, von Mering C (2010) MLTreeMap – accurate maximum likelihood placement of environmental DNA sequences into taxonomic and functional reference phylogenies. BMC Genomics 5:11–461

Summers DK (1996) The biology of plasmids. Blackwell Science, Oxford

Takahashi K, Nei M (2000) Efficiencies of fast algorithms of phylogenetic inference under the criteria of maximum parsimony, minimum evolution, and maximum likelihood when a large number of sequences are used. Mol Biol Evol 17:1251–1258

Tatusov RL, Koonin EV, Lipman DJ (1997) A genomic perspective on protein families. Science 278:631–637

Tatusov RL, Fedorova ND, Jackson JD, Jacobs AR, Kiryutin B, Koonin EV, Krylov DM, Mazumder R, Mekhedov SL, Nikolskaya AN, Rao BS, Smirnov S, Sverdlov AV, Vasudevan S, Wolf YI, Yin JJ, Natale DA (2003) The COG database: an updated version includes eukaryotes. BMC Bioinformatics 4:41

Tenover FC, Arbeit RD, Goering RV (1997) How to select and interpret molecular strain typing methods for epidemiological studies of bacterial infections: a review for healthcare epidemiologists. Infect Control Hosp Epidemiol 18:426–439

Thomas CM (2000) The horizontal gene pool: bacterial plasmids and gene spread. Harwood Academic, Amsterdam

Thompson JD, Gibson TJ, Higgins DG (2002) Multiple sequence alignment using ClustalW and ClustalX. Curr Protoc Bioinform. Chapter 2:Unit 2.3

Wan Y, Broschat SL, Call DR (2007) Validation of mixed-genome microarrays as a method for genetic discrimination. Appl Environ Microbiol 73:1425–1432

Wilmes P, Simmons SL, Denef VJ, Banfield JF (2009) The dynamic genetic repertoire of microbial communities. FEMS Microbiol Rev 33:109–132

Woese CR (1987) Bacterial evolution. Microbiol Mol Biol Rev 51:221–271

Yauk CL, Berndt ML (2007) Review of the literature examining the correlation among DNA microarray technologies. Environ Mol Mutagen 48:380–394

Zhou J (2003) Microarrays for bacterial detection and microbial community analysis. Curr Opin Microbiol 6:288–294

Chapter 7
Recent Advances and Perspectives in Metagenomic Studies of Soil Microbial Communities

E.V. Pershina, E.E. Andronov, A.G. Pinaev, and N.A. Provorov

Abstract Metagenomics is a modern and rapidly growing field of molecular genetics and ecology that studies the "collective" genome of the microbial community and is based on the analysis of environmental DNA, extracted directly from a variety of natural habitats. The advent of high-throughput sequencing techniques had opened the principally new opportunities in studies of the genetic structure of microbial communities, but at the same time highlighted significant difficulties, arising particularly during the investigation of the soil metagenome.

Soil is the most densely populated habitat on the planet, and can contain up to 1,000 Gbp of genetic information per gram suggesting the great misfortune during the analysis of the soil metagenome consisting in the preferential analysis of only a small fraction of the total soil metagenome with relatively low accuracy. We emphasize the necessity for structuring of the soil metagenome and identification of its main components. Considering that modern metagenomics should first be addressed to the "eternal" questions of soil microbiology, in this review we tried to identify the meaningful parts of soil metagenome primarily by the analysis of soil microbial communities. We discussed in detail the spatial organization of soil metagenome associated with micro- and macrostructure of the soil matrix, the structural organization of soil metagenome associated with the presence of heterogeneous pools of soil DNA, taxonomical and functional organization of the soil metagenome revealed from the investigation of global patterns of distribution of microbial communities in relation to the specific environmental factors.

We demonstrated that soil microbial communities are characterized by a presence of a restricted number of taxonomic organization types, which are based on the most powerful ecological factors such as soil pH and moisture. Comparing to more or less labile taxonomic structure, the functional structure of the soil metagenome is rather conservative and is maintained primarily by two factors – the

E.V. Pershina • E.E. Andronov • A.G. Pinaev • N.A. Provorov (✉)
All-Russia Research Institute for Agricultural Microbiology,
Saint-Petersburg, Russia
e-mail: provorov@newmail.ru

A. Malik et al. (eds.), *Management of Microbial Resources in the Environment*, 141
DOI 10.1007/978-94-007-5931-2_7, © Springer Science+Business Media Dordrecht 2013

microbial co-operation and the maintenance of high levels of genetic diversity. Finally we come to a conclusion that the soil metagenome represents an integrative hereditary system for maintenance of the basic soil functions under the variable ecological conditions.

Keywords Metagenomics • Microbiome • Soil microbial communities • Gene transfer • Soil DNA

1 What is the Metagenome?

It is quite difficult to give a precise definition of metagenomics. The term metagenomics usually means a number of molecular techniques involved in accession and analysis of the genetic information of microbial community derived directly from the environment. Thus we obtain a huge pool of genetic information that we call metagenome – the "collective" genome of the microbial community. Describing metagenomics we deal with DNA, that comparing to RNA molecules is more stable and heterogeneous and generally contain much more information not only about biologically-driven processes that currently take place in the environment but also about genetic potential of microbial community. All this information is stored in extremely large amount of environmental DNA (Vieites et al. 2010).

Therefore if we simply define the metagenome as a collection of the entire genetic information present in e.g. 1 g of soil (and it must be said that the majority of modern metagenomic projects is guided directly like this) we will deal, according to various estimates, with enormous 10^{14}–10^{16} b of DNA (Vogel et al. 2009; Trevors 2010; Hirsch et al. 2010). Sequencing of such a huge amount of soil DNA seems to be difficult and even unsolvable problem. Nowadays it leads to sometimes pessimistic forecasts that it is crucially impossible to analyse the whole soil metagenome (Baveye 2009; Tringe and Hugenholtz 2009).

Considering all this, we initiated the current review primarily to summarize all the present information about genetic diversity of bacteria in soil and find some reasonable traits for the future investigations. To run our work we must primarily give a narrow definition of soil metagenome. We define soil metagenome as a complex system of at least four pools of the genetic material: the DNA of the living cells, the DNA of the resting forms and finally the DNA of the dead organisms and extracellular DNA. All these genetic reservoirs are maintained in the dynamic equilibrium depending on the environmental conditions. It will be absolutely insufficient to restrict our scope to the living organisms alone. Giving credit to the importance of measuring metabolically active microbes in the soil we must stress the significance of the "non-active" soil DNA.

Soil is considered to be the most populated environment on the Earth, but at the same time, due to the highly changeable ecological conditions inherent to this habitat, the majority of microorganisms are present here in the various resting forms (the so called "microbiological pool" of soils). Instead of "microbiological pool" the extracellular DNA and potentially the DNA of the dead microbes and viruses can be treated as "genetic pool" of soils due to the well-known phenomenon of genetic transformation. This process plays a significant role particularly in soil due to its spatial and temporal heterogeneity. Investigation of the transformation or other processes involved in horizontal gene transfer could shed light on the rate of inter-species interaction within the microbial community as well as on the responsibility of soil microbes to act upon changing ecological conditions (adaptation potential of microbial community).

It is worth mentioning here that the traditional microbiological plating methods are basically irresponsible to gain any essential information either about soil "microbial pool" or about natural genetic reservoir. Contrarily, application of new high throughput molecular techniques that are sensitive to all the genetic material exposed in the environment opens radically new perspectives in studying soil microbial population. These new methods are applied not only to measure actual diversity of microorganisms and to answer the intriguing question of "who is doing what in the environment?" but even to elucidate the microbial and genetic potential hidden in soil, which makes it a highly integrated and continuously evolving ecosystem.

In this paper we consider the metagenome of soil community as an integral genetic system and give analysis of this system with respect to the taxonomic and functional structures of the soil microbiota. This approach will enable us to describe the correlation of the taxonomic structure of soil community and its functional structure, and thus to give account of the adaptive processes in soil microbiota responsible for soil sustainability and fertility.

2 Sources of Genetic Information in Soil

Before the great "molecular revolution" in microbial ecology, cultivation of microorganisms on artificial media was the main source of information about the microbial diversity. At the same time the data from microscopic studies testified that there were much more different bacterial species in the environment. Simultaneously, there was an opinion that cultivation of microorganisms on the reach medium (relatively to the naturally occurred) counted only for zymogenic soil microorganisms, which became alive in soil after the sharp changes in environment, particularly after enrichment of soil with organic matter.

Finally a plenty of methods for the analysis of environmental DNA molecules showed that by culturing we obtained less than 1% of actual biodiversity and the majority of microorganisms cannot be cultured in the laboratory. After the pioneering 16S rRNA based studies of microbial communities it was postulated that soil appeared to be the most populated environment on the Earth (Torsvik et al. 1990; Daniel 2005). A characteristic property of soil community structure is low number of dominant microbial species together with the "long tail" of rarely found species (Elshahed et al. 2008; Bent and Forney 2008). On the one hand, a lot of minor species in microbial community could be directly related with the highly heterogeneous nature of soil environment which produces many ecological niches, but, on the other hand, there is strong evidence that the vast majority of microorganisms in soil are staying in the resting forms due to the changeable surrounding conditions, and thus the temporary inactive microbes could comprise the great part of the "long tail". This cornerstone question in soil ecology is still unsolved, primarily due to the lack of DNA extraction techniques for the accurate separation of active cells from the resting microbial population. Anyway, after the first decade of molecular ecology studies it was shown, that microbial communities in natural environments possess an unprecedented levels of biodiversity and probably have their specific ecological characteristics.

The recent evidence of the stability of extracellular DNA (eDNA) in number of natural environments argues much for the specific integrative structure of microbial population in soil. Particularly it was shown that eDNA could comprise up to 60% of the total DNA (Agnelli et al. 2004).

Hence, now we are inclined to describe soil metagenome as a complex system with spatially and temporary separated and at the same time intimately related components, which play different roles in ecosystem functioning. First component includes living cells and closely related second component – resting forms of microorganisms; the third consists of dead cells, whose DNA supplements constantly the fourth pool of genetic information – eDNA (Levy-Bootha et al. 2007; Pietramellara et al. 2009). Finally, the fifth component (which also could be signed as eDNA, but has its own specific features) includes fragments of microbial DNA packed in virus particles. Of course, there are some more DNA sources in the soil, e.g. eukaryotic DNA, DNA, which comes in the soil from some other environments. All these DNA sources undoubtedly influence the soil microbial metagenome but they wouldn't be mentioned further in this paper.

Today we can only guess about the mechanisms of integration of soil microorganisms and their DNA in soil metagenome, because our knowledge of the microbial populations associated with each of the sources of soil DNA is still very poor.

Difficulties in studying of these main sources of soil DNA are closely connected to the problem of DNA extraction from soil. As opposed to other environments, e.g. marine ecosystem, the total and particularly differential extraction of soil DNA has a great obstacle mainly in the strong adhesion of cell as well as their DNA to different components of soil matrix (Nielsen et al. 2006; Saeki and Kunito 2010; Bakken and Frostegärd 2006). That is why beginning in 80-s and till now a lot of methods for DNA extraction were invented and reviewed (Bürgmann et al. 2001; Sagova-Mareckova et al. 2008; Martin-Laurent et al. 2001; Robe et al. 2003). The great dilemma of any method for soil DNA extraction is inability of "mild" procedures to extract representative portion of DNA and lack of specificity together with degradation of DNA in "strong" methods (Bakken and Frostegärd 2006). Thus today we are lacking in the appropriate method for separation of living (or dead) cells from bacterial resting stages. Partly it happens because very often the quantity and quality of extracted DNA come to the fore and lead us away from the biological sense of DNA extraction: too high percentage of soil DNA per gram is no better that too low, because the only important thing is to whom it belongs. A good theoretical basis for this is provided by the studies of physical and chemical parameters associated with the persistence of all mentioned DNA sources in the soil. From this point of view, the situation with eDNA seems to be relatively optimistic.

Not long ago there was a strong opinion that DNA coming from the cell to natural environment is immediately degraded by nucleases. But after a number of studies on the absorbance and stabilization of eDNA on mineral or organic materials and different soil components it became clear that DNA can maintain in soil for a long period of time (Wackernagel 2006; Cai et al. 2006) and probably could play significant role in functioning of soil ecosystem. The initial interest in studying free soil DNA was the process of natural genetic transformation of bacteria by eDNA and closely connected problem of dispersing the GMO genes in the environment. Another point of interest was eDNA distribution in soil and sorption on clay minerals which bridges soil DNA studies to the hypothetical processes of NA behaviour at the dawn of biological evolution (Saeki and Kunito 2010). Finally the very popular direction of studies of eDNA is formation of microbial biofilms (Böckelmann et al. 2007; Chiang and Tolker-Nielsen 2010), which might be very actual especially for soil microbial communities due to the uneven distribution of the polymeric organic compounds and pronounced ability of soil microorganisms to adhere on soil substances.

Together with these interdisciplinary topics, the knowledge about the amount and composition as well as the rate of the genetic transfer of eDNA play crucial role for understanding the organisation of soil metagenome. First of all, eDNA represents the genetic potential of the microbial community, that could be treated as connection between spatially and temporary isolated microbial populations and undoubtedly gives a great contribution to the soil adaptive potential. Presently we have many studies on the adsorption of DNA in soil (Saeki and Kunito 2010; Cai et al. 2006) and its movement through soil capillaries (Ascher et al. 2005; Ceccherini et al. 2007), the lifetime of soil eDNA (Romanowski et al. 1992), the number of studies on the horizontal gene transfer (Lu et al. 2010) with special focus on transition of genetic constructions from GMO to native soil microbiota (Paget et al. 1998; Wackernagel 2006).

As a result of these investigations we obtained many useful outcomes about the behavior of eDNA in the environment. It was shown that the amount of eDNA in soil can vary significantly depending on soil horizon and type of soil and ranges from 0.03 to 2 μg per gram (of dried soil). It was pointed by Nielsen et al. (2007) that a high number of bacterial isolates could produce eDNA in pure cultures, among them are bacteria which are commonly distributed in soil such as *Pseudomonas* and *Azotobacter*. Moreover this DNA release can be enhanced by presence of other bacteria or eukaryotes (Nielsen et al. 2007). These experimentally defined facts point on the possible integrative role of eDNA in interspecies interactions between prokaryotes as well as between prokaryotes and eukaryotes in soil microbial community.

Judging from the studies on adsorption of DNA molecules in soil it seems to be true that most of eDNA is associated with soil organic matter, primarily to humus components and clay-organic complexes, which significantly promote its stability

(Cai et al. 2006; Nielsen et al. 2007). Thus the vast majority of soil eDNA immediately after living cell envelope tightly bounds on organic or clay-organic particles in closest proximity to the metabolically active microorganisms. In the experiments of genetic transformation of *Azotobacter vinelandii* by naked eDNA and eDNA adsorbed on silica and on natural organic matter (NOM) Lu clearly showed detectable increase in transformation frequency in the last case: from $6*10^{-5}$ for naked eDNA to $2.5*10^{-4}$ for eDNA coupled with NOM (Lu et al. 2010).

Another fact that indicates the significant role of eDNA in microbial community functioning is stimulation of eDNA adhesion by low molecular weight organic acids (Pietramellara et al. 2009), which are known to be one of the general plant exudates. According to this we could assume that eDNA concentrates within rhizosphere – the most populated area in soil.

Summarizing all mentioned facts we may propose that eDNA constitutes the big portion in total soil DNA and concentrates primarily on the soil organic matter in the vicinity of cites with the highest microbial activity. These conditions seems to be quite favourable primarily for the "gene transfer", which naturally occurs in microbial populations and probably for many other functions that eDNA might play in soil e.g. the formation of biofilms, modulation of the symbiotic (or pathogenic) relationships between microorganisms, structuring of the soil matrix etc. The particular role that eDNA plays in regulation and maintaining the life of the soil microbial community is still unrated, but it seems to be significant.

Unfortunately until now we have much less information about the genetic composition of eDNA. There are only a few papers that showed the differences in the DGGE profiles of the microbial community in the total (tDNA) and eDNA (Agnelli et al. 2004; Ceccherini et al. 2009). From these papers the work of Agnelli et al. (2004), who analysed DGGE profiles from different horizons of forest soil deserves special attention. It has been shown that, along with a steadily decreasing amount of total DNA down through the soil profile, the eDNA dynamics was not linear. The maximum amount of eDNA was detected in humus-reach horizon A2 (60% of the total DNA), it was two-fold more than in the horizon A1 and a lot more than the eDNA concentration in the underlying horizons. Also Agneli with co-workers noted significant differences in DGGE profiles of A1 and A2 horizons: the bands associated with extracellular DNA were absent in total DNA at the same horizon and vice versa. At the same time in A2 horizon was discovered two additional intense bands which were absent in the total DNA in the A2 horizon, but present in the total DNA of A1 horizon. Additionally the A1 horizon was characterized by a great mismatch between total and extracellular DNA (only a small percentage of bands associated with the total DNA extracted from the A1 horizon were present in eDNA). In the underlying horizons the differences in the profiles of eDNA and tDNA were minimal (Agnelli et al. 2004). This local accumulation of eDNA in A2 could be connected both with the more active usage of eDNA by microorganisms in the A1 horizon and what is more likely with vertical migration of eDNA and it enhanced ability to bind to the humic substances. Thus, the accumulation of soil humic substances is closely connected with the storage of extracellular DNA, which makes the humus horizon a zone of intensive concentration of genetic information in soil.

Obviously, the interest in this problem will grow due to its practical and fundamental importance and future metagenomic studies should make the major contribution to determine the genetic composition of the eDNA.

3 Spatial Structure of the Soil Metagenome (Microbial Distribution and Soil Macro- and Micro-morphological Structure)

The important output for the metagenome research is dissection of the spatial structure of soil microbial community which is related to the microstructure of soil matrix and vertical stratification exposed in a number of specified horizons. The production of soil aggregates plays a crucial role in many typical soil processes, especially in degradation of organic matter and formation of humic substances (Chotte 2005). Soil sciences commonly divide soil aggregates on three groups depending on the their sizes: microaggregates <0.25 mm, mesoaggregates 0.25–7 (10) mm and macroaggregates >7 (10) mm. To investigate the role of microorganisms in maintenance of soil fertility the special attention must be given to the genesis and composition of the microbial community within individual aggregate to define differences microbial composition in micro, meso and macroaggregates.

The process of aggregate formation is mediated by the closest interaction between a wealth of different microorganisms and seems to be sequential. In the set of laboratory experiments it was shown that fungi and plant roots are mainly involved in the formation of big aggregates and thus determine the initial stages of the aggregation process, while bacteria produce smaller aggregates, which are further joined together by fungi and form an active centre within the soil aggregate (Chotte 2005).

Until now little is known about the taxonomic structure of microbial communities inhabiting soil microenvironments. There are only few studies that characterized taxonomic diversity of microbial communities in different aggregate fractions (Sessitsch et al. 2001; Davinic et al. 2012) and microorganisms living inside and on the surface of microaggregates (Mummey et al. 2006) using different molecular techniques.

Davinic et al. (2012) was the first who used pyrosequencing in studying microbial composition in three multi-scale soil fractions: macro-(>250 μm) and micro (53–250 μm)- soil aggregates and clay + silt particles (<53 μm). The composition of the microbial community, as well as simultaneously measured amounts of available C and N, of these fractions differed significantly. Macro-aggregates was characterized by high levels of available C and N and primarily contained *Actinobacteria*, *Bacteroidetes*, *Verucomicrobia* and δ-*Proteobacteria* as dominant bacterial divisions. Micro-aggregates showed the lowest levels of biodiversity and were formed by at least two big groups of bacteria – *Rubrobacteriales* and *Chloroflexi*. Clay-silt particles cared large number of minor soil phyla such as *Nitrospira*, OP10, WS3 together with the most abundant soil phyla – *Acidobacteria* (which corresponds with the result obtained by Sessitsch et al. 2001) and α-*Proteobacteria*. At the same time clay-salt particles had the great amount of recalcitrant organic C.

Following to the present data macro-aggregate community was in part similar to the whole bacterial diversity measurements, contrary, two other microenvironments, especially micro-aggregates, showed unique bacterial profiles (Davinic et al. 2012). It is very important to mention that Mummey and Stahl (2004) also defined *Rubrobacteriales* as the dominant group in the inner space of micro-aggregates of the disturbed (reclaimed) soil, in opposite to the undisturbed soil where the inner-micro-aggregate environment was populated by distinct groups of *Actinobacteria* – *Actinobacteridae*. Considering that Davinic with co-workers investigated to some extent extreme arid soil and actinobacteria belonging to the order *Rubrobacteriales* is reported to be radiation-tolerant extremophile bacteria this group could serve as a good predictor of the processes affecting soil health. Mummey and Stahl also showed that outer micro-aggregate space is largely occupied by *Proteobacteria*, which richness also correlates with the soil status and shows the highest value of 70% in reclaimed soil (Mummey and Stahl 2004). This fact is relative to the general observation of the increased number of proteobacteria in agricultural soil comparing with natural soil habitats (Spain et al. 2009). Thus these bacteria could also be the important indicators of anthropogenic soil treatment.

The aggregate formation studies is valuable not only for agricultural practice but for understanding the processes of biodegradation of the organic material. Considering the magisterial way of biodegradation of the plant matter in soil, where substances obtained on the previous stages are degraded by the microorganisms of the following levels, the formation of aggregates will be beneficial to concentrate the final products and thus to straighten out the whole process.

The formation of the isolated compartment within the soil aggregate is essential to offer saving the secretion of exoferments. Soil aggregate can be considered as specific and highly adaptive structure which is actively organized and maintained by different microorganisms (Young and Crawford 2004) integrated into the smallest ecological unit of soil microbiome – the aggregate-assosiated microbial community. The process of its formation somewhat resembles the interactions between fungus and photosynthetic microorganisms in lichens and tend to be substantially symbiotic. From all these facts it is reasonable to deduce that the aggregates are not only and perhaps not so much the structural units of the soil, as the structural units of the soil microbial community and its metagenome and thus must be put in the focus of close attention in the future metagenomic studies, especially for the success in functional metagenomics.

In spite of the apparent necessity in this knowledge (Grundmann and Gourbiere 1999; Girvan et al. 2003), and multiple evidences of biodiversity-forming role of soil texture (Ettema and Wardle 2002; Chau et al. 2011; Dechesne et al. 2003, Graundmann 2004; Nunan et al. 2003) there is a substantial lack of micro-scale studies, particularly in metagenomic approaches. Until now, soil sampling efforts present one of the most challenging questions of the modern molecular ecology, but sampling of 1 g of soil from the upper horizon is widely met. Considering all the given information about spatial structure of soil matrix the value of such mixed sample is at least questionable.

The aggregate structure of the soil matrix is one of the lowest levels in the morphological organisation of soils, moving upper we meet another important soil structures – the soil horizons that along with the differences in composition of the three types of soil aggregates are the main diagnostic features for the soil classification. In this review we concentrate our attention on the most populated soil humus-accumulative horizon. The process of soil genesis inherently represents the continuously evolving relationships between plants and microorganisms. The finishing point of this process is formation of two specified soil environments – the rhizosphere and humus. These two microbial habitats are closely connected to each other and define the fertility of soil, which tends to be one of the most important soil characteristic. That is why they usually call enormous scientific interest (Kuske et al. 2002; Andreetta et al. 2004). Here we just put the main ideas of these studies. The rhizospheral microbial community represent the example of direct interactions between plant and microorganisms mediated mainly by co-production of metabolites to the environment, while humus production represents the example of indirect interaction between plant and microorganisms. Now, this is generally recognized that microorganisms are metabolically involved in the production of specific humus substances (Chotte 2005) and that humus can't only be treated as the product of biodegradation of plant material.

Thereby, humus horizon is a specific and highly adaptive soil environment both for plants and microorganisms. Due to its porous structure, humus act as soil "sponge" accumulating water and other vital substances, it is also serves as reservoir of nutrients and vital elements like nitrogen (Voroney 2010). What is very important in soil management, humus is involved in maintaining the soil structure preventing the erosion processes. For microorganisms the formation of the humus horizon seems to be an adaptive response to the existence in heterogeneous, contrast, and frequently changing living conditions in soil. Because humus remains to be very stable environment it is inhabited mainly by the autochthonous microbial community (Chotte 2005). Relatively big proportion in studies of soil microbial communities in any case address to humus reach soil layers, but only small proportion of them treat humus-accumulative horizon as spatially and functionally isolated and continuously developing soil structure. Among them are the studies describing the microbial community changes along 130-years in the *Fagus silvatica* forest chronosequence (Trap et al. 2011). Measuring the great number of macromorphological characteristics of humus-accumulative horizon formed under the forest of different stand age classes Trap with co-authors showed significant changes in the diversity and activity of fungi and heterotrophic bacteria depending on the humus type. During the development of forest ecosystem from 15 to 130-years forest mull humus, characterized by high nutrient cycling gave the way to the moder humus with high number of recalcitrant substances and increased acidity. Despite of such unfavourable conditions moder humus was inhabited by more diverse microbial community. Trap et al. (2011) attributed this phenomenon to the presence of opportunistic bacteria that uses readably degradable substrates in mull humus and limits specialist bacteria which are capable to degrade the complex substances. In moder humus the high relative concentration of the recalcitrant substrates forms the selective pressure that favours the development of slow-growing specialized bacteria (Trap et al. 2011).

Unfortunately, in spite of the obvious significance of microbiological studies of humus-accumulative horizon and its development the research base on this topic is poor. But the interest to this aspect continuously growing up and there is an example of resent investigation of microbial community composition in upper soil horizons (A horizons) and subsurface horizons (B horizons) performed by use of pyrosequencing technique (Will et al. 2010). The microbial community in A horizon was enriched with the high-abundant soil phyla such *as Actinobacteria*, *Bacteroidetes*, *Cyanobacteria*, *Fibrobacteres*, *Firmicutes*, *Spirochaetes*, *Verrucomicrobia*, *Alphaproteobacteria*, *Betaproteobacteria*, and *Gammaproteobacteria*, whereas B horizons were dominated mostly by *Acidobacteria* and some minor soil phyla e.g. *Chloroflexi*, *Gemmatimonadetes*, *Nitrospira*, TM7, and WS3. Interestingly the microbial composition of upper horizons is quite similar to the composition of macro-aggregates, at the same time subsurface soil community remains that of the micro-aggregates and especially clay-silt particles. Mummey et al. (2006) also reported *Chloroflexi* as inner-microaggregate population, while *Bacteroidetes* were distributed mainly on aggregate surfaces (Mummey et al. 2006). These findings join the microstructure of soil matrix and vertical stratification of soil, separating the litter (or humus) associated bacteria involved primarily in the first steps of carbon cycling (*Bacteroidetes*, *Actinobacteria*) from the bacterial groups associated with other soil processes such as nitrification (*Nitrospira*), or bacterial groups that are attracted by microaerophylic or even anaerobic conditions (*Chloroflexi*, *Acidobacteria*).

We should stress here that analysed soils were characterized by nearly neutral values of pH (from 6.03 to 7.40) and relatively low water content that, as it will be shown in the next section, usually lead to the dominance of *Proteobacteria* and *Actinobacteria* instead of *Acidobacteria* in upper soil horizons (Lauber et al. 2009). In this case *Acidobacteria* seems to be "removed" to the deeper soil layers by competition for ecological niches with *Proteobacteria* more adapted to high pH rates.

4 Global Distribution of Soil Microbial Communities and Their Response to the Major Environmental Physicochemical Factors

Application of the molecular techniques for studying bacterial diversity in natural environment fundamentally altered our traditional taxonomic and ecological concepts, which were built primarily during the investigation of eukaryotic biodiversity in the past century. Analysis of the 16S rRNA gene sequences in different regions of the world, even in 1 g of soil revealed great microbial biodiversity of uncultured microorganisms within previously known phyla and opened the way to the discovering of many previously unknown prokaryotic taxa of higher ranks. Together with low abundant phyla containing at least some cultivable representatives, molecular ecology approaches succeed in discovering of *Acidobacteria* phylum which seems to have the same environmental characteristics and diverse physiology as it was proposed for the well-known ubiquitous *Proteobacteria* (Lee et al. 2008). The most common soil phyla are (according to their abundance in soil samples): *Proteobacteria, Acidobacteria, Actinobacteria, Verrucomicrobia, Bacteroidetes, Chloroflexi, Planctomycetes, Gemmatimonadetes* and *Firmicutes* (Jannsen 2006). The ratios between these phyla could change depending on the environmental conditions presented in a given soil and sometimes other phyla could be inserted in this list (such as *Nitrospira, Chlamydia* and some other low-abundant bacteria highly specialized to the local environmental conditions). Bacteria belonging to *Nitrospira, Spirochaetes*, TM7, OP10, WS3 etc. are usually considered to be minor soil phyla (Podar et al. 2007). Despite of mentioning at least nine phyla of soil abundant bacteria, practically four phyla (*Proteobacteria, Acidobacteria, Actinobacteria* and *Bacteroidetes*) form the great majority of bacteria inhabiting any kind of soil, these groups could comprise sometimes up to 90% of overall biodiversity in the environment (Tsai et al. 2009).

Making the description of global distribution of different groups of bacteria in soil we should address to the problem of bias occurring in molecular studies. The first source of the mistakes in measuring microbial diversity in soils is unequal number of studies of upper and deeper soil horizons with the significant superiority of the first; the second is PCR induced bias, which causes a disproportion in detectable bacteria phyla due to the sequence-primer mismatches. These theses were illustrated by Bergmann et al. (2011) in studies of the subsurface soil environments. Using taxon-specific primers for the phylum *Verrucomicrobia*, they showed that the proportion of these microorganisms had been underestimated in previous studies and could in fact comprise up to 23% of overall community composition (Bergmann et al. 2011). Finally, the relative abundance of several bacterial phyla could be influenced by the DNA extraction bias. By making repeating extractions of soil DNA from the soils which differed in clay, sand and organic composition, Feinstein with co-workers showed a dominance of *Acidobacteria, Gemmatimonadetes* and *Verrucomicrobia* in the first (mild) extracts, while *Proteobacteria* and *Actinobacteria* were dominant in samples obtained from the sixth extraction (Feinstein et al. 2009).

In exceptional cases the Firmicutes can be added to the “big four”. This happens, for example, in saline ecosystems, such as coasts of salt lakes, where *Firmicutes* “takes the place” of *Acidobacteria* or *Actinobacteria.* And as a gradient of salinity decreases the last two bacterial phyla reobtain their dominant position. These data from our last studies corresponds well with other investigations of saline soils (Hollister et al. 2010). Considering these facts one could suggest that there should be a strictly limited number of ecological factors that may influence the phylogenetic diversity of bacteria on the phylum level.

This idea was profoundly developed by Fierer and Lauber in a number of studies (Fierer and Jackson 2006; Fierer et al. 2007; Lauber et al. 2008, 2009), where they tried to find the main patterns of bacterial distribution in soils with the broad range of physical and geographical characteristics. The principal possibility of these studies was largely supplied by the improved molecular new generation techniques (Roesch et al. 2007) which provided simultaneous and rapid analysis of as much soil samples as it was needed for the nearly complete description of the microbial community changes induced by action of a given ecological factor. Now, we can summarize data from these studies and make an effort to determine the major soil biodiversity-forming factors.

Recently, primarily by using pyrosequencing techniques compiled with factor analysis, a plenty of soil characteristics such as soil pH, texture, vegetation, amount of acceptable C etc. were tested in their ability to influence soil biodiversity (Brockett et al. 2012; Singh et al. 2009b; Castro et al. 2010; King et al. 2010; Lauber et al. 2008). Among them at least two widespread factors – soil pH and soil water content were proposed to be most important for soil microorganisms (Bru et al. 2011; Lauber et al. 2009).

Soil pH strongly influences the relative abundance of two main soil phyla: *Acidobacteria* is favoured with the acidic values of soil pH, whereas *Proteobacteria* dominates at neutral or alkaline conditions (Lauber et al. 2009; Fierer and Jackson 2006).

Soil water content determines the abundance of *Actinobacteria*, which reaches its maximum value in arid soils (Connon et al. 2007; Brockett et al. 2012; Chowdhury et al. 2009; Gomez-Silva et al. 2008). This fact seems to be obvious due to certain resistance of this group to the drought conditions. But the substantial impact of water content to microbial diversity may be caused not only by the dominance of ecologically adapted actinobacteria but also by the spatial reorganization of microbial community. It is worth to mention here the studies addressed to influence of the soil texture on the microbial community composition (Carson et al. 2010; Chau et al. 2011). During the comparison of clay and sandy-reach soils, it was postulated that soil texture as well as soil water content determined the connectivity between microhabitats. Weak connectivity promotes the increase of biodiversity levels and vice versa (Torsvik and Ovreas 2002). Thus the highest levels of microbial diversity were detected in the sandy soils, in the contrast to the clay-based environments (Chau et al. 2011).

Surprisingly, soil biodiversity was not (or less) influenced by such factors as geographical location, latitude, temperature and vegetation that usually are very important in modelling the diversity of plants and animals (Bru et al. 2011; Chong et al. 2012; Fierer and Jackson 2006). This was proved by the investigation of antarctic soils and soils from high mountains (Chu et al. 2010; King et al. 2010). Despite of the extreme nature of these habitats it was shown that microbial composition there was generally the same as in the typical soil of temperate region. Investigating bacterial diversity in Colorado Mountains King tried to find spatial patterns in distribution of the dominant bacterial taxa and found the most powerful factors that might form bacterial diversity – these were pH, plant abundance and snow depth. These factors determined the presence and proportion of the four most abundant orders: *Rhodospirillales*, *Rhizobiales*, *Acidobacteria* G4 and *Saprospirales*. The first two orders showed high correlation with the density of vegetation: *Rhizobiales* was mostly associated with different plants (as they also do in Antarctic soils (Chu et al. 2010) and virtually in all types of soils all over the world) whereas phototrophic *Rhodospirillales* dominated exclusively in the areas with poor vegetation. Interestingly, *Rhodospirillales* was reported previously to appear in the marine environments and also was among dominant groups, discussed in the studies of the terrestrial saline sites (Hollister et al. 2010). Practically the same results were obtained by Bru et al. (2011) during the studies of bacterial and archaeal communities involved in N-cycling. It was shown that among five categories of environmental factors such as spatial distribution, land use, climate, soil physics and soil chemistry the last factor (especially soil pH) played the main role in prediction of bacterial abundance. On the second place Bru et al. (2011) put the land use category, while the role of the climate changes or geographical location was minimal (Bru et al. 2011). Authors also showed latitude-dependent variability in microbial communities with the predominance of bacterial diversity in north regions comparing to the south soils. They associated these differences with the soil parent material (reported to be a strong factor in determination of bacterial distribution) which was formed by limestone in the north and crystalline rocks in the south. Besides, there was one group of bacteria (ammonia oxidizers) that possessed unique pattern of abundance, which wasn't triggered by soil chemistry (Bru et al. 2011). So, to estimate the bacterial global distribution, it is important to assess not only the total biodiversity, but also the distribution of the main functional groups of microorganisms. Thus, this study, giving the suggestion to at least 43–85% of variables in bacterial community distribution (Bru et al. 2011), was the first to stress clearly the importance of geostatistical methods in microbial ecology.

A widespread distribution of bacterial taxons not even on the order level but also on bacterial genera level seems to be the most important ecological characteristic of microbial communities. If we compare differences in taxonomic structure among two different soils (for example, Waseca farm soil and rain forest soil from Puerto Rico, available at MG-RAST website: http://metagenomics.anl.gov/metagenomics.cgi?page=Home) we will find that there are too few differences in the number of observed taxons among them. The differences between high Andean soil and Waseca farm soil or rain forest soil are much more significant, but the tendency in the presence of the same taxa in these extremely different environments (especially in the phyla *Proteobacteria* and *Acidobacteria*) is saved, in the contrast to the Eukaryotic community portrait of these habitats which becomes more and more individual when moving from the typical to xtreme soils. More than that, comparing a great number of extreme soil habitats one could find that there are a lot of omnipresent microorganisms, and the champions among them are bacterial genera *Pseudomonas* (*Gammaproteobacteria*), *Arthrobacter* (*Actinobacteria*), *Sphingomonas* (*Alphaproteobacteria*), *Bacillus* (*Firmicutes*), *Rhodococcus* (*Actinobacteria*), *Flavobacterium* (*Bacteroidetes*) and some others (Lucas et al. 2008; Wagner 2008; Ruberto et al. 2008). All these bacteria tend to be the most common members of the soil microbial communities all over the world, exhibiting the highest levels of functional diversity.

5 Main Outputs from the Modern Studies of the Soil Metagenome

The topic pointed in the name of this section is the most popular among soil microbiologists and was reviewed many times (Raes et al. 2007; Hirsch et al. 2010; Rondon et al. 2000; Simon and Daniel 2011; Handelsman 2004; Daniel 2005). The main successes as well as the main problems of soil metagenome analysis were discussed in detail as well. Here we just put the general questions and give some idea for the future research.

The first "hot spot" to discuss is the size of soil metagenome. According to Vogel et al. (2009) there is as much as 1,000 Gbp of genetic information presented in 1 g of soil, which is about three orders of magnitude greater than in the more or less completely studied environments such as Sargasso Sea or Acid mine drainage (Vogel et al. 2009). It is also important to note that members of microbial community have different relative abundances with low number of dominant bacteria and a plenty of low-populated groups. Tringe and co-workers (2005) reported that during the investigation of soil metagenomic library, they obtained totally 150,000 reads and it was enough to get overlapping sequences in less than 1% of cases. Authors also predicted that eightfold coverage (the traditional number in genome assembly procedures) of at least one genome belonging to the most abundant species requires from 2 to 5 billions bp of sequence (Tringe et. al. 2005). And it will be necessary to analyse as much as 2 million clones to get more or less full information about genomes of all members of microbial community (Singh et al. 2009a). Considering the maximal volume of reads which could be obtained using the most resent sequencing machines built by Illumina or Applied Biosystems and estimated in average as 100Gb per run, sequencing of the whole soil metagenome seems to be very complicated and probably even unsolvable problem. That is the reason why modern soil metagenomic studies tend to result in complete (usually near-complete) investigation of at least one or two dominant species.

The second problem arises due to some technical restrictions and, in turn, is closely connected with the peculiarities of soil metagenome. It is the problem of assemblage of the whole bacterial genomes or even parts of genomes from the raw sequencing reads. The metadata processing includes a number of consequent steps, starting with assemblage of sequences to contigs and following by gene "calling" and gene "binning" procedures, which finally bind sequences to the phylogenetic groups (Kunin et al. 2008; Wooley and Ye 2009). Today, assemblage seems to be the most complicated and ambiguous procedure due to several reasons. The first is still short length of sequencing reads, produced by the most precise sequencing machines, the second is lack of repeated sequencing reads that minimizes the accuracy of the analysis, and in the case of soil metagenome it leads to the practically total absence of overlapping sequences (Chaisson and Pevzner 2008). These reasons, together with the presence of polymeric repeats and accidence of horizontal gene transfer between unrelated groups of microorganisms, lead to the imperfect and sometimes chimeric assemblages (Chaisson and Pevzner 2008; Kunin et al. 2008). But probably the main failure of assemblage procedures is inability to reconstruct complete bacterial genomes. So, entirely all modern internet resources such as CARMA (Gerlach et al. 2009) or MG-RAST (Meyer et al. 2008) or the most popular program for metadata processing – MEGAN (Huson et al. 2007) operate with single genes, more rarely with sequences containing multiple genes, and practically never with genomes.

The last problem to mention is the problem of "binning" nucleotide sequences to the specific phylogenetic groups, which origins from the biased and incomplete databases (Kunin et al. 2008). This problem again seems to be the most important in the case of soil, which has the lowest number of genes with known function, while the majority of sequences form the soil "black box" (Mocali and Benedetti 2010). It seems to be true that precisely these genes are involved in integration processes such as adhesion, biofilm formation, signalling etc. According to their specificity to soil components (adhesion) or to the specific partner (symbiotic genes) they were poorly studied as well as deposited in the databases. But these genes in particular should be put on the forefront to discover the most fundamental soil integrative processes such as aggregates formation, synthesis/degradation of the specific humic substances etc.

Functional-based approaches (Lämmle et al. 2010; Lorenz et al. 2003; Voget et al. 2003; Demanéche et al. 2009; Daniel 2004) are called to solve these mismatches. Lombard with co-workers described the main outputs in this field of metagenomics and determined the main classes of environmental gene products that attract special interest in the soil – these are broad range of enzymes and antibiotics, gene products involved in bioremediation, as well as in common biosynthetic and catabolic pathways (Lorenz et al. 2002). Studies of soil functional genes revealed some principally new applications for industry and medicine such as lactonase family proteins that tend to decrease biofilm formation by pathogenic *Pseudomonas aeruginosa* (Schipper et al. 2009), anticancer drug production (Pettit 2004) and some others (Lorenz et al. 2002).

Still the unsolved problems of functional analysis are low number of clones containing the gene of interest and restricted number of expression systems that don't fit the enormous biodiversity of functions present in soil (Wexler and Johnston 2010). Interesting findings were obtained by Craig et al. (2010), it was shown that expression of genes from the same metagenomic library in different host proteobacteria results in different range of gene products (Craig et al. 2010). Moreover, it is reasonable to deduce that gene expression will be different if single bacteria host strains and in multispecies (e.g. sin-trophic) systems. During the investigation of organization of soil microbial communities a large data was compiled, which proves that microorganisms exist in soil primarily in metabolically and genetically integrated associations (Garbeva and de Boer 2009; Schink 2002). The last experiments in co-cultivation of microorganisms performed by Burmølle et al. 2009 revealed the significant increase in culturability of soil bacteria while plating with model strains such as *Pseudomonas putida* and *Arthrobacter globiformis*. In the first case gram negative pseudomonades stimulate growth of colonies of different gram-positive bacteria belonging to the class *Bacilli*, while gram-positive *Arthrobacter* stimulates primarily gram-negative bacteria such as *Proteobacteria* and *Bacteroidetes* (Burmølle et al. 2009). Thus growth and the main physiological features of soil bacteria are highly variable and apparently require building the specific gene expression systems for the investigation of functional diversity in a given soil. While describing the global distribution of the soil microbial communities (see Sect. 4) we pointed that soil taxonomic diversity is restricted to the presence of only few bacterial phyla, with prevalence of *Proteobacteria* and *Acidobacteria*. Creating the effective expression systems one must rely greatly on the taxonomic composition of the microbial community to choose the most appropriate hosts for each type of soil. It might help to intensify findings of new gene products. In this aspect the development of new cultivation techniques to investigate uncultured microorganisms such as fluorescence-activated cell sorting, FACS proposed by Zengler with co-workers (2002) seems to be very promising. Studying the uncultivable *Acidobacteria*, as the second after proteobacteria most widespread bacterial phyla is of profound importance in having a glance on a hidden soil functions (Lee and Cho 2009).

Sequence and functional based studies contribute to the systematic view on soil metagenome that tends to be the main goal of existing metagenomic projects. There are currently three major consortia – Microbial Genomics Program, specifically the section MEP (Microbial Earth Project) sited on http://genome.jgi.doe.gov/programs/bacteria-archaea/MEP/index.jsf, EMP (Earth Microbiome project) (http://www.earthmicrobiome.org/), which respectively have sections devoted to soil metagenome. And a special project, directed exclusively to the study of soil microcosm – the TerraGenome (Vogel et al. 2009). Among the main goals of TerraGenome project are identification of new prokaryotic producers of biocatalysts, microorganisms involved in bioremediation and other substantial soil processes. The subject of the investigation is the Park Grass at Rothamsted, UK – an internationally recognized agroecology field experiment that has been running for more than 150 years (Vogel et al. 2009).

TerraGenome was called mostly to join the forces of international community of soil microbiologists; but immediately after announcement came under criticism (Baveye 2009; Singh et al. 2009a) mainly due to the problems associated with the analysis of metagenomic data, which were briefly discussed above.

After the last decade, that was greatly optimistic in adapting metagenomic surveys to almost all natural environments (Gilbert and Dupont 2011; Riesenfeld et al. 2004; Gill et al. 2006), many scientists came to understanding of the importance of traditional methods, like cultivation or 16S rRNA gene profiling and gained the qualitatively new level of the "way back" (Tyson and Banfield 2005; Kakirde et al. 2010; Lombard et al. 2011; Forney et al. 2004; Tringe and Hugenholtz 2009).

The latest reviews made by Lombard et al. (2011) and Kakirde et al. (2010) referred soil metagenome researchers to the analysis of more simple ecosystems such as microaggregates (Kakirde et al. 2010; Lombard et al. 2011) and extreme soils (Lombard et al. 2011). The authors point out the exceptional value of studying the microbiological component in relation to soil texture and vertical stratification. But not only these approaches could substantially decrease the value of the soil metagenome analysed at a time. The separation of the five previously described (see Sect. 2) sources of environmental DNA could contribute greatly to the future metagenomic surveys making the procedure of analysis not even simpler but more precise. Inattention to these problems induced great complexity in databases and associated computational studies (Bietz and Lee 2009), which became forced to work with enormous amounts of information not even enriched with any biological sense. All mentioned perspectives require integration of specialists from many scientific fields not only from biology and bioinformatics but especially from the soil sciences. It is worth recalling here the strong system of characterization of soil microstructure as well as soil vertical stratification that was made during the development of soil sciences and led to specific understanding of soil as a unique "natural body" that might have structural and functional modules (Young and Crawford 2004).

Together with rising interest to the investigation of soil on discrete micro-scale levels a noticeable tendency appears in intensification of 16S rRNA-based approaches (Roesch et al. 2007; Steven et al. 2012), primarily in the studies of global and local distribution of soil microorganisms (Fierer and Jackson 2006; Bru et al. 2011). Despite of probably the main drawback of these studies – the bias associated with choosing specific primers for amplification – 16S rRNA studies have essential benefits and until now serve as the main source of information on community structural and even functional diversity (Tringe and Hugenholtz 2009).

6 Concluding Remarks and Opening Perspectives in Soil Metagenomics

This review was initiated due to the growing understanding of the uniqueness of soil metagenomics in comparison to data from the other environments. We were trying to understand the main features of soil microbial community in order to identify key areas for the further research of its metagenome.

The first most obvious problem that differ soil from the other environments is enormous size of metagenome, which inevitably led researchers to the necessity of dividing it into some reasonable parts. Here metagenomics should move into close contact with soil science to accept already existent strict system for describing soil micro- and macro-structure. This interdisciplinary partnership is maintaining increasingly, and now we have some examples of beneficial combinations of pedological and molecular techniques (Young and Crawford 2004; Mummey et al. 2006; Eickhorst and Tippkötter 2008; Ceccherini et al. 2007). Particularly, Eickhorst and Tippkötter (2008) combined standard FISH techniques for visualising bacteria in their natural environment with micropedological technology of ultrathin dissections of soil matrix. Thus they have obtained the noise-free images of microbial colonies and determined their spatial distribution within soil microenvironments (Eickhorst and Tippkötter 2008). The microbial composition associated with different types of soil aggregates as well as studies of horizon-specific microbiota also induced a great interest and recently was supplied with modern sequencing techniques (Davinic et al. 2012; Will et al. 2010).

Another important issue was discovery of the great pool of extracellular DNA, persisting for a long time in stable complexes with different soil substances and in some cases forming up to 60% of the total soil DNA (Agnelli et al. 2004). The exclusive role of eDNA in microbial community functioning was proved by the studies of the natural genetic transformation in different groups of soil bacteria (Lu et al. 2010; Nielsen et al. 2007; Mercier et al. 2006). These findings highlighted the exceptional role of HGT among other biodiversity-forming factors particularly in soil ecosystem.

Finally, there is a growing body of research evaluating the contribution of different environmental factors to the global distribution of bacterial taxa. These studies revealed the priority of edaphic factors such as pH (Lauber et al. 2009) and humidity (Brockett et al. 2012) as well as soil texture (Carson et al. 2010; Girvan et al. 2003) in regulation of soil microbial diversity. Such factors as vegetation, land use, climate and some others were considered to have lesser influence on microbial community composition. The most intriguing feature of soil microbial communities was their sufficient independence from such traditional global characteristics as landscape and latitude. Thus, in contrast to eukaryotic organisms, microbial community of tropical soils was characterized by relatively low levels of biodiversity (apparently due to under-developed humus horizon) than the community of the temperate zone, which in turn does not differ much from the communities of polar regions such as antarctic soils (Chu et al. 2010).

Despite of significant progress in the application of geostatistical methods in the studies of microbial communities, scientists have failed to define a core set of biodiversity-forming factors. Perhaps this happened due to the multiple interactions between described ecological factors in their influence on the soil metagenome. For example, the soil texture (which along with the previously listed factors is one of the most powerful determinants of biodiversity), interacts with the moisture and with the soil nutrient status (Girvan et al. 2003). At the same time, the value of soil pH depends greatly on vegetation and land use strategies (Lauber et al. 2008). Therefore, further studies require precise determination and diversification of the major and minor biodiversity-shaping factors that respectively have strong or mild effects on the structure of soil microbial communities.

Nevertheless, even on the basis of existing data it can be concluded that there are a number of "architectural types" of microbial community organisation (and corresponding taxonomic structures of metagenome) defined by several strong edaphic factors such as pH and moisture (either could be treated as water potential that seems to be more physiologically significant characteristic). According to pH values, microbial community could be divided into two architectural types – the one with the dominance of *Proteobacteria* (neutral or alkaline conditions), and another one with prevalence of *Acidobacteria* (sites with low pH values). The situation around moisture is not as clear as for soil pH, but it was proved many times that the arid habitats was dominated by phylum *Actinobacteria* (Connon et al. 2007; Brockett et al. 2012), contrary sites with increase moisture rates were populated mostly by *Firmicutes*, *Bacteroidetes*. In the mentioned article the majority of *Firmicutes* and *Bacteroidetes* was found in the shore of saline lakes in US (Hollister et al. 2010), but, as it was noted by the authors themselves, salinity probably is not the main factor determining the structure of the microbial community in this area, because the biggest correlation in performed statistical analysis was found for soil moisture and organic C content (Hollister et al. 2010). Recently we have obtained practically the same results during the investigation of salt lake shore in Kazakhstan (Pershina et al. 2012).

"Architectural types", in turn include a number of "functional modules", characterizing tradeoffs between taxonomic and functional structures of metagenome. These functional modules could be related to the specific taxa, or could be spread among different groups of microorganisms. The last is caused by the paradox nature of bacterial taxa, as well as developed role of horizontal gene transfer in microbial populations, which together result in the existence of "common" pool of genes that may be at a time carried by phylogenetically unrelated groups of microorganisms. The process of denitrification is one of the examples. The ability of nitrate reduction is common for many typical soil bacteria such as *Pseudomonas* and *Bacillus*. This process is multistep and includes several consequent stages, which involve a number of genes that can persist in the genomes of taxonomically different but metabolically interchangeable microorganisms (Ettema and Wardle 2002).

As an example of taxon-specific functional modules we may call the processes of nitrification, or symbiotic nitrogen fixation which genes are presented in the genomes of the strictly limited range of soil bacteria. For the first process it was shown that phylogeny constructed on the basis of the amoA gene sequences was practically the same with 16S rRNA based phylogeny (Ettema and Wardle 2002). The second fact is partly proved by the investigation of the arid soil, where the dominance of *Actinobacteria* and *Acidobacteria* were noted. In spite of the known ability of some actinobacteria species to fix nitrogen this process was delegated not for the dominant species but for the well-known groups of proteobacteria such as *Rhizobium*, *Pseudomonas* and *Azospirillum*, which usually formed a minor part of the microbial community (Chowdhury et al. 2009).

From these two examples as well as from the data presented for extreme soils in Sect. 4 (Gomez-Silva et al. 2008; Wagner 2008; Lucas et al. 2008), we could guess that there are truly ubiquitious bacterial genera in soils, which show unprecedented physiological adaptation. The most probable sources of this adaptation are "labile", highly variable nature of bacterial genomes, coupled with the large pool of entire genetic information, stored in soil (eDNA). In this respect the most abundant soil bacterial phyla *Proteobacteria* reveals an example of evolutionary young and to greatest extent "paradoxal" bacterial taxon with the most variable genomes within single species (Ussery et al. 2009; Spain et al. 2009).

The other source of the enormous ecological potential of soil bacteria may be represented by microbial co-operation. It can be classified as a "protective symbiosis" between dominant (e.g. acidity-resistant soil *Acidobacteria*) and some specialized groups of microorganisms, that for example might arose during the formation of aggregates with more or less stable inner space.

Establishment of the correlation between "architectural models" of microbial community, and their constituent "functional modules" represent the fundamental goal of soil metagenomics.

Beside, by analogy with the genomes of individual species of microorganisms (Ussery et al. 2009), metagenome may contain the "core" and "accessory" parts (core and pan-metagenome). Obviously, it is impossible to determine the composition of these parts, without an exhaustive comparison procedure performed for at least two metagenomes. Therefore, at present time we can only speculate about what genes are "common" for the communities of different soils, and which are specific to a particular soil.

As it was revealed from the comparison of soil metagenomes deposited to MG-RAST server, the taxonomic composition of seemingly unrelated Waseca farm soil and rain forest soil from Puerto Rico (see Sect. 4) appeared to be very similar, even at the genus level. The same can be said about the functional diversity of these two metagenomes. This tells us that all the discovered functions belong to the "core metagenome". Then, what genes can constitute the accessory part of the soil metagenome and thereby can reveal the main differences between soils types? As noted in the study made by Tringe and co-workers (2005) the soil metagenome, in contrast to metagenomes from other habitats is characterized by the wide variety of genes associated with processes involved in genomic variability and particularly in HGT (Tringe et al. 2005). Because of the immense value of these genes in the processes of integration and ecological maintenance of the soil microbial communities one could expect the high levels of genetic diversity among them. Apparently, this is the particular group of genes that could be joined to the main biodiversity-forming factors. For example, the HGT between active soil bacteria could be an example of spatial integration of soil communities, while external DNA uptake (e.g. genetic transformation by eDNA) could contribute to even more significant process of temporal integration of soil communities (DNA remaining in the soil from the previous generation of microorganisms can "come alive" in the cells of living bacteria).

Other candidates for the role of "soil-specific" genes may be genes that provide above-described processes of microbial co-operation and formation of stable interspecies relationships. The facts revealing the exceptional importance of "interspecies dialog" in life of microbial community was reviewed many times (Burmølle et al. 2006; Schink 2002) and partly discussed in Sects. 2 and 3. Integration of individual organisms as well as their genomes can be made by genes involved in interspecies and intraspecies signalling, specific adhesion, etc., which are still poorly represented in soil libraries, because of significant prevalence of studies on pure cultures instead of multispecies systems of soil microorganisms. The main task for the future research here is to increase the knowledge about these genes by developing co-cultivation techniques, which undoubtedly will increase the percent of uncultured microorganisms, such as for example, *Acidobacteria* that undergo cultivation (Janssen et al. 2002)

Apparently, these genes together with a number of genes associated with adaptation to the major edaphic (pH, humidity) or structural (texture) factors, belong to the accessory part of soil metagenome, and may form the frame for identification of the main types of organization (or "architectural models") of soil microbial communities. These types are not numerous and represent something like "plans of construction" that are built on the basis of practically one or two bacterial phyla, among them are first of all *Proteobacteria*, *Acidobacteria* and *Actinobacteria* followed by less abundant *Bacteroidetes*, *Firmicutes*.

In summary, our review suggests that soil metagenome represents the integral genetic system, which possesses more or less labile taxonomic structure combined with a conservative functional structure. The metagenomic analysis may be used to reveal the adaptive processes in soil ecosystems based mostly on reformatting the phylogenetic composition of soil community and on the re-assortment of the ecologically important genes among different microbial genotypes. Following A.P. Kostychev who considered the soil as a biological system based mostly on the activities of resident microbial communities (Kostychev 1937), one can address the metagenomes of these communities as the hereditary systems of soil microbial communities, and of soil itself, responsible for the sustainability (reproducibility) of its basic functions including fertility, as well as for the evolutionary potential and maintenance of the global genetic resources in the soil ecosystem.

References

Agnelli A, Ascherb J, Cortia G, Ceccherini MT, Nannipieri P, Pietramellara G (2004) Distribution of microbial communities in a forest soil profile investigated by microbial biomass, soil respiration and DGGE of total and extracellular DNA. Soil Biol Biochem 36:859–868

Andreetta A, Macci C, Ceccherini MT, Cecchini G, Masciandaro G, Pietramellara G, Carnicelli S (2004) Microbial dynamics in Mediterranean Moder humus. Biol Fertil Soils 48:259–270

Ascher J, Ceccherini MT, Nannipieri P, Pietramellara G (2005) Extracellular DNA rise up in soil by water capillarity. Geophys Res Abstr 7:07946

Bakken L, Frostegård Å (2006) Nucleic acid extraction from soil. In: Nannipieri P, Smalla K (eds) Nucleic acids and proteins in soil. Springer, Berlin

Baveye PC (2009) To sequence or not to sequence the whole-soil metagenome? Nat Rev Microbiol 7(10):756

Bent SJ, Forney LJ (2008) The tragedy of the uncommon: understanding limitations in the analysis of microbial diversity. ISME 2:689–695

Bergmann GT, Bates ST, Eilers KG, Lauber CL, Caporaso JG, Walters WA, Knight R, Fierer N (2011) The under-recognized dominance of Verrucomicrobia in soil bacterial communities. Soil Biol Biochem 43:1450–1455

Bietz MJ, Lee CP (2009) Collaboration in metagenomics: sequence databases and the organization of scientific work. In: Balka E, Ciolfi L, Simone C, Tellioğlu H, Wagner I (eds) ECSCW 2009: Proceedings of the 11th European conference on computer supported cooperative work. Springer, London

Böckelmann U, Lünsdorf H, Szewzyk U (2007) The detection of extracellular DNA as a structural component in the EPS of bacterial strains. Geophys Res Abstr 9:01325

Brockett BFT, Prescott CE, Grayston SJ (2012) Soil moisture is the major factor influencing microbial community structure and enzyme activities across seven biogeoclimatic zones in Western Canada. Soil Biol Biochem 44:9–20

Bru D, Ramette A, Saby NP, Dequiedt S, Ranjard L, Jolivet C, Arrouays D, Philippot L (2011) Determinants of the distribution of nitrogen-cycling microbial communities at the landscape scale. ISME J 5:532–542

Bürgmann H, Pesaro M, Widmer F, Zeyer J (2001) A strategy for optimizing quality and quantity of DNA extracted from soil. J Microbiol Methods 45:7–20

Burmølle M, Hansen LH, Sørensen SJ (2006) Reporter gene technology in soil ecology: detection of bioavailability and microbial interactions. In: Nannipieri P, Smalla K (eds) Nucleic acids and proteins in soil. Springer, Berlin

Burmølle M, Johnsen K, Abu Al-Soud W, Hansen LH, Sørensen SJ (2009) The presence of embedded bacterial pure cultures in agar plates stimulate the culturability of soil bacteria. J Microbiol Methods 79:166–173

Cai P, Huang Q, Zhang X (2006) Interactions of DNA with clay minerals and soil colloidal particles and protection against degradation by DNase. Environ Sci Technol 40:2971–2976

Carson JK, Gonzalez-Quiñones V, Murphy DV, Hinz C, Shaw JA, Gleeson DB (2010) Low pore connectivity increases bacterial diversity in soil. Appl Environ Microbiol 76:3936–3942

Castro HF, Classen AT, Austin EE, Norby RJ, Schadt CW (2010) Soil microbial community responses to multiple experimental climate change drivers. Appl Environ Microbiol 76:999–1007

Ceccherini MT, Aschera J, Pietramellara G, Vogel TM, Nannipieri P (2007) Vertical advection of extracellular DNA by water capillarity in soil columns. Soil Biol Biochem 39:158–163

Ceccherini MT, Ascher J, Agnelli A, Borgogni F, Pantani OL, Pietramellara G (2009) Experimental discrimination and molecular characterization of the extracellular soil DNA fraction. Antonie Van Leeuwenhoek 96:653–657

Chaisson MJ, Pevzner PA (2008) Short read fragment assembly of bacterial genomes. Genome Res 18:324–330

Chau JF, Bagtzoglou AC, Willig MR (2011) The effect of soil texture on richness and diversity of bacterial communities. Environ Forensics 12:333–341

Chiang W, Tolker-Nielsen T (2010) Extracellular DNA as matrix component in microbial biofilms. In: Kikuchi Y, Rykova EY (eds) Extracellular nucleic acids. Springer, Berlin

Chong CW, Pearce DA, Convey P, Yew WS, Tan IKP (2012) Patterns in the distribution of soil bacterial 16S rRNA gene sequences from different regions of Antarctica. Geoderma 181–182:45–55

Chotte J (2005) Importance of microorganisms for soil aggregation. In: Buscot F, Varma A (eds) Microorganisms in soils: roles in genesis and functions. Springer, Berlin

Chowdhury SP, Schmid M, Hartmann A, Tripathi AK (2009) Diversity of 16S-rRNA and *nif*H genes derived from rhizosphere soil and roots of an endemic drought tolerant grass, *Lasiurus sindicus*. Eur J Soil Biol 45:114–122

Chu H, Fierer N, Lauber CL, Caporaso JG, Knight R, Grogan P (2010) Soil bacterial diversity in the Arctic is not fundamentally different from that found in other biomes. Environ Microbiol 12:2998–3006

Connon SA, Lester ED, Shafaat HS, Obenhuber DC, Ponce A (2007) Bacterial diversity in hyperarid Atacama Desert soils. J Geophys Res 112:G04S17. doi:10.1029/2006JG000311

Craig JW, Chang F, Kim JH, Obiajulu SC, Brady SF (2010) Expanding small-molecule functional metagenomics through parallel screening of broad-host-range cosmid environmental DNA libraries in diverse *Proteobacteria*. Appl Environ Microbiol 76(5):1633–1641. doi:10.1128/AEM.02169-09

Daniel R (2004) The soil metagenome – a rich resource for the discovery of novel natural products. Curr Opin Biotechnol 15:199–204

Daniel R (2005) The metagenomics of soil. Nat Rev Microbiol 3:470–478

Davinic M, Fultz LM, Acosta-Martinez V, Calderón FJ, Cox SB, Dowd SE, Allen VG, Zak JC, Moore-Kucera J (2012) Pyrosequencing and mid-infrared spectroscopy reveal distinct aggregate stratification of soil bacterial communities and organic matter composition. Soil Biol Biochem 46:63–72

Dechesne A, Pallud C, Debouzie D, Flandrois JP, Vogel TM, Gaudet JP, Grundmann GL (2003) A novel method for characterizing the microscale 3D spatial distribution of bacteria in soil. Soil Biol Biochem 35:1537–1546

Demanèche S, David MM, Navarro E, Simonet P, Vogel TM (2009) Evaluation of functional gene enrichment in a soil metagenomic clone library. JMicrobiol Methods 76:105–107

Eickhorst T, Tippkötter R (2008) Detection of microorganisms in undisturbed soil by combining fluorescence in situ hybridization (FISH) and micropedological methods. Soil Biol Biochem 40:284–1293

Elshahed MS, Youssef NH, Spain AM, Sheik C, Najar FZ, Sukharnikov LO, Roe BA, Davis JP, Schloss PD, Bailey VL, Krumholz LR (2008) Novelty and uniqueness patterns of rare members of the soil biosphere. Appl Environ Microbiol 74:5422–5428

Ettema CH, Wardle DA (2002) Spatial soil ecology. Trends Ecol Evol 17:177–183

Feinstein LM, Sul WJ, Blackwood CB (2009) Assessment of bias associated with incomplete extraction of microbial DNA from soil. Appl Environ Microbiol 75:5428–5433

Fierer N, Jackson RB (2006) The diversity and biogeography of soil bacterial communities. Proc Natl Acad Sci USA 103:626–631

Fierer N, Bradford MA, Jackson RB (2007) Toward an ecological classification of soil bacteria. Ecology 88(6):1354–1364

Forney LJ, Zhou X, Brown CJ (2004) Molecular microbial ecology: land of the one-eyed king. Curr Opin Microbiol 7:210–220

Garbeva P, de Boer W (2009) Inter-specific interactions between carbon-limited soil bacteria affect behavior and gene expression. Microb Ecol 58:36–46

Gerlach W, Jünemann S, Tille F et al (2009) WebCARMA: a web application for the functional and taxonomic classification of unassembled metagenomic reads. BMC Bioinformatics 10:430

Gilbert JA, Dupont CL (2011) Microbial metagenomics: beyond the genome. Annu Rev Mar Sci 3:347–371

Gill SR, Pop M, DeBoy RT, Eckburg PB, Turnbaugh PJ, Samuel BS, Gordon JI, Relman DA, Fraser-Liggett CM, Nelson KE (2006) Metagenomic analysis of the human distal gut microbiome. Science 312:1355–1359

Girvan MS, Bullimore J, Pretty JN, Osborn AM, Ball AS (2003) Soil type is the primary determinant of the composition of the total and active bacterial communities in arable soils. Appl Environ Microbiol 69:1800–1809

Gomez-Silva B, Rainey FA, Warren-Rhodes KA, McKay CP, Navarro-Gonzalez R (2008) Atacama desert soil microbiology. In: Dion P, Nautiyal CS (eds) Microbiology of extreme soils. Springer, Berlin

Grundmann GL (2004) Spatial scales of soil bacterial diversity – the size of a clone. FEMS Microbiol Ecol 48:119–127

Grundmann LG, Gourbiere F (1999) A micro-sampling approach to improve the inventory of bacterial diversity in soil. Appl Soil Ecol 13:123–126

Handelsman J (2004) Metagenomics: application of genomics to uncultured microorganisms. Microbiol Mol Biol Rev 68:669–685

Hirsch PR, Mauchline TH, Clark IM (2010) Culture-independent molecular techniques for soil microbial ecology. Soil Biol Biochem 42:878–887

Hollister EB, Engledow AS, Hammett AJM, Provin TL, Wilkinson HH, Gentry TL (2010) Shifts in microbial community structure along an ecological gradient of hypersaline soils and sediments. ISME 4:829–838

Huson DH, Auch AF, Qi J, Schuster SC (2007) MEGAN analysis of metagenomic data. Genome Res 17:377–386

Janssen PH (2006) Identifying the dominant soil bacterial taxa in libraries of 16S rDNA and 16S rRNA genes. Appl Environ Microbiol 72:1719–1728

Janssen PH, Yates PS, Grinton BE, Taylor PM, Sait M (2002) Improved culturability of soil bacteria and isolation in pure culture of novel members of the divisions acidobacteria, actinobacteria, proteobacteria, and verrucomicrobia. Appl Environ Microbiol 68:2391–2396

Kakirde KS, Parsley LC, Liles MR (2010) Size does matter: application-driven approaches for soil metagenomics. Soil Biol Biochem 42:1911–1923

King AJ, Freeman KR, Mccormick KF, Lynch RC, Lozupone C, Knight R, Schmidt SK (2010) Biogeography and habitat modelling of high-alpine bacteria. Nature Commun 1:53. doi:10.1038/ncomms1055

Kostychev PA (1937) The soils of chernozem region of Russia: their origin, composition and properties. Selchozgiz, Moscow-Leningrad

Kunin V, Copeland A, Lapidus A, Mavromatis K, Hugenholtz P (2008) A bioinformatician's guide to metagenomics. Microbiol Mol Biol Rev 72:557–578

Kuske CR, Ticknor LO, Miller ME, Dunbar JM, Davis JA, Barns SM, Belnap J (2002) Comparison of soil bacterial communities in rhizospheres of three plant species and the interspaces in an arid grassland. Appl Environ Microbiol 68(4):1854–1863. doi:10.1128/AEM.68.4.1854-1863.2002

Lämmle K, Zipper H, Breuer M, Hauer B, Buta C, Brunner H, Rupp S (2010) Identification of novel enzymes with different hydrolytic activities by metagenome expression cloning. J Biotechnol 127:575–592. doi:10.1016/j.jbiotec.2006.07.036

Lauber CL, Strickland MS, Bradford MA, Fierer N (2008) The influence of soil properties on the structure of bacterial and fungal communities across land-use types. Soil Biol Biochem 40:2407–2415

Lauber CL, Hamady M, Knight R, Fierer N (2009) Pyrosequencing-based assessment of soil pH as a predictor of soil bacterial community structure at the continental scale. Appl Environ Microbiol 75(15):5111–5120. doi:10.1128/AEM.00335-09

Lee S, Cho J (2009) Distribution patterns of the members of phylum Acidobacteria in global soil samples. J Microbiol Biotechnol 19:1281–1287

Lee S, Ka J, Cho J (2008) Members of the phylum Acidobacteria are dominant and metabolicallyactive in rhizosphere soil. FEMS Microbiol Lett 285:263–269

Levy-Bootha DJ, Campbell RG, Gulden RH, Harta MM, Powellc JR, Klironomos JN, Pauls KP, Swanton CJ, Trevorsa JT, Dunfieldd KE (2007) Cycling of extracellular DNA in the soil environment. Soil Biol Biochem 39:2977–2991

Lombard N, Prestat E, Elsas JDV, Simonet P (2011) Soil-specific limitations for access and analysis of soil microbial communities by metagenomics. FEMS Microbiol Ecol 78:31–49

Lorenz P, Liebeton K, Niehaus F, Eck J (2002) Screening for novel enzymes for biocatalytic processes: accessing the metagenome as a resource of novel functional sequence space. Curr Opin Biotechnol 13:572–577

Lorenz P, Liebeton K, Niehaus F, Schleper C, Eck J (2003) The impact of non-cultivated biodiversity on enzyme discovery and evolution. Biocatal Biotransfor 21(2):87–91

Lu N, Zilles JL, Nguyen TH (2010) Adsorption of extracellular chromosomal dna and its effects on natural transformation of *Azotobacter vinelandii*. Appl Environ Microbiol 76(13):4179–4184. doi:10.1128/AEM.00193-10

Martin-Laurent F, Philippot L, Hallet S, Chaussod R, Germon JC, Soulas G, Catroux G (2001) DNA extraction from soils: old bias for new microbial diversity analysis methods. Appl Environ Microbiol 67:2354–2359

Mercier A, Kay E, Simonet P (2006) Horizontal gene transfer by natural transformation in soil environment. In: Nannipieri P, Smalla K (eds) Nucleic acids and proteins in soil. Springer, Berlin

Meyer F, Paarmann D, D'Souza MD, Olson R, Glass EM, Kubal M, Paczian T, Rodriguez A, Stevens R, Wilke A, Wilkening J, Edwards RA (2008) The metagenomics RAST server – a public resource for the automatic phylogenetic and functional analysis of metagenomes. BMC Bioinformatics 9:386

Mocali S, Benedetti A (2010) Exploring research frontiers in microbiology: the challenge of metagenomics in soil microbiology. Res Microbiol 161:497–505. doi:10.1016/j.resmic.2010.04.010

Mummey DL, Stahl PD (2004) Analysis of soil whole- and inner-microaggregate bacterial communities. Microbial Ecol 48:41–50

Mummey D, Holben W, Six J, Stahl P (2006) Spatial stratification of soil bacterial populations in aggregates of diverse soils. Microbial Ecol 51:404–411

Nielsen KM, Calamai L, Pietramellara G (2006) Stabilization of extracellular DNA and proteins by transient binding to various soil components. In: Nannipieri P, Smalla K (eds) Nucleic acids and proteins in soil. Springer, Berlin

Nielsen KM, Johnsen PJ, Bensasson D, Daffonchio D (2007) Release and persistence of extracellular DNA in the environment. Environ Biosafety Res 6:37–53. doi:10.1051/ebr:2007031

Nunan N, Wu K, Young IM, Crawford JW, Ritz K (2003) Spatial distribution of bacterial communities and their relationships with the micro-architecture of soil. FEMS Microbiol Ecol 44:203–215

Paget E, Lebrun M, Freyssinet G, Simonet P (1998) The fate of recombinant plant DNA in soil. Eur J Soil Biol 34(2):81–88

Pershina EV, Tamazyan GS, Dolnik AS, Pinaev AG, Sergaliev NH, Andronov EE (2012) Studying the structure of soil microbial community in saline soils by high-throughput pyrosequencing. Ecol Genet 2(10):31–38 (in russian)

Pettit RK (2004) Soil DNA libraries for anticancer drug discovery. Cancer Chemother Pharmacol 54:1–6. doi:10.1007/s00280-004-0771-8

Pietramellara G, Ascher J, Borgogni F, Ceccherini MT, Guerri G, Nannipieri P (2009) Extracellular DNA in soil and sediment: fate and ecological relevance. Biol Fertil Soils 45:219–235

Podar M, Abulencia CB, Walcher M, Hutchison D, Zengler K, Garcia JA, Holland T, Cotton D, Hauser L, Keller M (2007) Targeted access to the genomes of low-abundance organisms in complex microbial communities. Appl Environ Microbiol 73(10):3205–3214. doi:10.1128/AEM.02985-06

Raes J, Foerstner UK, Bork P (2007) Get the most out of your metagenome: computational analysis of environmental sequence data. Curr Opin Microbiol 10:1–9

Riesenfeld CS, Schloss PD, Handelsman J (2004) Metagenomics, genomic analysis of microbial communities. Annu Rev Genet 38:525–552

Robe P, Nalin R, Capellano C, Vogel TM, Simonet P (2003) Extraction of DNA from soil. Eur J Soil Biol 39:183–190

Roesch LFW, Fulthorpe RR, Riva A, Casella G, Hadwin AKM, Kent AD, Daroub SH, Camargo FAO, Farmerie WG, Triplett EW (2007) Pyrosequencing enumerates and contrasts soil microbial diversity. ISME J 1:283–290

Romanowski G, Lorenz MG, Sayler G, Wackernagel W (1992) Persistence of free plasmid DNA in soil monitored by various methods, including a transformation assay. Appl Environ Microbiol 58:3012–3019

Rondon MR, August PR, Bettermann AD, Brady SF, Grossman TH, Liles MR, Loiacono KA, Lynch BA, MacNeil IA, Minor C, Tiong CL, Gilman M, Osburne MS, Clardy J, Handelsman J, Goodman RM (2000) Cloning the soil metagenome: a strategy for accessing the genetic and functional diversity of uncultured microorganisms. Appl Environ Microbiol 66:2541–2547

Ruberto LAM, Vazquez SC, Mac Cormack WP (2008) Bacteriology of extremely cold soils exposed to hydrocarbon pollution. In: Dion P, Nautiyal CS (eds) Microbiology of extreme soils. Springer, Berlin

Saeki K, Kunito T (2010) Adsorptions of DNA molecules by soils and variable-charged soil constituents. In: Méndez-Vilas A (ed) Current research technology and education topics in applied microbiology and microbial biotechnology. FORMATEX, Badajoz

Sagova-Mareckova M, Cermak L, Novotna J, Plhackova K, Forstova J, Kopecky J (2008) Innovative methods for soil DNA purification tested in soils with widely differing characteristics. Appl Environ Microbiol 74:2902–2907

Schink B (2002) Synergistic interactions in the microbial world. Antonie Van Leeuwenhoek 81:257–261

Schipper C, Hornung C, Bijtenhoorn P, Quitschau M, Grond S, Streit WR (2009) Metagenome-derived clones encoding two novel lactonase family proteins involved in biofilm inhibition in *Pseudomonas aeruginosa*. Appl Environ Microbiol 75(1):224–233. doi:10.1128/AEM.01389-08

Sessitsch A, Weilharter A, Gerzabek MH, Kirchmann H, Kandeler E (2001) Microbial population structures in soil particle size fractions of a long-term fertilizer field experiment. Appl Environ Microbiol 67:4215–4224

Simon C, Daniel R (2011) Metagenomic analyses: past and future trends. Appl Environ Microbiol 77(4):1153–1161. doi:10.1128/AEM.02345-10

Singh BK, Campbell CD, Sorenson SD, Zhou J (2009a) Soil genomics. Nat Rev Microbiol 7:756. doi:10.1038/nrmicro2119-c1

Singh BK, Dawson LA, Macdonald CA, Buckland SM (2009b) Impact of biotic and abiotic interction on soil microbial communities and functions: a field study. Appl Soil Ecol 41:239–248. doi:10.1016/j.apsoil.2008.10.003

Spain AM, Krumholz LR, Elshahed MS (2009) Abundance, composition, diversity and novelty of soil *Proteobacteria*. ISME J 3:992–1000

Steven B, Gallegos-Graves LV, Starkenburg SR, Chain PS, Kuske SR (2012) Targeted and shotgun metagenomic approaches provide different descriptions of dryland soil microbial communities in a manipulated field study. Environ Microbiol Rep 4(2):248–256

Torsvik V, Ovreas L (2002) Microbial diversity and function in soil, from genes to ecosystems. Curr Opin Microbiol 5:240–245

Torsvik V, Goksøyr J, Daae FL (1990) High diversity in DNA of soil bacteria. Appl Environ Microbiol 56(3):782–787

Trap J, Laval K, Akpa-Vinceslas M, Gangneux C, Bureau F, Decaëns T, Aubert M (2011) Humus macro-morphology and soil microbial community changes along a 130-yr-old *Fagus sylvatica* chronosequence. Soil Biol Biochem 43:1553–1562. doi:10.1016/j.soilbio.2011.04.005

Trevors JT (2010) One gram of soil: a microbial biochemical gene library. Antonie Van Leeuwenhoek 97:99–106. doi:10.1007/s10482-009-9397-5

Tringe SG, Hugenholtz P (2009) A renaissance for the pioneering 16S rRNA gene. Curr Opin Microbiol 11:442–446

Tringe SG, von Mering C, Kobayashi A, Salamov AA, Chen K, Chang HW, Podar M, Short JM, Mathur EJ, Detter JC, Bork P, Hugenholtz P, Rubin EM (2005) Comparative metagenomics of microbial communities. Science 308:554–557

Tsai S, Selvam A, Chang Y, Yang S (2009) Soil bacterial community composition across different topographic sites characterized by 16S rRNA gene clones in the Fushan Forest of Taiwan. Bot Stud 50:57–68

Tyson GW, Banfield JF (2005) Cultivating the uncultivated: a community genomics perspective. Trends Microbiol 13(9):411–415

Ussery DW, Wassenaar TM, Borini S (2009) Computing for comparative microbial genomics. Springer, Berlin

Vieites JM, Guazzaroni ME, Beloqui A, Golyshin PN, Ferrer M (2010) Molecular methods to study complex microbial communities. In: Streit WR, Daniel R (eds) Metagenomics. Methods and protocols. Humana Press, c/o Springer Science+Business Media, New York

Vogel TM, Simonet P, Jansson JK, Hirsh PR, Tiedje JM, Van Elsas JD, Bailey MJ, Nalin R, Philippot L (2009) TerraGenome: a consortium for the sequencing of a soil metagenome. Nat Rev Microbiol 7:252

Voget S, Leggewie C, Uesbeck A, Raasch C, Jaeger KE, Streit WR (2003) Prospecting for novel biocatalysts in a soil metagenome. Appl Environ Microbiol 69(10):6235–6242. doi:10.1128/AEM.69.10.6235-6242.2003

Voroney RP (2010) The soil habitat. In: Paul EA (ed) Soil microbiology, ecology, and biochemistry, 3rd edn. Academic, Burlington

Wackernagel W (2006) The various sources and the fate of nucleic acids in soil. In: Nannipieri P, Smalla K (eds) Nucleic acids and proteins in soil. Springer, Berlin

Wagner D (2008) Microbial communities and processes in arctic permafrost environments. In: Dion P, Nautiyal CS (eds) Microbiology of extreme soils. Springer, Berlin

Wexler M, Johnston AWB (2010) Wide host-range cloning for functional metagenomics. In: Streit WR, Daniel R (eds) Metagenomics. Methods and protocols. Humana Press, c/o Springer Science+Business Media, New York

Will C, Thürmer A, Wollherr A, Nacke H, Herold N, Schrumpf M, Gutknecht J, Wubet T, Buscot F, Daniel R (2010) Horizon-specific bacterial community composition of german grassland soils, as revealed by pyrosequencing-based analysis of 16S rRNA genes. Appl Environ Microbiol 76(20):6751–6759. doi:10.1128/AEM.01063-10

Wooley JC, Ye Y (2009) Metagenomics: facts and artifacts, and computational challenges. J Comp Sci Technol 25:71–81

Young IM, Crawford JW (2004) Interactions and selforganization in the soil-microbe complex. Science 304:1634–1637

Zengler K, Toledo G, Rappe M, Elkins J, Mathur EJ, Short JM, Keller M (2002) Cultivating the uncultured. Proc Natl Acad Sci USA 99:15681–15686

Chapter 8
Mobile Genetic Elements (MGEs) Carrying Catabolic Genes

Masaki Shintani and Hideaki Nojiri

Abstract Mobile genetic elements (MGEs) are important "vehicles" of diverse genes in the microbial genetic pool. Exchange of MGEs in the microbial community confers new traits to their hosts and promotes their rapid adaptation to various environments. For decades, a variety of bacteria capable of degrading "xenobiotic" compounds have been isolated for their potential importance in the removal of these compounds from contaminated environments. The genes responsible for the catabolic turnover of xenobiotics are sometimes located on MGEs such as plasmids, transposons, and integrative and conjugative elements (ICEs). This chapter summarizes our current knowledge of major MGEs that carry catabolic genes, and briefly describes their features. Recent works focused on the behavior of MGEs in natural environmental samples have also been described here.

Keywords Mobile genetic elements (MGEs) • Horizontal gene transfer (HGT) • Transposons • Catabolic plasmids • Xenobiotics

---

M. Shintani (✉)
Present address: Department of Materials Science and Chemical Engineering,
Faculty of Engineering, Shizuoka University, 3-5-1 Johoku, Naka-ku, Hamamatsu,
432-8561, Shizuoka, Japan

Japan Collection of Microorganisms, RIKEN BioResource Center
3-1-1 Kouyadai, Tsukuba, Ibaraki 305-0074, Japan
e-mail: tmshint@ipc.shizuoka.ac.jp

H. Nojiri
Biotechnology Research Center, The University of Tokyo,
1-1-1 Yayoi, Bunkyo-ku, Tokyo 113-8657, Japan

Agricultural Bioinformatics Research Unit, The University of Tokyo,
1-1-1 Yayoi, Bunkyo-ku, Tokyo 113-8657, Japan

A. Malik et al. (eds.), *Management of Microbial Resources in the Environment*,
DOI 10.1007/978-94-007-5931-2_8,

1 Introduction

Horizontal gene transfer (HGT) is one of the important mechanisms for rapid bacterial evolution and adaptation. HGT proceeds mainly by conjugation (Frost et al. 2005; Heuer and Smalla 2007; Aminov 2011) and is mediated by mobile genetic elements (MGEs), which are DNA segments that can move between bacterial cells (intercellular mobility) (Frost et al. 2005). The elements carry various kinds of genes, such as antibiotic resistance genes, virulence genes, and catabolic genes, and thus, MGEs are important "vehicles" of pathogenically- and environmentally-relevant traits. Plasmids, integrative and conjugative elements (ICEs), and transposons are the important MGEs. Since the 1960s, various bacteria capable of degrading "xenobiotic" compounds have been isolated because of their potential importance in the removal of these compounds from contaminated environments. In this review, we use the term "xenobiotic" compounds in a broad sense to signify compounds that are not natural to the environment, but are rather "guest" chemicals, as defined by Leisinger (1983). The genes involved in catabolic turnover of xenobiotic compounds are sometimes identified on MGEs, especially on plasmids, ICEs, and transposons.

Plasmids are circular or linear extrachromosomal replicons, which are often transmissible by conjugation (Sota and Top 2008; Frost et al. 2005). ICEs are also self-transmissible conjugative elements, but they are generally integrated into the host chromosome (Burrus and Waldor 2004; Wozniak and Waldor 2010). Conjugation can spread genetic elements among bacteria effectively (Guglielmini et al. 2011), and therefore, it is one of the most important mechanisms for rapid evolution and adaptation of bacteria. On the other hand, transposons are genetic elements that are mobilized and transferred between replicons by the activity of a transposase (Mahillon and Chandler 1998). Once transposons integrate into plasmids or ICEs, they can also be transferred into other cells (Frost et al. 2005). Insertion sequences (IS) are a transposons that carry only the transposase gene, and homologous recombination between multiple copies of the same IS element can promote genomic rearrangements (Mahillon and Chandler 1998).

Although many reviews have been published on MGEs that carry catabolic genes for xenobiotic compounds (Tan 1999; Top et al. 2000, 2002; Top and Springael 2003; van der Meer and Sentchilo 2003; Nojiri et al. 2004; Dennis 2005), a large number of new catabolic MGEs have since been reported due to the recent revolution in nucleotide sequencing technology. This chapter summarizes recent studies of major and/or new MGEs that carry catabolic genes, and briefly describes their features.

2 Catabolic Plasmids

Plasmids have been classified into incompatibility (Inc) groups on the basis of their replication and partition systems. When two different plasmids cannot be maintained in the same bacterial cell line, these two plasmids are called "incompatible" and are considered to belong to the same "Inc" group. There are 27 Inc groups for the *Enterobacteriaceae* (Carattoli 2009), at least 14 groups for the *Pseudomonas* (Thomas and Haines 2004), and around 18 groups for the gram-positive bacteria (Frost et al. 2005;

Sota and Top 2008), although these groupings do not include all the identified plasmids such as plasmids in *Sphingomonas*. Recently, a new classification of plasmids was proposed, which is based on their transfer systems generally composed of two sets of proteins for mating pair formation (MPF) and mobilization (MOB) (Smillie et al. 2010; Garcillán-Barcia et al. 2009, 2011). Combination of four types of MPFs (MPF_F, MPF_I, MPF_G, and MPF_T) and six classes of MOBs (MOB_F, MOB_H, MOB_Q, MOB_C, MOB_P, and MOB_V) enables us to classify a larger number of plasmids whose sequences have been deposited in DNA databases.

Features of major catabolic plasmids, such as host, growth substrate of host, Inc groups, MOB classes and MPF types, and transferability, are listed in Tables 8.1, 8.2, 8.3 and 8.4. They have been identified in bacteria of the phylum *Proteobacteria*, such as *Pseudomonas* (*γ-proteobacteria*), *Achromobacter* (*β-proteobacteria*), and *Sphingomonas* (*α-proteobacteria*), and in gram-positive bacteria such as *Arthrobacter*, *Flavobacterium*, and *Rhodococcus*, since the 1970s. Because detailed features of IncP-1, IncP-7, and IncP-9 group plasmids have been already described in our previous review (Shintani et al. 2010), we focused especially on the catabolic plasmids in sphingomonads or gram-positive bacteria in this chapter.

2.1 Catabolic Plasmids from Genus Pseudomonas and Those Belonging to Pseudomonas Incompatibility Groups

Many catabolic plasmids are classified into the IncP-1, IncP-2, IncP-7, and IncP-9 groups, which carry genes involved in the degradation of various xenobiotic compounds, such as those for toluene/xylene (*xyl*), (chloro)benzoate (*cba*), (chloro)aniline (*dca*), 2,4-dichlorophenoxyacetic acid (2,4-D) (*tfd*), naphthalene (*nah*), and carbazole (*car*), amongst others (Table 8.1). The complete nucleotide sequences of several plasmids in these groups, except for the IncP-2 plasmids, have been determined, and an Inc group-specific plasmid backbone was proposed by comparative analyses (Fig. 8.1). Dennis (2005) compared the genetic organization of IncP-1 plasmids and showed that most catabolic genes (or other genes, such as antibiotic resistance genes) of IncP-1 plasmids were inserted between the *trfA* and *oriV* regions and the *parA* and *tra* operons (Fig. 8.1a; Dennis 2005). Sota et al. (2007) showed that the structural similarity of IncP-1 plasmids was a result of both the region-specific insertion of transposons and the selective pressure for maintaining transferability and stability of the plasmids. Based on the comparisons of the nucleotide sequences of plasmids, conserved regions of IncP-9 and IncP-7 plasmids (i.e., a plasmid backbone) were also proposed (Fig. 8.1b, c; Sota et al. 2006; Yano et al. 2010). One important difference between these plasmids is their host range. IncP-1 plasmids are known to be broad host range plasmids that can transfer among bacteria belonging to different classes, such as *α*-, *β*-, and *γ-proteobacteria*. Indeed, the host range of IncP-1 catabolic plasmids is broad, as listed in Table 8.1. As for the IncP-7 and IncP-9 plasmids, their host ranges are narrower than that of the IncP-1 plasmids, and most of their hosts belong to *γ-proteobacteria*, and in particular, to the genus *Pseudomonas* (Table 8.1).

Table 8.1 Catabolic plasmids from genus *Pseudomonas* and those belonging to *Pseudomonas* incompatibility groups[a]

Plasmid	Host	Substrate[b]	Size (kb)	Inc group	MOB class[c]	T4SS type[c]	Genes	Transferability	References
pA81[d]	*Achromobacter xylosocidans* A8	Chlorobenzoate	98	P1	MOB_P	MPF_T	*mocpRAB CDhybRABCD*	+	Jencova et al. (2008)
pAC25	*Pseudomonas putida* AC858	3-Chlorobenzoate	117	P1	NA	NA	NA	+	Chatterjee et al. (1981)
pADP-1[d]	*Pseudomonas* sp. ADP	Atrazine	109	P1	MOB_P	MPF_T	*atzABCDEF*	+	de Souza et al. (1998); Martinez et al. (2001)
pBRC60	*Alcaligenes* sp. BR60	3-Chlorobenzoate	75	P1	NA	NA	*cba*	+	Fulthorpe and Wyndham (1991)
pC1	*Delftia acidovorans* CA28	3-Chloroaniline	100	P1	NA	NA	*tdnQ*	+	Boon et al. (2001)
pCNB1[d]	*Comamonas* sp. CNB-1	4-chloronitro-benzene	91	P1	MOB_P	MPF_T	*cnb*, *cat*	NA	Wu et al. (2005, 2006); Ma et al. (2007)
pEMT3	Unknwon soil bacterium	2,4-D, 3-Chlorobenzoate	60	P1	NA	NA	*tfdABC*	+	Top et al. (1995); Gstalder et al. (2003)
pENH91	*Ralstonia eutropha* NH9	3-Chlorobenzoate	78	P1	NA	NA	*cbnABCD*	NA	Ogawa and Miyashita (1995)
pEST4011[d]	*Achromobacter xylosoxidans* subsp. *denitrificans* EST4002	2,4-D	70	P1	MOB_P	MPF_T	*tfdCEBKA*, *tfdF*, *mdc*	+	Mäe et al. (1993); Vedler et al. (2000, 2004)
pIJB1[d]	*Burkholderia cepacia* 2a	2,4-D, malonate	102	P1	NA		*tfd*, *mdc*, *bph*	NA	Xia et al. (1998); Poh et al. (2002)
pJP4[d]	*Ralstonia eutropha* JMP134	2,4-D, 3-Chlorobenzoate	80	P1	MOB_P	MPF_T	*tfdA*, *tfdB*, *tfdCDEF*	+	Don and Pemperton (1981); Don et al. (1985); Trefault et al. (2004)

pNB2	*Comamonas testosteroni* I2	3-Chloroaniline	60	P1	NA	NA	*tdnQ*	+	Boon et al. (2000, 2001); Bathe (2004)
pNB8c[d]	*Delftia acidovorans* B8c	3-Chloroaniline	60	P1	NA	NA	*dca*	+	Boon et al. (2001); Dejonghe et al. (2002); Król et al. (2012)
pPS12-1	*Burkholderia* sp. PS12	1,2,4,5-Tetrachlorobenzene	85	P1	NA	NA	*tecAB*	NA	Beil et al. (1999)
pSS50	*Alcaligenes* sp. A5	4-Chlorobenzoate	53	P1	NA	NA	*bph*	+	Shields et al. (1985); Hooper et al. (1989); Layton et al. (1992)
pSS60	*Achromobacter* sp. LBS1C1	4-chlorobenzoate	63	P1	NA	NA	*bph*	+	Burlage et al. (1990)
pTSA	*Comamonas testosteroni* T-2	*p*-Toluenesulfonic acid	85	P1	NA	NA	*tsaMBCDR*, *psbAC*	+	Junker and Cook (1997); Tralau et al. (2001)
pUO1[d]	*Delftia acidovorans* B	Haloacetates	65	P1	MOB_P	MPF_T	*dehA1*, *dehA2*	+	Kawasaki et al. (1981); Sota et al. (2003)
pWDL7[d]	*Comamonas testosteroni* WDL2	3-Chloroaniline		P1	NA	NA	*dcaRBA2A1TQ*	+	Król et al. (2012)
CAM	*Pseudomonas putida* PpG1	Camphor	500	P2	NA		*cam*	NA	Chakrabarty (1973); Rheinwald et al. (1973); Tan (1999)
OCT	*Pseudomonas olevorans* PpG6	Camphor	500	P2	NA		*alkBFGHJKL*, *alkST*	+	Chakrabarty (1973)
pVI150	*Pseudomonas* sp. CF600	Phenol	NA	P2	NA		*dmp*	+	Bartilson et al. (1990)
pAK5	*Pseudomonas putida* AK5	Naphthalene	115	P7	NA		NA	NA	Izmalkova et al. (2005)
pCAR1[d]	*Pseudomonas resinovorans* CA10	Carbazole	199	P7	MOB_H	MPF_F	*carABCDEF* *antABC*	+	Nojiri et al. (2001); Maeda et al. (2003); Takahashi et al. (2009)

(continued)

Table 8.1 (continued)

Plasmid	Host	Substrate[b]	Size (kb)	Inc group	MOB class[c]	T4SS type[c]	Genes	Transferability	References
pDK1[d]	*Pseudomonas putida* HS1	Xylene, toluene	180	P7	NA	NA	*xyl*	+	Kunz and Chapman (1981); Yano et al. (2010)
pFME	*Pseudomonas fluorescens* FME4	Naphthalene	77	P7	NA	NA	NA	NA	Izmalkova et al. (2005)
pME5	*Pseudomonas fluorescens* FME5	Naphthalene	80	P7	NA	NA	NA	NA	Izmalkova et al. (2005)
pND6-1[d]	*Pseudomonas* sp. ND6	Naphthalene	102	P7	–	–	*nah*	–	Li et al. (2004)
pNK33	*Pseudomonas fluorescens* NK33	Naphthalene	100	P7	NA	NA	NA	NA	Izmalkova et al. (2005)
pNK43	*Pseudomonas fluorescens* NK43	Naphthalene	123	P7	NA	NA	NA	NA	Izmalkova et al. (2005)
pOS18	*Pseudomonas fluorescens* OS18P	Naphthalene	135	P7	NA	NA	NA	NA	Izmalkova et al. (2005)
pOS19	*Pseudomonas fluorescens* OS19P	Naphthalene	122	P7	NA	NA	NA	NA	Izmalkova et al. (2005)
pWW53[d]	*Pseudomonas putida* MT53	Xylene, Toluene	107	P7	–	–	*xyl*	–	Keil et al. (1985, 1987); Tsuda and Genka (2001); Yano et al. (2007)

NAH7[d]	*Pseudomonas putida* G7	Naphthalene, Phenanthrene, Anthracene	83	P9	MOB_F	MPF_T	*nah*	+	Dunn and Gunsalus (1973); Connors and Barnsley (1980); Sota et al. (2006)
NPL-1	*Pseudomonas putida* BS 202	Naphthalene	100	P9	NA	NA	NA	NA	Sevastsyanovich et al. (2008)
p15C	*Pseudomonas* sp. 15C	Naphthalene	110	P9	NA	NA	NA		Sevastsyanovich et al. (2008)
p8C	*Pseudomonas* sp. 8C	Naphthalene	110	P9	NA	NA	NA		Sevastsyanovich et al. (2008)
pBS1141	*Pseudomonas putida* BS 3701	Naphthalene	120	P9	NA	NA	NA	NA	Sevastsyanovich et al. (2008)
pBS1181	*Pseudomonas putida* BS 3750	Naphthalene	120	P9	NA	NA	NA	NA	Sevastsyanovich et al. (2008)
pBS1191	*Pseudomonas putida* BS 3790	Naphthalene	100	P9	NA	NA	NA	NA	Sevastsyanovich et al. (2008)
pBS240	*Pseudomonas putida* BS 639	Naphthalene	160	P9	NA	NA	NA	NA	Sevastsyanovich et al. (2008)
pBS243	*Pseudomonas putida* BS 638	Naphthalene	160	P9	NA	NA	NA	NA	Sevastsyanovich et al. (2008)
pBS265	*Pseudomonas putida* BS 394	ε-Caprolactam	130	P9	NA	NA	NA	NA	Sevastsyanovich et al. (2008)
pBS267	*Pseudomonas putida* BS 394	ε-Caprolactam	130	P9	NA	NA	NA	NA	Sevastsyanovich et al. (2008)
pBS268	*Pseudomonas putida* BS 394	ε-Caprolactam	85	P9	NA	NA	NA	NA	Sevastsyanovich et al. (2008)
pDTG1[d]	*Pseudomonas putida* NCBI9816-4	Naphthalene	81	P9	MOB_F	MPF_T	*nah*	–	Simon et al. (1993); Stuart-Keil et al. (1998); Dennis and Zylstra (2004)
pFKY1[e]	Unidentified soil bacterium	Naphthalene	200	P9	NA	NA	*nah*	+	Ono et al. (2007)

(continued)

Table 8.1 (continued)

Plasmid	Host	Substrate[b]	Size (kb)	Inc group	MOB class[c]	T4SS type[c]	Genes	Transferability	References
pFKY4[e]	Unidentified soil bacterium	Naphthalene	80	P9	NA	NA	*nah*	+	Ono et al. (2007)
pNAH20[d]	*Pseudomonas fluorescens* PC20	Naphthalene	83	P9	MOB_F	MPF_T	*nag*	+	Heinaru et al. (2009)
pNL22	*Pseudomonas fluorescens* 41a	Naphthalene	100	P9	NA	NA	NA	NA	Sevastsyanovich et al. (2008)
pNL25	*Pseudomonas putida* 21a	Naphthalene	75	P9	NA	NA	NA	NA	Sevastsyanovich et al. (2008)
pNL29	*Pseudomonas* sp. 58	Naphthalene	NA	P9	NA	NA	NA	NA	Sevastsyanovich et al. (2008)
pNL31	*Pseudomonas aeruginosa* 56	Naphthalene	NA	P9	NA	NA	NA	NA	Sevastsyanovich et al. (2008)
pNL4	*Pseudomonas putida* 10a	Naphthalene	75	P9	NA	NA	NA	NA	Sevastsyanovich et al. (2008)
pNL60	*Pseudomonas fluorescens* 18d	Naphthalene	120	P9	NA	NA	NA	NA	Sevastsyanovich et al. (2008)
pOV17	*Pseudomonas aureofaciens*OV17	Naphthalene	85	P9	NA	NA	NA	NA	Sevastsyanovich et al. (2008)
pSAH	*Alcaligenes* sp. O-1	2-Aminobenzene-sulfonate	180	P9	NA	NA	*abs*, *scm*	+	Jahnke et al. (1990); Ruff et al. (2010)
pSN11	*Pseudomonas putida* SN11	Naphthalene	83	P9	NA	NA	NA		Sevastsyanovich et al. (2008)
pSVS15	*Pseudomonas putida* SVS15	Toluene, Xylene	90	P9	NA	NA	NA		Sevastsyanovich et al. (2008)
pWW0[d]	*Pseudomonas putida* mt-2	Xylene, Toluene	117	P9	MOB_F	MPF_T	*xyl*	+	Williams and Murray (1974); Greated et al. (2002)

SAL1	*Pseudomonas putida* R1	Salicylate	85	P9	NA	NA	*sal*	+	Chakrabarty (1972)
NIC	*Pseudomonas convexa* Pc1	Nicotine, Nicotinate	NA	NA	NA	NA	NA	NA	Thacker et al. (1978)
pAC27	*Pseudomonas putida* AC867	3-Chlorobenzoate	110	NA	NA	NA	*clcABD*, *tfdC*	+	Chatterjee and Chakrabarty (1982, 1984); Ghosal et al. (1985); Frantz and Chacrabarty (1987); Ghosal and You (1988)
pAM10.6	*Pseudomonas fluorescens* biotype F Cb36	Phenol	11	NA	NA	NA	*pheBA*	NA	Peters et al. (1997)
pCg1	*Pseudomonas putida* Cg1	Naphthalene	86	NA	NA	NA	*nah*	+	Park et al. (2003)
pCINNP	*Pseudomonas putida* CINNP	Cinnamate	75	NA	NA	NA	NA	+	Andreoni and Bestetti (1986)
pCIT1	*Pseudomonas* sp. CIT1	Aniline	100	NA	NA	NA	NA	NA	Anson and Mackinnon (1984)
pCS1	*Pseudomonas diminuta*	Parathion	66	NA	NA	NA	*opd*	NA	Serdar et al. (1982); Mulbry et al. (1986)
pCT14[d]	*Pseudomonas* sp. CT14	Toluene	55	NA	MOB_F	–	*cbz*, *bphK*	NA	Bramucci et al. (2006)
pDBT2	*Pseudomonas alcaligenes* DBT2	Dibenzothiophene	80	NA	NA	NA	NA	NA	Foght and Westlake (1990); Top et al. (2000)
pEMT1[e]	Unidentified soil bacterium	2,4-D, 3-Chlorobenzoate	84	NA	NA	NA	*tfdARCGEFB*	+	Top et al. (1995)
pEMT8[e]	Unidentified soil bacterium	–[f]	75	NA	NA	NA	*tfdA*	+	Top et al. (1996)

(continued)

Table 8.1 (continued)

Plasmid	Host	Substrate[b]	Size (kb)	Inc group	MOB class[c]	T4SS type[c]	Genes	Transferability	References
pEST1026	*Pseudomonas putida* EST1020	Phenol	109	NA	NA	NA	*pheBA*	+	Kivisaar et al. (1990, 1991)
pHMT112	*Pseudomonas putida* ML2	Benzene	NA	NA	NA	NA	*bedC1C2BA*, *bedD*	NA	Tan and Mason (1990); Fong et al. (1996)
pKA4	*Pseudomonas pikettii 712*	2,4-D	41	NA	NA	NA	*tfdA*	+	Ka and Tiedje (1994)
pKJ1	*Pseudomonas* sp. TA8	Toluene	225	NA	NA	NA	NA	+	Yano and Nishi (1980)
pKW1	*Pseudomonas putida* GJ31	Chlorocatechol	180	NA	NA	NA	*cbz*	NA	Kunze et al. (2009)
pMH1	*Pseudomonas*sp. HF-1	Nicotine	21	NA	NA	NA	*hsp*	–	Wang et al. (2009)
pNB1	*Pseudomonas putida* HS12	Nitrobenzene	59	NA	NA	NA	*nbzA*, *nbzCDE*	NA	Park and Kim (2000)
pNB2	*Pseudomonas putida* HS12	Nitrobenzene	44	NA	NA	NA	*nbzB*	NA	Park and Kim (2000)
pP51	*Pseudomonas* sp. P51	Chlorinated Benzene	110	NA	NA	NA	*tcbCDEF*, *tcbAB*	+	van der Meer et al. (1991b)
pPGH1	*Pseudomonas putida* H	Pheonol	220	NA	NA	NA	*phlABCDEFGH*	NA	Herrmann et al. (1988)
pPOB	*Pseudomonas pseudoalcaligenes* POB310	Carboxyldiphenyl ethers	–	NA	NA	NA	*pabAB*	NA	Dehmel et al. (1995)
pRA500	*Pseudomonas putida* NCIB 9869	3,5-Xylenol	500	NA	NA	NA	*pchACXFHG*	+	Hopper and Kemp (1980); Top et al. (2000)
pRE4	*Pseudomonas putida* RE204	Isopropyl benzene	105	NA	NA	NA	*ipbABCEGFHD*	+	Eaton and Timmis (1986)

pTDN1	*Pseudomonas putida* UCC22	Aniline	NA	NA	NA	NA	*tdnQTA1A2BR*	NA	Saint et al. (1990)
pTKO	*Pseudomonas putida* PPK1	Toluene	150	NA	NA	NA	NA	NA	Keshavarz et al. (1985)
pUU204	*Pseudomonas* sp. E4	2-Chloropropionic acid	294	NA	NA	NA	Dehalogenase	–	Hardman et al. (1986)
pWW100	*Pseudomonas* sp. CB406	Biphenyl, Benzoate	200	NA	NA	NA	*bph*	–	Lloyd-Jones et al. (1994)
pZWL0	*Pseudomonas* sp. WBC-3	Methyl parathion, *p*-Nitrophenol	~70	NA	NA	NA	*mph*	+	Liu et al. (2005)

[a]Several original plasmid hosts of IncP-1 and P-9 groups are not necessarily of the genus *Pseudomonas*
[b]2,4-D represents 2,4-Dichlorophenoxyacetic acid
[c]Classification of MOB and MPF classes is based on the report of Smillie et al. (2010). NA means not available
[d]Whole nucleotide sequences are available
[e]This plasmid was captured by performing exogenous plasmid isolation from soil bacteria, and its original host was therefore unidentified
[f]This plasmid was able to complement the deficiency of 2,4-D metabolism of *tfdA*-disrupted-host

Table 8.2 Catabolic plasmids in sphingomonads

Plasmid	Host	Substrate[a]	Size (kb)	Rep type[b]	MOB class[c]	T4SS type[c]	Genes	Transferability	References
pCAR3[d]	*Novosphingobium* sp. KA1	Carbazole	240	pNL1	MOB_F	MPF_F	*car*	–	Habe et al. (2002); Shintani et al. (2007)
pCHQ1[d]	*Sphingobium japonicum* UT26	γ-HCH	191	pCHQ1	NA	NA	*linRED*	+	Nagata et al. (2006, 2010, 2011)
pISP1	*Sphingomonas* sp. MM-1	γ-HCH	200	NA	NA	NA	*lin*	NA	Tabata et al. (2011)
pISP3[d]	*Sphingomonas* sp. MM-1	γ-HCH	40	NA	NA	NA	*lin*	NA	Tabata et al. (2011)
pISP4	*Sphingomonas* sp. MM-1	γ-HCH	30	NA	NA	NA	*lin*	NA	Tabata et al. (2011)
pLA1[d]	*Novosphingobium pentaromativorans* US6-1	PAHs	188	pCHQ1	NA	NA	*bph*	NA	Luo et al. (2012)
pLA2[d,e]	*Novosphingobium pentaromativorans* US6-1	PAHs		pLB1	NA	NA	–	NA	Luo et al. (2012)
pLB1[d]	Unidentified soil bacterium	γ-HCH	66	pLB1	MOB_P	MPF_T	*linB*	+	Miyazaki et al. (2006)
pNL1[d]	*Novosphingobium aromaticivorans* DSM 12444	Biphenyl, Naphthalene	184	pNL1	MOB_F	MPF_F	*bph*, *xyl*	+	Stillwell et al. (1995); Romine et al. (1999)
pNL2[d,e]	*Novosphingobium aromaticivorans* DSM 12444	Biphenyl, Naphthalene	487	NA	–	–	–	NA	Fredrickson et al. (1991)

pSLGP[d,e]	*Sphingobium* sp. SYK-6	Lignin		pCHQ1	NA	NA	–	NA	Masai et al. (2010)
pSPHCH01[d,e]	*Sphingobium chlorophenolicum* L-1	Pentachlorophenol		pCHQ1	NA	NA	–	NA	Copley et al. (2012)
pSWIT01[d,e]	*Sphingomonas wittichii* RW1	Dibenzo-*p*-dioxin	310	Unclassified	NA	NA	–	NA	Miller et al. (2010)
pSWIT02[d]	*Sphingomonas wittichii* RW1	Dibenzo-*p*-dioxin	223	pNL1	MOB_P	MPF_T	*dxn*	NA	Miller et al. (2010)
pBN6	*Sphingomonas xenophaga* BN6	Naphthalenesulfonate	180	pNL1	NA	NA	*nsa*	+	Basta et al. (2004); Keck et al. (2006)
pCF01-05	*Sphingomonas* sp. CF06	Carbofuran	NA	NA	NA	NA	NA	+	Feng et al. (1997a, b)
pKS14	*Sphingomonas* sp. KS14	Phenanthrene, naphthalene	>500	NA	NA	NA	NA	NA	Cho and Kim (2001)
pZL	*Sphingomonas* sp. ZL5	PAHs	~60	NA	NA	NA	NA	+	Liu et al. (2004)
Megaplasmid	*Sphingopyxis* sp. 113P3	Polyvinyl alcohol	NA	NA	NA	NA	*pvaA*	NA	Hu et al. (2008)
Large plasmid	*Sphingopyxis terrae*	Polyethilene glycol	NA	NA	NA	NA	*pegB*	NA	Tani et al. (2007)
Plasmid	*Sphingomonas paucimobilis* TNE12	Fluoranthene	240	NA	NA	NA	NA	NA	Shuttleworth et al. (2000)
Plasmid a	*Sphingobium francense* SP+	γ-HCH	<32	NA	NA	NA	*linB*	NA	Cérmémonie et al. (2006)

(continued)

Table 8.2 (continued)

Plasmid	Host	Substrate[a]	Size (kb)	Rep type[b]	MOB class[c]	T4SS type[c]	Genes	Transferability	References
Plasmid b	*Sphingobium francense* SP+	γ-HCH	<32	NA	NA	NA	*linE*	NA	Cérmémonie et al. (2006)
Plasmid e	*Sphingobium francense* SP+	γ-HCH	~214	NA	NA	NA	*linA*, *linX*	NA	Cérmémonie et al. (2006)
Plasmid	*Sphingobium indicum* B40	γ-HCH	~214	NA	NA	NA	*linA*	NA	Cérmémonie et al. (2006)
Plasmid	*Sphingomonas* sp. HH69	Dibenzofuran	240	pNL1	NA	NA	*dxnA*	NA	Basta et al. (2004)

[a]γ-HCH indicates γ-hexachlorocyclohexane and PAHs indicate polycyclic aromatic hydrocarbons

[b]Rep type is classified on the basis of the amino acid sequence identity (>70%) of putative Rep genes of each sequenced plasmid. As for pBN6 and the plasmid of *Sphingomonas* sp. HH69, the classification is based on the Southern blot analysis of Basta et al. (2005)

[c]Classifications of MOB classes and MPF types are based on the report of Smillie et al. (2010). NA means not available

[d]Whole nucleotide sequences are available

[e]No catabolic genes have been reported in the plasmid so far

Table 8.3 Catabolic plasmids in other gram-negative bacteria

Plasmid	Host	Substrate	Size (kb)	MOB class[a]	T4SS type[a]	Genes	Transferability	References
pWW174	*Acinetobacter calcoaceticus* RJE174	Benzene	200	NA	NA	*cat*	+	Winstanley et al. (1987)
pYA1	*Acinetobacter* sp. YAA	Aniline	NA	NA	NA	*atdA*	NA	Fujii et al. (1997)
pCPE3	*Alcaligenes* sp. CPE3	Chlorobenzoates	16	NA	NA	*cbaABC*	+	Di Gioia et al. (1998)
pKA2	*Alcaligens paradoxus* 2811P	2,4-D[b]	43	NA	NA	*tfdA*	+	Ka and Tiedje (1994)
pCMS1	*Brevundimonas diminuta* MG	Organophospate	66	NA	NA	*opd*	+	Mulbry et al. (1987); Pandeeti et al. (2011)
pTOM	*Burkholderia cepacia* G4	Toluene	108	NA	NA	*tom*	+	Shields et al. (1995)
pNF1	*Burkholderia* sp. NF100	Methylhydroquinone	105	NA	NA	*mhq*	+	Hayatsu et al. (2000); Tago et al. (2005)
pOPH1	*Comamonas acidovorans* UCC61	Phtalate	70	NA	NA	*pht*	NA	Dutton et al. (1995)
pBS1010	*Comamonas testosteroni* BS1310	*p*-Toluenesulfonate	130	NA	NA	NA	NA	Top et al. (2000)

(continued)

Table 8.3 (continued)

Plasmid	Host	Substrate	Size (kb)	MOB class[a]	T4SS type[a]	Genes	Transferability	References
pMC1	*Delftia acidovorans* MC1	Dichlorprop[c]	NA	NA	NA	*rdpA* *sdpA*	NA	Schleinitz et al. (2004)
pBRX1	*Klebsiella ozaenae*	Bromoxynil	82	NA	NA	*bxn*	NA	Stalker and McBride (1987)
pPNAP01[d]	*Polaromonas naphthalenivorans* CJ2	Naphthalene	353	MOB_H, MOB_P	MPF_T	*bph*, *pht*[e]	NA	Jeon et al. (2003, 2006); Yagi et al. (2009)
pPNAP04[d]	*Polaromonas naphthalenivorans* CJ2	Naphthalene	144	–	–	*pht*[e]	NA	Jeon et al. (2003, 2006); Yagi et al. (2009)
pAC200	*Rhizobium* sp. AC100	Carbaryl[f]	25	BA	NA	*cehA*	NA	Hashimoto et al. (2002)

[a]Classifications of MOB classes and MPF types are based on the report of Smillie et al. (2010). NA means not available
[b]2,4-D represents 2,4-Dichlorophenoxyacetic acid
[c]Dichlorprop represents 2(2,4-dichlorophenoxy)propionate
[d]Whole nucleotide sequences are available
[e]Putative biphenyl- and phthalate-degradative genes were located on pPNAP01 and pPNAP04, although naphthalene degradative genes were not detected
[f]Carbaryl represents 1-naphtyl-N-methylcarbamate

Table 8.4 Catabolic plasmids in gram-positive bacteria

Plasmid	Host	Substrate[a]	Size (kb)	Linear/circular	MOB class[b]	T4SS type[b]	Genes	Transferabllity	References
pRE1	*Arthrobacter keyseri* 12B	Phthalate	130	NA	NA	NA	*pht*, *pcm*	NA	Eaton (2001)
pAO1[c]	*Arthrobacter nicotinovorans*	Nicotine	165	Circular	NA	NA	*ndh*	+	Baitsch et al. (2001); Igloi and Brandsh (2003)
Plasmid	*Arthrobacter nicotinovorans* HIM	Atrazine	96	NA	NA	NA	*atzABC*	NA	Aislabie et al. (2005)
pAL1[c]	*Arthrobacter nitroguajacolicus* Rü61a	2-Methylquinoline	113	linear	NA	NA	*qox*, *moq*,*hod*, *amq*	+	Overhage et al. (2005); Parschat et al. (2007)
Plasmid	*Arthrobacter* sp. DNS10	Atragine	NA	NA	NA	NA	NA	NA	Zhang et al. (2011b)
pRC1	*Arthrobacter* sp. RC100	Carbaryl, 1-naphthol	130	NA	NA	NA	NA	+	Hayatsu et al. (1999)
pRC2	*Arthrobacter* sp. RC100	Carbaryl, 1-naphthol	120	NA	NA	NA	NA	+	Hayatsu et al. (1999)
Plasmid	*Bacillus licheniformis*	Dimethoate	54	NA	NA	NA	NA	+	Mandel et al. (2005)
pPDL2	*Flavobacterium* sp. ATCC27551	Organophosphate	39	NA	NA	NA	*opd*	NA	Mulbry et al. (1987); Siddavattam et al. (2003)
pOAD2[c]	*Flavobacterium* sp. KI723T1	Nylon	46	Circular	NA	NA	*nylABC*	NA	Negoro et al. (1980); Negoro and Okada (1982); Kato et al. (1995)

(continued)

Table 8.4 (continued)

Plasmid	Host	Substrate[a]	Size (kb)	Linear/circular	MOB class[b]	T4SS type[b]	Genes	Transferablity	References
pLW1071[c]	*Geobacillus thermodentrificans* NG80-2	Long-chain alkane	58	Circular	MOB_Q	–	*ladA*	NA	Feng et al. (2007)
p174	*Gordonia polyiso prenivorans* VH2	Rubber	174	Circular	NA	NA	*lcp2*	NA	Hiessl et al. (2012)
pGKT2[c]	*Gordonia* sp. KTR9	Hexahydro-1,3,5-trinitro-1,3,5-trazine	182	Circular	NA	NA	*xplABglnA-xplB*, *xplA*, *xplR*	NA	Indest et al. (2010)
pKB1[c]	*Gordonia westfalica* Kb1	Poly (*cis*-1,4-isoprene)	101	Circular	MOB_F	–	*cad*	+	Bröker et al. (2004, 2008)
Small plasmid	*Gordonia* sp. CC-NAPH129-6	Naphthalene	97	NA	NA	NA	*nar*	NA	Lin et al. (2012)
Plasmid	*Nocardioides* sp. DF412	Dibenzofuran	NA	NA	NA	NA	*dfdA*	NA	Miyauchi et al. (2008)
pNC30	*Rhodococcus carallinus* B-276	Propene	185	Linear	NA	NA	*amoABC*	NA	Saeki et al. (1999)
pBD2[c]	*Rhodococcus erythropolis* BD2	Isopropylbenzene	210	Linear	NA	NA	*ipb*	+	Darbrock et al. (1994); Kesseler et al. (1996); Stecker et al. (2003)
pREL1[c]	*Rhodococcus erythropolis* PR4	Alkane	272	Linear	NA	NA	*alk*	NA	Sekine et al. (2006)
pREC1[c]	*Rhodococcus erythropolis* PR4	Alkane	104	Circular	MOB_F	–	β-oxydation enzymes	NA	Sekine et al. (2006)
pTSA421	*Rhodococcus erythropolis* TA421	Biphenyl/PCBs	560	Linear	NA	NA	*bph*	NA	Kosono et al. (1997)

pLP6	*Rhodococcus globerulus* P6	Biphenyl/PCBs	650	Linear	NA	NA	*bphC2*	NA	Kosono et al. (1997)
pRHL1[c]	*Rhodococcus jostii* RHA1	Biphenyl/PCBs, ethylbenzene, limonene, carveol	1100	Linear	NA	NA	*bph*, *etb*	NA	Masai et al. (1997); Shimizu et al. (2001)
pRHL2[c]	*Rhodococcus jostii* RHA1	Biphenyl/PCBs, ethylbenzene, limonene, carveol	450	Linear	NA	NA	*bph*, *etb*	+	Masai et al. (1997); Shimizu et al. (2001)
pRHL3[c]	*Rhodococcus jostii* RHA1	Biphenyl/PCBs, ethylbenzene, limonene, carveol	330	Linear	NA	NA	Limonene mono-oxygenase	NA	Warren et al. (2004)
p1CP	*Rhodococcus opacus* 1CP	Chloroaromatic compounds	740	Linear	NA	NA	*macA*, *clc*	NA	König et al. (2004)
pNUO1	*Rhodococcus opacus* M213	–	ca 750	Linear	NA	NA	*edoD*	NA	Uz et al. (2000)
Plasmid	*Rhodococcus rhodochrous* K37	PCBs	200	Linear	NA	NA	*bphC*	NA	Taguchi et al. (2004)
pRTL1	*Rhodococcus rhodochrous* NCIMB 13064	1-Chloroalkane	100	NA	NA	NA	*dhaA*, *adhA*, *aldA*	+	Kulakova et al. (1995, 1997)

(continued)

Table 8.4 (continued)

Plasmid	Host	Substrate[a]	Size (kb)	Linear/circular	MOB class[b]	T4SS type[b]	Genes	Transferabllity	References
Plasmid	*Rhodococcus* sp. I24	Naphthalene, toluene	50	NA	NA	NA	*nid*	+	Priefert et al. (2004)
Plasmid	*Rhodococcus* sp. I24	Naphthalene, toluene	340	NA	NA	NA	Toluene inducible dioxygenase	–	Priefert et al. (2004)
pDBF1	*Terrabacter* sp. DBF63	Dibenzofuran, fluorene	160	Linear	NA	NA	*dbf-fln*, *pht*, *pca*	NA	Nojiri et al. (2002); Habe et al. (2005)
pDBF2	*Terrabacter* sp. DBF63	Dibenzofuran, fluorene	190	Linear	NA	NA	*dbf-fln*, *pht*, *pca*	NA	Nojiri et al. (2002); Habe et al. (2005)
pYK3	*Terrabacter* sp. YK3	Dibenzofuran	NA	NA	NA	NA	*dfdA*	NA	Iida et al. (2002)

[a]Carbaryl represents 1-naphtyl-*N*-methylcarbamate. PCBs represents polychlorinated biphenyls
[b]Classifications of MOB classes and MPF types were based on the report of Smillie et al. (2010). NA means not available
[c]Whole nucleotide sequences are available

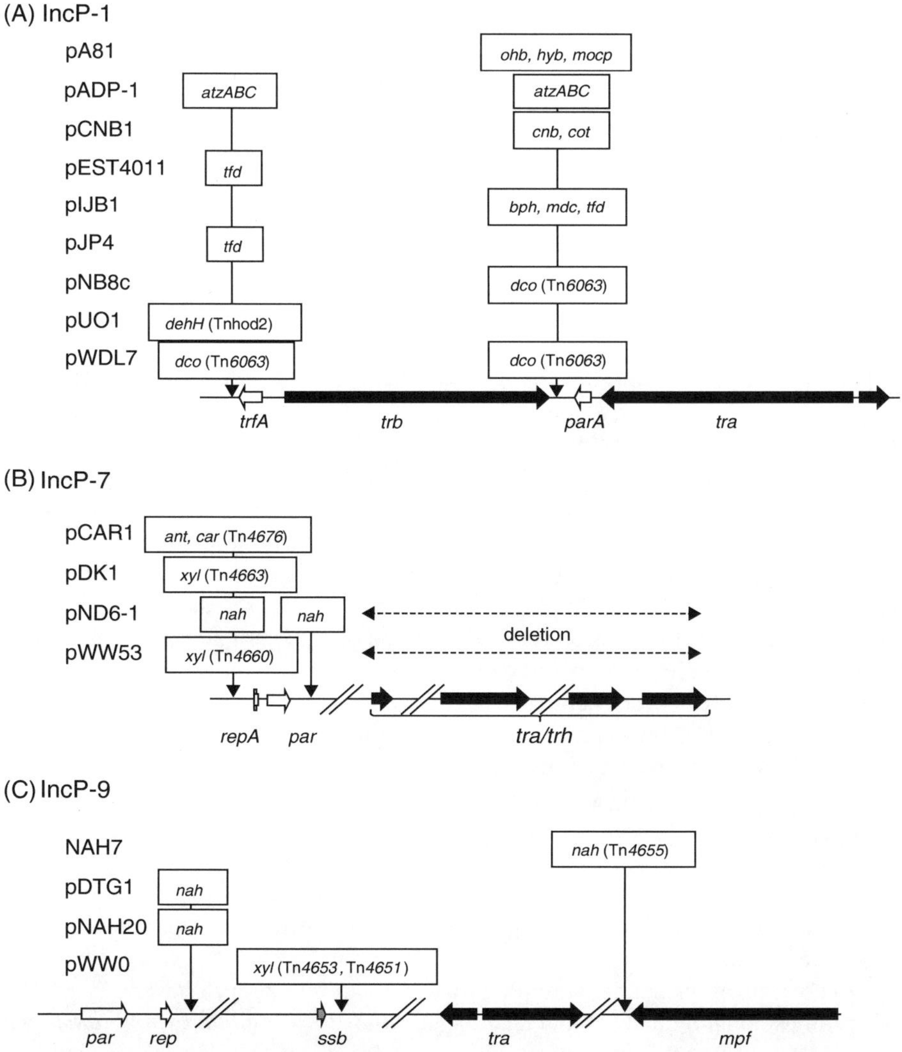

Fig. 8.1 Proposed IncP-1 (**a**), IncP-7 (**b**), and IncP-9 (**c**) backbones in conjunction with the insertion sites (*vertical arrows*) of catabolic genes on each plasmid (see Table 8.1). *Horizontal white arrows* indicate genes for replication and stability of the plasmid, and those in *black* indicate genes for conjugative transfer. The *ssb* gene of the IncP-9 backbone is shown in *gray*. The *tra*/*trh* genes of the IncP-7 backbone were deleted in pND6-1 and pWW53

On the other hand, many other catabolic plasmids have been isolated from *Pseudomonas*. However, the nucleotide sequences of replication or transfer regions for these plasmids are not available, and therefore, it is difficult to classify these plasmids. One exception is pCT14, which carries several genes for a *meta* cleavage pathway for aromatic rings, including *cbzTEXG*, *bphK*, and *tdnG* (Bramucci et al. 2006). Although the gene encoding its replication protein and the *oriV* region were proposed, there are no genes of similar sequence in the GenBank/EMBL/DDBJ database; this plasmid is predicted to be of the MOB_F class (Table 8.1).

2.2 Catabolic Plasmids of Sphingomonads

Over the past decade, many catabolic plasmids from xenobiotic-degrading sphingomonads (genera *Sphingomonas*, *Sphingobium*, *Novosphingobium*, and *Sphingopyxis*) belonging to the class *α-proteobacteria*, have been identified (Table 8.2). pNL1 was isolated from *Novosphingobium aromaticivorans* DSM 12444 (its previous name was *N. aromaticivorans* F199), and it is the first catabolic plasmid in sphingomonads whose 184-kb nucleotide sequence has been reported (Romine et al. 1999). Some xenobiotic-degrading sphingomonads carry multiple plasmids in one cell (Basta et al. 2004; Cérémonie et al. 2006; Tabata et al. 2011).The strain DSM 12444 also carries another plasmid of 487 kb, pNL2 (Fredrickson et al. 1991). Basta et al. (2005) compared plasmids from 16 sphingomonad strains that degrade various polycyclic aromatic hydrocarbons (PAHs). Based on Southern blot analyses, a plasmid of the naphthalenesulfonate-degrader *Sphingomonas xenophaga* BN6 and a plasmid of the dibenzofuran-degrader *Sphingomonas* sp. HH69 were shown to possess a pNL1-type Rep (replication initiation protein) gene (Basta et al. 2005). Nucleotide sequence comparisons revealed that similar Rep genes were also found in pCAR3, which also carries *car* genes, in the carbazole-degrader *Novosphingobium* sp. KA1 (its previous name was *Sphingomonas* sp. KA1, Shintani et al. 2007), and in pSWIT02, which also carries *dxn* genes, in the dibenzo-*p*-dioxin degrader *Sphingomonas wittichii* RW1 (Miller et al. 2010). The Rep type is classified based on the amino acid sequence identity (>70%) of putative Rep gene products of each sequenced plasmid.

Notably, many plasmids were identified in γ-hexachlorocyclohexane (γ-HCH)-degrading sphingomonads (Table 8.2, Nagata et al. 2007). *Sphingobium japonicum* UT26 is an archetypal γ-HCH-degrading bacterium, and its whole genome sequence has been determined (Nagata et al. 2010, 2011). This strain has three plasmids, and one of them is the 191-kb pCHQ1, which carries *linRDEB* (Nagata et al. 2007, 2010, 2011). No Inc groups have been suggested for plasmids from sphingomonads; however, several types of Rep genes are known to be conserved among these bacteria. Indeed, there are other plasmids in sphingomonads that contain genes which show high identities with the Rep gene of pCHQ1 (Table 8.2): pLA1, which was identified in a PAHs-degrader, *Novosphingobium pentaromativorans* US6-1, and carries *bph* and *xyl* genes involved in biphenyl and toluene/xylene degradation (Luo et al. 2012); pSLGP in a lignin-degrader, *Sphingobium* sp. SK-6 (Masai et al. 2012); and pSPHCH01, in a pentachlorophenol-degrader, *Sphingobium chlorophenolicum*

L-1 (Copley et al. 2012). The last two plasmids, however, do not carry catabolic genes. pLB1 also carries the *linB* gene, which was identified by performing an exogenous plasmid isolation technique from γ-HCH-contaminated soil using a *linB*-disrupted UT26 mutant. The original host of pLB1 was unidentifiable, but the plasmid can transfer to *Sphingobium japonicum* UT26 (Miyazaki et al. 2006). The Rep type of pLB1 is different from that of pCHQ1 because it shows compatibility to pCHQ1 (Miyazaki et al. 2006). Similarly, in addition to pLA1 (pCHQ1-type), *N. pentaromativorans* US6-1 harbors another plasmid, pLA2, which carries the pLB1-type Rep gene but has no catabolic genes. The conservation of the Rep genes suggests that many plasmids in sphingomonads may be self-transmissible, although this property has been experimentally proved to exist in only a few (Table 8.2).

Plasmids belonging to the same *Pseudomonas* incompatibility groups always have the same types of genes for conjugative transfer (Table 8.1). In contrast, plasmids in sphingomonads have different types of genes for conjugative transfer, whereas they have the same Rep genes, suggesting that they have "mosaic" genetic structures. While the Rep gene of pSWIT02 is pNL1-like, the genes for plasmid transfer show higher similarity to those of pCHQ1 than to those of pNL1. On the other hand, putative plasmid transfer genes of pLA1 are more similar to those of pNL1 than those of pCHQ1, while its Rep gene is more similar to that of pCHQ1 (Luo et al. 2012). In addition, several catabolic genes, such as *bph* on pNL1, *car* on pCAR1, or *lin* on pCHQ1 are not organized in a single operon but dispersed on the plasmid or host chromosome in sphingomonads (Romine et al. 1999; Shintani et al. 2007; Nagata et al. 2011). The varied distribution of similar genes and dispersed organization of genes indicate that catabolic plasmids in sphingomonads might have been transferred among the genus, and might have undergone DNA rearrangements with other plasmids and host chromosomes, resulting in the "mosaic" structure.

2.3 Catabolic Plasmids of Other Gram-Negative Bacteria

Catabolic plasmids have also been observed in other gram-negative bacteria belonging to classes *α-*, *β-*, and *γ-proteobacteria*, as listed in Table 8.3, although they have not been investigated in detail. The whole genome sequence of the naphthalene-degrading *Polaromonas naphthalenivorans* CJ2 has been determined (Jeon et al. 2003, 2006; Yagi et al. 2009). This strain possesses eight plasmids, and at least two of them, pPNAP01 and pPNAP04, carry putative aromatic hydrocarbon-degradative genes (Yagi et al. 2009). The partial sequence of pCMS1, the organophosphate degradative plasmid of *Brevundimonas diminuta* MG, revealed that its putative transfer genes showed 67–74% identity with those of the IncP-1 plasmid pEST4011 (Pandeeti et al. 2011). This fact implied an evolutionary relationship between pCMS1 and IncP-1 plasmids. Analysis of the nucleotide sequences and identification of open reading frames on these plasmids will be important for elucidating the steps in the evolution of these plasmids in gram-negative bacteria.

2.4 Catabolic Plasmids of Gram-Positive Bacteria

Several plasmids have been identified in xenobiotic-degrading gram-positive bacteria belonging to classes *Actinobacteria*, *Bacilli*, and *Flavobacteriia* (Table 8.4). Some of these bacteria carry circular plasmids and others harbor linear plasmids (Table 8.4). The linear plasmids belong to a class of genetic elements called invertrons, which carry terminal inverted repeats (TIRs) that are covalently bound to terminal proteins at both 5′ termini (Sakaguchi 1990). Linear plasmids have been proposed to have evolved from bacteriophages (Hinnebusch and Tilly 1993). The details of the mechanisms of plasmid transfer between gram-positive bacteria are still unclear (Grohmann et al. 2003).

Rhodococcus is one of the most important genera among gram-positive degraders of alkanes, PCBs, and naphthalene, and many plasmids have been identified in the *Rhodococcus* species (Table 8.4). pBD2 is a conjugative linear plasmid that carries *ipb* genes for the catabolism of isopropylbenzene, and it was detected in *R. erythropolis* BD2 (Dabrock et al. 1994; Stecker et al. 2003). pREL1 and pREC1 were identified in *R. erythropolis* PR4, an alkane-degrader (Sekine et al. 2006). Several DNA regions in pREL1 and pBD2 are conserved, including genes that encode for terminal protein, lipoproteins, and heavy metal resistance. However, the degradative genes for alkane (pREL1) and for isopropylbenzene (pBD2) are not conserved (Sekine et al. 2006).

R. jostii RHA1 can degrade polychlorinated biphenyls (PCBs) (Seto et al. 1995), and its complete genome sequence has been determined (McLeod et al. 2006). This strain harbors three linear plasmids, pRHL1, pRHL2, and pRHL3 (Shimizu et al. 2001; Masai et al. 1997), and most of the genes involved in the biphenyl degradative pathway are located on the two larger plasmids, pRHL1 and pRHL2 (Shimizu et al. 2001). Notably, many catabolic isozyme genes are distributed throughout the RHA1 genome (Kitagawa et al. 2001; Sakai et al. 2002; McLeod et al. 2006). The four replicons of RHA1, including the three plasmids and its linear chromosome, were suggested to be similar types of linear elements, because their TIRs are highly similar (McLeod et al. 2006).

Arthrobacter utilizes a wide and varied range of xenobiotic compounds and several catabolic plasmids have been identified in this genus (Table 8.4). pAL1 is a linear catabolic plasmid that was detected in the 2-methylquinoline-degrading *Arthrobacter nitorguajacolicus* Rü61a strain (Parschat et al. 2007; Overhage et al. 2005). The replication region of pAL1 was analyzed in detail, and it revealed that this plasmid carries a novel Rep gene (Kolkenbrock et al. 2010; Wagenknecht and Meinhardt 2011). Parschat et al. (2007) showed that several regions of pAL1 are conserved in pAL1 and the pBD2, pREL1, and pRHL2 plasmids mentioned above, and also in the dibenzofuran-degradative plasmid pDBF1 from *Terrabacter* sp. DBF63 (Nojiri et al. 2002; Habe et al. 2005). One of the regions includes putative genes for a secretion system possibly involved in conjugation (Parschat et al. 2007). Similarly, 2,3-dihydroxybiphenyl dioxygenase BphC genes are conserved on pLP6 and pTSA421 found in *R. globerulus* P6 and *R. erythropolis* TA421 (Kosono et al. 1997).

Other types of catabolic plasmids have also been reported (Table 8.4). pLW1071 is a circular plasmid from *Geobacillus thermodentrificans* NG80-2 that carries degradative genes for long-chain alkanes (Feng et al. 2007). This plasmid is unique in comparison to other sequenced plasmids, except for a plasmid from *Geobacillus* sp., G11MC16 (accession no. NZ_ABVH01000017). The putative Rep gene of the plasmid of G11MC16 was similar to that of NG80-2. pGKT2 is a 182-kb circular plasmid carrying *xplAB* genes found in the hexahydro-1,3,5-trinitro-1,3,5-triazine degrader *Gordonia* sp. KTR9 (Indest et al. 2010). *Gordonia* spp. are a metabolically diverse group, with regards to their ability to degrade xenobiotic compounds, and recently, two other catabolic plasmids have been reported in this genus (Table 8.4). Catabolic genes in gram-positive bacteria may also be spread by self-transmissible plasmids (listed in Table 8.4), similar to that observed in the case of gram-negative plasmids, and have an important role in their HGT, although their host range remains unclear.

3 Catabolic Transposons

In some cases, catabolic genes are flanked by two copies of the same or highly-identical insertion sequences (ISs). These elements are known as composite transposons. Tn*5280* (van der Meer et al. 1991a), Tn*HadI* (Kawasaki et al. 1985); Sota et al. 2002), and *DEH* (Weightman et al. 2002) are composite transposons whose transposition ability has been experimentally validated (Table 8.5). As for Tn-*DhaI*, it encodes *pcrABCT* which is involved in reductive dechlorination of tetrachloroethene in *Desulfitobacterium hafniense* TCE1, and detection of the circular form of the transposon strongly indicated that it could transpose (Maillard et al. 2005). As genome sequences of an increasing number of xenobiotic-degrading bacteria are determined, many composite transposon-like genetic structures are being discovered (Table 8.5). Homologous recombination events among several copies of the identical ISs located on regions surrounding catabolic genes possibly increase the plasticity of the genome. There are two kinds of ISs, IS*6100* and IS*1071*, which were frequently associated with various catabolic genes. IS*6100* was originally isolated as part of the composite transposon Tn*6100* from *Mycobacterium fortuitum* (Martin et al. 1990), and was found in a wide range of host bacteria, such as *Sphingomonas* (Dogra et al. 2004), *Arthrobacter* (Kato et al. 1994), *Pseudomonas* (Hall et al. 1994), *Xanthomonas* (Sundin and Bender 1995), *Salmonella* (Boyd et al. 2000), and *Corynebacterium* (Tauch et al. 2002). The IS elements were also found in many kinds of xenobiotic-degrading bacteria, and some of them form composite transposon-like structures (Table 8.5). IS*6100* was found in many γ-HCH-degrading sphingomonads in the region flanking the *lin* genes involved in γ-HCH-degradation, suggesting that this IS may have played a key role in the recruitment of the *lin* genes in these bacteria (Nagata et al. 2011).

IS*1071* was originally identified in a chlorobenzoate-catabolic transposon, Tn*5271*, from *Comamonas testosteroni* BR60 (Nakatsu et al. 1991). IS*1071* belongs

Table 8.5 Catabolic transposons

Elements (plasmid)	Substrates[a]	Host	Size (kb)	Transpos-ability[b]	IS	Gene	References
Class I composite transposons							
Tn*5542*[c] (pHMT112)	Benzene	*Pseudomonas putida* ML2	12	NA	IS*1489*	*bed*	Fong et al. (2000)
Tn*5280* (pP51)	Chlorobenzene	*Pseudomonas* sp. P51	9	+	IS*1066*, IS*1067*	*tcbAaAbAcAdB*	van der Meer et al. (1991a)
Tn*5707* (pENH91)	3-Chlorobenzoate	*Alcaligenes eutrophus* NH9	15	NA	IS*1600*	*cbnRABCD*	Ogawa and Miyashita (1999)
Tn*Ppu-alk1*	Pentane	*Pseudomonas putida* P1	22	NA	IS*Ppu4*	*alkST*, *alkBFGHJKLN*	van Beilen et al. (2001)
Tn*5271* (pBRC60)	Chlorobenzoates	*Comamonas testosteroni* BR60	17	NA	IS*1071*	*cbaABC*	Nakatsu et al. (1991)
Tn*5271*-like	Chlorobenzoates	*Alcaligenes* sp. CPE3	16	NA	IS*1071*	*cbaABC*	Di Gioia et al. (1998)
Tn*Had1* (pUO1)	Haloacetate	*Delftia acidovorans* B	9	+	IS*1071*	*dehH1*	Sota et al. (2002)
DEH[c]	α-Halocarboxylic acids	*Pseudomonas putida* PP3	10	+	IS*Ppu12*	*dehI*, *dehR*	Weightman et al. (2002)
–[c] (pWW0)	Xylene, Toluene	*Pseudomonas putida* mt-2	40	NA	IS*1246*	*xyl*	Tsuda and Iino (1987); Greated et al. (2002)
- (pTDN1)	Aniline	*Pseudomonas putida* UCC22	26	NA	IS*1071*	*tdnQA1A2B*	Saint et al. (1990); Fukumori and Saint (1997, 2001)
- (pTSA)	*p*-Toluenesulphonate	*Comamonas testosteroni* T2	21	NA	IS*1071*	*tsaMBCD*	Junker and Cook (1997); Tralau et al. (2001)
–[c] (pADP-1)	Atrazine	*Pseudomonas* sp. ADP	13	NA	IS*1071*	*atzA*	Martinez et al. (2001)

–[c] (pADP-1)	Atrazine	*Pseudomonas* sp. ADP	15	NA	IS*1071*	*atzB*	Martinez et al. (2001)
- (pPOB)	4-Carboxydiphenyl ether	*Pseudomonas pseudoalcaligenes* POB310	NA	NA	IS*1071*	*pobAB*	Dehmel et al. (1995)
Tn-*Dha1*[c]	Tetrachloroehene	*Desulfitobacterium hafniense* TCE1	10	+[d]	IS*Dha1*	*pcrABCT*	Maillard et al. (2005)
Tn*6063*[c] (pWDL2)	3-Chloroaniline	*Comamonas testosteroni* WDL2	22	NA	IS*1071*	*dcaQTA1A2BR*	Król et al. (2012)
Tn*6063*[c] (pNB8c)	3-Chloroaniline	*Delftia acidovorans* B8c	22	NA	IS*1071*	*dcaQTA1A2BR*	Król et al. (2012)
Tn*CNB1*	4-Chloronitrobenzene	*Comamonas* sp. CNB-1	45	NA	IS*1071*	*cnb*	Ma et al. (2007)
–[c]	Aniline	*Delftia tsuruhatensis* AD9	~25	NA	IS*1071*	*tad*	Liang et al. (2005)
–[c] (pJP4)	2,4-D	*Ralstonia eutropha* JMP134	~44	NA	IS*1071*	*tfd-I, tfd-II*	Trefault et al. (2004)
–[c] (pJP4)	2,4-D	*Ralstonia eutropha* JMP134	~10	NA	IS*JP4*	*tfd-II*	Trefault et al. (2004)
–[c] (pEST4011)	2,4-D	*Achromobacter xylosoxidans* subsp. *denitrificans* EST4002	48	NA	IS*1071*:: IS*1471*	*tfd*	Vedler et al. (2004)
Tn*AxI* (pA81)	Chlorobenzoate	*Achromobacter xylosocidans* A8	39	NA	IS*Ax1a*, IS*Ax1b*	*mocpRABCD*, *hybRABCD*	Jencova et al. (2008)
–[c] (pSAH)	2-Aminobenzenesulfonate	*Alcaligenes* sp. O-1	12	NA	IS*1240*-like	*abs*	Ruff et al. (2010)
–[c]	Carbazole	*Sphingobium yanoikuyae* XLDN2-5	8	NA	IS*6100*	*carRAaBaBbCAc*	Gai et al. (2010)

(continued)

Table 8.5 (continued)

Elements (plasmid)	Substrates[a]	Host	Size (kb)	Transpos-ability[b]	IS	Gene	References
–[c]	Carbazole	*Sphingobium yanoikuyae* XLDN2-5	4	NA	IS*6100*	*antRACAdAbAa*	Gai et al. (2010)
–[c]	Carbazole	*Sphingobium yanoikuyae* XLDN2-5	7	NA	IS*6100*	*fdr*	Gai et al. (2010)
–[c]	2-Chloronitrobenzene	*Pseudomonas stutzeri* ZWLR2-1	9	NA	IS*6100*	*cnbCEFAbAaD*	Liu et al. (2011)
–[c]	2-Chloronitrobenzene	*Pseudomonas stutzeri* ZWLR2-1	5	NA	IS*6100*	*cnbAcAd*	Liu et al. (2011)
–[c]	2-Chloronitrobenzene	*Pseudomonas stutzeri* ZWLR2-1	12	NA	IS*6100*	*cnbCEFAbAaD*, *cnbAcAd*	Liu et al. (2011)
Tn*mph*[c] (pZWL0)	Methyl parathion	*Pseudomonas* sp. WBC-3	4	+	IS*6100*	*mph*	Wei et al. (2009)
Tn*opdA*	Organophosphate	*Agrobacterium radiobacter* P230	6	+	IS*6100*	*opdA*	Horne et al. (2003)
–[c] (pLB1)	γ-HCH	Unidentified soil bacterium	4	NA	IS*6100*	*linB*	Miyazaki et al. (2006)
–[c] (pOAD2)	Nylon oligomers	*Flavobacterium* sp. K172	15	NA	IS*6100*	*nylABC*	Kato et al. (1994, 1995)
–[c] (pCAR1)	Carbazole	*Pseudomonas resinovorans* CA10	6	NA	IS*Pre1* IS*Pre2*	*antABC*	Nojiri et al. (2001); Maeda et al. (2003); Takahashi et al. (2009)
–[c] (pCAR1)	Carbazole	*Pseudomonas resinovorans* CA10	16	NA	IS*Pre1*	*carABCD*	Nojiri et al. (2001); Maeda et al. (2003); Takahashi et al. (2009)
–	Carbazole	*Pseudomonas stutzeri* OM1	55	NA	IS*Pst3*	*carABCDEF*, *antABC*	Shintani et al. (2003)

–[c] (pCAR3)	Carbazole	*Novosphingobium* sp. KA1	82	NA	ISSsp1	*car*, *and*, *cat*	Shintani et al. (2007)
–	Monobromoacetate	*X. autotrophicus* GJ10	NA	+	IS*1247*	*dhlB*	van der Ploneg et al. (1995)
Class II transposon							
Tn*4651*[c] (pWW0)	Xylene, tolene	*Pseudomonas putida* mt-2	56	+	–	*xyl*	Tsuda and Iino (1987); Tsuda et al. (1989)
Tn*4653*[c] (pWW0)	Xylene, toluene	*Pseudomonas putida* mt-2	70	+	–	*xyl*	Tsuda and Iino (1988); Tsuda et al. (1989)
Tn*4656*[c] (pWW53)	Xylene, toluene	*Pseudomonas putida* MT53	37	+	–	*xyl*	Tsuda and Genka (2001)
Tn*4657*[c] (pWW53)	Xylene, toluene	*Pseudomonas putida* MT53	86	+	–	*xyl*	Yano et al. (2007)
Tn*4660*[c] (pWW53)	Xylene, toluene	*Pseudomonas putida* MT53	62	–	–	*xyl*	Yano et al. (2007)
Tn*4663*[c] (pDK1)	Xylene, toluene	*Pseudomonas putida* HS1	41	+	–	*xyl*	Yano et al. (2010)
Tn*4655*[c] (NAH7)	Naphthalene	*Pseudomonas putida* G7	37	–[e]	–	*xyl*	Tsuda and Iino (1990); Sota et al. (2006)
Tn*Had2*[c] (pUO1)	Haloacetate	*Delftia acidovorans* B	16	+	–	*dehH1*, *dehH2*	Sota et al. (2002)
Tn*4676*[c] (pCAR1)	Carbazole	*Pseudomonas resinovorans* CA10	73	+	–	*carABCDEF*,	Maeda et al. (2003); Shintani et al. (2005, 2011)
–[c] (pKW1)	Chlorobenzene	*Pseudomonas putida* GJ31	15	NA	–	*cbz*	Kunze et al. (2009)

[a]2,4-D represents 2,4-Dichlorophenoxyacetic acid
[b]NA means not available
[c]Whole nucleotide sequences are available
[d]Strong indications for the transposition activity of IS*Dha1* were observed by PCR amplification and sequencing of the intervening sequence located between both IRs of IS*Dha1* (Maillard et al. 2005)
[e]Tn*4655* did not carry the *tnpA* gene but was able to form a cointegrate when the *tnpA* gene from Tn*4653* was supplied in *trans* (Tsuda and Iino 1990; Sota et al. 2006)

to the class II transposons, which generally carry the genes for their transposition (*tnpA*, *tnpR*, and *res*) and one or more phenotypic traits between their terminal inverted repeats (Grindley 2002). This type of transposon generates a cointegrate of donor and target molecules, and the cointegrate is then resolved at the resolution (*res*) sites by TnpR (resolvase). This resolution function, however, is lacking in IS*1071*. The copy number of class II transposons doubles after their transposition by means of a mechanism known as "copy and paste" transposition (Grindley 2002). Many IS*1071* sequences have been identified in close proximity to various xenobiotic-degradative genes on self-transmissible plasmids from environmental bacteria (Table 8.5). These data indicate that IS*1071* might have been involved in the recruitment of catabolic genes to these plasmids and in the dissemination of these genes among various host strains.

It should be noted that some class II transposons (Grindley 2002) that carry catabolic genes are found in various xenobiotic-degrading bacteria (Table 8.5). In addition to the extensively characterized Tn*4651*/Tn*4653* in the toluene/xylene-degradative plasmid pWW0 (IncP-9) (Tsuda and Iino 1987, 1988; Tsuda et al. 1989), these types of transposons are found in two other toluene/xylene-degradative plasmids, namely pWW53 (IncP-7) and pDK1 (IncP-7), the carbazole degradative plasmid pCAR1 (IncP-7), and the naphthalene degradative plasmid NAH7 (IncP-9). Notably, the transposition function of most of these transposons has been experimentally verified (Table 8.5, Yano et al. 2007, 2010; Shintani et al. 2005, 2011). Although Tn*4655* in NAH7 lacks the *tnpA* gene (Sota et al. 2006), it is able to form a cointegrate when the *tnpA* gene of Tn*4653* is supplied *in trans* (Tsuda and Iino 1990; Sota et al. 2006). These class II transposons might have been efficiently spread among bacterial replicons via their "copy and paste" transposition, and they can carry longer DNA regions than class I composite transposons can.

4 Catabolic ICEs

ICEs are self-transmissible MGEs that are integrated in the chromosome. These elements carry genes for conjugative transfer and also excision systems to excise from the chromosome (Burrus and Waldor 2004; Wozniak and Waldor 2010). They are replicated as a part of the chromosome, they excise from the chromosome, circularize and then transfer to new hosts, sometimes leading to the integration into these new host chromosomes (Burrus and Waldor 2004; Wozniak and Waldor 2010). ICEs are difficult to identify experimentally, because they are usually physically linked to the host chromosome (Wozniak and Waldor 2010). ICE_{clc} (Ravatn et al. 1998a), *bph-sal* element (Nishi et al. 2000), and ICE_{KKS}*4677* (Ohtsubo et al. 2003, 2006, 2012) are the ICEs that have been verified experimentally (Table 8.6). Among these, the most in-depth analyses, such as on the mechanisms for excision, transfer, and impact on the host cell, have been performed for ICE_{clc} (Ravatn et al. 1998a, b; Gaillard et al. 2006, 2008, 2010; Sentchilo et al. 2009; Miyazaki and van der Meer 2011a, b).

Table 8.6 Catabolic ICEs

ICE	Host	ICE family[a]	Substrate	Size (kb)	Transferability[b]	Gene	References
ICE_{KKS102}*4677*[c]	*Acirovorax* sp. KKS102	ICE_{Tn4371}	Biphenyl	62	+	*bph*	Ohtsubo et al. (2003, 2006, 2012)
ICE-GI1[c]	*Bordetella petrii* DSM 12804	ICE_{clc}	–	255	NA	Putative monooxygenase	Lechner et al. (2009)
ICE-GI2[c]	*Bordetella petrii* DSMZ 12804	ICE_{clc}	–	143	NA	*ben*, *cat*	Lechner et al. (2009)
ICE-GI3[c]	*Bordetella petrii* DSM 12804	ICE_{clc}	–	102	+	*cat*	Lechner et al. (2009)
ICE*Bxe*LB400-1[c]	*Burkholderia xenovorans* LB400	ICE_{clc}	Biphenyl	123	NA	*clc*	Cain et al. (2006); Gaillard et al. (2006)
phn-island[c]	*Delftia* sp. Cs1-4	Unclassified	Phenanthrene	232	NA	*phn*, *oph*	Hickey et al. (2012)
bph-sal element	*P. putida* KF715	Unclassified	Biphenyl/salicylate	90	+	*bph*, *nah*	Nishi et al. (2000)
ICE_{Tn4371}*6065*[c]	*Palaromonas naphthalenivorans* CJ2 (pPNAP01)	ICE_{Tn4371}	Naphthalene	70	NA	*bph*, *nah*	Ryan et al. (2009)
ICE_{clc}(B13)[c]	*Pseudomonas knackmussii* B13	ICE_{clc}	Chlorocatechol	105	+	*clc*	Ravatn et al. (1998a)
ICE_{clc}(JS705)	*Ralstonia eutropha* JS705	ICE_{clc}	Chlorocatechol	115	NA	*clc*, *mcb*	Müller et al. (2003)
ICE_{Tn4371}[c]	*Ralstonia oxalatica* A5	ICE_{Tn4371}	Biphenyl	55	NA	*bph*	Springael et al. (1993); Merlin et al. (1999); Toussaint et al. (2003)

[a]Classification based on ICEberg (http://db-mml.sjtu.edu.cn/ICEberg/) and Bi et al. (2012)
[b]NA means not available
[c]Whole nucleotide sequences are available

Recently, *in silico* analyses of complete bacterial genomes have identified putative ICEs in several *β*- and *γ-proteobacteria*. Indeed, such analyses of many complete bacterial genomes showed that ICEs are spread among various bacterial subdivisions, and more than 400 putative ICEs are listed in ICEberg (http://db-mml.sjtu.edu.cn/ICEberg/) (Bi et al. 2012). Ryan et al. (2009) reported that an ICE_{Tn4731}-related ICE was found in several bacterial genome sequences, and one of them, ICE_{Tn4371}*6065*, carrying the *bph* gene, was found in a naphthalene degrader, *Polaromonas naphthalenivorans* CJ2. Interestingly, *Bordetella petrii* DSM 12804 possesses at least seven large ICEs mostly encoding metabolic functions involved in the degradation of aromatic compounds and detoxification of heavy metals (Lechner et al. 2009). Four of them, ICE-GI1, ICE-GI2, ICE-GI3, and ICE-GI6, are closely related to ICE_{clc}, and the first three carry putative catabolic genes (Table 8.6). It should be noted that their circular intermediates have been detected, and that transmissibility of ICE-GI3 has been confirmed (Lechner et al. 2009). Hichey et al. found a new ICE in the genome of the PAHs-degrader, *Delftia* sp. Ds1-4, which carries all of the required phenanthrene catabolic genes (Hickey et al. 2012). Because ICEs are not necessarily replicated as circular forms after their integration into the host chromosome, host ranges of ICEs are not dictated by whether the ICEs can be replicated in the host cells. Therefore, their host ranges are likely to be wider than that of other MGEs.

5 Behaviors of Catabolic MGEs

Bioaugmentation by inoculation of highly efficient xenobiotic degraders into polluted sites has been studied as an attractive approach to remove pollutants. However, it is difficult to maintain the high levels of degradative ability of these inoculants, because they are not necessarily able to compete or survive in natural environments (Top et al. 2002). The catabolic MGEs, especially conjugative elements, can be used in alternative bioaugmentation by utilizing the transferability of MGEs into the indigenous bacteria in the polluted sites. In bioaugmentation via inoculation with degraders harboring MGEs, known as "gene bioaugmentation" or "plasmid-mediated bioaugmentation," the survival of the inoculated degraders is not needed (Bathe 2004; Bathe et al. 2005; Dejonghe et al. 2000; Pepper et al. 2002). There are still, however, large gaps between laboratory conditions and natural systems, and the basic features of MGEs in laboratory conditions do not necessarily reflect their actual behavior in natural systems. Many trials have been conducted to bridge the differences between these conditions by using artificial model environments, which model natural habitats such as soil, plants, and water. While the behaviors of the IncP-1, P-7, and P-9 group plasmids have been summarized recently (Shintani et al. 2010), those of other plasmids, which belong to unknown Inc groups, have been also reported. Detailed analyses have been performed to analyze the effect of conjugative transfer of two kinds of 2,4-D degradative plasmids in soil by using pEMT1 and IncP-1 plasmid pEMT3 in different donors (Top et al. 1995; Dejonghe et al. 2000; Goris et al. 2002). Top et al. (2002) concluded that these catabolic plasmids were most often transferred to, and their genes expressed in, strains that belong to the genera *Burkholderia*, *Ralstonia*, and *Pseudomonas*. Transfer of the plasmid pTOM carrying constitutively transcribed toluene-degradative genes (*tom*) was shown from *Burkholderia cepacia* to different endogenous endophytic bacteria in yellow lupine (Barac et al. 2004) or poplar cuttings (Taghavi et al. 2005). Springael et al. reported that ICE_{clc} (B13) of *P. putida* BN210 was transferred to different bacteria belonging to the class of *β-proteobacteria* in biofilm reactors under non-sterile conditions (Springael et al. 2002).

These studies, together with those of IncP-1, P-7 and P-9 plasmids, strongly indicate that HGT by means of catabolic MGEs generally occurs in natural environments. Nevertheless, it is still difficult to predict how the catabolic plasmids or their hosts behave in these environments. A more in-depth understanding of HGT of MGEs will be required for practical application of plasmid-mediated bioaugmentation. Behaviors of the MGEs should be analyzed in microbial communities that include uncultivated and non-cultivable bacteria in natural environments. Several cultivation-independent methods to monitor the behavior of environmental bacteria have been reported. Metagenomic analysis combined with reverse-transcriptase real-time PCR analysis revealed the changes in the bacterial community and in abundant functional genes in contaminated environments (Yergeau et al. 2012). Ishii et al. (2011) identified the active N_2O reducers in rice paddy soil using stable isotope probing and functional single-cell isolation by micromanipulation. In another study, fluorescence-activated cell sorting (FACS) and micromanipulation enabled the identification and cultivation of independent plasmid transconjugants (Musovic et al. 2006, 2010). The combinations of these cultivation-independent and cultivation-dependent methods will shed light on HGT in microbial communities in various natural environments.

6 Conclusion and Perspectives

As an increasing number of whole genome sequences of bacteria capable of degrading various kinds of xenobiotic compounds are analyzed, a large number of catabolic MGEs have been discovered and studied recently. *In silico* analyses of the genome sequences of these bacteria enable us to detect new ISs and ICEs; however, experimental confirmation of their ability to mobilize is still required to further our understanding of how they are transmitted among bacteria or replicons. On the other hand, nucleotide sequence information on other Inc group plasmids from *Pseudomonas*, such as IncP-2 or other plasmids not affiliated to any Inc group (Table 8.1), is also required for further classification of the newly-identified plasmids.

Jones and Marchesi (2007) developed a method for transposon-aided capture of plasmids to discover novel plasmids in various bacterial habitats. This method allowed them to identify plasmids that did not rely on the plasmids' own replication and transfer systems. Indeed, many novel MGEs have been identified in various sites by the method mentioned above and by metagenomic analyses, such as in activated sludge (Zhang et al. 2011a), river or sea sediments (Elsaied et al. 2011; Kristiansson et al. 2011), wastewater treatment plants (Szczepanowski et al. 2008), human dental plaque (Warburton et al. 2011), and human gut (Jones et al. 2010). These reports suggest that a huge number of unidentified MGEs exist in the environment. Detection and analyses of new catabolic MGEs will help us to understand the mechanism by which MGEs spread and also determine which MGEs are capable of spreading in natural bacterial communities, including those that contain uncultivated and non-cultivable bacteria. These MGEs can possibly be used as new tools for genetic analysis of unidentified bacteria.

Acknowledgement The writing of this book chapter was supported by the Special Postdoctoral Researcher Program of Riken and by JSPS KAKENHI Grant Number 24780087 to M.S.

References

Aislabie J, Bej AK, Ryburn J, Lloyd N, Wilkins A (2005) Characterization of *Arthrobacter nicotinovorans* HIM, an atrazine-degrading bacterium, from agricultural soil New Zealand. FEMS Microbiol Ecol 52:279–286

Aminov RI (2011) Horizontal gene exchange in environmental microbiota. Front Microbiol 2:158

Andreoni V, Bestetti G (1986) Comparative analysis of different *Pseudomonas* strains that degrade cinnamic acid. Appl Environ Microbiol 52:930–934

Anson JG, Mackinnon G (1984) Novel pseudomonas plasmid involved in aniline degradation. Appl Environ Microbiol 48:868–869

Baitsch D, Sandu C, Brandsch R, Igloi GL (2001) Gene cluster on pAO1 of *Arthrobacter nicotinovorans* involved in degradation of the plant alkaloid nicotine: cloning, purification, and characterization of 2,6-dihydroxypyridine 3-hydroxylase. J Bacteriol 183:5262–5267

Barac T, Taghavi S, Borremans B, Provoost A, Oeyen L, Colpaert JV, Vangronsveld J, van der Lelie D (2004) Engineered endophytic bacteria improve phytoremediation of water-soluble, volatile, organic pollutants. Nat Biotechnol 22:583–588

Bartilson M, Nordlund I, Shingler V (1990) Location and organization of the dimethylphenol catabolic genes of *Pseudomonas* CF600. Mol Gen Genet 220:294–300

Basta T, Keck A, Klein J, Stolz A (2004) Detection and characterization of conjugative degradative plasmids in xenobiotic-degrading *Sphingomonas* strains. J Bacteriol 186:3862–3872

Basta T, Buerger S, Stolz A (2005) Structural and replicative diversity of large plasmids from sphingomonads that degrade polycyclic aromatic compounds and xenobiotics. Microbiology 151:2025–2037

Bathe S (2004) Conjugal transfer of plasmid pNB2 to activated sludge bacteria leads to 3-chloroaniline degradation in enrichment cultures. Lett Appl Microbiol 38:527–531

Bathe S, Schwarzenbeck N, Hausner M (2005) Plasmid-mediated bioaugmentation of activated sludge bacteria in a sequencing batch moving bed reactor using pNB2. Lett Appl Microbiol 41:242–247

Beil S, Timmis KN, Pieper DH (1999) Genetic and biochemical analyses of the *tec* operon suggest a route for evolution of chlorobenzene degradation genes. J Bacteriol 181:341–346

Bi D, Xu Z, Harrison EM, Tai C, Wei Y, He X, Jia S, Deng Z, Rajakumar K, Ou HY (2012) ICEberg: a web-based resource for integrative and conjugative elements found in Bacteria. Nucleic Acids Res 40:D621–D626

Boon N, Goris J, De Vos P, Verstraete W, Top EM (2000) Bioaugmentation of activated sludge by an indigenous 3-chloroaniline-degrading *Comamonas testosteroni* strain, I2*gfp*. Appl Environ Microbiol 66:2906–2913

Boon N, Goris J, De Vos P, Verstraete W, Top EM (2001) Genetic diversity among 3-chloroaniline- and aniline-degrading strains of the *Comamonadaceae*. Appl Environ Microbiol 67:1107–1115

Boyd DA, Peters GA, Ng L, Mulvey MR (2000) Partial characterization of a genomic island associated with the multidrug resistance region of *Salmonella enterica Typhymurium* DT104. FEMS Microbiol Lett 189:285–291

Bramucci M, Chen M, Nagarajan V (2006) Genetic organization of a plasmid from an industrial wastewater bioreactor. Appl Microbiol Biotechnol 71:67–74

Bröker D, Arensköter M, Legatzki A, Nies DH, Steinbüchel A (2004) Characterization of the 101-kilobase-pair megaplasmid pKB1, isolated from the rubber-degrading bacterium *Gordonia westfalica* Kb1. J Bacteriol 186:212–225

Bröker D, Arensköter M, Steinbüchel A (2008) Transfer of megaplasmid pKB1 from the rubber-degrading bacterium *Gordonia westfalica* strain Kb1 to related bacteria and its modification. Appl Microbiol Biotechnol 77:1317–1327

Burlage RS, Bemis LA, Layton AC, Sayler GS, Larimer F (1990) Comparative genetic organization of incompatibility group P degradative plasmids. J Bacteriol 172:6818–6825

Burrus V, Waldor MK (2004) Shaping bacterial genomes with integrative and conjugative elements. Res Microbiol 155:376–386

Carattoli A (2009) Resistance plasmid families in *Enterobacteriaceae*. Antimicrob Agents Chemother 53:2227–2238

Cérémonie H, Boubakri H, Mavingui P, Simonet P, Vogel TM (2006) Plasmid-encoded gamma-hexachlorocyclohexane degradation genes and insertion sequences in *Sphingobium francense* (ex-*Sphingomonas paucimobilis* Sp+). FEMS Microbiol Lett 257:243–252

Chain PS, Denef VJ, Konstantinidis KT, Vergez LM, Agulló L, Reyes VL, Hauser L, Córdova M, Gómez L, González M, Land M, Lao V, Larimer F, LiPuma JJ, Mahenthiralingam E, Malfatti SA, Marx CJ, Parnell JJ, Ramette A, Richardson P, Seeger M, Smith D, Spilker T, Sul WJ, Tsoi TV, Ulrich LE, Zhulin IB, Tiedje JM (2006) *Burkholderia xenovorans* LB400 harbors a multi-replicon, 9.73-Mbp genome shaped for versatility. Proc Natl Acad Sci USA 103:15280–15287

Chakrabarty AM (1972) Genetic basis of the biodegradation of salicylate in *Pseudomonas*. J Bacteriol 112:815–823

Chakrabarty AM (1973) Genetic fusion of incompatible plasmids in *Pseudomonas*. Proc Natl Acad Sci USA 70:1641–1644

Chatterjee DK, Chakrabarty AM (1982) Genetic rearrangements in plasmids specifying total degradation of chlorinated benzoic acids. Mol Gen Genet 188:279–285

Chatterjee DK, Chakrabarty AM (1984) Restriction mapping of a chlorobenzoate degradative plasmid and molecular cloning of the degradative genes. Gene 27:173–181

Chatterjee DK, Kellogg ST, Hamada S, Chakrabarty AM (1981) Plasmid specifying total degradation of 3-chlorobenzoate by a modified *ortho* pathway. J Bacteriol 146:639–646

Cho JC, Kim SJ (2001) Detection of mega plasmid from polycyclic aromatic hydrocarbon-degrading *Sphingomonas* sp. strain KS14. J Mol Microbiol Biotechnol 3:503–506

Connors MA, Barnsley EA (1980) Metabolism of naphthalene by pseudomonads: salicylaldehyde as the first possible inducer in the metabolic pathway. J Bacteriol 141:1052–1054

Copley SD, Rokicki J, Turner P, Daligault H, Nolan M, Land M (2012) The whole genome sequence of *Sphingobium chlorophenolicum* L-1: insights into the evolution of the pentachlorophenol degradation pathway. Genome Biol Evol 4:184–198

Dabrock B, Kesseler M, Averhoff B, Gottschalk G (1994) Identification and characterization of a transmissible linear plasmid from *Rhodococcus erythropolis* BD2 that encodes isopropylbenzene and trichloroethene catabolism. Appl Environ Microbiol 60:853–860

de Souza ML, Wackett LP, Sadowsky MJ (1998) The *atzABC* genes encoding atrazine catabolism are located on a self-transmissible plasmid in *Pseudomonas* sp. strain ADP. Appl Environ Microbiol 64:2323–2326

Dehmel U, Engesser KH, Timmis KN, Dwyer DF (1995) Cloning, nucleotide sequence, and expression of the gene encoding a novel dioxygenase involved in metabolism of carboxydiphenyl ethers in *Pseudomonas pseudoalcaligenes* POB310. Arch Microbiol 163:35–41

Dejonghe W, Goris J, El Fantroussi S, Höfte M, De Vos P, Verstraete W, Top E (2000) Effect of dissemination of 2,4-dichlorophenoxyacetic acid (2,4-D) degradation plasmids on 2,4-D degradation and on bacterial community structure in two different soil horizons. Appl Environ Microbiol 66:3297–3304

Dejonghe W, Goris J, Dierickx A, De Dobbeleer V, Crul K, De Vos P, Verstraete W, Top EM (2002) Diversity of 3-chloroaniline and 3,4-dichloroaniline degrading bacteria isolated from three different soils and involvement of their plasmids in chloroaniline degradation. FEMS Microbiol Ecol 42:315–325

Dennis JJ (2005) The evolution of IncP catabolic plasmids. Curr Opin Biotechnol 16:291–298

Dennis JJ, Zylstra GJ (2004) Complete sequence and genetic organization of pDTG1, the 83 kilobase naphthalene degradation plasmid from *Pseudomonas putida* strain NCIB 9816-4. J Mol Biol 341:753–768

Di Gioia D, Peel M, Fava F, Wyndham RC (1998) Structures of homologous composite transposons carrying *cbaABC* genes from Europe and North America. Appl Environ Microbiol 64:1940–1946

Dogra C, Raina V, Pal R, Suar M, Lal S, Gartemann KH, Holliger C, van der Meer JR, Lal R (2004) Organization of *lin* genes and IS*6100* among different strains of hexachlorocyclohexane-degrading *Sphingomonas paucimobilis*: evidence for horizontal gene transfer. J Bacteriol 186:2225–2235

Don RH, Pemberton JM (1981) Properties of six pesticide degradation plasmids isolated from *Alcaligenes paradoxus* and *Alcaligenes eutrophus*. J Bacteriol 145:681–686

Don RH, Weightman AJ, Knackmuss HJ, Timmis KN (1985) Transposon mutagenesis and cloning analysis of the pathways for degradation of 2,4-dichlorophenoxyacetic acid and 3-chlorobenzoate in *Alcaligenes eutrophus* JMP134(pJP4). J Bacteriol 161:85–90

Dunn NW, Gunsalus IC (1973) Transmissible plasmid coding early enzymes of naphthalene oxidation in *Pseudomonas putida*. J Bacteriol 114:974–979

Dutton JR, Venables WA, Saint CP (1995) *Comamonas acidovorans* UCC61 catabolizes *o*-phthalate via a 4,5-oxygenation pathway that is encoded on a 70 kbp section of plasmid pOPH1 bounded by directly repeated sequences. Microbiology 141:1673–1682

Eaton RW (2001) Plasmid-encoded phthalate catabolic pathway in *Arthrobacter keyseri* 12B. J Bacteriol 183:3689–3703

Eaton RW, Timmis KN (1986) Characterization of a plasmid-specified pathway for catabolism of isopropylbenzene in *Pseudomonas putida* RE204. J Bacteriol 168:123–131

Elsaied H, Stokes HW, Kitamura K, Kurusu Y, Kamagata Y, Maruyama A (2011) Marine integrons containing novel integrase genes, attachment sites, *attI*, and associated gene cassettes in polluted sediments from Suez and Tokyo Bays. ISME J 5:1162–1177

Feng X, Ou LT, Ogram A (1997a) Cloning and sequence analysis of a novel insertion element from plasmids harbored by the carbofuran-degrading bacterium, *Sphingomonas* sp. CFO6. Plasmid 37:169–179

Feng X, Ou LT, Ogram A (1997b) Plasmid-mediated mineralization of carbofuran by *Sphingomonas* sp. strain CF06. Appl Environ Microbiol 63:1332–1337

Feng L, Wang W, Cheng J, Ren Y, Zhao G, Gao C, Tang Y, Liu X, Han W, Peng X, Liu R, Wang L (2007) Genome and proteome of long-chain alkane degrading *Geobacillus thermodenitrificans* NG80-2 isolated from a deep-subsurface oil reservoir. Proc Natl Acad Sci USA 104:5602–5607

Foght JM, Westlake DW (1990) Expression of dibenzothiophene-degradative genes in two *Pseudomonas* species. Can J Microbiol 36:718–724

Fong KP, Goh CB, Tan HM (1996) Characterization and expression of the plasmid-borne *bedD* gene from *Pseudomonas putida* ML2, which codes for a NAD+-dependent *cis*-benzene dihydrodiol dehydrogenase. J Bacteriol 178:5592–5601

Fong KP, Goh CB, Tan HM (2000) The genes for benzene catabolism in *Pseudomonas putida* ML2 are flanked by two copies of the insertion element IS*1489*, forming a class-I-type catabolic transposon, Tn*5542*. Plasmid 43:103–110

Frantz B, Chakrabarty AM (1987) Organization and nucleotide sequence determination of a gene cluster involved in 3-chlorocatechol degradation. Proc Natl Acad Sci USA 84:4460–4464

Fredrickson JK, Brockman FJ, Workman DJ, Li SW, Stevens TO (1991) Isolation and characterization of a subsurface bacterium capable of growth on toluene, naphthalene, and other aromatic compounds. Appl Environ Microbiol 57:796–803

Frost LS, Leplae R, Summers AO, Toussaint A (2005) Mobile genetic elements: the agents of open source evolution. Nat Rev Microbiol 3:722–732

Fujii T, Takeo M, Maeda Y (1997) Plasmid-encoded genes specifying aniline oxidation from *Acinetobacter* sp. strain YAA. Microbiology 143(Pt 1):93–99

Fukumori F, Saint CP (1997) Nucleotide sequences and regulational analysis of genes involved in conversion of aniline to catechol in *Pseudomonas putida* UCC22(pTDN1). J Bacteriol 179:399–408

Fukumori F, Saint CP (2001) Complete nucleotide sequence of the catechol metabolic region of plasmid pTDN1. J Gen Appl Microbiol 47:329–333

Fulthorpe RR, Wyndham RC (1991) Transfer and expression of the catabolic plasmid pBRC60 in wild bacterial recipients in a freshwater ecosystem. Appl Environ Microbiol 57:1546–1553

Gai Z, Wang X, Liu X, Tai C, Tang H, He X, Wu G, Deng Z, Xu P (2010) The genes coding for the conversion of carbazole to catechol are flanked by IS*6100* elements in *Sphingomonas* sp. strain XLDN2-5. PLoS One 5:e10018

Gaillard M, Vallaeys T, Vorhölter FJ, Minoia M, Werlen C, Sentchilo V, Pühler A, van der Meer JR (2006) The *clc* element of *Pseudomonas* sp. strain B13, a genomic island with various catabolic properties. J Bacteriol 188:1999–2013

Gaillard M, Pernet N, Vogne C, Hagenbüchle O, van der Meer JR (2008) Host and invader impact of transfer of the *clc* genomic island into *Pseudomonas aeruginosa* PAO1. Proc Natl Acad Sci USA 105:7058–7063

Gaillard M, Pradervand N, Minoia M, Sentchilo V, Johnson DR, van der Meer JR (2010) Transcriptome analysis of the mobile genome ICE*clc* in *Pseudomonas knackmussii* B13. BMC Microbiol 10:153

Garcillán-Barcia MP, Francia MV, de la Cruz F (2009) The diversity of conjugative relaxases and its application in plasmid classification. FEMS Microbiol Rev 33:657–687

Garcillán-Barcia MP, Alvarado A, de la Cruz F (2011) Identification of bacterial plasmids based on mobility and plasmid population biology. FEMS Microbiol Rev 35:936–956

Ghosal D, You IS (1988) Nucleotide homology and organization of chlorocatechol oxidation genes of plasmids pJP4 and pAC27. Mol Gen Genet 211:113–120

Ghosal D, You IS, Chatterjee DK, Chakrabarty AM (1985) Genes specifying degradation of 3-chlorobenzoic acid in plasmids pAC27 and pJP4. Proc Natl Acad Sci USA 82:1638–1642

Goris J, Dejonghe W, Falsen E, De Clerck E, Geeraerts B, Willems A, Top E, Vandamme P, De Vos P (2002) Diversity of transconjugants that acquired plasmid pJP4 or pEMT1 after inoculation of a donor strain in the A- and B-horizon of an agricultural soil and description of *Burkholderia hospita* sp. nov. and *Burkholderia terricola* sp. nov. Syst Appl Microbiol 25:340–352

Greated A, Lambertsen L, Williams PA, Thomas CM (2002) Complete sequence of the IncP-9 TOL plasmid pWW0 from *Pseudomonas putida*. Environ Microbiol 4:856–871

Grindley NDF (2002) The movement of Tn*3*-like elements: transposition and cointegrate resolution. In: Craig NL, Craigie R, Gellert M, Lambowitz AM (eds) Mobile DNA II. American Society for Microbiology, Washington, DC, pp 272–302

Grohmann E, Muth G, Espinosa M (2003) Conjugative plasmid transfer in gram-positive bacteria. Microbiol Mol Biol Rev 67:277–301, table of contents

Gstalder ME, Faelen M, Mine N, Top EM, Mergeay M, Couturier M (2003) Replication functions of new broad host range plasmids isolated from polluted soils. Res Microbiol 154:499–509

Guglielmini J, Quintais L, Garcillán-Barcia MP, de la Cruz F, Rocha EP (2011) The repertoire of ICE in prokaryotes underscores the unity, diversity, and ubiquity of conjugation. PLoS Genet 7:e1002222

Habe H, Ashikawa Y, Saiki Y, Yoshida T, Nojiri H, Omori T (2002) *Sphingomonas* sp. strain KA1, carrying a carbazole dioxygenase gene homologue, degrades chlorinated dibenzo-*p*-dioxins in soil. FEMS Microbiol Lett 211:43–49

Habe H, Chung JS, Ishida A, Kasuga K, Ide K, Takemura T, Nojiri H, Yamane H, Omori T (2005) The fluorene catabolic linear plasmid in *Terrabacter* sp. strain DBF63 carries the beta-ketoadipate pathway genes, *pcaRHGBDCFIJ*, also found in proteobacteria. Microbiology 151:3713–3722

Hall RM, Brown HJ, Brookes DE, Stokes HW (1994) Integrons found in different locations have identical 5′ ends but variable 3′ ends. J Bacteriol 176:6286–6294

Hardman DJ, Gowland PC, Slater JH (1986) Large plasmids from soil bacteria enriched on halogenated alkanoic acids. Appl Environ Microbiol 51:44–51

Hashimoto M, Fukui M, Hayano K, Hayatsu M (2002) Nucleotide sequence and genetic structure of a novel carbaryl hydrolase gene (*cehA*) from *Rhizobium* sp. strain AC100. Appl Environ Microbiol 68:1220–1227

Hayatsu M, Hirano M, Nagata T (1999) Involvement of two plasmids in the degradation of carbaryl by *Arthrobacter* sp. strain RC100. Appl Environ Microbiol 65:1015–1019

Hayatsu M, Hirano M, Tokuda S (2000) Involvement of two plasmids in fenitrothion degradation by *Burkholderia* sp. strain NF100. Appl Environ Microbiol 66:1737–1740

Heinaru E, Vedler E, Jutkina J, Aava M, Heinaru A (2009) Conjugal transfer and mobilization capacity of the completely sequenced naphthalene plasmid pNAH20 from multiplasmid strain *Pseudomonas fluorescens* PC20. FEMS Microbiol Ecol 70:563–574

Herrmann H, Janke D, Krejsa S, Roy M (1988) *In vivo* generation of R68.45-pPGH1 hybrid plasmids conferring a Phl+ (meta pathway) phenotype. Mol Gen Genet 214:173–176

Heuer H, Smalla K (2007) Horizontal gene transfer between bacteria. Environ Biosafety Res 6:3–13

Hickey WJ, Chen S, Zhao J (2012) The *phn* island: a new genomic island encoding catabolism of polynuclear aromatic hydrocarbons. Front Microbiol 3:125

Hiessl S, Schuldes J, Thürmer A, Halbsguth T, Bröker D, Angelov A, Liebl W, Daniel R, Steinbüchel A (2012) Involvement of two latex-clearing proteins during rubber degradation and insights into the subsequent degradation pathway revealed by the genome sequence of *Gordonia polyisoprenivorans* strain VH2. Appl Environ Microbiol 74:2874–2887

Hinnebusch J, Tilly K (1993) Linear plasmids and chromosomes in bacteria. Mol Microbiol 10:917–922

Hooper SW, Dockendorff TC, Sayler GS (1989) Characteristics and restriction analysis of the 4-chlorobiphenyl catabolic plasmid, pSS50. Appl Environ Microbiol 55:1286–1288

Hopper DJ, Kemp PD (1980) Regulation of enzymes of the 3,5-xylenol-degradative pathway in *Pseudomonas putida*: evidence for a plasmid. J Bacteriol 142:21–26

Horne I, Qiu X, Russell RJ, Oakeshott JG (2003) The phosphotriesterase gene *opdA* in *Agrobacterium radiobacter* P230 is transposable. FEMS Microbiol Lett 222:1–8

Hu X, Mamoto R, Fujioka Y, Tani A, Kimbara K, Kawai F (2008) The *pva* operon is located on the megaplasmid of *Sphingopyxis* sp. strain 113P3 and is constitutively expressed, although expression is enhanced by PVA. Appl Microbiol Biotechnol 78:685–693

Igloi GL, Brandsch R (2003) Sequence of the 165-kilobase catabolic plasmid pAO1 from *Arthrobacter nicotinovorans* and identification of a pAO1-dependent nicotine uptake system. J Bacteriol 185:1976–1986

Iida T, Mukouzaka Y, Nakamura K, Kudo T (2002) Plasmid-borne genes code for an angular dioxygenase involved in dibenzofuran degradation by *Terrabacter* sp. strain YK3. Appl Environ Microbiol 68:3716–3723

Indest KJ, Jung CM, Chen HP, Hancock D, Florizone C, Eltis LD, Crocker FH (2010) Functional characterization of pGKT2, a 182-kilobase plasmid containing the *xplAB* genes, which are involved in the degradation of hexahydro-1,3,5-trinitro-1,3,5-triazine by *Gordonia* sp. strain KTR9. Appl Environ Microbiol 76:6329–6337

Ishii S, Ohno H, Tsuboi M, Otsuka S, Senoo K (2011) Identification and isolation of active N_2O reducers in rice paddy soil. ISME J 5:1936–1945

Izmalkova TY, Sazonova OI, Sokolov SL, Kosheleva IA, Boronin AM (2005) The P-7 incompatibility group plasmids responsible for biodegradation of naphthalene and salicylate in fluorescent *Pseudomonads*. Microbiology (Pleiades publish) 74:342–348

Jahnke M, El-Banna T, Klintworth R, Auling G (1990) Mineralization of orthanilic acid is a plasmid-associated trait in *Alcaligenes* sp. O-1. J Gen Microbiol 136:2241–2249
Jencova V, Strnad H, Chodora Z, Ulbrich P, Vlcek C, Hickey WJ, Paces V (2008) Nucleotide sequence, organization and characterization of the (halo)aromatic acid catabolic plasmid pA81 from *Achromobacter xylosoxidans* A8. Res Microbiol 159:118–127
Jeon CO, Park W, Padmanabhan P, DeRito C, Snape JR, Madsen EL (2003) Discovery of a bacterium, with distinctive dioxygenase, that is responsible for *in situ* biodegradation in contaminated sediment. Proc Natl Acad Sci USA 100:13591–13596
Jeon CO, Park M, Ro HS, Park W, Madsen EL (2006) The naphthalene catabolic (*nag*) genes of *Polaromonas naphthalenivorans* CJ2: evolutionary implications for two gene clusters and novel regulatory control. Appl Environ Microbiol 72:1086–1095
Jones BV, Marchesi JR (2007) Transposon-aided capture (TRACA) of plasmids resident in the human gut mobile metagenome. Nat Methods 4:55–61
Jones BV, Sun F, Marchesi JR (2010) Comparative metagenomic analysis of plasmid encoded functions in the human gut microbiome. BMC Genomics 11:46
Junker F, Cook AM (1997) Conjugative plasmids and the degradation of arylsulfonates in *Comamonas testosteroni*. Appl Environ Microbiol 63:2403–2410
Ka JO, Tiedje JM (1994) Integration and excision of a 2,4-dichlorophenoxyacetic acid-degradative plasmid in *Alcaligenes paradoxus* and evidence of its natural intergeneric transfer. J Bacteriol 176:5284–5289
Kato K, Ohtsuki K, Mitsuda H, Yomo T, Negoro S, Urabe I (1994) Insertion sequence IS*6100* on plasmid pOAD2, which degrades nylon oligomers. J Bacteriol 176:1197–1200
Kato K, Ohtsuki K, Koda Y, Maekawa T, Yomo T, Negoro S, Urabe I (1995) A plasmid encoding enzymes for nylon oligomer degradation: nucleotide sequence and analysis of pOAD2. Microbiology 141:2585–2590
Kawasaki H, Yahara H, Tonomura K (1981) Isolation and characterization of plasmid pUO1 mediating dehalogenation of haloacetate and mercury resistance in *Moraxella* sp. B. Agric Biol Chem 45:1477–1481
Kawasaki H, Takao M, Koiso A, Tonomura K (1985) Genetic rearrangement of plasmids: *in vivo* recombination between a dehalogenation plasmid and multiple resistance plasmid RP4 in *Pseudomonas* sp. Appl Environ Microbiol 49:1544–1546
Keck A, Conradt D, Mahler A, Stolz A, Mattes R, Klein J (2006) Identification and functional analysis of the genes for naphthalenesulfonate catabolism by *Sphingomonas xenophaga* BN6. Microbiology 152:1929–1940
Keil H, Keil S, Pickup RW, Williams PA (1985) Evolutionary conservation of genes coding for *meta* pathway enzymes within TOL plasmids pWW0 and pWW53. J Bacteriol 164:887–895
Keil H, Saint CM, Williams PA (1987) Gene organization of the first catabolic operon of TOL plasmid pWW53: production of indigo by the *xylA* gene product. J Bacteriol 169:764–770
Keshavarz T, Lilly MD, Clarke PH (1985) Stability of a catabolic plasmid in continuous culture. J Gen Microbiol 131:1193–1203
Kesseler M, Dabbs ER, Averhoff B, Gottschalk G (1996) Studies on the isopropylbenzene 2,3-dioxygenase and the 3-isopropylcatechol 2,3-dioxygenase genes encoded by the linear plasmid of *Rhodococcus erythropolis* BD2. Microbiology 142:3241–3251
Kitagawa W, Suzuki A, Hoaki T, Masai E, Fukuda M (2001) Multiplicity of aromatic ring hydroxylation dioxygenase genes in a strong PCB degrader, *Rhodococcus* sp. strain RHA1 demonstrated by denaturing gradient gel electrophoresis. Biosci Biotechnol Biochem 65:1907–1911
Kivisaar M, Hõrak R, Kasak L, Heinaru A, Habicht J (1990) Selection of independent plasmids determining phenol degradation in *Pseudomonas putida* and the cloning and expression of genes encoding phenol monooxygenase and catechol 1,2-dioxygenase. Plasmid 24:25–36
Kivisaar M, Kasak L, Nurk A (1991) Sequence of the plasmid-encoded catechol 1,2-dioxygenase-expressing gene, *pheB*, of phenol-degrading *Pseudomonas* sp. strain EST1001. Gene 98:15–20

Kolkenbrock S, Naumann B, Hippler M, Fetzner S (2010) A novel replicative enzyme encoded by the linear *Arthrobacter* plasmid pAL1. J Bacteriol 192:4935–4943
König C, Eulberg D, Gröning J, Lakner S, Seibert V, Kaschabek SR, Schlömann M (2004) A linear megaplasmid, p1CP, carrying the genes for chlorocatechol catabolism of *Rhodococcus opacus* 1CP. Microbiology 150:3075–3087
Kosono S, Maeda M, Fuji F, Arai H, Kudo T (1997) Three of the seven *bphC* genes of *Rhodococcus erythropolis* TA421, isolated from a termite ecosystem, are located on an indigenous plasmid associated with biphenyl degradation. Appl Environ Microbiol 63:3282–3285
Kristiansson E, Fick J, Janzon A, Grabic R, Rutgersson C, Weijdegård B, Söderström H, Larsson DG (2011) Pyrosequencing of antibiotic-contaminated river sediments reveals high levels of resistance and gene transfer elements. PLoS One 6:e17038
Król JE, Penrod JT, McCaslin H, Rogers LM, Yano H, Stancik AD, Dejonghe W, Brown CJ, Parales RE, Wuertz S, Top EM (2012) Role of IncP-1β plasmids pWDL7::*rfp* and pNB8c in chloroaniline catabolism as determined by genomic and functional analyses. Appl Environ Microbiol 78:828–838
Kulakova AN, Stafford TM, Larkin MJ, Kulakov LA (1995) Plasmid pRTL1 controlling 1-chloroalkane degradation by *Rhodococcus rhodochrous* NCIMB13064. Plasmid 33:208–217
Kulakova AN, Larkin MJ, Kulakov LA (1997) The plasmid-located haloalkane dehalogenase gene from *Rhodococcus rhodochrous* NCIMB 13064. Microbiology 143:109–115
Kunz DA, Chapman PJ (1981) Isolation and characterization of spontaneously occurring TOL plasmid mutants of *Pseudomonas putida* HS1. J Bacteriol 146:952–964
Kunze M, Zerlin KF, Retzlaff A, Pohl JO, Schmidt E, Janssen DB, Vilchez-Vargas R, Pieper DH, Reineke W (2009) Degradation of chloroaromatics by *Pseudomonas putida* GJ31: assembled route for chlorobenzene degradation encoded by clusters on plasmid pKW1 and the chromosome. Microbiology 155:4069–4083
Layton AC, Sanseverino J, Wallace W, Corcoran C, Sayler GS (1992) Evidence for 4-chlorobenzoic acid dehalogenation mediated by plasmids related to pSS50. Appl Environ Microbiol 58:399–402
Lechner M, Schmitt K, Bauer S, Hot D, Hubans C, Levillain E, Locht C, Lemoine Y, Gross R (2009) Genomic island excisions in *Bordetella petrii*. BMC Microbiol 9:141
Leisinger T (1983) Microorganisms and xenobiotic compounds. Experientia 39:1183–1220
Li W, Shi J, Wang X, Han Y, Tong W, Ma L, Liu B, Cai B (2004) Complete nucleotide sequence and organization of the naphthalene catabolic plasmid pND6-1 from *Pseudomonas* sp. strain ND6. Gene 336:231–240
Liang Q, Takeo M, Chen M, Zhang W, Xu Y, Lin M (2005) Chromosome-encoded gene cluster for the metabolic pathway that converts aniline to TCA-cycle intermediates in *Delftia tsuruhatensis* AD9. Microbiology 151:3435–3446
Lin CL, Shen FT, Tan CC, Huang CC, Chen BY, Arun AB, Young CC (2012) Characterization of *Gordonia* sp. strain CC-NAPH129-6 capable of naphthalene degradation. Microbiol Res 167(7):395–404
Liu Y, Zhang J, Zhang Z (2004) Isolation and characterization of polycyclic aromatic hydrocarbons-degrading *Sphingomonas* sp. strain ZL5. Biodegradation 15:205–212
Liu H, Zhang JJ, Wang SJ, Zhang XE, Zhou NY (2005) Plasmid-borne catabolism of methyl parathion and *p*-nitrophenol in *Pseudomonas* sp. strain WBC-3. Biochem Biophys Res Commun 334:1107–1114
Liu H, Wang SJ, Zhang JJ, Dai H, Tang H, Zhou NY (2011) Patchwork assembly of *nag*-like nitroarene dioxygenase genes and the 3-chlorocatechol degradation cluster for evolution of the 2-chloronitrobenzene catabolism pathway in *Pseudomonas stutzeri* ZWLR2-1. Appl Environ Microbiol 77:4547–4552
Lloyd-Jones G, de Jong C, Ogden RC, Duetz WA, Williams PA (1994) Recombination of the *bph* (Biphenyl) catabolic genes from plasmid pWW100 and their deletion during growth on benzoate. Appl Environ Microbiol 60:691–696

Luo YR, Kang SG, Kim SJ, Kim MR, Li N, Lee JH, Kwon KK (2012) Genome sequence of benzo(a)pyrene-degrading bacterium *Novosphingobium pentaromativorans* US6-1. J Bacteriol 194:907

Ma YF, Wu JF, Wang SY, Jiang CY, Zhang Y, Qi SW, Liu L, Zhao GP, Liu SJ (2007) Nucleotide sequence of plasmid pCNB1 from *Comamonas* strain CNB-1 reveals novel genetic organization and evolution for 4-chloronitrobenzene degradation. Appl Environ Microbiol 73:4477–4483

Mäe A, Marits R, Ausmees N, Kõiv V, Heinaru A (1993) Characterization of a new 2,4-dichlorophenoxyacetic acid degrading plasmid pEST4011: physical map and localization of catabolic genes. J Gen Microbiol 139:3165–3170

Maeda K, Nojiri H, Shintani M, Yoshida T, Habe H, Omori T (2003) Complete nucleotide sequence of carbazole/dioxin-degrading plasmid pCAR1 in *Pseudomonas resinovorans* strain CA10 indicates its mosaicity and the presence of large catabolic transposon Tn*4676*. J Mol Biol 326:21–33

Mahillon J, Chandler M (1998) Insertion sequences. Microbiol Mol Biol Rev 62:725–774

Maillard J, Regeard C, Holliger C (2005) Isolation and characterization of Tn-*Dha1*, a transposon containing the tetrachloroethene reductive dehalogenase of *Desulfitobacterium hafniense* strain TCE1. Environ Microbiol 7:107–117

Mandal MD, Mandal A, Pal NK (2005) Plasmid-mediated dimethoate degradation by *Bacillus licheniformis* isolated from a fresh water fish *Labeo rohita*. J Biomed Biotechnol 3:280–286

Martin C, Timm J, Rauzier J, Gomez-Lus R, Davies J, Gicquel B (1990) Transposition of an antibiotic resistance element in mycobacteria. Nature 345:739–743

Martinez B, Tomkins J, Wackett LP, Wing R, Sadowsky MJ (2001) Complete nucleotide sequence and organization of the atrazine catabolic plasmid pADP-1 from *Pseudomonas* sp. strain ADP. J Bacteriol 183:5684–5697

Masai E, Sugiyama K, Iwashita N, Shimizu S, Hauschild JE, Hatta T, Kimbara K, Yano K, Fukuda M (1997) The *bphDEF* meta-cleavage pathway genes involved in biphenyl/polychlorinated biphenyl degradation are located on a linear plasmid and separated from the initial *bphACB* genes in *Rhodococcus* sp. strain RHA1. Gene 187:141–149

Masai E, Kamimura N, Kasai D, Oguchi A, Ankai A, Fukui S, Takahashi M, Yashiro I, Sasaki H, Harada T, Nakamura S, Katano Y, Narita-Yamada S, Nakazawa H, Hara H, Katayama Y, Fukuda M, Yamazaki S, Fujita N (2012) Complete genome sequence of *Sphingobium* sp. strain SYK-6, a degrader of lignin-derived biaryls and monoaryls. J Bacteriol 194:534–535

McLeod MP, Warren RL, Hsiao WW, Araki N, Myhre M, Fernandes C, Miyazawa D, Wong W, Lillquist AL, Wang D, Dosanjh M, Hara H, Petrescu A, Morin RD, Yang G, Stott JM, Schein JE, Shin H, Smailus D, Siddiqui AS, Marra MA, Jones SJ, Holt R, Brinkman FS, Miyauchi K, Fukuda M, Davies JE, Mohn WW, Eltis LD (2006) The complete genome of *Rhodococcus* sp. RHA1 provides insights into a catabolic powerhouse. Proc Natl Acad Sci USA 103:15582–15587

Merlin C, Springael D, Toussaint A (1999) Tn*4371*: a modular structure encoding a phage-like integrase, a *Pseudomonas*-like catabolic pathway, and RP4/Ti-like transfer functions. Plasmid 41:40–54

Miller TR, Delcher AL, Salzberg SL, Saunders E, Detter JC, Halden RU (2010) Genome sequence of the dioxin-mineralizing bacterium *Sphingomonas wittichii* RW1. J Bacteriol 192:6101–6102

Miyauchi K, Sukda P, Nishida T, Ito E, Matsumoto Y, Masai E, Fukuda M (2008) Isolation of dibenzofuran-degrading bacterium, *Nocardioides* sp. DF412, and characterization of its dibenzofuran degradation genes. J Biosci Bioeng 105:628–635

Miyazaki R, van der Meer JR (2011a) A dual functional origin of transfer in the ICE*clc* genomic island of *Pseudomonas knackmussii* B13. Mol Microbiol 79:743–758

Miyazaki R, van der Meer JR (2011b) How can a dual *oriT* system contribute to efficient transfer of an integrative and conjugative element? Mobile Genet Elem 1:82–84

Miyazaki R, Sato Y, Ito M, Ohtsubo Y, Nagata Y, Tsuda M (2006) Complete nucleotide sequence of an exogenously isolated plasmid, pLB1, involved in gamma-hexachlorocyclohexane degradation. Appl Environ Microbiol 72:6923–6933

Mulbry WW, Karns JS, Kearney PC, Nelson JO, McDaniel CS, Wild JR (1986) Identification of a plasmid-borne parathion hydrolase gene from *Flavobacterium* sp. by southern hybridization with *opd* from *Pseudomonas diminuta*. Appl Environ Microbiol 51:926–930

Mulbry WW, Kearney PC, Nelson JO, Karns JS (1987) Physical comparison of parathion hydrolase plasmids from *Pseudomonas diminuta* and *Flavobacterium* sp. Plasmid 18:173–177

Müller TA, Werlen C, Spain J, Van Der Meer JR (2003) Evolution of a chlorobenzene degradative pathway among bacteria in a contaminated groundwater mediated by a genomic island in *Ralstonia*. Environ Microbiol 5:163–173

Musovic S, Oregaard G, Kroer N, Sørensen SJ (2006) Cultivation-independent examination of horizontal transfer and host range of an IncP-1 plasmid among gram-positive and gram-negative bacteria indigenous to the barley rhizosphere. Appl Environ Microbiol 72:6687–6692

Musovic S, Dechesne A, Sørensen J, Smets BF (2010) Novel assay to assess permissiveness of a soil microbial community toward receipt of mobile genetic elements. Appl Environ Microbiol 76:4813–4818

Nagata Y, Kamakura M, Endo R, Miyazaki R, Ohtsubo Y, Tsuda M (2006) Distribution of gamma-hexachlorocyclohexane-degrading genes on three replicons in *Sphingobium japonicum* UT26. FEMS Microbiol Lett 256:112–118

Nagata Y, Endo R, Ito M, Ohtsubo Y, Tsuda M (2007) Aerobic degradation of lindane (gamma-hexachlorocyclohexane) in bacteria and its biochemical and molecular basis. Appl Microbiol Biotechnol 76:741–752

Nagata Y, Ohtsubo Y, Endo R, Ichikawa N, Ankai A, Oguchi A, Fukui S, Fujita N, Tsuda M (2010) Complete genome sequence of the representative γ-hexachlorocyclohexane-degrading bacterium *Sphingobium japonicum* UT26. J Bacteriol 192:5852–5853

Nagata Y, Natsui S, Endo R, Ohtsubo Y, Ichikawa N, Ankai A, Oguchi A, Fukui S, Fujita N, Tsuda M (2011) Genomic organization and genomic structural rearrangements of *Sphingobium japonicum* UT26, an archetypal γ-hexachlorocyclohexane-degrading bacterium. Enzyme Microb Technol 49:499–508

Nakatsu C, Ng J, Singh R, Straus N, Wyndham C (1991) Chlorobenzoate catabolic transposon Tn*5271* is a composite class I element with flanking class II insertion sequences. Proc Natl Acad Sci USA 88:8312–8316

Negoro S, Okada H (1982) Physical map of Nylon oligomer degradative plasmid pOAD2 harbored in *Flavobacterium* sp. KI72. Agric Biol Chem 46:745–750

Negoro S, Shinagawa H, Nakata A, Kinoshita S, Hatozaki T, Okada H (1980) Plasmid control of 6-aminohexanoic acid cyclic dimer degradation enzymes of *Flavobacterium* sp. KI72. J Bacteriol 143:238–245

Nishi A, Tominaga K, Furukawa K (2000) A 90-kilobase conjugative chromosomal element coding for biphenyl and salicylate catabolism in *Pseudomonas putida* KF715. J Bacteriol 182:1949–1955

Nojiri H, Sekiguchi H, Maeda K, Urata M, Nakai S, Yoshida T, Habe H, Omori T (2001) Genetic characterization and evolutionary implications of a *car* gene cluster in the carbazole degrader *Pseudomonas* sp. strain CA10. J Bacteriol 183:3663–3679

Nojiri H, Kamakura M, Urata M, Tanaka T, Chung JS, Takemura T, Yoshida T, Habe H, Omori T (2002) Dioxin catabolic genes are dispersed on the *Terrabacter* sp. DBF63 genome. Biochem Biophys Res Commun 296:233–240

Nojiri H, Shintani M, Omori T (2004) Divergence of mobile genetic elements involved in the distribution of xenobiotic-catabolic capacity. Appl Microbiol Biotechnol 64:154–174

Ogawa N, Miyashita K (1995) Recombination of a 3-chlorobenzoate catabolic plasmid from *Alcaligenes eutrophus* NH9 mediated by direct repeat elements. Appl Environ Microbiol 61:3788–3795

Ogawa N, Miyashita K (1999) The chlorocatechol-catabolic transposon Tn*5707* of *Alcaligenes eutrophus* NH9, carrying a gene cluster highly homologous to that in the 1,2,4-trichlorobenzene-degrading bacterium *Pseudomonas* sp. strain P51, confers the ability to grow on 3-chlorobenzoate. Appl Environ Microbiol 65:724–731

Ohtsubo Y, Shimura M, Delawary M, Kimbara K, Takagi M, Kudo T, Ohta A, Nagata Y (2003) Novel approach to the improvement of biphenyl and polychlorinated biphenyl degradation

activity: promoter implantation by homologous recombination. Appl Environ Microbiol 69:146–153
Ohtsubo Y, Goto H, Nagata Y, Kudo T, Tsuda M (2006) Identification of a response regulator gene for catabolite control from a PCB-degrading beta-proteobacteria, *Acidovorax* sp. KKS102. Mol Microbiol 60:1563–1575
Ohtsubo Y, Ishibashi Y, Naganawa H, Hirokawa S, Atobe S, Nagata Y, Tsuda M (2012) Conjugal transfer of PCB/biphenyl degradation genes in *Acidovorax* sp. KKS102 located on integrative and conjugative element. J Bacteriol 194(16):4237–4248
Ono A, Miyazaki R, Sota M, Ohtsubo Y, Nagata Y, Tsuda M (2007) Isolation and characterization of naphthalene-catabolic genes and plasmids from oil-contaminated soil by using two cultivation-independent approaches. Appl Microbiol Biotechnol 74:501–510
Overhage J, Sielker S, Homburg S, Parschat K, Fetzner S (2005) Identification of large linear plasmids in *Arthrobacter* spp. encoding the degradation of quinaldine to anthranilate. Microbiology 151:491–500
Pandeeti EV, Chakka D, Pandey JP, Siddavattam D (2011) Indigenous organophosphate-degrading (*opd*) plasmid pCMS1 of *Brevundimonas diminuta* is self-transmissible and plays a key role in horizontal mobility of the *opd* gene. Plasmid 65:226–231
Park HS, Kim HS (2000) Identification and characterization of the nitrobenzene catabolic plasmids pNB1 and pNB2 in *Pseudomonas putida* HS12. J Bacteriol 182:573–580
Park W, Jeon CO, Hohnstock-Ashe AM, Winans SC, Zylstra GJ, Madsen EL (2003) Identification and characterization of the conjugal transfer region of the pCg1 plasmid from naphthalene-degrading *Pseudomonas putida* Cg1. Appl Environ Microbiol 69:3263–3271
Parschat K, Overhage J, Strittmatter AW, Henne A, Gottschalk G, Fetzner S (2007) Complete nucleotide sequence of the 113-kilobase linear catabolic plasmid pAL1 of *Arthrobacter nitroguajacolicus* Rü61a and transcriptional analysis of genes involved in quinaldine degradation. J Bacteriol 189:3855–3867
Pepper IL, Gentry TJ, Newby DT, Roane TM, Josephson KL (2002) The role of cell bioaugmentation and gene bioaugmentation in the remediation of co-contaminated soils. Environ Health Perspect 110(Suppl 6):943–946
Peters M, Heinaru E, Talpsep E, Wand H, Stottmeister U, Heinaru A, Nurk A (1997) Acquisition of a deliberately introduced phenol degradation operon, *pheBA*, by different indigenous *Pseudomonas* species. Appl Environ Microbiol 63:4899–4906
Poh RP, Smith AR, Bruce IJ (2002) Complete characterisation of Tn*5530* from *Burkholderia cepacia* strain 2a (pIJB1) and studies of 2,4-dichlorophenoxyacetate uptake by the organism. Plasmid 48:1–12
Priefert H, O'Brien XM, Lessard PA, Dexter AF, Choi EE, Tomic S, Nagpal G, Cho JJ, Agosto M, Yang L, Treadway SL, Tamashiro L, Wallace M, Sinskey AJ (2004) Indene bioconversion by a toluene inducible dioxygenase of *Rhodococcus* sp. I24. Appl Microbiol Biotechnol 65:168–176
Ravatn R, Studer S, Springael D, Zehnder AJ, van der Meer JR (1998a) Chromosomal integration, tandem amplification, and deamplification in *Pseudomonas putida* F1 of a 105-kilobase genetic element containing the chlorocatechol degradative genes from *Pseudomonas* sp. Strain B13. J Bacteriol 180:4360–4369
Ravatn R, Zehnder AJ, van der Meer JR (1998b) Low-frequency horizontal transfer of an element containing the chlorocatechol degradation genes from *Pseudomonas* sp. strain B13 to *Pseudomonas putida* F1 and to indigenous bacteria in laboratory-scale activated-sludge microcosms. Appl Environ Microbiol 64:2126–2132
Rheinwald JG, Chakrabarty AM, Gunsalus IC (1973) A transmissible plasmid controlling camphor oxidation in *Pseudomonas putida*. Proc Natl Acad Sci USA 70:885–889
Romine MF, Stillwell LC, Wong KK, Thurston SJ, Sisk EC, Sensen C, Gaasterland T, Fredrickson JK, Saffer JD (1999) Complete sequence of a 184-kilobase catabolic plasmid from *Sphingomonas aromaticivorans* F199. J Bacteriol 181:1585–1602
Ruff J, Smits TH, Cook AM, Schleheck D (2010) Identification of two vicinal operons for the degradation of 2-aminobenzenesulfonate encoded on plasmid pSAH in *Alcaligenes* sp. strain O-1. Microbiol Res 165:288–299

Ryan MP, Pembroke JT, Adley CC (2009) Novel Tn*4371*-ICE like element in *Ralstonia pickettii* and genome mining for comparative elements. BMC Microbiol 9:242

Saeki H, Akira M, Furuhashi K, Averhoff B, Gottschalk G (1999) Degradation of trichloroethene by a linear-plasmid-encoded alkene monooxygenase in *Rhodococcus corallinus* (*Nocardia corallina*) B-276. Microbiology 145:1721–1730

Saint CP, McClure NC, Venables WA (1990) Physical map of the aromatic amine and *m*-toluate catabolic plasmid pTDN1 in *Pseudomonas putida*: location of a unique *meta*-cleavage pathway. J Gen Microbiol 136:615–625

Sakaguchi K (1990) Invertrons, a class of structurally and functionally related genetic elements that includes linear DNA plasmids, transposable elements, and genomes of adeno-type viruses. Microbiol Rev 54:66–74

Sakai M, Masai E, Asami H, Sugiyama K, Kimbara K, Fukuda M (2002) Diversity of 2,3-dihydroxybiphenyl dioxygenase genes in a strong PCB degrader, *Rhodococcus* sp. strain RHA1. J Biosci Bioeng 93:421–427

Schleinitz KM, Kleinsteuber S, Vallaeys T, Babel W (2004) Localization and characterization of two novel genes encoding stereospecific dioxygenases catalyzing 2(2,4-dichlorophenoxy) propionate cleavage in *Delftia acidovorans* MC1. Appl Environ Microbiol 70:5357–5365

Sekine M, Tanikawa S, Omata S, Saito M, Fujisawa T, Tsukatani N, Tajima T, Sekigawa T, Kosugi H, Matsuo Y, Nishiko R, Imamura K, Ito M, Narita H, Tago S, Fujita N, Harayama S (2006) Sequence analysis of three plasmids harboured in *Rhodococcus erythropolis* strain PR4. Environ Microbiol 8:334–346

Sentchilo V, Czechowska K, Pradervand N, Minoia M, Miyazaki R, van der Meer JR (2009) Intracellular excision and reintegration dynamics of the ICE*clc* genomic island of *Pseudomonas knackmussii* sp. strain B13. Mol Microbiol 72:1293–1306

Serdar CM, Gibson DT, Munnecke DM, Lancaster JH (1982) Plasmid involvement in parathion hydrolysis by *Pseudomonas diminuta*. Appl Environ Microbiol 44:246–249

Seto M, Kimbara K, Shimura M, Hatta T, Fukuda M, Yano K (1995) A novel transformation of polychlorinated biphenyls by *Rhodococcus* sp. strain RHA1. Appl Environ Microbiol 61:3353–3358

Sevastsyanovich YR, Krasowiak R, Bingle LE, Haines AS, Sokolov SL, Kosheleva IA, Leuchuk AA, Titok MA, Smalla K, Thomas CM (2008) Diversity of IncP-9 plasmids of *Pseudomonas*. Microbiology 154:2929–2941

Shields MS, Hooper SW, Sayler GS (1985) Plasmid-mediated mineralization of 4-chlorobiphenyl. J Bacteriol 163:882–889

Shields MS, Reagin MJ, Gerger RR, Campbell R, Somerville C (1995) TOM, a new aromatic degradative plasmid from *Burkholderia* (*Pseudomonas*) *cepacia* G4. Appl Environ Microbiol 61:1352–1356

Shimizu S, Kobayashi H, Masai E, Fukuda M (2001) Characterization of the 450-kb linear plasmid in a polychlorinated biphenyl degrader, *Rhodococcus* sp. strain RHA1. Appl Environ Microbiol 67:2021–2028

Shintani M, Nojiri H, Yoshida T, Habe H, Omori T (2003) Carbazole/dioxin-degrading *car* gene cluster is located on the chromosome of *Pseudomonas stutzeri* strain OM1 in a form different from the simple transposition of Tn*4676*. Biotechnol Lett 25:1255–1261

Shintani M, Yoshida T, Habe H, Omori T, Nojiri H (2005) Large plasmid pCAR2 and class II transposon Tn*4676* are functional mobile genetic elements to distribute the carbazole/dioxin-degradative car gene cluster in different bacteria. Appl Microbiol Biotechnol 67:370–382

Shintani M, Urata M, Inoue K, Eto K, Habe H, Ornori T, Yamane H, Nojiri H (2007) The *Sphingomonas* plasmid pCAR3 is involved in complete mineralization of carbazole. J Bacteriol 189:2007–2020

Shintani M, Takahashi Y, Yamane H, Nojiri H (2010) The behavior and significance of degradative plasmids belonging to Inc groups in *Pseudomonas* within natural environments and microcosms. Microbes Environ 25:253–265

Shintani M, Matsumoto T, Yoshikawa H, Yamane H, Ohkuma M, Nojiri H (2011) DNA rearrangement has occurred in the carbazole-degradative plasmid pCAR1 and the chromosome of its unsuitable host, *Pseudomonas fluorescens* Pf0-1. Microbiology 157:3405–3416

Shuttleworth KL, Sung J, Kim E, Cerniglia CE (2000) Physiological and genetic comparison of two aromatic hydrocarbon-degrading *Sphingomonas* strains. Mol Cells 10:199–205

Siddavattam D, Khajamohiddin S, Manavathi B, Pakala SB, Merrick M (2003) Transposon-like organization of the plasmid-borne organophosphate degradation (*opd*) gene cluster found in *Flavobacterium* sp. Appl Environ Microbiol 69:2533–2539

Simon MJ, Osslund TD, Saunders R, Ensley BD, Suggs S, Harcourt A, Suen WC, Cruden DL, Gibson DT, Zylstra GJ (1993) Sequences of genes encoding naphthalene dioxygenase in *Pseudomonas putida* strains G7 and NCIB 9816-4. Gene 127:31–37

Smillie C, Garcillán-Barcia MP, Francia MV, Rocha EP, de la Cruz F (2010) Mobility of plasmids. Microbiol Mol Biol Rev 74:434–452

Sota M, Top EM (2008) Horizontal gene transfer mediated by plasmids. In: Lipps G (ed) Plasmids: current research and future trends. Horizon Scientific Press, Norfolk

Sota M, Endo M, Nitta K, Kawasaki H, Tsuda M (2002) Characterization of a class II defective transposon carrying two haloacetate dehalogenase genes from *Delftia acidovorans* plasmid pUO1. Appl Environ Microbiol 68:2307–2315

Sota M, Kawasaki H, Tsuda M (2003) Structure of haloacetate-catabolic IncP-1beta plasmid pUO1 and genetic mobility of its residing haloacetate-catabolic transposon. J Bacteriol 185:6741–6745

Sota M, Yano H, Ono A, Miyazaki R, Ishii H, Genka H, Top EM, Tsuda M (2006) Genomic and functional analysis of the IncP-9 naphthalene-catabolic plasmid NAH7 and its transposon Tn*4655* suggests catabolic gene spread by a tyrosine recombinase. J Bacteriol 188:4057–4067

Sota M, Tsuda M, Yano H, Suzuki H, Forney LJ, Top EM (2007) Region-specific insertion of transposons in combination with selection for high plasmid transferability and stability accounts for the structural similarity of IncP-1 plasmids. J Bacteriol 189:3091–3098

Springael D, Kreps S, Mergeay M (1993) Identification of a catabolic transposon, Tn*4371*, carrying biphenyl and 4-chlorobiphenyl degradation genes in *Alcaligenes eutrophus* A5. J Bacteriol 175:1674–1681

Springael D, Peys K, Ryngaert A, Van Roy S, Hooyberghs L, Ravatn R, Heyndrickx M, van der Meer JR, Vandecasteele C, Mergeay M, Diels L (2002) Community shifts in a seeded 3-chlorobenzoate degrading membrane biofilm reactor: indications for involvement of *in situ* horizontal transfer of the *clc*-element from inoculum to contaminant bacteria. Environ Microbiol 4:70–80

Stalker DM, McBride KE (1987) Cloning and expression in *Escherichia coli* of a *Klebsiella ozaenae* plasmid-borne gene encoding a nitrilase specific for the herbicide bromoxynil. J Bacteriol 169:955–960

Stecker C, Johann A, Herzberg C, Averhoff B, Gottschalk G (2003) Complete nucleotide sequence and genetic organization of the 210-kilobase linear plasmid of *Rhodococcus erythropolis* BD2. J Bacteriol 185:5269–5274

Stillwell LC, Thurston SJ, Schneider RP, Romine MF, Fredrickson JK, Saffer JD (1995) Physical mapping and characterization of a catabolic plasmid from the deep-subsurface bacterium *Sphingomonas* sp. strain F199. J Bacteriol 177:4537–4539

Stuart-Keil KG, Hohnstock AM, Drees KP, Herrick JB, Madsen EL (1998) Plasmids responsible for horizontal transfer of naphthalene catabolism genes between bacteria at a coal tar-contaminated site are homologous to pDTG1 from *Pseudomonas putida* NCIB 9816-4. Appl Environ Microbiol 64:3633–3640

Sundin GW, Bender CL (1995) Expression of the *strA-strB* streptomycin resistance genes in *Pseudomonas syringae* and *Xanthomonas campestris* and characterization of IS*6100* in *X. campestris*. Appl Environ Microbiol 61:2891–2897

Szczepanowski R, Bekel T, Goesmann A, Krause L, Krömeke H, Kaiser O, Eichler W, Pühler A, Schlüter A (2008) Insight into the plasmid metagenome of wastewater treatment plant bacteria showing reduced susceptibility to antimicrobial drugs analysed by the 454-pyrosequencing technology. J Biotechnol 136:54–64

Tabata M, Endo R, Ito M, Ohtsubo Y, Kumar A, Tsuda M, Nagata Y (2011) The *lin* genes for γ-hexachlorocyclohexane degradation in *Sphingomonas* sp. MM-1 proved to be dispersed across multiple plasmids. Biosci Biotechnol Biochem 75:466–472

Taghavi S, Barac T, Greenberg B, Borremans B, Vangronsveld J, van der Lelie D (2005) Horizontal gene transfer to endogenous endophytic bacteria from poplar improves phytoremediation of toluene. Appl Environ Microbiol 71:8500–8505

Tago K, Sato J, Takesa H, Kawagishi H, Hayatsu M (2005) Characterization of methylhydroquinone-metabolizing oxygenase genes encoded on plasmid in *Burkholderia* sp. NF100. J Biosci Bioeng 100:517–523

Taguchi K, Motoyama M, Kudo T (2004) Multiplicity of 2,3-dihydroxybiphenyl dioxygenase genes in the Gram-positive polychlorinated biphenyl degrading bacterium *Rhodococcus rhodochrous* K37. Biosci Biotechnol Biochem 68:787–795

Takahashi Y, Shintani M, Yamane H, Nojiri H (2009) The complete nucleotide sequence of pCAR2: pCAR2 and pCAR1 were structurally identical IncP-7 carbazole degradative plasmids. Biosci Biotechnol Biochem 73:744–746

Tan HM (1999) Bacterial catabolic transposons. Appl Microbiol Biotechnol 51:1–12

Tan HM, Mason JR (1990) Cloning and expression of the plasmid-encoded benzene dioxygenase genes from *Pseudomonas putida* ML2. FEMS Microbiol Lett 60:259–264

Tani A, Charoenpanich J, Mori T, Takeichi M, Kimbara K, Kawai F (2007) Structure and conservation of a polyethylene glycol-degradative operon in sphingomonads. Microbiology 153:338–346

Tauch A, Götker S, Pühler A, Kalinowski J, Thierbach G (2002) The 27.8-kb R-plasmid pTET3 from *Corynebacterium glutamicum* encodes the aminoglycoside adenyltransferase gene cassette *aadA9* and the regulated tetracycline efflux system Tet 33 flanked by active copies of the widespread insertion sequence IS*6100*. Plasmid 48:117–129

Thacker R, Rørvig O, Kahlon P, Gunsalus IC (1978) NIC, a conjugative nicotine-nicotinate degradative plasmid in *Pseudomonas convexa*. J Bacteriol 135:289–290

Thomas CM, Haines AS (2004) Plasmids of the genus Pseudomonas. In: Ramos J-L (ed) *Pseudomonas*. Kluwer Academic/Plenum Publishers, New York

Top EM, Springael D (2003) The role of mobile genetic elements in bacterial adaptation to xenobiotic organic compounds. Curr Opin Biotechnol 14:262–269

Top EM, Holben WE, Forney LJ (1995) Characterization of diverse 2,4-dichlorophenoxyacetic acid-degradative plasmids isolated from soil by complementation. Appl Environ Microbiol 61:1691–1698

Top EM, Maltseva OV, Forney LJ (1996) Capture of a catabolic plasmid that encodes only 2,4-dichlorophenoxyacetic acid:alpha-ketoglutaric acid dioxygenase (TfdA) by genetic complementation. Appl Environ Microbiol 62:2470–2476

Top EM, Moënne-Loccoz Y, Pembroke T, Thomas CM (2000) Phenotypic traits conferred by plasmids. In: Thomas CM (ed) The horizontal gene pool: bacterial plasmid and gene spread. Harwood, New York, pp 249–285

Top EM, Springael D, Boon N (2002) Catabolic mobile genetic elements and their potential use in bioaugmentation of polluted soils and waters. FEMS Microbiol Ecol 42:199–208

Toussaint A, Merlin C, Monchy S, Benotmane MA, Leplae R, Mergeay M, Springael D (2003) The biphenyl- and 4-chlorobiphenyl-catabolic transposon Tn*4371*, a member of a new family of genomic islands related to IncP and Ti plasmids. Appl Environ Microbiol 69:4837–4845

Tralau T, Cook AM, Ruff J (2001) Map of the IncP1beta plasmid pTSA encoding the widespread genes (*tsa*) for *p*-toluenesulfonate degradation in *Comamonas testosteroni* T-2. Appl Environ Microbiol 67:1508–1516

Trefault N, De la Iglesia R, Molina AM, Manzano M, Ledger T, Pérez-Pantoja D, Sánchez MA, Stuardo M, González B (2004) Genetic organization of the catabolic plasmid pJP4 from *Ralstonia eutropha* JMP134 (pJP4) reveals mechanisms of adaptation to chloroaromatic pollutants and evolution of specialized chloroaromatic degradation pathways. Environ Microbiol 6:655–668

Tsuda M, Genka H (2001) Identification and characterization of Tn*4656*, a novel class II transposon carrying a set of toluene-degrading genes from TOL plasmid pWW53. J Bacteriol 183:6215–6224

Tsuda M, Iino T (1987) Genetic analysis of a transposon carrying toluene degrading genes on a TOL plasmid pWW0. Mol Gen Genet 210:270–276

Tsuda M, Iino T (1988) Identification and characterization of Tn*4653*, a transposon covering the toluene transposon Tn*4651* on TOL plasmid pWW0. Mol Gen Genet 213:72–77

Tsuda M, Iino T (1990) Naphthalene degrading genes on plasmid NAH7 are on a defective transposon. Mol Gen Genet 223:33–39

Tsuda M, Minegishi K, Iino T (1989) Toluene transposons Tn*4651* and Tn*4653* are class II transposons. J Bacteriol 171:1386–1393

Uz I, Duan YP, Ogram A (2000) Characterization of the naphthalene-degrading bacterium, *Rhodococcus opacus* M213. FEMS Microbiol Lett 185:231–238

van Beilen JB, Panke S, Lucchini S, Franchini AG, Röthlisberger M, Witholt B (2001) Analysis of *Pseudomonas putida* alkane-degradation gene clusters and flanking insertion sequences: evolution and regulation of the *alk* genes. Microbiology 147:1621–1630

van der Meer JR, Sentchilo V (2003) Genomic islands and the evolution of catabolic pathways in bacteria. Curr Opin Biotechnol 14:248–254

van der Meer JR, Zehnder AJ, de Vos WM (1991a) Identification of a novel composite transposable element, Tn*5280*, carrying chlorobenzene dioxygenase genes of *Pseudomonas* sp. strain P51. J Bacteriol 173:7077–7083

van der Meer JR, van Neerven AR, de Vries EJ, de Vos WM, Zehnder AJ (1991b) Cloning and characterization of plasmid-encoded genes for the degradation of 1,2-dichloro-, 1,4-dichloro-, and 1,2,4-trichlorobenzene of *Pseudomonas* sp. strain P51. J Bacteriol 173:6–15

van der Ploeg J, Willemsen M, van Hall G, Janssen DB (1995) Adaptation of *Xanthobacter autotrophicus* GJ10 to bromoacetate due to activation and mobilization of the haloacetate dehalogenase gene by insertion element IS*1247*. J Bacteriol 177:1348–1356

Vedler E, Kõiv V, Heinaru A (2000) Analysis of the 2,4-dichlorophenoxyacetic acid-degradative plasmid pEST4011 of *Achromobacter xylosoxidans* subsp. denitrificans strain EST4002. Gene 255:281–288

Vedler E, Vahter M, Heinaru A (2004) The completely sequenced plasmid pEST4011 contains a novel IncP1 backbone and a catabolic transposon harboring *tfd* genes for 2,4-dichlorophenoxyacetic acid degradation. J Bacteriol 186:7161–7174

Wagenknecht M, Meinhardt F (2011) Replication-involved genes of pAL1, the linear plasmid of *Arthrobacter nitroguajacolicus* Rü61a – phylogenetic and transcriptional analysis. Plasmid 65:176–184

Wang M, Yang G, Min H, Lv Z (2009) A novel nicotine catabolic plasmid pMH1 in *Pseudomonas* sp. strain HF-1. Can J Microbiol 55:228–233

Warburton PJ, Allan E, Hunter S, Ward J, Booth V, Wade WG, Mullany P (2011) Isolation of bacterial extrachromosomal DNA from human dental plaque associated with periodontal disease, using transposon-aided capture (TRACA). FEMS Microbiol Ecol 78:349–354

Warren R, Hsiao WW, Kudo H, Myhre M, Dosanjh M, Petrescu A, Kobayashi H, Shimizu S, Miyauchi K, Masai E, Yang G, Stott JM, Schein JE, Shin H, Khattra J, Smailus D, Butterfield YS, Siddiqui A, Holt R, Marra MA, Jones SJ, Mohn WW, Brinkman FS, Fukuda M, Davies J, Eltis LD (2004) Functional characterization of a catabolic plasmid from polychlorinated-biphenyl-degrading *Rhodococcus* sp. strain RHA1. J Bacteriol 186:7783–7795

Wei M, Zhang JJ, Liu H, Wang SJ, Fu H, Zhou NY (2009) A transposable class I composite transposon carrying mph (methyl parathion hydrolase) from *Pseudomonas* sp. strain WBC-3. FEMS Microbiol Lett 292:85–91

Weightman AJ, Topping AW, Hill KE, Lee LL, Sakai K, Slater JH, Thomas AW (2002) Transposition of *DEH*, a broad-host-range transposon flanked by IS*Ppu12*, in *Pseudomonas putida* is associated with genomic rearrangements and dehalogenase gene silencing. J Bacteriol 184:6581–6591

Williams PA, Murray K (1974) Metabolism of benzoate and the methylbenzoates by *Pseudomonas putida* (*arvilla*) mt-2: evidence for the existence of a TOL plasmid. J Bacteriol 120:416–423

Winstanley C, Taylor SC, Williams PA (1987) pWW174: a large plasmid from *Acinetobacter calcoaceticus* encoding benzene catabolism by the beta-ketoadipate pathway. Mol Microbiol 1:219–227

Wozniak RA, Waldor MK (2010) Integrative and conjugative elements: mosaic mobile genetic elements enabling dynamic lateral gene flow. Nat Rev Microbiol 8:552–563

Wu JF, Sun CW, Jiang CY, Liu ZP, Liu SJ (2005) A novel 2-aminophenol 1,6-dioxygenase involved in the degradation of *p*-chloronitrobenzene by *Comamonas* strain CNB-1: purification, properties, genetic cloning and expression in *Escherichia coli*. Arch Microbiol 183:1–8

Wu JF, Jiang CY, Wang BJ, Ma YF, Liu ZP, Liu SJ (2006) Novel partial reductive pathway for 4-chloronitrobenzene and nitrobenzene degradation in *Comamonas* sp. strain CNB-1. Appl Environ Microbiol 72:1759–1765

Xia XS, Aathithan S, Oswiecimska K, Smith AR, Bruce IJ (1998) A novel plasmid pIJB1 possessing a putative 2,4-dichlorophenoxyacetate degradative transposon Tn*5530* in *Burkholderia cepacia* strain 2a. Plasmid 39:154–159

Yagi JM, Sims D, Brettin T, Bruce D, Madsen EL (2009) The genome of *Polaromonas naphthalenivorans* strain CJ2, isolated from coal tar-contaminated sediment, reveals physiological and metabolic versatility and evolution through extensive horizontal gene transfer. Environ Microbiol 11:2253–2270

Yano K, Nishi T (1980) pKJ1, a naturally occurring conjugative plasmid coding for toluene degradation and resistance to streptomycin and sulfonamides. J Bacteriol 143:552–560

Yano H, Garruto C, Sota M, Ohtsubo Y, Nagata Y, Zylstra G, Williams P, Tsuda M (2007) Complete sequence determination combined with analysis of transposition/site-specific recombination events to explain genetic organization of IncP-7 TOL plasmid pWW53 and related mobile genetic elements. J Mol Biol 369:11–26

Yano H, Miyakoshi M, Ohshima K, Tabata M, Nagata Y, Hattori M, Tsuda M (2010) Complete nucleotide sequence of TOL plasmid pDK1 provides evidence for evolutionary history of IncP-7 catabolic plasmids. J Bacteriol 192:4337–4347

Yergeau E, Sanschagrin S, Beaumier D, Greer CW (2012) Metagenomic analysis of the bioremediation of diesel-contaminated Canadian high arctic soils. PLoS One 7:e30058

Zhang T, Zhang XX, Ye L (2011a) Plasmid metagenome reveals high levels of antibiotic resistance genes and mobile genetic elements in activated sludge. PLoS One 6:e26041

Zhang Y, Jiang Z, Cao B, Hu M, Wang Z, Dong X (2011b) Chemotaxis to atrazine and detection of a xenobiotic catabolic plasmid in *Arthrobacter* sp DNS10. Environ Sci Pollut Res Int 19(7):2951–8, Epub 2012 Feb 17

Chapter 9
Conjugative Plasmids in Anthropogenic Soils

Elisabeth Grohmann

Abstract Horizontal transfer of mobile genetic elements such as plasmids and bacteriophages and their associated hitchhiking elements such as transposons, integrative and conjugative elements and insertion sequences shapes bacterial chromosomes and enables their adaptation to changing environmental conditions. In soils, a diversity of conjugative plasmids involved in microbial metabolism, coping with stress factors or providing their carriers with traits outcompeting competitors in the ecological niche have been detected by classical and molecular techniques. The rhizosphere displays a hot spot for conjugative plasmid transfer in soil. Plasmid transfer has been studied in a variety of soil habitats from pristine to heavily polluted soils, such as heavy metal and radionuclide contaminated ones. In all of these environments, plasmid transfer has been demonstrated to contribute to the spread of genes that provide their recipients with ecological traits conferring adaptation to contaminant stress, to cope with the presence of toxic pharmaceuticals (e.g. antibiotics) or enable them to degrade xenobiotic compounds. Transconjugants harbouring the traits acquired by horizontal transfer were shown to survive in soil and to spread the acquired beneficial factors to both indigenous related and phylogenetically distant microorganisms. PCR screens have been developed to detect conjugative plasmids of different incompatibility groups and virulence/pathogenicity-encoded traits in a rapid, reliable and sensible way. This chapter summarizes the state of the art of mobile genetic elements and their transmission in soils.

Keywords Conjugative plasmids • Mobilizable plasmids • Antibiotic resistance • Molecular tools • Rhizosphere

E. Grohmann (✉)
Division of Infectious Diseases, University Medical Centre Freiburg,
Hugstetter Strasse 55, 79106 Freiburg, Germany
e-mail: elisabeth.grohmann@uniklinik-freiburg.de

A. Malik et al. (eds.), *Management of Microbial Resources in the Environment*,
DOI 10.1007/978-94-007-5931-2_9, © Springer Science+Business Media Dordrecht 2013

1 Introduction

Soil is a very complex habitat dominated by the soil solid phase. In contrast to aquatic systems, it is relatively recalcitrant to mixing, but soluble components of the solid soil matrix may dissolve in soil water and reprecipitate at other sites (Daniel 2004). The soil microorganisms are localized in close association with soil particles, such as complexes of clay-organic matter (Foster 1988). Microorganisms can be found as single cells or microcolonies, which are often enclosed by a polysaccharide matrix usually known as EPS (extracellular polymeric substances). Their metabolism and interactions with other organisms and with soil particles is dependent on the conditions at the microhabitat level, which often differ between microhabitats even at very small distances. The microhabitats in soil include micropores and the surfaces of soil aggregates of various compositions and sizes (Banjard and Richaume 2001; Torsvik et al. 2002; Daniel 2004). Thus, soil can be regarded as very heterogeneous with respect to conditions for microbial growth and for the distribution of microorganisms and matrix substances. This heterogeneity results in multiple local chemical and physical microgradients, a wide variety of microbial niches and a high diversity of soil microorganisms. This microbial diversity exceeds that of other environments and is far greater than that of eukaryotic organisms: 1 g of soil contains up to 10 billion microorganisms of possibly thousands of different species (Roselló-Mora and Amann 2001; Daniel 2004). The genetic complexity of microbial soil communities has been estimated by re-association of community DNA. This type of analyses has shown that the soil community size is equivalent to 6,000–10,000 *Escherichia coli* genomes (Torsvik et al. 1998; Øvreås 2000). Re-association kinetics of the total bacterial DNA in a 30 g soil sample revealed that it contained more than 500,000 species (Doolittle 1999). Thus, the genetic diversity of the soil microbiota is a rich and widely unexplored resource for new enzymes and bioactive compounds (Daniel 2004).

Anthropogenic soils can be defined as human-altered and/or -transported soils (ICOMANTH 2007). The International Committee for Anthropogenic Soils (ICOMANTH) suggests dividing them into six different categories, namely, Mine and Dredge Soils, Urban Soils, Farmed/Altered Soils, Wet Soils, Polluted Soils, and Other Anthropogenic Soils. In all these different types of soils mobile genetic elements (MGE) have been found: MGEs can be divided into the following major groups: plasmids (subdivided into conjugative plasmids, mobilizable plasmids and nonmobilizable plasmids), transposons (classical transposons and conjugative transposons or integrative and conjugative elements (ICE)) and bacteriophages (e.g. reviewed in Wozniak and Waldor 2010; Skippington and Ragan 2011). Out of those, of particular interest are the conjugative and mobilizable plasmids which are able to self-transfer their genetic information (conjugative plasmids) or mobilize it with the help of a conjugative helper plasmid (mobilizable plasmids) to closely and also distantly related microorganisms via horizontal gene transfer (HGT). These plasmids can encode a variety of genetic information additionally to the genes required for self-transfer or their conjugative mobilization. This information can include, among others (i) antibiotic resistance genes (ii) heavy metal resistance genes, (iii) degradative traits enabling the host harbouring the respective plasmid to degrade xenobiotic substances, such as herbicides, fungicides or pesticides and (iv) "spiteful" traits, like the ones governing the production of bacteriocins to outcompete bacterial neighbours.

This chapter will summarize the current knowledge on plasmids, in particular mobilizable and conjugative ones from various anthropogenic soils and discuss their potential applications in bioremediation and soil sanitation.

2 Horizontal Gene Transfer and Plasmid Types

HGT is a major source of phenotypic innovation and adaptation among bacteria. Determinants for antibiotic resistance and other adaptive traits can spread rapidly, particularly by conjugative plasmids (Skippington and Ragan 2011). Plasmids encode a range of phenotypic features and are important agents of HGT among bacteria (Frost et al. 2005; Thomas and Nielsen 2005; Schlüter et al. 2007; Barlow 2009). As a class of MGEs, plasmids are defined by three key features: the capacity to exist and replicate extra chromosomally, the ability to be transferred between distinct hosts and the absence of genes essential to their hosts (Skippington and Ragan 2011). Plasmids are highly diverse in size, structure, transmission, evolutionary history and accessory phenotypes (Slater et al. 2008; Carattoli 2009). This diversity is due in part to the succession of HGT events, resulting in size variation and mosaic structures (Mellata et al. 2009). Plasmids are typically mobilized by conjugative transfer. While not all plasmids encode the functions essential for cell-to cell DNA transfer, mobilizable plasmids can be mobilized by co-resident conjugative plasmids. Unlike conjugative plasmids, which can be maintained and replicate autonomously in their host, most ICEs can be maintained only through integration into the host replicon (Wozniak and Waldor 2010). The chromosome itself can be partially or completely transferred by conjugation, provided it contains a conjugative element and the interbacterial junction is stable long enough-up to an hour or more (Thomas and Nielsen 2005). The remarkable ability of conjugative transfer to mediate plasmid exchange between taxonomically and genetically unrelated bacterial hosts facilitates gene sharing within broad genetic exchange communities (Ochman et al. 2000). Conjugative transfer commonly crosses species and genus boundaries (Davison 1999) and can extend across biological domains (Buchanan-Wollaston et al. 1987; Heinemann and Sprague 1989). Due to general mechanisms such as exclusion, which restrain the conjugative transfer of plasmids, not all strains or species within a community are equally efficient as transfer donors. Some subpopulations of bacteria, including bacteria harbouring plasmids carrying antibiotic resistance genes, have high donor activity; these so-called amplifiers (Dionisio et al. 2002) can accelerate the spread of plasmids within their genetic exchange community (Skippington and Ragan 2011).

Successful conjugative transfer requires donor and recipient cells to be compatible, as determined by surface proteins on the recipient (Thomas and Nielsen 2005). Donor cells encode a specialized multi-protein complex, termed the conjugation apparatus. This conjugation apparatus is encoded by the so-called type IV secretion systems. An important prerequisite for conjugative transfer is an intimate association between the cell surfaces of the interacting donor and recipient cells. In gram-negative bacteria, this physical contact is first established by complex extracellular filaments, designated "sex pili". For the majority of gram-positive bacteria, the means to achieve this intimate cell-cell contact have not yet been identified but do not apparently involve sex pili (Grohmann et al. 2003; Alvarez-Martinez and Christie 2009).

Likely, the frequency of conjugation, however, depends primarily on the donor bacterium; recipient *E. coli* cells, for example, contain no gene indispensable for conjugation (Perez-Mendoza and de la Cruz 2009).

Plasmids can be divided into incompatibility (Inc) groups (Novick 1987). Incompatibility has been described as a manifestation of relatedness: plasmids that utilize common mechanisms for replication or stability cannot proliferate in the same cell line (Carattoli et al. 2005). Inc groups thus constrain plasmid host range. Plasmids must nonetheless adapt to unfavourable hosts if they are to persist long term within a Genetic Exchange Community (Skippington and Ragan 2011). Some plasmids can be maintained only in one or a few bacterial hosts (narrow-host-range plasmids), others replicate in diverse bacterial genera (broad-host-range plasmids). The latter may not be equally stable in all hosts, particularly as their ability to persist in a bacterial population is determined in part by host-encoded traits (De Gelder et al. 2007). Under certain selective conditions, plasmids can extend their host range, often via a relatively small number of genetic changes (De Gelder et al. 2008). Plasmids in the same plasmid family can show very different host ranges (Wu and Tseng 2000), suggesting that broad-host-range plasmids can probably arise selectively from those of narrower host ranges (Thomas and Nielsen 2005). Fondi et al. (2010) introduced the concept of the panplasmidome, the set of all plasmids harboured by members of a taxonomic group. Based on analysis of plasmids from the genus *Acinetobacter*, they concluded that plasmids likely mediate preferential flow of genetic information within and between Genetic Exchange Communities.

3 Traits Transferred via Horizontal Gene Transfer

Prokaryotic genomes are known to vary in their rates of gene gain and loss. Most bacteria are highly dynamic, with high rates of gene gain and loss, whereas some, typically those of obligatory endosymbionts, have stable or shrinking genomes (Rankin et al. 2011). Within genomes, genes differ in their aptitude to be mobile. In addition to the core genome (approximately 2,000 genes in *E. coli* that are present in the first 20 sequenced strains), non-core genes contribute significantly to the overall diversity of gene repertoires in a species, which together with the core are designated the pan-genome (Rankin et al. 2011). In a study of *E. coli*, these non-core genes made up 90% of the pan-genome when 20 strains were put together (Touchon et al. 2009). There is increasing information on important separations of functions between the vertically transmitted core genome, which encodes fundamental cellular processes, and the horizontally transmissible accessory genome, which encodes for a variety of secondary metabolites conferring resistance to specific toxins or antibiotics or the ability to exploit a specific niche (Hacker and Carniel 2001; Norman et al. 2009). The accessory genome contains acquired functions, MGEs, non-expressed genes and genes expressed under particular modes of selection (Rankin et al. 2011). Despite the large interest in research into the molecular mechanisms of HGT, the ecological and evolutionary forces that drive the basic divisions of mobility and function are still poorly understood (see for example, Slater et al. 2008).

Horizontally transferred genes can confer a variety of beneficial and negative effects on their bacterial hosts. The fitness costs of MGEs differ significantly with the element. The introduction of plasmids, ICEs, mobilizable islands or transposable elements into the bacterial genome, does not pose the same life-or-death dilemma as temperate and virulent phages respectively exert on their hosts (Rankin et al. 2011). However, these elements are most likely costly (Diaz Ricci and Hernandez 2000). MGE replication requires the synthesis of proteins, RNA and DNA, which incur in a fitness cost. This cost can be low in general and can be easily exceeded by other adaptive or addictive traits. However, HGT often involves the creation of proteinaceous structures, such as conjugative pili, that can be costly (Rankin et al. 2011). MGEs can also be costly because of the genes they carry compete with other genetic elements. For instance, F-like plasmids use exclusion proteins to prevent super-infection by other closely related plasmids that are very highly expressed (Achtman 1975). MGEs may also carry genes coding for adaptive traits. Thus, they have been denominated "agents of open source evolution" (Frost et al. 2005), suggesting that they facilitate their host access to a vast genetic resource that can then be improved upon and made available to other organisms. As already mentioned, many beneficial traits for the receiving organisms are carried by plasmids, including virulence factors, antibiotic resistance, detoxifying agents and enzymes for secondary metabolism (Philppon et al. 2002; Yates et al. 2006; Martínez 2008, 2009; Rankin et al. 2011). Additionally to having an effect on the bacterial host, many horizontally transferred genes also code for traits that can affect the fitness of a host's neighbours (Rankin et al. 2011). This can be either in a positive way, by producing proteins that can have a beneficial effect on the degrading enzymes of the host's neighbours (Livermore 1995), or in a negative way, by producing compounds that harm the host's neighbours, such as bacteriocins (for instance, van der Ploeg 2005; Brown et al. 2006). In spite of their potentially positive effects on host fitness, it is important to mention that MGEs do not necessarily share the same interest as that of the host genome and can thus be considered first and foremost as infectious agents; they infect their hosts much in the same way as parasites infect their host (through cell to cell contact for plasmids), and can thus persist in spite of potential costs they may impose on the host (Rankin et al. 2011). Whether they confer beneficial or detrimental effects on a host will depend on the selective forces exerting an effect on both the MGE and the host chromosome.

Rankin et al. (2011) classified in an excellent way the behaviour of MGEs by their net influence on their host cell (from parasitic to mutualistic, see for example, Ferdy and Godelle 2005) and additionally by their net influence on neighbouring cells (whether they turn their host cell into "helpers" or "harmers" of neighbouring cells). The classification scheme commonly used in social evolution theory divides social behaviours into two types, depending on their effects on the fitness of a target individual and that of its neighbour. Social behaviours are divided into "mutualism", "selfishness", "altruism" and "spite" (West et al. 2007b). Thus, Rankin et al. (2011) distinguished traits that are kept within the bacterial cell ("private" traits, governing the parasitism–mutualism axis) from those secreted outside the cell ("public" traits, which govern the helping-harming axis). Some of the "private" and "public" traits will be discussed in the following sections.

3.1 Private Traits: Genomic Parasites and Mutualists

3.1.1 MGEs as Parasites

As many prokaryotes have very large effective population sizes, any gene conferring a deleterious effect on their host will be selected against and purged from the population. However, many MGEs do exert costs on their hosts (Diaz Ricci and Hernandez 2000; Fox et al. 2008) raising the question as to how costly and poorly transmissible MGEs can persist within a genome (see for instance, Bergstrom et al. 2000; Lili et al. 2007). The persistence of any parasite is fundamentally determined by its rate of horizontal transmission (Anderson and May 1992), which can be easily measured for plasmids and has been found to vary over eight orders of magnitude among various *E. coli* plasmids (Dionisio et al. 2002). Although highly mobile plasmids can readily persist as so called molecular "parasites" (Bahl et al. 2007), the persistence of elements with extremely low rates of horizontal transfer cannot be readily explained in this way, given any reduction in vertical transmission imposed by the plasmid (see Bergstrom et al. 2000; Dionisio et al. 2002). However, one of the unsolved issues in this context is that HGT rates have until now been primarily determined under laboratory conditions, and are still difficult to estimate for natural populations (Slater et al. 2008).

Infectious elements such as plasmids are likely to face a trade off between horizontal transfer and vertical transmission, mediated by the costs that they impose on their hosts (Haft et al. 2009; Turner 2004). Studies on conjugative plasmids demonstrated that the cost of MGEs generally decreases under selection (Turner et al. 1998; Bouma and Lenski 1988); plasmids that exert lower costs on their hosts have a higher representation in daughter cells. However, as an increase in vertical transfer generally comes at a cost to HGT, there will be an optimum in which gene transfer through some mix of both horizontal and vertical mechanisms is maximized (Paulsson 2002; Rankin et al. 2011).

There is a large variance in the transmissibility of bacterial plasmids (Dionisio et al. 2002), and selection is likely to exert an effect upon hosts to reduce the frequency of conjugation in the case the transfer is costly. Thus, in many plasmids the expression of the conjugation machinery depends on cell density (Kozlowicz et al. 2006), or is repressed after some lag upon acquisition of the plasmid (Polzleitner et al. 1997), presumably because by then most recipient cells already contain the plasmid and the costs of conjugation offset the gains in further transfer.

Considering MGEs as infectious agents highlights the selection pressures driving them to optimize their own parasite function, balancing gains through transfer (vertical and horizontal) with losses through burden and eventually death of the host. If the spatial density of hosts is high, favouring HGT, MGEs will be selected to increase their horizontal transmission, even at a high cost to the host. However, if HGT is reduced to very low levels, then MGEs can only increase their fitness by coding for traits enhancing vertical transmission and for traits that are beneficial to the host (Ferdy and Godelle 2005). Costs of MGEs are likely to be reduced by coevolution with their bacterial hosts. If plasmid-chromosome co-evolution results in the evolution of smaller plasmids, which impose little or no costs on their hosts, there would be no additional fitness cost from a gene that is carried on the plasmid compared with the chromosome. Very small plasmids may not exert a visible fitness effect on their host if they do not have large copy numbers. If there are no or really low costs to a MGE, it will be readily able to spread in a population through HGT (Rankin et al. 2011). Actually, reduced costs of plasmid carriage may have been the driving force for the creation of secondary chromosomes in many bacteria. Such chromosomes often have plasmid-like features and are ubiquitous in some bacterial clades (Egan and Waldor 2003; Slater et al. 2009; Rankin et al. 2011). Accordingly, recent studies showed that the very large plasmids have lost mobility and acquired essential genes (Smillie et al. 2010).

3.1.2 MGEs as Mutualists

MGEs have been associated with the spread of a wide range of adaptive traits that enhance the fitness of their hosts. For example, plasmids commonly carry resistance genes that allow bacteria to grow in the presence of antibiotics or heavy metals, which usually show spatially or temporarily variable distributions (Eberhard 1990). Plasmids also play important roles in the establishment of antagonisms or mutualisms between prokaryotes and eukaryotes, such as virulence traits (Buchrieser et al. 2000), nitrogen fixation by genes encoded on rhizobia symbiotic plasmids (Nuti et al. 1979) or amino acid production, for instance in the insect endosymbiont *Buchnera* (Gil et al. 2006). Genes carried on plasmids may be readily transferred to the bacterial chromosome. Theoretical models suggest that constant selection pressure should favour transfer of beneficial plasmid genes to the chromosome to avoid the costs of plasmid carriage (Eberhard 1990; Bergstrom et al. 2000). This raises the question: Why do beneficial genes remain on MGEs and do not integrate themselves into the chromosome? If selection pressures differ over time or space, genes that are beneficial in some environments, but not others, will be able to persist on

MGEs. In a spatially structured environment, MGEs help to facilitate the transmission of beneficial traits that have previously evolved in local populations to other subpopulations (Bergstrom et al. 2000). Parasites and symbionts usually lie on a continuum between harming (in the case of parasites) and helping (in the case of mutualists) their partners. MGEs that encode beneficial genes can be seen as mutualists and should face the same evolutionary dilemma (Sachs and Simms 2007; Rankin et al. 2011). When should a MGE provide a benefit and when should it harm its host, and how will these evolutionary decisions in turn influence mobility? The more a MGE transmits vertically, the more it will depend on the reproduction of the host, and selection processes will therefore favour genes that are beneficial to the host, whereas higher HGT frequencies will have a negative impact on the host. Over long timescales, elements that harm the host are likely to be lost, and those that are beneficial will probably be integrated into the host genome. Despite the potential benefits of some plasmids to the bacteria involved, sharing beneficial DNA with other members of the population can be seen as a social dilemma. The reason for this is that a bacterium that theoretically does not suppress the transmission of a plasmid will be benefiting its potential competitors. If plasmid transfer is costly, then this may be seen as an altruistic act (Rankin et al. 2011).

3.1.3 MGEs as Drivers of Bacterial Sociality

Social behaviours can be categorized by their net lifetime direct effect on the fitness of a focal bacterium (the actor) and on neighbouring bacteria (the recipients) as described by Hamilton (1964) and West et al. (2007b, c). The resulting four behaviours are "selfishness", which confers a benefit on the actor while exerting a cost on the social partner, "altruism", "spite" and "mutually beneficial behaviours". Over the past 10 years many research efforts have been devoted to microbial sociality, and all four of these social behaviours can be found in microbial populations (West et al. 2006, 2007a, b, c; Xavier and Foster 2007; Rankin et al. 2011). Rankin et al. (2011) argued that all of these four behaviours can be observed to be encoded by horizontally transferred genes. Due to their effect on the local genetic structure, MGEs lend themselves to promoting cooperative social traits (Rankin et al. 2011).

3.2 *Public Traits*

3.2.1 Cooperation: Altruism and Mutually Beneficial Traits

Bacteria are remarkably cooperative organisms, producing a diversity of shared, secreted products (public goods) that can enhance growth in a diversity of challenging environments (West et al. 2006, 2007a). However, similar to any social organism, cooperative bacteria are always vulnerable to exploitation by non-producing cheats. In the absence of any "family" or spatial structure, any individual in a population that does not produce a public good will have an advantage over individuals that do, a situation referred to as the tragedy of the commons (Hardin 1968; Rankin et al. 2007, 2011). As non-producing individuals have a fitness advantage over those that produce public goods, non-producing individuals will invade over time. Many traits that have been shown to be involved in bacterial cooperation and virulence are encoded on MGEs, such as traits that can degrade toxins in the local environment (Philppon et al. 2002; Lee et al. 2006; and secreted toxins (Waldor and Mekalanos 1996; Ahmer et al. 1999).

As plasmids can spread horizontally within a local population, they can change the genetic structuring among individuals in local populations. Relatedness is an influential measure of population genetic structure, used in particular to decipher the direction of selection on social traits (Hamilton 1964; Griffin et al. 2004; Rankin et al. 2011), with higher relatedness between individuals tending to favour the evolution of cooperative traits (Hamilton 1964). Relatedness at mobile loci can increase as a result of HGT. Therefore, it is expected that many genes involved in the production of cooperative traits, or public goods, will be carried by MGEs, in particular by conjugative plasmids (Rankin et al. 2011). Genes encoding proteins that are secreted outside of the cell can be referred to as being part of the secretome. As proteins are costly to produce, secreting them outside of a cell has a clear cost to the producing individuals. Nogueira et al. (2009) analyzed the genomes of 20 *Escherichia* and *Shigella* lineages, determined where proteins were likely to be expressed within a cell, and identified the genes that coded for secreted proteins. The genomes were further analyzed for transfer hotspots (areas in the genome more likely to be transmitted horizontally) and for determining whether the genes coding for secreted proteins were more likely to be located on transfer hotspots or plasmids. They found that 8% of chromosomal hotspot genes coded for extracellular proteins. On the contrary, 15% of the plasmid genes code for extracellular proteins. This suggests that HGT has a high impact on the evolution of social behaviours in bacteria (Nogueira et al. 2009).

A well-studied plasmid influencing the social environment is the tumour-inducing T_i plasmid in the plant pathogen, *Agrobacterium tumefaciens*. For the bacterium to be virulent, it must carry a copy of the T_i plasmid, parts of it are then inserted via HGT into the plant cells. Once in the plant, the plasmid induces cell division, which creates a crown gall (Zupan et al. 2000). The gall releases opines that can then be used as an important source of nitrogen and energy for the bacterium (Zupan et al. 2000; White and Winans 2007). Interestingly, only bacteria that are infected with the plasmid can use the opines meaning that cooperation by plasmids only favours other individuals that harbour the plasmid, making the trait a so-called "greenbeard" (Rankin et al. 2011).

3.2.2 Spiteful Traits

The flipside of altruism is spite, in which an individual pays a cost to impose another cost on another individual (Hamilton 1970; Gardner and West 2004a, b, 2006; Rankin et al. 2011). This is especially common in bacteria, in which individual cells may produce bacteriocins, or other toxins, to kill their conspecifics (Riley and Wertz 2002; Gardner et al. 2004; Ackermann et al. 2008). Spite has long been observed as a puzzling evolutionary phenomenon, as a given actor pays a net cost to impose a net cost on another member of the population (Foster et al. 2001). Spite occurs in various systems (for example, Foster et al. 2000; Gardner et al. 2004, 2007; Rankin et al. 2011) but has most often been invoked regarding anti-competitor behaviours in bacteria (West et al. 2006, 2007a; Dionisio 2007; Brown and Buckling 2008; Rankin et al. 2011). Spiteful extracellular products, which are costly for the producer and cause harm to other members of the population, are frequently found on MGEs, especially on plasmids and phages (see, for example, Brown et al. 2006). In the case of bacteriocins, plasmids can encode both toxins and the corresponding genes for resistance to the toxin (Riley and Wertz 2002). Spite can evolve if the individual exerting an effect spitefully and the recipient of the spiteful behaviour are negatively related. Usually this occurs under small population sizes (Gardner and West 2004b) or if there is a way of recognizing unrelated individuals (Keller and Ross 1998; Brown and Buckling 2008; Rankin et al. 2011). If bacteriocins are encoded on plasmids, then it is easier; if a cell encoding bacteriocins lyses, it kills all members of the neighbourhood that do not carry the plasmid, and this favours individuals that carry the plasmid (as they also carry resistance to the bacteriocin). Therefore, one may regard such genes as being "greenbeard", as bacteriocinogenic individuals (carriers of the toxin-immunity gene complex) preferentially help other individuals carrying the exactly same complex to survive (Rankin et al. 2011). One should expect bacteriocins to be carried by plasmids with intermediate levels of HGT; if transfer is too high, and all individual cells in a local neighbourhood carry a plasmid, there are no non-carriers to kill. In contrast, if plasmids are too rare, then there would be insufficient toxin and subsequent killing to compensate for the fixed costs of toxin production (Chao and Levin 1981).

4 Soil Types and Plasmids Encountered

It is increasingly being recognized that conjugative plasmid transfer across species boundaries plays a vital role in the adaptability of bacterial populations to varying soil conditions. There are specific driving forces for and constraints to plasmid transfer within bacterial communities in soils (Heuer and Smalla 2012). Plasmid-mediated genetic variation enables bacteria to respond rapidly via adaptive responses to challenges such as variable antibiotic or metal concentrations, or the opportunity to utilize xenobiotic compounds as the sole carbon source. Plasmids seem to be an ancient and successful strategy to ensure survival of a bacterial soil population in spatial and temporal heterogeneous conditions with differing environmental stresses or opportunities that occur irregularly or as a novel challenge in soil (Heuer and Smalla 2012).

In all types of soil, for instance, in polluted, mine, urban, farmed, and fertilized soils self-transmissible and mobilizable plasmids have been found and some of them have been studied in detail. However, most of the studies up to now have been only performed in polluted and fertilized/manured soils.

4.1 Plasmids in Polluted and Mine Soils

Most work has been dedicated to studies of plasmids encoding traits that enable the degradation of xenobiotics.

Lin et al. (2012) isolated a *Gordonia* strain capable to utilize naphthalene as a sole carbon source from a diesel-oil industrial park site located in Kaohsiung County, Taiwan. The naphthalene catabolic genes in the *Gordonia* strain CC-NAPH129-6 are organized in an operon-like gene cluster and were found to be located on a 97-kb plasmid harboured by the strain. Furthermore, a partial transposase sequence containing an insertion sequence (IS) element structure was found in the plasmid, which was flanked by direct repeats downstream of *narC,* one of the structural genes of the naphthalene catabolic operon. Comparative analyses of the naphthalene catabolic genes, the 16S rRNA and *gyrB* gene present in strain CC-NAPH129-6 and naphthalene-degrading *Rhodococcus* species implied that the naphthalene catabolic genes in strain CC-NAPH129-6 might be horizontally transferred from *Rhodococcus* spp. Thus, this is the first report on the presence of naphthalene catabolic genes with an operon-like structure in *Gordonia*, and it might provide evidence of the importance of this actinobacterial lineage in the bioremediation of oil-contaminated soils (Lin et al. 2012).

Over the past 40 years, the s-triazine herbicide, atrazine (2-chloro-4-ethylamino-6-isopropylamino-s-triazine), has been widely used in agriculture to control grassy and broadleaf weeds (Udiković-Kolić et al. 2010). To characterize the atrazine-degrading potential of bacterial communities enriched from an agrochemical factory area, upper soil layer samples were collected within the Herbos factory area, Zagreb, Croatia, from three different locations exposed to long-term contamination with atrazine and other s-triazine compounds. The bacterial communities enriched from these three different sites of varying atrazine contamination mineralized 65–80% of ^{14}C ring-labelled atrazine within 4 days. The presence of the atrazine degradative genes, in different combinations, *trzN-atzBC-trzD*, *trzN-atzABC-trzD* and *trzN-atzABC-DEF-trzD*, was confirmed by PCR. In all enriched communities, the *trzN-atzBC* genes were located on a putatively conjugative 165-kb plasmid, while the *atzBC* or *atzC* genes were located on separate plasmids (Udiković-Kolić et al. 2010). Quantitative real-time PCR revealed that catabolic genes were present in up to 4% of the bacterial community. Sequencing of selected clones identified members belonging to the *Proteobacteria* (α-, β- and γ-subclasses), *Actinobacteria* and *Bacteroidetes*. The presence of catabolic genes in a small proportion of the microbial community suggests that only a subset of the community has the ability to degrade atrazine (Udiković-Kolić et al. 2010).

Gene bioaugmentation is a bioremediation strategy that enhances the biodegradative potential via dissemination of degradative genes from introduced microorganisms to indigenous microorganisms. Inoue et al. (2011) performed bioremediation studies with 2,4-dichlorophenoxy-acetic acid (2,4-D)-contaminated soil slurry and *Pseudomonas putida* and *E. coli* strains harbouring the conjugative 2,4-D degradative plasmid pJP4 in microcosms to assess possible effects of gene bioaugmentation on the overall microbial community structure and ecological functions (carbon source utilization and nitrogen transformation potentials). Although exogenous bacteria decreased rapidly, 2,4-D degradation was stimulated in bioaugmented microcosms, probably due to the conjugative transfer of pJP4 to the indigenous soil community. Terminal restriction fragment length polymorphism analysis revealed that, although the bacterial community structure was changed immediately after introducing the exogenous bacteria to the inoculated microcosms, that community gradually approached the one of the uninoculated microcosm (Inoue et al. 2011). In this study transconjugants appeared to be limited to *Burkholderia* spp., which have been frequently found to be 2,4-D-degrading transconjugants of plasmid pJP4 in soil (DiGiovanni et al. 1996; Goris et al. 2002). Furthermore, persistence of transconjugant population even without the selective pressure (for instance, 2,4-D) argues that plasmid pJP4 will be relatively stable in the soil microbial community.

Inoue et al. (2011) concluded that the impact of pJP4 dissemination among the indigenous microbial community would be less pronounced than the introduction of exogenous bacteria, and the influence of gene bioaugmentation with *P. putida* and *E. coli* strains harbouring pJP4 on the indigenous soil microbial populations and their carbon and nitrogen transformation capabilities are not irretrievable and long-lasting. The results of Inoue et al. (2011) suggest that gene bioaugmentation can be effective for the remediation of 2,4-D contaminated soil without drastic impacts on the indigenous microbial community. Ikuma et al. (2012) have recently shown in soil slurry batch reactors that efficient TOL plasmid-mediated bioaugmentation could be further improved by minimal altering of environmental conditions, such as the addition of nutrient.

Heavy-metal contamination of the environment is now wide-spread. Soils may become contaminated from a variety of anthropogenic sources, such as smelters, mining, industry, and application of metal-containing pesticides and fertilizers. Soil microorganisms are very sensitive to moderate heavy-metal concentrations. Pereira Almeida et al. (2006) designed a study to screen for possible mechanisms involved in cadmium resistance of *Rhizobium* isolates, originating from an area in close proximity to heavy metal industries that have been in operation for nearly 30 years. The cadmium tolerance levels of several isolates derived from sites with different heavy-metal contamination were determined. Unexpectedly, extremely cadmium tolerant isolates accumulated higher levels of metal, suggesting the presence of intracellular agents that prevent metal interfering with important metabolic pathways. Plasmid profiles also showed differences; most tolerant isolates harboured two plasmids, with sizes of 485 and 415 kb, indicating that extra chromosomal DNA may be involved in cadmium resistance (Pereira Almeida et al. 2006). One of the characteristics encoded by these plasmids that can interfere with heavy metal tolerance is the synthesis of lipopolysaccharides (LPSs) which can function as a barrier to the entry of metals into the cell (Pereira Almeida et al. 2006). Plasmid profiles showed variations between *Rhizobium* isolates from differently contaminated soils. Most of the moderately tolerant and tolerant isolates harboured three or four plasmids; however, plasmid profiles were more similar between tolerant isolates. Plasmids of 485 and 415 kb were detected only in tolerant and extremely tolerant isolates surviving in conditions containing between 250 and 500 μM of cadmium chloride. Pereira Almeida and co-workers showed that there is a resistance mechanism common to all isolates, which is characterized by the presence and/or induction of a LPS quantity that sequestrates most of the metal extracellularly, acting as a first-defence barrier against heavy-metal stress. However, LPSs failed to reach the highest levels of stress imposed. Pereira Almeida et al. (2006) concluded that rhizobia response to heavy-metal stress is a complex phenomenon, where different mechanisms, some of them carried on plasmids, act together to confer the levels of tolerance observed. Mohamed and Abo-Amer have recently isolated a *P. aeruginosa* RA65 strain from car-traffic impacted Egyptian soil. RA65 harboured a single plasmid of about 9.5 kb which mediated heavy metal resistance (for instance to lead, cadmium and zinc). Consequently, this putatively mobilizable small plasmid could be efficiently used in bioremediation of heavy metal-contaminated soils (Mohamed and Abo-Amer 2012).

4.2 Plasmids in Manured/Wastewater-Fertilized Soils

Binh and co-workers investigated the prevalence and types of self-transmissible antibiotic resistance plasmids in piggery manure, which is applied as crop fertilizer on fields in Germany. Samples from manure storage tanks of 15 farms were analysed, representing diverse sizes of herds, meat or piglet production. Antibiotic resistance plasmids from manure bacteria were captured in *gfp*-tagged rifampicin-resistant *E. coli* recipients and characterized (Binh et al. 2008). The presence of different plasmid types was also investigated in total community DNA by PCR and Southern blot. Two hundred and twenty-eight transconjugants were recovered from the 15 manures using selective media supplemented with the antibiotics, amoxicillin, sulfadiazine or tetracycline. The restriction patterns of 81 plasmids representing different resistance profiles or originating from different samples clustered into seven groups. Replicon typing revealed that most of the plasmids belonged to the incompatibility group IncN, followed by members of the recently discovered pHHV216-like plasmids and the IncP-1 type. The amoxicillin resistance gene *bla*-TEM was found on 44 plasmids, and the sulphonamide resistance genes *sul1*, *sul2* and/or *sul3* on 68 plasmids (Binh et al. 2008). Hybridization of replicon-specific sequences amplified from community DNA revealed that IncP-1 and pHHV216-like plasmids were present in all manures. The authors concluded that "field-scale" piggery manure is a reservoir of broad-host range plasmids conferring multiple antibiotic resistance genes.

The transfer frequencies of MGEs from the manure bacteria to *E. coli* CV601 *gfp*$^+$ in this study were in the same range (10^{-4} to 10^{-8}) as the plasmid transfer frequencies observed in a similar study (Smalla et al. 2000a). Most of the captured plasmids conferred multiple antibiotic resistance, 4 of them even encoded resistance towards 7 of the 8 antibiotics investigated and 40 towards 6 of the 8 antibiotics tested. It is documented that the selection by one antibiotic may co-select for other antibiotic resistances (Normark and Normark 2002). Therefore, it is not surprising that, although Binh et al. (2008) used only one antibiotic for the selection of putative transconjugants, plasmids with multiple resistances were captured. Sulfadiazine resistance was remarkably frequent among the transconjugants (204 out of 228), although only one third of them had originally been isolated from sulfadiazine-supplemented plates. The abundance of sulfadiazine resistance among the transconjugants might reflect the use of sulfadiazine in animal husbandries (Boxall et al. 2003). The *sul2* gene was the most frequently detected *sul* gene conferring resistance to sulfadiazine (46 out of 81) captured from 12 manures. Interestingly, *sul2* was detected on all 19 pHHV216-like plasmids, a new plasmid type with low% G+C content. These plasmids are self-transmissible, encode multiple antibiotic resistances and have been recently identified as a major vector of *sul2* in manure and soil (Heuer et al. 2009). Thus, it may be argued that Low-G+C%-type plasmids like pHHV216 play an important role in the spread of *sul2* genes in manure. Binh and co-workers (2008) demonstrated that the majority of the amoxicillin resistant transconjugants carried the *bla-TEM* gene, confirming the results by Binh et al. (2007) on the abundance of *bla-TEM* genes on MGEs captured from manure and amoxicillin-treated soils. Furthermore, Binh et al. (2008) confirmed that IncN plasmids originating from manure are important vehicles for the *bla*-TEM gene dissemination among manure bacteria. Thus, the use of manure as a soil fertilizer will introduce not only bacteria harbouring MGEs with antibiotic resistance genes but additionally nutrients and antibiotics excreted un-metabolized by farm animals (Witte 2000) that may increase the abundance of transmissible antibiotic resistance in soil- and plant-associated bacterial communities (Heuer and Smalla 2007; Binh et al. 2008).

The impact of wastewater irrigation on the abundance of antibiotics and their corresponding putatively transmissible resistance genes was investigated in soils of the Mezquital Valley (Mexico), the world's largest sewage field, by Dalkmann et al. (2012). Irrigation with untreated wastewater releases among others pharmaceuticals, pathogenic bacteria, and resistance genes into the environment, but little is known about the growing accumulation of these contaminants when wastewater is applied for decades. The researchers sampled a chronosequence of soils having received from zero ("rain-fed soil") to 100 years of wastewater irrigation in the Mezquital Valley and investigated the accumulation of six different antibiotics, as well as the occurrence of *Enterococcus* spp., and *sul* (*sul1*, *sul2*) and *qnr* (*qnrA*, *qnrB*, *qnrS*) genes, conferring resistance to fluorquinolones.

Total concentrations of the antibiotics ciprofloxacin, sulfamethoxazole, and carbamazepine increased with irrigation duration reaching 95% of their upper limit after 19–28 years, in the Mezquital Valley soils. Accumulation was soil-type-specific, with largest accumulation rates in Leptosols and no time-trend in Vertisols. It is to note that *qnrA* genes were not detected, but *qnrS* and *qnrB* were found in two of the wastewater-irrigated soils. Relative concentrations of *sul1* genes in wastewater-irrigated soils were two orders of magnitude larger (3.15×10^{-3} copies/16S rDNA) than in rain-fed soils (4.35×10^{-5}copies/16S rDNA), while those of *sul2* exceeded those in rain-fed soils by a factor of 22. The concentration of *sul* genes continued to increase with long-lasting irrigation together with *Enterococcus* spp. 23S rDNA and total 16S rDNA contents. However, increasing total concentrations of antibiotics in soil were not accompanied by increasing relative abundances of resistance genes. Dalkmann and co-workers assumed that increasing total concentrations of resistance genes with longer time of irrigation could be related to increasing microbial biomass by wastewater irrigation and probably to longer survival of wastewater-borne bacteria between irrigation events. Moreover, increasing absolute concentrations of resistance genes suggest that the genetic information is preserved for longer periods than the compounds themselves.

Heuer and Smalla (2012) studied the role of the broad-host range IncP-1ε plasmids in the dissemination of antibiotic resistance in manure samples and arable soils. IncP-1ε plasmids were detected in total DNA from all tested manure samples and in arable soil via a novel 5′-nuclease assay for real-time PCR. The authors reported a correlation between IncP-1ε plasmid abundance and antibiotic usage (application of sulfadiazine) in a soil microcosm experiment. Fifty IncP-1ε plasmids that were captured in *E. coli* CV601*gfp* from bacterial communities of manure and arable soil were further characterized. All plasmids contained class 1 integrons with highly variable sizes of the gene cassette region and the *sul1* gene (Heuer and Smalla 2012). Three IncP-1ε plasmids captured from soil bacteria and one from manure were completely sequenced. The backbones were nearly identical to that of the previously described IncP-1ε plasmid pKJK5 (Binh et al. 2008). Diverse *Beta-* and *Gamma-Proteobacteria* were found as hosts of the IncP-1ε plasmids in the soil microcosms. The remarkable diversity of antibiotic resistance gene cassettes encoded on these plasmids, the ability to efficiently transfer under soil conditions and the broad host range of IncP-1ε plasmids strongly suggest that these plasmids are important vehicles for the transmission of antibiotic resistances in agricultural, in particular manured soils (Heuer and Smalla 2012).

4.3 Plasmids in Urban Soils

Malik et al. (2008) used PCR typing methods to assess the presence of plasmids of the Inc groups IncP, IncN, IncW and IncQ in total DNA extracts from anthropogenic soils from India and Germany. Soils from two different locations in Germany, (the urban park Tiergarten in Berlin and the abandoned irrigation field Berlin-Buch), and from four different polluted locations in India (for instance an urban road which has been irrigated with wastewater, mainly from lock manufacturing and steel industries, for more than two decades and soil from the Jajmau area of Kanpur which receives partially treated effluents from the tannery industries) were analyzed. PCR amplification of the total DNA extracts revealed the prevalence of IncP-specific sequences in Berlin-Buch and Indian soil samples. The IncP plasmids contained at least one conjugative transfer function, in contrast, IncQ, IncN and IncW-specific sequences were never detected (Malik et al. 2008). The prevalence of the antibiotic resistance genes *ampC*, *tet (O)*, *ermB*, *bla-SHV-5*, *mecA*, and *vanA* was also studied. Three Indian soil samples irrigated with wastewater contained the *ampC* gene, whereas the other resistance genes were not found in any of the samples.

Detection of IncP replicon-specific (*trfA*) and transfer-specific sequences (*oriT*) by PCR amplification and Southern hybridization was a clear indication that IncP plasmids with the potential to disseminate resistance genes and other virulence factors are prevalent in these polluted urban habitats. Furthermore, through exogenous plasmid isolation conjugative plasmids belonging to the IncPβ group encoding resistance to ampicillin were isolated from all the urban sites investigated (Malik et al. 2008).

Contamination of soil and water with antibiotic-resistant bacteria may create reservoirs of antibiotic resistance genes that have the potential to negatively impact future public health through HGT. Cummings et al. (2011) studied the presence of plasmid-mediated quinolone resistance genes – many of the plasmids are self-transmissible – in surface sediments of the Tijuana River Estuary, a sewage-impacted coastal wetland along the U.S.-Mexico border and sediments of Famosa Slough, a nearby urban wetland that is largely unaffected by sewage. *qnrA*, *qnrB*, *qnrS*, *qepA*, and *aac(6′)-Ib-cr* quinolone resistance genes were detected by PCR amplification of metagenomic DNA from the sewage-impacted wetland, while the almost unaffected wetland contained only *qnrB*, *qnrS*, and *qepA*. The number of PCR-positive sites increased in both wetlands after rainfall (Cummings et al. 2011). *qnrA* amplicons were cloned, and nucleotide sequences of most of the *qnrA* amplicons were affiliated with *qnrA* genes found on plasmids of clinical isolates.

Urban stormwater in coastal cities is notorious for sewage contamination (Gersberg et al. 2006; Coulliette and Noble 2008) that results in beach closing and costs municipalities millions of dollars each year (Given et al. 2006; Cummings et al. 2011). The problems are more evident in developing countries where wastewater infrastructure is lacking. Acute public health threats from bacteria and viruses are usually alleviated within 72 h, probably as a result of dilution, osmotic stress due to seawater salinity and photo-inactivation due to exposure to direct sunlight (Boehm et al. 2009). The chronic threats due to repeated inoculation in the area with biological pollutants, however, are largely unknown (Cummings et al. 2011).

It is well documented that human and farm animal faeces contain elevated levels of antibiotic resistant bacteria and resistance genes ("antibiotic resistance pollutants") (Chen et al. 2007; Sommer et al. 2009) as does wastewater within treatment systems (Jindal et al. 2006; Schlüter et al. 2007). It is reasonable to speculate that antibiotic resistance pollutants may find their way back into the human community via vectors, such as shore birds, insects, and marine animals or through direct contact with the water by swimming or fishing (Bonnedahl et al. 2009; Cummings et al. 2011).

It is believed that antibiotic resistance genes evolved in the natural environment, for instance in the microbial antibiotic producers, and subsequently moved into clinically important bacteria through HGT (Martinez 2009). Indeed, the origin of many of the *qnr* resistance genes has been traced to marine organisms (for instance, Poirel et al. 2005; Cattoir et al. 2007). Thus, the contamination of natural environments with plasmid-encoded antibiotic resistance genes may have significant future public health consequences (Cummings et al. 2011).

4.4 Plasmids in Farmed Soils

Umamasheswari and Murali (2010) studied three crop fields in India, namely paddy, sugarcane and tomato crops which have been exposed to the pesticides bavistin, monocrotophos and kinado plus, for the composition of the bacterial population and degradation of pesticides. Pesticide resistant bacteria, affiliated to the human nosocomial pathogens *Staphylococcus aureus*, *E. faecalis* and *Pseudomonas aeruginosa* were found. All the resistant isolates studied harboured a plasmid of 240 kb. Plasmid curing experiments demonstrated that all the isolates lost their bavistin resistance phenotype after plasmid curing. Umamasheswari and Murali (2010) concluded that the bavistin resistance phenotype in these isolates was plasmid borne. However, whether the 240-kb plasmid conferring bavistin resistance is conjugative, remains to be elucidated.

4.5 Plasmids in Pristine Soils

Gentry et al. (2004) performed a study to determine the diversity of 2-, 3-, and 4-chlorobenzoate (CB) degraders in two pristine soils with similar physical and chemical characteristics. Surface soils were collected from forested sites of the Coronado National Forest in Madera Canyon and Oversite Canyon, Arizona, USA, and amended with 500 μg of 2-, 3-, or 4-CB g^{-1} soil. CB degraders were isolated, grouped by DNA fingerprints, identified via 16S rDNA sequences, and screened for the presence of plasmids. Nearly all of the 2-CB degraders originating from the Madera soil were identified as a *Burkholderia* spp. encoding chromosomally located degradative genes. In contrast, several different 3-CB degraders were isolated from the Oversite soil, most of them were identified as *Burkholderia* spp. that encoded the degradative genes on large plasmids (Gentry et al. 2004).

Knowledge of the microbial diversity of the CB degrader community along with an understanding of the microbial ecology of the degrader community that develops when a pristine soil is first exposed to high levels of CBs may help elucidate the mechanisms of microbial adaptation and degrader development, such as HGT, that occur as result of the contamination event (Gentry et al. 2004). Southern hybridization of selected CB degraders revealed details on the encoded genes: the genes from the modified ortho-3-CB cleavage pathway (*tfdC* and *tfdD*) that hybridized to plasmids from the Oversite 3-CB degraders are commonly encoded on conjugative plasmids (Perkins et al. 1990; Gentry et al. 2004). Other 3-CB degradative genes are also found on transmissible plasmids or transposons (Nakatsu et al. 1997). Gentry and co-workers (2004) concluded that even though the Madera and Oversite Canyon soils were from similar locations and were chemically and physically comparable, the bacterial communities that developed resulted in varying abilities to degrade 2-, 3-, and 4-CB. Furthermore, the origin of degradative genes in many of the isolated 3-CB degraders may have been due to gene transfer events, as several of the 3-CB degraders harboured large plasmids, and at least two of them appeared to encode homologous 3-CB degradative genes (Gentry et al. 2004).

5 Techniques Applied to Detect and Characterize Plasmids in Soil

There is a broad spectrum of culture-dependent and culture-independent techniques in use to characterize plasmids in and from soils.

The methods can be divided in those that result in the isolation of the MGE and/or are dependent on the availability of the isolated plasmid and in those independent of plasmid extraction.

The classical methods (including isolation of the MGE) are:

(i) Bacteria isolation from soil and isolation of the plasmid(s) harboured by the strain: provided their sizes permit isolation (large plasmids are prone to physical breakage), plasmid isolation can be performed by diverse commercially available kits for plasmid isolation or by the traditional caesium chloride/ethidium bromide gradient (Sambrook and Russel 2001). The clear separation of the plasmid DNA from the genomic DNA has the great advantage that the plasmid features and traits encoded by the plasmid can be studied separately from the core genome.
(ii) Plasmid Genome Sequencing: in the last 10 years several plasmid sequencing projects have been carried out which revealed an overwhelming amount of interesting data on plasmids from very diverse habitats, including from different soils.

Genome-sequencing data and comparative genomics support the assumption that the horizontal acquisition of genetic modules is a major driver in bacterial evolution and adaptability (de la Cruz and Davies 2000; Smillie et al. 2010; Heuer and Smalla 2012). The increasing number of sequenced plasmids from soil also confirmed this assumption for plasmids from soil bacteria and enabled new insights into the diversity of "plasmid backbone genes" required for replication, maintenance and transfer of plasmids.

A major finding of the plasmid genome sequencing projects is the modular character of most of the plasmids. The sequences of the genes coding for replication, maintenance and transfer provided a far more reliable basis for plasmid grouping than the classical incompatibility testing, enabling insights into the evolution of plasmids (Sevastsyanovich et al. 2008; Smillie et al. 2010; Heuer and Smalla 2012). Sequencing of plasmids from soil bacterial isolates and exogenously isolated plasmids broadened our view of various plasmid groups and suggests that plasmids consist of core (backbone) genes and flexible (accessory) genes as indicated before. In soils various different types of plasmids have been detected such as broad-host range plasmids, PromA-like plasmids, conjugative narrow host range plasmids, IncP-9 plasmids, low GC plasmids, and plasmids of the *B. cereus* group.

5.1 Broad Host Range (BHR) Plasmids

Plasmids with a BHR are putatively important for many taxa in soil, as they can transfer useful traits such as antibiotic, heavy metal- or UV-resistance genes, efflux pumps, restriction modification systems, and toxin–antitoxin systems between different members of the bacterial communities in soil (Heuer and Smalla 2012). Several studies have demonstrated that the transfer range of plasmids is often much broader than their replication range (Musovic et al. 2006; Baharoglu et al. 2010). The ability of plasmids to replicate in a wide range of hosts requires that plasmid replication is to a large extent independent of the host's replication machinery. Furthermore, a tight regulation of plasmid-encoded genes ensures a reduced metabolic burden for their host. Repression of transfer genes might represent a compromise between the benefits for the plasmid to transfer horizontally and the associated metabolic burden on the hosts (Haft et al. 2009; Heuer and Smalla 2012). To predict the evolutionary range of hosts in which plasmids evolved, Suzuki et al. (2010) recently explored the use of nucleotide composition for a range of plasmids. Based on the assumption that plasmids acquire the genomic signature of their hosts, the trinucleotide composition of a variety of plasmids was compared with all sequenced bacterial genomes (Heuer and Smalla 2012). Indeed, the signatures of several typical BHR plasmids such as IncP-1, PromA and IncU showed signatures that were dissimilar to any chromosomal signature, suggesting that these plasmids did not fix in a particular host (Suzuki et al. 2010). BHR plasmids most often reported from soil bacteria belonged either to the IncP-1 group or the recently proposed PromA group (Van der Auwera et al. 2009). Sequence-based insights into the diversity of backbone genes of these main groups of BHR plasmids in soil revealed a strong modular organization of the plasmid cores, comprising genes for replication, maintenance and transfer and hot spots for acquisition of accessory genes.

Plasmids belonging to the IncP-1 group have attracted the attention of plasmid biologists for more than 40 years, as these plasmids transfer and stably replicate in a wide range of bacterial hosts and are considered BHR plasmids (Adamczyk and Jagura-Burdzy 2003). In the 1990s, the complete sequence of the two archetype plasmids of the IncP-1α, RP4 (Pansegrau et al. 1994) and the IncP-1β group, R751 (Thorsted et al. 1998) provided the first insights from comparative genomics. Nowadays, 28 published IncP-1 sequences are available. The sequenced plasmids originated from different geographic regions and environments, e.g. sewage, soils and river sediments. Genomic signature analysis data from Norberg et al. (2011) indicated that the different IncP-1 plasmids analyzed to date have adapted to hosts belonging to different species and that their backbones seem to originate from various parental plasmids. Accessory elements were inserted at various sites, the so-called hot spots of insertion (Heuer et al. 2004). Although IncP-1 plasmids were shown experimentally to have a wide host range, stable replication in the absence of selective pressure is strain-specific (De Gelder et al. 2007; Heuer et al. 2007; Sota et al. 2010). The complete sequences of nine IncP-1 plasmids isolated from soil bacteria of various geographic origins revealed that IncP-1plasmids are also important shuttles of beneficial traits in soil bacteria (Trefault et al. 2004; Smalla et al. 2006; Bahl et al. 2007; Heuer and Smalla 2012).

5.2 PromA-Like Plasmids

The complete sequence of plasmids pSB102 and pIPO2, captured directly from rhizosphere bacteria by exogenous plasmid isolation, revealed striking similarities (Schneiker et al. 2001; Tauch et al. 2002). Additionally, the overall organization of the plasmids was very similar to plasmid pXF51 from the plant pathogen *Xylella fastidiosa*. Given the similarities between these plasmids, a novel family of BHR plasmids was proposed because both pIPO2 and pSB102 transferred to a wide range of gram-negative bacteria but did not belong to any of the known BHR plasmid groups revealed by PCR and Southern hybridization probing, and confirmed by sequencing. Comparative genomics based on their complete sequences enabled Tauch et al. (2002) to discover that plasmids pSB102, pIPO2 and pXF51 had a strikingly similar overall genetic organization and led to the proposal of a new family of environmental BHR plasmids. Van der Auwera et al. (2009) proposed naming BHR-plasmids belonging to the pIPO2 family PromA in accordance with the traditional BHR plasmids PromN (IncN), PromP (IncP-1), PromU (IncU) and PromW (IncW).

5.3 Conjugative Narrow Host Range Plasmids

Conjugative plasmids that stably replicate only in a restricted number of taxonomically related species or those that are found only in a limited number of hosts are termed narrow host range plasmids. Narrow host range plasmids might tend to transfer modules that require more integration into cellular networks for their function. For example, plasmid-transferred genes amending upper degradative pathways of aromatic compounds are only beneficial for particular hosts with a corresponding lower pathway (Heuer and Smalla 2012).

5.4 IncP-9 Plasmids

Plasmids belonging to the IncP-9 group were most often detected in *Pseudomonas* isolates from polluted soils and are supposed to play an important role in the adaptation of *Pseudomonas* populations (Heuer and Smalla 2012). In contrast to IncP-1 plasmids their biology is far less studied and the host range of IncP-9 plasmids seems to be narrow, as no transfer to recipients other than *Pseudomonas* has been shown (Krasowiak et al. 2002). The complete sequences of four IncP-9 plasmids are available, revealing an approximately 35-kb IncP-9 core, with genes involved in replication, partitioning and transfer (Heuer and Smalla 2012). On the basis of the *oriV* and *rep* sequences of 28 IncP-9 plasmids, a novel primer system was recently developed and applied to detect IncP-9 plasmids in total community DNA in soils from various geographic regions (Heuer and Smalla 2012). Cloning and sequencing of PCR amplicons revealed a surprisingly high diversity of IncP-9 amplicons, suggesting that, similar to IncP-1 plasmids, various subgroups of IncP-9 plasmids might co-occur in the same environmental niche and contribute to a rapid adaptability of *Pseudomonas* populations.

5.5 Low GC Plasmids

Studies on the effect of veterinary drugs introduced into soil via manure on the abundance of transmissible antibiotic resistance recently led to the discovery of another novel plasmid group. Plasmids belonging to this new group were the plasmids most frequently captured directly from manure-treated soil in *E. coli* in several independent experiments (Heuer and Smalla 2012). These plasmids did not hybridize with any of the previously described plasmid probes or primers and displayed a huge diversity, based on their plasmid restriction patterns and the antibiotic resistance genes encoded (Binh et al. 2008; Heuer et al. 2009). The complete sequences of three representatives of this group – pHHV35, pHHV216 and pHH1107 – were recently determined and showed that all three of them shared virtually an identical plasmid backbone of approximately 30 kb and had an unusually low GC content of about 36%. The accessory regions of pHHV35, pHHV216 and pHH1107 showed a mosaic structure with multiple antibiotic resistance genes and were similar in size (27, 28.3, and 28 kb), respectively. Real-time PCR revealed a surprisingly high abundance of these replicons in several soils (Heuer and Smalla 2012). Considering the huge diversity of their accessory gene load, Heuer et al. (2009) concluded that this plasmid group might play an important role in disseminating antibiotic resistance genes among soil bacterial populations.

5.6 Plasmids of the B. cereus Group

Recent whole genome sequencing projects of several isolates of the *Bacillus cereus* group revealed that the isolates previously classified as *B. cereus*, *Bacillus thuringiensis* or *Bacillus anthracis* are only distinguished by the content of plasmids and their accessory genes. The availability of the pXO1 and pXO2 plasmid sequences provided the basis for the development of PCR primer systems targeting different plasmid backbone genes and genes coding for toxins (Heuer and Smalla 2012). The application of these primer systems for screening environmental isolates belonging to the *B. cereus* group provided evidence that pXO1- and pXO2-like replicons can be detected in *B. cereus* isolates from soils (Hu et al. 2009a, b). Self-transfer and plasmid mobilization was observed among pXO2 plasmids carrying transfer (*tra*) genes. Interestingly, components (VirB4, VirB11, VirD4) of the type IV secretion-like transfer system of gram-negative bacteria were found encoded on several pXO2-like plasmids, suggesting that the three key proteins also form a fundamental core in *Bacillus* plasmids (Hu et al. 2009b). A model for a type IV secretion-like transfer system was recently proposed for the enterococcal BHR plasmid pIP501 based on a protein–protein interaction mapping of all pIP501-encoded Tra proteins (Abajy et al. 2007). The 350-kb pXO16 plasmid from *B. thuringiensis* conferring an aggregation phenotype DNA transfer was shown not only to conjugate efficiently, but also mobilize and retro-mobilize (Timmery et al. 2009).

The culture-independent methods can be divided into different categories. The most important techniques are listed below:

(i) Polymerase Chain Reaction (PCR)-based methods. The PCR methods are usually applied on total genomic DNA that has been isolated from the microbial soil community. They target the origin of conjugative transfer, *oriT* (Malik et al. 2008), *tra* genes, such as the conjugative relaxase (Anjum et al. 2012; Schiwon et al. unpublished) and the type IV secretion system signature genes, *virB1*, *virB4*, and *virD4* (Schiwon et al. unpublished) or antibiotic resistance genes or gene cassettes encoded on the plasmids (Malik et al. 2008; Anjum et al. 2011, 2012; Kopmann et al. 2013; Dalkmann et al. 2012). These PCRs are performed in the qualitative way to look for the absence or presence of these genes and/or to compare different samples with each other or as quantitative real-time PCR to determine the concentration of the respective gene(s) in the soil (Walsh et al. 2011; Dalkmann et al. 2012). The qualitative PCRs are frequently followed by dot blot or Southern hybridization on the respective genes to confirm the PCR results obtained or to detect *tra* or antibiotic resistance genes which are present in low copy-number and therefore cannot be unambiguously detected by qualitative PCR (Ansari et al. 2008; Malik et al. 2008; Anjum et al. 2012).

(ii) Conjugative transfer experiments and exogenous plasmid isolation. Through conjugative transfer experiments between different soil bacteria or between a labelled donor bacterium and the microbial soil community plasmid transfer frequencies in soil can be estimated and transferred or mobilized plasmids can be isolated from the transconjugants and further characterized. As label, fluorescence based methods, such as *gfp* or *rfp* are frequently applied (for instance, Arends et al. 2012). These labels have the advantage that they can be easily detected in situ through fluorescence microscopy, thus soil bacteria which have acquired a *gfp*-marked plasmid can be visualized and enumerated by epifluorescence microscopy, fluorescence-activated cell sorting (FACS) or even spatially localized in the sample by confocal microscopy. Exogenous plasmid isolation usually applies the addition of a marked recipient strain (often encoding, resistance to a specific antibiotic-in the classical exogenous plasmid isolation protocol, a rifampin resistant *E. coli* strain is used) to samples of the microbial soil community. Capture of conjugative plasmids by the introduced recipient strain is visualized by growth on double selective plates, amended with both the antibiotics selecting for the recipient and for potential self-transmissible plasmids conferring another resistance phenotype to the transconjugant (Bale et al. 1988; Hill et al. 1992).

(iii) Soil column/soil microcosm experiments. Many researchers apply soil microcosms as model systems for their studies on microbial soil communities, as they enable simulating a soil ecosystem in the laboratory. Application of soil columns for the understanding of processes taking place within and among soil microbial populations includes among others studies on conjugative plasmid transfer from a labelled donor bacterium introduced into the microcosm to the indigenous bacterial populations. Studies of this type using *gfp* and *rfp*-labelled *E. faecalis* donor bacteria have been performed by Broszat et al. (unpublished data) on wastewater-irrigated and rain-fed soils from the Mezquital Valley (Mexico). Furthermore, soil columns are well-suited to test degradative capabilities of bacteria, which are frequently encoded on plasmids, e.g. on pJP4 plasmid coding for 2,4-D degradation (Quan et al. 2011). In this type of soil microcosms, the substrate to be degraded is frequently applied in radiolabelled form and the incorporation of radiolabelled carbon in bacterial 16S rRNA genes is monitored by stable isotope probing (SIP). For instance, Uhlik et al. (2012) identified bacteria utilizing biphenyl, benzoate, and naphthalene in microcosms with contaminated soil and amended with ^{13}C-labelled biphenyl, benzoate, or naphthalene by SIP and sequence analysis of 16S rRNA gene pools via amplicon pyrosequencing (Pilloni et al. 2012).

(iv) "Omic" technologies, such as transcriptomics and metagenomics. Transcriptomics or microarray studies allow comparisons of gene expression in soil microbial communities which have been exposed to varying physiological conditions, subjected to environmental stresses, such as water stress or amendment with toxic compounds. Metagenomics studies offer the possibility to obtain complete genomic information on the microbial habitat of interest. If 16S rDNA amplified from a habitat or total DNA isolated from the habitat is applied to Illumina® sequencing platforms, the output is a complete data set of the core genome and accessory genome (including the genomes of MGEs such as conjugative plasmids) of the soil of interest (for a review refer to Degnan and Ochman 2012). First metagenomics studies on wastewater-irrigated soils *versus* rain-fed soils from the Mezquital Valley, Mexico have been carried out by Broszat et al. (unpublished data).

Heuer and Smalla have recently published an excellent review on the role of conjugative plasmids in soil bacteria, on the bacterial soil populations and the repertoire of tools to identify them (Heuer and Smalla 2012). The knowledge on plasmids in soil bacteria is still biased towards those carried by taxa that are accessible through cultivation techniques. Furthermore, plasmids are also assumed to be present only in a small proportion of a given bacterial population and thus their abundance might often be below the detection limit. However, under conditions of selective pressure they might become detectable (Heuer and Smalla 2012). An overview of the experimental tools available at present to study plasmids in soil bacteria is given in Fig. 9.1. Direct detection of plasmid backbone genes in total community DNA was already suggested in the 1990s to obtain insights into the occurrence of conjugative plasmids in soil, and PCR detection systems were provided for the known BHR plasmids at the time (Götz et al. 1996). The increasing number of newly sequenced

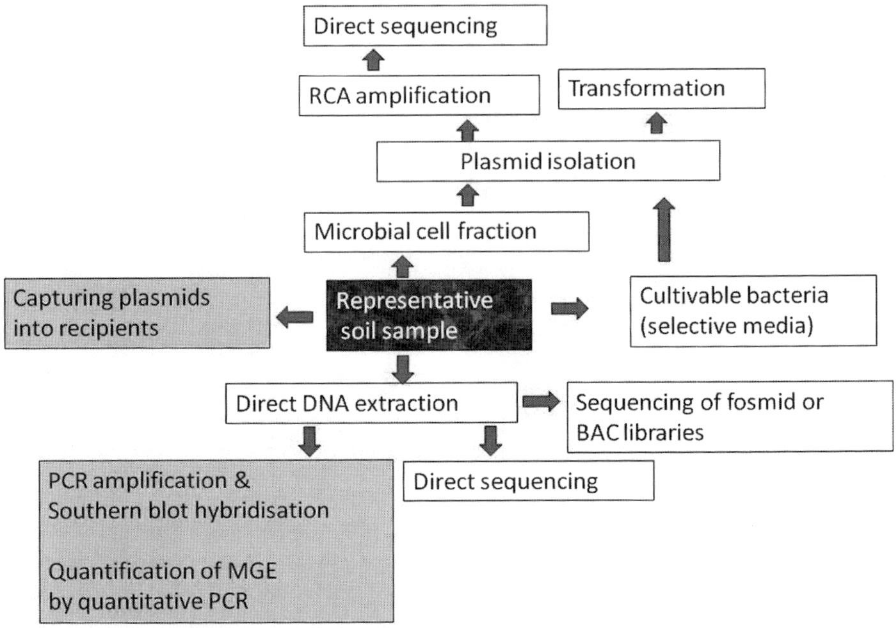

Fig. 9.1 Cultivation-dependent and -independent approaches to detect and analyze plasmids from soil bacteria. *RCA* Rolling circle amplification, *BAC* Bacterial artificial chromosome (With permission from Heuer and Smalla (2012))

plasmids revealed that primer systems designed on the basis of only a few sequences available at the time might fail to detect the huge diversity of different plasmid groups; for instance, the originally proposed primer systems targeting the *trfA* gene of IncP-1 plasmids will not efficiently amplify plasmid-specific sequences from the newly discovered groups of IncP-1γ, ε and δ from community DNA (Heuer and Smalla 2012). This limitation has to be taken into consideration when interpreting previously published studies (for instance, Heuer et al. 2002; van Overbeek et al. 2002; Binh et al. 2008). If PCR amplicons are obtained from soil DNA, the diversity of PCR-amplified plasmid sequences can be analyzed by either cloning and sequencing (fosmid or BAC libraries, see Fig. 9.1) or direct pyrosequencing.

Although PCR-Southern blot detection is at best semi-quantitative, the rather high specificity of the hybridization probes enabled the assessment of the presence of different plasmid groups or subgroups (Heuer and Smalla 2012). Quantitative real-time PCR targeting plasmid-specific genes provides the chance to quantify plasmids in total community DNA and to determine their relative abundance (plasmid copy number related to 16S rRNA gene copy number). It enables the correlation between plasmid abundance as well as plasmid-encoded functions and environmental pollution. The main limitation of PCR-based screening is that the genetic context of the plasmid-specific sequences amplified, as well as of the host, remain unknown (Heuer and Smalla 2012). To explore the diversity and the types of accessory genes carried, the plasmids have to be either isolated from the culturable fraction or captured directly from soil bacteria into suitable recipients (see above).

New insights into plasmids from soil bacteria will come from metagenomics studies. Pyrosequencing of DNA from rivers highly polluted with antibiotics revealed the occurrence of different plasmids and enabled the assembly of four small plasmids encoding different antibiotic resistance genes (Kristiansson et al. 2011). The main advantage of the pyrosequencing of total or plasmid DNA extracted from soils is that novel plasmid sequences will be discovered. A major limitation is that the sequencing depth affects the detection of plasmids in less abundant populations and thus will cover only a fraction of the entire soil metagenome (Kristiansson et al. 2011). Additionally, assembly, particularly of large plasmids, will be difficult due to the low copy number or the use of random circle amplification (Heuer and Smalla 2012).

6 Conclusions and Perspectives

At present it is known that plasmids are found in all types of soils and in virtually all known bacterial taxa. They play an important role in shaping the physiological capabilities of the soil microbial community, and have primordial importance in adaptation processes of soil populations to environmental changes and stress situations. In the last decade a great number of plasmid genomes have been fully sequenced, plasmid sequencing projects continue and will further broaden our knowledge on the plethora of genetic information carried on the genomes of mobile genetic elements. Of particular interest are conjugative plasmids that encode beneficial traits for the microbial soil community, such as the catabolic plasmids, which are frequently used in bioaugmentation experiments or plasmids encoding for heavy metal resistance which are applied in first attempts in soil sanitation. Another promising approach that involves plasmids is rhizostabilization: it is based on the introduction of rhizosphere bacteria that tolerate increased concentrations of heavy metals. The beneficial plant-microbe interactions on the plant root surfaces help improving phytoextraction and phytostabilization in heavily contaminated soils.

Continuous methodological improvements such as the successful application of fluorescence microscopy and FACS in soils will enable to estimate transfer frequencies of introduced self-transmissible or mobilizable plasmids in soils, independent of long term survival of the introduced donor strain. Such approaches with labelled mobilizable broad-host-range plasmids are under progress in polluted soils. Continuous efforts in plasmid genome sequencing projects as well as in metagenomics approaches on various soil types will further extend our knowledge on the wealth of beneficial traits carried by plasmids that await our exploitation. At the end of the decade we will hopefully reach a level of knowledge on plasmid diversity and mosaic accessory genome structures comparable to that on bacterial genomes.

Acknowledgements I sincerely thank Miquel Salgot and Jacques Mahillon for critical reading of the manuscript. I regret that not all valuable contributions of colleagues in the field could have been included in this chapter due to space limitation.

References

Abajy MY, Kopec J, Schiwon K, Burzynski M, Doring M, Bohn C, Grohmann E (2007) A type IV-secretion-like system is required for conjugative DNA transport of broad-host-range plasmid pIP501 in gram-positive bacteria. J Bacteriol 189:2487–2496

Achtman M (1975) Mating aggregates in *Escherichia coli* conjugation. J Bacteriol 123:505–515

Ackermann M, Stecher B, Freed NE, Songhet P, Hardt W-D, Doebeli M (2008) Self-destructive cooperation mediated by phenotypic noise. Nature 454:987–990

Adamczyk M, Jagura-Burdzy G (2003) Spread and survival of promiscuous IncP1 plasmids. Acta Biochim Pol 50:425–453

Ahmer BMM, Tran M, Heron F (1999) The virulence plasmid of *Salmonella typhimurium* is self-transmissible. J Bacteriol 181:1364–1368

Alvarez-Martinez CE, Christie PJ (2009) Biological diversity of prokaryotic type IV secretion systems. Microbiol Mol Biol Rev 73:775–808

Anderson RM, May RM (1992) Infectious diseases of humans: dynamics and control. Oxford University Press, Oxford

Anjum R, Grohmann E, Malik A (2011) Molecular characterization of conjugative plasmids in pesticide tolerant and multi-resistant bacterial isolates from contaminated alluvial soil. Chemosphere 84:175–181

Anjum R, Grohmann E, Malik A (2012) Exogenous isolation of conjugative plasmids from pesticide contaminated soil. World J Microbiol Biotechnol 28:567–574

Ansari MI, Grohmann E, Malik A (2008) Conjugative plasmids in multi-resistant bacterial isolates from Indian soil. J Appl Microbiol 104:1774–1781

Arends K, Schiwon K, Sakinc T, Huebner J, Grohmann E (2012) A GFP-labeled monitoring tool to quantify conjugative plasmid transfer between gram-positive and gram-negative bacteria. Appl Environ Microbiol 78:895–899

Baharoglu Z, Bikard D, Mazel D (2010) Conjugative DNA transfer induces the bacterial SOS response and promotes antibiotic resistance development through integron activation. PLoS Genet 6(10):e1001165

Bahl MI, Hansen LH, Goesmann A, Sorensen SJ (2007) The multiple antibiotic resistance IncP-1 plasmid pKJK5 isolated from a soil environment is phylogenetically divergent from members of the previously established alpha, beta and delta sub-groups. Plasmid 58:31–43

Bale MJ, Day MJ, Fry JC (1988) Novel method for studying plasmid transfer in undisturbed river epilithon. Appl Environ Microbiol 54:2756–2758

Banjard L, Richaume A (2001) Quantitative and qualitative microscale distribution of bacteria in soil. Res Microbiol 152:707–716

Barlow M (2009) What antimicrobial resistance has taught us about horizontal gene transfer. In: Gogarten MB, Gogarten JP, Olendzenski LC (eds) Horizontal gene transfer – genomes in flux. Humana Press/Springer, New York, pp 397–411

Bergstrom CT, Lipsitch M, Levin BR (2000) Natural selection, infectious transfer and the existence conditions for bacterial plasmids. Genetics 155:1505–1519

Binh CT, Heuer H, Gomes NC, Kotzerke A, Fulle M, Wilke BM, Schloter M, Smalla K (2007) Short-term effects of amoxicillin on bacterial communities in manured soil. FEMS Microbiol Ecol 62:290–302

Binh TCT, Heuer H, Kaupenjohann M, Smalla K (2008) Piggery manure used for soil fertilization is a reservoir for transferable resistance plasmids. FEMS Microbiol Ecol 66:25–37

Boehm AB, Yamahara KM, Love DC, Peterson BM, McNeill K, Nelson KL (2009) Covariation and photoinactivation of traditional and novel indicator organisms and human viruses at a sewage-impacted marine beach. Environ Sci Technol 43:8046–8052

Bonnedahl J, Drobni M, Gauthier-Clerc M, Hernandez J, Granholm S, Kayser Y, Melhus A, Kahlmeter G, Waldenström J, Johansson A, Olsen B (2009) Dissemination of *Escherichia coli* with CTX-M-type ESBL between humans and yellow-legged gulls in the south of France. PLoS One 4(6):e5958

Bouma JE, Lenski RE (1988) Evolution of a bacteria plasmid association. Nature 335:351–352

Boxall ABA, Fogg LA, Kay P, Blackwell PA, Pemberton EJ, Croxford A (2003) Prioritisation of veterinary medicines in the UK environment. Toxicol Lett 142:207–218

Brown SP, Buckling A (2008) A social life for discerning microbes. Cell 135:600–604

Brown SP, Le Chat L, De Paepe M, Taddei F (2006) Ecology of microbial invasions: amplification allows virus carriers to invade more rapidly when rare. Curr Biol 16:2048–2052

Buchanan-Wollaston V, Passiatore JE, Cannon F (1987) The *mob* and *oriT* mobilization functions of a bacterial plasmid promote its transfer to plants. Nature 328:170–175

Buchrieser C, Glaser P, Rusniok C, Nedjari H, D'Hauteville H, Kunst F et al (2000) The virulence plasmid pWR100 and the repertoire of proteins secreted by the type III secretion apparatus of *Shigella flexneri*. Mol Microbiol 38:760–771

Carattoli A (2009) Resistance plasmid families in Enterobacteriaceae. Antimicrob Agents Chemother 53:2227–2238

Carattoli A, Bertini A, Villa L, Falbo V, Hopkins KL, Threlfall EJ (2005) Identification of plasmids by PCR-based replicontyping. J Microbiol Methods 63:219–228

Cattoir V, Poirel L, Mazel D, Soussy CJ, Nordmann P (2007) *Vibrio splendidus* as the source of plasmid-mediated *qnrS*-like quinolone resistance determinants. Antimicrob Agents Chemother 51:2650–2651

Chao L, Levin BR (1981) Structured habitats and the evolution of anticompetitor toxins in bacteria. Proc Natl Acad Sci USA 78:6324–6328

Chen J, Yu Z, Michel FC Jr, Wittum T, Morrison M (2007) Development and application of real-time PCR assays for quantification of erm genes conferring resistance to macrolides lincosamides-streptogram in livestock manure and manure management systems. Appl Environ Microbiol 73:4407–4416

Coulliette AD, Noble RT (2008) Impacts of rainfall on the water quality of the Newport River Estuary (Eastern North Carolina, USA). J Water Health 6:473–482

Cummings DE, Archer KF, Arriola DJ, Baker PA, Faucett KG, Laroya JB, Pfeil KL, Ryan CR, Ryan KR, Zuill DE (2011) Broad dissemination of plasmid-mediated quinolone resistance genes in sediments of two urban coastal wetlands. Environ Sci Technol 45:447–454

Dalkmann P, Broszat M, Willaschek E, Sakinc T, Huebner J, Siebe C, Amelung W, Grohmann E, Siemens J (2012) Accumulation of pharmaceuticals, Enterococcus, and resistance genes in soils after up to 100 years of wastewater irrigation in the Mezquital Valley, Mexico. PLoS ONE 7(9):e45397

Daniel R (2004) The soil metagenome – a rich resource for the discovery of novel natural products. Curr Opin Biotechnol 15:199–204

Davison J (1999) Genetic exchange between bacteria in the environment. Plasmid 42:73–91

De Gelder L, Ponciano J, Joyce P, Top EM (2007) Stability of a promiscuous plasmid in different hosts: no guarantee for a long-term relationship. Microbiology 153:452–463

De Gelder L, Williams JJ, Ponciano JM, Sota M, Top EM (2008) Adaptive plasmid evolution results in host-range expansion of a broad-host-range plasmid. Genetics 178:2179–2190

de la Cruz F, Davies J (2000) Horizontal gene transfer and the origin of species: lessons from bacteria. Trends Microbiol 8:128–133

Degnan PH, Ochman H (2012) Illumina-based analysis of microbial community diversity. ISME J 6:183–194

Diaz Ricci JC, Hernandez ME (2000) Plasmid effects on *Escherichia coli* metabolism. Crit Rev Biotechnol 20:79–108

DiGiovanni GD, Neilson JW, Pepper IL, Sinclair NA (1996) Gene transfer of *Alcaligenes eutrophus* JMP134 plasmid pJP4 to indigenous soil recipients. Appl Environ Microbiol 62:3297–3304
Dionisio F (2007) Selfish and spiteful behaviour through parasites and pathogens. Evol Ecol Res 9:1199–1210
Dionisio F, Matic I, Radman M, Rodrigues OR, Taddei F (2002) Plasmids spread very fast in heterogeneous bacterial communities. Genetics 162:1525–1532
Doolittle W (1999) Phylogenetic classification and the universal tree. Science 284:2124–2128
Eberhard WG (1990) Evolution in bacterial plasmids and levels of selection. Q Rev Biol 65:3–22
Egan ES, Waldor MK (2003) Distinct replication requirements for the two *Vibrio cholerae* chromosomes. Cell 114:521–530
Ferdy JB, Godelle B (2005) Diversification of transmission modes and the evolution of mututalism. Am Nat 166:613–627
Fondi M, Bacci G, Brilli M, Papaleo MC, Mengoni M, Vaneechoutte M, Dijkshoorn L, Fani R (2010) Exploring the evolutionary dynamics of plasmids: the *Acinetobacter* panplasmidome. BMC Evol Biol 10:59
Foster RC (1988) Microenvironments of soil microorganisms. Biol Fertil Soils 6:189–203
Foster KR, Ratnieks FLW, Wenseleers T (2000) Spite in social insects. Trends Ecol Evol 15:469–470
Foster KR, Wenseleers T, Ratnieks FLW (2001) Spite: Hamilton's unproven theory. Ann Zool Fennici 38:229–238
Fox RE, Zhong X, Krone SM, Top EM (2008) Spatial structure and nutrients promote invasion of IncP-1 plasmids in bacterial populations. ISME J 2:1024–1039
Frost LS, Leplae R, Summers AO, Toussaint A (2005) Mobile genetic elements: the agents of open source evolution. Nat Rev Microbiol 3:722–732
Gardner A, West SA (2004a) Spite among siblings. Science 305:1413–1414
Gardner A, West SA (2004b) Spite and the scale of competition. J Evol Biol 17:1195–1203
Gardner A, West SA (2006) Spite. Curr Biol 16:R662–R664
Gardner A, West SA, Buckling A (2004) Bacteriocins, spite and virulence. Proc R Soc Lond B 271:1529–1535
Gardner A, Hardy ICW, Taylor PD, West SA (2007) Spiteful soldiers and sex ratio conflict in polyembryonic parasitoid wasps. Am Nat 169:519–533
Gentry TJ, Wang G, Rensing C, Pepper IL (2004) Chlorobenzoate-degrading bacteria in similar pristine soils exhibit different community structures and population dynamics in response to anthropogenic 2-, 3-, and 4-chlorobenzoate levels. Microb Ecol 48:90–102
Gersberg RM, Rose MA, Robles-Sikisaka R, Dhar AK (2006) Quantitative detection of hepatitis A virus and enteroviruses near the United States-Mexico border and correlation with levels of fecal indicator bacteria. Appl Environ Microbiol 72:7438–7444
Gil G, Sabater-Muñoz B, Perez-Brocal V, Silva FJ, Latorre A (2006) Plasmids in the aphid endosymbiont *Buchnera aphidicola* with the smallest genomes. A puzzling evolutionary story. Gene 370:17–25
Given S, Pendleton LH, Boehm AB (2006) Regional public health cost estimates of contaminated coastal waters: a case study of gastroenteritis at southern California beaches. Environ Sci Technol 40:4851–4858
Goris J, Dejonghe W, Falsen E, De Clerck E, Geeraerts B, Willems A, Top EM, Vandamme P, De Vos P (2002) Diversity of transconjugants that acquired plasmid pJP4 or pEMT1 after inoculation of a donor strain in the A- and B-horizon of an agricultural soil and description of *Burkholderia hospital* sp. nov. and *Burkholderia terricola* sp. nov. Syst Appl Microbiol 25:340–352
Götz A, Pukall R, Smit E, Tietze E, Prager R, Tschäpe H, van Elsas JD, Smalla K (1996) Detection and characterization of broad-host-range plasmids in environmental bacteria by PCR. Appl Environ Microbiol 62:2621–2628
Griffin AS, West SA, Buckling A (2004) Cooperation and competition in pathogenic bacteria. Nature 430:1024–1027
Grohmann E, Muth G, Espinosa M (2003) Conjugative plasmid transfer in gram-positive bacteria. Microbiol Mol Biol Rev 67:277–301

Hacker J, Carniel E (2001) Ecological fitness, genomic islands and bacterial pathogenicity. A Darwinian view of the evolution of microbes. EMBO Rep 2:376–381

Haft RJF, Mittler JE, Traxler B (2009) Competition favours reduced cost of plasmids to host bacteria. ISME J 3:761–769

Hamilton WD (1964) The genetical evolution of social behaviour. I & II. J Theor Biol 7:1–52

Hamilton WD (1970) Selfish and spiteful behaviour in an evolutionary model. Nature 228:1218–1220

Hardin G (1968) The tragedy of the commons. Science 162:1243–1248

Heinemann JA, Sprague JF Jr (1989) Bacterial conjugative plasmids mobilize DNA transfer between bacteria and yeast. Nature 340:205–209

Heuer H, Smalla K (2007) Manure and sulfadiazine synergistically increased bacterial antibiotic resistance in soil over at least two months. Environ Microbiol 9:657–666

Heuer H, Smalla K (2012) Plasmids foster diversification and adaptation of bacterial populations in soil. FEMS Microbiol Rev 36:1083–1104

Heuer H, Kroegerrecklenfort E, Wellington EMH et al (2002) Gentamicin resistance genes in environmental bacteria: prevalence and transfer. FEMS Microbiol Ecol 42:289–302

Heuer H, Szczepanowski R, Schneiker S, Pühler A, Top EM, Schlüter A (2004) The complete sequences of plasmids pB2 and pB3 provide evidence for a recent ancestor of the IncP-1β group without any accessory genes. Microbiology 150:3591–3599

Heuer H, Fox RE, Top EM (2007) Frequent conjugative transfer accelerates adaptation of a broad-host-range plasmid to an unfavorable *Pseudomonas putida* host. FEMS Microbiol Ecol 59:738–748

Heuer H, Kopmann C, Binh CTT, Top EM, Smalla K (2009) Spreading antibiotic resistance through spread manure: characteristics of a novel plasmid type with low %G+C content. Environ Microbiol 11:937–949

Hill KE, Weightman AJ, Fry JC (1992) Isolation and screening of plasmids from the epilithon which mobilize recombinant plasmid pD10. Appl Environ Microbiol 58:1292–1300

Hu X, Swiecicka I, Timmery S, Mahillon J (2009a) Sympatric soil communities of *Bacillus cereus* sensu lato: population structure and potential plasmid dynamics of pXO1- and pXO2-like elements. FEMS Microbiol Ecol 70:344–355

Hu X, Van der Auwera G, Timmery S, Zhu L, Mahillon J (2009b) Distribution, diversity, and potential mobility of extrachromosomal elements related to the *Bacillus anthracis* pXO1 and pXO2 virulence plasmids. Appl Environ Microbiol 75:3016–3028

ICOMANTH (International Committee for Anthropogenic Soils) Report No. 2 – Version 2.0, August, 2007

Ikuma K, Holzem RM, Gunsch CK (2012) Impacts of organic carbon availability and recipient bacteria characteristics on the potential for TOL plasmid genetic bioaugmentation in soil slurries. Chemosphere 89:158–163

Inoue D, Yamazaki Y, Tsutsui H, Sei K, Soda S, Fujita M, Ike M (2011) Impacts of gene bioaugmentation with pJP4-harboring bacteria of 2,4-D-contaminated soil slurry on the indigenous microbial community. Biodegradation 23:263–276

Jindal A, Kocherginskaya S, Mehboob A, Robert M, Mackie RI, Raskin L, Zilles JL (2006) Antimicrobial use and resistance in swine waste treatment systems. Appl Environ Microbiol 72:7813–7820

Keller L, Ross KG (1998) Selfish genes: a green beard in the red fire ant. Nature 394:573–575

Kopmann C, Jechalke S, Rosendahl I, Groeneweg J, Krögerrecklenfort E, Zimmerling U, Weichelt V, Siemens J, Amelung W, Heuer H, Smalla K (2013) Abundance and transferability of antibiotic resistance as related to the fate of sulfadiazine in maize rhizosphere and bulk soil. FEMS Microbiol Ecol 83:125–134

Kozlowicz BK, Shi K, Gu ZY, Ohlendorf DH, Earhart CA, Dunny GM (2006) Molecular basis for control of conjugation by bacterial pheromone and inhibitor peptides. Mol Microbiol 62:958–969

Krasowiak R, Smalla K, Sokolov S, Kosheleva I, Sevastyanovich Y, Titok M, Thomas CM (2002) PCR primers for detection and characterisation of IncP-9 plasmids. FEMS Microbiol Ecol 42:217–225

Kristiansson E, Fick J, Janzon A, Grabic R, Rutgersson C, Weijdegard B, Soederstroem H, Larsson DG (2011) Pyrosequencing of antibiotic-contaminated river sediments reveals high levels of resistance and gene transfer elements. PLoS One 6:e17038

Lee K, Lee M, Shin JH, Lee MH, Kang SH, Park AJ et al (2006) Prevalence of plasmid-mediated AmpC β-Lactamases in *Escherichia coli* and *Klebsiella pneumoniae* in Korea. Microb Drug Resist 12:44–49

Lili LN, Britton NF, Feil EJ (2007) The persistence of parasitic plasmids. Genetics 177:399–405

Lin CC, Shen FT, Tan CC, Huang CC, Chen BY, Arun AB, Young CC (2012) Characterization of *Gordonia* sp. strain CC-NAPH129-6 capable of naphthalene degradation. Microbiol Res 167(7):395–404

Livermore DM (1995) Beta-Lactamases in laboratory and clinical resistance. Clin Microbiol Rev 8:557–584

Malik A, Celik EK, Bohn C, Böckelmann U, Knobel K, Grohmann E (2008) Detection of conjugative plasmids and antibiotic resistance genes in anthropogenic soils from Germany and India. FEMS Microbiol Lett 279:207–216

Martínez JL (2008) Antibiotics and antibiotic resistance genes in natural environments. Science 321:365–367

Martinez JL (2009) The role of natural environments in the evolution of resistance traits in pathogenic bacteria. Proc R Soc Lond B 276:2521–2530

Mellata M, Touchman JW, Curtiss R III (2009) Full sequence and comparative analysis of the plasmid pAPEC-1 of avian pathogenic *E. coli* w7122 (O78:K80:H9). PLoS One 4:e4232

Mohamed RM, Abo-Amer AE (2012) Isolation and characterization of heavy-metal resistant microbes from roadside soil and phylloplane. J Basic Microbiol 52:53–65

Musovic S, Oregaard G, Kroer N, Sorensen SJ (2006) Cultivation-independent examination of horizontal transfer and host range of an IncP-1 plasmid among gram-positive and gram-negative bacteria indigenous to the barley rhizosphere. Appl Environ Microbiol 72:6687–6692

Nakatsu CH, Providenti M, Wyndham RC (1997) The cis-diol dehydrogenase *cbaC* gene of Tn*5271* is required for growth on 3-chlorobenzoate but not 3,4-dichlorobenozate. Gene 196:209–218

Nogueira T, Rankin DJ, Touchon M, Taddei F, Brown SP, Rocha EP (2009) Gene mobility drives the evolution of bacterial cooperation and virulence. Curr Biol 19:1683–1691

Norberg P, Bergström M, Jethava V, Dubhashi D, Hermansson M (2011) The IncP-1 plasmid backbone adapts to different host bacterial species and evolves through homologous recombination. Nat Commun 2:268

Norman A, Hansen LH, Sorensen SJ (2009) Conjugative plasmids: vessels of the communal gene pool. Philos Trans R Soc Lond B 364:2275–2289

Normark BH, Normark S (2002) Evolution and spread of antibiotic resistance. J Intern Med 252:91–106

Novick RP (1987) Plasmid incompatibility. Microbiol Rev 51:381–395

Nuti MP, Lepidi AA, Prakash RK, Schilperoort RA, Cannon FC (1979) Evidence for nitrogen fixation (*nif*) genes on indigenous *Rhizobium* plasmids. Nature 282:533–535

Ochman H, Lawrence JG, Groisman EA (2000) Lateral gene transfer and the nature of bacterial innovation. Nature 405:299–304

Øvreås L (2000) Population and community level approaches for analysing microbial diversity in natural environments. Ecol Lett 3:236–251

Pansegrau W, Lanka E, Barth PT, Figurski DH, Guiney DG, Haas D, Helinski DR, Schwab H, Stanisich VA, Thomas CM (1994) Complete nucleotide sequence of Birmingham IncP-alpha plasmids – compilation and comparative analysis. J Mol Biol 239:623–663

Paulsson J (2002) Multileveled selection on plasmid replication. Genetics 161:1373–1384

Pereira SI, Lima AI, Figueira EM (2006) Screening possible mechanisms mediating cadmium resistance in *Rhizobium leguminosarum bv. viciae* isolated from contaminated Portuguese soils. Microb Ecol 52:176–186

Perez-Mendoza D, de La Cruz F (2009) *Escherichia coli* genes affecting recipient ability in plasmid conjugation: are there any? BMC Genomics 10:71

Perkins EJ, Gordon MP, Caceres O, Lurquin PF (1990) Organization and sequence analysis of the 2,4-dichlorophenol hydroxylase and dichlorocatechol oxidative operons of plasmid pJP4. J Bacteriol 172:2351–2359
Philppon A, Arlet G, Jacoby GA (2002) Plasmid-determined ampC-type beta-lactamases. Antimicrob Agents Chemother 46:1
Pilloni G, Granitsiotis MS, Engel M, Lueders T (2012) Testing the limits of 454 pyrotag sequencing: reproducibility, quantitative assessment and comparison to T-RFLP fingerprinting of aquifer microbes. PLoS One 7(7):e40467
Poirel L, Rodriguez-Martinez J-M, Mammeri H, Liard A, Nordmann P (2005) Origin of plasmid-mediated resistance determinant QnrA. Antimicrob Agents Chemother 49:3523–3525
Polzleitner E, Zechner EL, Renner W, Fratte R, Jauk B, Högenauer G et al (1997) TraM of plasmid R1 controls transfer gene expression as an integrated control element in a complex regulatory network. Mol Microbiol 25:495–507
Quan X, Tang H, Ma J (2011) Effects of gene augmentation on the removal of 2,4-dichlorophenoxyacetic acid in a biofilm reactor under different scales and substrate conditions. J Hazard Mater 185:689–695
Rankin DJ, Bargum K, Kokko H (2007) The tragedy of the commons in evolutionary biology. Trends Ecol Evol 22:643–651
Rankin DJ, Rocha EPC, Brown SP (2011) What traits are carried on mobile genetic elements, and why? Heredity 106:1–10
Riley MA, Wertz JE (2002) Bacteriocins: evolution, ecology and application. Annu Rev Microbiol 56:117–137
Roselló-Mora R, Amann R (2001) The species concept for prokaryotes. FEMS Microbiol Rev 25:39–67
Sachs JL, Simms EL (2007) Pathways to mutualism breakdown. Trends Ecol Evol 21:585–592
Sambrook J, Russel DW (2001) Purification of closed circular DNA by equilibrium centrifugation in CsCl-Ethidium Bromide gradients: continuous gradients. In: Irwing N (ed) Molecular cloning. A laboratory manual. Cold Spring Harbor Laboratory Press, New York
Schlüter A, Szczepanowski R, Pühler A, Top EM (2007) Genomics of IncP-1 antibiotic resistance plasmids isolated from wastewater treatment plants provides evidence for a widely accessible drug resistance gene pool. FEMS Microbiol Rev 31:449–477
Schneiker S, Keller M, Dröge M, Lanka E, Pühler A, Selbitschka W (2001) The genetic organization and evolution of the broad host range mercury resistance plasmid pSB102 isolated from a microbial population residing in the rhizosphere of alfalfa. Nucleic Acids Res 29:5169–5181
Sevastsyanovich YR, Krasowiak R, Bingle LEH, Haines AS, Sokolov SL, Kosheleva IA, Leuchuk AA, Titok MA, Smalla K, Thomas CM (2008) Diversity of IncP-9 plasmids of *Pseudomonas*. Microbiology 154:2929–2941
Skippington E, Ragan MA (2011) Lateral genetic transfer and the construction of genetic exchange communities. FEMS Microbiol Rev 35:707–735
Slater FR, Bailey MJ, Tett AJ, Turner SL (2008) Progress towards understanding the fate of plasmids in bacterial communities. FEMS Microbiol Ecol 66:3–13
Slater SC, Goldman BS, Goodner B, Setubal JC, Farrand SK, Nester EW et al (2009) Genome sequences of three *Agrobacterium* biovars help elucidate the evolution of multichromosome genomes in bacteria. J Bacteriol 191:2501–2511
Smalla K, Heuer H, Goetz A, Niemeyer D, Kroegerrecklenfort E, Tietze E (2000a) Exogenous isolation of antibiotic resistance plasmids from piggery manure slurries reveals a high prevalence and diversity of IncQ-like plasmids. Appl Environ Microbiol 66:4854–4862
Smalla K, Haines AS, Jones K, Kroegerrecklenfort E, Heuer H, Schloter M, Thomas CM (2006) Increased abundance of IncP-1β plasmids and mercury resistance genes in mercury polluted river sediments: first discovery of IncP-1β plasmids with a complex *mer* transposon as the sole accessory element. Appl Environ Microbiol 72:7253–7259
Smillie C, Garcillán-Barcia MP, Francia MV, Rocha EP, de la Cruz F (2010) Mobility of plasmids. Microbiol Mol Biol Rev 74:434–452

Sommer MOA, Dantas G, Church GM (2009) Functional characterization of the antibiotic resistance reservoir in the human microflora. Science 325:1128–1131

Sota M, Yano H, Hughes JM, Daughdrill GW, Abdo Z, Forney LJ, Top EM (2010) Shifts in the host range of a promiscuous plasmid through parallel evolution of its replication initiation protein. ISME J 4:1568–1580

Suzuki H, Yano H, Brown CJ, Top EM (2010) Predicting plasmid promiscuity based on genomic signature. J Bacteriol 192:6045–6055

Tauch A, Schneiker S, Selbitschka W et al (2002) The complete nucleotide sequence and environmental distribution of the cryptic, conjugative, broad-host-range plasmid pIPO2 isolated from bacteria of the wheat rhizosphere. Microbiology 148:1637–1653

Thomas CM, Nielsen KM (2005) Mechanisms of, and barriers to, horizontal gene transfer between bacteria. Nat Rev Microbiol 3:711–721

Thorsted PA, Macartney DP, Akhtar P et al (1998) Complete sequence of the IncP beta plasmid R751: implications for evolution and organisation of the IncP backbone. J Mol Biol 282:969–990

Timmery S, Modrie P, Minet O, Mahillon J (2009) Plasmid capture by the *Bacillus thuringiensis* conjugative plasmid pXO16. J Bacteriol 191:2197–2205

Torsvik V, Daa FL, Sandaa RA, Øvreås L (1998) Novel techniques for analysing microbial diversity in natural and perturbed environments. J Biotechnol 64:53–62

Torsvik V, Daa FL, Sandaa RA, Øvreås L (2002) Microbial diversity and function in soil: from genes to ecosystems. Curr Opin Microbiol 5:240–245

Touchon M, Hoede C, Tenaillon O, Barbe V, Baeriswyl S, Bidet P et al (2009) Organised genome dynamics in the *Escherichia coli* species results in highly diverse adaptive paths. PLoS Genet 5:e1000344

Trefault N, De la Iglesia R, Molina AM, Manzano M, Ledger T, Perez-Pantoja D, Sanchez MA, Stuardo M, Gonzalez B (2004) Genetic organization of the catabolic plasmid pJP4 from *Ralstonia eutropha* JMP134 (pJP4) reveals mechanisms of adaptation to chloroaromatic pollutants and evolution of specialized chloroaromatic degradation pathways. Environ Microbiol 6:655–668

Turner PE (2004) Phenotypic plasticity in bacterial plasmids. Genetics 167:9–20

Turner PE, Cooper VS, Lenski RE (1998) Tradeoff between horizontal and vertical modes of transmission in bacterial plasmids. Evolution 52:315–329

Udikovi -Koli N, Hršak D, Devers M, Klepac-Ceraj V, Petri I, Martin-Laurent F (2010) Taxonomic and functional diversity of atrazine-degrading bacterial communities enriched from agrochemical factory soil. J Appl Microbiol 109:355–367

Uhlik O, Wald J, Strejcek M, Musilova L, Ridl J, Hroudova M, Vlcek C, Cardenas E, Mackova M, Macek T (2012) Identification of bacteria utilizing biphenyl, benzoate, and naphthalene in long-term contaminated soil. PLoS One 7(7):e40653

Umamaheswari S, Murali M (2010) Prevalence of plasmid mediated pesticide resistant bacterial assemblages in crop fields. J Environ Biol 31:957–964

Van der Auwera GA, Krol JE, Suzuki H, Foster B, Van Houdt R, Brown CJ, Mergeay M, Top EM (2009) Plasmids captured in *C. metallidurans* CH34: defining the PromA family of broad-host-range plasmids. Antonie Van Leeuwenhoek 96:193–204

van der Ploeg JR (2005) Regulation of bacteriocin production in *Streptococcus mutans* by the quorum-sensing system required for development of genetic competence. J Bacteriol 187:3980–3989

van Overbeek LS, Wellington EMH, Egan S, Smalla K, Heuer H, Collard JM, Guillaume G, Karagouni AD, Nikolakopoulou TL, van Elsas JD (2002) Prevalence of streptomycin-resistance genes in bacterial populations in European habitats. FEMS Microbiol Ecol 42:277–288

Waldor MK, Mekalanos J (1996) Lysogenic conversion by a filamentous phage encoding cholera toxin. Science 272:1910–1914

Walsh F, Ingenfeld A, Zampicolli M, Hilber-Bodmer M, Frey JE, Duffy B (2011) Real-time PCR methods for quantitative monitoring of streptomycin and tetracycline resistance genes in agricultural ecosystems. J Microbiol Methods 86:150–155

West SA, Griffin AS, Gardner A, Diggle SP (2006) Social evolution theory for microorganisms. Nat Rev Microbiol 4:597–607

West SA, Diggle SP, Buckling A, Gardner A, Griffin AS (2007a) The social lives of microbes. Ann Rev Ecol Evol Syst 38:53–77

West SA, Griffin AS, Gardner A (2007b) Evolutionary explanations for cooperation. Curr Biol 17:R661–R672

West SA, Griffin AS, Gardner A (2007c) Social semantics: altruism, cooperation, mutualism, strong reciprocity and group selection. J Evol Biol 20:415–432

White CE, Winans SC (2007) Cell-cell communication in the plant pathogen *Agrobacterium tumefaciens*. Philos Trans R Soc Lond B Biol Sci 362:1135–1148

Witte W (2000) Selective pressure by antibiotic use in livestock. Int J Antimicrob Agents 16(Suppl 1):S19–24

Wozniak RAF, Waldor MK (2010) Integrative and conjugative elements: mosaic mobile genetic elements enabling dynamic lateral gene flow. Nat Rev Microbiol 8:552–563

Wu LT, Tseng YH (2000) Characterization of the IncW cryptic plasmid pXV2 from *Xathomonas campetris* pv. Vesicatoria. Plasmid 44:163–172

Xavier JB, Foster KR (2007) Cooperation and conflict in microbial biofilms. Proc Natl Acad Sci USA 104:876–881

Yates CM, Shaw DJ, Roe AJ, Woolhouse MEJ, Amyes SGB (2006) Enhancement of bacterial competitive fitness by apramycin resistance plasmids from non-pathogenic *Escherichia coli*. Biol Lett 2:463–465

Zupan J, Muth TR, Draper O, Zambryski P (2000) The transfer of DNA from *Agrobacterium tumefaciens* into plants: a feast of fundamental insights. Plant J 23:11–28

Chapter 10
Potential Eco-friendly Soil Microorganisms: Road Towards Green and Sustainable Agriculture

Surinder Kaur, Gurpreet Singh Dhillon, Satinder Kaur Brar, Vijay Bahadur Chauhan, Ramesh Chand, and Mausam Verma

Abstract Rapidly increasing human population, industrialization and urbanization has led to agricultural land shrinkage resulting in food crisis. The widespread use of agrochemicals to enhance the food production has worsened the situation and jeopardized the environment, human and animal health. This mandates the need and importance of sustainable agriculture. In this context, microorganisms play a vital and vibrant role in sustainable agriculture. Microorganisms are ubiquitous and participate in an inimitable role in maintaining the dynamism and integrity of the sustainable biosphere. Microorganisms mediated transformation, degradation and recycling of organic matter ensures the sustained existence of life in an eco-friendly manner. They constitute the largest group of living creatures with varying potential in biochemical, physiological and nutritional modes. They play an integral role in many fields, such as bioremediation, biomedicine, environmental and agricultural sectors. Microorganisms leverage the agricultural sector through mutual symbiosis in agricultural plants, biological control agents against several plant pathogens, plant growth promotion and nutrient cycling.

S. Kaur
Department of Mycology & Plant Pathology, Institute of Agricultural Sciences, Banaras Hindu University (BHU), Varanasi, Uttar Pradesh 221005, India

INRS-ETE, Université du Québec, 490, Rue de la Couronne, QC, Québec G1K 9A9, Canada

G.S. Dhillon • S.K. Brar (✉)
INRS-ETE, Université du Québec, 490, Rue de la Couronne, QC, Québec G1K 9A9, Canada
e-mail: satinder.brar@ete.inrs.ca

V.B. Chauhan • R. Chand
Department of Mycology & Plant Pathology, Institute of Agricultural Sciences, Banaras Hindu University (BHU), Varanasi, Uttar Pradesh 221005, India

M. Verma
Institut de Recherche et de Développement en Agroenvironnement Inc. (IRDA), 2700 rue Einstein, Québec, QC G1P 3W8, Canada

A. Malik et al. (eds.), *Management of Microbial Resources in the Environment*,
DOI 10.1007/978-94-007-5931-2_10, © Springer Science+Business Media Dordrecht 2013

With the advancement in microbial technology, these microbes have been successfully exploited with significant achievements in recent years. An appreciation and utilization of such agriculturally important microorganisms (AIMs) have the potential to improve living standards by enhancing the nation's economy and wealth creation. This chapter discusses various aspects of microorganisms and their potential role in agriculture. The chapter highlights the contribution of efficient soil microbes in sustainable agriculture and management of soil fertility for sustainable agriculture. The chapter also addresses the potential plant growth promoting rhizobacteria (PGPR) along with dynamics and significance of other soil microbial communities and their valuable and obliging roles in agricultural productivity and sustainable agriculture development. Various microorganisms that contribute to the production of agricultural antibiotics and compost production are also discussed. The importance of soil microorganisms towards sustaining agro-ecosystem, it's functioning and the directions for future research are also discussed. Finally, the chapter sheds light on the health and safety issues related to the utilization of these microorganisms including their non-target effects and persistence in the environment and potential for horizontal gene transfer.

Keywords Agriculturally important microorganisms (AIMs) • Biocontrol agents (BCAs) • Health and safety issues • Plant growth promoting rhizobacteria (PGPR) • Sustainable agriculture

Abbreviations

ACC	1-aminocyclopropane-1-carboxylate
a.i.	active ingredient
AIMs	agriculturally important microorganisms
BCAs	biocontrol agents
Bt	*Bacillus thuringensis*
GABA	γ-aminobutyric acid
IAA	indole acetic acid
ISR	induced systemic resistance
mRNA	messenger ribonucleic acid
PGPR	plant growth promoting rhizobacteria
PSB	phosphate solubilising bacteria

1 Introduction

The rapid growth and development of human population has led to advancement in agriculture and industrial revolution which in turn has invited several negative effects, such as environmental pollution, excessive soil erosion, surface run-off of chemical fertilizers and pesticides to surface and groundwater contamination throughout the globe. The disproportionate erosion of topsoil from farmland caused by agricultural practices such as intensive tillage and row-crop production has led to extensive soil degradation and both surface waters and groundwater pollution. The organic wastes from animal production, agriculture, marine processing industries and municipal wastes, such as sewage and garbage have become a major concern of environmental safety in both developed and developing countries. An area that appears to hold the greatest promise for technological advancement in crop production, crop protection, and natural resource conservation is that of beneficial and efficient microorganisms applied as soil, plant and environmental inoculants (Higa 1995).

In this context, the concept of sustainable agriculture emerged that aims for the development of productive, profitable and eco-friendly farming systems. It should also conserve the natural resources as well as guarantee the food safety and quality. An enormous number of microbes have been explored and used in the past to function in various different areas , such as food processing, safety and quality, medical technology, genetic engineering, biocontrol agents, agriculture, municipal waste treatments, kitchen waste management, biofertilizers, waste water treatment, composting, metal remediation, nitrogen fixation, phytohormone and vitamin production, bioremediation and phytoremediation, overcoming abiotic stress and many others (Figueiredo et al. 2008; Sandhya et al. 2009; Singh et al. 2011a). Considering the potential of microorganisms against several pathogens and in improving the plant health, they are now being used in nature farming as well as organic agriculture (Daly and Stewart 1999; Maeder et al. 2002; Khaliq et al. 2006). The concept of Effective Microorganisms (EMs) was developed by Professor Teruo Higa, University of the Ryukyus, Okinawa, Japan (Higa 1991; Higa and Wididana 1991). Efficient microorganisms play an important role in nitrogen fixation, decomposition of organic wastes, detoxification of many chemicals, disease suppression, disease control and plant growth promotion (Birkhofer et al. 2008). They also play a diverse role in agro-ecosystem and maintain sustainable agriculture. Hence these microorganism are designated as Agriculturally Important Microorganisms (AIMs) throughout the chapter.

2 Microbial Technology and Agriculture

The utilization of microorganisms in an efficient manner requires potential and efficient technology. The development and advancement of microbiology, agricultural microbiology, genetic engineering, molecular tools, bioinformatics and metagenomics has led to the efficient utilization of many AIMs for agricultural intensification. Microbial technology is being used with considerable success in various sectors, such as, industrial fermentations, antibiotics production, biocontrol agents and plant growth hormones, among others. However, they are often not accepted by the scientific community owing to the inconsistency and non-reproducible results. This could be explained on the basis that microorganisms require optimum growth conditions in order to metabolize their substrates and yield effective results.

In the meantime, various types of microbial cultures and inoculants are available in the market and functioning in various sectors.

In order to fulfil the increasing demands of food and agricultural commodities, microbial technology and agriculture must interact in close proximity to yield effective and feasible results. This could be achieved by the development and execution of conceptual designs and new technologies for the efficient utilization of AIMs for sustainable agriculture and environment. Conceptual design relates to envisioning a principle or model and then to formulate a strategy and technique to achieve the authenticity. It also necessitates the careful coordination of the materials, environment and the technology comprising the method. Likewise, an idealistic approach must be adopted while validating microbial technologies to agricultural production and conservation systems.

3 Soil Microbes for Sustainable Agriculture

Several microorganisms, such as photosynthetic bacteria, actinomycetes, wood degrading fungi, nematophagous fungi, fungal parasites, nitrogen fixers function together in the same natural environment by producing several bioactive compounds, such as vitamins, hormones and enzymes that stimulate plant growth and suppress harmful pathogens. Therefore, these microbes can open new avenues for shifting the soil microbiological equilibrium in ways that can improve soil quality, enhance crop production and protection, conserve natural resources, and ultimately create a more sustainable agriculture and environment, crop residue recycling, and biocontrol of pests (Stockdale et al. 2002). Crop growth and development are affected by the microbes in close proximity of the plant roots i.e., the rhizospheric microflora (Raaijmakers et al. 2009). It thus suggests the importance of managing and exploring soil microflora. An array of biological activities is influenced by AIMs.

4 Ecological Benefits of Managing Soil Microflora

Profitable cropping systems must be designed to realize a nutrient balance through continuous nutrient replenishment. Nutrients are replenished from rocks, processes of decomposition and mineralization of organic materials and atmospheric fixation by soil microorganisms. Under natural conditions, the release rate of nutrients from minerals to soils and plants are not readily available. Increasing land pressure has led to either reduced or even missing fallow periods, hence culminating nutrient mining from soil (Shannon et al. 2002). In most cases, the rate of replenishment is slow and cannot surpass nutrient mining by crops and loss of nutrients by soil mismanagement. Soil replenishment of nutrients is therefore a necessity for sustainable production systems. Only a small proportion of applied fertilizers are available to crops, as nutrients may be lost through leaching, washing away, immobilization and volarization (Wells et al. 2000; Ghorbani et al. 2009). Mechanisms to combat loss of nutrients and enhance nutrient use efficiency can be achieved efficiently by microbiological processes (Richardson et al. 2009).

Nutrient cycling processes are mediated by soil organisms, hence forming a major component of the soil system and therefore greatly contributing to soil health (Shannon et al. 2002). Amongst the keystone microorganisms, soil microorganisms that contribute to soil health and subsequent plant growth are bacteria (symbiotic and free living), such as *Rhizobium*, *Bradyrhzobium*, *Azotobacter*, *Azospirillum sp.*, *Bacillus sp.* and *Pseudomonas sp.* and fungi in symbiosis with plants (Arbuscular Mycorrhizae), fungi in the rhizosphere (*Trichoderma sp.*) and other plant growth promoting organisms (Rodríguez and Fraga 1999; Singh et al. 2011a).

5 Role of Soil Microflora in Sustainable Agriculture

Microbial population can be increased by the addition of organic amendments to the soil. Generally, the soil microorganisms are heterotrophic in nature i.e. they require carbon and nitrogen for metabolism and biosynthesis. Applications of organic materials, such as seaweed, fish meal and chitin from crustacean shells not only helps to balance the micronutrient content of a soil but also increases the population of beneficial antibiotic-producing actinomycetes. They help to improve the soil texture and transform it into a disease-suppressive state (Sullivan 2001). Nevertheless, ecosystem and environmental conditions will affect the probability that a particular beneficial microorganism will become predominant with organic farming or nature farming methods (Peacock et al. 2001).

6 Plant Growth Promoting Rhizobacteria (PGPR) in Sustainable Agriculture

Plant growth-promoting rhizobacteria are free-living bacteria that exert beneficial effects in and around the plant rhizosphere that directly or indirectly affect the plant system and agriculture. In 1904, Hiltner described the concept of rhizosphere as, "the narrow zone of soil surrounding the roots where microbial population is stimulated by root activities" (Saharan and Nehra 2011). Several microbes, such as bacteria, fungi, protozoa and algae exist and interact in the rhizosphere. Among all microbial groups, bacteria are the most frequent in soils. However, it has been estimated that bacteria account for less than half of the total biomass in agricultural soils due to their small size (1–10 μm) (Alexander 1977). Approximately, bacterial cells exist at 10^4–10^9 colony forming units/g soil. The

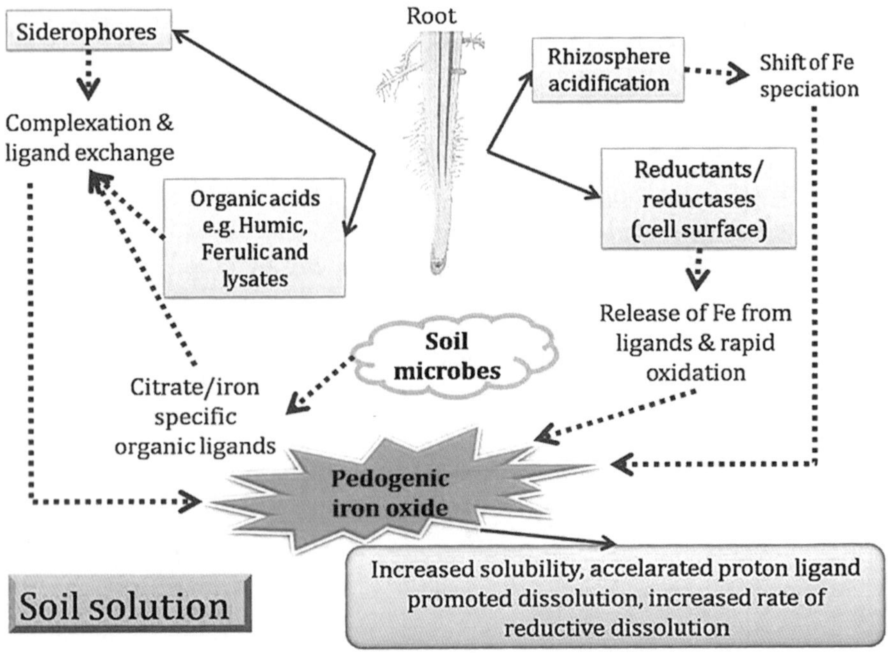

Fig. 10.1 Different mechanisms of plant growth promoting rhizobacteria in plant rhizosphere

abundance of bacterial cells in the rhizosphere probably suggests their major role in influencing plant physiology, especially considering their competitiveness in root colonization (Antoun and Kloepper 2001; Barriuso et al. 2008). Some of the most important PGPR belong to the genera *Acetobacter*, *Acinetobacter*, *Alcaligenes*, *Arthrobacter*, *Azoarcus*, *Azospirillum*, *Azotobacter*, *Bacillus*, *Beijerinckia*, *Burkholderia*, *Derxia*, *Enterobacter*, *Gluconacetobacter*, *Klebsiella*, *Ochrobactrum*, *Pseudomonas*, *Rhodococcus*, *Serratia*, *Zoogloea* etc. and have been subject of extensive research for decades (Singh et al. 2011a).

PGPR function in diverse ways to enhance plant growth. In nutshell, different mechanisms involved in plant growth are: production of root exudates, repression of soil-borne pathogens (by the production of hydrogen cyanide, antibiotics, and/or competition for nutrients), siderophore production, nitrate reduction, nitrogen fixation, phosphate solubilization, production of organic acids, and phytohormones (indole acetic acid or IAA), release of enzymes (dehydrogenase, phosphatase, nitrogenase, 1-aminocyclopropane-1-carboxylate (ACC) deaminase), and the induction of systemic resistance (ISR) (Kohler et al. 2006; Figueiredo et al. 2010) as described in Fig. 10.1. Regardless of the mechanism(s) of plant growth promotion, the most important factor for PGPR functioning is the rhizosphere/rhizoplane (root surface) colonization or the root itself (within root tissue) (Singh et al. 2011b).

6.1 Plant Growth-Promoting Rhizobacteria as Biofertilizers

In the state of affairs after "Green revolution", sustainable advances in agricultural production, productivity and profitability can be attained only by a quantum jump in productivity per unit of land, water and energy and even per capita, without compromising environmental health. Resource flow to the agriculture sector is declining and indebtedness of small and marginal farmers is rising. Input costs are increasing, while factor productivity is declining. The current scenario suggests the need of cheaper available alternatives to chemical fertilizers, such as *biofertilizers* and *organic amendments*. Biofertilizers and organic amendments are cost-effective substitutes to costly and harmful chemical fertilizers to improve soil quality, fertility, biology and agricultural productivity (Saleem et al. 2007; Ray et al. 2008; Babalola 2010).

Biofertilizers are composed of microorganisms that efficiently convert micronutrients to a readily available form from an inaccessible form through biological processes in soil. Microbial inoculants and biofertilizers are an important component of organic farming accounting for about 65% of the nitrogen supply to crops worldwide (Singh et al. 2011 b). In comparison with chemical pesticides and fertilizers, microbial inoculants or biofertilizers have several advantages, such as no health hazards, eco-friendly, targeted activity, effective at small quantity, self-multiplication while being controlled by the plant as well as by the indigenous microbial populations, no residual effect or bio-magnification and can be incorporated in conventional or integrated pest management systems (Berg 2009).

Physical, chemical and biological changes in properties of rhizospheric soil significantly influence growth and health of plants. Both the structural and functional characteristics of roots are vital factors for nutrient acquisition. They determine the capacity of plants to utilize available essential nutrients from soil while assuaging the toxic ones (Marschner 1995; Hinsinger 1998). Efficient utilization of nutrients from soil is an imperative issue for plants in nutrient limiting environments. Additionally, nutrient uptake depends on a wide range of physico-chemical parameters, such as root architecture, environmental and seasonal factors, biological interactions and competition (Lynch 2005). According to Tinker and Nye (2000) the importance of different root traits and rhizosphere-mediated processes is dependent on the nutrient in question and other factors, such as plant species and soil type. For instance, nutrients present at low concentrations in soil solution and/or with poor diffusivity (e.g. P as either HPO_4^{2-} or $H_2PO_4^-$, and micronutrients, such as Fe and Zn), root growth and proliferation and the release of root exudates are of meticulous significance (Barber 1995). In contrast, nutrients with greater diffusion coefficients (e.g. NO^{3-}, SO^{4-} and Ca^{2+}), or those present in higher concentrations (e.g. K^+, NH^{4+}), move more freely towards the root probably via mechanism of mass flow. In such cases, characteristics, such as the root architecture and its distribution that facilitates water uptake are of immense importance (Barber 1995; Tinker and Nye 2000; Lynch 2005).

As biofertilizers, PGPR stimulate the plant growth directly through increase in nutrition acquisition, such as phosphate solubilization and nitrogen fixation or by

Table 10.1 Different functions of plant growth promoting rhizobacteria in agro-ecosystem

Group	Microorganisms	Mechanism	References
Biofertilizers	*Pseudomonas*, *Azospirillum*, *Azotobacter*, *Bacillus*, *Burkholdaria*, *Enterobacter*, *Rhizobium*, *Erwinia* and *Flavobacterium*	Enhanced nitrogen and zinc uptake, plant growth promotion, increased yield, phosphate solubilization, phytohormone production, iron chelation, antibiotics production and induction of plant systemic resistance	Zaidi and Mohammad (2006); Turan et al. (2006); Kohler et al. (2006)
Bio-control agents	*Pseudomonas fluorescens*, *Kluyvera cryocrescens*, *Bacillus pumilus*, *B. amyloliquefaciens*, *B. subtilis*, *Burkholderia*, *Azospirillum*	Various antibiotic compounds, iron chelators and exoenzymes such as proteases, lipases, chitinases, and glucanases, competitive root colonization	Jiang et al. (2006); Saravanakumar and Samiyappan (2007); Hernandez-Rodriguez et al. (2008); Naureen et al. (2009)
Abiotic stress tolerance	*P. fluorescens*, *P. mendocina*, *P. alcaligenes*, *Bacillus polymyxa*, *Mycobacterium phlei*, *Azospirillum*, *Achromobacter piechaudii*, *Variovorax paradoxus*, *Piriformaspora indica*	Osmolytes production and over-expression of salt-stress proteins	Egamberdiyeva (2007); Kohler et al. (2009); Paul and Nair (2008)
Nutrient recycling			
Nitrogen cycle	*Rhizobium*, *Nitrosomonas*, *Nitrobacter*, *Pseudomonas*, *Bacillus*, *Micrococcus*	Nitrogen fixation, ammonification, nitrification, denitrification	Egamberdiyeva (2005); Tilak et al. (2005)
Phosphorus cycle	*Pseudomonas*, *Bacillus*, *Streptomyces*, *Mycobacterium*	Mineralization, immobilization, solubilization	Rivas et al. (2006); Ramachandran et al. (2007)

Microbial transformation of micronutrients			
Iron	*Thiobacillus thiooxidans*, *Pseudomonas*, *Streptomyces Bacillus*, *Klebsiella*	Oxidation	Meziane et al. (2005); De Vleesschauwer et al. (2006); Ongena et al. (2005)
Manganese	*Bacillus*, *Klebsiella*, *Pseudomonas*, *Athrobacter*	Oxidation	Sidhu (1998)
Biodegradation/bioremediation	*Pseudomonas*, *Actinobacter*, *Flavobacterium*, *Klebsiella*, *Moraxella* and *Arthrobacter*, *Rhodococcus*, *Stenotrophomonas maltophilia*, *Alcaligenes denitrificans*, *B. subtilis*, *Enterobacter gergoviae*, *Flavimonas oryzihabitans*	Hydrolysis and degradation by bacterial enzyme-esterase and amidase	Lee et al. (2002); Turan et al. (2008)

rendering the inaccessible nutrients available to the plants (Persello et al. 2003). Biofertilizers containing nitrogen (N_2) fixers (*Rhizobium spp.*, *Bradyrhizobium spp.*, *Azotobacter chroococcum*), P-solubilizer (*Bacillus megaterium*) and K-solubilizer (*B. mucilaginous*) have been developed for commercial applications. PGPR play diverse roles in soil for plant growth promotion as described in Table10.1.

Although soils are generally rich in total phosphorus albeit only a small proportion is available for uptake by the plants. However, in spite of lower concentration (μM) of phosphorus present in soil, plants efficiently acquire it from low concentration soil solution (Jungk 2001; Khan et al. 2006). It might be due to the root exudations or microbial processes. N_2-fixing and phosphate solubilizing bacteria (PSB) are important for crops as they increase N and P uptake and play a fundamental role as PGPR in the biofertilization (Zahir et al. 2004; Zaidi and Mohammad 2006). PGPR utilized in the treatment of arable soils have significantly increased the agronomic yields (Harish et al. 2009a, b; Yazdani et al. 2009).

Phosphate-solubilizing bacteria have already been applied in the agronomic practices as potential bio-inoculants to increase the crop productivity (Kumar et al. 2011). "Phosphobacterin" a biofertilizer product is formulated from *Bacillus megaterium* var. *Phosphaticum*. Studies have suggested that the co-inoculation of N_2 fixing bacteria and PSB has proved more efficient and effective than the single microbe in providing a more balanced nutrition to agriculture crops, such as sorghum, barley, black gram, soybean and wheat (Abdalla and Omer 2001; Tanwar et al. 2002; Galal 2003). Microbial consortia developed with PGPR and rhizobia have efficiently increased plant nodulation and N_2-fixation (Figueiredo et al. 2010). Consortia of PGPR strains, *P. alcaligenes* PsA15, *B. polymyxa* BcP26 and *Mycobacterium phlei* MbP18, have found to exert pronounced stimulatory effects on plant growth and uptake of N, P and K by maize in nutrient-deficient calcisol soils (Egamberdiyeva 2007). PGPRs can also be utilized in combination with biopolymers, such as chitin and chitosan. *P. fluorescens* strain, CHA0 in combination with chitin increased plant growth, leaf nutrient contents and banana yield under perennial cropping systems (Kavino et al. 2010). Therefore, it might be implicated that PGPR might be symbolized as the potential soil microflora in the lieu of sustainable and eco-friendly agriculture. The application of such microbes as biofertilizers also contributes to minimize the use of expensive phosphate fertilizers.

6.2 *Plant Growth-Promoting Rhizobacteria as Biocontrol Agents (BCA)*

Soil microbial diversity is one of the most dynamic emerging areas that have been explored in various fields, such as improving soil structure, soil health, plant disease management among others. Soil microbes have the potential to suppress plant pathogens in their natural environment through different mechanisms. Various microbes have been discovered and emerged with the potential to mitigate the undesirable effects of phytopathogenic microbes on agriculture and forestry. The use of

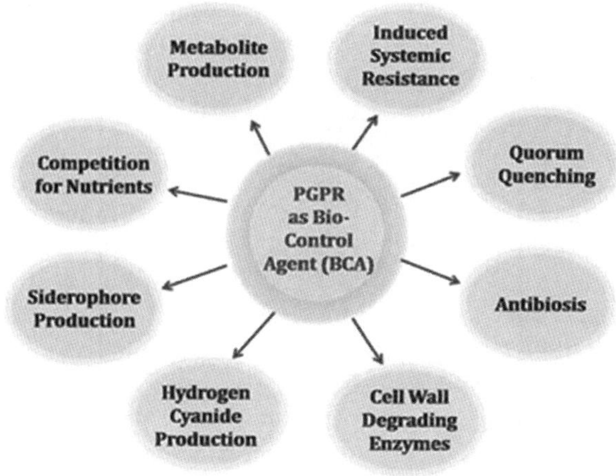

Fig. 10.2 Diagrammatic representation of mechanism of plant growth promoting rhizobacteria as biocontrol agents

biocontrol agents (BCAs) along with the current practices with some specifications has been able to bring down their impact below threshold level. BCAs have been harmonizing or even substituting the chemical counterparts.

Traditional methods used to protect crops from diseases have been largely based on the use of chemical pesticides. Applications of fungicides, fumigants, herbicides and insecticides can have drastic effects on the environment and consumers. Chemical methods may not be economical in the long run because they pollute the atmosphere, damage the environment, leave harmful residues and can lead to the development of resistant strains among the target organisms with repeated use. Therefore, a reduction or elimination of synthetic pesticide applications in agriculture is highly desirable. One of the most promising means to achieve this goal is by the use of BCAs for disease control alone, or by integrating them with lower concentration of chemicals in the control of plant pathogens resulting in alleviated impact of the chemicals on the environment. Biocontrol of pests in agriculture is a method of controlling pests including insects, mites, weeds and soil-borne pathogens. A number of BCAs have been registered and are available commercially, including strains belonging to bacterial genera, such as *Agrobacterium*, *Pseudomonas*, *Streptomyces* and *Bacillus* and fungal genera, such as *Gliocladium*, *Trichoderma*, *Ampelomyces*, *Candida* and *Coniothyrium*.

PGPR play an important role against various phytopathogens through different mechanisms, such as siderophore production (iron chelating), synthesis of anti-fungal metabolites, antibiotic and hydrogen cyanide production, phenazines, pyrrolnitrin, pyoluteorin, metabolites, phytohormones, competition for nutrition, production of exoenzymes (e.g. proteases, lipases, chitinases and glucanases) as well as the competitive root colonization and induction of systemic resistance (Bloemberg and Lugtenberg 2001; Lugtenberg et al. 2001; Persello et al. 2003) as depicted in Fig. 10.2.

Among the various PGPRs identified, *P. fluorescens* is one of the most extensively studied rhizobacteria owing to its antagonistic actions against several plant pathogens (Kavino et al. 2007; Saravanakumar and Samiyappan 2007; Harish et al. 2009a). *P. fluorescens* MSP-393 has been proved as biocontrol agent for many of the crops grown in saline agricultural soils (Paul and Nair 2008). As discussed earlier, the efficacy of microbes functioning in diverse areas can often be enhanced by developing microbial consortia or amending certain adjuvants, such as different biopolymers (chitin and chitosan etc.). The biocontrol efficacy of PGP fluorescent pseudomonads has been increased by mixing two or more strains of *Pseudomonas* spp. and with amendment of chitin or other substances (Radjacommare et al. 2002; Saravanakumar et al. 2009). *B. thuringiensis* has also been used to control most of the economically important insect pests, including American bollworm sp., *Earias* spp., *Spodoptera* sp. and *Plutella* sp. strains of *B. subtilis*, and *P. fluorescens* are used to control bacterial as well as fungal pathogens.

Pseudomonas fluorescens strain CHA0 in combination with chitin based bioformulations significantly reduced the banana bunchy top virus (BBTV) incidence in hill banana variety under greenhouse and field conditions in Western Ghats, India (Kavino et al. 2008, 2010). PGPR play an important role in PGPR-mediated insect management. PGPR induce chitinases production which hydrolyze chitin, the main structural component of the gut linings of insects (Harish et al. 2009b). Chitinases also degrade fungal cell walls and cause cell lysis (Radjacommare et al. 2002). Exploration of such PGPR strains will aid in maintaining the concept of sustainable agriculture. However at the same time, a better understanding of PGPR and their ecology is needed before biological control can be implemented (Bull et al. 1992).

6.3 Plant Growth-Promoting Rhizobacteria Mediated Alleviation of Abiotic Stress

Agricultural crops face many stresses induced by both biotic as well as abiotic factors affecting crop yield. Such stresses impose serious challenges in introducing agricultural crops into affected areas. Apart from affecting agricultural crops, different stresses persuade the occurrence and activity of soil microorganisms. Abiotic stress factors include heavy metals, high and low temperature, salinity, drought, flood, ultraviolet light and air pollution (ozone). Vegetable and other crops in many semi-arid and arid regions of the world face serious loss in yield due to increasing salinity of irrigation water as well as soil salinity (Parida and Das 2005; Saharan and Nehra 2011). Salt stress of agricultural land is rising every year across the globe and has become a matter of serious concern (Paul and Nair 2008). Among various abiotic stresses, soil salinity is one of the most severe limiting factors affecting nodulation, yield and physiological response in plants. Salt stress affects leaf growth rate due to decreased water uptake that results in reduced photosynthesis (Saharan and Nehra 2011). High salt stress also inhibits the synthesis and activity of nitrogenases in plants as reported in *Azospirillum brasilense* (Tripathi et al. 2002). Similarly, Han and Lee (2005) also reported physiological disorder in lettuce plants affected by high salt stress. Salt and water stress (drought) adversely affect a number of metabolic and plant physiological processes, such as plant growth and nutritional uptake (Hu 2005). In this context, the role of PGPR in plant health and alleviating osmo-tolerance mechanisms is imperative. Catabolic versatility and efficient root colonizing habit has gained paramount importance to PGPR mainly, soil-borne pseudomonads. Pseudomonads are known to produce an array of enzymes and metabolites that facilitate the plant to endure varied biotic as well as abiotic stress (Vessey 2003). Egamberdiyeva (2007) has reported three PGPR isolates, *P. alcaligenes* PsA15, *Bacillus polymyxa* BcP26 and *Mycobacterium phlei* MbP18 to survive in arid and saline soils, such as calcisol. These strains were tolerant to high temperature and salt concentrations.

Rhizobium, *Bradyrhizobium*, *Azotobacter*, *Azospirillum*, *Pseudomonas* and *Bacillus* has been reported from abiotic stressed ecosystems (Venkateswarlu et al. 2008; Selvakumar et al. 2009; Upadhyay et al. 2009). Most of the rhizobacteria modulate their cytoplasmic osmolarity by producing osmo-protectants such as, potassium ions, glutamate, trehalose, proline, glycine beatine, proline betaine and ectoine etc. (Grover et al. 2011). *Azospirillum* also reduces drought induced stress in sorghum plants by improving water content, increasing water potential and lowering canopy temperature of foliage. Arbuscular Mycorrhizal Fungi (AMF) are also known to improve plant growth under salt stress (Cho et al. 2006). A PGPR, *P. mendocina*, alone or in combination with an AMF, *Glomus intraradices* or *G. mosseae* improved plant growth, enhanced plant growth, nutrient uptake and other physiological activities of *Lactuca sativa* affected by salt stress (Kohler et al. 2009). *De novo* synthesis of osmolytes and over-expression of salt-stress proteins by *P. fluorescens* MSP-393 have been reported to overcome the harsh effects of high osmolarity (Paul and Nair 2008). High salt, pH and temperature under alkaline soil conditions affect the activity of PSB (Gaind and Gaur 1991). Phosphate solubilization occurs efficiently under acidic soil conditions as a result of organic acids released from bacterial metabolism (Seshadri et al. 2007). Phosphate solubilising bacteria lower the soil pH by secreting organic acids thereby improving phosphate solubilization (Chen et al. 2006). Therefore, it might be implicated that PGPR-PSB can be utilized to improve natural ecosystems, such as savannas that predominate in tropics, and have low fertile acid soils poor in minerals, particularly phosphorus. Therefore, all-embracing investigations are required for further exploitation of PGPR and their interaction with other soil microflora might open new avenues to facilitate sustainable agriculture in salt affect soils.

6.4 Plant Growth-Promoting Rhizobacteria Mediated Phytoremediation

Phytoremediation is a cheap and energy proficient detoxification method. It influences plant metabolism by concentrating the heavy metals in their shoot biomass thereby reducing their bioavailability (Saharan and Nehra 2011). Soil microbes alleviate toxic effects of heavy metals on the plants via different mechanisms, such as acid secretion, over-expression of proteins, phyto-antibiotic production, and other chemicals (Denton 2007). The metal resistant PGPR can act efficiently in carbon sequestering and enhancing plant growth under the conditions of metal stress (Rajkumar and Freitas 2008). Besides PGPR, mycorrhizal fungi also play an important role in phytoremediation (Tam 1995; Belimov et al. 2005; Denton 2007). According to Khan (2005), PGPR, PSB, mycorrhizal-helping bacteria and AMF in the plant rhizosphere play an important role in phytoremediation. Cadmium induces plant-stress-induced ethylene biosynthesis and the accumulation of ACC in the roots (Pennasio and Roggero 1992). PGPR help in overcoming the inhibitory effects of metals, drought, flooding and salt stress (Burd et al. 2000; Grichko and Glick 2001; Nie et al. 2002; Mayak et al. 2004a, b; Amico et al. 2008). Govindasamy et al. (2009) detected and characterized ACC in PGPR in wheat seedlings treated with ACC deaminase under *in vitro* conditions by measuring ethylene production under cadmium stress. *Pseudomonas putida* and *P. fluorescens* are also potent heavy metal tolerant PGPR that have been evaluated successfully under the conditions of hyperosmolarity and contaminated soils (Schnider-Keel et al. 2001; Chacko et al. 2009). *Streptomyces acidiscabies* E13 strain exerts positive growth promoting effects in cowpea in nickel contaminated soil probably by binding iron and nickel through the production of hydroxamate siderophores (Dimkpa et al. 2008).

6.5 Plant Growth-Promoting Rhizobacteria Mediated Ethylene Stress Alleviation

Plant growth-promoting rhizobacteria producing ACC deaminase protect plants under unfavourable environmental conditions, such as flooding, heavy metals, phytopathogens, drought and high salt. ACC deaminase-containing PGPR also protect the crops from adverse effects of salt stress ethylene (Belimov et al. 2001). PGPR containing ACC deaminase help in alleviating the stress induced by ethylene and mitigate the negative effect on plants. It has been comprehensively studied in numerous PGPR, such as *Agrobacterium genomovars*, *Azospirillum lipoferum*, *Alcaligenes*, *Bacillus*, *Burkholderia*, *Enterobacter*, *Methylobacterium fujisawaense*, *Pseudomonas* sp., *Ralstonia solanacearum*, *Rhizobium*, *Rhodococcus*, *Sinorhizobium meliloti*, *Variovorax paradoxus* and *Enterobacter cloacae* (Hontzeas et al. 2004a; Uchiumi et al. 2004; Belimov et al. 2005; Pandey et al. 2005a, b; Blaha et al. 2006; Stiens et al. 2006).

Ethylene is an important phytohormone; however, its overproduction under stressful conditions inhibits plant growth or might result in plant mortality, particularly in seedlings. PGPR contain an enzyme, that lowers the ethylene concentration in developing seedlings or stressed plants by producing an enzyme ACC deaminase. ACC deaminasedegrades ACC and utilize it as a carbon source thereby lowering the ACC uptake and hence ethylene (Van Loon 2007; Saleem et al. 2007). Seed treatment with ACC deaminase containing PGPR acts as a sink for ACC and maintains plant ethylene levels that might prejudice root growth (Glick et al. 1999). PGP *Enterobacter cloacae* UW4 strain has been reported to induce root elongation and proliferations in canola (*Brassica rapa*) by minimising ethylene levels (Hontzeas et al. 2004b). PGPR containing ACC-deaminase impart resistance to plants against several forms of stress probably by lowering ethylene levels resulting in enhanced root systems (Safronova et al. 2006; Shaharoona et al. 2006). Utilization of PGPR containing ACC deaminase activity in plant growth promotion and gene expression analysis under both normal and stresses conditions has gained sincere attention (Sergeeva et al. 2006). The advanced biotechnological tools such as metagenomics, proteomics, cDNA microarrays, gene expression studies in conjuction with bioinformatics can further aid in better understanding and efficient utilization of PGPR strains and hence attaining agricultural sustainability.

7 Nutrient Recycling

7.1 Nitrogen Cycle Processes Mediated by Microorganisms

Nitrogen cycles are strongly mediated by microorganisms and comprise four major pathways: (1) Biological nitrification: Microbial-mediated mineralization of organic forms of N to ammonium (NH_4^+) and its subsequent nitrification to nitrite (NO_2^-) and nitrate (NO_3^-) mediated by *Nitrosococcus* and *Nitrobacter* bacteria is of major significance to N availability; (2) symbiotic nitrogen (N_2) fixation which is mediated by *Rhizobium* and *Bradyrhizobium* associated with legumes and *Frankia*, *Nostoc* and *Azolla* associated with non legume; (3) the free-living nitrogen fixers, such as the rhizosphere nitrogen fixing *Azotobacter*, *Azospirillum* and *Bacillus*, free-living nitrogen fixing Cyanobacteria and *Clostridium* and; (4) denitrification, the conversion of nitrate to gaseous nitrogen mediated by facultative heterotrophic bacteria. Arbuscular Mycorrhizal Fungi (AMF) are also known to play an important role in nitrogen cycling and nutrition in soil by increasing plant nitrate uptake and efficient utilization. The major supply of nitrogen in leguminous plants occurs via symbiosis, a process that is highly associated with the phosphorus cycle.

Microorganisms play an important role for the breakdown of organic material and nutrient cycling with few compounds becoming recalcitrant (Dorn et al. 1974). Diversity of bacterial functions contributes a major role in nutrient recycling and plant litter decomposition, thereby shaping the execution of a system (Lynch 1983; Collins et al. 1992). They represent a vibrant source and sink of nutrients in all ecosystems and occupy a significant place in the ecosystem and soil food web. Approximately, 90–95% of all nutrient cycling is passed through them towards higher trophic levels.

Bacterial cells break down several intricate compounds into simpler organic forms. Therefore, they improve the soil organic content, soil structure and physical properties, increase water retention capacity, nutrient availability as well as act as a cementing agent by holding soil particles together. The symbiotic relationship between N-fixing bacteria and legumes is one of the most dynamic plant–bacterial interactions (Sprent 1979). However, environmental conditions and the host determine the distribution, diversity and interaction of specific N-fixing bacteria (Turco and Bezdicek 1987; Hirsch et al. 1993). Another important role that bacteria play in the environment is the sulphur cycle. However, a limited number of genera of facultative and obligate anaerobic bacteria is accountable for the process within a narrow range of environmental conditions. Sulfate reduction and oxidation is carried out by a variety of facultative and obligate anaerobic and a limited number of aerobic autotrophic bacteria respectively (Tiedje et al. 1984; Bock et al. 1989). According to Jones (1991), more than 50 species of methanotrophs are responsible for the production or catabolism of methane. However, the process of production and catabolism of methane in agricultural soils is generally meagre due to the lack of anaerobic conditions. A limited number of genera of obligate aerobic bacteria widely distributed in soils carry out methane oxidation in agricultural soil and are involved in ammonium oxidation and oxygen depletion with subsequent N_2O production (Topp and Hansen 1991).

7.2 Phosphorus Cycle Processes Mediated by Microorganisms

Phosphorus is one of the most important minerals abundantly present in the soil. However, it is also the least accessible nutrient in the plant rhizosphere. Phosphorus is known to play a crucial role in symbiotic nitrogen fixation in many nodulating legumes, such as soy beans.

The capability of the extra radical mycelium of AMF to extend the phosphate depletion zones has an added advantage to the plants by increasing the uptake of comparatively immobile phosphate ions and sometimes increasing phosphorous use efficiency (Koide et al. 2000). The plants strategies for phosphorous uptake are embraced by modification of their rhizosphere conditions, hasty root growth and elongation and root hair proliferation. Modification of the rhizosphere is mediated by the formation of symbiotic association with AMF; releasing bio-available phosphorous by chelating agents secreted by mycorrhizal roots in acid and calcareous soils; mobilization of organic P by phosphate enzymes and acidification via increased proton efflux and pCO_2 enhancement in mycorhizosphere in high-P-fixing soils. AMF besides increasing the uptake and efficient utilization of phosphorous also enhance the uptake and utilization of other macronutrients, such as N, potassium, magnesium (Clark and Zeto 2000; Hodge et al. 2001) and some micronutrients, such as zinc, copper, iron, nitrogen, potassium, calcium and manganese (Clark and Zeto 2000). The availability of these nutrients may sometimes direct the pattern of symbiotic association (Ryan and Angus 2003). The availability of nutrients enhances the plant growth and development, increases vigor and increased resistance to many biotic as well as abiotic stresses and finally the yield (Borowicz 1997; Feng et al. 2002; Karagiannidis et al. 2002). Significant decrease in yield has been observed due to the disruption in AMF probably due to lower phosphorous uptake (Thompson 1994). On the other hand, there are also some examples that suggest the negative effects of AM colonization. Under the conditions of higher phosphorous availability, the colonization of roots by AMF may reduce the crop growth (Kahiluoto et al. 2001) while the plants may not counter native AMF even under low phosphorous availability (Ryan et al. 2002). In some cases, the inoculation of roots with AMF may also not result in a positive response (Sainz et al. 1998). Degree of selection between the host and the AMF and their interaction with different hosts have been accounted for such contradictory reports.

8 Microbes Mediated Agricultural Antibiotics/Metabolites: Biopesticides

With the advancement in human civilization, development of technology, trade and globalization has invited certain serious issues. The agro-ecosystem is also being affected by this advancement owing to the extensive use of chemical pesticides in agriculture. While the use of chemical pesticides is on the rise, the pest induced losses are also increasing. Fortunately, realization of the negative effects of these chemicals on nature and natural resources has forced us to search for alternatives to chemical pesticides. In the current scenario, the focus is more on reliable, sustainable and environment friendly agents of pest control, i.e. biopesticides. Biopesticides have been gaining increasing interest among those concerned with developing environment friendly, safe and Integrated Crop Management (ICM)-compatible approaches and tactics for pest management (Copping and Menn 2000).

Pesticides of biological origin i.e. viruses, bacteria, pheromones, plant, animal compounds and certain minerals are termed as biopesticides. Biopesticides are generally grouped into three major categories: (1) pesticide of microbial origin (e.g. a bacterial, fungal, viral or protozoan pesticide) that contains microorganisms as the active ingredient (a.i). Such pesticides are relatively specific for their target pest(s); (2) microorganisms derived molecules that act as a.i against pests. The most widely used biopesticides in this category are obtained from subspecies and strains of *B. thuringiensis* (popularly called Bt); (3) Plant originated pesticidal substances termed as Plant-Incorporated-Protectants (PIPs). Plant-Incorporated-Protectants are produced by plants through genetic manipulations (http://www.epa.gov/oecaagct/tbio.html). The key feature of biopesticides is their eco-friendly nature and biodegradability which reduces their residual effects and environmental pollution hence they play a role in sustainable agriculture. Different types of biopesticides belonging to these two categories are described in Fig. 10.3 (Table 10.2).

8.1 Bacteria Derived Molecules

8.1.1 Fungicide/Bactericide

(a) **Blasticidin-S**
Blasticidin-S was isolated from the soil actinomycete *Streptomyces griseochromogenes* in 1955 by Fukunaga and later by Takeuchi in 1958 while its fungicidal properties were first described by Misato in 1959 (Copping and Duke 2007).

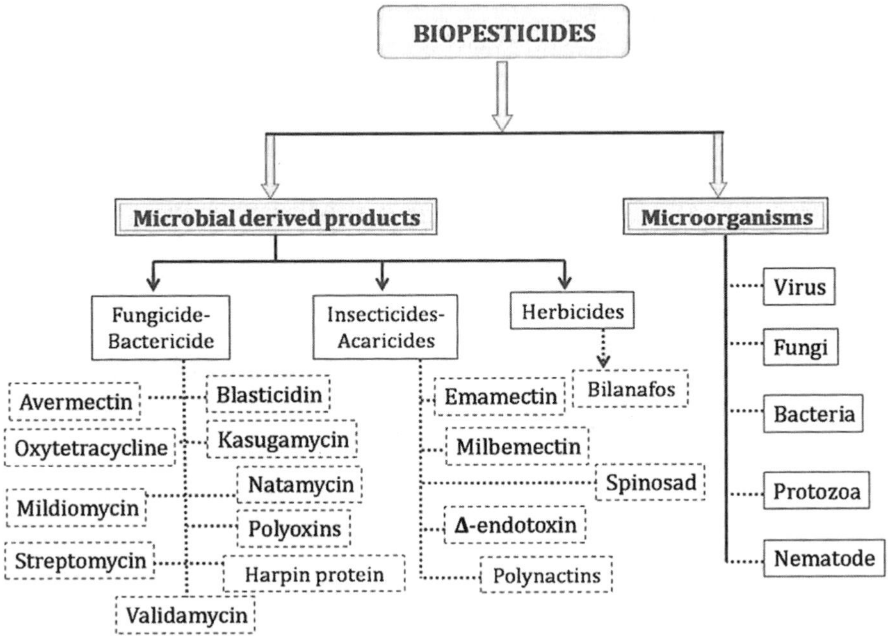

Fig. 10.3 Different types of biopesticides

Blasticidin-S functions by inhibiting peptidyl transfer and protein chain elongation by binding to the 50S ribosome in prokaryotes and blocking protein biosynthesis (Huang et al. 1964). Blasticidin-S inhibits spore germination and mycelial growth of *Pyricularia oryzae* at a concentration <1 mg/ml under *in vitro* conditions. Foliar application (100–300 g/ha) of the fungicide protects the plants from the rice blast, however excessive application could result in yellow spots on rice leaves. Blasticidin-S could be phytotoxic to alfalfa, aubergines, clover, potatoes, soybean, tobacco and tomatoes. More recently, its use has decreased following the introduction of new, less toxic, pathogen-specific, synthetic rice blast products (Copping and Duke 2007).

(b) **Harpin protein**

Harpin protein is produced by *Erwinia amylovora* (Burrill) Winslow, causal organism of fire blight disease in apples and pears. Harpin is an acidic, heat-stable, cell envelope-associated protein with a molecular mass of about 40 kDa, comprising 403 amino acid residues, lacking cysteine. A weakened strain of *Escherichia coli* (Migula) has been used to produce commercial harpin and harpin $\alpha\beta$ by transfer of the DNA fragment encoding harpin protein from *E. amylovora* to *E. coli* K-12 strain (Wei et al. 1992). Harpin proteins are known to enhance seed germination, plant vigour, flowering, fruit setting and increase yields as well as suppress a wide range of phytopathogens. These proteins activate innate defence mechanisms in plants, known as systemic acquired resistance

Table 10.2 Microbes mediated agricultural antibiotics/metabolites

	Isolation	Sold as/chemical composition	Used against	Formulation	Side effects
Blasticidin S	*Streptomyces griseochromogenes*	Benzylaminobenzenesulfonic acid salt	Rice blast (*Pyricularia oryzae*)	CF, DP, EC, and WP	Toxic to mammal, Rodents (>100 $mgkg^{-1}$), Rats (L.D >500 $mgkg^{-1}$); eye irritant
Harpin protein	*Erwinia amylovora* (Burrill) Winslow	403 amino acid residues lacking cysteine	Bacterial leaf spot, Bacterial speck, Bacterial wilt, Fusarium wilt, Phytophthora root rot, Rice stem rot, Sheath blight, Apple scab, Fire blight, Botrytis bunch rot, Black rot, Black leaf spot, Cucumber mosaic virus, Root-knot nematode, Tobacco cyst nematode and Tobacco mosaic virus (TMV)	ST, SD, FA, WG	Harpin is not considered to be a toxic pesticide. Harpin protein is classified as a toxicity category IV product by the US-EPA via the oral, dermal and inhalation route, and a toxicity category IV eye and skin irritant
Kasugamycin	*Streptomyces kasugaensis*	Kasugamycin hydrochloride hydrate	Rice blast (*P. oryzae*), Leaf spot in sugar beet and Celery (*Cercospora* spp.), Bacterial disease in rice and vegetables and Scab (*Venturia* spp.) in apples and pears	SF, protectants, WP, DP, SC, GF, FS, ST	Non-toxic to mammals, fish and non-target organisms; biodegradable; safe to non-target organism; environmentally safe
Mildiomycin	*Streptoverticillium rimofaciens* strain B-98 891		Powdery mildews (*Erysiphe* spp., *Uncinula necator*), *Podosphaera* spp. and *Sphaerotheca* spp.)	Eradicant, with some systemic activity, WP	Biodegradable and Non-toxic to mammals, fish and non-target organisms
Natamyci-N	*Streptomyces natalensis* and *S. chattanoogensis*		Basal rots on ornamental bulbs such as daffodils (caused by *Fusarium oxysporum*).	DT, WP	Biodegradable and Non-toxic to mammals, fish and non-target organisms

(continued)

Table 10.2 (continued)

	Isolation	Sold as/chemical composition	Used against	Formulation	Side effects
Oxytetracycline	*Streptomyces rimosus*	Hydrochloride	Fireblight (*E. amylovora*) and diseases caused by *Pseudomonas* and *Xanthomonas*; mycoplasma like organisms; stone and pome fruit and in turf.	WSP, FS	Biodegradable and Non-toxic to mammals, fish and non-target organisms
Polyoxin	*Streptomyces cacoai* var *asoensis* Isono	Polyoxins D	Powdery mildews, Rice sheath blight, Apple and Pear canker, Pear black spot and Apple cork spot, Grey mould. It is ineffective against bacteria and yeasts	FS, WP, EC, WSG,	Biodegradable and Non-toxic to mammals, fish and non-target organisms
Streptomycinn	*Streptomyces griseus*	Sesquisu-lfate	Bacterial shot-hole, bacterial rots, wilts, fire blight and other bacterial diseases. It is recommended for use in pome fruit, stone fruit, citrus fruit, olives, vegetables, potatoes, tobacco, cotton and ornamentals	Systemic, WP, FS	Very low mammalian toxicity; no adverse effects on non-target organisms or the environment
Validamycin	*Streptomyces hygroscopicus*		*Rhizoctonia solani* and other *Rhizoctonia* species in rice, potatoes, vegetables, strawberries, tobacco, ginger, cotton, rice and sugar beet	NS, FS, SD, ST, DP	Very low mammalian toxicity (sufficient to be used as a pharmaceutical), and it has not been shown to have any adverse effects on non-target organisms or the environment

Bilanafos	*Streptomyces hygroscopicus; Streptomyces viridochromeogenes*	Tripeptide phosphinothricin	Post-emergence in vines, apples, brassicas, cucurbits, mulberries, azaleas, annual weeds and at higher rates for control of perennial weeds rubber and annual weeds	LP	Relatively non-toxic to mammals and other non-target organisms; they have very little activity in soil, mainly owing to rapid microbial degradation. Low environmental impact herbicides
The avermectin	**ABAMECTIN:** *Streptomyces avermitilis*	Mixture of two avermectins, avermectin B1a and avermectin B1b	Mites, leaf miners, suckers, beetles and other insects, and it is also used for control of fire ants (*Solenopsis* spp.).	EC, contact and stomach action	Highly toxic to mammals, insects and mites, bio degradable
	EMAMEC-TIN *Streptomyces avermitilis*	Emamec-tin benzoate (emamect-in B1a and emamect-in B1b)	Caterpillar pests (Lepidoptera), with suppressive activity against mites, leaf miners and thrips, Pinewood nematodes	Contact and stomach action, WG, EC	Lower mammalian toxicity; toxic to beneficial insects such as honey bees
Milbemyctin	*Streptomyces hygroscopicus* subsp. *aureolacrimosus*	Milbemycin A3 and milbemycin A4 in the ratio 3:7	Citrus red mites, pink citrus rust mites, Kanzawa spider mites and other spider mites, leaf miners in citrus fruit, tea, aubergines and protected ornamentals	Contact and stomach action, EC, WP	Moderate oral mammalian toxicity but is much less toxic via the dermal route; non-persistent in the environment; relatively non-toxic to non-target organisms, although some beneficial insects are susceptible

(continued)

Table 10.2 (continued)

	Isolation	Sold as/chemical composition	Used against	Formulation	Side effects
Polynactin	*Streptomyces aureus*	A mixture of tetranactin, trinactin and dinactin	Spider mites, such as carmine spider mite, two-spotted mite and European red mite in orchard fruit trees, spider mites	EC	Non toxic to mammals and beneficial insects, high toxicity to aquatic organisms
Spinosad	*Saccharopolyspora spinosa*	Mixture of spinosyn A and spinosyn D	Caterpillars, Thrips, Flies, Beetles, drywood termites, fire ants, grasshoppers, chewing and sucking lice. Spinosad may be used on row crops (including cotton), vegetables, fruit trees, turf, vines and ornamentals	Water-based suspension concentrate formulation	Very low mammalian toxicity; non-toxic to birds but slightly to moderately toxic to fish; highly toxic to honey bees; dry residues are non-toxic
Myrothecium verrucaria	Isolated from a nematode		Plant parasitic nematodes, including root-knot (*Meloidogyne* spp.), cyst, sting and burrowing nematodes. It is used in turf, tobacco, grapes, citrus, brassicae and bananas. Controls both adult and juvenile nematodes on contact and inhibits the hatching and development of eggs	DP, LF	There is no evidence of allergic reactions on non-target organisms and environment. Possible toxicity to aquatic organisms

CF contact fungicide, *DT* dip treatment, *DP* dustable powder, *EC* emulsifiable concentrate, *FA* Foliar applications, *FS* Foliar spray, *GF* granule formulation, *LF* liquid formulations, *ST* seed treatment, *SD* soil drench, *SC* soluble concentrate, *SF* systemic fungicide, *WSG* water soluble granule, *WSP* water-soluble powder, *WG* wettable granule, *WP* wettable powder

(SAR) (Wei et al. 1992). Since harpin does not show any direct inhibitory or toxic effect on plant pathogens, it is claimed that it does not impose any selection pressure on pathogens (Copping and Duke 2007). Harpin is likely to be an important tool in sustainable agriculture by decreasing the use of conventional pesticides and resistance management programmes. It degrades rapidly and does not impose comprehensible adverse effects on birds, fish, aquatic invertebrates, honey bees, non-target plants, algae or ground water contamination, therefore, it is considered eco-friendly (Copping and Duke 2007).

(c) **Kasugamycin**
Kasugamycin and kasugamycin hydrochloride hydrate were isolated from the soil actinomycete *Streptomyces kasugaensis* Hamad (Hamad and biological activity of these compounds were first described by Hamada in 1965 (Copping and Duke 2007). Kasugamycin functions by inhibiting protein biosynthesis by interfering with the binding of aminoacyl-tRNA to both the mRNA-30S and the mRNA-70S ribosomal subunit complexes, thus preventing the incorporation of amino acids into proteins (Tanaka et al. 1966). Kasugamycin is used in combination with other fungicides with different modes of action. It is slightly phytotoxic to crops, such as peas, beans, soybeans, grapes, citrus and apples while it is comparatively safe on rice, tomatoes, sugar beet, potatoes and many other vegetables. The use of kasugamycin has declined due to resistance development and the release of new, disease-specific chemical fungicides for rice blast control.

(d) **Mildiomycin**
Mildiomycin was first reported by Harada and Kishi (1978) and its fungicidal properties were reported by Kusaka et al. (1979). Mildiomycin functions by blocking peptidyl-transferase, hereby inhibits protein biosynthesis in fungi (Om et al. 1984). It is effective at concentrations of 50–100 mg/l. Mildiomycin has not been widely used for disease control outside Japan. Mildiomycin has very low mammalian toxicity and it has not been shown to have any adverse effects on non-target organisms or the environment.

(e) **Natamycin**
Natamycin is also known as myprozine, pimaricin and tennectin. The structure of natamycin was elucidated by Golding et al. (1966) and Meyer and Pimaricin (1968). The mode of action is still unknown.

(f) **Oxytetracycline**
Oxytetracycline was first described by Finlay in 1950 (Copping and Duke 2007). Oxytetracycline is a potent inhibitor of protein biosynthesis in bacteria and binds to the 30S and 50S bacterial ribosomal subunits thereby inhibiting the binding of aminoacyl-tRNA and the termination factors, RF1 and RF2 to the A site of the bacterial ribosome (Caskey 1973). It is rapidly taken up by plant leaves, particularly through stomata, and is readily translocated to other plant tissues. However, its use as a crop protection agent has been declining. It is used to treat several mammalian diseases, such as acne, spirochaetal infections,

clostridial wound infection, anthrax, skin, ear, eye, respiratory and urinary tract infections and gonorrhoea.

(g) **The Polyoxins**

The fungicidal polyoxins are polyoxin B and polyoxorim. Polyoxin was first isolated by Isono et al. (1965), Polyoxorim (polyoxin D) was isolated by Suzuki et al. (1965) and Isono et al. (1967). Polyoxins functions through inhibition of cell wall biosynthesis or chitin synthesis in fungal cell wall. It causes abnormal swelling on the germ tubes of spores and hyphal tips of the pathogen (Eguchi et al. 1968; Isono and Suzuki 1979). It is effective against bacteria, yeasts and phycomycetes.

(h) **Streptomycin**

Streptomycin was isolated by Schatz in 1944 (Dorothy et al. 1945). Streptomycin binds to the 30S ribosomal subunit thereby inhibiting protein biosynthesis that leads to misreading of the genetic code. Development of resistance to streptomycin has reduced its use in crop protection markets. Recommended use rates can cause chlorosis to rice, grapes, pears, peaches and some ornamentals, but these symptoms can be relieved by the addition of iron chloride or iron citrate to the spray tank (Copping and Duke 2007).

(i) **Validamycin**

Validamycin is also known as validamycin A. It is not systemic but possesses fungistatic properties. In *Rhizoctonia solani*, it does not exhibit fungicidal action but causes abnormal branching of the tips of the pathogen followed by a cessation of further development and inhibits trehalase (Matsuura 1983; Shigemoto et al. 1989). High concentrations (1 g/l) showed no phytotoxicity to over 150 different target crops. Validamycin continues to find wide usage in a wide variety of crops, particularly in Japan.

8.1.2 Herbicides

(a) **Bilanafos**

The tripeptide herbicide bilanafos is commonly known as bilanafos and bialaphos. Bilanafos induces toxicity only after its metabolic conversion to a potential irreversible inhibitor of glutamine synthetase (GS), phosphinothricin {4- [hydroxy (methyl) phosphinoyl]-L-homoalanine} (Lydon and Duke 1999). Glutamine synthetase inhibition results in accumulation of ammonium ions and inhibition of photorespiration. The rapid toxic effects of the herbicide can be reversed by supplying the plant with glutamine. However, this does not reduce the levels of ammonium ions. In case of C3 plants, toxicity results from rapid cessation of photorespiration due to the inhibition of glutamine synthetase, resulting in accumulation of glyoxylate in the chloroplast and rapid inhibition of ribulose bisphosphate (RuBP) carboxylase (Lydon and Duke 1999).

8.1.3 Insecticides and Acaricides

(a) **The Avermectins: Abamectin and Emamectin**
Abamectin is an insecticide and acaricide. It is a mixture of two avermectins, avermectin B1a and avermectin B1b, and was introduced as an insecticide/acaricide. It has the approved common names abamectin and abamectine, but it is also known as avermectin B1. Abamectin acts on γ–aminobutyric acid (GABA) receptor in the peripheral nervous system (Fisher 1993). The compound stimulates the release of GABA from nerve endings and enhances the binding of GABA to receptor sites on the post-synaptic membrane of inhibitory motor neurons of nematodes and on the post-junction membrane of muscle cells of insects and other arthropods. This enhanced GABA binding results in an increased flow of chloride ions into the cell, with consequent hyperpolarization and elimination of signal transduction, resulting in an inhibition of neurotransmission (Jansson and Dybas 1996). Emamectin, another avermectin acts as insecticide/acaricide. It targets GABA receptors in the peripheral nervous system in lepidopteran (Fisher 1993).

(b) **Milbemycin**
Milbemycin (also known as milbemectin) acts as an insecticide and acaricide by stimulating the release of GABA from nerve endings. It enhances the binding of the molecule to the receptor sites on the post-synaptic membrane of inhibitory motor neurons of mites and other arthropods. This results in an increased flow of chloride ions into the cell, which subsequently causes hyperpolarization and elimination of signal transduction and finally inhibits neurotransmission (Lankas and Gordon 1989).

(c) **Polynactins**
Polynactins are secondary metabolites of *Streptomyces aureus* Waksman & Henrici isolate S-3466. Polynactins act on the lipid layer of the mitochondrial membrane resulting in the leakage of basic cations such as, potassium ions (Ando et al. 1974).

(d) **Spinosad**
Spinosad is a secondary metabolite from the soil actinomycetes. It is target specific and functions by activating nicotinic acetylcholine receptor, but at a different site than nicotine or the neonicotinoids. It is also known to affect GABA receptors, however their role is unclear. The activity of this herbicide results in rapid death of target phytophagous insects. It has been recommended to use this herbicide with sturdy, proactive resistance management strategy, however its moderate residual activity reduces the possibility of the onset of resistance (Thompson et al. 2000). It has been an important component of conventional and integrated farming systems. It has high biodegradability in soil via photolysis and by soil microorganisms below the soil surface (Saunders and Bret 1997).

8.2 Fungi Derived Molecules

8.2.1 Fungicides and Bactericides

Yeast extract hydrolysate, a by-product of the fermentation of brewer's yeast (*Saccharomyces cerevisiae* Meyer exHansen) is the only fungi derived product that is sold as a fungicide/bactericide (www.epa.gov/pesticides/biopesticides/ingredients/index.htm). It is sold under the trade name KeyPlex 350, however its composition is not known.

8.2.2 Herbicides

Till date, no herbicide of fungal origin is available on the market (Duke et al. 2002; Copping and Duke 2007).

8.2.3 Insecticides, Acaricides and Nematicides

(a) **Myrothecium verrucaria**
Myrothecium verrucaria (Alb. & Schwein.) Ditmar isolate AARC-0255 is a potent nematicide belonging to order Hypocreales. The dried mycelium of the fungus is used as a nematicide and the a.i is amixture of substances that are in suspension and in solution. The mode of action of the fungus is still unidentified. Although the living fungus causes plant disease, its a.i (heat killed) acts as a herbicide.

9 Microbes Mediated Composting

Some of the most common organic wastes and amendments used for soil application such as, animal manure, kitchen waste, agriculture waste, pulp and paper industrial sludge, municipal wastewater sludge can be converted into compost. The unknown composition of the raw organic wastes and uncertainty in terms of pathogens, toxic compounds, weed seeds, heavy metals and foul odours render them unsuitable for field application (Singh et al. 2011 b). Untreated or unprocessed wastes often pose a serious threat to the agro-ecosystem as well as human health and cause toxicity to beneficial soil microflora. Large quantities of organic waste, high cost of treatment and unavailability of sufficient dumping sites and associated environmental hazards raises questions on incineration and land filling practice. On the contrary, organic waste treatment has emerged as an attractive and cost-effective strategy being used for land application to supply plant nutrients and organic matter for improved crop production. Most commonly used organic amendments are animal and green manures, compost, nematicidal plants and proteinaceous wastes (Singh et al. 2011 b). Organic soil amendments have been used as a traditional cultural practice to improve soil fertility and texture. Organic amendments also help in plant disease management by inhibiting soil-borne pathogens, such as *Aphanomyces euteiches*, *Gaeumannomyces graminis*, *Macrophomina phaseolina*, *Rhizoctonia solani*, *Thielaviopsis basicola*, *Verticillium dahlia*, etc. (Bonanomi et al. 2010). Organic amendments also help in suppression of various phytopathogenic nematodes (Oka 2010).

Composting is one of the most appropriate methods for the management of solid organic waste for agricultural use while minimizing the constraints associated with it. Composting improves soil biophysical properties, organic matter and provides plant nutrients and increases the crops yield. According to an estimate, nearly 827 million tons of compostable materials are generated every year, mainly by agriculture, municipalities, and industry while only 140 million tons, or 17%, of those are collected for composting (Ahmad et al. 2007; Singh et al. 2011 b). Composting is a biological process of conversion of heterogeneous organic wastes (manure, sludge, yard wastes, leaves, fruits, vegetables, and food wastes) into humus-like substances. Technologies have been developed to fasten the process by utilizing consortia of microorganisms, such as bacteria, actinomycetes and fungi under controlled conditions of moisture, temperature and aeration. These microbes occur widely in nature and are indigenous to soil, dust, fruit and vegetable matter and wastes. Apart from benefiting agro ecosystem, composts also mitigate the adverse effects of greenhouse gases, such as carbon dioxide released from agricultural soil and methane gas from livestock and their manures. This is done principally by carbon sequestration i.e., locking up carbon in organic matter and organisms within the soil due to the stable nature of composts.

10 Non-target Effects of Microorganisms/Biological Control Agents/Health and Safety Issues

Soil microbes or in particular AIMs, such as PGPR or BCAs that are extensively used in agriculture might possess some under-estimated or neglected non-targeted effects. Four safety issues were described by Burges (1981) and Cook et al. (1996): (1) competitive displacement; (2) allergenicity; (3) toxicity and; (4) pathogenicity. The potential risk of BCAs could be *Gene transfer* from BCAs to other microorganisms that could result in the development of new genotypes. Risk associated with the probability of gene transfer depends on many factors such as biology of the recipient organism, nature of the trait transferred as well as the environment. However, according to Cook et al. (1996), there are no safety issues associated with the gene transfer, whether it occurs naturally or is done intentionally.

Competitive displacement embraces when BCAs eliminate or replace native non-target species through temporal and spatial struggle for nutrients i.e., microbe-microbe interaction. Targeted effect of *Competitive displacement* can be explained by some of the following examples: (1) spore suspension of saprophytic fungus *Phlebia gigantea* protects the pine stump from annosus root rot of pine caused by *Heterobasidion annosum* (Rishbeth 1975); (2) yeasts and bacteria have been used in post-harvest disease management in fruits during storage (Roberts 1990; Wisniewski et al. 1991). *Candida oleophila* Montrocher (Aspire) and two isolates of *Pseudomonas syringae* van Hall with this ability (Bio-Save 10 and Bio-Save 11) were registered with the U.S. Environmental Protection Agency in 1995 (Cook et al. 1996); (3) non-pathogenic strain of *P. syringae* (Sno-Max) protects the frost-sensitive plants, such as tomato and potato to prevent the subsequent natural establishment of ice-nucleation-active bacteria (Lindow 1983). The wide use of BCAs might reduce the diversity and/or abundance of other microorganisms in an ecosystem that might significantly affect other natural processes occurring in the eco-system, such as nutrient recycling by affecting their ability to release enzymes that permit the metabolism of complex organic molecules that are unavailable to many organisms, such as lignin or symbiotic relationships (mycorrhizae) with fungi (Orth et al. 1993). The interaction of AIMs in close proximity with other organisms, such as insects or fungi or bacteria or among themselves might be affected.

Spores of AIMs might develop in patients with weak or suppressed immunity resulting in *allergic reactions* are the non targeted effects (Latge and Paris 1991). Allergenicity is a potential safety concern to the workers engaged in production facilities exposed repeatedly to high concentrations of spores of fungi, such as *Beauveria* or *Metarhizium* due to the development of hypersensitive reactions. An intradermal skin testing confirmed the allergenic potential of *Beauveria bassiana* due to the presence of abundant IgE reactive proteins that are also cross-reactive with allergens from other fungi. (Westwood et al. 2005). Although no evidence suggests that the toxicity of *Metarhizium anisopliae* to humans or other mammals, there are some case studies that documents the infection by *M. anisopliae* in immune-compromised and immune-competent individuals as well as a case of invasive rhinitis in a cat (Ward et al. 2011).

Antibiosis and the toxic alkaloids produced against aphids or other insect herbivores are an example of *Toxigenicity* (Cook and Baker 1983; Siegel et al. 1987). For example *P. fluorescens*, *G. virens*, *B. thuringiensis*, *A. radiobacter*, *Acremonium sp*. (Siegel et al. 1987; Lumsden et al. 1992). However, it has also been reported that the alkaloids might cause toxicosis in livestock grazing on treated grass/leaves and the several Bt toxins might affect arthropods (Rogoff 1982; Siegel et al. 1987; Laird et al. 1990). Some plant-associated microorganisms used for seed treatments produce toxic metabolites during seed germination and exert injury to the germinating seedlings resulting in stand failure or stunted plants. Such types of non-targeted effects should be recognised early during the research and development phase. Potential risk of these microorganisms to affect organisms other than the target pests is of most important concern from a risk management perspective. Simberloff and Stiling (1996) reviewed the probability of risk associated with microorganisms, particularly BCAs. The prospective reproducing and spreading habit of BCAs to non-target environments might produce unforeseen effects on native organisms. BCAs, being perceived as 'natural' and 'low

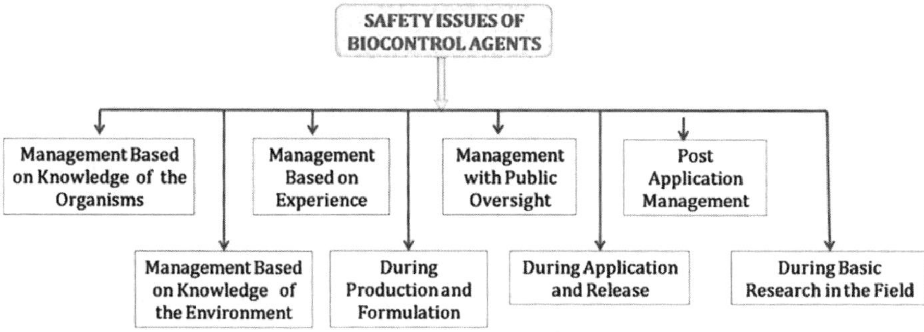

Fig. 10.4 Safety issues pertinent to biocontrol agents

risk', are often exempted from the rigorous testing as in the case of chemical pesticides (Lumsden and Walter 1995; Anonymous 2001). Some of the microorganisms used in plant disease management are hypo-virulent strains of plant pathogens that increase the possibility of gene transfer between isolates that might result in gain of virulence or loss of biological control abilities (Gullino et al. 1995). The ecosystem exhibits an intricate environment and complex interactions between biotic and abiotic factors. It renders difficult to investigate the interaction of BCAs within their environment. However, the live behaviour of these microorganisms implicates the importance of understanding their functioning and to anticipate possible hitches and alleviate any adverse effects. According to Wapshere (1974) and Briese (2003), centrifugal phylogeny scheme used to verify any unfavourable effects of BCAs in weed management on non-target plants can also be used to determine the host range of plants for BCAs for plant diseases. Nevertheless due to variable susceptibility within a single plant species, phylogeny testing might not be helpful in identifying vulnerable plant species. This implicates the need of studying genetic relationships between hosts and pathogens for determining host range (Weidemann and Tebeest 1990).

The movement of BCAs in an ecosystem needs to be monitored regularly. Molecular techniques can be used to monitor the integrity as well as stability of strains released in the field (Hermosa et al. 2000, 2001; Avis et al. 2001). According to Teng and Yang (1993), the analysis of risks and impacts associated with release of BCAs can be accessed through risk determination, data and information generation, prediction of impact, risk and benefit evaluation. However, at the same time, it is practically unfeasible to envisage all impacts or interactions of BCAs with non-target organisms. In order to utilize the maximum potential of BCAs, some of the safety issues pertinent to the safe use of microorganisms for biological control of pests and diseases and to avoid the risk associated with the production and utilization of BCAs are described in Fig. 10.4.

Identification, assessment and management of the risks or potential risks would lead to the "safe use" of BCAs. Policies should be formulated to study the potential risks of BCAs.

11 Concluding Remarks

In the scenario of "*Green Revolution*" or sustainable advances in agricultural production, productivity, profitability and enhanced job opportunities can be attained only by a substantial progress in productivity per unit of land, water, and energy and even per capita, without causing any harm to the environment. Agricultural fields vary with space and time and hence require site-specific crop management for maximising the efficiency of the input resources. Productivity and profitability enhancement should be our key to success. In this endeavour, cutting-edge technologies will play a pivotal role. Sustained growth in agricultural productivity will depend upon continued improvements in genotypes, germplasms, improved nutritional value of crops, resistance to abiotic stress and disease and pest management yield stability. The increasing cost of chemical pesticides and fertilizers and the low purchasing power of most of the farmers suggest the importance of cheaper, cost-effective and easily accessible alternatives for practicing sustainable agriculture.

Decision making for agricultural sustainability requires an active participation of not only at the field and farm levels but also on a large scale including the end users, stakeholders, researchers together on one stage. Agricultural sustainability can be attained by attaining flexible understanding and anticipation of social, economic and biological factors. The wealth of soil microorganisms can be exploited for the cause of mankind in a safe and effective manner. Intensive future research is required in this area, predominantly on the field evaluation and application of potential microorganisms. Unfavourable interactions between farming and conservation occur due to the lack of understanding between plant-microbe and microbe-microbe interactions. Therefore, well designed farming practices and successful bio-diversification using these microorganisms could lead in the right direction of sustainable agriculture. Such understanding is also of utmost importance in biological control systems where wildlife is to stay a major component of agricultural systems.

Considerable research in the field of microbial technology has proved to be an effective and potential means for enriching soil fertility and crop health. However, the technology needs further improvement for its better exploitation under sustainable agriculture development programs. In this direction, AIMs particularly PGPR have emerged as an excellent model system with novel genetic constituents and bioactive chemicals with multifaceted use in agriculture and environmental sustainability. Our understanding of PGPR diversity, colonization ability and mechanisms of interactions, formulation and application could assist there better utilization in agricultural ecosystem. At the same time, the exploration of live microbial creatures raises health and safety issues and their non-target effects on other organisms, such as toxigenicity, allergenicity and pathogenicity, persistence in the environment and potential for horizontal gene transfer.

Bio-augmentation and bio-stimulation approaches using microbial inoculants, biofertilizers, bio(chemicals) and organic amendments help in improving and maintaining soil biology, fertility, crop productivity, biocontrol methods for soil borne diseases and plant–parasitic nematodes as well as soil remediation. However, the knowledge of microbial diversity and function in soils is limited due to the lack of taxonomic and efficient methodology. The recent rapid advances in the field of genomics in particular, metagenomics, proteomics and functional genomics have led to a better understanding of the microbial diversity.

The functional aspects of agro-ecosystems and soil biodiversity, important to farmers are numerous and beyond the scope of the chapter. The full potential of soil microorganisms in sustainable agriculture could be achieved by proper amalgamation with above-ground bio-diversity to sustain ecosystem functioning. The acquaintance gained through in-depth experimentation and testing will reap benefits only, if inspired and combined with farmer's knowledge, problem perception and opportunities for application. At the same time, the role of society, in understanding the importance of agro-biodiversity, at large cannot be neglected.

Acknowledgements The authors acknowledge ICAR, New Delhi, India for providing Senior Research Fellowship for doctorate studies and Canadian Government for providing Canadian Commonwealth Scholarship (CCSP, 2010–2011) to Surinder Kaur. The views or opinions expressed in this article are those of the authors.

References

Abdalla MH, Omer SA (2001) Survival of *Rhizobia/Bradirhizobia* and a rock phosphate solubilizing fungus *Aspergillus niger* on various carriers from some agro-industrial wastes and their effect on nodulation and growth of fababean and soybean. J Plant Nutr 24:261–272

Ahmad R, Jilani G, Arshad M, Zahir ZA, Khalid A (2007) Bio-conversion of organic wastes for their recycling in agriculture: an overview of perspectives and prospects. Ann Microbiol 57:471–479

Alexander M (1977) Introduction to soil microbiology, 2nd edn. Academic, New York, p 467

Amico ED, Cavalca L, Andreoni V (2008) Improvement of *Brassica napus* growth under cadmium stress by cadmium resistant Rhizobacteria. Soil Biol Biochem 40(1):74–84

Ando K, Sagawa T, Oishi H, Suzuki K, Nawata T (1974) Tetranactin, a pesticidal antibiotic. Proceedings of the 1st intersect congr IAMS. Sci Counc Jpn 3:630

Anonymous (2001) Guidelines for the registration of microbial pest control agents and products. Pest Management Regulatory Agency, Ottawa

Antoun H, Kloepper JW (2001) Plant growth promoting rhizobacteria (PGPR). In: Brenner S, Miller JH (eds) Encyclopedia of genetics. Academic, New York, pp 1477–1480

Avis TJ, Hamelin RC, Bélanger RR (2001) Approaches to molecular characterization of fungal biocontrol agents: some case studies. Can J Plant Pathol 23:8–12

Babalola OO (2010) Beneficial bacteria of agricultural importance. Biotechnol Lett 32:1559–1570

Barber SA (1995) Soil nutrient bioavailability: a mechanistic approach, 2nd edn. Wiley, New York

Barriuso J, Solano BR, Lucas JA, Lobo AP, Villaraco AG, Mañero FJG (2008) Ecology, genetic diversity and screening strategies of plant growth promoting rhizobacteria (PGPR). In: Ahmad I, Pichtel J, Hayat S (eds) Plant-bacteria interactions. Wiley-VCH, Weinheim, pp 1–13

Belimov AA, Safronova VI, Sergeyeva TA, Egorova TN et al (2001) Characterization of plant growth promoting rhizobacteria isolated from polluted soils and containing 1-aminocyclopropane-1-carboxylate deaminase. Can J Microbiol 47(7):642–652

Belimov AA, Hontzeas N, Safronova VI, Demchinskaya SV, Piluzza G, Bullitta S, Glick BR (2005) Cadmium-tolerant plant growth-promoting bacteria associated with the roots of Indian mustard (*Brassica juncea* L Czern.). Soil Biol Biochem 37(2):241–250

Berg G (2009) Plant–microbe interactions promoting plant growth and health: perspectives for controlled use of microorganisms in agriculture. Appl Microbiol Biotechnol 84:11–18

Birkhofer KT, Bezemer M, Bloem J, Bonkowski M, Christensen S, Dubois D, Ekelund F, Fliebach A, Gunst L, Hedlund K, Mader P, Mikola J, Robin C, Setala H, Tatin-Froux F, Wim H, der Putten V, Scheu S (2008) Long-term organic farming fosters below and aboveground biota: implications for soil quality, biological control and productivity. Soil Biol Biochem 40(9):2297–2308

Blaha D, Combaret CP, Mirza MS, Loccoz YM (2006) Phylogeny of the 1-aminocyclopropane-1-carboxylic acid deaminase encoding gene acdS in phytobeneficial and pathogenic Proteobacteria and relation with strain biogeography. FEMS Microbiol Ecol 56:455–470

Bloemberg GV, Lugtenberg BJJ (2001) Molecular basis of plant growth promotion and biocontrol by rhizobacteria. Curr Opin Plant Biol 4:343–350

Bock E, Koops HP, Harms H (1989) Nitrifying bacteria. In: Schlegel H, Bowein B (eds) Autotrophic bacteria. Science Tech Publishers, Madison, pp 81–96

Bonanomi G, Antignani V, Capodilupo M, Scala F (2010) Identifying the characteristics of organic soil amendments that suppress soilborne plant diseases. Soil Biol Biochem 42:136–144

Borowicz VA (1997) A fungal root symbiont modifies plant resistance to an insect herbivore. Oecologia 112:534–542

Briese DT (2003) The centrifugal phylogenetic method used to select plants for host-speciWcity testing of weed biological control agents: can and should it be modernized? In: SpaVord-Jacob HS, Briese DT (eds) Improving the selection, testing and evaluation of weed biological control agents, Tech. ser. 7. CRC for Australian Weed Management, Adelaide, pp 23–34

Bull AT, Goodfellow M, Slater JH (1992) Biodiversity as a source of innovation in biotechnology. Annu Rev Microbiol 46:219–252

Burd GI, Dixon DG, Glick BR (2000) Plant growth-promoting bacteria that decrease heavy metal toxicity in plants. Can J Microbiol 46:237–245

Burges HD (1981) Safety, safety testing and quality control of microbial pesticides. In: Burges HD (ed) Microbial control of pests and plant diseases 1970–1980. Academic, London, pp 739–767, 949 pp

Caskey CT (1973) Inhibitors of protein biosynthesis. In: Hochster RM, Kates M, Quastel JH (eds) Metabolic inhibitors, vol IV. Academic, New York, p 131

Chacko S, Ramteke PW, John SA (2009) Amidase from plant growth promoting rhizobacterium. J Bacteriol Res 1(4):046–050

Chen YP, Rekha PD, Arun AB, Shen FT, Lai WA, Young CC (2006) Phosphate solubilizing bacteria from subtropical soil and their tricalcium phosphate solubilizing abilities. Appl Soil Ecol 34(Suppl 1):33–41

Cho K, Toler H, Lee J, Owenley B, Stutz JC, Moore JL, Auge RM (2006) Mycorrhizal symbiosis and response of sorghum plants to combined drought and salinity stresses. J Plant Physiol 163:517–528

Clark RB, Zeto SK (2000) Mineral acquisition by arbuscular mycorrhizal plants. J Plant Nutr 23:867–902

Collins HP, Rasmussen PE, Douglas CL Jr (1992) Crop rotation and residue management effects on soil carbon and bacterial dynamics. Soil Sci Soc Am J 56:783–788

Cook RJ, Baker KF (1983) The nature and practice of biological control. APS Press, St. Paul

Cook RJ, Bruckart WL, Coulson JR, Goettel MS, Humber RA, Lumsden RD, Maddox JV, McManus ML, Moore L, Meyer SF, Quimby PC, Stack JP, Vaughn JL (1996) Safety of microorganisms intended for pest and plant disease control: a framework for scientific evaluation. Biol Control 7:333–351

Copping GL, Duke OS (2007) Natural products that have been used commercially as crop protection agents. Pest Manag Sci 63:524–554

Copping GL, Menn JJ (2000) Biopesticides: a review of their action, applications and efficacy. Pest Manag Sci 56:651–676

Daly MJ, Stewart DPC (1999) Influence of "effective microorganisms" (EM) on vegetable production and carbon mineralization–a preliminary investigation. J Sustain Agric 14:2–3

De Vleesschauwer D, Cornelis P, Höfte M (2006) Redox-active pyocyanin secreted by *Pseudomonas aeruginosa* 7NSK2 triggers systemic resistance to *Magnaporthe grisea* but enhances *Rhizoctonia solani* susceptibility in rice. Mol Plant Microbe Interact 19:1406–1419

Denton B (2007) Advances in phytoremediation of heavy metals using plant growth promoting bacteria and fungi. MMG 445 Basic Biotechnol 3:1–5

Dimkpa C, Svatoš A, Merten D, Büchel G, Kothe E (2008) Hydroxamate siderophores produced by *Streptomyces acidiscabies* E13 bind nickel and promote growth in cowpea (*Vigna unguiculata* L) under nickel stress. Can J Microbiol 54(3):163–172

Dorn E, Hellwig M, Reineke W, Knackmuss HJ (1974) Isolation and characterization of a 3-chlorobenzoate degrading pseudomonad. Arch Microbiol 99:61–70

Dorothy GS, Harry JR, Dorothy MC (1945) The influence of streptomycin and streptothricin on the intestinal flora of mice. J Bacteriol 50(6):613–621

Duke SO, Dayan FE, Rimando AM, Schrader KK, Aliotta G, Oliva A et al (2002) Chemicals from nature for weed management. Weed Sci 50:138–151

Egamberdiyeva D (2005) Plant-growth-promoting rhizobacteria isolated from a Calcisol in a semi-arid region of Uzbekistan: biochemical characterization and effectiveness. J Plant Nutr Soil Sci 168(Suppl 1):94–99

Egamberdiyeva D (2007) The effect of plant growth promoting bacteria on growth and nutrient uptake of maize in two different soils. Appl Soil Ecol 36:184–189

Eguchi J, Sasaki S, Ohta N, Akashiba T, Tsuchiyama T, Suzuki S (1968) Studies on polyoxins, antifungal antibiotics. Mechanism of action on the diseases caused by *Alternaria* spp. Ann Phytopathol Soc Jpn 34:280

Feng G, Zhang FS, Li XL, Tian CY, Tang C, Rengel Z (2002) Improved tolerance of maize plants to salt stress by arbuscular mycorrhiza is related to higher accumulation of soluble sugars in roots. Mycorrhiza 12:185–190

Figueiredo MVB, Burity HA, Martinez CR, Chanway CP (2008) Alleviation of drought stress in common bean (*Phaseolus vulgaris* L.) by co-inoculation with *Paenibacillus polymyxa* and *Rhizobium tropici*. Appl Soil Ecol 40:182–188

Figueiredo MVB, Seldin L, Araujo FF, Mariano RLR (2010) Plant growth promoting rhizobacteria: fundamentals and applications. In: Maheshwari DK (ed) Plant growth and health promoting bacteria, microbiology monographs 18. Springer, Berlin

Fisher MH (1993) Recent progress in avermectin research. In: Duke SO, Menn JJ, Plimmer JR (eds) Pest control with enhanced environmental safety, ACS symposium series no. 524. American Chemical Society, Washington, DC, pp 169–182

Gaind S, Gaur AC (1991) Thermotolerant phosphate solubilizing microorganisms and their interaction with mungbean. Plant Soil 133(1):141–149

Galal YGM (2003) Assessment of nitrogen availability to wheat (*Triticum aestivum* L.) from inorganic and organic N sources as affected by *Azospirillum brasilense* and *Rhizobium leguminosarum* inoculation. Egypt J Microbiol 38:57–73

Ghorbani R, Wilcockson S, Koocheki A, Leifert C (2009) Soil management for sustainable crop disease control: a review. In: Lichtfouse E (ed) Organic farming, pest control and remediation of soil pollutants, vol 1, Sustainable agriculture reviews. Springer, Dordrecht/New York

Glick BR, Patten CL, Holguin G, Penrose DM (1999) Biochemical and genetic mechanisms used by plant growth promoting bacteria. Imperial College Press, London

Golding BT, Rickards RW, Meyer WE, Patrick JB, Barber M (1966) The structure of the macrolide antibiotic pimaricin. Tetrahedron Lett 30:3551–3557

Govindasamy V, Senthilkumar M, Mageshwaran V, Annapurna K (2009) Detection and characterization of ACC in plant growth promoting rhizobacteria. J Plant Biochem Biotechnol 18(1):71–76

Grichko VP, Glick BR (2001) Amelioration of flooding stress by ACC deaminase containing plant growth-promoting bacteria. Plant Physiol Biochem 39:11–17

Grover M, Ali Z, Sandhya V, Rasul A, Venkateswarlu B (2011) Role of microorganisms in adaptation of agriculture crops to abiotic stresses. World J Microbiol Biotechnol 27:1231–1240

Gullino ML, Migheli Q, Mezzalama M (1995) Risk analysis in the release of biological control agents. Plant Dis 79:1193–1201

Han HS, Lee KD (2005) Plant growth promoting rhizobacteria effect on antioxidant status, photosynthesis, mineral uptake and growth of lettuce under soil salinity. Res J Agric Biol Sci 1(3):210–215

Harada S, Kishi T (1978) Isolation and characterisation of mildiomycin, a new nucleoside antibiotic. J Antibiot 31:519–524

Harish S, Kavino M, Kumar N, Balasubramanian P, Samiyappan R (2009a) Induction of defense-related proteins by mixtures of plant growth promoting endophytic bacteria against Banana bunchy top virus. Biol Control 51:16–25

Harish S, Kavino M, Kumar N, Samiyappan R (2009b) Differential expression of pathogenesis-related proteins and defense enzymes in banana: interaction between endophytic bacteria banana bunchy top virus and *Pentalonia nigronervosa*. Biocontrol Sci Technol 19:843–857

Hermosa MR, Grondona I, Iturriaga EA, Diaz-Minguez JM, Castro C, Monte E, Garcia-Acha I (2000) Molecular characterization and identification of biocontrol isolates of *Trichoderma* spp. Appl Environ Microbiol 66:1890–1898

Hermosa MR, Grondona I, Diaz-Minguez JM, Iturriaga EA, Monte E (2001) Development of a strain-specific SCAR marker for the detection of *Trichoderma atroviride* 11, a biological control agent against soilborne fungal plant pathogens. Curr Genet 38:343–350

Hernandez-Rodriguez A, Heydrich Perez M, Acebo Guerrero Y, Velazquez del Valle MG, Hernandez Lauzardo AN (2008) Antagonistic activity of Cuban native rhizobacteria against *Fusarium verticillioides* (Sacc.) Nirenb. In maize (*Zea mays* L.). Appl Soil Ecol 39:180–186

Higa T (1991) Effective microorganisms: a biotechnology for mankind. In: Parr JF, Hornick SB, Whitman CE (eds) Proceedings of the first international conference on Kyusei nature farming. U.S. Department of Agriculture, Washington, D.C., pp 8–14

Higa T (1995) Effective microorganisms: their role in Kyusei nature farming and sustainable agriculture. In: Parr JF, Hornick SB, Simpson ME (eds) Proceedings of the third international conference on Kyusei nature farming. U.S. Department of Agriculture, Washington. D.C

Higa T, Wididana GN (1991) The concept and theories of effective microorganisms. In: Parr JF, Hornick SB, Whitman CE (eds) Proceedings of the first international conference on Kyusei nature farming. U.S. Department of Agriculture, Washington, DC, pp 118–124

Hinsinger P (1998) How do plant roots acquire mineral nutrients? Chemical processes involved in the rhizosphere. Adv Agron 64:225–265

Hirsch PR, Jones MJ, McGrath SP, Giller KE (1993) Heavy metals from past applications of sewage sludge decrease the genetic diversity of *Rhizobium leguminosarum* biovar *trifolii* populations. Soil Biol Biochem 25:1485–1590

Hodge A, Campbell CD, Fitter AH (2001) An arbuscular mycorrhizal fungus accelerates decomposition and acquires nitrogen directly from organic material. Nature 413:297–299

Hontzeas N, Saleh SS, Glick BR (2004a) Changes in gene expression in canola roots induced by ACC-deaminase-containing plant-growth-promoting bacteria. Mol Plant Microbe Interact 17:865–871

Hontzeas N, Zoidakis J, Glick BR, Abu Omar MM (2004b) Expression and characterization of 1-aminocyclopropane-1-carboxylate deaminase from the rhizobacterium *P. putida* UW4: a key enzyme in bacterial plant growth promotion. Biochim Biophys Acta Mol Cell Biol Lipids 1703:11–19

Hu C (2005) Induction of growth promotion and stress tolerance in *Arabidopsis* and tomato by plant growth promoting rhizobacteria. Auburn University, Dissertation. http://hdl.handle.net/10415/769

Huang KT, Misato T, Suyama H (1964) Effect of blasticidin S on protein biosynthesis of *Pyricularia oryzae*. J Antibiot Ser A 17:65

Isono K, Suzuki S (1979) The polyoxins: pyrimidine nucleoside peptide antibiotics inhibiting cell wall biosynthesis. Heterocycles 13:333

Isono K, Nagatsu J, Kawashima Y, Suzuki S (1965) Studies on polyoxins, antifungal antibiotics. Part 1. Isolation and characterisation of polyoxins A and B. J Antibiot Ser A 18:115

Isono K, Nagatsu J, Kobinata K, Sasaki K, Suzuki S (1967) Studies on polyoxins, antifungal antibiotics. Part V. Isolation and characterisation of polyoxins C, D, E, F, G, H and I. Agric Biol Chem 31:190

Jansson RK, Dybas RA (1996) Avermectins: biochemical mode of action, biological activity and agricultural importance. In: Ishaaya I (ed) Insecticides with novel modes of action: mechanisms and application. Springer, New York

Jiang ZQ, Guo Y-H, Li S-M, Qi H-Y, Guo J-H (2006) Evaluation of biocontrol efficiency of different *Bacillus* preparations and field application methods against Phytophthora blight of bell pepper. Biol Control 36:216–223

Jones JW (1991) Diversity and physiology of methanogens. In: Rogers JE, Whitman WB (eds) Bacterial production and consumption of greenhouse gases: methane, nitrogen oxides, and halomethanes. American Society for Microbiology, Washington DC, pp 3–55

Jungk A (2001) Root hair and acquisition of plant nutrients from soils. J Plant Nutr Soil Sci 164:121–129

Kahiluoto H, Ketoja E, Vestberg M, Saarela I (2001) Promotion of AM utilization through reduced P fertilization. Plant Soil 231:65–79

Karagiannidis N, Bletsos F, Stavropoulos N (2002) Effect of Verticillium wilt (*Verticillium dahliae* Kleb.) and mycorrhiza (*Glomus mosseae*) on root colonization, growth and nutrient uptake in tomato and eggplant seedlings. Sci Hortic 94:145–156

Kavino M, Harish S, Kumar N, Saravanakumar D, Damodaran T, Soorianathasundaram K, Samiyappan R (2007) Rhizosphere and endophytic bacteria for induction of systemic resistance of banana plantlets against bunchy top virus. Soil Biol Biochem 39:1087–1098

Kavino M, Harish S, Kumar N, Saravanakumar D, Samiyappan R (2008) Induction of systemic resistance in banana (*Musa* spp.) against banana bunchy top virus (BBTV) by combining chitin with root-colonizing *Pseudomonas fluorescens* strain CHA0. Eur J Plant Pathol 120:353–362

Kavino M, Harish S, Kumar N, Saravanakumar D, Samiyappan R (2010) Effect of chitinolytic PGPR on growth, yield and physiological attributes of banana (*Musa* spp.) under field conditions. Appl Soil Ecol 45:71–77

Khaliq A, Abbasi MK, Hussain T (2006) Effects of integrated use of organic and inorganic nutrient sources with effective microorganisms (EM) on seed cotton yield in Pakistan. Bioresour Technol 97(8):967–972

Khan AG (2005) Role of soil microbes in the rhizospheres of plants growing on trace metal contaminated soils in phytoremediation. J Trace Elem Med Biol 18(4):355–364

Khan MS, Zaidi A, Wani PA (2006) Role of phosphate solubilizing microorganisms in sustainable agriculture-a review. In: Lichtfouse E, Navarrete M, Debaeke P, Veronique S, Alberola C (eds) Sustainable agriculture, vol 5. Springer, Netherlands, pp 551–570

Kohler J, Caravaca F, Carrasco L, Roldan A (2006) Contribution of *Pseudomonas mendocina* and *Glomus intraradices* to aggregates stabilization and promotion of biological properties in rhizosphere soil of lettuce plants under field conditions. Soil Use Manag 22:298–304

Kohler J, Hernandez JA, Caravaca F, Roldan A (2009) Induction of antioxidant enzymes is involved in the greater effectiveness of a PGPR versus AM fungi with respect to increasing the tolerance of lettuce to severe salt stress. Environ Exp Bot 65:245–252

Koide RT, Goff MD, Dickie IA (2000) Component growth efficiencies of mycorrhizal and nonmycorrhizal plants. New Phytol 148:163–168

Kumar A, Prakash A, Johri BN (2011) *Bacillus* as PGPR in crop ecosystem. In: Maheshwari DK (ed) Bacteria in agrobiology: crop ecosystems. Springer, Heidelberg. ISBN 3642183565, 9783642183560

Kusaka T, Suetomi K, Iwasa T, Harada S (1979) TF-138: a new fungicide. In: Proceedings of the 1979 British crop protection conference – pests and diseases, BCPC, Farnham, Surrey, 1979, pp 589–595

Laird M, Lacey LA, Davidson EW (eds) (1990) Safety of microbial insecticides. CRC Press, Boca Raton

Lankas G, Gordon LR (1989) Toxicology. In: Campbell WC (ed) Ivermectin and abamectin. Springer, New York, pp 89–112

Latge JP, Paris S (1991) The fungal spore reservoir of allergens. In: Cole GT, Hoch, HC (eds) The fungal spore and disease Initiation in plants and animals. Plenum Press, New York, pp 379–401

Lee EY, Jun YS, Cho KS, Ryu HW (2002) Degradation characteristics of toluene, benzene, ethylbenzene, and xylene by *Stenotrophomonas maltophilia* T3-c. J Air Waste Manage Assoc 52:400–406

Lindow S (1983) Methods of preventing frost injury caused by epiphytic ice nucleation active bacteria. Plant Dis 67:327–333

Lugtenberg BJJ, Dekkers L, Bloemberg GV (2001) Molecular determinants of rhizosphere colonization by *Pseudomonas*. Ann Rev Phytopathol 39:461–490

Lumsden RD, Walter JF (1995) Development of the biocontrol fungus *Gliocladium virens:* risk assessment and approval for horticultural use. In: Hokkanen MT, Lynch JM (eds) Biological control: benefits and risks. Cambridge University Press, Cambridge, pp 263–269

Lumsden R, Locke JC, Adkins ST, Walter JF, Ridout CJ (1992) Isolation and localization of the antibiotic gliotoxin produced by *Gliocladium virens* from alginate prill in soil and soilless media. Phytopathology 82:230–235

Lydon J, Duke SO (1999) Inhibitors of glutamine synthesis. In: Singh BK (ed) Plant amino acids. Marcel Dekker, New York, pp 445–464

Lynch JM (1983) Soil biotechnology, microbiological factors in crop productivity. Blackwell Scientific Publications, Oxford, 191 pp

Lynch JP (2005) Root architecture and nutrient acquisition. In: BassiriRad H (ed) Nutrient acquisition by plants: an ecological perspective. Springer, Berlin, pp 147–183

Maeder P, Fliessbach A, Dubois D, Gunst L, Fried P, Niggli U (2002) Soil fertility and biodiversity in organic farming. Science 296(5573):1694–1697

Marschner H (1995) Mineral nutrition of higher plants, 2nd edn. Academic, London

Matsuura K (1983) Characteristics of validamycin A in controlling rhizoctonia diseases. In: Miyamoto J, Kearney PC (eds) IUPAC pesticide chemistry, vol 2. Pergamon Press, Oxford, p 301

Mayak S, Tirosh T, Glick BR (2004a) Plant growth-promoting bacteria that confer resistance to water stress in tomato and pepper. Plant Sci 166:525–530

Mayak S, Tirosh T, Glick BR (2004b) Plant growth-promoting bacteria that confer resistance in tomato and pepper to salt stress. Plant Physiol Biochem 42:565–572

Meyer WE (1968) Pimaricin: the glycosidic form of mycosamine. Chem Commun (Lond) 8:470

Meziane H, Van der Sluis I, Van Loon LC, Höfte M, Bakker PAHM (2005) Determinants of *Pseudomonas putida* WCS358 involved in inducing systemic resistance in plants. Mol Plant Pathol 6:177–185

Naureen ZP, Hafeez AH, Roberts FY, Michael R (2009) Identification of rice blast disease-suppressing bacterial strains from the rhizosphere of rice grown in Pakistan. Crop Prot 28:1052–1060

Nie L, Shah S, Rashid A, Burd GI, Dixon DG, Glick BR (2002) Phytoremediation of arsenate contaminated soil by transgenic canola and the plant growth promoting bacterium *Enterobacter cloacae* CAL2. Plant Physiol Biochem 40:355–361

Oka Y (2010) Mechanisms of nematode suppression by organic soil amendments: a review. Appl Soil Ecol 44:101–115

Om Y, Yamaguchi I, Misado T (1984) Inhibition of protein biosynthesis by mildiomycin, an anti-mildew substance. J Pestic Sci 9:317–323

Ongena M, Jourdan E, Schäfer M, Kech C, Budzikiewicz H, Luxen A, Thonart P (2005) Isolation of an N-alkylated benzylamine derivative from *Pseudomonas putida* BTP1 as elicitor of induced systemic resistance in bean. Mol Plant Microbe Interact 18:562–569

Orth AB, Royse DJ, Tien M (1993) Ubiquity of lignin degrading peroxidases among various wood-degrading fungi. Appl Environ Microbiol 59:4017–4023

Pandey KD, Shukla PN, Giri DD, Kashyap AK (2005a) Cyanobacteria in alkaline soil and the effect of cyanobacteria inoculation with pyrite amendments on their reclamation. Biol Fertil Soils 41:451–457

Pandey P, Kang SC, Maheshwari DK (2005b) Isolation of endophytic plant growth promoting *Burkholderia sp.* MSSP from root nodules of *Mimosa pudica*. Curr Sci 89:170–180

Parida AK, Das AB (2005) Salt tolerance and salinity effects on plants: a review. Ecotoxicol Environ Saf 60:324–349

Paul D, Nair S (2008) Stress adaptations in a plant growth promoting rhizobacterium (PGPR) with increasing salinity in the coastal agricultural soils. J Basic Microbiol 48:378–384

Peacock AD, Mullen MD, Ringelberg DB, Tyler DD, Hedrick DB, Gale PM, White DC (2001) Soil microbial community responses to dairy manure or ammonium nitrate applications. Soil Biol Biochem 33:1011–1019

Pennasio S, Roggero P (1992) Effect of cadmium and nickel on ethylene biosynthesis in soybean. Plant Biol 34(3–4):345–349

Persello CF, Nussaume L, Robaglia C (2003) Tales from the underground: molecular plant–rhizobacteria interactions. Plant Cell Environ 26:189–199

Raaijmakers JM, Paulitz TC, Steinberg C, Alabouvette C, Loccoz YM (2009) The rhizosphere: a playground and battlefield for soil borne pathogens and beneficial microorganisms. Plant Soil 321:341–361

Radjacommare R, Nandakumar R, Kandan A, Suresh S, Bharathi M, Raguchander T, Samiyappan R (2002) *Pseudomonas fluorescens* based bio-formulation for the management of sheath blight disease and leaf folder insect in rice. Crop Prot 21:671–677

Rajkumar M, Freitas H (2008) Influence of metal resistant-plant growth-promoting bacteria on the growth of *Ricinus communis* in soil contaminated with heavy metals. Chemosphere 71(5):834–842

Ramachandran K, Srinivasan V, Hamza S, Anandaraj M (2007) Phosphate solubilizing bacteria isolated from the rhizosphere soil and its growth promotion on black pepper (*Piper nigrum* L) cuttings. Dev Plant Soil Sci 102:324–331

Ray RC, Ward OP, Singh A, Isobe S (2008) Commercialization of microbial biotechnology in horticulture: summary, outlook of achievements, constraints and prospects. In: Ray RC, Ward OP (eds) Microbial biotechnology in horticulture, vol 3. Science, New Hampshire, pp 341–365

Richardson AE, Barea JM, McNeill AM, Combaret CP (2009) Acquisition of phosphorus and nitrogen in the rhizosphere and plant growth promotion by microorganisms. Plant Soil 321:305–339

Rishbeth J (1975) Stump inoculation: a biological control of *Fomes annosus*. In: Bruehl GW (ed) Biology and control of soil-borne plant pathogens. Am. Phytopathol. Soc, St. Paul, pp 158–162

Rivas R, Peix A, Mateos PF, Trujillo ME, Martinez-Molina E, Velazqueze E (2006) Biodiversity of populations of phosphate solubilizing rhizobia that nodulates chickpea in different spanish soils. Plant Soil 287(Suppl 1–2):23–33

Roberts RG (1990) Biological control of mucor rot of pear by *Cryptococcus laurentii, C. flavus, and C. albidius*. Phytopathology 80:526–530

Rodríguez H, Fraga R (1999) Phosphate solubilizing bacteria and their role in plant growth promotion. Biotechnol Adv 17:319–339

Rogoff MH (1982) Regulatory safety data requirements for registration of microbial pesticides. In: Kurstak E (ed) Microbial and viral pesticides. Marcel-Dekker, New York, pp 645–679

Ryan MH, Angus JF (2003) Arbuscular mycorrhizae in wheat and field pea crops on a low P soil, increased Zn-uptake but no increase in Puptake or yield. Plant Soil 250:225–239

Ryan MH, Norton RM, Kirkegaard JA, McCormick KM, Knights SE, Angus JF (2002) Increasing mycorrhizal colonisation does not improve growth and nutrition of wheat on vertosols in south-eastern Australia. Aust J Agric Res 53:1173–1181

Safronova VI, Stepanok VV, Engqvist GL, Alekseyev YV, Belimov AA (2006) Root-associated bacteria containing 1-minocyclopropane-1-carboxylate deaminase improve growth and nutrient uptake by pea genotypes cultivated in cadmium supplemented soil. Biol Fertil Soils 42:267–272

Saharan BS, Nehra V (2011) Plant growth promoting rhizobacteria: a critical review. Life Sci Med Res 2011:LSMR-21. http://astonjournals.com/lsmr

Sainz MJ, Taboada-Castro MT, Vilarino A (1998) Growth, mineral nutrition and mycorrhizal colonization of red clover and cucumber plants grown in a soil amended with composted urban wastes. Plant Soil 205:85–92

Saleem M, Arshad M, Hussain S, Bhatti AS (2007) Perspective of plant growth promoting rhizobacteria (PGPR) containing ACC deaminase in stress agriculture. J Ind Microbiol Biotechnol 34:635–648

Sandhya V, Ali SZ, Grover M, Kishore N, Venkateswarlu B (2009) *Pseudomonas sp.* strain P45 protects sunflowers seedlings from drought stress through improved soil structure. J Oilseed Res 26:600–601

Saravanakumar D, Samiyappan R (2007) ACC deaminase from *Pseudomonas fluorescens* mediated saline resistance in groundnut (*Arachis hypogea*) plants. J Appl Microbiol 102:1283–1292

Saravanakumar D, Lavanya N, Muthumeena K, Raguchander T, Samiyappan R (2009) Fluorescent pseudomonad mixtures mediate disease resistance in rice plants against sheath rot (*Sarocladium oryzae*) disease. Biocontrol 54:273–286

Saunders DG, Bret BL (1997) Fate of spinosad in the environment. Down Earth 52:21–28

Schnider-Keel U, Lejbø lle KB, Bachler E, Haas D, Keel C (2001) The sigma factor Algu (Algt) controls exopolysaccharides production and tolerance towards desiccation and osmotic stress in the biocontrol agent, *Pseudomonas fluorescens* CHAO. Appl Environ Microbiol 67(2):5683–5693

Selvakumar G, Joshi P, Mishra PK, Bisht JP, Gupta HS (2009) Mountain aspects influence the genetic clustering of psychrotolerant phosphate solubilizing Pseudomonads in the Uttarkhand Himalayas. Curr Microbiol 59:432–438

Sergeeva E, Shah S, Glick BR (2006) Growth of transgenic canola (*Brassica napus* cv, Westar) expressing a bacterial 1-aminocyclopropane-1-carboxylate (ACC) deaminase gene on high concentrations of salt. World J Microbiol Biotechnol 22:277–282

Seshadri S, Ignacimuthu S, Vadivelu M, Lakshminarasimhan C (2007) Inorganic phosphate solubilization by two insect pathogenic *Bacillus* sp. Dev Plant Soil Sci 102:351–355

Shaharoona B, Arshad M, Zahir ZA (2006) Performance of *Pseudomonas spp.* containing ACC-deaminase for improving growth and yield of maize (*Zea mays* L.) in the presence of nitrogenous fertilizer. Soil Biol Biochem 38:2971–2975

Shannon D, Sen AM, Johnson DB (2002) A comparative study of the microbiology of soils managed under organic and conventional regimes. Soil Use Manag 18:274–283

Shigemoto R, Okuno T, Matsuura K (1989) Effect of validamycin A on the activity of trehalase of *Rhizoctonia solani* and several sclerotial fungi. Ann Phytopathol Soc Jpn 55:238

Sidhu GS (1998) Role of microorganisms in soil fertility. Utra Gro Plant Food Company. 1–10 pp 1–6

Siegel MR, Latch GCM, Johnson MC (1987) Fungal endophytes of grasses. Annu Rev Phytopathol 25:293–315

Simberloff D, Stiling P (1996) How risky is biocontrol. Ecology 77:1965–1974

Singh A, Parmar N, Kuhad RC (2011a) Bioaugmentation, biostimulation and biocontrol. In: Singh A, Parmar N, Kuhad RC, Ward OP (eds) Soil biology, vol 28. Springer, Heidelberg. ISBN 364219768X, 9783642197680

Singh JS, Pandey VC, Singh DP (2011b) Efficient soil microorganisms: a new dimension for sustainable agriculture and environmental development. Agric Ecosyst Environ 140:339–353

Spores of AIMs might develop in patients with weak or suppressed immunity resulting in *allergic reactions* are the non targeted effects (Latge and Paris 1991)

Sprent JL (1979) The biology of nitrogen-fixing organisms. McGraw-Hill, London, 196 pp

Stiens M, Schneiker S, Keller M, Kuhn S, Puhler A, Schluter A (2006) Sequence analysis of the 144-kilobase accessory plasmid psmesm11a, isolated from a dominant *Sinorhizobium meliloti* strain identified during a long-term Weld release experiment. Appl Environ Microbiol 72:3662–3672

Stockdale EA, Shepherd MA, Fortune S, Cuttle SP (2002) Soil fertility in organic farming systems – fundamentally different? Soil Use Manag 18:301–308

Sullivan P (2001) Sustainable management of soil-born plant diseases. ATTRA, USDA's Rural Business Cooperative Service. www.attar.org. Accessed 12 June 2012

Suzuki S, Isono K, Nagatsu J, Mizutani T, Kawashima Y, Mizuno T (1965) A new antibiotic, polyoxin A. J Antibiot Ser A 20:109

Tam PCF (1995) Heavy metal tolerance by ectomycorrhyzal fungi and metal amelioration by *Pisolithus tinctorium*. Mycorrhiza 5:181–187

Tanaka N, Yamaguchi H, Umezawa H (1966) Mechanism of kasugamycin action on polypeptide synthesis. J Biochem 60:429–434

Tanwar SPS, Sharma GL, Chahar MS (2002) Effect of phosphorus and biofertilizers on the growth and productivity of black gram. Annl Agric Res 23:491–493

Teng PS, Yang XB (1993) Biological impact and risk assessment in plant pathology. Annu Rev Phytopathol 31:495–521

Thompson JP (1994) Inoculation with vesicular-arbuscular mycorrhizal fungi from cropped soil overcomes long-fallow disorder of linseed (*Linum usitatissimum* L.) by improving P and Zn uptake. Soil Biol Biochem 26:1133–1143

Thompson GD, Dutton R, Sparks TC (2000) Spinosad – a case study: an example from a natural products discovery programme. Pest Manag Sci 56:696–702

Tiedje JM, Sextone AJ, Parkin TB, Revsbech NP, Shelton DR (1984) Anaerobic processes in soil. Plant Soil 76:197–212

Tilak KVBR, Ranganayaki N, Pal KK, De R, Saxena AK, Nautiyal CS, Mittal S, Tripathi AK, Johri BN (2005) Diversity of plant growth and soil health supporting bacteria. Curr Sci 89(1):136–150

Tinker PB, Nye PH (2000) Solute movement in the rhizosphere. Oxford University Press, New York

Topp E, Hansen RS (1991) Metabolism of radiatively important trace gases by methane-oxidizing bacteria. In: Rogers JE, Whitman WB (eds) Microbial production and consumption of greenhouse gases: methane, nitrogen oxides, and halomethanes. American Society for Microbiology, Washington, D.C., pp 71–90

Tripathi AK, Nagarajan T, Verma SC, Le Rudulier D (2002) Inhibition of biosynthesis and activity of nitrogenase in *Azospirillum brasilense* Sp7 under salinity stress. Curr Microbiol 44(5):363–367

Turan M, Ataoglu N, Sahin F (2006) Evaluation of the capacity of phosphate solubilising bacteria and fungi on different forms of phosphorus in liquid culture. Sustain Agric 28:99–108

Turan K, Levent K, Muhittin S, Songul K (2008) Bacterial biodegradation of aldicarb and determination of bacterium which has the most biodegradative effect. Turk J Biochem 33:209–214

Turco RF, Bezdicek DF (1987) Diversity within two serogroups of *Rhizobium leguminosarum* native to soils in the Palouse of eastern Washington. Ann Appl Biol 111:103–114

Uchiumi T, Oowada T, Itakura M, Mitsui H, Nukui N, Dawadi P, Kaneko T et al (2004) Expression islands clustered on symbiosis island of *Mesorhizobium loti* genome. J Bacteriol 186:2439–2448

Upadhyay SK, Singh DP, Saikia R (2009) Genetic diversity of plant growth promoting rhizobacteria from rhizospheric soil of wheat under saline conditions. Curr Microbiol 59(5):489–496

Van Loon LC (2007) Plant responses to plant growth-promoting rhizobacteria. Eur J Plant Pathol 119:243–254

Venkateswarlu B, Desai S, Prasad YG (2008) Agriculturally important microorganisms for stressed ecosystems: challenges in technology development and application. In: Khachatourians GG, Arora DK, Rajendran TP, Srivastava AK (eds) Agriculturally important microorganisms, vol 1. Academic World, Bhopal, pp 225–246

Vessey JK (2003) Plant growth promoting rhizobacteria as biofertilizers. Plant Soil 255:571–586

Wapshere AJ (1974) A strategy for evaluating the safety of organisms for biological weed control. Ann Appl Biol 77:210–211

Ward DWM, Chung YJ, Copeland LB, Doerfler LD (2011) Allergic responses induced by a fungal biopesticide M*etarhizium anisopliae* and house dust mite are compared in a mouse model. J Toxicol 2011:360805, 13 pages

Wei Z-M, Laby RJ, Zumoff CH, Bauer DW, He SY, Collmer A et al (1992) Harpin, elicitor of the hypersensitive response produced by the plant pathogen *Erwinia amylovora*. Sci (Wash) 257:85–88

Weidemann GJ, Tebeest DO (1990) Biology of host range testing for biocontrol of weeds. Weed Technol 4:465–470

Wells AT, Chan KY, Cornish PS (2000) Comparison of conventional and alternative vegetable farming systems on the properties of a yellow earth in New South Wales. Agric Ecosyst Environ 80:47–60

Westwood GS, Huang SW, Keyhani NO (2005) Allergens of the entomopathogenic fungus *Beauveria bassiana*. Clin Mol Allergy 3:1

Wisniewski M, Biles C, Droby S, McLaughlin R, Wilson C, Chalutz E (1991) Mode of action of the postharvest biocontrol yeast. *Pichia guilliermondii*. I. Characterization of attachment to *Botrytis cinerea*. Physiol Mol Plant Pathol 39:245–258

Yazdani M, Bahmanyar MA, Pirdashti H, Esmaili MA (2009) Effect of phosphate solubilization microorganisms (PSM) and plant growth promoting rhizobacteria (PGPR) on yield and yield components of corn (*Zea mays* L.). World Acad Sci Eng Technol 49:90–92

Zahir A, Arshad ZM, Frankenberger WF (2004) Plant growth promoting rhizobacteria. Adv Agron 81:97–168

Zaidi A, Mohammad S (2006) Co-inoculation effects of phosphate solubilising microorganisms and *Glomus fasciculatum* on green gram bradyrhizobium symbiosis. Agric Sci 30:223–230

Chapter 11
Current Aspects of Metal Resistant Bacteria in Bioremediation: From Genes to Ecosystem

Farhana Masood and Abdul Malik

Abstract Global industrialization has resulted in a widespread contamination of the environment with persistent addition of organic and inorganic wastes. The contaminants enter the environment either by natural processes or through human activity. The natural contamination originates from excessive withering of minerals from rocks or displacement from the groundwater or subsurface layers of the soil. Disposal of industrial effluents, sewage sludge, deposition of air-borne industrial wastes, military operations, mining, land-fill operations, industrial solid-waste disposal and the growing use of agricultural chemicals such as pesticides, herbicides and fertilizers are sources of human-assisted contamination of the environment. Heavy metals exert some important roles in some biochemical reactions, being essential to the growth and development of microorganisms, plant and animals. However, in high concentrations they can form unspecific compounds, creating cytotoxic effects. They exhibit a range of toxicities towards microorganisms, depending on physico-chemical factors, speciation etc., while toxic effects can arise from natural processes in the soil, and on microbial communities are more commonly associated with anthropogenic contamination or redistribution of toxic metals in terrestrial ecosystems. A variety of mechanisms have been implicated in the adaptation, tolerance, and resistance of microorganisms to a metal pollutant: precipitation of metals as phosphates, carbonates, and/or sulfides, volatilization via methylation or ethylation, physical exclusion of electronegative components in membranes and extracellular polymeric substances (EPS), energy-dependent metal efflux systems, and intracellular sequestration with low molecular weight, cysteine-rich proteins. The efficiency of these mechanisms depends on many parameters, among which the metal itself, the species studied, time, temperature, pH, presence of plant communities near the microfauna, interactions of the metal with other compounds. Most microorganisms are known

F. Masood (✉) • A. Malik
Department of Agricultural Microbiology, Faculty of Agricultural Sciences, Aligarh Muslim University, Aligarh 202002, India
e-mail: farhanamasud4@gmail.com; ab_malik30@yahoo.com

A. Malik et al. (eds.), *Management of Microbial Resources in the Environment*, 289
DOI 10.1007/978-94-007-5931-2_11, © Springer Science+Business Media Dordrecht 2013

to have specific genes for resistance to toxic ions of heavy metals. This chapter summarizes the recent progress in the field of molecular microbial ecology of metal resistant bacteria with emphasis that how the genetic capacity of the organisms can be exploited for the remediation of heavy metal pollution. Genetic improvement may help to develop the field of existing methodologies to decontamination processes are also discussed in the chapter.

Keywords Bioremediation • Genetic engineering • Heavy metals • Metal resistance • Ecosystem

1 Introduction

Heavy metal pollution is one of the most important environmental problems today. Various industries produce and discharge wastes containing different heavy metals into the environment. Mining, electroplating, metal processing, textile, battery manufacturing, tanneries, petroleum refining, paint manufacture, pesticides, pigment manufacture, printing and photographic industries are the main sources of heavy metals (Williams et al. 1998; Kadirvelu et al. 2001; Lesmana et al. 2009). Thus, metal as a kind of resource is becoming shortage and also brings about serious environmental pollution, threatening human health and ecosystem. Three kinds of heavy metals are of concern, including toxic metals (such as Hg, Cr, Pb, Zn, Cu, Ni, Cd, As, Co, Sn, etc.), precious metals (such as Pd, Pt, Ag, Au, Ru etc.) and radionuclides (such as U, Th, Ra, Am, etc.) (Wang and Chen 2006). Unlike organic wastes, heavy metals are non-biodegradable and they can be accumulated in living tissues, causing various diseases and disorders; therefore they must be removed before discharge (Ansari et al. 2011). Heavy metal toxicity can result in damage or reduced mental and central nervous function, lower energy levels and damage to blood composition, lungs, kidneys, liver and other vital organs (Lesmana et al. 2009).

Rapid industrialization and accelerating global development over the past two centuries have greatly increased the rate at which trace metals are released into the global environment and as a result many of the fresh water bodies are becoming greatly altered. Heavy metals in aquatic system can be naturally produced by the slow leaching from soil/rock to water, which are usually at low levels, causing no serious deleterious effects on human health (Kadirvelu et al. 2001). The development of industry and agriculture promotes the rapid increase of environmental metal pollution. Aquatic heavy metal pollution usually represents high levels of Hg, Cr, Pb, Cd, Cu, Zn, Ni etc. in water system (Tunali et al. 2006). The wastewater mainly originates from mining, mill run, metallurgy, plating, chemical plant, curry and paper making industry. Some of the metallic compounds may adsorbed on the suspended particles and sediments however, under favourable conditions i.e. pH and Eh values they are again released into the water. Some heavy metals including Hg, Cr, Cd, Ni, Cu, Pb etc. introduced into environmental water system may pose high toxicities on the aquatic organisms (Ahmaruzzaman 2011).

Soil is a major sink for heavy metals released into the environment. Many soils in industrialized countries are affected by acid deposition, mine waste and organic refuses, such as sewage sludge that introduce pollutants to the soil (Moral et al. 2005; Akinola et al. 2011). According to Moral et al. (2005), the level of pollution of soils by heavy metals depends on the retention capacity of soil, especially on physico-chemical properties (mineralogy, grain size, organic matter) affecting soil particle surfaces and also on the chemical properties of the metal. These metals may be retained by soil components in the near surface soil horizons or may precipitate or co-precipitate as sulphides, carbonates, oxides or hydroxides with Fe, Mn, Ca etc. (van Hullebusch et al. 2005). The mobility of trace metals reflects their capacity to pass from one soil compartment to another where the element is bound less energetically, the ultimate compartment being soil solution, which determines the bioavailability. The distribution of metals among various compartments or chemical forms can be measured by sequential extraction procedures. Knowledge of how contaminants are partitioned among various chemical forms allows a better insight into degradation of soil and water quality following the input of metals around mining and metallurgical plants. Therefore, soil pollution by heavy metals occurring both on surface and in deeper layers of soil is of great concern for environmental quality control. The pattern of pollutant content is the synergistic result of mixed processes, including diffusion of deposited air-borne particulate matter, fluvial deposition of contaminated sediments and irregular leaching of soil layers, assisted by rainwater, even down to groundwater (Frentiu et al. 2009). The impact of heavy metals resulting from mining and ore roasting on soil is attenuated by several processes such as adsorption, precipitation and complex formation with soil compounds. Soil pollution with heavy metals is multidimensional. Upon entering the soil in large amounts, heavy metals primarily affect biological characteristics: the total content of microorganism changes, their species diversity reduced, and the intensity of basic microbiological processes and the activity of soil enzymes decreases. In addition, heavy metals also change humus content, structure, and pH of soils. These processes ultimately lead to the partial or, in some cases complete loss of soil fertility. Any increase in contamination emission adversely affect crop productivity (Zhu et al. 2011).

2 Impacts of Heavy Metals

Metals play an integral role in the life processes of organisms. Some metals, such as, cobalt, chromium, copper, iron, potassium, magnesium, manganese, sodium, nickel and zinc, are essential, serve as micronutrients used for- (i) redox-processes (ii) to stabilize molecules through electrostatic interactions (iii) acting as components of various enzymes and (iv) regulation of osmotic pressure (Bruins et al. 2000). Other metals, like, silver, cadmium, gold, lead and mercury, have no biological function, and are nonessential and potentially toxic to organisms. Nonessential metals may displace essential metals from their native binding sites or interact with ligands leading to their toxicity (Nies 1999; Bruins et al. 2000). For example, Hg^{2+}, Cd^{2+} and Ag^{2+} can bind to SH groups of proteins, and thus inhibit the activity of enzymes (Nies 1999). Moreover, both essential and nonessential metals can damage (i) cell membranes (ii) alter enzyme specificity (iii) disrupt cellular functions and (iv) damage the structure of DNA at high concentration (Bruins et al. 2000). To be toxic most metal ions have to enter into the cell. Many divalent metal cations (e.g. Mn^{2+}, Fe^{2+}, Co^{2+}, Ni^{2+}, Cu^{2+} and Zn^{2+}) resemble structurally. Also, the structure of oxy-anions such as chromate resembles that of sulfate, and the same is true for arsenate and phosphate. Thus, to be able to differentiate between structurally very similar metal ions, the organism's uptake systems have to be tightly regulated.

3 Metal Resistance Mechanisms in Bacteria

The major problem of heavy metal concentration is ion imbalance. Microorganisms have evolved mechanisms to solve this problem by using two types of uptake systems for metal ions. One is fast, unspecific, and driven by the chemiosmotic gradient across the cytoplasmic membrane of bacteria. Since this mechanism is used by a variety of substrates, it is constitutively expressed (Nies 1999). The second type of uptake system has high substrate specificity. It is slower, often uses ATP hydrolysis as the energy source and is only produced by the cell in times of need, starvation or a special metabolic situation (Nies and Silver 1995). Even though microorganisms have specific uptake systems, high concentrations of nonessential metals may be transported across the cell by a constitutively expressed unspecific system.

Because metal ions cannot be degraded or modified like toxic organic compounds, there are six possible mechanisms for a metal resistance system- (i) metal exclusion by permeability barriers (ii) active transport of the metal away from the cell organism (iii) intracellular sequestration of the metal by protein binding (iv) extracellular sequestration (v) enzymatic detoxification of the metal to a less toxic form and (vi) reduction in the sensitivity of cellular targets to metal ions (Nies and Silver 1995; Bruins et al. 2000; Agrawal et al. 2011). The detoxification mechanisms may be directed against one metal or a group of chemically related metals. Furthermore, the detoxification mechanisms may vary depending on the type of microorganism (Nies and Silver 1995; Agrawal et al. 2011). Most microorganisms are known to have specific genes for resistance to toxic ions of heavy metals. Mostly, the resistance genes are found on plasmids or on chromosomes (Nies 1999; Spain and Alm 2003). There are differences between chromosomal and plasmid-based metal resistance systems. Essential metal resistance systems are usually chromosome-based and more complex than plasmid systems. Plasmid-encoded systems, on the other hand, are usually toxic-ion efflux mechanism. Plasmid-encoded metal resistance determinants have been reported to be inducible (Silver et al. 1981; Rosen 2002; Poole 2005).

Bacterial plasmids have resistance genes to many toxic metals and metalloids, e.g. Ag^{+}, AsO^{-2} , AsO_3^{-4} , Cd^{2+}, Co^{2+}, CrO_2^{-4} , Cu^{2+}, Hg^{2+}, Ni^{2+}, Sb^{3+}, TeO_2^{-3} , T^{1+} and

Zn^{2+}. Related systems are also frequently located on bacterial chromosomes, e.g. Hg^{2+} resistance in *Bacillus*, Cd^{2+} efflux in *Bacillus*, arsenic efflux in *E. coli* (Silver and Phung 1996). Generalizations regarding bacterial metal resistance may include that (1) plasmid-determined resistances are highly specific, (2) resistance systems have been found on plasmids in all bacterial groups tested and (3) resistance mechanisms generally involve efflux from the cells or enzymatic detoxification (Silver and Phung 1996; Nies 1999, 2003). However, other less-specific interactions, e.g. sorption, may contribute to the overall response. Efflux pumps, determined by plasmid and chromosomal systems, are either ATPases or chemiosmotic systems, with mechanisms often showing similarity in different types of bacteria. Cd^{2+} resistance may involve (1) an efflux ATPase in Gram-positive bacteria, (2) cation-H^+ antiport in Gram-negative bacteria, and (3) intracellular metallothionein in cyanobacteria (Silver and Phung 1996, 2005). Arsenic-resistant Gram-negative bacteria have an arsenite efflux ATPase and an arsenate reductase (which reduces arsenate [As(V)] to arsenite [As(III)] which comprise the underlying biochemical mechanism) (Oremland et al. 2002; Macur et al. 2004). Similar systems for Hg^{2+} resistance occur on plasmids from Gram-positive and Gram- negative bacteria with component genes being involved in transport of Hg^{2+} to the detoxifying enzyme, mercuric reductase, which reduces Hg^{2+} to elemental Hg^0 (Silver and Phung 1996; Kannan and Krishnamoorthy 2006). The enzyme organomercurial lyase can break the C-Hg bond in organomercurials (Schottel et al. 1974). It appears that contaminating mercury selects for higher frequencies of Hg^{2+}-resistant bacteria in polluted habitats (Silver and Phung 1996; Barkay et al. 2003). Plasmid-determined chromate resistance appears unconnected with chromate [Cr(VI)] reduction to Cr(III) (Ohtake et al. 1987; Cervantes et al. 2001; Cervantes and Campos-Garcia 2007), resistance depending on reduced CrO_2^{-4} uptake (Cervantes and Campos-Garcia 2007). A Cd^{2+} efflux ATPase is widely found in Gram-positive bacteria including soil *Bacillus* spp. (Nies 2003). The large plasmids of *Alcaligenes eutrophus* have numerous toxic metal resistance determinants, e.g. three for Hg^{2+}, one for Cr^{6+}, and two for divalent cations, *czc* (Cd^{2+}, Zn^{2+} and Co^{2+} resistance) and *cnr* (Co^{2+} and Ni^{2+} resistance; Silver and Phung 1996; Rensing et al. 2002; Mergeay et al. 2003). Czc functions as a chemiosmotic divalent cation/H^+ antiporter (Nies and Silver 1995). In *Enterococcus hirae* (previously *Streptococcus faecalis*), copper resistance is determined by two genes, *copA* and *copB* which respectively determine uptake and efflux P-type ATPases (Solioz and Stoyanov 2003). Plasmid-determined Cu^{2+} resistance has been described in *Pseudomonas* (Cooksey 1994; Unaldi et al. 2003), *Xanthomonas* (Lee et al. 1994; Voloudakis et al. 2005) and *Escherichia coli* (Lee et al. 2002). Chromosomal genes also affect Cu^{2+} transport and resistance by determining functions such as uptake, efflux and intracellular Cu^{2+} binding (Gadd 2005).

It is clear that a more lucid understanding of the mechanisms of metal toxicity on living cells may lead to the development of novel technologies to mitigate such toxicity. To this end, microorganisms are useful models for the study of various aspects of oxidative stress at the biochemical, molecular, and cellular levels. This is because the natural stress factors, as well as the damage caused by oxidative stress to nucleic acids, proteins, lipids, and other cell components, are very similar in all types of organisms.

This is also true because at all levels of cell organization, the principles of cellular defense against oxidative stress are similar, for example, the nature and role of antioxidants and antioxidative enzymes that act to decrease ROS concentrations, the repair of damaged macromolecules, and the elimination of irreparable proteins (Poljsak et al. 2010). Many microorganisms have also evolved other complex mechanisms to counteract the toxic effects of metals (Silver and Phung 1996). Thus bacteria, yeasts, algae, and fungi have been located in and isolated from sites contaminated with heavy metals, and are now under study as possible bioremediators of environmental contamination (Machado et al. 2008; Ray and Ray 2009; Ruta et al. 2010). Despite apparent toxicity, many microbes grow and even flourish in apparently metal-polluted locations, and a variety of mechanisms, both active and incidental, contribute to resistance (Avery 2001; Holden and Adams 2003; Fomina et al. 2007; Ansari and Malik 2007; Masood and Malik 2011). Microbial resistance to toxic metals is widespread, with frequencies ranging from a few per cent in pristine environments to nearly 100% in heavily polluted environments (Silver and Phung 2009).

4 Conventional Technologies for Treating Environments Contaminated by Heavy Metals

The removal of metals from the environment is extremely important. Methods for removing metal ions from aqueous solution mainly consist of physical, chemical and biological technologies. Conventional methods for removing metal ions from aqueous solution have been suggested, such as chemical precipitation, filtration, ion exchange, electrochemical treatment, membrane technologies, adsorption on activated carbon, evaporation etc. However, chemical precipitation and electrochemical treatment are ineffective, especially when metal ion concentration in aqueous solution is among 1–100 mg/L, and also produce large quantity of sludge required to treat with great difficulty. Ion exchange, membrane technologies and activated carbon adsorption process are extremely expensive when treating large amount of water and wastewater containing heavy metal in low concentration, they cannot be used at large scale. On the other hand, bioremediation is increasingly gaining importance as an alternative technology, due to the advantages it offers: simplicity, efficiency and low cost (Goyal et al. 2003; Tabak et al. 2005; Hameed 2006; Machado et al. 2008; Wang and Chen 2009).

5 Bioremediation

Bioremediation is often considered a cost effective and environmental friendly method and is gradually making inroads for environmental clean-up applications, which rely on immobilization or mobilization of the toxic metals. A combination of these approaches is often used by industry to treat metal-contaminated sites; combining approaches can be more cost-effective than using just one. The ability of a microorganism to survive and grow in a metal-contaminated habitat can depend on genetic and/or physiological adaptation. Such physiological changes in the microbial cells reduce the rate of metal uptake and intracellular metal toxicity, while genetic changes result in the reduced intracellular and extracellular concentrations of the toxic metal species. However, these natural transformations are limited by the relative slow rates. Development of new genetic tools and a better understanding of microorganism's natural transformation ability at the genetic level are essential to accelerate the progress of designer microbes for improved hazardous waste removal. Several attempts have been made recently to enhance biotransformation and bioaccumulation of toxic wastes by microorganisms. In the last three decades extensive research has been performed on basic and applied aspects of microbial interaction with metals with a bioremediation perspective. These studies include mainly isolation of superior metal resistant and/or accumulating or metal transforming microorganisms, their identification, biochemical and genetic characterization, elucidation of microbe-metal interaction mechanisms and all related studies relevant for bioremediation.

5.1 *Microbial Transformations of Metals*

5.1.1 Mobilization

Microorganisms can mobilize metals through autotrophic and heterotrophic leaching, chelation by microbial metabolites and siderophores, and methylation, which can result in volatilization. All these processes result in metal compounds and minerals (oxides, phosphates, sulphides and more complex mineral ores) dissolution, and desorption of metal species from exchange sites on, e.g. clay minerals or organic matter. Microorganisms can acidify their environment by proton efflux via plasma membrane H^+-ATPases, maintenance of charge balance or as a result of respiratory carbon dioxide accumulation. Acidification can lead to metal release via a number of obvious routes, e.g. competition between protons and the metal in a metal–anion complex or in a sorbed form, resulting in the release of free metal cations. Organic acids can supply both protons and metal complexing anions (Gadd 1999; Gadd and Sayer 2000; Huang et al. 2004; Lian et al. 2008a, b). For example, citrate and oxalate can form stable complexes with a large number of metals. Many metal citrates are highly mobile and not readily degraded (Francis et al. 1992). Oxalic acid can also act as a leaching agent for those metals that form soluble oxalate complexes, including Al and Fe (Strasser et al. 1994). Organic acid production is also an important agent of mineral deterioration, playing a role in both biogenic chemical weathering and soil formation (Gadd 1999). Such solubilisation phenomena can also have consequences for mobilization of metals from toxic metal-containing minerals, e.g. pyromorphite ($Pb_5(PO_4)_3Cl$) which can form in urban and industrially contaminated soils. Pyromorphite can be solubilized by phosphate-solubilizing fungi, with concomitant production of lead oxalate (Sayer et al. 1999; Fomina et al. 2004, 2005a, b). Most chemolithotrophic (autotrophic) leaching is carried out by acidophilic bacteria which fix CO_2 and obtain energy from the oxidation of Fe(II) or reduced sulphur compounds which causes the solubilisation of metals because of the resulting production of Fe(III) and H_2SO_4 (Rawlings 1997; Schippers and Sand 1999). The microorganisms involved include sulphur-oxidizing bacteria, e.g. *Thiobacillus thiooxidans*, iron and sulphur-oxidizing bacteria, e.g. *Thiobacillus ferrooxidans* and iron oxidizing bacteria, e.g. *Leptospirillum ferrooxidans* (Siddiqui et al. 2009). As a result of sulphur- and iron-oxidation, metal sulphides are solubilized concomitant with the pH of their immediate environment being decreased, therefore resulting in solubilization of other metal compounds including metals sorbed to soil and mineral constituents (Rawlings 2002; Huang 2008).

Functional groups such as hydroxamates and catecholates, siderophores can also be observed in mobilizing metal complexation (Nair et al. 2007). Although they serve a specific purpose for the organism in obtaining iron, they are nonspecific for the range of metals that they can bind. Siderophores have been shown to bind Al, As, Cd, Co, Cr, Cu, Mn, Ni, Pb, Sn, and Zn, in addition to Fe (Nair et al. 2007). Siderophore pyridine-2, 6-bis (thiocarboxylic acid) produced by *Pseudomonas stutzeri* KC precipitated As, Cd, Hg, and Pb, conferring resistance to the bacterium (Zawadzka et al. 2007). Similarly, microbially-produced biosurfactants including rhamnolipid produced by *Pseudomonas aeruginosa*, surfactin by *Bacillus subtilis*, and sophorolipid from the yeast *Torulopsis bombicola* have been implicated in metal resistance (Sandrin et al. 2000; Mulligan et al. 2001).

Microorganisms can also mobilize metals, metalloids and organometallic compounds by reduction and oxidation processes (Lovley 2001; Lloyd 2003. For example, solubilities of Fe and Mn increase on reduction of Fe(III) to Fe(II) and Mn(IV) to Mn(II) (Lovley 2001; McLean et al. 2002). Most iron reduction is carried out by specialized anaerobic bacteria that use Fe(III) as a terminal electron acceptor. Dissimilatory metal-reducing bacteria can use a variety of metal(loid)s with an appropriate redox couple, including Fe(III), Mn(IV), Se(IV), Cr(VI) and U(VI) (DiChristina et al. 2005; Geets et al. 2008). While Fe and Mn increase their solubility upon reduction, the solubility of other metals such as U(VI) to U(IV) and Cr(VI) to Cr(III) decreases, resulting in immobilization (Geets et al. 2008). Reduction of Hg(II) to Hg(0) by bacteria and fungi results in diffusion of elemental Hg out of cells (Hobman et al. 2000; Barkay and Wagner-Dobler 2005). *Bacillus* and *Streptomyces* spp. can oxidize Hg(0) to Hg(II) and, therefore participate in the oxidative phase of the global Hg cycle (Barkay and Wagner-Dobler 2005; Ehrlich and Newman 2009). Fe(III) and Mn(IV) oxides absorb metals strongly and this may hinder metal mobilization. Microbial reduction of Fe(III) and Mn(IV) may be one way for releasing such metals and this process may be enhanced with the addition of humic materials, or related compounds. Such compounds may also act as electron shuttles for, e.g. U(VI) and Cr(VI), converting them to less soluble forms, especially if located in tight pore spaces where microorganisms cannot enter (Lovley and Coates 1997; Lloyd 2003). Ferric iron, Fe(III), can be enzymatically reduced to ferrous iron, Fe(II), with a suitable electron donor. Many Fe(III) reducers are heterotrophs and in some anaerobic environments such Fe(III) respiration may be a more important mechanism of carbon source decomposition than sulfate reduction (Ehrlich and Newman 2009). Some Fe(III) reduction can be the result of metabolic products such as H_2S or formate, or other secondary metabolites. Naturally occurring microbially produced metal chelators that may solubilize Fe(III) include oxalate, citrate, humic acids and tannins.

5.1.2 Immobilization

A number of processes lead to immobilization of metals. Although immobilization reduces the external free metal species, it may also promote solubilization in some circumstances by shifting the equilibrium to release more metal into solution. Biosorption can be defined as the microbial uptake of organic and inorganic metal species, both soluble and insoluble, by physico-chemical mechanisms such as adsorption. In living cells, metabolic activity may also influence this process because of changes in pH, E*h*, organic and inorganic nutrients and metabolites. Biosorption can also provide nucleation sites for the formation of stable minerals (McLean et al. 2002; Gadd 2005; Violante et al. 2010 as in continuation well as sorption to cellular surfaces, cationic metal species can be accumulated within cells via membrane transport systems of varying affinity and specificity (Ansari and Malik 2007; Ansari et al. 2011; Masood and Malik 2011). Once inside cells, metal species may be bound (e.g. to metallothioneins, phytochelatins), precipitated (e.g. as reduced forms), localized within intracellular structures or organelles (e.g. fungal vacuoles), or translocated to specific structures (e.g. fungal fruiting bodies) depending on the element concerned and the organism (Gadd 1996; White et al. 1997; Gadd and Sayer 2000). Peptidoglycan carboxyl groups are the main binding site for cations in Gram-positive bacterial cell walls with phosphate groups contributing significantly in Gram-negative species (McLean et al. 2002; Wang and Chen 2009). It seems likely that microbial binding and biomineralization (mineral formation) reactions have a more significant role in metal speciation and mobility in the terrestrial environment than has previously been supposed (McLean et al. 2002; Gadd 2009).

A range of specific and non-specific metal-binding compounds are produced by microorganisms. Non-specific metal-binding compounds range from simple organic acids and alcohols to macromolecules such as polysaccharides, humic and fulvic acids (Gadd 2005). Exopolymers, e.g., exopolysaccharides (EPS), and cell walls made up of lipopolysaccharides, proteins and carbohydrates with various functional groups, create sorption sites for metal binding and metal immobilization. The EPS produced by the bacterium *Paenibacillus jamilae* complexed up to 230 mg Pb per g EPS (Morillo et al. 2006). Similarly, the EPS from the cyanobacterium *Nostoc spongiaeforme* is a highly effective sorbent of zinc (Singh et al. 2011). Extracellular polysaccharides can also adsorb or entrap particulate matter such as precipitated metal sulphides and oxides (Gadd 2005, 2010) and these processes may be particularly important in microbial biofilms (White and Gadd 1998, 2000). Where microbial reduction of a metal to a lower redox state occurs, mobility and toxicity may be reduced (Lovley 2001; Finneran et al. 2002; Holden and Adams 2003; Wall and

Krumholz 2006). Such processes may also accompany other indirect reductive metal precipitation mechanisms, e.g. in sulphate-reducing bacterial systems where reduction of Cr(VI) can be a result of indirect reduction by Fe^{2+} and the produced sulphide. Aerobic or anaerobic reduction of Cr(VI) to Cr(III) is widespread in microorganisms (Smith and Gadd 2000; McLean and Beveridge 2001). U(VI) can be reduced to U(IV) by certain Fe(III)-dissimilatory microorganisms and this reduction in solubility can be the basis of U removal from contaminated waters and leachates (Lovley 2001; Finneran et al. 2002; Landa 2005; Lloyd and Renshaw 2005). Sulphur and sulphate-reducing bacteria are geochemically important in reductive precipitation of toxic metals, e.g. U(VI), Cr(VI), Tc(VII), Pd(II) (Lloyd 2003; Lloyd and Renshaw 2005). Some sulphate-reducing bacteria like *Desulfotomaculum reducens* share physiological properties of both sulphate- and metal-reducing groups of bacteria, and can use Cr(VI), Mn(IV), Fe(III) and U(IV) as sole electron acceptors (Haveman and Pedersen 2002). Sulphate-reducing bacteria (SRB) oxidize organic compounds or hydrogen coupled with the reduction of sulphate, producing sulphide (Muyzer and Stams 2008). The solubility products of most heavy metal sulphides are very low, so that even a moderate output of sulphide can remove metals (Gadd 2005; Violante et al. 2010). Sulphate-reducing bacteria can also create extremely reducing conditions which can chemically reduce species such as U(VI) (Hietala and Roane 2009; Du et al. 2011).

Bacterial Fe oxidation is ubiquitous in environments with sufficient Fe^{2+} and conditions to support bacterial growth such as drainage waters and tailings piles in mined areas, pyritic and hydric soils (bogs and sediments), drain pipes and irrigation ditches, and plant rhizospheres. Iron-oxidizers found in acidic soil environments are acidophilic chemolithotrophs, such as *Thiobacillus ferrooxidans*, significant for its role in generating acid mine drainage (Johnson 2003). Fungi also oxidize metals in their environment. Desert varnish is an oxidized metal layer (patina) a few millimetres thick found on rocks and in soils of arid and semi-arid regions, and is believed to be of fungal and bacterial origin.

5.2 Metalloid Transformations

Main microbial transformations carried out in the soil are reduction and methylation which can lead to alterations in bioavailability and toxicity. For selenium, some bacteria can use SeO_2^{-4} as a terminal *e*– acceptor in dissimilatory reduction and also reduce and incorporate Se into organic components, e.g. selenoproteins (assimilatory reduction). Methylation and subsequent volatilization of methylated selenium derivatives is also a widely found property of soil bacteria and fungi and may be an important process in Se transport from terrestrial to aquatic environments (Dungan and Frankenberger 1999). Selenate (SeO_2^{-4}) and selenite (SeO_2^{-3}) can be reduced to Se^0, with SeO_2^{-3} reduction appearing more ubiquitous than SeO_2^{-4} reduction. However, only SeO_2^{-4} can support bacterial growth under anaerobic conditions: SeO_2^{-4} reduction to Se^0 is a major sink for Se oxyanions in anoxic sediments (Stolz and Oremland 1999; Oremland and Stolz 2000). Anaerobic sulphate-reducing bacteria like *Desulfovibrio desulfuricans* can reduce selenate/selenite to Se^0, but neither oxyanion could be used for respiratory growth (Tomei et al. 1995). Reduction to Se^0 can be considered a detoxification mechanism (Dungan and Frankenberger 1999; Yee and Kobayashi 2008). The opposite process of Se^0 oxidation can occur in soils and sediments (Dowdle and Oremland 1998; Losi and Frankenberger 1998). It is possible that Se^0 oxidation is a similar process to S oxidation, and may be mediated by heterotrophs and autotrophs (Losi and Frankenberger 1998). In aerobic soil slurries, Se^{4+} was the main product with lower amounts of Se^{6+} being produced: heterotrophic and autotrophic thiobacilli were believed to be the active organisms (Dowdle and Oremland 1998).

Methylation of Se is a ubiquitous property of microorganisms and can occur in soils, sediments and water (Gadd 2005). Bacteria and fungi are the most important Se-methylaters in soil (Karlson and Frankenberger 1988) with the most frequently produced volatile being dimethyl selenide (DMSe; Karlson and Frankenberger 1988, 1989; Thompson-Eagle et al. 1989). Other volatiles produced in smaller amounts include dimethyl diselenide (DMDSe; Dungan and Frankenberger 1999). Like reduction, volatilization can be considered a detoxification mechanism since volatile Se derivatives are lost from the soil system. Those environmental factors that affect microbial activity can markedly affect Se methylation, e.g. pH, temperature, organic amendments, Se speciation etc., addition of organic amendments can stimulate methylation (Dungan and Frankenberger 1999). The opposite process of demethylation can also occur in soil and water systems. Anaerobic demethylation may be mediated by methylotrophic bacteria (Oremland et al. 1989). Tellurium may also be transformed by analogous mechanisms as selenium, i. e. reduction and methylation (Chasteen and Bentley 2003). Reduction of tellurite to Te^0 results in a grey to black colouration of microbial colonies with extracellular and intracellular precipitation being observed (Gharieb et al. 1999). Dimethyl telluride (DMTe) is the main product of Te methylation (Chasteen and Bentley 2003). Arsenic methylation can be mediated by many organisms with compounds having the general structure $(CH_3)_nAsH_{3-}n$ and mono-, di- and trimethylarsine (n = 1, 2, 3 respectively) being major volatile compounds (Bentley and Chasteen 2002). Reduction of arsenic oxyanions by reductase enzymes is also frequent and a determinant of As resistance. However, there appears to be no involvement of such reductases in biomethylation (Bentley and Chasteen 2002; Paez-Espino et al. 2009).

5.3 Biomineralization

Microorganisms can form minerals by a process called biomineralization, which offers an interesting bioremediation strategy for capturing pollutants within relatively stable solid phases. Biomineralization occurs naturally in very diverse environments but may be artificially stimulated for bioremediation purposes as proposed in several past studies (e.g., Senko et al. 2002; Beazley et al. 2007; Yabusaki et al. 2007). Biomineralization refers to biologically induced mineralization where an organism modifies the local microenvironment creating conditions that promote chemical precipitation of extracellular mineral phases (Hamilton 2003; Dupraz et al. 2009). Three different types of bacterially mediated biomineralization processes have been distinguished in the literature (Dupraz et al. 2009): (1) biologically controlled biomineralization, referring to cases in which a specific cellular activity directs the nucleation, growth, morphology, and final location of a mineral; the emblematic example for that process in the case of bacteria is the formation of intracellular chemically pure magnetite crystals in magnetotactic bacteria (Blakemore 1975; Bazylinski and Frankel 2004; Komeili 2007), but no significant impact on immobilization of metal pollutants has been reported so far; (2) biologically induced biomineralization resulting from indirect modification of chemical conditions, such as a pH shift or redox transformations, in the environment by biological activity; many examples are reported in the literature for this process and can lead to efficient immobilization of almost any known inorganic pollutant (e.g., Borch et al. 2010; Lowenstam 1981); and (3) biologically influenced biomineralization, which is defined as passive mineral precipitation in the presence of organic matter, such as cell surfaces or extracellular polymeric substances (EPS), whose properties influence crystal morphology and composition. The term organomineralization encompasses biologically influenced and biologically induced biomineralization (Dupraz et al. 2009). Although by definition organomineralization involves processes that are not supposed to involve specific genes, they may have a negative or positive impact on cell viability. An increasing number of studies have shown that microbial extracellular polymers can be involved in biomineralization, hence inorganic pollutant trapping (Benzerara et al. 2008; Chan et al. 2009; Miot et al. 2009). Calcium oxalate is the most common form of oxalate encountered in the environment, mostly occurring as the dihydrate (weddellite) or the more stable monohydrate (Gadd 1999). Calcium oxalate crystals are commonly associated with free-living, pathogenic and plant symbiotic fungi and are formed by the precipitation of solubilized calcium as the oxalate (Gharieb et al. 1999; Gadd 1999, 2010). This has an important influence on biogeochemical processes in soils, acting as a reservoir for calcium, and also influencing phosphate availability. The importance of biomineralization in the sequestration of inorganic pollutants is obvious, especially at field sites where the metabolic activity of microorganisms has been artificially stimulated for bioremediation purposes (Barkay and Schaefer 2001; Adriano et al. 2004). A group of studies have, for example, clearly shown that the secretion of phosphate groups due to phosphatase activity can induce a significant immobilization of uranium in the form of autunite at acidic and alkaline pH (Macaskie et al. 2000; Merroun and Selenska-Pobell 2008; Nilgiriwala et al. 2008; Beazley et al. 2007). Phosphatase activity can be stimulated by injection of glycerol-3-phosphate as the sole carbon and phosphorus source.

6 Gene Transfer and Genetic Engineering of Metal-Resistance Genes

The genetic capacity of bacteria can be exploited for the remediation of heavy metal pollution. Genes associated with metal resistance/detoxification mechanisms have long been studied in microorganisms (Silver and Phung 2005). The *cad* operon responsible for cadmium efflux is well-characterized in *Staphylococcus aureus* (Nies and Silver 1995). The *mer* operon in mercury resistance is well-understood in a variety of microorganisms (Nies 1999). The *czc* operon has been elucidated in numerous bacteria (Abou-Shanab et al. 2007). Such metal-resistance genes have been introduced into other bacteria and plants to help establish metal-tolerant populations within contaminated systems. For example, numerous tree species have been genetically altered to produce the MT-related glutathione for metal sequestration (Merkle 2006). Gene bioaugmentation is defined as the process of obtaining enhanced biological activity after gene transfer from an introduced donor organism into a member of the indigenous soil population (Maier 2000). In a metal-contaminated environment, enhanced metal detoxification activity could be achieved if metal-resistant/detoxifying genes were transferred to bacterial populations within the soil (Pepper et al. 2002). Gene bioaugmentation has primarily been studied for mitigation of organic pollutants (Urgun-Demirtas et al. 2006; Jussila et al. 2007), and while not widely used gene bioaugmentation can potentially establish large, diverse metal-resistant populations within a soil.

In situ bioremediation uses naturally occurring non engineered microorganisms and is often enhanced (biostimulation) by the addition of nutrients, such as N and P, surfactants and oxygen during the treatment (Watanabe 2001). In such treatments, nature of the microbial ecological niches is unpredictable. Another possible method to improve the bioremediation efficiency is the bioaugmentation, where the indigenously isolated bacteria are injected into the contaminated site, which may include genetically engineered (GE) bacterial strains, but it is rarely done. A major difficulty with in situ bioremediation by using transgenic bacteria is the unpredictable end result on account of various environmental factors that may interfere. However, bioremediation based on GE bacteria is an emerging technology and has been receiving more attention as an ecofriendly and efficient way of cleaning up toxic-metal contaminations (Cases and de Lorenzo 2005a, b; Shukla et al. 2010; Liu et al. 2011).

Recent developments in bioremediation technology, which includes the utilization of protein engineering, metabolic engineering, whole-transcriptome profiling, and proteomics are also considered significant for removal of obstinate pollutants such as heavy metals (Thomas 2008). Cell surface appearance of precise proteins in the GE bacteria is known to contribute in detoxification of heavy metals as well as other recalcitrant compounds (Muhammad et al. 2008). Now, it is widely considered that molecular biology has potential application in designing the bacteria for remediative tasks. The GE bacteria have higher degradative capacity and have been demonstrated successfully for the degradation of various pollutants under defined conditions (Barac et al. 2004). Various strains of *Ralstonia eutropha, Pseudomonas putida*, *Mycobacterium marinum*, *Escherichia coli*, *Sphingomonas desiccabilis*, *Bacillus idriensis*, etc., have genes inserted into their genomes, which empowers them to specific bioremediation of toxic metal compounds in contaminated environment (Kube et al. 2005; Parnell et al. 2006; Schue et al. 2009; Liu et al. 2011). Hu et al. (2005) using whole-genome transcriptional analysis, have identified the pathways responding to heavy metal toxicity in *Caulobacter crescentus*, a bacterium known for its ability to live in environments with characteristics of those heavy metal contaminated sites. The bacterium is able to tolerate uranium, cadmium and chromium, possibly due to protection against oxidative stress and multiple efflux pumps. In order to identify genes induced by exposure to heavy metals, Moore et al. (2005) used a transcriptional profiling assay of *Bacillus subtilis*, discovering that many of the genes affected by metal stress were controlled by metalloregulatory proteins known as Fur, MntR, PerR, ArsR and CueR. Recently Hasin et al. (2010) reported a well-characterized model methanotroph *Methylococcus capsulatus* able to bioremediate chromium(VI) pollution over a wide range of concentrations (1.4–1,000 mg/L of Cr^{6+}), thus extending the bioremediation potential via successful genetic manipulation of this major group of microorganisms.

The genetic engineering offers the advantage of constructing the bacterial strains which can withstand the adverse stressful conditions and can be deliberately employed as bioremediators under complex environmental conditions. Thus, identification and genetic engineering of indigenous bacteria, which are fully adapted to such complex environmental conditions, is largely expected to offer a most viable future remediation technology. Considering the huge selection pressure on the free living GE bacteria imposed by the environmental factors, both biotic (antagonism, competition and predation) and abiotic (temperature, pH, moisture, adsorption, etc.), it is difficult and a discouraging task to derive a competent modelling scheme for survival of engineered bacteria. It is considered that the indigenous microbial flora, if taken for constructing the recombinant bacteria, will have natural advantage over the "exotic strain" of GE bacteria. Since the rate of bioremediation is not the only function of degradative genes but is also dependent on total population and innate capabilities of the bacteria to withstand the existing complex stressful environmental conditions. Therefore, choosing and engineering the right bacterial strain with rapid growth (largest potential population and greatest nutrient responses) and more efficient bioremediation capabilities without environmental risk will be a crucial step for achieving a safe and sustainable environment (Singh et al. 2011).

Application of GE bacteria for bioremediation of heavy metals has come in the forefront due to its selective nature of transforming the toxicants and thus, posing less health hazards as compared to physico-chemical methods. Bacterial metal detoxification and removal can be an efficient strategy due to its low cost, high efficiency and eco-friendly nature. Recent advances in the field of molecular biology have enabled us to understand the metal-microbe interaction and their application for bioremediation of metal in the environment (Rajendran et al. 2003). The detoxification machineries in GE bacteria have been considered useful for metal bioremediation, and a comparison of their efficiency of remediation with the previously isolated metal-resistant bacteria may yield interesting results.

7 Nanotechnology in Bioremediation

The area of nanotechnology encompasses the synthesis of nanoscale materials, the understanding and the utililization of their often exotic physicochemical and optoelectronic properties, and the organization of nanoscale structures into predefined superstructures. Thus, nanotechnology promises to play an increasingly important role in many key technologies of the new millenium. Recently, the emerging field of nanotechnology has also contributed significantly in remediating these common soil and water pollutants in environment friendly manner. Biological synthesis of metal nanoparticles using microbes, such as bacteria, yeasts, algae, actinomycetes and fungi, is gaining momentum due to the ecofriendly nature of the organisms which reduce toxic chemicals (Muralisastry et al. 2003). Metal-microbe interaction is very important in several biotechnological applications, including in the fields of biomineralization, bioremediation, bioleaching, and microbial corrosion (Joerger et al. 2001). Nano materials, besides providing new research challenges, form the basis of a new class of atomically engineered materials. Confluence of environmental biotechnology and nanotechnology will lead to the most exciting progress in the development of nanodevices having bio-capabilities in novel metal remediation strategies.

Biological systems provide many examples of specifically tailored, nanostructured molecules with highly optimized properties and characteristics. Thus biological materials are considered as a nanophase system in its own right and as the starting point for producing other novel nanophase systems. Nanomaterial *in vivo* biosynthesis is the best option for metal bioremediation, since a biologically controlled mineralization process produces materials with well-defined characteristics. The biominerals are composite materials and consist of an inorganic component and a special organic matrix; the organic matrix has a vital influence on the morphology of the inorganic compound. Li et al. (2003) have reported the sorption of Pb(II), Cu(II) and Cd(II) onto multiwalled carbon nanotubes (MWCNTs). The maximum sorption capacities of 97.08 mg/g for Pb(II), 24.49 mg/g for Cu(II) and 10.86 mg/g for Cd(II) at room temperature, pH 5.0 and metal ion equilibrium concentration of 10 mg/L. It was also found that the metal-ion sorption capacities of the MWCNTs were 3–4 times higher than that of powder activated carbon and granular activated carbon which are two commonly used sorbents in water purification.

8 Bioinformatics Use in Bioremediation

Advances in the molecular biology technologies are making a global gene expression profile possible; genome wide analysis of DNA (genomics), RNA expression (transcriptomics) and protein expression (proteomics) as well as exploring complexes of protein aggregation such as protein-protein interaction (interactomics) create the opportunity to systematically study the physiological expressions of such organisms. Attempts are made to interpret some of the areas of genomics and proteomics which have been employed for bioremediation studies. The bioinformatic data obtained viz. various sources are interpreted by combining these techniques and applying them towards future studies of active bioremediation to develop environmental cleanup technologies. Bioinformatic techniques have been developed to identify and analyze various components of cells such as gene and protein function, interactions, and metabolic and regulatory pathways. The next decade will belong to the understanding of cellular mechanism and manipulation using the integration of bioinformatics.

9 New Frontiers in Bioremediation

The biological treatment of metal-contaminated systems is a continuing field of research. While much need for elucidation remains, the field of microbial metal remediation is making great strides towards applicable technologies. There are many new and exciting proposed uses of microorganisms and their products in soil metal remediation. With our increased understanding of multiple microbial metal resistance mechanisms, treatment of sites with fluctuating metal concentrations and multiple metal and organic contaminants may be possible. The use of microbial products, the microbial enhancement of phytoremediation technologies, in combination with the genetic engineering of plants and microorganisms for enhanced metal uptake, represent continuing, exciting directions in the use of microorganisms in soil metal remediation. Molecular approaches including '-omics' tools were used relatively recently in several studies to get better insight in to the bacterial interaction with toxic metals. Such extensive studies have shown beyond doubt that microbial interaction with metals and minerals have potential for treatment of environmental pollution (Lloyd and Renshaw 2005; Gadd 2010).

10 Conclusions

The pollution of soil and water with heavy metals is an environmental concern today. Metals and other inorganic contaminants are among the most prevalent forms of contamination found at waste sites, and their remediation in soils and sediments is among the most technically difficult. The high cost of existing cleanup technologies led to the search for new cleanup strategies that have the potential to be low cost, low impact, visually benign, and environmentally sound. Some microorganisms act in the biosphere as geochemical agents promoting precipitations, transformations, or dissolutions of minerals. The use of these microorganisms for bioremediation offers new tools to degrade or transform toxic contaminants. Various molecular, genetic, and metabolic engineering tools have accelerated the progress toward bioremediation and have led to specifically designed microorganisms for various bio-based clean-up processes. However, these genetic modifications should be understood in full and any research must always determine the actual risks and benefits involved.

References

Abou-Shanab RAI, van Berkum P, Angle JS (2007) Heavy metal resistance and genotypic analysis of metal resistance genes in gram-positive and gram-negative bacteria present in Ni-rich serpentine soil and in the rhizosphere of *Alyssum murale*. Chemosphere 68:360–367

Adriano DC, Wenzel WW, Vangronsveld J, Bolan NS (2004) Role of assisted natural remediation in environmental cleanup. Geoderma 122:121–142

Agrawal J, Sherameti I, Varma A (2011) Detoxification of heavy metals: state of art. In: Sherameti I, Varma A (eds) Detoxification of heavy metals. Springer, Heidelberg

Ahmaruzzaman M (2011) Industrial wastes as low-cost potential adsorbents for the treatment of wastewater laden with heavy metals. Adv Colloid Interface Sci 166:36–59

Akinola OM, Njoku LKL, Ifitezue VN (2011) Assessment of heavy metals (lead and cadmium) concentration in *Paspalum orbiculare* near municipal refuse dumpsites in Lagos State, Nigeria. J Ecol Nat Environ 3:509–514

Ansari MI, Malik A (2007) Biosorption of nickel and cadmium by metal resistant bacterial isolates from agricultural soil irrigated with wastewater. Bioresour Technol 98:3149–3153

Ansari MI, Masood F, Malik A (2011) Bacterial biosorption: a technique for remediation of heavy metals. In: Ahmad I, Ahmad F, Pichtel J (eds) Microbes and microbial technology: agricultural and environmental applications. Springer, New York

Avery SV (2001) Metal toxicity in yeast and the role of oxidative stress. Adv Appl Microbiol 49:111–142

Barac T, Taghavi S, Borremans B, Provoost A, Oeyen L, Colpaert JV, Vangronsveld J, van der Lelie D (2004) Engineered endophytic bacteria improve phytoremediation of water-soluble, volatile, organic pollutants. Nat Biotechnol 22:583–588

Barkay T, Schaefer JJ (2001) Metal and radionuclide bioremediation: issues, considerations and potentials. Curr Opin Microbiol 4:318–323

Barkay T, Wagner-Dobler I (2005) Microbial transformations of mercury: potentials, challenges, and achievements in controlling mercury toxicity in the environment. Adv Appl Microbiol 57:1–52

Barkay TS, Miller M, Summers AO (2003) Bacterial mercury resistance from atoms to ecosystems. FEMS Microbiol Rev 27:355–384

Bazylinski DA, Frankel RB (2004) Magnetosome formation in prokaryotes. Nat Rev Microbiol 2:217–230

Beazley MJ, Martinez RJ, Sobecky PA, Webb SM, Taillefert M (2007) Uranium biomineralization as a result of bacterial phosphatase activity: insights from bacterial isolates from a contaminated subsurface. Environ Sci Technol 41:5701–5707

Bentley R, Chasteen TG (2002) Microbial methylation of metalloids: arsenic, antimony and bismuth. Microbiol Mol Biol Rev 66:250–271

Benzerara K, Morin G, Yoon TH, Miot J, Tyliszczak T, Casiot C, Bruneel O, Farges F, Brown GE Jr (2008) Nanoscale study of as biomineralization in an acid mine drainage system. Geochim Cosmochim Acta 72:3949–3963

Blakemore R (1975) Magnetotactic bacteria. Science 190:377–379

Borch T, Kretzschmar R, Kappler A, Van Cappellen P, Ginder-Vogel M, Voegelin A, Campbell K (2010) Biogeochemical redox processes and their impact on contaminant dynamics. Environ Sci Technol 44:15–23

Bruins MR, Kapil S, Oehme FW (2000) Microbial resistance to metals in the environment. Ecotoxicol Environ Saf 45:198–207

Cases I, de Lorenzo V (2005a) Genetically modified organisms for the environment: stories of success and failure and what we have learned from them. Int Microbiol 8:213–222

Cases I, de Lorenzo V (2005b) Promoters in the environment: transcriptional regulation in its natural context. Nat Rev Microbiol 3:105–118

Cervantes C, Campos-Garcia J (2007) Reduction and efflux of chromate by bacteria. In: Nies DH, Silver S (eds) Molecular microbiology of heavy metals. Springer, Berlin

Cervantes C, Campos-Garcia J, Devars S, Gutierrez-Corona F, Loza-Tavera H, Torres-Guzman JC, Moreno-Sanchez R (2001) Interactions of chromium with microorganisms and plants. FEMS Microbiol Rev 25:335–347

Chan CS, Fakra SC, Edwards DC, Emerson D, Banfield JF (2009) Iron oxihydroxide mineralization on microbial extracellular polysaccharides. Geochim Cosmochim Acta 73:3807–3818

Chasteen TG, Bentley R (2003) Biomethylation of selenium and tellurium: microorganisms and plants. Chem Rev 103:1–26

Cooksey DA (1994) Molecular mechanisms of copper resistance and accumulation in bacteria. FEMS Microbiol Rev 14:381–386

DiChristina TJ, Fredrickson JK, Zachara JM (2005) Enzymology of electron transport: energy generation with geochemical consequences. Rev Miner Geochem 59:27–52

Dowdle PR, Oremland RS (1998) Microbial oxidation of elemental selenium in soil slurries and bacterial cultures. Environ Sci Technol 32:749–3755

Du X, Boonchayaanant B, Wu W, Fendorf S, Bargar J, Criddle CS (2011) Reduction of uranium(VI) by soluble iron(II) conforms with thermodynamic predictions. Environ Sci Technol 45:4718–4725

Dungan RS, Frankenberger WT (1999) Microbial transformations of selenium and the bioremediation of seleniferous environments. Bioremediation J 3:171–188

Dupraz C, Reid RP, Braissant O, Decho AW, Norman RS, Visscher PT (2009) Processes of carbonate precipitation in modern microbial mats. Earth Sci Rev 96:141–162

Ehrlich HL, Newman DK (2009) Geomicrobiology, 5th edn. CRC Press/Taylor & Francis, Boca Raton

Finneran KT, Anderson RT, Nevin KP, Lovley DR (2002) Bioremediation of uranium-contaminated aquifers with microbial U(VI) reduction. Soil Sediment Contam 11:339–357

Fomina M, Alexander IJ, Hillier S, Gadd GM (2004) Zinc phosphate and pyromorphite solubilization by soil plant-symbiotic fungi. Geomicrobiol J 21:351–366

Fomina M, Hillier S, Charnock JM, Melville K, Alexander IJ, Gadd GM (2005a) Role of oxalic acid over-excretion in toxic metal mineral transformations by *Beauveria caledonica*. Appl Environ Microbiol 71:371–381

Fomina M, Alexander IJ, Colpaert JV, Gadd GM (2005b) Solubilization of toxic metal minerals and metal tolerance of mycorrhizal fungi. Soil Biol Biochem 37:851–866

Fomina M, Podgorsky VS, Olishevska SV, Kadoshnikov VM, Pisanska IR, Hillier S, Gadd GM (2007) Fungal deterioration of barrier concrete used in nuclear waste disposal. Geomicrobiol J 24:643–653

Francis AJ, Dodge CJ, Gillow JB (1992) Biodegradation of metal citrate complexes and implications for toxic metal mobility. Nature 356:140–142

Frentiu T, Ponta M, Levei E, Cordos EA (2009) Study of partitioning and dynamics of metals in contaminated soil using modified four-step BCR sequential extraction procedure. Chem Pap 63:239–248

Gadd GM (2005) Microorganisms in toxic metal-polluted soils. In: Buscot F, Varma A (eds) Microorganisms in soils: roles in genesis and functions. Springer, Heidelberg

Gadd GM (1996) Influence of microorganisms on the environmental fate of radionuclides. Endeavour 20:150–156

Gadd GM (1999) Fungal production of citric and oxalic acid: importance in metal speciation, physiology and biogeochemical processes. Adv Microb Physiol 41:4792

Gadd GM (2009) Biosorption: critical review of scientific rationale, environmental importance and significance for pollution treatment. J Chem Technol Biotechnol 84:13–28

Gadd GM (2010) Minerals and microbes: geomicrobiology and bioremediation. Microbiology 156:609–643

Gadd GM, Sayer JA (2000) Fungal transformations of metals and metalloids. In: Lovley DR (ed) Environmental microbe-metal interactions. American Society of Microbiology, Washington, DC

Geets J, Vangronsveld J, Diels L, Taghavi S, van der Lelie D (2008) Microbial activities, monitoring and application as part of a management strategy for heavy metal-contaminated soil and

ground water. In: Naidu R (ed) Developments in soil science. Elsevier B.V., London, pp 521–559
Gharieb MM, Kierans M, Gadd GM (1999) Transformation and tolerance of tellurite by filamentous fungi: accumulation, reduction and volatilization. Mycol Res 103:299–305
Goyal N, Jain SC, Banerjee UC (2003) Comparative studies on the microbial adsorption of heavy metals. Adv Environ Res 7:311–319
Hameed MSA (2006) Continuous removal and recovery of lead by alginate beads free and alginate-immobilized *Chlorella vulgaris*. Afr J Biotechnol 5:1819–1823
Hamilton WA (2003) Microbially influenced corrosion as a model system for the study of metal–microbe interactions: a unifying electron transfer hypothesis. Biofouling 19:65–76
Hasin AA, Gurman SJ, Murphy LM, Perry A, Smith TJ, Gardiner PE (2010) Remediation of chromium (VI) by a methane-oxidizing bacterium. Environ Sci Technol 44:400–405
Haveman SA, Pedersen K (2002) Microbially mediated redox processes in natural analogues for radioactive waste. J Contam Hydrol 55:161–174
Hietala KA, Roane TM (2009) Microbial remediation of metals in soils. In: Singh A et al. (eds) Advances in applied bioremediation, Soil biology, vol 17, Springer, Berlin/Heidelberg, pp 201–220
Holden JF, Adams MWW (2003) Microbe–metal interactions in marine hydrothermal vents. Curr Opin Chem Biol 7:160–165
Hobman JL, Wilson JR, Brown NL (2000) Microbial mercury reduction. In: Lovley DR (ed) Environmental microbe-metal interactions. Am Soc Microbiol Washington, pp 177–197
Hu P, Brodie EO, Suzuki Y, McAdams HH, Andersen GL (2005) Whole-genome transcriptional analysis of heavymetal stresses in *Caulobacter crescentus*. J Bacteriol 187:8437–8449
Huang PM (2008) Impacts of physicochemical–biological interactions on metal and metalloid transformations in soils: an overview. In: Violante A, Huang PM, Gadd GM (eds) Biophysico-chemical processes of heavy metals and metalloids in soil environments. John Wiley and Sons, Wiley, Chichester
Huang PM, Wang MC, Wang MK (2004) Mineral–organic–microbial interactions. In: Hillel D, Rosenzweig C, Powlson DS, Scow KM, Singer MJ, Sparks DL, Hatfield J (eds) Encyclopedia of soils in the environment. Elsevier, Amsterdam
Joerger K, Joerger R, Olsson E, Granqvist CG (2001) Bacteria as workers in the living factory: metal accumulating-bacteria and their potential for material science. Trends Biotechnol 19:15–20
Johnson DB (2003) Chemical and microbiological characteristics of mineral spoils and drainage waters at abandoned coal and metal mines. Water Air Soil Pollut 3:47–66
Jussila MM, Zhao J, Suominen L, Lindström K (2007) TOL plasmid transfer during bacterial conjugation in vitro and rhizoremediation of oil compounds in vivo. Environ Pollut 146:510–524
Kadirvelu K, Thamaraiselvi K, Namasivayam C (2001) Removal of heavy metals from industrial wastewaters by adsorption onto activated carbon prepared from an agricultural solid waste. Bioresour Technol 76:63–65
Kannan SK, Krishnamoorthy R (2006) Isolation of mercury resistant bacteria and influence of abiotic factors on bioavailability of mercury – a case study in Pulicat Lake North of Chennai, South East India. Sci Total Environ 367:341–353
Karlson U, Frankenberger WT (1988) Effects of carbon and trace element addition on alkylselenide production by soil. Soil Sci Soc Am J 52:1640–1644
Karlson U, Frankenberger WT (1989) Accelerated rates of selenium volatilization from California soils. Soil Sci Soc Am J 53:749–753
Komeili A (2007) Molecular mechanisms of magnetosome formation. Annu Rev Biochem 76:351–366
Kube M, Beck A, Zinder SH, Kuhl H, Reinhardt R, Adrian L (2005) Genome sequence of the chlorinated compound respiring bacterium *Dehalococcoides* species strain CBDB1. Nat Biotechnol 23:1269–1273
Landa ER (2005) Microbial biogeochemistry of uranium mill tailings. Adv Appl Microbiol 57:113–130

Lee Y-A, Hendson M, Panopoulus NJ, Schrott MN (1994) Molecular cloning, chromosomal mapping, and sequence analysis of copper resistance genes from *Xanthomonas campestris pv. juglandis*: homology with blue copper proteins and multicopper oxidase. J Bacteriol 176:173–188

Lee SM, Grass G, Rensing C, Barrett SR, Yates CJD, Stoyanov JV, Brown NL (2002) The Pco proteins are involved in periplasmic copper handling in *Escherichia coli*. Biochem Biophys Res Commun 295:616–620

Lesmana SO, Febriana N, Soetaredjo FE, Sunarso J, Ismadji S (2009) Studies on potential applications of biomass for the separation of heavy metals from water and wastewater. Biochem Eng J 44:19–41

Li YH, Ding J, Luan ZK, Di ZC, Zhu YF, Xu C, Wu DH, Wei BQ (2003) Competitive adsorption of Pb^{2+}, Cu^{2+} and Cd^{2+} ions from aqueous solutions by multiwalled carbon nanotubes. Carbon 41:2787–2792

Lian B, Chen Y, Zhu L, Yang R (2008a) Effect of microbial weathering on carbonate rocks. Earth Sci Front 15:90–99

Lian B, Wang B, Pan M, Liu C, Teng HH (2008b) Microbial release of potassium from K-bearing minerals by thermophilic fungus *Aspergillus fumigatus*. Geochim Cosmochim Acta 72:87–98

Liu S, Zhang F, Chen J, Sun GX (2011) Arsenic removal from contaminated soil via biovolatilization by genetically engineered bacteria under laboratory conditions. J Environ Sci 23:1544–1550

Lloyd JR (2003) Microbial reduction of metals and radionuclides. FEMS Microbiol Rev 27:411–425

Lloyd JR, Renshaw JC (2005) Bioremediation of radioactive waste: radionuclide-microbe interactions in laboratory and field-scale studies. Curr Opin Biotechnol 16:254–260

Losi ME, Frankenberger WT (1998) Microbial oxidation and solubilization of precipitated elemental selenium in soil. J Environ Qual 27:836–843

Lovley DR (2001) Anaerobes to the rescue. Science 293:1444–1446

Lovley DR, Coates JD (1997) Bioremediation of metal contamination. Curr Opin Biotechnol 8:285–289

Lowenstam HA (1981) Minerals formed by organisms. Science 211:1126–1131

Macaskie LE, Bonthrone KM, Yong P, Goddard DT (2000) Enzymically mediated bioprecipitation of uranium by a *Citrobacter* sp.: a concerted role for exocellular lipopolysaccharide and associated phosphatase in biomineral formation. Microbiology 146:1855–1867

Machado MD, Santos MSF, Gouveia C, Soares HMVM, Soares EV (2008) Removal of heavy metals using a brewer's yeast strain of *Saccharomyces cerevisiae*: the flocculation as a separation process. Bioresour Technol 99:2107–2115

Macur RE, Jackson CR, Botero LM, Mcdermott TR, Inskeep WP (2004) Bacterial populations associated with the oxidation and reduction of arsenic in an unsaturated soil. Environ Sci Technol 38:104–111

Maier RM (2000) Microorganisms and organic pollutants. In: Maier RM, Pepper IL, Gerba CP (eds) Environmental microbiology. Academic, San Diego

Masood F, Malik A (2011) Biosorption of metal ions from aqueous solution and tannery effluents by *Bacillus* sp. FM1. J Environ Sci Health Part A 46:1667–1674

McLean J, Beveridge TJ (2001) Chromate reduction by a pseudomonad isolated from a site contaminated with chromate copper arsenate. Appl Environ Microbiol 67:1076–1084

McLean JS, Lee JU, Beveridge TJ (2002) Interactions of bacteria and environmental metals, fine-grained mineral development, and bioremediation strategies. In: Huang PM, Bollag J-M, Senesi N (eds) Interactions between soil particles and microorganisms. Wiley, New York, pp 227–261

Merkle SA (2006) Engineering forest trees with heavy metal resistance genes. Silvae Genetica 55:263–268

Merroun ML, Selenska-Pobell S (2008) Bacterial interactions with uranium: an environmental perspective. J Contam Hydrol 102:285–295

Mergeay M, Monchy S, Vallaeys T, Auquier V, Benotmane A, Bertin P, Taghavi S, Dunn J, van der Lelie D, Wattinez R (2003) Ralstonia metallidurans, a bacterium specifically adapted to toxic metals: towards a catalogue of metal-responsive genes. FEMS Microbiol Rev 27:385–410

Miot J, Benzerara K, Morin G, Kappler A, Bernard S, Obst M, Ferard C, Skouri-Panet F, Guigner JM, Posth N, Galvez M, Brown GE Jr, Guyot F (2009) Iron biomineralization by neutrophilic iron-oxidizing bacteria. Geochim Cosmochim Acta 73:696–711
Moore CM, Gaballa A, Hui M, Ye RW, Helmann JD (2005) Genetic and physiological responses of *Bacillus subtilis* to metal ion stress. Mol Microbiol 57:27–40
Moral R, Gilkes RJ, Jordan MM (2005) Distribution of heavy metals in calcareous and non-calcareous soils in Spain. Water Air Soil Pollut 162:127–142
Morillo JA, Aguilera M, Ramos-Cormenzana A, Monteoliva-Sánchez M (2006) Production of a metal-binding exopolysaccharide by *Paenibacillus jamilae* using two-phase olive-mill waste as fermentation substrate. Curr Microbiol 53:189–193
Muhammad A, Muhammad S, Sarfraz H (2008) Perspectives of bacterial ACC deaminase in phytoremediation. Trends Biotechnol 25:356–362
Mulligan CN, Yong RN, Gibbs BF (2001) Heavy metal removal from sediments by biosurfactants. J Hazard Mater 85:111–125
Muralisastry, Ahmad A, Khan MI, Rajivkumar (2003) Biosynthesis of metal nanoparticles using fungi and actinnomycetes. Curr Sci 85:162–170
Muyzer G, Stams AJM (2008) The ecology and biotechnology of sulphate-reducing bacteria. Nat Rev Microbiol 6:441–454
Nair A, Juwarkar AA, Singh SK (2007) Production and characterization of siderophores and its application in arsenic removal from contaminated soil. Water Air Soil Pollut 180:199–212
Nies DH (1999) Microbial heavy-metal resistance. Appl Microbiol Biotechnol 51:730–750
Nies DH (2003) Efflux-mediated heavy metal resistance in prokaryotes. FEMS Microbiol 27:313–339
Nies DH, Silver S (1995) Ion efflux systems involved in bacterial metal resistances. Ind J Microbiol 14:186–199
Nilgiriwala KS, Alahari A, Rao AS, Apte SK (2008) Cloning and overexpression of alkaline phosphatase PhoK from *Sphingomonas* sp. strain BSAR-1 for bioprecipitation of uranium from alkaline solutions. Appl Environ Microbiol 74:5516–5523
Ohtake H, Cervantes C, Silver S (1987) Decreased chromate uptake in *Pseudomonas fluorescens* carrying a chromate resistance plasmid. J Bacteriol 169:3853–3856
Oremland R, Stolz J (2000) Dissimilatory reduction of selenate and arsenate in nature. In: Lovley DR (ed) Environmental microbe–metal interactions. American Society for Microbiology, Washington, DC
Oremland RS, Hollibaugh JT, Maest AS, Presser TS, Miller LG, Culbertson CW (1989) Selenate reduction to elemental selenium by anaerobic bacteria in sediments and culture: biogeochemical significance of a novel sulfate-independent respiration. Appl Environ Microbiol 55:2333–2343
Oremland RS, Hoeft SE, Santini JM, Bano N, Hollibaugh RA, Hollibaugh JT (2002) Anaerobic oxidation of arsenite in Mono Lake water and by facultative, arsenite-oxidizing chemoautotroph, strain MLHE-1. Appl Environ Microbiol 68:4795–4802
Paez-Espino D, Tamames J, de Lorenzo V, Canovas D (2009) Microbial responses to environmental arsenic. Biometals 22:117–130
Parnell JJ, Park J, Denef V, Tsoi T, Hashsham S, Quensen JI, Tiedje JM (2006) Coping with polychlorinated biphenyl (PCB) toxicity: physiological and genome wide responses of *Burkholderia xenovorans* LB400 to PCB-mediated stress. Appl Environ Microbiol 72:6607–6614
Pepper IL, Gentry TJ, Newby DT, Roane TM, Josephson KL (2002) The role of cell bioaugmentation and gene bioaugmentation in the remediation of co-contaminated soils. Environ Health Perspect 110:943–946
Poljsak B, Pócsi I, Raspor P, Pesti M (2010) Interference of chromium with biological systems in yeast and fungi: a review. J Basic Microbiol 50:21–36
Poole K (2005) Efflux-mediated antimicrobial resistance. J Antimicrob Chemother 56:20–51
Rajendran P, Muthukrishnan J, Gunasekaran P (2003) Microbes in heavy metal remediation. Indian J Exp Biol 41:935–944

Rawlings DE (1997) Mesophilic, autotrophic bioleaching bacteria: description, physiology and role. In: Rawlings DE (ed) Biomining: theory, microbes and industrial processes. Springer, Berlin, pp 229–245

Rawlings DE (2002) Heavy metal mining using microbes. Annu Rev Microbiol 56:65–91

Ray S, Ray MK (2009) Bioremediation of heavy metal toxicity-with special reference to chromium. Al Ameen J Med Sci 2:57–63

Rensing C, Newby DT, Pepper IL (2002) The role of selective pressure and selfish DNA in horizontal gene transfer and soil microbial community adaptation. Soil Biol Biochem 34:285–296

Rosen BP (2002) Biochemistry of arsenic detoxification. FEBS Lett 529:86–92

Ruta L, Paraschivescu C, Matache M, Avramescu S, Farcasanu IC (2010) Removing heavy metals from synthetic effluents using "kamikaze" *Saccharomyces cerevisiae* cells. Appl Microbiol Biotechnol 85:763–771

Sandrin TR, Chech AM, Maier RM (2000) A rhamnolipid biosurfactant reduces cadmium toxicity during naphthalene biodegradation. Appl Environ Microbiol 66:4585–4588

Sayer JA, Cotter-Howells JD, Watson C, Hillier S, Gadd GM (1999) Lead mineral transformation by fungi. Curr Biol 9:691–694

Schippers A, Sand W (1999) Bacterial leaching of metal sulphides proceeds by two indirect mechanisms via thiosulfate or via polysulfides and sulphur. Appl Environ Microbiol 65:319–321

Schottel J, Mandal A, Clark D, Silver S, Hedges RW (1974) Volatilization of mercury and organomercurials determined by inducible R-factor systems in enteric bacteria. Nature 251:335–337

Schue M, Dover LG, Besra GS, Parkhill J, Brown NL (2009) Sequence and analysis of a plasmid encoded mercury resistance operon from *Mycobacterium marinum* identifies MerH, a new mercuric ion transporter. J Bacteriol 19:439–444

Senko JM, Istok JD, Suflita JM, Krumholz LR (2002) In-situ evidence for uranium immobilization and remobilization. Environ Sci Technol 36:1491–1496

Shukla KP, Singh NK, Sharma S (2010) Bioremediation: developments, current practices and perspectives. Genet Eng Biotechnol J 3:1–20

Siddiqui MH, Kumar A, Kesari KK, Arif JM (2009) Biomining-a useful approach toward metal extraction. Am Eurasian J Agron 2:84–88

Silver S, Phung LT (1996) Bacterial heavy metal resistance: new surprises. Annu Rev Microbiol 50:753–789

Silver S, Phung LT (2005) A bacterial view of the periodic table: genes and proteins for toxic inorganic ions. J Ind Microbiol Biotechnol 32:587–605

Silver S, Phung LT (2009) Heavy metals, bacterial resistance. In: Schaechter M (ed) Encyclopedia of microbiology. Elsevier, Oxford

Silver S, Budd K, Leahy KM, Shaw WV, Hammond D, Novick RP, Willsky GR, Malamy MH, Rosenberg H (1981) Inducible 92 plasmid-determined resistance to arsenate, arsenite and antimony(III) in *Escherichia coli* and *Staphylococcus aureus*. J Bacteriol 146:983–996

Singh JS, Abhilash PC, Singh HB, Singh RP, Singh DP (2011) Genetically engineered bacteria: an emerging tool for environmental remediation and future research perspectives. Gene 480:1–9

Smith WL, Gadd GM (2000) Reduction and precipitation of chromate by mixed culture sulphate-reducing bacterial biofilms. J Appl Microbiol 88:983–991

Solioz M, Stoyanov JV (2003) Copper homeostasis in *Enterococcus hirae*. FEMS Microbiol Rev 27:183–195

Spain A, Alm E (2003) Implications of microbial heavy metal tolerance in the environment. Rev Undergrad Res 2:1–6

Stolz JF, Oremland RS (1999) Bacterial respiration of arsenic and selenium. FEMS Microbiol Rev 23:615–627

Strasser H, Burgstaller W, Schinner F (1994) High yield production of oxalic acid for metal leaching purposes by *Aspergillus niger*. FEMS Microbiol Lett 119:365–370

Tabak HH, Lens P, van Hullebusch ED, Dejonghe W (2005) Developments in bioremediation of soils and sediments polluted with metals and radionuclides–1. Microbial processes and mechanisms affecting bioremediation of metal contamination and influencing metal toxicity and transport. Rev Environ Sci Bio Technol 4:115–156

Thomas KW (2008) Molecular approaches in bioremediation. Curr Opin Biotechnol 19:572–578
Thompson-Eagle ET, Frankenberger WT, Karlson U (1989) Volatilization of selenium by *Alternaria alternata*. Appl Environ Microbiol 55:1406–1413
Tomei FA, Barton LL, Lemanski CL, Zocco TG, Fink NH, Sillerud LO (1995) Transformation of selenate and selenite to elemental selenium by *Desulfovibrio desulfuricans*. J Ind Microbiol 14:329–336
Tunali S, Cabuk A, Akar T (2006) Removal of lead and copper ions from aqueous solutions by bacterial strain isolated from soil. Chem Eng J 115:203–211
Unaldi MN, Korkmaz H, Ankan B, Coral G (2003) Plasmid-encoded heavy metal resistance in *Pseudomonas* sp. Bull Environ Contam Toxicol 71:1145–1150
Urgun-Demirtas M, Stark B, Pagilla K (2006) Use of genetically engineered microorganisms (GEMS) for the bioremediation of contaminants. Crit Rev Biotechnol 26:145–164
Van Hullebusch ED, Lens PNL, Tabak HH (2005) Developments in bioremediation of soils and sediments polluted with metals and radionuclides. 3. Influence of chemical speciation and bioavailability on contaminants immobilization/mobilization bio-processes. Rev Environ Sci Biotechnol 4:185–212
Violante A, Cozzolino V, Perelomov L, Caporale AG, Pigna M (2010) Mobility and bioavailability of heavy metals and metalloids in soil environments. J Soil Sci Plant Nutr 10:268–292
Voloudakis AE, Reignier TM, Cooksey DA (2005) Regulation of resistance to copper in Xanthomonas axonopodis pv. vesicatoria. Appl Environ Microbiol 71:782–789
Wall JD, Krumholz LR (2006) Uranium reduction. Annu Rev Microbiol 60:149–166
Wang JL, Chen C (2006) Biosorption of heavy metals by *Saccharomyces cerevisiae*: a review. Biotechnol Adv 24:427–451
Wang J, Chen C (2009) Biosorbents for heavy metals removal and their future. Biotechnol Adv 27:195–226
Watanabe ME (2001) Can bioremediation bounce back? Nat Biotechnol 19:1111–1115
White C, Gadd GM (1998) Accumulation and effects of cadmium on sulphate-reducing bacterial biofilms. Microbiology 144:1407–1415
White C, Gadd GM (1997) An internal sedimentation bioreactor for laboratory-scale removal of toxic metals from soil leachates using biogenic sulphide precipitation. J Ind Microbiol 18:414–421
White C, Gadd GM (2000) Copper accumulation by sulphate reducing bacterial biofilms and effects on growth. FEMS Microbiol Lett 183:313–318
Williams CJ, Aderhold D, Edyvean GJ (1998) Comparison between biosorbents for the removal of metal ions from aqueous solution. Water Res 32:216–224
Yabusaki SB, Fang Y, Long PE, Resch CT, Peacock AD, Komlos J, Jaffe PR, Morrison SJ, Dayvault RD, White DC, Anderson RT (2007) Uranium removal from groundwater via in situ biostimulation: field-scale modeling of transport and biological processes. J Contam Hydrol 93:216–235
Yee N, Kobayashi DY (2008) Molecular genetics of selenate reduction by *Enterobacter cloacae* SLD1a-1. Adv Appl Microbiol 64:107–123
Zawadzka AM, Crawford RL, Paszczynski AJ (2007) Pyridine-2, 6-bis (thiocarboxylic acid) produced by *Pseudomonas stutzeri* KC reduces chromium(VI) and precipitates mercury, cadmium, lead and arsenic. Biometals 20:145–158
Zhu R, Wu M, Yang J (2011) Mobilities and leachabilities of heavy metals in sludge with humus soil. J Environ Sci 23:247–254

Chapter 12
Anaerobic Digestion of the Organic Fraction of Municipal Solid Waste

Muna Albanna

Abstract Municipal solid waste (MSW) is the waste generated from residential sources, such as households, and from institutional and commercial sources such as offices, schools, hotels and other sources. The main components of MSW are food, garden waste, paper, board, plastic, textile, metal, and glass waste. The composition of MSW varies depending on a range of factors; the household waste reflects population density and economic prosperity, seasonality, housing standards and the presence of waste minimization initiatives. The MSW consists of a high proportion of organic fraction resulting from scraps, food residues, paper and garden waste. The organic fraction of municipal solid waste (OFMSW) represents 70% of the waste composition with moisture content around 85–90%.

The uncontrolled decomposition of OFMSW can contribute to global warming and result in large-scale contamination of soil, water, and air. Macias-Corral et al. (Bioresour Technol 99:8288–8293, 2008) indicated that the decomposition of one metric ton of OFMSW can potentially release 50–110 m^3 of carbon dioxide (CO_2) and 90–140 m^3 of methane (CH_4). In addition, the high levels of moisture content make this type of waste ineffectual for incineration. Therefore, the anaerobic digestion of OFMSW can be an environmentally sustainable technology to reduce the harmful effects of MSW, reduce the volume and toxicity of this waste, in addition to many advantages including potential for energy recovery, production of an end-product suitable for soil conditioning, and decreased dependency on landfills.

Anaerobic biodegradation of the organic material proceeds in the absence of oxygen and presence of anaerobic microorganisms. Anaerobic Digestion (AD) is an engineered biological process by which complex organic materials are first hydrolyzed and fermented by acid bacteria into volatile fatty acids that are consumed by methanogenic bacteria and converted into biogas afterwards. The biogas generated

M. Albanna (✉)
Water and Environmental Engineering Department, School of Natural Resources Engineering, German Jordanian University, P.O. Box: 35247, Amman 11180, Jordan
e-mail: muna.albanna@gju.edu.jo

A. Malik et al. (eds.), *Management of Microbial Resources in the Environment*,
DOI 10.1007/978-94-007-5931-2_12, © Springer Science+Business Media Dordrecht 2013

can be used as a renewable source of energy, and the solid compost material can be used as soil fertilizer. For all these advantages, the AD technology has been supported by legislation in many countries around the world and encouraged as sustainable solid waste management option.

This chapter presents in depth the AD technology to determine its economic and environmental competitiveness, as one of the options for processing the biodegradable organic materials in MSW. The chapter also discusses the stages of the digestion of the waste such as; the pretreatment, the separation processes, waste digestion, gas recovery and residue treatment. The AD operating parameters are illustrated, such as – but not limited to –: waste composition, temperature, organic content, residence time, pH, carbon/nitrogen ratio, and compost quality.

Keywords Municipal solid waste (MSW) • Anaerobic digestion • Sustainable technology • Renewable energy

Abbreviations

AD	Anaerobic Digestion
EIONET	European Environmental Information and Observation Network
EU	European Union
HRT	Hydraulic Retention Time (days)
LCA	Life Cycle Assessment
MSW	Municipal Solid Waste
OFMSW	Organic Fraction of Municipal Solid Waste
RCRA	Resource Conversion and Recovery Act
SRT	Solid Retention Time (days)
UASB	Upflow Anaerobic Sludge Bed
USEPA	United States Environmental Protection Agency
UNESCAP	United Nations Economic and Social Commission for Asia and the Pacific
VFA	Volatile Fatty Acids

1 Introduction

The economic development of societies, which started after the industrial revolution, had major effects on the environment: the natural resources are used extensively, and pollution and waste are produced as results of many activities related to the human civilization. Solid waste treatment, disposal and management are huge challenges for all municipalities, industries and businesses around the world. Global societies produce enormous quantities of waste as a result of many activities (municipal, agricultural, industrial and commercial). The crucial challenge is that the waste volumes produced are rapidly increasing due to the increase in population and modernized lifestyles. The recent statistics declare that the annual amount of the world's waste production is approximately four billion tonnes; only one quarter is recovered or recycled (Chalmin and Gaillochet 2009). Future projections estimate that the world's waste production could reach up annually to 27 billion tonnes by 2050; a third of which may be generated in Asia (Shanghai Manual 2011). In addition, the substances in the different types of waste are increasing in complexity and diversity. Affordable and secure solid waste management is indeed one of the key challenges of the twenty-first century as declared by many international forums around the world.

Many societies recognized that some of the current waste management methods are no longer acceptable due to the problems associated with the treatment paths, such as the effects on the environment and human health. These specifics forced different governments to look for improved practices that may lead to improved quality of services and reduce the related negative impacts. Also, many countries around the world started taking several measures to control the growing quantities of waste. One of these measures is to use it as a resource since the entire waste management system can be designed to maximize the benefits from the discarded and disposed materials using different modern practices and techniques. The anaerobic digestion (AD) of the organic fraction of municipal solid waste (OFMSW) is one of the simple technologies with a low energy requirement that may be used to convert the OFMSW, agricultural waste and other organic biomass into biogas that can be used as alternative and sustainable source of energy.

1.1 Definition of Solid Waste

Defining waste is not easy due to the fact that the waste contains numerous substances that are increasing in complexity and variety. One of the broad definitions was recognized by the Resource Conservation and Recovery Act (RCRA) which is the major federal statute that governs solid waste in the United States. This act defines solid waste as "any garbage, refuse, sludge from a waste treatment plant, water supply treatment plant or air pollution control facility and other discarded or salvageable materials, including solid, liquid, semisolid, or contained gaseous materials resulting from industrial, commercial, mining and agricultural operations, and from community activities" (USEPA 2008).

1.2 Composition of Solid Waste

There are different types of generated non-hazardous waste some of which are municipal, commercial, agriculture, non-hazardous industrial, mining, construction, and sewage wastes. The types of waste and the generation rates of each type of waste vary over time and location, taking into consideration the industrial development and human activities. EIONET (2009) reported that over 1.8 billion tonnes of wastes are generated annually in Europe, and this waste is mainly made up of municipal solid waste (MSW), commercial, agriculture, non-hazardous industrial and construction and demolition. The composition of solid waste reported by the latter author may describe the waste stream for most of the countries around the world, yet with different percentages.

The type of waste that takes much attention is the MSW, mainly due to the fact that management of this waste stream is expensive for all the municipalities around the world. The dilemma of MSW management is significant as it is visible to the communities and directly related to the national environmental concerns. Add to this the fact that the quantities of MSW are still increasing, even with all the advanced technologies for the waste recovery and treatment, and the more stringent laws and regulations. The MSW generation rate in the USA increased from 242.5 million tones in 2000 to around 250 million tonnes in 2010 (USEPA 2011). Strange (2002) stated that the United Kingdom generates more than 400 million tonnes/year of non-hazardous solid waste and 7% of this waste is MSW (calculated to be equivalent to 28 million tonnes).

The MSW consists of a non-biodegradable portion and a biodegradable organic fraction. The composition of MSW varies depending on a range of factors; the household waste reflects population density and economic prosperity, seasonality, housing standards and the presence of waste minimization initiatives (Albanna 2011). Figure 12.1 illustrates the composition of MSW in (a) the United States

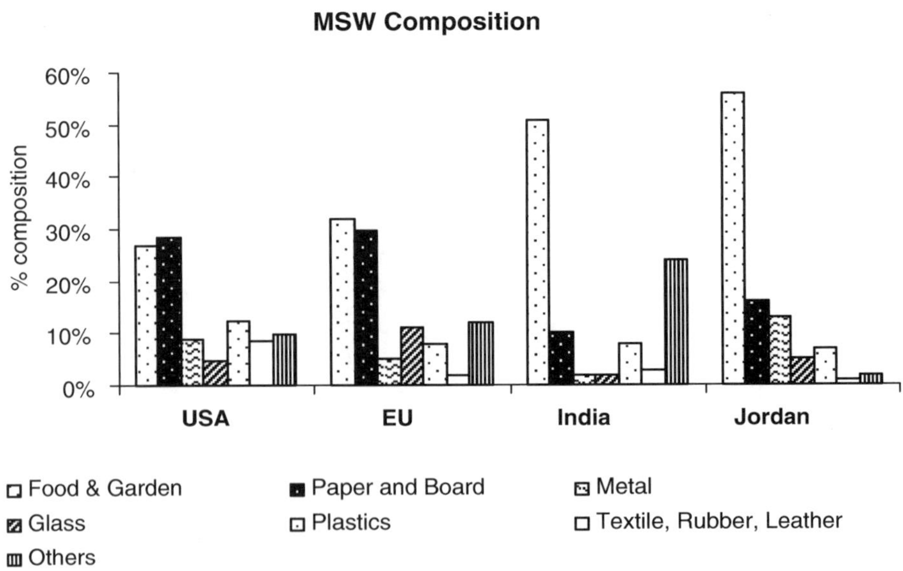

Fig. 12.1 Composition of MSW (**a**) in the USA (Adapted from USEPA 2011), (**b**) European Community (Adapted from European Commission Report 2001), (**c**) in India – Thiruvananthapuram City (Adapted from Narayana 2009), and (**d**) in Jordan (Adapted from Marayyan 2004)

(adapted from USEPA 2011), (b) EU countries (adapted from the European Commission report 2001), (c) India (adapted from Narayana 2009), and (d) Jordan (adapted from Marayyan 2004).

As demonstrated in Fig. 12.1, food and garden wastes constitute the largest proportion of MSW in most of the countries. Lardinois and van de Klundert (1993) illustrated that the vegetables and putrescible waste (organic fraction) forms 40–80% of the MSW in low income countries, 20–65% of the MSW in middle income countries and 20–50% of the MSW generated in the industrialized countries.

In addition to the MSW and agricultural waste; the animal manure and sewage sludge are two significant solid waste streams that usually contain high organic content. Macias-Coral et al. (2008) identified the animal manure as a major environmental problem in the United States as the quantities produced are 130 times greater than the amount of human waste. Eskicioglu et al. (2008) indicated that approximately 670,000 t of dry sewage sludge or biosolids are produced annually in Canada. According to the latter authors, these quantities of biosolids are subjected to increase in future since many wastewater treatment plants are expanding to serve the demands of growing communities.

As mentioned above; it is evident that huge quantities of waste which are rich in organic content are produced annually around the world. The municipal, sewage, and agricultural waste streams have potential for use as a substrate for anaerobic bacteria that are able to degrade the organic matter and produce biogas as a viable and recognized alternative source of energy.

2 Solid Waste Treatment Options of OFMSW

Throughout historical developments of human civilization; the acceptable treatment and disposal options of waste created a huge challenge. The Greek civilization was the first to recognize the problems associated with the waste disposal and consequently forced a law that controlled the waste dumping (Williams 2005). However, since early ages and until recent decades; landfilling used to be the dominant waste disposal method and incineration used to be the main method for waste treatment. As mentioned earlier and according to estimates by the USEPA (2011), the amount of MSW generated in the USA in 2010 was 250 million tonnes. From this amount, only 85 million tonnes were recycled or composted. Analysis of the illustrated statistics shows that the percentage of recovered MSW in 2010 in the USA was 34.1% of the generated waste, while 11.7% of the wastes were combusted for energy recovery, and the rest of the generated waste (54.2%) were landfilled. Nevertheless, due to the environmental risks associated with different disposal and treatment options, and mainly landfilling, many communities realized that integrated waste management systems are one of the global top priorities. Figure 12.2 presents the disposal methods for MSW in several countries (the data for USA adapted from USEPA 2011, and data for countries in Asia adapted from UNESCAP 2000).

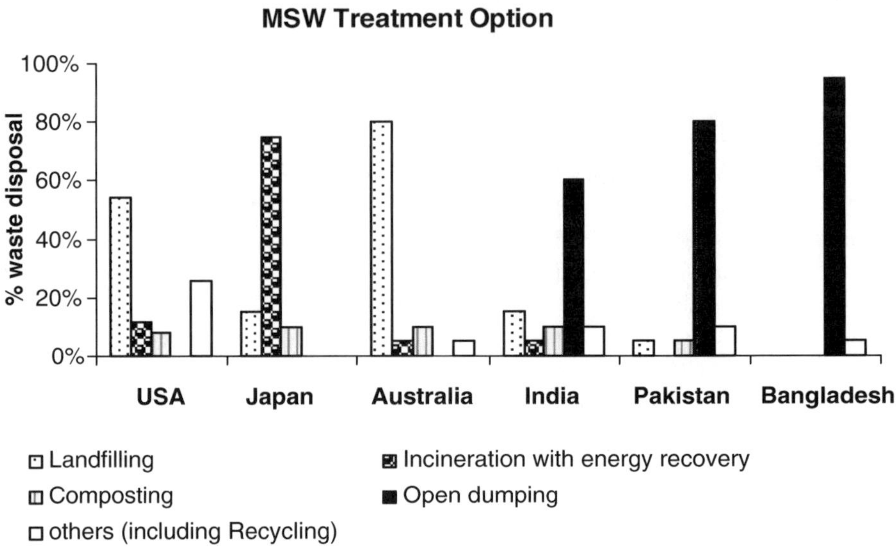

Fig. 12.2 Disposal methods for MSW in USA (Adapted from USEPA 2011), Japan, Australia, India, Pakistan and Bangladesh (Adapted from UNESCAP 2000)

As illustrated in Fig. 12.2, the solid waste disposal methods are different amongst countries, though typically the main solid waste disposal method is either landfilling or open dumping, which have negative implications on the environment and on human health.

If the MSW is landfilled or dumped, unquestionably large amounts of landfill gas as well as leachate will be produced as a result of the anaerobic biodegradation inside the landfills or open dumps. The fugitive emissions of landfill gas – that is composed mainly of methane (CH_4) and carbon dioxide (CO_2) – generated from the anaerobic biodegradation of the OFMSW will escape into the atmosphere, pollute the environment and add to the anthropogenic greenhouse gases emissions. Decomposition of one metric ton of organic solid waste can potentially release 50–110 m^3 of CO_2 and around 90–140 m^3 of CH_4 into the atmosphere (Macias-Corral et al. 2008). It is worth mentioning that landfilling of the combustible waste has been outlawed in many countries in Europe since a decade ago. In these countries, the OFMSW has to be recycled, treated by anaerobic digestion or incinerated (Eurostat 2012; Bond and Templeton 2011).

Despite the fact that landfilling is still the dominant way of MSW disposal in many countries around the world, there are still some other treatment options that take advantage of the high organic fraction that the MSW contains. The primary treatment and management options for the OFMSW may include: composting, waste to energy technologies, incineration and landfilling. Yet, the high moisture content levels of the OFMSW may hinder the incineration processes, also composting of this type of waste may produce odors, ammonia and large amounts of residuals that eventually have to be landfilled. The different treatment technologies of the waste are based on the processes that change the physical, chemical, or biological characteristics of the waste to reduce their negative impacts and threat on the environment, and/or to provide opportunities of using the waste as alternative resources. From this perspective, the AD will be a better treatment option for OFMSW. The engineered biological processes in controlled anaerobic reactors can offer a sustainable approach to address the problems associated with OFMSW treatment and provide useful products such as biogas and organic fertilizer. This chapter focuses on the AD of the OFMSW, as one of the significant treatment options and promising waste to energy recovery techniques. Within the identified treatment option; the waste will be neutralized, energy can be recovered and used for electricity generation, and other clean by products can be reclaimed as soil conditioner.

3 Anaerobic Digestion of the Organic Fraction of Solid Waste

The AD is the treatment of the organic fraction of solid waste in the absence of air for a specific mean cell residence time at a specific temperature. Anaerobic fermentation processes are among the oldest processes for the stabilization of solids and biosolids, and have been identified and used primarily for the treatment of waste sludge and high strength organic wastes (Metcalf et al. 2004). The latter authors illustrated that the stabilization of the organic fraction of the solid waste is accomplished biologically using a variety of microorganisms which convert the colloidal and dissolved carbonaceous organic matter into various gases and into protoplasm. Also, many toxic and recalcitrant organic compounds such as non-halogenated aromatic and aliphatic compounds can serve as a growth substrate and are degraded under anaerobic conditions.

Davis (2011) classified the diverse microorganisms involved in the environmental processes by the following: energy and carbon source, oxygen relationship, and temperature. The relationship between the source of carbon and the source of energy for the microorganisms is significant, since the carbon is the basic building block for cell synthesis. A source of energy must be obtained from outside the cell to enable synthesis to proceed. The bacteria that use the organic matter as a supply of carbon are called heterotrophic bacteria and the ones that use the CO_2 as a supply for the carbon needs are autotrophs. Also, bacteria can be classified by their ability or inability to utilize oxygen in oxidation-reduction reactions. Obligate aerobes are microorganisms that must have oxygen, while obligate anaerobes are microorganisms that cannot survive in the presence of oxygen. Facultative anaerobes can use oxygen in oxidation/reduction reactions, and under certain conditions, they can also grow in the absence of oxygen. Under anoxic conditions, a group of facultative anaerobes called denitrifiers utilize nitrites (NO_2^-) and nitrates (NO_3^-) instead of oxygen. Nitrate nitrogen is converted to nitrogen gas in the absence of oxygen in the anoxic de-nitrification process.

In addition, the bacteria can also be classified according to temperature, since each species of bacteria reproduces best within a limited range of temperatures. There are four temperature ranges used to classify bacteria. The psychrophile species grow best at temperatures below 20°C, while the mesophiles grow best at temperatures between 20 and 45°C. The third and fourth groups; the thermophiles grow at temperatures between 45 and 60°C, and the hyperthermophiles grow best above 60°C to near boiling. It is important to note that the facultative thermophiles growth usually extends from the thermophilic ranges into the mesophilic range (Hahn and Hoffstede 2010). Usually, the AD processes are designed to operate in either the mesophilic (20–45°C) or thermophilic (45–60°C) temperature ranges. Most of the studies reported in literature have been confined to biogas production at mesophilic and thermophilic temperatures. The psychrophilic digestion has not been as extensively explored due to the fact that low temperatures significantly decrease the bacterial kinetics and consequently the biogas yield. Balasubramaniyam et al. (2008) reported that it is not expected to design competitive systems using the psychrophilic digestion. In addition, De Mes et al. (2003) reported that the psychrophilic digestion requires longer retention time, which will result in larger reactors volume. The same authors reported that the mesophilic digestion is an advantageous option since it requires less reactor volume, and the thermophilic digestion can be viable when the waste is discharged at high temperature range or when the pathogen removal is considered crucial.

4 Phases of Anaerobic Biodegradation

As any other biological degradation process; the AD of the organic waste includes a wide variety of microbial communities. The conversion of the complex organic compounds into biogas involves a series of metabolic reactions and requires different groups of microorganisms that convert complex macromolecules into low molecular weight compounds (Wilkinson 2011; Hahn and Hoffstede 2010; Metcalf et al. 2004). The AD metabolism is carried out in a sequence of four stages: hydrolysis, acidogenesis, acetogenesis and methanogenesis.

Hydrolysis is the first step of the fermentation process, and it is essential since the methanogenic and acetogenic bacteria are not able to use polymers. In this phase, hydrolytic bacteria release extracellular enzymes that break down the particulate materials or the biodegradable non-dissolved organic solids and convert them into soluble compounds that can be hydrolyzed (or loosened with water) further to simple monomers. In this context, carbohydrates are broken down to simple sugars, fat into fatty acids and glycerol, and proteins into amino acids by the actions of hydrolytic bacteria (Wilkinson 2011). These monomers and particulates can be used as substrates in subsequent reactions and are used by the bacteria that perform the fermentation. In the AD of the OFMSW; hydrolysis is considered the rate limiting step, since it is relatively slow and it depends on the temperature and the complexity of the waste (Park and Ahnn 2011; De Mes et al. 2003).

The hydrolysis phase is followed by the acidogenesis phase, where two groups of fermentative bacteria transform the monomers produced in the previous phase into many intermediary products. The amino acids, sugars, and some fatty acids are degraded further in this phase. Metcalf et al. (2004) indicated that the organic substrates serve as both the electron donors and acceptors. Acidogenic bacteria convert the monomers resulting from the hydrolysis phase into alcohols, volatile fatty acids – such as butyric, propionic, and lactic acids-, in addition to keytones – such as methanol, ethanol, acetone and glycerol. In the acetogenesis phase, the carbohydrates fermentation continues and acetate is formed as the main end product. The latter authors illustrated that the main products that result from the fermentation phases are acetate, hydrogen (H_2), carbon dioxide (CO_2), propionate and butyrate. The latter products are fermented further to produce more H_2, CO_2 and acetate. Therefore, the main final products of the fermentation up to this phase are H_2, CO_2 and acetate, and they are the predecessors of the next phase; the methanogenesis phase.

The methanogenesis phase is carried out by a diverse group of microorganisms known as methanogens. In the beginning of this phase, the aceticlastic methanogens split the acetate into CH_4 and CO_2. While, the second group of methanogens – the hydrogen-utilizing methanogens use the H_2 as electron donor and CO_2 as the electron acceptor to produce CH_4. The acetogens in this phase use CO_2 to oxidize the H_2 and form the acetic acid, and this acetic acid is also converted to CH_4. Figure 12.3 illustrates the anaerobic digestion phases as adapted from Hahn and Hoffstede (2010).

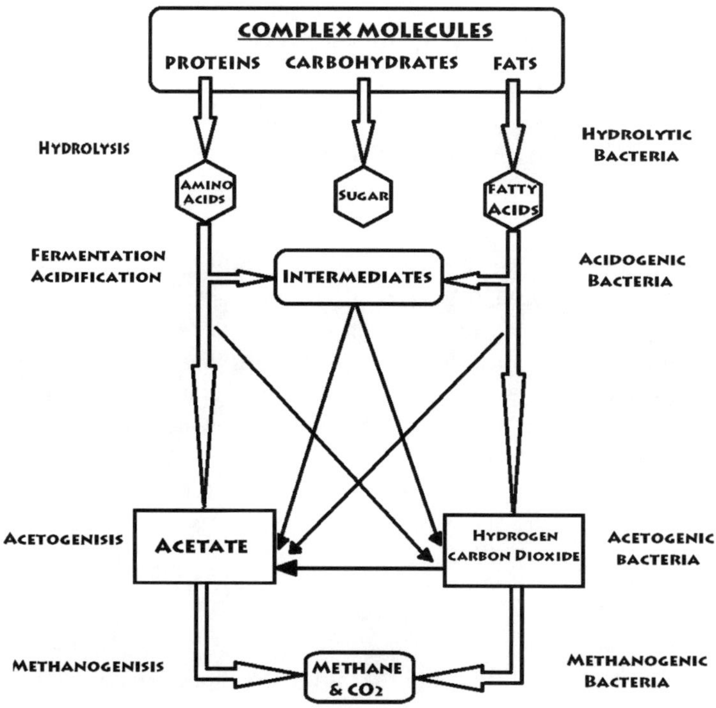

Fig. 12.3 Phases of anaerobic biodegradation (Adapted from Hahn and Hoffstede 2010)

De Mes et al. (2003) explained that in order to maintain stable digestion and to prevent the accumulation of intermediate compounds, it is important that the different biological conversions remain sufficiently coupled during the AD process.

The stoichiometry of the anaerobic fermentation phases can be summarized as following, and as adapted from Metcalf et al. (2004):

$$4H_2 + CO_2 \rightarrow CH_2 + 2H_2O \tag{12.1}$$

$$4HCOO^- + 4H^+ \rightarrow CH_4 + 3CO_2 + 2H_2O \tag{12.2}$$

$$4CO + 2H_2O \rightarrow CH_4 + 3CO_2 \tag{12.3}$$

$$4CH_3OH \rightarrow 3CH_4 + CO_2 + 2H_2O \tag{12.4}$$

$$4(CH_3)_3N + H_2O \rightarrow 9CH_4 + 3CO_2 + 6H_2O + 4NH_3 \tag{12.5}$$

$$CH_3COOH \rightarrow CH_4 + CO_2 \tag{12.6}$$

The kinetics of the microbial process of AD may be divided into kinetics of growth and kinetics of substrate utilization. Methanogenesis is the rate controlling process as methanogens have much slower growth rates than acidogenic and acetogenic bacteria. The rate of increase of the biomass or microorganism concentration is modeled as a first-order process (Nwabanne et al. 2009).

5 Factors Affecting Anaerobic Biodegradation of the Organic Fraction of Solid Waste

As the complex digestion processes involve different degradation steps, it is worth noting that the interdependent bacterial consortium must continue to exist in equilibrium, as discussed in the previous section. If one of the environmental parameters needed for the bacterial consortium involved in the processes changes, then the system will shift away from equilibrium and will result in disturbing the processes, and consequently the operating system. Thus, the following environmental parameters must be monitored and controlled to ensure stability and functionality of biological processes.

5.1 Temperature

Temperature is one of the most significant environmental parameters that affect the digestion process, since it affects the bacterial growth kinetics and stability, and subsequently the biogas production. The temperature does not affect the microbial kinetics only, but also has an extreme effect on gas transfer rates and settling characteristics of biosolids (Metcalf et al. 2004).

At temperatures below their optimum for growth, microorganisms are unable to attack substrates from their environment because of lowered affinity. High temperatures may lower the biogas production rates due to the production of volatile gases that may include ammonia which also affect the bacterial activities at the methanogenesis phase (Khalid et al. 2011).

The biogas production may be carried out at different temperatures through the thermophilic, mesophilic and psychrophilic digestion processes. Balasubramaniyam et al. (2008) reported- based on case studies- that the daily biogas production rates per m^3 of OFMSW at different temperatures varied from 0.05 m^3 of biogas at temperature range from 6–10°C, to 0.1–0.33 m^3 of biogas at temperature range from 16–22°C, and to 0.2–0.33 m^3 of biogas at temperature range higher than 22°C. This information may show that the biodigester system can function under different temperatures, but the biogas production is less at lower temperature range.

Many researchers illustrated that the AD is usually carried out at the mesophilic temperatures due to the fact that the processes require smaller energy expenses (Metcalf et al. 2004). Khalid et al. (2011) demonstrated that a temperature range between 35 and 37°C is considered right for the production of biogas in the mesophilic digestion, and temperatures below 65°C are considered appropriate in the thermophilic digestion. Several research groups (Saha et al. 2011; Eskicioglu et al. 2009; Metcalf et al. 2004) stated that the thermophilic digestion is usually more efficient at volatile solids reductions and CH_4 production than mesophilic digestion, in addition to many other advantages that include: increased solid destruction, higher loading rates, as well as increased bacterial destruction. However, thermophilic digestion has some disadvantages such as: need of energy and more sensitivity to toxicity and environmental changes. The mesophilic bacteria are more robust to environmental changes, and demonstrated considerable stability in many different applications, and therefore, are more reliable in most AD facilities (Wilkinson 2011)

Studying the effect of temperature is important for the design and operation of biologically active AD systems. It is necessary to maintain a stable temperature range due to the fact that methanogenic bacteria are sensitive to temperature changes, therefore, fluctuations in temperature will result in disturbing the bacterial activities which will result in fluctuations in biogas production rates.

5.2 Moisture Content

The high moisture content will enhance the AD due to the fact that higher water contents affect the rate of dissolving degradable organic matter, while lower moisture content levels will result in decrease in the bacterial activities, presumably due to physiological response to water stress. From their experiments; many researchers reported that the highest CH_4 production rates occurred at moisture content between 60 and 80% (Khalid et al. 2011).

It is difficult to maintain the same availability of water throughout the digestion cycle. Therefore, moisture levels are one of the important parameters to be monitored and controlled during the AD operations in order to avoid fluctuations in the biogas production rates.

5.3 pH Value and Alkalinity

The pH plays a significant role in the phases of AD. The different stages of AD are pH dependent, since each bacterial group involved in the AD has a specific pH range for optimal growth. Hydrolytic and acidogenic bacteria's optimal pH range is between 5.2 and 6.3, while the pH range for methanogenic bacteria is between 6.8 and 7.4.

Acidogenesis and acetogenic bacteria will result in producing large amounts of volatile fatty acids, and consequently the pH will drop. Since the optimal pH for methanogenesis is between 6.8 and 7.4; adding the bicarbonate alkalinity to the operating system is important to buffer the system.

It can be concluded that the first phases of AD may occur at a wide range of pH values, but the methanogenesis phase will only proceed at neutral pH, therefore, it is expected to get lower CH_4 production rates when the pH values are outside the range of 6.5–7.5 (De Mes et al. 2003). Hahn and Hoffstede (2010) demonstrated that in the one-step AD process, the pH is usually maintained at the optimal pH for methanogenic bacteria to prevent the dominance of acidogenic bacteria, which can cause the accumulation of organic acids and digester instability. Ward et al. (2008) reported that controlling the pH range between 6.8 and 7.2 is most suitable for the overall AD processes.

5.4 Substrate

The substrate that provides the carbon source for the bacterial activities may vary in type and complexity. The OFMSW can be classified as follows (Tchobanoglous et al. 1993)

- Water-soluble constituents, such as sugars, starches, amino acids, in addition to other organic acids,
- Hemicellulose,
- Cellulose,
- Fats, lipids, and oils, which are long chain fatty acids and esters of alcohols,
- Lignin which is a polymeric material containing aromatic rings with methoxyl groups ($-OCH_3$) , present in some paper products,
- Lignocellulose, a combination of lignin and cellulose,
- Proteins, which are composed of chains of amino acids.

The most suitable substrate can be characterized for carbohydrate, lipid, protein and fiber contents (Khalid et al. 2011). They stated that carbohydrates are the most important organic source of the MSW for biogas production, and starch is an effective cheap substrate compared to sucrose and glucose. The degradation of carbohydrates provides around 600–700 L of biogas per kg organic dry matter with 50% or more CH_4 content (Hahn and Hoffstede 2010).

Fat is an excellent substrate and can produce the highest CH_4 yield; however, it is hardly soluble in water and can be difficult to blend into the digestion process (Hahn and Hoffstede 2010). From this perspective, the pre-treatment of this substrate (mainly heating) may allow the fat to mix with the other substrate and become more readily accessible to the bacteria. Hahn and Hoffstede observed that the degradation of fat provides around 1,000–1,250 L of biogas per kg organic dry matter with 68–73% CH_4 content. The same authors explained that problems can occur when using proteins as a substrate due to the nitrogen and sulphur content. Nitrogen is released in form of ammonia when a protein is degraded, which will cause the pH to rise. Also sulphur is released in form of hydrogen sulphide. Hahn and Hoffstede (2010) demonstrated that in the integrated process and during the degradation of proteins, part of CO_2 produced in the biogas dissolves in the liquid phase and forms hydrogen carbonate which buffers the ammonia released. Hahn and Hoffstede stated that the degradation of protein yields around 600–700 L of biogas per kg organic dry matter with the highest CH_4 content of 70–75%.

5.5 Nutrients

In addition to a carbon source, bacteria need other elements as nutrients for their cellular metabolism. The available level of nutrients affects significantly the AD and consequently the biogas production. To improve the nutrients levels and the available C/N ratios; co-digestion of different organic mixtures was applied by several researchers. The recommended carbon/nitrogen ratio for the anaerobic bacteria is 40. Khalid et al. (2011) reported that nutrients levels of the elements carbon:nitrogen:phosphorus:sulphur (C:N:P:S) at 600:15:5:3 are adequate for the methanogenesis. If the nitrogen concentration is high, ammonia will accumulate in the system and will cause inhibition of the biological process and will raise the pH (above 8.5). If the carbon concentration is high, this will cause rapid consumption of nitrogen by methanogenic bacteria and consequently this will lower the biogas production. Inhibition of AD of waste with a high organic content is usually also caused by high ammonia concentrations produced by the degradation of proteins from nitrogen-rich waste (Cuetos et al. 2008).

Generally, the bacteria need balanced macro-nutrients (such as carbon, nitrogen, phosphorus, potassium, and sulfur) and micro-nutrients such as (cobalt, iron, copper, nickel, zinc and others) (Wilkinson 2011). The macro-nutrients are essential as cell building material, for microbial synthesis, and as primary source of energy. Most of the micro-nutrients are required in the enzymes that play important roles in the microbial activities.

There are some types of waste, such as the olive mill waste, that have special toxic characteristics such as: acid pH, low alkalinity, and low nitrogen content. Sampaio et al. (2011) reported that several synthetic nutrients and additions of chemicals and pretreatment may be included in the AD process in order to overcome these problems and enable the AD of such waste. However, these pre-treatments involve inputs which will raise the cost-benefit ratio and may lead to organic load reduction and alternatively reduce the biogas production.

5.6 *Hydraulic and Solid Retention Times*

The retention time represents the time the feedstock remains in the digester and it is important for the bacterial growth. There are two measures of retention time: hydraulic retention time (HRT) and solid retention time (SRT). The HRT – also known as the residence time- is the theoretical period of time that the material spends in the digester, and is calculated based on the volumetric loading rate at which a digester is operated. The HRT is important for bacterial growth because the material has to remain in the digester for a sufficient time so that the microorganisms have enough time to grow and multiply. Wilkinson (2011) stated that the HRT will differ for each substrate, with a range of 14–30 days for dry processes, and for wet processes HRT can be around 3 days. The HRT is the most important factor in determining the volume of the digester which in turn determines the cost of the plant (Abu-Hamatteh et al. 2010).

The HRT can be calculated as follows:

$$HRT(day) = \frac{Digester\ Volume\ (m^3)}{Feed\ Stock\ Mass\ Flow\ (m^3\ /\ day)} \tag{12.7}$$

The SRT is the amount of time solid material is retained in the digester. It is also a measurement of the concentration of bacteria maintained within the system with time. High STR indicates that larger populations of biomass are retained within the system, while low levels of SRT indicate that bacterial growth cannot efficiently replace the bacteria lost with the effluent. Hahn and Hoffstede (2010) indicated that the process where bacterial dilution is faster than its growth is called "wash out".

The organic loading rate is an important design parameter and describes the amount of degradable substrate or volatile solids that enter the system over a period of time. Higher loading rate means greater amount of degradable substrate per unit of time, which if too high can cause stress in the digester due to increased bacterial activities and consequently acid production, decrease in pH and harm effects to the methanogenic bacteria.

6 Advantages of Anaerobic Digestion

In addition to the biogas production as energy source and the transformation of organic waste into high quality fertilizer; the waste treatment option through the anaerobic digestion of the OFMSW has many advantages that added to the popularity of this option, such as the low energy required, small reactor volume, fewer nutrients required, as well as improvement of hygienic conditions through reduction of pathogens.

However, and as for the disadvantages; the processes may have longer start-up and retention time compared to other biological processes, and it may also be sensitive to the adverse effect of lower temperatures on the bacterial kinetics. Moreover, the chemical composition and structure of lignocellulose materials may hinder the rate of biodegradation of the OFMSW (Khalid et al. 2011). Consequently, the compost that results from the biological degradation of the organic matter may need further aerobic treatment to be safely used as compost.

6.1 Biogas

Traditionally, the AD has been employed to stabilize wastewater sludge and agricultural waste. However, due to the global pressure to shift towards renewable and alternative energy sources, the AD has become not only a way to treat the waste, but also to recover the useable energy. The feedstock and biomass have extended to include MSW and many types of waste that contain high percentages of organic content. Recent attention has been given to the AD of the OFMSW for the production of biogas. The various types of waste streams that can be digested with special intention to recover the energy may include: domestic wastes such as vegetables, fruits, and yard wastes; agricultural wastes; manure; and non-hazardous industrial and domestic wastewater.

The biogas generated from the anaerobic digestion processes is generally composed of 48–75% CH_4, 30–45% CO_2, and traces of other gases (Hahn and Hoffstede 2010; Ward et al. 2008). The energy content depends mainly on the CH_4 content and is between 5.0 and 6.5 kWh/m^3 raw biogas, and ignition temperature is about 650–750°C (Hahn and Hoffstede 2010). De Mes et al. (2003) reported that the fuel value of the biogas that contains 55–75% (v/v) CH_4 ranges between 22 and 30 MJ/Nm^3 in its higher heating value, and 19–26 MJ/Nm^3 in its lower heating value.

Several reports indicate that the AD of OFMSW yields promising amounts of biogas (Khalid et al. 2011). The latter authors indicated that the biogas yield may vary from 200 to 850 L of CH_4/kg of volatile solids, by using several types of organic waste such as MSW, fruit and vegetable waste, manure, and oil mill waste. It is worth mentioning that the lowest yield values were obtained from the MSW and lignin rich organic waste, and the higher yield values from oil mill waste. Figure 12.4 illustrates the different types of organic waste and the biogas yield as reported by Abu-Hamatteh et al. (2010).

Zhou et al. (2011) indicated that the energy production by means of AD of the municipal and agricultural wastes globally could reach up to 200 billion kWh/year. Crolla et al. (2007) demonstrated the results of the team's research to evaluate the production of biogas and electricity from anaerobic digestion of dairy manure in

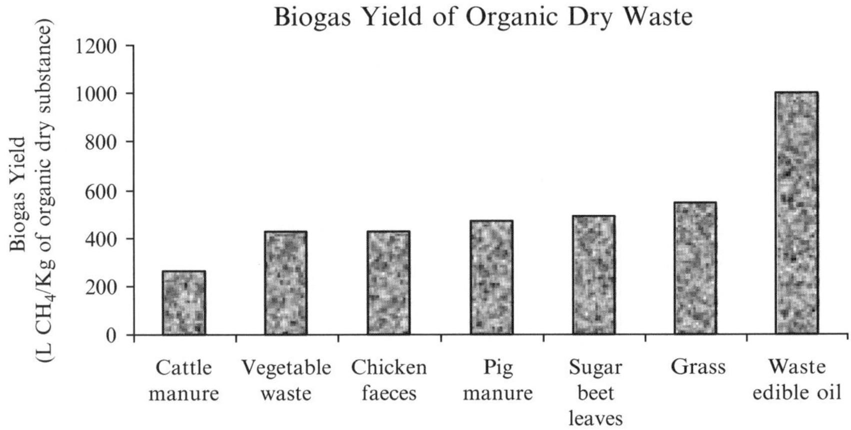

Fig. 12.4 Biogas yield of different organic dry waste (Adapted from Abu-Hamatteh et al. 2010)

two digesters built in two farms (160 and 105 dairy cows respectively) in Ontario, Canada and operating under mesophilic conditions. The authors reported that the biogas produced (419 and 357 m^3/day respectively) was used for electricity generation, heat production to heat the digesters and the farm's utilities around, and for electricity selling to the grid. They confirmed that the odors were reduced significantly as they observed 95% reduction of volatile fatty acids (VFAs). The results also showed 70–95% reduction in pathogens. The data confirmed the effect of the digestion processes to reduce greenhouse gases from the waste, since the emissions from storage without digestion were reduced from 245 to 56 kg CH_4/head/year from storage with digestion. Energy recovery by means of anaerobic digestion has some benefits in most of the categories; the anaerobic digestion showed saved potential impacts in global warming category. Benefits in acidification and nutrient enrichment categories are mainly due to saved NOx emissions from avoiding coal-based electricity production.

The biogas yield is not only affected by the type and composition of waste used as the substrate, but is also affected by several factors, including the microbial composition, temperature, moisture and bioreactor design. Eskicioglu et al. (2011) studied the AD of dry stillage since this biomass is rich in nutrients such as fiber, protein, lipids and starch. Their study showed that anaerobic digestion processes of the corn stillage produced 88 ± 8 L of CH_4/L of stillage from mesophilic digestion, and 96 ± 19 L of CH_4/L of stillage from thermophilic digestion, and at the end of the experiment the organic removal efficiencies were similar. This observation was explained by the fact that this substrate has a highly biodegradable nature and the percentage of organic matter is 93% of the solid waste. In addition the results of these experiments showed that the thermophilic digestion showed slightly higher biogas yield than the mesophilic digestion.

6.2 *Water Recycle*

The re-use of the liquid effluents generated from the AD digestion process or the digestate has significant economic benefits. If the MSW is subjected to source separation, then the digestate from the digestion process of the organic fraction can comply with the governing quality standards. De Mes et al. (2003) indicated that the digestate resulting from the thermophilic digestion may be easily used as fertilizer without much treatment, and in this case the digestate is a bio-fertilizer that can be distributed through the network of local farmers.

It should be mentioned that all the remaining liquid effluents resulting from the AD process must be disposed off. Therefore, the discharge to the wastewater treatment plants will involve considerable costs for transport and treatment charges depending on the effluent quality. From this perspective, the on-site reuse systems may be advantageous, such as the use of water from the digestion process for substrate dilution to reduce fresh water consumption. In addition, Alkan-Ozkaynak and Karthikeyan (2011) illustrated that the digestate from anaerobic treatment of thin stillage would be of suitable quality for recycling as process water. The digestate resulting from the AD of thin stillage may offset part of fresh water requirement for corn-ethanol production.

6.3 *Combined High Solids Anaerobic Digestion/Aerobic Composting*

The recovery of the nutrients-rich compost to be used as soil conditioner or fertilizer depends on the compliance with the governing quality standards and mainly the concentration of the heavy metals and pathogens (De Mes et al. 2003). Therefore, the combined (high solids anaerobic digestion/aerobic composting) process is one of the promising AD technologies. Tchobanoglous et al. (1993) illustrated the high solids AD/aerobic composting process as a two stage process according to Fig. 12.5 below.

The first stage (according to Tchobanoglous et al. 1993) of the two stage process involves the AD of the high solids, which will degrade 25–30% of the OFMSW to produce biogas. The AD operates under thermophilic conditions (54–59°C) and HRT

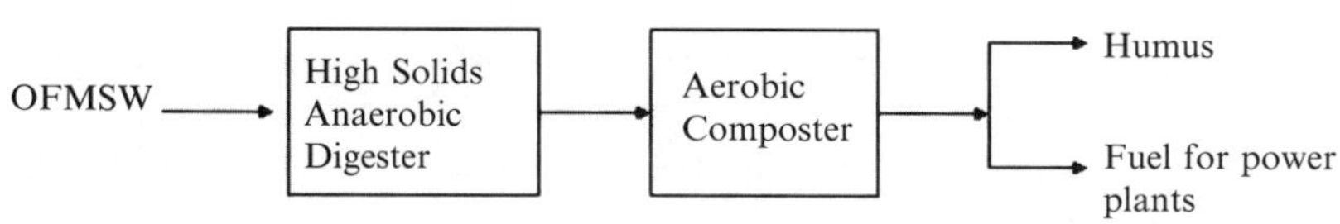

Fig. 12.5 Flow diagram for high-solids anaerobic digestion/aerobic composting process (Adapted from Tchobanoglous et al. 1993)

of 30 days. The second stage involves the aerobic composting of the anaerobically digested solids to increase the solids content from 25 to 65% and more. The output is fine humus like material that can be used as soil amendment, with specific weight of about 35 lb/ft^3 (560.65 kg/m^3), and thermal content of about 6,000–6,400 Btu/lb.

Anaerobic digestion preserves the nutrient content of the manure, so that it can be land applied as a fertilizer. Anaerobic digestion also greatly reduces the pathogen content of the manure, and greatly reduces the odor.

6.4 Weight and Volume Reductions of Waste

Macias-Corral et al. (2008) reported that the AD of OFMSW showed weight and volume reductions. The AD of cow manure showed mean weight and volume reductions of 43 and 20% respectively. The same authors also stated the co-digestion had improved not only the biogas yield, but also the weight and volume reductions of the waste; the co-digestion of the OFMSW and cow manure showed weight and volume reductions of 78 and 98% respectively, and the co-digestion of the cotton gin waste and cow manure showed weight and volume reductions of 52 and 58% respectively.

7 Optimization of Anaerobic Digestion Processes

Previously, the AD process has been extensively used for sewage sludge treatment. Nowadays, there is a great interest worldwide to use this process for biogas production from different organic substrates, due to the increasing concerns on energy from renewable and alternative resources. The biological treatment process aims to create optimal degradation conditions, accelerate the AD process and reduce the process time, reduce the volume of digestate and to enhance the biogas production. From this perspective; co-digestion and pre-treatment of waste were introduced.

7.1 Co-digestion

Co-digestion is a waste treatment method where different types of waste are mixed and treated together. It is used to improve the yield of the AD of the organic wastes. The benefits of co-digestion are dilution of toxic compounds, improved balance of nutrients needed for bacteria, enhancement of the biodegradation of the organic fraction of the waste, stabilization of the feed and improved nitrogen to carbon ratio. Also, it was confirmed that the co-substrate addition effectively decreased the lag phase of the organic biodegradation which explains the increase in CH_4 production

(Li et al. 2011). All of these processes help improve the biogas yields resulting from the biodegradation processes.

Several researchers conducted laboratory scale experiments for co-digestion of OFMSW and animal manure or sludge. Many reported that the co-digestion of two wastes can reduce inhibition of methanogenesis and increase CH_4 yields (Macias-Corral et al. 2008). The latter authors confirmed that the co-digestion of OFMSW and other agricultural waste with cow manure utilized intrinsic cellulose degrading bacteria, and the additional nutrients in the manure optimized the digestion of the fiber in the agricultural waste as well as the paper fraction of the MSW.

Table 12.1 demonstrates the CH_4 yield from the co-digestion of different types of solid waste.

As illustrated in Table 12.1, the co-digestion of different types of solid waste enhanced the biogas production by a range of 40–250% of the biogas generated from the biodegradation processes of a single type of waste.

Zhu and Yao (2011) proposed a logistics model for the biomass-to-bioenergy industry with multiple types of biomass feed stocks. The numerical study confirmed that the use of multiple types of biomass may have several advantages, including increase of the supply of biomass, smoothing the biogas production, and consequently increasing the unit profit of the generated biogas.

It is concluded that the co-digestion can affect positively the digestion process due to the clear synergistic effect which overcomes the imbalance in nutrients, reduces inhibition, and improves biodegradation of the organic fraction of the waste.

7.2 Pretreatment

Khalid et al. (2011) have indicated that the hydrolysis of the complex organic matter to soluble compounds is the rate-limiting step of anaerobic processes for wastes with a high solid content; therefore, various physical, chemical and enzymatic pretreatments may be required to increase substrate solubility and accelerate the biodegradation rate of OFMSW. The pretreatment methods include mechanical treatment, ultrasound, chemical treatment, thermal hydrolysis and thermo-chemical pretreatment in addition to microwave pretreatment (Shahriari et al. 2011). The latter authors reported that all of the reported methods have increased biodegradability of biomass and other organic solid wastes with varying degrees of success. Eskicioglu et al. (2009) explained that these methods could disrupt the extracellular polymeric substances (EPS) and divalent cation network and increase the extent of waste activated sludge biodegradability through an enhanced hydrolysis phase.

Previous studies reported that the pretreatment of waste activated sludge can enhance the CH_4 production for both mesophilic AD by 43–145%, and thermophilic AD by 4–58% (Saha et al. 2011). It was noted by Saha and coworkers that the thermophilic digestion is usually more efficient at volatile solids reductions and CH_4 production than the mesophilic, which explains the reduced benefits of pretreatment in thermophilic systems.

Table 12.1 Biogas yield from the co-digestion of different types of solid waste

Substrate	Co-substrate	CH_4 yield (L of biogas/kg of total VS)	Percentage of improvement	Reference
Waste activated sludge	–	117 ± 2.02		Li et al. (2011)
Waste activated sludge	Fat, oil and grease	418 ± 13.7	+257%	Li et al. (2011)
Waste activated sludge	Kitchen waste	324 ± 4.11	+177%.	Li et al. (2011)
Cow manure	–	62		Macias-Corral et al. (2008)
OFMSW	–	37		Macias-Corral et al. (2008)
Cow manure	OFMSW	173	+179%	Macias-Corral et al. (2008)
Cow manure	Cotton gin waste	87	+40%	Macias-Corral et al. (2008)
83% Olive mill waste	17% Piggery effluent	340		Sampaio et al. (2011)
Cattle excreta	Olive mill waste	179	+337	adapted from Khalid et al. (2011), and Goberna et al. (2010)
Fruit and vegetable waste	Abattoir wastewater	611	+51.5%	adapted from Khalid et al. (2011), and Bouallagui et al. (2009)
Potato waste	Sugar beet waste	680	+62%	adapted from Khalid et al. (2011), and Parawira et al. (2004)
Slaughter house waste	MSW	500	+100%	Cuetos et al. (2008)

The positive effects of different pre-treatment processes on the bacterial digestion of the OFMSW may vary. Saha et al. (2011) studied the effects of different pretreatment methods that were applied to pulp mill waste sludge to enhance CH_4 production and reduce digester sludge retention time. Their experiments showed that microwave pretreatment proved to be the most effective in terms of biodegradation rate, since this pretreatment method increased the specific CH_4 yield by 90% compared to samples without treatment in mesophilic digestion processes. They confirmed that although the chemo-mechanical pretreatment was the least method to enhance the CH_4 yield, this method was the least energy intensive for both mesophilic and thermophilic digestion of the waste activated sludge. As an example, they showed that the input energy of the chemo-mechanical pretreatment in mesophilic digestion was 1,111 kWh/t of total solids and the CH_4 output was 1,644 kWh/t of total solids. For the ultrasound pretreatment in the same conditions (ultrasound for 90 min), the input energy was 32,700 kWh/t of total solids and the CH_4 output was 1,780 kWh/t of total solids. As for the microwave pretreatment (microwave temperature 175°C); the input energy was 218,000 kWh/t of total solids and the CH_4 output was 2,206 kWh/t of total solids. These results lead to the conclusion that neither the microwave nor the ultrasound treatment were energy economical due to much smaller CH_4 yields compared to the high energy inputs.

Park and Ahnn (2011) investigated the effect of thermal and microwave pretreatments on the AD of mixtures of primary municipal and secondary municipal sludge in mesophilic digesters at different retention times. Their results showed that the biogas production was 33% higher when the samples were thermally pretreated and 53% when the samples were microwaved. Also, the results showed that the microwave pretreatment is more effective than thermal pretreatment in increasing the solublization degree and mesophilic anaerobic biodegradation of the different sewage sludge.

Several studies (Marin et al. 2011; Saha et al. 2011; Park and Ahnn 2011) have reported that the microwave treatment is a promising pretreatment option to accelerate the hydrolysis in anaerobic digestion processes. However, Solyom et al. (2011) concluded that in all microwave treatment experiments, the absorbed energy shall be reported as an important operating variable, which may affect significantly the process operating conditions.

8 Bioreactors and Digesters Types

There are many ways in which the AD can occur. The modern advancements in the design of bioreactors improved the use of AD techniques for the treatment of OFMSW at a much higher rate than the conventional landfilling processes (Khalid et al. 2011). There are different types of bioreactor designs that have been developed in the past decades targeting the optimization of the reaction rates of the biological processes. The operational concerns associated with the anaerobic processes may include long start-up time, sensitivity to possible toxic compounds present in the waste, operational stability, and corrosiveness of the generated gas (Metcalf et al. 2004). The AD bio-processes are almost the same; however, they may vary depending on: the composition of the substrate, the volume of the waste stream, the complexity of design as well as the construction and operation. The main differences concerning the design of the different digesters vary according to the operating parameters, such as: the biogas potential of the substrate, the dry and wet processes, the mesophilic and thermophilic processes, the mechanically mixed and no-mixed reactors, the batch and continuous reactors and one-stage or multi-stage processes.

Khalid et al. (2011) stated that there are three main groups of bioreactors, and as follows:

- Batch reactors: are considered the simplest in design. The digestion processes are quick, simple and cheap. The operations are based on filling the reactors with the feed stock that is kept inside the reactor for a designed retention time. Yet, the limitations of this main type of reactors may include; high fluctuations in gas quantity and quality, looses in quantities during emptying, as well as limitation in the designed height.
- One stage continuously fed system: in this type all the biochemical reactions take place in one bioreactor.
- Two-stage or multi-stage continuously fed systems: the biochemical processes that include hydrolysis, acidification, acetogenesis and methanogenesis, take place separately. In this context, Shahriari (2011) explained that the multi-stage or multi phase AD process is based on the principle of optimizing the conditions for the individual biological reactions of AD, with the hydrolytic bacteria in one reactor and the methanogens in another reactor in order to increase gas production and waste stabilization rates. The system is based on separation of the bacterial consortia based on hydraulic separation with a short residence time in the acid phase digester followed by a longer residence time in the methanogenic phase reactor. The two stage system is the most promising and optimized process for the biological treatment of the OFMSW as well as the biogas production, due to the fact that it allows the selection and enrichment of different bacteria involved in each phase. Macias-Corral et al. (2008) reported that their experiments showed that the biogas in two-phase AD had a higher CH_4 content (72% or more) than the conventional single-phase systems, which usually produce a gas that contains 60% CH_4.

The main three types of bioreactors mentioned above in addition to a variety of methanizers such as continuously stirred tank rectors (CSTR), tubular bioreactor, anaerobic sequencing batch reactor, up flow anaerobic sludge blanket (UASB) and anaerobic filters have been used successfully for the biological treatment of the OFMSW (Khalid et al. 2011).

Hahn and Hoffstede (2010) and De Mes et al. (2003) illustrated digester types used in a wide range of different substrates as follows:

- High and low rates systems: the low rate systems are usually operated with long HRT and are usually used to treat slurry and solid wastes that necessitate a long time for appropriate anaerobic biodegradation, while high rate systems are operated with short HRT and are used for wastewater. The high-rate systems have a mechanism either to retrain bacterial sludge mass in the digester or to separate bacterial sludge from the effluent and return it to the digester. The high rate systems are recognized to be cost effective and efficient. High rate systems can be batch, plug flow and CSTR, while the low rate systems can be fluidized bed and UASB.

– Wet and dry fermentation systems: The wet fermentation systems are usually used to digest low solid content and the digester type is usually the CSTR. The dry fermentation systems are usually used to digest solid feed stock with higher solid content (30% and more) and usually require less process water and have lower heating costs. In this context, Khalid et al. (2011) demonstrated that the wet bioreactors may contain 16% or less total solids, while the dry bioreactors may contain 22–40% total solids.

De Mes et al. (2003) demonstrated that the wet digestion has been carried out in a number of commercial and pilot scale plants, such as: AVECON in Finnland; VAGRON in the Netherlands; and Bigadan process in Denmark and Sweden. The dry fermentation systems have been developed and used in a number of commercial and pilot scale plants around the world, namely: the Valorga process in France; the Dranco or Dry Anaerobic Composting system in Belgium; the Komogas process in Switzerland; and the Biocel system in the Netherlands.

– Single and double stage systems: Single stage reactors use one reactor for both the acidogenic phase and the methanogenic reactions of AD. The single stage AD processes are currently used at the majority of the existing facilities due to simplicity and lower costs. It is based on the fact that the digestion process can basically be divided into the main steps hydrolysis and methanogenesis, and these processes occur in the same digester. In order to ensure successful digestion the pH value has to be kept around neutral.

The two- and multi-stage AD process is based on the principle of optimizing the conditions for the individual biological reactions of AD, with the hydrolytic bacteria in one reactor and the methanogens in another reactor in order to increase gas production and waste stabilization rates. De Mes et al. (2003) indicated that there are two types of two-phase digestion systems; the different stages are separated based on a wet fermentation in the first one, and separated based on a dry fermentation in the second one. The two or multi-stage digestion systems have been developed and used in a number of commercial and pilot scale plants around the world, namely: the BTA process in Germany which is a three-phase liquid system for digestion of the OSMSW; and the BRV process in Switzerland which is based on aerobic/anaerobic conversion systems.

Mesophilic and thermophilic systems: The mesophilic AD processes require longer HRT; however they are not efficient in killing pathogens. Thermophilic AD provides higher growth rates of the bacteria involved and higher loading rates of organic materials as well. These processes are adapted usually for AD of industrial organic wastes, manure and sewage sludge.

9 Economy of Anaerobic Digestion

The AD of the OFMSW has a very important role as an alternative source of energy, and may help to face the crucial energy challenges in the coming years. Fox and van Kalles (2010) estimated that the biodegradable waste in the United Kingdom is around 100 million tonnes, all of which is suitable as a feedstock for the AD. They added that if this waste is anaerobically digested, the potential renewable energy generation could be 10–20 Tera Watt per hour (TWh) of heat and electricity, and this renewable energy amount could form up to around 10% of the energy demand for the country. Unlike the fossil fuel or coal; the biogas generated does not contribute either to the greenhouse gas effect or to the ozone depletion.

Under the right conditions a biogas plant will yield several benefits for the end-users, the main benefits are (Bond and Templeton 2011; Balasubramaniyam et al. 2008; De Mes et al. 2003):

1. Production of energy through CH_4 recovery for lighting, heat, and electricity,
2. Improved sanitation through reduction of pathogens, worm eggs and flies, in addition to raw waste stabilization,
3. Environmental benefits through the reduction of greenhouse gas emissions from the degradation of the OFMSW in landfills and open dumps,
4. Retention of the fertilizer nutrients,
5. The AD processes usually have little energy needs or requirements. De Mes at al. (2003) reported that the energy requirements for the processes are in the range of 0.05–0.1 kWH/m^3, depending on the need for pumping and recycling effluents,
6. Economic benefits (a substitute to spending on expensive fuels and fertilizers), in addition to relatively low construction costs.

The AD of OFMSW has become a well-established technology mostly in Europe. Boldrin et al. (2011) reported that as the biological treatment is becoming a common option for management of the OFMSW, there are around 2,000 composting facilities in Europe, as well as 185 AD plants. Many technological developments have taken place recently in the biogas sector; nevertheless there is still great potential for technical developments and improvements to solve the pending issues that affect the optimization of the biological processes.

Fox and van Kalles (2010) and De Mes et al. (2003) stated that although the potential benefits of the AD are evident, this technology has some disadvantages and suffered a lack of wide spread implementation due to the following reasons: the variability of substrates which affects the process stability; the high sensitivity of methanogenic bacteria that produce the biogas to a large number of chemical compounds that may be toxic; the added time and costs associated with the source separation and collection systems of the OFMSW; the lack of understanding of the technology options and methods to optimize the biogas yield; the lack of confidence in the quality of digestate for market use; and finally the inference of multiple governmental associations where each have their own goals and ideas for end product use.

In order to balance the advantages and disadvantages of the AD process for any industry or community, life cycle assessment (LCA) is a useful systematic approach and viable standardized methodology for reporting and assessing the effects associated with the biological treatment (Boldrin et al. 2011). Boldrin stated that a complete LCA study is iteratively carried out through four phases: goal and scope definition, life cycle inventory and impact. In order to assess the life cycle impacts of AD technology, it is important to consider: the emissions needed to transport the waste and/or to manage the system; the emissions resulting or mitigated from the treatment; the use of the outputs; and the emissions mitigated as a result of displacement from other sources (Fox and van Kalles 2010).

De Mes et al. (2003) indicated that the costs for electricity produced from biogas in the year 2000 were between 0.1 and 0.22 Euro/kWh, while the electricity from landfill gas was produced for 0.04–0.07 Euro/kWh. Despite this cost difference, financial incentives are encouraged for enhancing the competitiveness of the AD process and the biogas plants, and will have a positive impact on the expansion of the AD processes.

10 Conclusions

The concept of anaerobic biodegradation of OFMSW as engineered systems for solid waste treatment and management is viable and may extend the biogas production to its maximum potential. It is concluded that the AD has become increasingly popular in recent years as a sustainable technology producing clean energy and green byproducts. In the developing countries, the AD of OFMSW shall be encouraged as a promising option for the treatment of the MSW to face the elevated challenges of solid waste management. The biogas yield varies with the types of the substrates/feedstock available and the process conditions. It was reported by several studies that the OFMSW's biogas yield may vary between 80 and 200 m^3/t of waste. Co-digestion is a significant factor for improving the reactors efficiencies which will result in economic feasibility of the AD systems. The AD process can be engineered to optimize the biogas yield by controlling the operating conditions such as temperature, moisture content, bacterial activities and waste properties. It is worth noting that the different government's policies and regulations are vital to encourage the investment in the AD process which contributes to the alternative energy production. The recent developments of the AD techniques have shown that increasingly more feed stocks can be used in this process, and then, the AD technique can be used as an effective global treatment option for many types of generated wastes. The AD of the OFMSW can be described as relatively new in many countries and its application is still in the initial phase. In this context, financial and legislative incentives are key elements for economic feasibility of the biogas digesters and plants.

The optimized engineering designs that will assure lower operating and maintenance costs and efforts, improved efficiencies, and increased functionality will definitely promote the AD as an attractive solid waste treatment option and viable technique to generate a sustainable alternative source of energy.

References

Abu-Hamatteh ZSH, Al-Jufout S, Abbassi B, Besieso MS (2010) Biogas energy: unexplored source of a renewable energy in Jordan. European Association for the Development of Renewable Energies, Environment and Power Quality (EA4EPQ). In: International conference on renewable energies and power quality (ICREPQ'10) Granada (Spain), The European Association for the Development of Renewable Energies, 23–25 Mar 2010

Albanna M (2011) Solid waste management options and their impacts on climate change and human health. In: Malik A, Grohmann E (eds) Environmental protection strategies for sustainable development. Springer Dordrecht Heidelberg London New York, pp 498–524

Alkan-Ozkaynak A, Karthikeyan KG (2011) Anaerobic digestion of thin stillage for energy recovery and water reuse in corn-ethanol plants. Bioresour Technol 102:9891–9896

Balasubramaniyam U, Zisengwe L.S, Meriggi N, Buysman E, Zeeman G (2008) Biogas production in climates with long cold winters. Wageningen University, Wageningen. http://www.susana.org/docs_ccbk/susana_download/2-1502-biogascoldclimatesweb-wecf0608.pdf. Cited 31 May 2012

Boldrin A, Neidel TL, Damgaard A, Bhande G, Moller J, Christensen TH (2011) Modelling of environmental impacts for biological treatment of organic municipal waste in EASEWASTE. Waste Manag 31:619–630

Bond T, Templeton M (2011) History and future of domestic biogas plants in the developing world. Energy Sustain Dev 15:347–354

Bouallagui H, Lahdheb H, Romdan E, Rachdi B, Hamdi M (2009) Improvement of fruit and vegetable waste anaerobic digestion performance and stability with co-substrates addition. J Environ Manag 90:1844–1849

Chalmin P, Gaillochet C (2009) From waste to resource. http://www.uncrd.or.jp/env/spc/docs/plenary3/PS3-F-Veolia_Hierso-Print%20abstract.pdf. Cited on 31 May 2012

Crolla A, Kinsley C, Desilets E, Pattey E, Desjardins R, Ball-Coelho B, Sahota T (2007) Assessment of the production and land application of anaerobically digested manure from medium sized livestocks farms. In: Growing the margins: energy conservation and generation for farms and food processes, London Convention Center, London, Ontario, Canada, 11–14 April 2007

Cuetos MJ, G'omez X, Otero M, Mor'an A (2008) Anaerobic digestion of solid slaughterhouse waste (SHW) at laboratory scale: influence of co-digestion with the organic fraction of municipal solid waste (OFMSW). Biochem Eng J 40:99–106

Davis M (2011) Water and wastewater engineering- design principles and practices. McGraw-Hill Inc. International Edition, New York. ISBN 978-007-128924-5

De Mes TZD, Stams AJM, Reith JH, Zeeman G (2003) Methane production by anaerobic digestion of wastewater and solid wastes. In: Reith JH, Wijffels RH, Barten H (eds) Bio-methane and bio-hydrogen- status and perspectives of biological methane and hydrogen production. Dutch Biological Hydrogen Foundation, Netherlands. ISBN 90-9017165-7

Eskicioglu C, Prorot A, Marin J, Droste RL, Kennedy KJ (2008) Synergetic pretreatment of sewage sludge by microwave irradiation in presence of H_2O_2 for enhanced anaerobic digestion. Water Res 42:4674–4682

Eskicioglu C, Kennedy KJ, Droste RL (2009) Enhanced disinfection and methane production from sewage sludge by microwave irradiation. Desalination 248:279–285

Eskicioglu C, Kennedy K, Marin J, Strehler B (2011) Anaerobic digestion of whole stillage from dry-grind corn ethanol plant under mesophilic and thermophilic conditions. Bioresour Technol 102:1079–1086

EU- European Commission (2001) Waste management options and climate change. Office for official publications of the European Communities, Luxembourg, ISBM 92-894-1733-1. http://ec.europa.eu/environment/waste/studies/pdf/climate_change.pdf. Cited 31 May 2012

Eurostat European Commission (2012) Municipal waste statistics. http://epp.eurostat.ec.europa.eu/statistics_explained/index.php/Municipal_waste_statistics. Cited 10 June 2012

Fox N, van Kalles M (2010) Anaerobic digestion of municipal food waste on Merseyside: a practical assessment. Masters thesis, Cranfield University

Goberna M, Schoen MA, Sperl D, Wett W, Insam H (2010) Mesophilic and thermophilic co-fermentation of cattle excreta and olive mill wastes in pilot anaerobic digesters. Biomass Bioenergy 34:340–346

Hahn H, Hoffstede U (2010) Assessment report on operational experience. Project supported by the European commsion under RTD contract: 019795. http://www.biogasmax.eu/media/d2_11_biogasmax_iwes_vfinal_nov2010__095398400_1109_10022011.pdf. Cited on 31 May 2012

Khalid A, Arshad M, Anjum M, Mahmood T, Dawson L (2011) The anaerobic digestion of solid organic waste. Waste Manag 31:1737–1744

Lardinois I, van de Klundert A (1993) Organic waste options for small-scale resource recovery, Urban solid waste series 1. TOOL, Amsterdam and Waste Consultants, Gouda. ISBN 90-70857-33-2

Li C, Champagne P, Anderson B (2011) Evaluating and modeling biogas production from municipal fat, oil, and grease and synthetic kitchen waste in anaerobic co-digestions. Bioresour Technol 102:9471–9480

Macias-Corral M, Samani Z, Hanson A, Smith G, Funk P, Yu H, Longworth J (2008) Anaerobic digestion of municipal solid waste and agricultural waste and the effect of co-digestion with dairy cow manure. Bioresour Technol 99:8288–8293

Marayyan B (2004) Methane gas emissions from landfill: opportunities and constraints – the case of Al-Russifa city. J Environ Assess Policy Manag 6(3):367–384

Marin J, Kennedy KJ, Eskicioglu C (2011) Enhanced solubilization and anaerobic biodegradability of source-separated kitchen waste by microwave pre-treatment. Waste Manag Res 29(2):208–218

Metcalf M, Eddy Metcalf & Eddy, Inc., Tchobanoglous G, Burton F, Stensel HD (2004) Water and wastewater engineering-design principles and practices. McGraw-Hill International Edition, New York, ISBN 978-007-128924-5

Narayana H (2009) Municipal solid waste management in India: from waste disposal to recovery of resources? Waste Manag 29:1163–1166

Nwabanne JT, Onukwuli OD, Ifeakandu CM (2009) Bio-kinetics of anaerobic digestion of municipal waste. Int J Environ Resour 3(4):511–516

EIONET – European Environmental Information and Observation Network (2009) What is waste?. http://scp.eionet.europa.eu/themes/waste/#10. Cited 31 May 2012

Parawira W, Murto M, Zvauya R, Mattiasson B (2004) Anaerobic batch digestion of solid potato waste alone and in combination with sugar beet leaves. Renew Energy 29:1811–1823

Park WJ, Ahnn JH (2011) Effects of microwave pretreatment on mesophilic anaerobic digestion for mixture of primary and secondary sludges compared with thermal pretreatment. Environ Eng Res 16(2):103–109

Saha M, Eskicioglu C, Marin J (2011) Microwave, ultrasonic and chemo-mechanical pretreatments for enhancing methane potential of pulp mill wastewater treatment sludge. Bioresour Technol 102:7815–7826

Sampaio MA, Goncalves MR, Marques IP (2011) Anaerobic digestion challenges of raw olive mill wastewater. Bioresour Technol 102:10810–10818

Shahriari H (2011) Enhancement of anaerobic digestion of organic fraction of municipal solid waste by microwave pretreatment. Ph.D. of applied sciences in environmental engineering, University of Ottawa, Canada

Shahriari H, Warith M, Kennedy K (2011) Effect of microwave temperature, intensity and moisture content on solubilisation of organic fraction of municipal solid waste. Int J Environ Technol Manag 14:67–83

Shanghai Manual (2011) A guide for sustainable urban development of 21sr century, United Nations Bureau International des Expositions, Shanghai 2010 World Exposition Executive Committee

Solyom K, Mato R, Perez-Elvira SI, Cocero MJ (2011) The influence of the energy absorbed from microwave pretreatment on biogas production from secondary wastewater sludge. Bioresour Technol 102:10849–10854

Strange K (2002) Overview of waste management options; their efficacy and acceptability. In: Hester RE, Harrison RM (eds) Environmental and health impact of solid waste management activities, Issues in environmental science and technologies no. 18. Royal Society of Chemistry, Cambridge, pp 1–52

Tchobanoglous G, Theisen H, Vigil S (1993) Integrated solid waste management – engineering principles and management issues. McGraw-Hill Inc. International Edition, New York. ISBN 0-07-112865-4

UNESCAP- United Nations Economic and Social Commission for Asia and the Pacific (2000) Sustainable Asia – waste. http://www.unescap.org/esd/environment/soe/2000/documents/CH08.PDF. Cited 31 May 2012

USEPA - United States Environmental Protection Agency (2011) Municipal solid waste generation, recycling, and disposal in the United States: facts and figures for 2010. Solid Waste and Emergency Response (5306P), Washington, DC

USEPA – United States Environmental Protection Agency (2008) Waste management resources. http://www.epa.gov/tribalcompliance/wmanagement/wmwastedrill.html. Cited 12 Dec 2012

Ward AJ, Hobbs PJ, Holliman PJ, Jones DL (2008) Optimization of the anaerobic digestion of agricultural resources. Bioresour Technol 99:7928–7940

Wilkinson A (2011) Anaerobic digestion of corn ethanol thin stillage for biogas production in batch and down-flow fixed film reactor., Master of applied sciences in environmental engineering, University of Ottawa, Canada

Williams P (2005) Waste treatment and disposal. John Wiley and Sons Ltd., West Sussex. ISBN 0-470-84912-6

Zhou H, Loffler D, Kranert M (2011) Model-based predictions of anaerobic digestion of agricultural substrates for biogas production. Bioresour Technol 102:10819–10828

Zhu XA, Yao Q (2011) Logistics system design for biomass-to-bioenergy industry with multiple types of feed stocks. Bioresour Technol 102:10936–10945

Chapter 13
Microbial Insecticides: Food Security and Human Health

M. Shafiq Ansari, Salman Ahmad, Nadeem Ahmad, Tufail Ahmad, and Fazil Hasan

Abstract Sustainable agricultural systems must be adopted to increase the food and fiber production keeping in view of human health and increase in population; the number of undernourished has increased to almost 20% (The state of food insecurity in the world economic crises – impacts and lessons learned. In Food and Agriculture Organization of the United Nations, Rome). Insect pests have been causing serious damage in the fields and stored grains and their products. Interventions is required to limit the losses, therefore, synthetic insecticides have played a significant role in their management for more than 60 years. Indiscriminate use of insecticides have left undesirable residues in the environment, which are toxic to human beings and non target organisms as well as insects have developed resistance against them and resurgence of pests that lead to find a suitable, sustainable and efficient method of management. Microorganisms: bacteria, viruses, fungi and protozoa form the most abundant and diverse groups, which offer a vast resource for exploitation to use in the management program. *Bacillus thuringiensis* is a gram positive, occurs in soil and ubiquitous in distribution. It produces parasporal crystalline body which contains one or more *cry* proteins that can be toxic to a number of insects. *cry* proteins are encoded by *cry* genes and 200 of them are identified. Similarly, a number of insect pests are also vulnerable to viral diseases. Nuclear Polyhedrosis and Granulosis Viruses are commonly used against the Lepidoptera. They are highly species specific and safer to human beings. Fungi, often act as important natural control agents that limit the insect population. Promising results are obtained by *Beauveria bassiana* and *Metarhizium anisopliae* against many insect pests. *B. bassiana* grows naturally in soil throughout the world and causes white muscardine disease. Therefore, intensive work is required to improve the efficacy of microbial insecticides through molecular biology and genetic engineering to enhance their role in the insect management for better food security.

M.S. Ansari (✉) • S. Ahmad • N. Ahmad • T. Ahmad • F. Hasan
Department of Plant Protection, Faculty of Agricultural Sciences,
Aligarh Muslim University, Aligarh 202002, India
e-mail: mohdsansari@yahoo.com; salmanalig.alig@gmail.com;
nadeem777in@yahoo.com; fazilento10@gmail.com

A. Malik et al. (eds.), *Management of Microbial Resources in the Environment*,
DOI 10.1007/978-94-007-5931-2_13, © Springer Science+Business Media Dordrecht 2013

Keywords Microbial insecticides • δ -endotoxin • Entomopathogenic fungi • Food security • Biocontrol

1 Introduction

Agriculture is the backbone of economy of many countries and approximately, two billion people in the world depend on agriculture for their livelihood. Majority of them are farming families with small holdings, small-scale breeders and entrepreneurs, laborers and nomadic groups. Food security is a major concern because of increase in human population and simultaneously increasing food demand. Efficient and sustainable agricultural systems would be adopted to meet the demand of humans who are living below poverty line. Food and fiber crops are damaged by over 10,000 species of insects, with an estimated annual loss of 13.6% globally (Benedict 2003). Despite the annual investment of US $35,000 for the application of 3 million tonnes of pesticides and the use of various biological and other non-chemical methods of control worldwide, global crop losses remain a matter of concern (Pimentel 2007, 2009). The insect pests are polyphagous and migratory in nature, with high fecundity and short life span and diapausing under adverse conditions. They are the most successful group of animals that exist in a countless of environment and nearly 645 species of insects and mites have already developed resistance against insecticides including *B. thuringiensis* (Narayanan 2004; Sayyed et al. 2004; Sarfraz and Keddie 2005). In order to reduce the losses caused by the insects and to meet the demands of increasing population, synthetic chemical insecticides have played an important role in the management of insect pests for nearly 60 years (Smith 1970; Kumar et al. 2008). Despite of these credentials, much use of insecticides has been ecologically unsound. Therefore, alternate method of control strategy may be adopted.

Insect pathogens have great potential and can be used in pest management in at least four ways by: (1) Utilizing naturally occurring diseases (2) Introduction of insect pathogen in insect pest population (3) Application of insect pathogens (or their products) as microbial insecticides (4) Novel manipulation of insect pathogen genes, usually involving recombinant DNA technology. This article explores the role of microbial genetic resources for food security without deterioration of environment and human health.

2 Types of Insect Pathogens

(a) Bacteria
(b) Viruses
(c) Fungi
(d) Protozoa

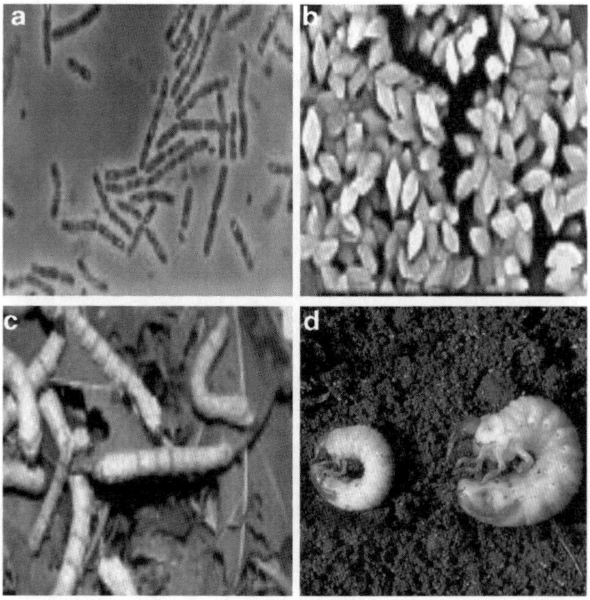

Fig. 13.1 *Bacillus thuringiensis*'s Spores (**a**), *cry* proteins (**b**), *Bt* infected silkworm larvae (**c**), *B. popillae* infected larvae of Japanese beetle (**d**)

2.1 Bacteria

Bacillus thuringiensis was recognized as living insecticides by Steinhaus (1954) that occurs naturally in the soil and is distributed worldwide. It is a gram-positive, aerobic, rod-shaped and produces a proteinaceous parasporal crystal during sporulation that consists of α, β-exotoxins and δ- endotoxin, which are highly insecticidal even at very low concentrations (Xavier et al. 2007). The molecular weight of proteins of the bipyramidal crystal (5×15.5 nm) is of 120,000 Da. The insecticidal activity is attributed to the parasporal crystal, δ-endotoxin which is toxic to the larvae of Lepidoptera, Diptera, and Coleoptera but harmless to human being and non-target organisms (de Maagd et al. 2000; Kumar et al. 2008).

The insecticidal crystal proteins are encoded by *cry* genes and nearly 200 genes are classified on the basis of amino acid homology and insecticidal activity (Hernandez and Ferre 2005; XU Jian et al. 2006). The proteins toxic for Lepidopteran insects belong to *cry*1, *cry*2, and *cry* 9; for Coleopteran *cry*3, *cry*7 and *cry*8 (Fig 13.1a–c) and for Dipteran *cry*2, *cry*4, *cry*10, *cry*11, *cry*16, *cry*17, and *cry*19 genes (Crickmore et al. 1998).

B. popillae is used as a biocontrol agent that causes milky disease in the larvae of Japanese beetle (Fig 13.1d). The bacterium is applied to turf grass as a dust which only controls the grubs effectively and virtually has no adverse effect on other soil borne insects.

2.1.1 Mode of Entry

Bacterial pathogens must be ingested by a susceptible insect. The enzymes produced by bacteria viz. lecithinase, proteinase and chitinase within the digestive tract that act on the midgut cells and enable them to enter the hemocoel of insect. Toxins are produced in the initial stages and play a significant role in the invasion of bacteria through the digestive tract. In some cases the bacteria multiply in the gut before invading the hemocoel, when the larva is placed under stressed conditions, such as abnormal nutrition, unfavorable temperature and humidity or other microbial infections.

2.1.2 Mode of Action

The primary action of *cry* toxins is to lyse midgut epithelial cells in the target insect by forming pores in the cells of apical microvilli membrane (Aronson and Shai 2001; de Maagd et al. 2000; Bravo et al. 2004). Each bacterial cell forms a spore at one end and a crystal at the other. Insecticidal activity of *Bt* is due to the presence of crystalline protein body called parasporal crystal or δ -endotoxin formed during sporulation (Heimpel 1967) that contains protoxin (Herrero et al. 2004). The crystal is composed of large protein with molecular weight of 130–135 kDa. After ingestion, it is hydrolyzed into active toxin of 60–65 kDa in the presence of alkaline conditions of midgut (pH 10–12) and proteases (Gill et al. 1992; Knowles 1994). Finally, activated toxins bind to specific receptors present in the larval midgut epithelial cells. As a result of binding, ATPs are rapidly lost from the cell stimulating respiration and glucose uptake and also create ion channels or pores eventually. The cells burst spilling their cytoplasmic content into lumen. Goblet cells may be the site of original lesion. The pore formation causes osmotic shock upsetting the gut's ion balance. As a result of this process, the cell membrane lyses, paralysis occurs and consequently, the insect stops feeding and dies from starvation (Knowles 1994).

2.1.3 Symptoms of Infection

Bacterial infections are broadly classified as;

1. **Bacteremia:** Bacteremia occurs when the bacteria multiply in insect's hemolymph without production of toxins. Bacteremia occurs in case of bacterial symbionts occurring in the body of insects and rarely occurs with bacterial pathogens.
2. **Septicemia:** Septicemia occurs most frequently with pathogenic bacteria, which invade the hemocoel, multiply, produce toxins and kills the insects.
3. **Toxemia:** Toxemia occurs when the bacteria is usually confined to the gut of lumen and produce toxins there. Insects are usually killed by these toxins as in case of brachytosis of the tent caterpillar.

2.1.4 Application of *Bt*

Three major applications of *Bt* toxins have been achieved: (i) in the control of insect defoliators in agricultural crops, (ii) control of vectors e.g. mosquitoes for transmission of human diseases, and (iii) in the development of transgenic plants. *B. thuringiensis* formulations have been used to control insect pests since1920s (Lemaux 2008) and its commercial formulation was first marketed by USA in 1958. These formulations have the characteristics of being eco-friendly, safe, easy to use and cheap (Carlton 1988), as compared to chemical pesticides. They are now used under trade names such as Dipel and Thuricide as specific insecticides. In 1985, Plant Genetic Systems, Belgium was the first company to develop genetically engineered (tobacco) plants with insect tolerance by expressing *cry* genes from *B. thuringiensis* (Hofte et al. 1986; Vaeck et al. 1987). In 1995, *Bt* toxins expressing in the potato plants were approved safe by the Environmental Protection Agency, USA as a first *Bt* crop. Transgenic cotton often referred to as *Bt* cotton, is grown in a number of countries. *Bt* cotton with a single *cry*1A gene and stacked with *cry*2A gene has offered better protection against cotton bollworm, *Helicoverpa armigera* (Gujar et al. 2007). Liquid formulation of *Bt* is applied on the crops through overhead irrigation systems or in a granular form for control of European corn borer (Cranshaw 2008). A synergistic effect was obtained when *Spodoptera litura* ingested *B. thuringiensis* subsp. *kurstaki* (*Btk*) with plant extracts (Rajguru et al. 2011). *B. thuringiensis* subsp. *israelensis* is used as solid or granules with slow release rings or brickettes on the standing water to control the mosquito larvae. The persistence of *Bt.* in water is more than on sun-exposed plant surface.

2.1.5 Disadvantages of *Bt*

Bt. is effective only against immature stages of target insects which are defoliators and not effective against sap suckers and borers. Repeated applications of *Bt* are required to achieve successful results but used judiciously, not indiscriminately because resistance has also been reported in a number of economically important insects like; *Plutella xylostella*, *Plodia interpunctella*, *Heliothis virescens*, *H. armigera*, *Spodoptera exigua*, *S. litura* (Narayanan 2004). High level of resistance to α-endotoxin of *Bt.* subspecies *kurstaki* is also recorded against the Indian meal moth, *P. interpunctella* (McGaughey 1985).

2.2 Viruses

The entomopathogenic viruses offer a promising option of microbial control second only to bacteria in terms of adoption, and success in the field conditions. Viral diseases are commonly occurring among the insects and one of the most widely investigated infections. The advanced techniques in biotechnology, biochemistry, recombinant DNA technology, serology, pathology and tissue culture have provided the opportunity to study the viral diseases in insects. With these advancements, the applied virology had extended beyond pest management to the field of genetic engineering, where virus serves as a vector for expression of foreign gene to form biochemically important products.

2.2.1 Classification of Insect Viruses

Insect viruses are divided into, the occluded and the non-occluded viruses. After virion formation in infected cell, it is occluded within a protein matrix, forming paracrystalline body that is referred to as occlusion body that contributes to stability and persistence in the environment. In non-occluded, the virion occurs freely or occasionally forms paracrystalline arrays of virions, known as inclusion body and has no occlusion body protein scattered in the virions.

Types of insect virus:

(a) Iridoviruses
(b) Cytoplasmic Polyhedrosis Viruses
(c) Entomopoxviruses
(d) Ascoviruses
(e) Baculoviruses

Iridoviruses

Iridoviruses (Family- Iridoviridae) are non-occluded with a linear double-stranded DNA genome and icosahedral in shape (125–200 nm) which replicate in cytoplasm in the infected hosts. The virion is made up of three domains; a central core containing DNA-protein complexes, an intermediate lipid membrane and an outer proteinaceous capsid. The iridoviruses have been most commonly reported from larval stages of Diptera, Lepidoptera and Coleoptera. Nearly 44 species of Diptera are known to be infected by iridoviruses (Williams 2008). Invertebrates are, generally infected by iridoviruses and chloriridoviruses especially insects and terrestrial isopods, and also living in aquatic and damp habitats, and are both called as invertebrate iridescent viruses (IIVs) because of the opalescent hues observed in heavily infected hosts (Fig 13.2a). The infection is invariably lethal, but there is now growing evidence that sublethal infections may be common in certain host species (Tonka and Weiser 2000). The IIV infections have little potential to control the medically and agriculturally important insect pests due to broad host range shown in laboratory tests and often low prevalence of patent disease (Henderson et al. 2001; Jakob et al. 2002).

Cytoplasmic Polyhedrosis Viruses

Cytoplasmic polyhedrosis viruses (Family- Reoviridae) consist of double-stranded RNA with 10 segments. The CPV particle is icosahedral in shape and formed in the cytoplasm of midgut epithelium of lepidopterous larvae. Inclusion body of CPV is also

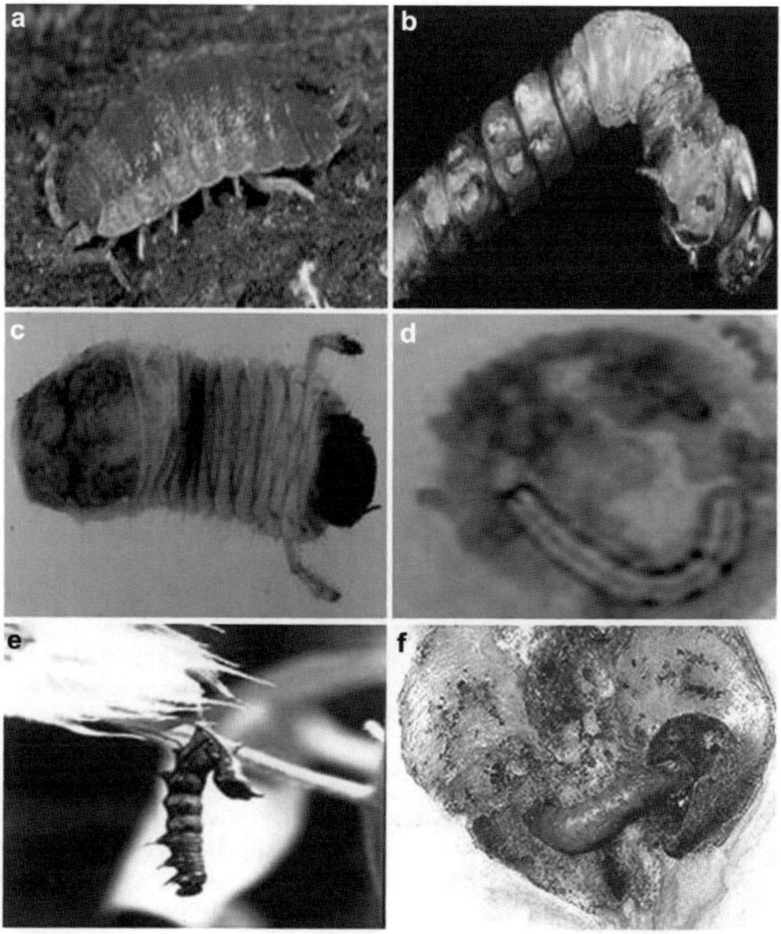

Fig 13.2 Woodlouse (*Porcelio scaber*) infected with Iridovirus – *blue colour* is due to Iridovirus infection (**a**), a larvae infected with Cytoplasmic polyhedrosis viruses – the skin of larvae rupture and polyhedral bodies are released (**b**), Soil Scarab (*Othnonius batses*) infected with Entomopox virus (**c**), Larvae of *Spodoptera* spp. Infected with Ascovirus (**d**), larvae of *Helocoverpa armigera* infected with Baculovirus NPV (**e**), larvae of *Cydia pomonella* infected with Baculovirus GV (**f**)

called polyhedron, which is stable, protecting the virions from adverse environmental factors and serving as vehicle for transmission of virus from one host to another. The infection occurs when the insect ingests polyhedra contaminated food (foliage of a plant). As the infection progresses, the synthesis of the polyhedron matrix proteins intensifies resulting in an increase in the inclusion body. The inclusion body may also increase in size by the uniting of one or more bodies to form a large body. Proteolytic enzymes may function in the dissolution of ingested polyhedra by the insect and the liberated virions enter the midgut lumen and become attached to the migut columnar cells by the spikes on the viral surface (Asai et al. 1973) where RNA is released leaving an empty shell on the cell surface. During severe infection, polyhedra are released with feces of lepidopterous caterpillar, thus contaminating more foliage responsible for the transmission of CPV when a healthy insect consumes the leaves. The disease is often chronic, while Lepidopterous with lethal infection may retard their development and are smaller, lighter in weight and may have an increased number of larval moults as compared with uninfected larvae. In later stages, the infected midgut is turned yellow to white rather than translucent brown because of accumulation of larger number of polyhedra (Fig 13.2b). The CPV is relatively common among the Lepidoptera and Diptera (suborder Nematocera) (mosquitoes, blackflies and midges).

Entomopoxviruses

Entomopoxviruses (EPVs) (Family – Poxviridae) consist of double stranded DNA with virions (150×300 nm) that are brick-shaped or oval. EPVs replicate in the cytoplasm in most of the insects, causing an acute and fatal disease. They are commonly reported from Lepidoptera, Coleoptera, Diptera, Hymenoptera and Orthoptera (Fig 13.2c). These are DNA viruses which have naturally infecting grasshoppers. Occlusion bodies of EPVs are varied from spindle to oval shaped and occluding nearly 100 or more virions.

Entomopoxviruses form two types of proteinaceous crystalline bodies: spindles and spheroids. The spheroid occludes virions, whereas the spindle does not; thus, spheroids are main agent of infection. The major constituent proteins of the spheroid are spheroidin and that of spindle are fusolin and both of them are most abundant proteins of EPVs (Mitsuhashi et al. 2007). It is easily transmitted by feeding, with relatively narrow host range generally being restricted to closely related species. EPV of insect are related to vertebrate poxviruses, such as the variola virus (causative agent of smallpox), and they may be the evolutionary source of the vertebrate poxviruses.

Ascoviruses

The ascoviruses (Family – Ascoviridae) consisting of an envelope, an internal lipid membrane associated with the inner particle and a core. Enveloped virions are large, approximately 130×400 nm. The virions are bacilliform, ovoid and allantoid, containing a linear double-stranded DNA. They are reported to infest Noctuids (*Spodoptera frugiperda, S. exigua, Trichoplusia ni, Helicoverpa virescens*) of Lepidoptera (Fig 13.2d). Ascovirus causes stunting of growth of larvae and large numbers of virion containing vesicles are formed in the cytoplasmic membrane of the cell. As the infection progresses, the hemolymph of an infected larva becomes opaque white as large numbers of retractile vesicles are released after breaking down of cell membrane into hemolymph (Tillman et al. 2004). Infected caterpillars are generally stopped feeding and retarded growth rate, concomitantly increased the larval longevity as compared to uninfected larvae. Ascovirus is transmitted mechanically from one lepidopteran host to another by female endoparasitic wasps but rarely through feeding.

Baculoviruses

The genus Baculovirus (Baculoviridae) constitutes one of the largest groups of insect pathogenic viruses. It has two prominent members, namely, Nuclear Polyhedrosis virus (NPV) and Granulovirus (GV), which show the greatest potential due to their exclusive pathogenicity to insects spanning over 400 species of Lepidoptera, Hymnoptera, Coleoptera, Diptera, and Decapoda (Gupta et al. 2007). Baculoviruses are large, enveloped; rod-shaped nucleocapsid in which an amorphous but a definite layer exists between the nucleocapsid and the envelope. The baculovirus has a single molecule of circular, supercoiled, double-stranded DNA. NPV contains many virions occluded within occlusion body called polyhedra, while, GV contains only one or, rarely, two or more virons in occlusion body called granules (Rohrmann 2008). Both NPVs and GVs are highly infectious, and in some insects periodically cause epizootics (outbreaks of disease). Baculoviruses primarily infect the larval stage through feeding upon the plant foliage, in soil or eggs contaminated with occlusion bodies (Blissard et al. 2000). The dissolution occurred both by the high pH and the presence of alkaline proteinases in the midgut lumen of most Lepidoptera. The virions are released then passed through the peritrophic membrane and fused with the midgut epithelial cell plasma membrane. The nucleocapsids are then released into the cytoplasm and migrate to the nucleus, where transcription of viral genes and replication of viral genome takes place. The BVs produced in the midgut epithelial cells spread via hemolymph into all tissues of the insect causing secondary infection, but the predominant target is the fat body cell. At the end of infection, the cells and tissues of dead insects are disintegrated and occlusion bodies are released into the environment, thus named as "wilting disease".

Commercially, the Baculovirus products are mostly produced in the form of concentrated wettable powders apart from liquid or granular forms. The virus is inactivated by ultra violet (UV) rays of wavelength 290–320 nm; UV protectants such as metallic oxides are used besides the addition of anti-evaporants and spreaders/wetting agents to the virus formulations. In field conditions, they are generally failed, if not applied at proper time and right place; thus, information regarding the insect behaviour and distribution within the crop in each instar, and the area of foliage ingested per instar all should be well known for the effective use of the virus (Simon et al. 2008). Genetic engineering could be performed using genes of various origins such as *Bt* gene, scorpion toxin (BelT/AaIT) gene, straw itch mite toxin (TxP1) gene, and insect hormone genes such as diuretic hormone from *Manduca sexta* that alters the larval fluid metabolism, eclosion hormone associated with ecdysis from *M. sexta* that causes initiation of the eclosion in the inoculated larvae, and ecdysteroid UDP-glucosyltransferase (EGT), juvenile hormone esterase (JHE) expression in Baculovirus (Lasa et al. 2009) that inhibit the larval feeding and molting (Bonning and Hammock 1996).

Nuclear Polyhedrosis Virus (NPV)

The NPV is the first virus to be detected in the insects but most commonly reported from Lepidoptera (Fig 13.2e). Most NPVs have been isolated from Lepidoptera (88%), Hymenoptera (6%) and Diptera (5%) (Saufi 2008). The virions are large

(80–200 × 280 nm) consisting of one or more nucleocapsids with double-stranded circular DNA enclosed in an envelope. Most of the enveloped nucleocapsids in the nucleus are occuluded in polyhedra. The number of enveloped virions in polyhedra may be as high as 200 virions. NPVs replicate in the nuclei of epidermal cells, blood cells, fat body and trachea. When ingested by a susceptible insect, the polyhedra dissolve and virions enter into midgut cells, replicate, and then pass to tissues more commonly associated with NPV infections. The larva infected with NPV becomes discolored (brown and yellow), hemolymph turns cloudy and milky, becomes less active and stops feeding; the whole body decomposes and becomes liquefied and, moves to the top of the plant and hangs upward down by its abdominal and caudal prolegs. After death the larva rapidly darkens and the older infected larvae which die in the fifth or late instars gradually turn slightly pale with pinkish tinge several days before death, but otherwise remain nearly as dark brown as the healthy larvae. Currently, high virus production costs make the viral treatments uncompetitive compared with the chemical treatments, but more economical than *B. thuringiensis* treatments (Bhargava et al. 2008; Suryawanshi et al. 2008).

Granulosis Virus (GV)

The occlusion body of granulosis virus is granule or capsule-like and it is ovoid or ovocylindrical (0.3 × 0.5 μm) but the shapes may vary greatly. Each capsule has one, rarely two or more enveloped nucleocapsids. GV infections have been reported from more than 100 species of insects, limited to Lepidoptera (Murphy et al. 1995). Granuloviruses are generally considered to have a narrow host range (Federici 1997) but four of them: CpGV (*Cydia pomonella* GV) (Fig 13.2f), HaGV (*Helicoverpa armigera* – GV), SpfrGV (*Spodoptera frugiperda*-GV) and XecnGV (*Xestia c-nigrum* – GV) have relatively wide host range (Winstanley and O'Reilly 1999). The fat body is the primary site of infection, but the epidermis and tracheal matrix may occasionally be affected (Huger 1963). GVs are transmitted orally and via the egg. Latent infections also occur. The period between ingestion of the virus and the death of the host generally ranges between 4 and 25 days. External symptoms are not usually apparent in early stages of infection, but towards the later stages infected larvae frequently develop a lighter color. Liquefaction of the infected larvae occurred, similar to NPV infection, but when the epidermis is not involved; liquefaction does not take place (Martignoni and Iwai 1986). Within a few days after infection, the host larvae become unable to digest the food, and so weaken and die (Thakore 2006). The infected larvae hang upside down from the leaves and twigs in a characteristic way and brownish fluid oozes from them which is a highly infective and is readily disseminated amongst the healthy insect population.

2.2.2 Symptoms of Infection

The insects that are killed by viruses are shiny-oily appearance, and often seen hanging limply from vegetation. The infected larvae are extremely fragile, with ruptured cuticle and releasing fluid with infective virus particles. Infection to other insects will occur only when they eat upon leaves contaminated by virus killed larvae.

These viruses are highly species specific and beneficial insects and not affect to parasitoids and predators and are safe to humans and non target organisms. There has been a growing demand amongst the farmers for these bioagents because of their usefulness. NPV can be used on field crops, including chickpea, sorghum, cotton and maize etc.

2.3 Fungi

Most of the insects are infected by naturally occurring entomopathogenic fungi. 750 species of fungi representing 100 genera are known to be associated with insects (Benjamin et al. 2002; Zimmerman 2007). The epizootic potential of fungi is considerably high and may spread quickly through insect population. Fungi usually attached to the cuticle of the insects in the form of asexual spores (conidia) or sexual spores. Under favorable temperature and high moisture, they germinate as hyphae and colonize on to the cuticle; eventually they penetrate through it and reach the hemocoel. They are also called as mycoinsecticides and are being used on a limited scale to control insect pests.

2.3.1 Process of Infection

The fungal pathogens are unique among the bioagents which do not require the consumption by the insect through food to act up on and can invade their hosts directly through the exoskeleton or cuticle. Therefore, they can infect non-feeding stages such as eggs and pupae. The site of invasion is often at intersegmental folds, between the mouthparts or through the spiracles, where high humidity promotes germination and the hyphae penetrated easily (Hajek and Leger 1994; Clarkson and Charnley 1996). *Metarhizium* spp. and *Beauveria* spp. are opportunistic hemibiotrophs with a parasitic phase in the live host and saprotrophic phase during post-mortem growth on the cadaver (Charnley and Collins 2007). They may use toxins to overcome host defences, while, Entomophthorales are biotrophs with little or no saprotrophism; kill the insect by tissue colonisation without producing toxins (Charnley 2003; Freimoser et al. 2003).

2.3.2 Host Response to Fungal Attack

The host insects respond by different ways to the fungal attack. The cuticle is not only the major but also the first barrier to the host invasion. Structural features such as sclerotisation impede penetration, while enzyme inhibitors and tyrosinases, which generate antimicrobial melanins, are frontline defences against weak pathogens

(Charnley and Collins 2007). Virulent fungal pathogens are least affected by blood-borne defences. Phagocytosis by individual blood cells and cooperative behaviour between haemocyte subpopulations viz. encapsulation and granuloma formation are often not recorded after the initial incursion. This has been attributed to a failure of the insect's non-self recognition system, in some cases brought about by toxic fungal metabolites, in others due to the removal of immunogenic components from fungal cell walls in the blood of infected insects (Charnley 2003).

2.3.3 Mode of Action

The fungal spore adheres first on to the cuticle and, under favorable environmental conditions it germinates in the form of germ tube which further penetrates the cuticle of the host. The process of penetration through the cuticle involves chemical (enzymes) and physical forces. The mechanical force is noticeable at the tip of invading hypha where the cuticular layers are distorted from pressure. The enzymes detected on germ tube are proteases, aminopeptidases, lipase, esterase and N-acetyl-glucosamidase (chitinase). These enzymes liberate monomers that can be metabolized by the germ tube in order to continue to grow into the integument. Vegetative structures, hyphae multiply by budding in the hemocoel and filled with hyphae, the insect usually dies, and the fungus develops saprophytically. The cells and tissues of an infected insect may begin to disintegrate prior to insect's death or they may break down after death. The fungal hyphae continue to grow usually resulting mummification, and dead insects retain their shape and form. Finally, mycelia emerged from the cadaver after death. In Lepidopteran larvae infected with certain Entomophthorales, they become flaccid with watery contents and fragile integument that may appear dark in color.

2.3.4 Groups of Entomopathogenic Fungi

Many common entomopathogenic fungi belong to order, Hypocreales of the Ascomycota: the asexual (anamorph) phases; *Beauveria*, *Metarhizium*, *Nomuraea*, *Paecilomyces, Hirsutella* (Fig 13.3) and sexual (teleomorph) state *Cordyceps*; others (*Entomophthora*, *Zoophthora*, *Pandora*, *Entomophaga*) are from the order Entomophthorales of the Zygomycota.

Beauveria spp.

Beauveria grows naturally in the soil throughout the world and acts as parasite on various arthropod species, causing white muscardine disease in insects. *B. bassiana* is the most biologically active species, discovered by Agostino Bassi in 1835 as the cause of the muscardine disease in silkworm. *B. bassiana* (formerly as *Tritirachium shiotae*) is the anamorph (asexually reproducing form) of *Cordyceps*

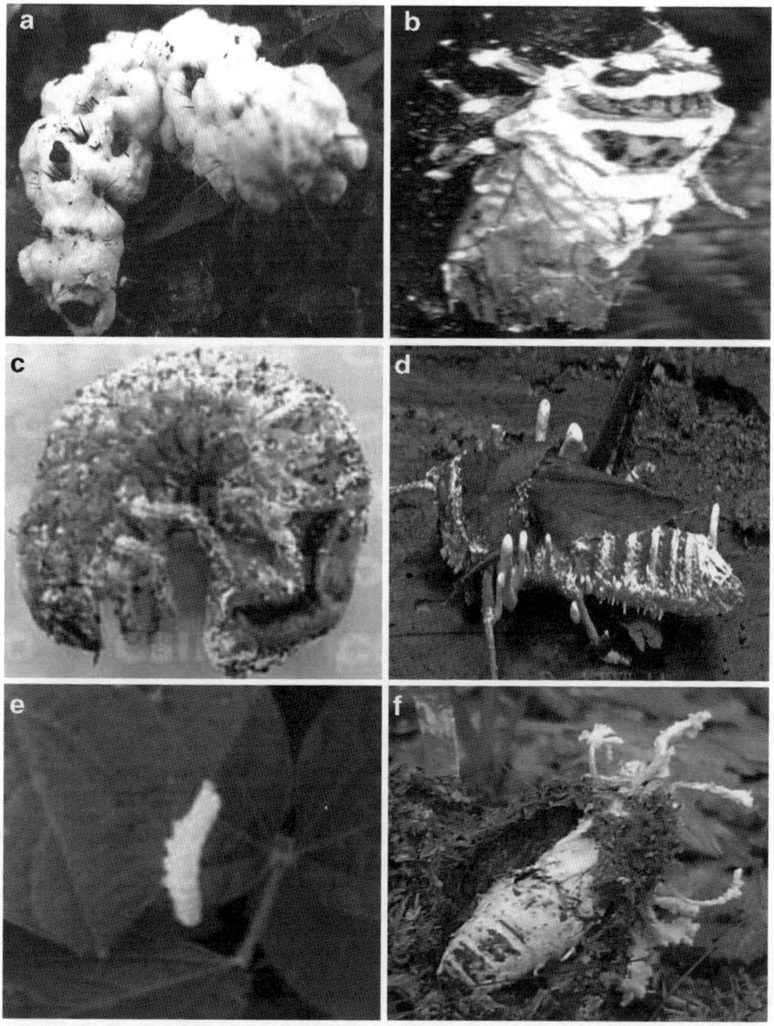

Fig 13.3 Hairy caterpillar and Cicada infected with *Beauveria bassiana* (**a** and **b**), larvae of *white* grub infected with *Metarhizium anisopliae* (**c**), stroma of Cordyceps sp. emerging from a locust (**d**), infection of Nomuraea sp. on a lepidopterous larvae (**e**), a lepidopterous pupa infected with *Paecilomyces tenuipus*

bassiana. Teleomorph (the sexually reproducing form) was discovered in 2001. *B. bassiana* is one of the most commonly occurring entomopathogenic fungi and shows strong pathogenicity to Lepidoptera, Hymenoptera, and Coleoptera (Harris et al. 2000; Fernandez et al. 2001; Mannion et al. 2001; Phoofolo et al. 2001). It is used to control a number of agricultural insect pests such as whiteflies, thrips, termites, aphids and beetles on major agricultural crops and even mosquitoes. Conidia of *B. bassiana* are also effective in controlling of mosquito larvae when applied as conidia dust on to breeding sites (Prasad and Veerwal 2010). The use of *B. bassiana* is environmentally safe and generally non-toxic to beneficial insects. It grows as a white mold on culture media and produces many dry powdery conidia, white spore balls. Each spore ball is composed of cluster of conidiogenous cells, which are short, ovoid and terminate in rachis with is a narrow apical extension. The conidia are single-celled and hydrophobic in nature.

Mode of Action

The conidia when come into contact with cuticle of an insect, they germinate and penetrate into the hemocoel where they multiply until hemocoel is filled with mycelia. At this stage the infected insect is usually dead and has a rather firm consistency, like a small loaf of bread. Under favorable conditions the fungus continues to grow and produces structures that protrude through the cuticle and form conidia. Death is caused by destruction of tissues and the toxins produced by *Beauveria* like- bassianin, bassiacridin, beauvericin, bassianolide, beauverolides, tenellin and oosporein (Strasser et al. 2000; Vey et al. 2001; Romeis et al. 2006). *B. bassiana* can be used to control the insects of economically important crops.

Metarhizium spp.

Metarhizium anisopliae (*Entomophthora anisopliae)* is widely distributed and soil-inhabiting fungus, known as green muscardine fungus and was first isolated by Metchnikoff (1879) from the beetle, *Anisoplia austriaca*. The colony of *M. anisopliae* appears white at initial stage but dark green at later (Zimmerman 2007). Spores of *M. anisopliae* bind to the cuticle where germination occurred and the germ tubes penetrate in the hemocoel where lateral extension of hyphae are produced and continue to prolifrate until the insect is filled with mycelia. They sporulate in the hemocoel and the cuticle breaks down and makes the insect "fuzzy". Death may occur because of the destruction of tissues, nutritional deficiency and toxins produced by *M. anisopliae*; cyclodepsipeptides, destruxins A. B, C, D and E and desmethyldestruxin B (Tamura et al. 1964; Suzuki et al. 1970). Destruxins have been considered as a new generation insecticides. They cause titanic paralysis when inoculated into larvae of *Gallerria mellonella* (Romeis et al. 2006). *M. anisopliae* could be used to control locusts, grasshoppers, termites, curculionids, and scarabeids.

2.3.5 Application in the Insect Pest Management

Several strategies have been adopted for the use entomopathogenic fungi in the integrated pest management programs. *B. bassiana* (Mycotrol®) applied to seedlings grown in a nursery was effective in controlling DBM before they were transplanted into the field. Classical biological control is also known as permanent introduction of pathogens in nature. This involves the establishment of a disease in a population where it does not occur normally, in order to obtain long-term or permanent suppression of the pest. It may be promoted through introduction of natural fungal epizootics by adopting appropriate cultural and crop protection practices of harnessing entomopathogenic fungi for pest management. There are also new opportunities for developing biopesticides in IPM by combining ecological factors with post-genomic technologies (Chandler et al. 2011). Loc and Chi (2007) showed that the isolates of *B. bassiana* were isolated from naturally infected DBM exhibited the highest infectivity to *P. xylostella*.

2.4 Protozoa

Protozoans are diverse group of motile, unicellular eukaryotic organisms. Generally, they are referred to as animal-like protists because of movement (motile). However, both protozoa and protists are paraphyletic groups (not including all genetic relatives of the group). It is a potential microbial genetic resource for the management of insect pests as they have ability to cause diseases in the insects. They range from well known microsporida such as *Nosema locustae*, a pathogen of grasshoppers, *N. apis*, a pathogen of honey bee, and *N. pyrausta*, a pathogen of European corn borer. They are generally host specific and slow acting, most often producing chronic infections. Of some 14,000 described species of Protozoa, about 500 are pathogens of insects. Most of them are chronic pathogens that may debilitate a host without producing obvious disease symptoms but some species are extremely virulent, causing stunted growth, slow development, and early death (Tanada and Kaya 1993).

Protozoan diseases of insects are ubiquitous and comprised of an important regulatory role in insect populations (Lacey et al. 2001). They are generally host specific and slow acting, most often producing chronic infections. They develop only in living hosts and many species require an intermediate host. Microsporida are among the most commonly observed. Their main advantages are persistence and recycling in host populations and their devastating effect on reproduction and overall fitness of target insects (Solter and Becnel 2000). The grasshopper pathogen *N. locustae* Canning is the only species that has been registered and commercially available species of microsporidium, marketed under several labels for the control of grasshoppers and crickets. It is applied with insect-attractant bait. Because of its slow mode of action, this product is better suited to long-term management of rangeland pests than to the more intensive demands of commercial crop or even home garden production. Other *Nosema* species have been shown to infect spider mites and webworms, but have yet to be developed sufficiently for commercial use. Similarly, *N. pyrausta* (=*Perezia pyraustae*) infects several insect species, including European corn borer, for which it can be an important natural control. However, its commercial use is still in the developmental phase. Infection can spread from diseased to healthy larvae via contaminated frass, and by migration of infected larvae between plants. *Vairimorpha necatrix* is another microsporidium with commercial potential.

It has a wide host range among caterpillar pests, including corn earworm and European corn borer, various armyworms, fall webworm, and cabbage looper. It can be more virulent than other species and infected insects may die within 6 days of infection. The main disadvantages of the Protozoa as inundatively applied microbial control agents are the requirement for *in vivo* production and low levels of immediate mortality.

2.4.1 Mode of Action

To cause infection, most microsporidia must be ingested by an insect. However, there may also be some natural transmission within a pest population, for example by predators and parasitoids. The pathogen enters the insect body via the gut wall, spreads to various tissues and organs, and multiplies, sometimes causing tissue breakdown and septicemia.

2.4.2 Symptoms

Infected insects may become sluggish and smaller than normal, sometimes with reduced feeding and reproduction, and difficulty molting. Death may follow if the level of infection is high. One advantage of this type of infection is that the weakened insects are more likely to be susceptible to adverse weather and other mortality factors.

3 Conclusion

The perusal of literature shows that there are several ecofriendly ways available to reduce the pesticide usage in agriculture. Sustainable agriculture will rely increasingly on alternative interventions for insect pest management that are environment friendly and safer to non target organisms as well as to human beings. Effective microbial control agents that can fill the void of phased out chemicals exist, but their further development and implementation will require the following advances: improvements in the efficacy of pathogens, their production, and formulation; better understanding of how they will fit into integrated systems and their interaction with the environment and other biotic components; greater appreciation for their full advantages (efficacy, safety, selectivity, etc.), not simply their comparison with chemical pesticides; and acceptance by growers and the general public. Despite our optimistic appraisal of the future of entomopathogens as biological control agents, portions of the biopesticide industries which are currently facing financial setbacks. Although the market for microbial insecticides is growing, it represents only approximately 1–1.5% of the total crop protection market and most of this is due to sales of *B. thuringiensis* (Lacey et al. 2001). We believe that in the near future microbials will face even stiffer competition from new pesticide chemistry and transgenic plants. Improvement in microbial products, benefit to growers that microbial control offers, and the need to develop alternatives to conventional chemical insecticides should overcome many of the obstacles that it is now facing. However, if future development is only market driven, then there will be considerable delay in the implementation of microbial control agents that have good potential for use in insect management programs.

References

Aronson AI, Shai Y (2001) Why *Bacillus thuringiensis* insecticidal toxins are so effective? Unique features of their mode of action. FEMS Microbiol Lett 195:1–8

Asai J, Kawamoto F, Kawase S (1973) On the structure of cytoplasmic polyhedrosis virus of silkworm, *Bombyx mori*. J Invert Pathol 29:69–73

Benedict JH (2003) Strategies for controlling insect, mite and nematodes pests. In: Chrispeels MJ, Sadava DE (eds) Plants, genes, and crop biotechnology. Jones and Bartlett Publishers, Sudbury, pp 414–442

Benjamin MA, Zhioua E, Ostfeld RS (2002) Laboratory and field evaluation of the entomopathogenic fungus, *Metarhizium anisopliae* (Deuteromycetes) for controlling questing adult, *Ixodes scapularis* (Acari: Ixodidae). J Med Entomol 39:723–728

Bhargava MC, Choudhary RK, Jain PC (2008) Genetic engineering of plants for insect resistance. In: Jain PC, Bhargava MC (eds) Entomology: novel approaches. New India Publishing, New Delhi, pp 133–144

Blissard GW, Black B, Crook N, Keddie BA, Possee R, Rohrmann G, Theilmann D, Volkman L (2000) Family Baculoviridae. In: van Regenmortel MHV, Fauquet CM, Bishop DHL, Carstens EGB, Estate MK, Lemon SM, Mainloff J, Mayo MA, McGoech DJ, Pringle CR, Wickners RB (eds) Virus Taxonom. Academic, San Diego, pp 195–202

Bonning BC, Hammock BD (1996) Development of recombinant baculoviruses for insect control. Annu Rev Entomol 41:191–210

Bravo A, Gómez I, Conde J, Muñoz-Garay C, Sánchez J, Zhuang M, Gill SS, Soberón M (2004) Oligomerization triggers differential binding of a pore-forming toxin to a different receptor leading to efficient interaction with membrane microdomains. Biochem Biophys Acta 1667:38–46

Carlton BC (1988) Development of genetically improved strain of *Bacillus thuringiensis* – a biological insecticide. Am Chem Soc 9:260–297

Chandler D, Bailey AS, Mark G, Davidson TG, Greaves J, Grant WP (2011) The development, regulation and use of biopesticides for integrated pest management. Philos Trans R Soc Biol Sci. doi:10.1098/rstb.2010.0390

Charnley AK (2003) Fungal pathogens of insects: cuticle degrading enzymesandtoxins. Adv Bot Res 40:241–321

Charnley AK, Collins SA (2007) Entomopathogenic fungi and their role in pest control. In: Kubicek CP, Druzhinina IS (eds) Environmental and microbial relationships. The Mycota IV, 2nd edn. Springer, Berlin/Heidelberg

Clarkson JM, Charnley AK (1996) New insights into the mechanisms of fungal pathogenesis in insects. Trends Microbiol 4:197–203

Cranshaw WS (2008) *Bacillus thuringiensis*: home and garden, Insect series no 5.556. Colorado State University Cooperative Extension, Fort Collins

Crickmore N, Zeigler DR, Feitelson J, Schnepf E, Van Rie J, Lereclus D, Baum J, Dean DH (1998) Revision of nomenclature for the *Bacillus thuringiensis* pesticidal crystal proteins. Microbiol Mol Biol Rev 62:807–813 de Maagd RA, Weemen-Hendriks M, Stiekema W, Bosch D (2000) Domain III substitution in *Bacillus thuringiensis* delta-endotoxin Cry1C domain III can function as a specific determinant for *Spodoptera exigua* in different, but not all, Cry1-Cry1C hybrids. Appl Environ Microbiol 66:1559–1563

Federici BA (1997) Baculovirus pathogenesis. In: Miller LK (ed) The baculoviruses. Plenum Press, New York, pp 33–59

Fernandez S, Groden E, Vandenberg JD, Furlong MJ (2001) The effect of mode of exposure to *Beauveria bassiana* on conidia acquisition and host mortality of Colorado potato beetle, *Leptinotarsa decemlineata*. J Invert Pathol 77:217–226

Freimoser FM, Screen S, Hu G, St Leger R (2003) EST analysis of genes expressed by the zygomycete pathogen *Conidiobolus coronatus* during growth on insect cuticle. Microbiology 149:1893–1900

Gill SS, Cowles EA, Pietrantonio PV (1992) The mode of action of *Bacillus thuringiensis* endotoxins. Annu Rev Entomol 37:615–636

Gujar GT, Kalia V, Kumar A, Singh BP, Mittal A, Nair R, Mohan M (2007) *Helicoverpa armigera* baseline susceptibility to *Bacillus thuringiensis* (Bt) Cry toxins and resistance management for *Bt* cotton in India. J Invert Pathol 95:214–219

Gupta RP, Kalia A, Kapoor S (2007) Biopesticides. In: Bioinoculants: a step towards sustainable agriculture. In: Singh A, Parmar N, Kuhad RC (eds) Bioaugmentation, biostimulation and biocontrol soil biology. New India Publishing Agency, New Delhi, pp 223–256

Hajek AE, St Leger RJ (1994) Interactions between fungal pathogens and insect hosts. Annu Rev Entomol 39:293–322

Harris RS, Harcourt SJ, Glare TR, Rose EA, Nelson TJ (2000) Susceptibility of *Vespula vulgaris* (Hymenoptera: Vespidae) to generalist entomopathogenic fungi and their potential for wasp control. J Invert Pathol 75:251–258

Heimpel AM (1967) A taxonomic key proposed for the species of "crystaloferous" bacteria. J Invertebr Pathol 9:364–375

Henderson CW, Johnson CL, Lodhi SA, Bilimoria SL (2001) Replication of Chilo iridescent virus in the cotton boll weevil, *Anthonomus grandis*, and development of an infectivity assay. Arch Virol 146:767–775

Hernandez CS, Ferre J (2005) Common receptor for *Bacillus thuringiensis* toxins Cry 1Ac, Cry1Fa, and Cry1Ja in *Helicoverpa armigera*, *Helicoverpa zea*, and *Spodoptera exigua*. Appl Environ Microbiol 71:5627–5629

Herrero SJ, Cabrera G, Ferré J, Bakker PL, de Maagd RA (2004) Mutations in the *Bacillus thuringiensis* Cry1Ca toxin demonstrate the role of domains II and III in specificity towards *Spodoptera exigua* larvae. Biochem J 384:507–513

Höfte H, de Greve H, Seurinck J (1986) Structural and functional analysis of a cloned delta endotoxin of *Bacillus thuringiensis* Berliner. Eur J Biochem 161:273–280

Huger A (1963) Granulosis of insects. In: Steinhaus EA (ed) Insect pathology: an advanced treatise, vol 689. Academic, New York/London, pp 531–575

Jakob NJ, Kleespies RG, Tidona CA, Müller K, Gelderblom HR, Darai G (2002) Comparative analysis of the genome and host range characteristics of two insect iridoviruses: Chilo iridescent virus and a cricket iridovirus isolate. J Gen Virol 83:463–470

Jian X, Qin L, Xiang-dong Y, Shu-de Z (2006) A review of recent development of *Bacillus thuringiensis*. ICP genetically engineered microbes. Entomol J East China 15:53–58

Knowles BH (1994) Mechanism of action of *Bacillus thuringiensis* insecticidal δ-endotoxins. Adv Insect Physiol 24:275–308

Kumar S, Chandra A, Pandey KC (2008) *Bacillus thuringiensis* (*Bt*) transgenic crop: an environment friendly insect-pest management strategy. J Environ Biol 29:641–653

Lacey LA, Frutos R, Kaya HK, Vail P (2001) Insect pathogens as biological control agents: do they have a future? Biol Control 21:230–248

Lasa R, Moreno I, Caballero P, Williams T (2009) Application of juvenile hormone analogue and optical brightener technologies to the production of *Spodoptera frugiperda* multiple nucleopolyhedrovirus. IOBC Wprs Bull 45:153–156

Lemaux P (2008) Genetically engineered plants and foods: a scientist's analysis of the issues (part I). Annu Rev Plant Biol 59:771–812. doi:10.1146/annurev.arplant.58.032806.103840
Loc NT, Chi VTB (2007) Biocontrol potential of Metarhizium anisopliae and Beauveria bassiana against diamondback moth, Plutella xylostella. Omonrice 15:86–93
Mannion CM, McLane W, Klein MG, Moyseenko J, Oliver JB, Cowan D (2001) Management of early-instar Japanese beetle (Coleoptera: Searabaeidae) in field-grown nursery crops. J Econ Entomol 94:1151–1161
Martignoni ME, Iwai PJ (1986) A catalogue of viral diseases of insects, mites and ticks, 4th edn. USDA Forest Service PNW-195, Washington, DC
McGaughey WH (1985) Insect resistance to the biological insecticide *Bacillus thuringiensis*. Science 229:193–195
Metchnikoff E (1879) Diseases of the larva of the grain weevil. Insects harmful to agriculture (series). Issue III, The grain weevil. Published by the commission attached to the Odessa Zemstvo office for the investigation of the problem of insects harmful to agriculture, Odessa, 32 p
Mitsuhashi W, Kawakita H, Murakami R, Takemoto Y, Saiki T, Miyamoto K, Wada S (2007) Spindles of an entomopoxvirus facilitate its infection of the host insect by disrupting the peritrophic membrane. J Virol 18:4235–4243
Murphy FA, Fauquet CM, Bishop DHL, Ghabrial SA, Jarvis AW, Martelli GP, Mayo MA, Summers MD (1995) Virus taxonomy; classification and nomenclature of viruses. Sixth report of the international committee on taxonomy of viruses. Springer, Wien/New York. 586 pp
Narayanan K (2004) Insect defence: its impact on microbial control of insect pests. Curr Sci 86:6–25
Phoofolo MW, Obrycki JJ, Lewis LC (2001) Quantitative assessment of biotic mortality factors of the European corn borer (Lepidoptera: Crambidae) in field corn. J Econ Entomol 94:617–622
Pimentel D (2007) Area-wide pest management: environmental, economic and food issues. In: Vreysen MJB, Robinson AS, Hendrichs J (eds) Area-wide control of insect pests: from research to field implementation. Springer, Dordrecht, pp 35–47
Pimentel D (2009) Pesticides and pest control. In: Peshin R, Dhawan AK (eds) Integrated pest management: innovation- development. Springer, Dordrecht, pp 83–88
Prasad A, Veerwal B (2010) Biotoxicity of entomopathogenic fungus *Beauveria bassiana* (*balsamo*) *vuillemin*, against early larval instars of anopheline mosquitoes. J Herb Med Toxicol 4(2):181–188
Rajguru M, Sharma AN, Banerjee S (2011) Feasibility assessment of plant extracts fortified with *Bacillus thuringiensis* Berliner for management of *Spodoptera litura* Fab. Int J Trop Insect Sci 31(1–2):92–97
Rohrmann GF (2008) Baculovirus molecular biology. National Center for Biotechnol. Information (US), NCBI, Bethesda
Romeis J, Meissle M, Bigler F (2006) Transgenic crops expressing *Bacillus thuringiensis* toxins and biological control. Nat Biotechnol 24:63–71
Sarfraz M, Keddie BA (2005) Conserving the efficacy of insecticides against *Plutella xylostella* (L.) (Lep., Plutellidae). J Appl Entomol 129:149–157
Saufi AEM (2008) Characterization of an Egyptian *Spodoptera littoralis* nuclear polyhedrosis virus and a possible use of a highly conserved region from polyhedron gene nucleopolyhedrovirus detection. Virol J 5:13
Sayyed AH, Omar D, Wright DJ (2004) Genetics of spinosad resistance in a multi-resistant field selected population of *Plutella xylostella*. Pest Manag Sci 60:827–832
Simon O, Williams T, Lopez-Ferber M, Taulemesse JM, Caballero P (2008) Population genetic structure determines the speed of kill and occlusion body production in *Spodoptera frugiperda* multiple nucleopolyhedrovirus. Biol Control 44:321–330
Smith RF (1970) Pesticides: their use and limitations in the pest management. In: Labb RL, Guthrie FE (eds) Concepts of pest management., pp 102–113
Solter LF, Becnel JJ (2000) Entomopathogenic microsporida. In: Lacey LA, Kaya HK (eds) Field manual of techniques in invertebrate pathology: application and evaluation of pathogens for control of insects and other invertebrate pests. Kluwer, Dordrecht, pp 231–254

Steinhaus EA (1954) Further observations on *Bacillus thuringiensis* Berliner and other spore forming bacteria. Hilgardia 23:1

Strasser H, Vey A, Butt T (2000) Are there any risks in using entomopathogenic fungi for pest control, with particular reference to the bioactive metabolites of *Metarhizium*, *Tolypocladium* and *Beauveria* species? Biocontrol Sci Technol 10:717–735

Suryawanshi DS, Bhede BV, Bhosale SV, More DG (2008) Insecticide resistance in field population of American bollworm, *Helicoverpa armigera* Hub. (Lepidoptera: Noctuidae). Indian J Entomol 70:44–46

Suzuki A, Taguchi H, Tamura S (1970) Isolation and structure elucidation of three new insecticidal cyclodepsipeptides, destruxins C and D and desmethyldestruxin B, produced by *Metarrhizium anisopliae*. Agric Biol Chem 34:813–817

Tamura S, Kuyama S, Kodaira Y, Higashikawa S (1964) Studies on destruxin B, an insecticidal depsipeptide produced by *Oospora destructor*. Inst Appl Microbiol Symp Microbiol 6:127–140

Tanada Y, Kaya HK (1993) Insect pathology. Academic, San Diego, p 666

Thakore Y (2006) The biopesticide market for global agricultural use. Ind Biotechnol 2:194–208

Tillman PG, Styer EL, Hamm JJ (2004) Transmission of an ascovirus from Heliothis virescens (Lepidoptera, Noctuidae) and effects of the pathogen on survival of a parasitoid, Cardiochiles nigriceps (Hymenoptera, Braconidae). Environ Entomol 33:633–643

Tonka T, Weiser J (2000) Iridovirus infection in mayfly larvae. J Invertebr Pathol 76:229–231

Vaeck M, Reynaerts A, Hofte A (1987) Transgenic plants protected from insect attack. Nature 328:33–37. doi:10.1038/328033a0

Vey A, Hoagland RE, Butt TM (2001) Toxic metabolites of fungal biocontrol agents. In: Butt TM, Jackson C, Magan N (eds) Fungi as biocontrol agents. Progress, problems and potential. CABI Publishing, Oxford, pp 311–346

Williams T (2008) Natural invertebrate hosts of iridoviruses (Iridoviridae). Neotrop Entomol 37:615–632

Winstanley D, O'Reilly DR (1999) Granuloviruses. In: Webster R, Granoff A (eds) The encyclopedia of virology, 2nd edn. Academic, London, pp 140–146

Xavier R, Reena J, Sreeramanan S (2007) Environmental distribution and diversity of insecticidal proteins of *Bacillus thuringiensis* Berliner. Malays J Microbiol 3(2):1–6

Zimmerman G (2007) Review on safety of entomopathogenic fungus *Metarhizium anisopliae*. Biocontrol Sci Technol 17:879–920

Chapter 14
Plant Growth Promoting Rhizobacteria (PGPR): Microbes in Sustainable Agriculture

Jay Shankar Singh and D.P. Singh

Abstract Plant growth promoting rhizobacteria (PGPR) contain diverse type of plant growth promoting attributes that has many beneficial effects on crop productivity. The PGPR appear to promote the plant growth via – suppression of plant disease (bio-controls), enhanced nutrient achievement, or phytohormone production (bio-fertilizers). The PGPR protect plants from several biotic and abiotic stresses. Co-inoculation of PGPR can ease the adverse effects on crop plants due to various environmental stresses such as soil salinity, droughts, temperature and nutrient deficiency. During the last couple of decades given the negative environmental impact of artificial fertilizers and their increasing costs, the use of PGPR for sustainable environment and safe agriculture has increased globally. Thus, the PGPR offer an environmentally sustainable approach to increase crop production for future generation.

Keywords Agriculture • Biofertilizers • Bio-control • Environment • Rhizobacteria

1 Introduction

The bacteria colonizing the plant rhizosphere are known as plant growth promoting rhizobacteria (PGPR) (Kloepper et al. 1980). Most of the PGPR strains are found to be associated with plant growth indirectly, via alterations in the structure of rhizosphere soil

J.S. Singh (✉)
Department of Environmental Microbiology, Babasaheb Bhimrao Ambedkar (Central) University, Raibarelly Road, Lucknow 226025, Uttar Pradesh, India
e-mail: jayshankar_1@yahoo.co.in

D.P. Singh
Department of Environmental Science, Babasaheb Bhimrao Ambedkar (Central) University, Raibarelly Road, Lucknow 226025, Uttar Pradesh, India

A. Malik et al. (eds.), *Management of Microbial Resources in the Environment*,

DOI 10.1007/978-94-007-5931-2_14, © Springer Science+Business Media Dordrecht 2013

(Noel et al. 1996). Direct mechanisms include production of plant growth regulators, solubilization of mineral materials (Son et al. 2006; Chen et al. 2008) or fixation of atmospheric nitrogen. For example, *Bacillus* strains induce plant resistance against stress and produce various plant hormones for growth improvement (Rajendran et al. 2008). In addition it has been demonstrated that inoculation with plant growth promoting improves plant growth under a variety of salinity stress conditions (Han and Lee 2005).

In the last few years, the number of PGPR like *Pseudomonas, Azospirillum, Azotobacter, Klebsiella, Enterobacter, Alcaligenes, Arthrobacter, Burkholderia, Bacillus* and *Serratia* have been identified and have gained importance because their dominant role in the microcosm of rhizosphere (Kloepper et al. 1989; Joseph et al. 2007). Now many PGPR inoculants are commercialized which apparently promote the plant growth through at least one mechanism; improved nutrient supply, biofertilizers, bio-control agent and phytohormone production. (Rajendran et al. 2008; Singh et al. 2011; Upadhyay et al. 2012).

Environmental stresses are the major constraints, which hampers the crop production, causing loss of significant quantity of yield. Soil salinity is one of the major environmental constrains that significantly reduces plant nutrient uptake, particularly phosphate (P) ions which tends to get precipitated with Ca ions in saline soil (Grattan and Grieve 1999). Phosphate solubilizing bacteria can increase P availability to plants by solubilizing the insoluble P, and thereby improve the availability of nutrients including the phosphate (Gyaneshwar et al. 2002). Application of such beneficial soil microorganisms in agriculture has drawn the attention of scientists all over the world (Berg 2009; Weyens et al. 2009). Inoculation of PGPR strains, well adapted to rhizospheric conditions and exhibiting P solubilizing attributes, are considered more effective for balanced nutrition to plants and crop yields (Galal 2003). The PGPR can stimulate not only plant growth and yield, but can also alleviate the effects of biotic or abiotic stresses on plants (Lugtenberg and Kamilova 2009). An increase in various agronomic yields involving production of growth stimulating phytohormones as well as by improved phosphate nutrition of plants due to PGPR has been reported (Kohler et al. 2006). Therefore, PGPR inoculation to alleviate environmental stress may be considered as an innovative and cost effective alternative to overcome the various plants stresses (Bano and Fatima 2009). Biofertilizer is the most commonly referred term used for the beneficial soil microorganisms such as PGPR which increase the availability and uptake of nutrients for plants (Vessey 2003). Application of such beneficial soil microorganisms in sustainable agriculture has drawn the attention of scientists all over the world (Berg 2009; Weyens et al. 2009). This review describes the future perspectives of PGPR in sustainable agricultural productivity.

2 Perspective of PGPR in Stress Agriculture

The PGPR associated with plant roots play important role in enhancing the plant productivity and disease resistance. Recently several workers have demonstrated that PGPR provide protection to the plants against several biotic and abiotic stresses.

Table 14.1 Inoculation with PGPR containing ACC deaminase and subsequent physiological changes in plants

PGPR	Plant	Physiological changes	References
Pseudomonas cepacia	*Glycine max*	Rhizobacterium caused an early soybean growth	Cattelana et al. (1999)
Alcaligenes sp. *Bacillus pumilus* *Pseudomonas* sp. *Variovorax paradoxus*	*Brassica napus*	Inoculated plant demonstrated more vigorous growth than the control	Belimov et al. (2001)
*B. circulans*DUC1, *B. wrmus*DUC2, *B. globisporus*DUC3	*Brassica campestris*	Bacterial inoculation enhanced root and shoot elongation	Ghosh et al. (2003)
*Enterobactersakazakii*8MR5 *Pseudomonas* sp. MKS8 *Klebsiellaoxytoca*10MKR7	*Zea mays* L.	Inoculation increased agronomic parameters of maize	Babalola et al. (2003)
P. syringae; Mk20, *P. fluorescens*Mk25, and *P. fluorescens* biotype G	Mung bean	Improve seedling growth and nodulation	Ahmad et al. (2011)
Bacillus subtilis SU47 and *Arthrobacter* sp. SU18	Wheat	Increase in dry biomass, total soluble sugars and proline content	Upadhyay et al. (2012)

The PGPR dependent enhanced stress resistance in plants may be mediated by specific enzymes, inducing alterations at physiological and molecular level. Among these enzymes, bacterial 1-aminocyclopropane-1-carboxylate (ACC) deaminase is reported to play a well defined role in the regulation of level of ethylene- a plant growth hormone (Glick 2005). Ethylene is a gaseous plant growth hormone produced endogenously by almost all plants and is known to alter the growth and development of plants (Saleem et al. 2007). Apart from the role of ethylene as a plant growth regulator, it is an established stress hormone in plants. These PGPR boost plant growth particularly under stressed conditions by the regulation of accelerated ethylene production in response to a multitude of abiotic and biotic stresses like salinity, drought, waterlogging, temperature, pathogenicity and contaminants. Recently, Ahmad et al. (2011) showed that co-inoculation with PGPR containing ACC deaminase and *Rhizobium* spp. could be a useful approach for inducing salt tolerance and thus improving growth and nodulation in mung bean under salt-affected conditions. Inoculation of PGPR containing ACC deaminase and subsequent physiological changes in plants are given in Table 14.1.

2.1 Salinity Stress

Soil salinity constitutes a serious problem for vegetable as well as for other crops. An alteration in physiology of plants induced by salt stress leads to reduced nutritional uptake and also the plant growth (Singh et al. 2011). Salinity stress boosts endogenous ethylene production in plants, which in most cases serves as a stress hormone (O'Donnell et al. 1996; Blumwald 2000). It is very likely that reducing salinity-induced ethylene production by any mechanism could reverse the negative impact of salinity onto the plant growth. Since the PGPR is known to colonize the plant roots (Upadhyay et al. 2011), it can be an effective tool in developing the strategies to enhance the wheat production in salinity affected areas (Ashraf et al. 2004). We very recently during an experiment demonstrated that co-inoculation of *B. subtilis* and *Arthrobacter* sp. might ease the unfavorable influence of the soil salinity on wheat plants (Upadhyay et al. 2012). Studies have revealed that plants inoculated with PGPR containing ACC deaminase were better able to thrive through the salinity stress while demonstrating a normal growth pattern. In this direction, Mayak et al. (2004b) during an investigation showed that *Achromobacter piechaudii* having ACC deaminase activity considerably enhanced the tomato seedlings fresh and dry weights when treated with NaCl salt (up to 172 mM). The bacterium *Achromobacter piechaudii* has been found to reduce the generation of ethylene by salt stressed tomato seedlings. However, the sodium content of the plant was not decreased whereas the uptake of phosphorous and potassium were slightly increased, which might have contributed in part, to the activation of processes involved in the alleviation of the adverse effect of salt on plant growth. The bacterium also increased the water use efficiency (WUE) in saline environment and helped in alleviating the salt suppression of photosynthesis. Saravanakumar and Samiyappan (2007) reported that *Pseudomonas fluorescens* strain TDK1 containg ACC deaminase activity enhanced the saline resistance in groundnut plants and increased yield as compared with that inoculated with *Pseudomonas* strains lacking ACC deaminase activity. Cheng et al. (2007) have also confirmed that ACC deaminase bacteria conferred salt tolerance onto plants by lowering the synthesis of salt-induced stress ethylene and promoted the growth of canola in saline environment. Nadeem et al. (2006) have observed almost similar results in the case of maize growth under salt stress in response to inoculation with ACC deaminase PGPR.

2.2 Drought Stress

Drought is one of the major constraints to the yield of crop globally. Since the global population continues to increase and water resources for crop production decline, the development of drought-tolerant cultivars and water-use-efficient crops is a global concern. Drought affects virtually all climatic regions of the world (Wilhite 2000) and more than one half of the earth is susceptible to drought every year (Kogan 1997) Drought is one of the major environmental stresses that limit the growth of plants and the production of crops. Plants respond to drought stress at cellular and molecular levels (Bray 1997; Ingram and Bartels 1996; Shinozaki and Yamaguchi-Shinozaki 1999). Like many other environmental factors, drought also induces accelerated ethylene production in plant tissues which leads to abnormal growth of a plant (Mattoo and Suttle 1991). Mayak et al. (2004a) reported that ACC deaminase in *Achromobacter piechaudii* ARV8 significantly increased the fresh and dry weights of both tomato and pepper seedlings exposed to transient water stress. In addition, the bacterium also reduced the production of ethylene by tomato seedlings exposed to water stress. During water scarcity, the bacterium did not influence the water content of plants; however, it significantly improved the recovery of plants when watering was resumed. Interestingly, inoculation of tomato plants with the bacterium resulted in continued plant growth both during water stress as well as when watering was resumed. The bacterial effects were more pronounced and more consistent under controlled soil drying (moisture stress conditions). In short term experiments, the ACC deaminase producing bacteria showed positive effects on the root and shoot biomass, leaf area and plant transpiration. In case of long-term experiments, plants inoculated with ACC deaminase producing bacteria gave higher yield (25–41%); greater seed number and seed nitrogen accumulation than that in the uninoculated plants. Moreover, PGPR inoculation of drought affected pea plants restored the level of root nodulation when compared with well-watered uninoculated plants. It has been shown that the inoculation with ACC deaminase producing bacteria partially eliminated the effects of water scarcity on growth and yield of *Pisum sativum* L. both in pot and field conditions (Arshad et al. 2008).

Several investigations have been carried out under greenhouse and field situations by using mixtures of PGPR strains with symbiotic nitrogen-fixing rhizobia (Figueiredo et al. 2008) or with mycorrhizal fungi (Kohler et al. 2008). The co-inoculation of PGPR with *Rhizobium tropici* under low soil water content conditions has been found to show improved plant growth and enhanced number of root nodules in *Phaseolus vulgaris* L. (Figueiredo et al. 2008). Interestingly, the effect of co-inoculation of two strains of *P. polymyxa* strains on nodule number exhibited synergistic effects.

A survey on the effect of drought stress in relation to plant hormone revealed that how an increase in the abscisic acid (ABA) content in the plant leaves with simultaneous decrease in the endogenous level of cytokinin elicited a differential response of stomata closure (Cowan et al. 1999; Figueiredo et al. 2008). Earlier the cytokinin – ABA antagonism has been observed as they share a common biosynthetic origin (Cowan et al. 1999). It will be interesting to determine whether cytokinin produced by *P. polymyxa* affects ABA signaling in plants or rhizobia-elicited nodulation (Timmusk and Wagner 1999; Figueiredo et al. 2008) Co-inoculation of lettuce (*Lactuca sativa* L.) with PGPR *Pseudomonas mendocina* and arbuscular mycorrhizal fungi (*Glomus intraradices* or *G. mosseae*) is known to increase the level of antioxidant enzyme catalase under severe drought conditions, suggesting that these can be used as co-inoculants to alleviate the oxidative damage elicited by the drought stress (Kohler et al. 2008).

2.3 Temperature Stress

Changes in the global climate, notably spatial and temporal variations in temperature, are predicted to have important consequences for crop production. Transient or constantly high temperatures may cause an array of morpho-anatomical, physiological and biochemical changes in plants, which induce changes in the growth and development of plants, and ultimately account for severe decrease in the economic yield. Plants with ACC deaminase expression can successfully cope with the stressful situations by lowering the production of ethylene. Bensalim et al. (1998) reported that a plant growth promoting rhizobacterium *Burkholderia phytofirmans* strain PsJN was able to maintain normal growth of potato plant under heat stress. Barka et al. (2006) reported that ACC deaminase activity of the same bacterium (*Burkholderia phytofirmans* strain PsJN) enhanced plant growth and physiological activity of grapevine (*Vitis vinifera* L.) cv. Chardonnay explants at both ambient (26°C) and low (4°C) temperature conditions. Inoculation of PGPR was able to increase the root growth (11.8- and 10.7-fold at 26 and 4°C, respectively) and plantlet biomass (6- and 2.2-fold at 26 and 4°C, respectively). It was inferred that this bacterium can successfully improve the cold tolerance property of the plantlet when compared with that of the non-bacterized control. Cheng et al. (2007) demonstrated that ACC deaminase producing bacterium *P. putida* UW4 have also promoted the growth of psychrotolerant Canola plant at extremely low temperature under the salt stress. These studies have clearly demonstrated the potential of ACC deaminase enzyme in normalizing the plant growth exposed to temperature extremes by lowering the accelerated production of ethylene.

The various kinds of stresses in fact accentuate the biosynthesis of ethylene, which in the most cases retards the plant growth through several mechanisms at molecular level. In the present scenario, the application of PGPR containing ACC deaminase activity is considered vital in regulation of ethylene production in plants. Thus, application of PGPR containing ACC deaminase in agriculture might prove beneficial and could be a sound step towards sustainable crop production.

2.4 Nutrient Stress

Another crucial abiotic stress faced by the crop plants is supply of inadequate soil nutrients. Although soil fertilization is typically required for better agricultural production. However, external application of nitrate and phosphate fertilizers eventually contaminate the surface and ground waters. Phosphate and nitrate run-off is associated with eutrophication of surface water bodies, resulting in ecological nuisance in aquatic ecosystem. Application of fertilization of soil is attributable to low nutrient level, which is created due to mutual interaction of nutrients such as phosphorous interaction with highly reactive iron, aluminium and calcium minerals in soil, resulting into locking of up to 90% of the soil available phosphorous (Gyaneshwar et al. 2002) and rendering it unavailable to the plants. The PGPR have potential to maintain adequate plant nutrition and also reduce the negative environmental impact of fertilizers. Plant growth promotion by some PGPR has been associated with improved availability of soil nutrients as well as improved uptake of nutrients (Gyaneshwar et al. 2002). PGPR have been known to improve the nitrate uptake by plants (Mantelin and Touraine 2004; Adesemoye et al. 2008). It has been reported that a general increase in the plant growth and nutrient uptake by PGPR is due to profuse root development (Mantelin and Touraine 2004) and altered root structure due to production of phytohormones like indole acetic acid (IAA) (Adesemoye et al. 2008). An increase in root surface area and numbers of root tips can contribute to improved system tolerance and plant defense against pathogens. An increase in the nutrient uptake efficiency of plants due to PGPR is also suggested to be mediated by stimulated activity of proton pump ATPase (Mantelin and Touraine 2004).

Owing to the ever increasing price of chemical fertilizers and its adverse environmental impact, there is a growing need worldwide to reduce the application of chemical fertilizers. Hence, several studies are now testing the hypothesis that PGPR might enable us to maintain agricultural productivity with reduced application of fertilizers. The preliminary results are found to be promising. The use of PGPR isolates as inoculants is reportedly highly beneficial for rice cultivation as they can enhance the growth of rice by inducing other plant growth promoting traits (Ashrafuzzaman et al. 2009). During a field study with wheat plant (*Triticum aestivum* L.), the total yield of crop plants, given only 75% of the recommended doses of N-P-K fertilizer plus a PGPR strain was comparable to the yield for plants given full dose of fertilizer, but without PGPR (Shaharoona et al. 2008). In another study on tomato (Hernanndez and Chailloux 2004), the dry weight of tomato transplants grown with two PGPR strains and 75% fertilizer dose in the greenhouse was significantly higher than that with the full dose of fertilizer and without PGPR. After transplanting the plants in the field, yields with some combinations of PGPR and mycorrhizal fungi at 50% recommended field fertilization were greater than the yield in plants treated with full dose of fertilizer, but without microbes.

Another current hypothesis is that PGPR, used as components of integrated nutrient management systems, can help reduce the build-up of nutrients in fertilized soils. Support for this hypothesis was presented in a report (Adesemoye et al. 2008) of a 3-year field study on maize that evaluated PGPR with and without mycorrhizal fungi, manure and inorganic fertilizer, as well as with and without tillage. Significant increases in grain yield from microbial treatments were accompanied by increased nitrogen content per gram of grain tissue and removal of significantly higher amounts of nitrogen, phosphorous and potassium. Therefore, within the tested nutrient management system, PGPR contributed significantly to reducing nutrient build up in the soil. Many current studies are underway that will further define the utility of PGPR in nutrient management strategies aimed at reducing fertilizer application rates and nutrient runoff from agricultural sources.

3 Perspective of PGPR as Biofertilizers

The word biofertilizer may be defined as living microorganisms, which can promote growth of plants by increasing the availability of primary nutrients to the host plant. More recent research findings indicated that the treatment of agricultural soils with PGPR inoculation significantly increases agronomic yields when compared with that of uninoculated soil. There are many mechanisms by which the PGPR can promote the growth of plants. Some of the important mechanisms by which PGPR can enhance the growth and yield of crop plants are listed below:

(1) PGPR mediated N_2 fixation
(2) Availability of nutrients in the rhizosphere
(3) Root surface area enhancement
(4) Phytohormone synthesis.

3.1 N_2 Fixation

The PGPR as microbial inoculants have been accepted as an alternative source of N-fertilizers and they can promote plant growth and productivity. They are considered environment friendly and can be used to ensure a sustainable crop production. In the biofertilizer technology, efforts are being made to increase biological nitrogen fixation for cereals and other non-legumes, by introducing the N_2-fixing bacteria in plant roots (Cocking 2000). The plant growth promotion by means of nitrogen fixation is an important criterian used for selection of rhizobacteria as an effective biofertilizer. Application of both symbiotic and free-living N_2-fixing bacteria (PGPR) has shown a considerable beneficial effect on plant growth (Kloepper et al. 1980; Bashan and Holguin 1998). The bacterial species belonging to genera *Azospirillum*, *Alcaligenes*, *Arthrobacter*, *Acinetobacter*, *Bacillus*, *Burkholderia*, *Enterobacter*, *Erwinia*, *Flavobacterium*, *Pseudomonas*, *Rhizobium* and *Serratia* are known to be associated with the plant rhizosphere and are found to show beneficial effect on plant growth (Tilak et al. 2005; Egamberdiyeva 2005). There is an important role played by plant root exudates in selecting and enriching the compatible bacterial strains. Therefore, the type of bacteria colonizing the rhizosphere of plants depends upon the nature and concentrations of organic constituents of root exudates, but also on the ability and efficiency of the microorganism to utilize the organic source present in the root exudates (Curl and Truelove 1986). The rhizospheric bacteria, that have efficient systems for uptake and catabolism of organic constituents of root exudates (Barraquio et al. 2000), can get attached with the root surfaces and derive maximum benefit from root exudates. The associative interactions of plants with microorganisms must have evolved as a result of co evolution. The use of latter group as bio-inoculants, for a long-term successful and sustainable interaction, envisages that microbes must be pre-adapted (Chaiharn et al. 2008). The PGPR application as inoculants offers an attractive and environment friendly technology to replace the chemical fertilizers and pesticides (Ashrafuzzaman et al. 2009). The use of bio-fertilizer and bio-enhancer such as N_2 fixing bacteria would not only reduce the dependence of farmers on the use of chemical fertilizers applications, but also reduce the cost of agricultural production. Utilization of PGPR is a viable alternative to organic fertilizers, which will help in preserving the environment and reducing the pollution (Stefan et al. 2008). Thus, the PGPR or combinations of PGPR and AMF can be exploited for improving the nutrient use efficiency of plants (Adesemoye et al. 2009). Applying the combined inoculation of PGPR as biofertilizer affects beneficially the yield and growth of chickpea in field conditions (Rokhzadi et al. 2008). Application of PGPR strains, especially *Azospirillum* spp. was reported to fix N_2 in oil palm (*Elaeis guineensis*) and sweet potato (*Ipomoea batatas*). *Bacillus sphaericus* UPMB10 was observed to produce beneficial effects on oil palm (Amir et al. 2001). In sweet potato, PGPR inoculation along with 33% the total N-fertilizer requirement produced a plant biomass comparable to yield obtained with fully fertilized plants, which represents about 67% saving of N-fertilizer (Saad et al. 1999). Halimi et al. (2000) observed that the PGPR can supplement the nutrient requirement of tomato on soilless culture media under protected environment. *Azospirillum* inoculation process has reportedly increased the N_2 fixation, mineral nutrient content (P, K, Ca and Mg) and growth of maize (Rai and Hunt 1993).

3.2 *Enhanced Rhizospheric Nutrient Status*

There is ample evidence that the primary mode of PGPR action is related with enhanced availability of nutrients to plant rhizosphere (Glick 1995; Rodriguez and Fraga 1999). The method by which availability of nutrients increases involves solubilization of unavailable form of nutrients The solubilisation of phosphate and production of siderophores help in the availability and transport of certain nutrients (notably ferric iron).

3.2.1 Phosphate Solubilization

Phosphorus (P) is second to nitrogen in mineral nutrients which commonly limits the growth of terrestrial plants. Ironically, there is large reserve of total P in the soil, but the amount of phosphorous available to plants is usually a tiny fraction of the total amount (Stevenson and Cole 1999). The reduced availability of P to plants in the vast majority of soil is due to insoluble form of P and plants can only absorb P in two soluble forms i.e., the monobasic ($H_2PO_4^-$) and the dibasic (HPO_4^{2-}) ions (Glass 1989). P-solubilizing bacteria are commonly present in the rhizosphere secrete organic acids which converts the non-available form of P to soluble form. The phosphatases enzyme present in the P solubilising bacteria facilitates the conversion of insoluble forms of P to soluble form (Kim et al. 1998). The solubilization of P in the rhizosphere is considered as the most common action of PGPR in the direction of improving the P availability to the host plants (Richardson 2001). Few examples of such actions include associations between *Azotobacter chroococcum* with wheat, *Bacillus* sp. and five crop species, *Enterobacter agglomerans* with tomato, *Pseudomonas chlororaphis* and *P. putida* with soybean, *Rhizobium* sp. and *Bradyrhizobium japonicum* with radish, *Rhizobium leguminosarum* with *Phaseolus* (Antoun et al. 1998; Kim et al. 1998; Pal 1998; Chabot et al. 1998; Kumar and Narula 1999; Singh and Kapoor 1999; Cattelana et al. 1999).

Phosphate-solubilizing bacteria are common in rhizospheres (Nautiyal et al. 2000; Vazquez et al. 2000). However, their ability to solubilize P with no extra input suggested that a rhizospheric P solubilizing bacterium can be considered as PGPR. Cattelana et al. (1999) found only two out of five rhizospheric isolates positive for P solubilisation, which registered a positive impact on soybean seedling growth. Thus, not all P solubilizing PGPR are able to increase the plant growth by increasing the P availability to the hosts. de Freitas et al. (1997) isolated number of P-solubilizing *Bacillus* sp. and a *Xanthomonas maltophilia* isolate from canola (*Brassica napus* L.) rhizosphere, which had positive effects on plant growth, but no effects on P content of the host plants.

3.2.2 Iron Absorption by Siderophore Production

The PGPR are reported to secrete some extracellular metabolites called siderophores. For the first time, Kloepper et al. (1980) reported the significance of siderophores in plant growth promotion produced by certain genera of PGPR. Siderophores are commonly referred to as microbial chelating agent of Fe. The presence of siderophore producing PGPR in rhizosphere can enhance the Fe supply to plants and thereby, improve the plant growth and yield. Further, this compound after chelating Fe^{3+} make the soil Fe^{3+} deficient for other soil microbes and consequently inhibits the activity of other competing microbes.

Iron (Fe) is an essential nutrient of plants, but it is relatively insoluble in soil solutions. Plant roots prefer absorption of iron in its reduced form i.e., ferrous (Fe^{2+}) ion. The ferric (Fe^{3+}) ion is more common in well aerated soil and tends to get precipitated in the form of iron-oxide (Salisbury and Ross 1992). Plants normally excrete various soluble organic compounds (chelators and phytosiderophores) which tend to bind with Fe^{3+} form and prevent it precipitation. Chelators help in the transport the Fe^{3+} to the root surface where it is reduced to Fe^{2+} and is ready for absorption. Phytosiderophores, excreted by grasses, are absorbed with the Fe^{3+} across the plasmalemma (von Wiren et al. 2000).

There is evidence that a number of plant species can absorb bacterial Fe^{3+} siderophore complexes (Bar-Ness et al. 1992; Wang et al. 1993). However, the significance of bacterial Fe^{3+} siderophore being taken up by plants important mode of iron nutrition in plants (Duijff et al. 1994), it is more vital process especially in calcareous soils (Masalha et al. 2000). There is another school of thought which subscribes to the theory that contribution of bacterial siderophores to the overall iron requirements of plants is small (Glick 1995). Bar-Ness et al. (1992) had earlier supported the concept of bacterial siderophore uptake by plants (Bar-Ness et al. 1992), concluded that two bacterial siderophores (pseudobactin and ferrioxamine B) were inefficient as iron sources for plants and that rhizospheric siderophore-producing bacteria can be in competition with the plant for iron. A vast majority of the researchers believe that the microbial siderophores in the rhizosphere are closely associated with the bio-control activities in the region due to their ability to create an iron deficient condition for competing plant pathogens (Hiifte et al. 1994).

3.3 Root Surface Area Improvement

Despite wide ranging impact of PGPR on the solubility and availability of soil nutrients, they are known to affect the root density, surface area and morphoanatomy of the plant. More specifically, enhanced root surface area can have a huge influence on the nutrient uptake efficiency (Vessey 2003). The plant growth promoting attribute of the PGPR is mainly associated with morphological and physiological changes in the inoculated plant roots and which sufficiently improve the water and mineral intake (Sarig et al. 1988).

So far the accumulated evidences on the positive effects of biofertilizing-PGPR points to bacteria-mediated changes in root growth and morphology. Bacterial mediated increases in root weight are commonly reported in response to PGPR inoculations (Vessey and Buss 2002), particularly an increase in the root length and surface area (Galleguillos et al. 2000; German et al. 2000; Holguin and Glick 2001). Fallik et al. (1994) reported that maize plants inoculated with *Azospirillum brasilense* resulted in a proliferation of root hairs and exhibited profound impact on the root surface area. The root length and root surface area are considered important parameters more effective evaluation of PGPR potential, just mention of increase in the root weight i. e. not sufficient. For instance, the clipped soybean roots in the presence of *A. brasilense* Sp7 showed about 63% increase in the root dry weight, but there was about sixfold increase in specific root length (root length per unit root dry weight), and more than tenfold increase in total root length (Molla et al. 2001).

3.4 Production of Phytohormone by PGPR

Accumulating evidences indicated that PGPR influence the plant growth and development by bringing about changes in the level of phytohormones such as auxins, gibberellins, and cytokinins. The effects of auxins on plant growth are found to be concentration dependent, where lower concentrations stimulate the growth and higher concentrations reduce the overall growth (Arshad and Frankenberger 1991). However, different plant seedlings respond differently to variable level of auxin concentrations (Sarwar and Frankenberger 1994). The microorganisms (Ahmad et al. 2005) which produce the highest amount of auxins i.e. indole acetic acid (IAA) and indole acetamide (IAM) in non-sterilized soil, cause maximum increase in growth and yield of the wheat crop (Khalid et al. 2004). Even few strains showing low rate IAA production, if secreted continuously, also exhibit improved plant growth (Tsavkelova et al. 2007). It has been observed that addition of IAA to soil not only improves the root and shoot weight, but also ensures better survival of rhizosphere bacteria (Narula et al. 2006). The *Dendrobium moschatum*-originally isolated from the roots of the epiphytic orchid and the strains of *Rhizobium*, *Microbacterium*, *Sphingomonas*, and *Mycobacterium* genera are among the most active IAA producers (Tsavkelova et al. 2007). The species of *Pseudomonas* and *Bacillus* are also considered to promote the growth of plants, but the production of phytohormones or growth regulators is not characterized. But they are known to induce the greater amounts of fine roots which have the effect on the absorptive surface of plant roots and uptake of water and nutrients. Rhizobia were the first group of bacteria characterized for production of IAA which helps and promotes the growth and pathogen resistance in plants (Mandal et al. 2007; Basu and Ghosh 2001; Ghosh and Basu 2002; Roy and Basu 2004). Sridevi and Mallaiah (2007) also showed that all the strains of *Rhizobium* isolated from root nodules of *Sesbania sesban* (L) Merr. were able to produce IAA. The *Rhizobium* sp. isolated from the root nodules of plant *Vigna mungo* (L) Hepper showed high levels of IAA and was made available to young and healthy root nodules (Mandal et al. 2007). All the *Rhizobium* spp. isolated from *Crotalaria* sp. showed IAA production, but the isolates differ significantly in the level of auxin production, perhaps depends upon the cultural conditions. The studies conducted so far indicated that Rhizobia can be effectively used as both bioenhancer and biofertilizer for increasing the crop production as it can easily improve the nutrient uptake (N, P and K) by producing IAA and subsequently increases the plant root system (Etesami et al. 2009). Among all the isolates maximum amount of IAA is produced by isolate from *C. retusa* (Sridevi et al. 2008). Independent of the origin (rhizosphere vs. phyllosphere), the isolated bacterial strains produced IAA, which had positive impact on the growth of pea and wheat plants. Among them, the highest concentration of IAA was particularly produced by the bacterial strain *P. fluorescens* and *Kocuria varians* (Ahmad et al. 2005; Egamberdieva 2008). While working on chickpea, Joseph et al. (2007) observed that all the isolates of genus *Bacillus, Pseudomonas* and *Azotobacter* were able to produce IAA. On the other hand, about 85.7% isolates of genus *Rhizobium* exhibited IAA production. Choi et al. (2008) have reported that *Pseudomonas fluorescens* B16 is a plant growth-promoting rhizobacterium and it produces Pyrroloquinoline Quinone which is a plant growth promotion factor. Even the plant growth promoting attribute of nitrogen fixing *Azotobacter* strain is associated with production of phytohormone rather than to its diazotrophic activity. *Pseudomonas* bacteria, especially *P. fluorescens* and *P. putida* are the most important PGPR strains known to produce auxin and promote the plant yield.

In fact, a variety of auxin like substances such as indole-3-acetic acid (IAA), indole-3-pyruvic acid, indole-3-butyric acid and indole lactic acid (Costacurta et al. 1994; Martinez-Morales et al. 2003); cytokinins (Horemans et al. 1986; Cacciari et al. 1989) and gibberellins (Bottini et al. 1989) are known as growth promoters. But the production of auxin is comparatively and quantitatively most important for plants (Barassi et al. 2007). Khakipour et al. (2008) tried to evaluate the auxin production potential of *Pseudomonas* strains through chromatography, using HPLC devise; compared the methods of IAA synthesis in *Azospirillum brasilense* strain SM which has potential to trigger the IAA accumulation in Sorghum plants under nutrient stresses. Further, it also exhibited ability to promote the growth of number of other plants like Mung bean, Wheat and Maize (Malhotra and Srivastava 2008). Some of the P-solubilizing bacteria (PSB) and fungi (PSF) are also known as plant growth promoters due to their ability to produce IAA but there is a differential rate of IAA production among different PSB and PSF isolates

(Souchie et al. 2007). *Bacillus megaterium* isolated from tea plant rhizosphere was found to contribute positively to the plant growth due to its ability to produce IAA (Chakraborty et al. 2006). It has been suggested that the cytokinin receptors in *B. megaterium* also play a complimentary role in plant growth promotion (Ortiz-Castro et al. 2008).

4 PGPR as Bio-control Agents

The plant growth promoting bacteria colonizing the rhizosphere of plants produce substances, which not only increase the growth of plants, but also protect them against various diseases. They are known to suppress a broad spectrum of bacterial, viral, fungal and nematode diseases. The PGPR provide protection to the plants against pathogens by direct antagonistic interactions with pathogens and work as bio-control agents. As the PGPR are indigenous to soil and the plant rhizosphere, it is easier for them to play the role of bio-control agents against plant pathogens. On the other side, they can also induce the host resistance against plant disease. Most of the studies demonstrating the plant protection by PGPR have been carried out in laboratory and greenhouse. The results obtained so far under the field condition have been inconsistent. Recent progress in our understanding of their diversity, colonizing ability, and mechanism of action, formulation and application should be fully utilized to facilitate and devise the cost effective plant protection technology against plant pathogens. Some of these rhizobacteria may also be tried as a part of integrated pest management programmes. Greater the application of PGPR be encouraged in plant protection and biofertilization (Siddiqui 2006). The isolation of bacterial strains from the rhizosphere of *Lolium perenne* rhizosphere are found to be effective as plant growth promoters and as bio-control agents and subsequently resulting into enhanced yield (Shoebitz et al. 2007). The *Pseudomonades* – a major group of rhizobacteria with potential for biological control (Kremer and Kennedy 1996), are ubiquitous and are present in agricultural soils. A great deal of information is available regarding the process of root colonization by *pseudomonads* and also the biotic/abiotic factors regulating the colonization, bacterial traits and genes, which confer the special attributes to this soil bacteria for rhizosphere competence and suppression of pathogen (Weller 2007). *Pseudomonads* possess many more traits that make them well suited as bio-control and growth-promoting agents (Weller 1988). These include the ability to (i) grow rapidly *in vitro* and produce larger biomass; (ii) efficient utilization of seed and root exudates; (iii) ability to colonize and multiply in the rhizosphere environments and within the plant tissue; (iv) ability to produce a wide spectrum of bioactive metabolites (i.e., antibiotics, siderophores, volatiles, and growth-promoting substances); (v) compete aggressively with other microorganisms; and (vi) adapt to environmental stresses. In addition, pseudomonads are responsible for the natural suppression of growth of some soil borne pathogens (Weller et al. 2002). The major weakness of pseudomonads as bio-control agents is their inability to produce resting spores like many other *Bacillus* spp. which restricts the wide application of the bacteria for commercial use. *Fluorescent pseudomonas* spp. has been studied for decades for their plant growth-promoting effects through effective suppression of soil borne plant diseases. Among various bio-control agents, *Fluorescent pseudomonads*, are well equipped with multiple mechanisms of bio-control as well as plant growth promotion (Banasco et al. 1998; Dileep et al. 1998). They produce a wide spectrum of compounds like antibiotics, chitinolytic enzymes, P solubilising enzymes, growth promoting hormones, siderophores, HCN and catalase enzyme (Kraus and Loper 1995). *Pseudomonas fluorescens* MSP-393 has been tried as plant growth-promoting rhizobacterium as well as bio-control agent for rice crop grown in saline soils of coastal areas (Paul et al. 2006). Cold-tolerant fluorescent *Pseudomonas* isolated from Garhwal district of Uttarakhand has shown its potential both as plant growth promoter and bio-control agents for pea (Negi et al. 2005).

Bacillus subtilis, an endospore forming bacterium predominantly an inhabitant of soil, has been widely recognized as a powerful bio-control agent as it produces different biologically active compounds with a broad spectrum of activity (Nagorska et al. 2007). *Bacillus megaterium* – an isolate from tea plant rhizosphere can solubilize phosphate, produce IAA, siderophore and antifungal metabolites and thereby, it helps in the plant growth promotion and protection of plant diseases (Chakraborty et al. 2006). Two strains *Bacillus thuringiensis* (*kurstaki*) and *B. sphaericus* have also been reported to have the ability to solubilise phosphates and help in the control of the lepidopteron pests (Seshadri et al. 2007).

In recent years, role of siderophore producing PGPR have been implicated in bio-control of soil-borne plant pathogens. Now the Microbiologists have developed the techniques for introduction of siderophore producing PGPR in soil system through seed, soil or root system. The suppression of plant pathogens by PGPR, can indirectly contribute to enhancement in the plant growth/ yield via a variety of mechanisms. These include:

- The ability to produce siderophores (as discussed above) that chelate iron, making it unavailable to pathogens.
- The capacity to synthesize anti-fungal metabolites such as antibiotics, fungal cell wall-lysing enzymes, or hydrogen cyanide, which suppress the growth of fungal pathogens.
- The ability to successfully compete with pathogens for nutrients or specific niches on the root; and the ability to induce systemic resistance.

Among the various PGPRs identified, *Pseudomonas fluorescens* is considered as one of the most extensively used rhizobacterium because of its antagonistic action against several plant pathogens. Banana bunchy top virus (BBTV) is one of the deadly viruses which severely affects the yield of banana (*Musa* spp.) crop in Western Ghats, Tamil Nadu, India. It has been demonstrated that application of *P. fluorescens* strain significantly reduces the incidence of BBTV disease incidence in the banana under both greenhouse and field conditions. Different PGPR spp. as bio-control agents against various plant diseases has been given in Table 14.2.

Table 14.2 Bio-control behaviour of PGPR against various plant diseases

PGPR	Pathogens	Plant diseases	References
Fluorescent pseudomonas	Under gnotobiotic conditions	Black root-rot of tobacco	Voisard et al. (1989)
P. fluorescens CHA0	*Thielaviopsis basicola*	Black root rot of tobacco	Voisard et al. (1989)
F. pseudomonas EM85	*Rhizoctonia solani*	Damping-off of cotton	Pal et al. (2000)
P. oryzihabitans and *Xanthomonas nematophila* strains	*Pythium* and *Rhizoctonia* sp.	Damping-off of cotton	Kapsalis et al. (2008)
F. pseudomonads	*Rhizoctonia bataticola* and *Fusarium oxysporum*	Rice and sugarcane rhizosphere	Kumar et al. (2002)
Pseudomonas strains	*Xanthomonas oryzae* pv. *Oryzae* and *Rhizoctonia solani*	Bacterial leaf blight and sheathblight pathogens of rice (*Oryzasativa*)	Rangarajan et al. (2001)
F. pseudomonads	*Helminthosporium sativum*	Endo-rhizosphere of wheat	Gaur et al. (2004)
P. fluorescens CHA0	*Meloidogyne javanica*	Root-knot nematode	Siddiqui et al. (2005)
P. putida	*Macrophomina phaseolina*	Root-rot disease complex of chickpea	Saraf et al. (2008)
P. aeruginosa Sha8	*F.oxysporium* and *Helmithosporium* sp.	Antagonistic activities	Hassanein et al. (2009)
P. fluorescens CHA0	*Tetrahymena pyriformis*	Pathogenic against the ciliated protozoa	Jousset et al. (2009)

4.1 PGPR as Biological Fungicides

The PGPR and bacterial endophytes are known to play a vital role in the management of various fungal diseases. But one of the major hurdles experienced with bio-control agents is the lack of appropriate delivery system. Some PGPR can synthesize antifungal compounds, viz. synthesis of 2, 4-diacetyl phloroglucinol by *P.* fluorescens which inhibits the growth of pathogenic fungi (Nowak-Thompson et al. 1994). Certain PGPR can degrade fusaric acid produced by the fungi *Fusarium* sp. – a causative agent of wilt (Toyoda and Utsumi 1991). There are few PGPR strains which can produce enzymes that hydrolyses the fungal cell wall, for instance secretion of chitinase and laminarinase enzymes by the *Pseudomonas stutzeri* which can lyse the mycelia of *Fusarium solani* (Mauch et al. 1988). *Pseudomonas fluorescent* has been suggested not only as a plant growth promoter, but also as potential bio-control agent due to its ability to protect the plants against the incidence of a wide range of important fungal diseases such as black root-rot of tobacco (Voisard et al. 1989), root-rot of pea (Papavizas and Ayers 1974), root-rot of wheat (Garagulia et al. 1974), damping-off of sugar beet (Kumar et al. 2002). There is ample scope for genetic manipulation of these organisms to improve their efficacy s bio-control agents (Dowling and O'Gara 1994). A number of fluorescent Pseudomonads exhibit strong antifungal activity (Reddy and Rao 2009). *Pseudomonas fluorescent* spp. EM85 and *P. oryzihabitans* exhibit strong antagonistic interaction with *Rhizoctonia solani* – a causal agent of damping-off of cotton (Pal et al. 2000). The *X. nematophila* strain also produce secondary metabolites which can suppress the growth of *Pythium* and *Rhizoctonia* species causing damping-off in cotton (Kapsalis et al. 2008). The fluorescent Pseudomonads also exhibit strong antagonistic effect against *Rhizoctonia bataticola* and *Fusarium oxysporum* associated with the rhizosphere of rice and sugarcane plants (Kumar et al. 2002). *Xanthomonas oryzae* pv. *oryzae* and *Rhizoctonia solani* – the bacterial leaf blight (BB) and sheath blight (ShB) pathogens of rice (*Oryza sativa*) are suppressed by indigenously present *Pseudomonas* strains isolated from rice cultivated coastal agricultural fields, having saline soils (Rangarajan et al. 2001). *Pseudomonas fluorescens* isolated from rice plant rhizosphere are found to have strong antifungal activity against *P. oryzae* and *R. solani* mainly due to excess production of antifungal metabolites (Reddy et al. 2008). About 50–60% of several fluorescent pseudomonads isolated from the rhizosphere and endorhizosphere of wheat plant growing Indo-Gangetic plains exhibit antagonistic interactions with *Helminthosporium sativum* (Gaur et al. 2004). Zadeh et al. (2008) reported antagonistic potential of non-pathogenic rhizosphere isolates of fluorescent Pseudomonas in the bio-control of *Pseudomonas savastanoi* – a causative agent of Olive knot disease. The *P. corrugata*, which grows at 4°C under laboratory conditions (Pandey and Palni 1998), also produces antifungal compounds like diacetylphloroglucinol and/or phenazine. *Pseudomonas fluorescens* CHA0 is reported to suppresses the black root rot of tobacco plant caused by the fungus *Thielaviopsis basicola* (Voisard et al. 1989) and it also contributes to biological control of root-knot nematode disease caused by *Meloidogyne javanica* (Siddiqui et al. 2005). In addition, certain soils from Morens, Switzerland, were found to be natural suppressive agent for *Thielaviopsis basicola*-mediated black root rot of tobacco due to presence of f*luorescent Pseudomonads* populations (Pal et al. 2000). *P. putida* has potential for the bio-control of root-rot disease complex of chickpea due to its antagonistic interaction against *Macrophomina phaseolina* (Saraf et al. 2008). It has also been shown that anaerobic regulator ANR-mediated cyanogenesis has important role in the suppression of black root rot (Laville et al. 1998). The suppression of *Phytophthora capsici by fuorescent Pseudomonads* in all seasons of plant growth helps in control of foot rot disease (Paul and Sarma 2006). Some metabolites produced by *Pseudomons aeruginosa* Sha8 include toxic volatile compound which reduces the growth of both *F. oxysporium* and *Helmithosporium* sp. but not the growth of *A. niger* (Hassanein et al. 2009). *B. luciferensis* strain KJ2C12 controls the *Phytophthora blight* of

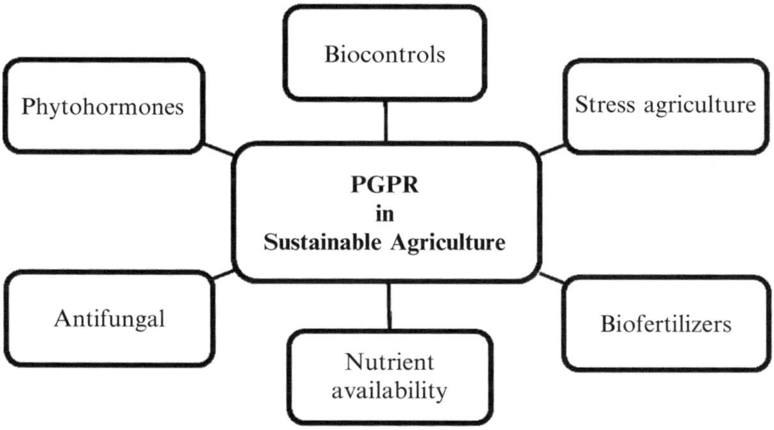

Fig. 14.1 A schematic diagram showing perspectives of PGPR in sustainable agriculture development

pepper by effective root colonization and enhanced production of protease enzyme and, also by increasing the soil microbial activity (Kim et al. 2009).

5 Conclusions

It has been demonstrated and proven that PGPR can be the very effective and potential microorganisms for enriching the soil fertility and enhancing the productivity in various agriculture yields. In the present scenario, application of PGPR is vital to overcome the problems of various environmental stresses such as soil salinity, drought, water logging, temperature and nutritional stresses to crop plants. Inoculation of plants under salinity stress with PGPR having ACC-deaminase activity is expected to mitigate the inhibitory effects of salinity on root growth by lowering the ethylene concentration in the plant. The PGPR is the most commonly used as biofertilizer which increase the availability and uptake of nutrients in plants. Current and future progress in our understanding of PGPR diversity, colonization ability, mechanisms of action, formulation, and application could facilitate their development as reliable components in the management of sustainable agricultural systems. On a commercial scale, application of PGPR in agriculture might prove beneficial and could be a sound step towards sustainable crop production and conservation. A schematic diagram showing role of PGPR in sustainable agriculture productivity is presented in Fig. 14.1.

Acknowledgements The authors wish to thank the Head of the Department of Environmental Microbiology and Department of Environmental Science, Babasaheb Bhimrao Ambedkar (Central) University, Lucknow for providing infrastructure facilities. Financial support from UGC, New Delhi is gratefully acknowledged.

References

Adesemoye AO, Torbert HA, Kloepper JW (2008) Enhanced plant nutrient use efficiency with PGPR and AMF in an integrated nutrient management system. Can J Microbiol 54:876–886

Adesemoye A, Torbert H, Kloepper J (2009) Plant growth-promoting rhizobacteria allow reduced application rates of chemical fertilizers. Microb Ecol 58:921–929

Ahmad F, Ahmad I, Khan MS (2005) Indole acetic acid production by the indigenous isolates of *Azotobacter* and *Fluorescent Pseudomonas* in the presence and absence of tryptophan. Turk J Biol 29:29–34

Ahmad M, Zahir ZA, Asghar HN, Asghar M (2011) Inducing salt tolerance in mung bean through coinoculation with rhizobia and plant-growth-promoting rhizobacteria containing 1-aminocyclopropane-1-carboxylate deaminase. Can J Microbiol 57(7):578–589

Amir HG, Shamsuddin ZH, Halimi MS, Ramlan MF, Marziah M (2001) Effects of Azospirillum inoculation on N2 fixation and growth of oil palm plantlets at nursery stage. J Oil Palm Res 13:42–49

Antoun H, Beauchamp CJ, Goussard N, Chabot R, Lalande R (1998) Potential of *Rhizobium* and *Bradyrhizobium* species as plant growth promoting rhizobacteria on non-legumes: effect on radishes (*Raphanus sativus* L.). Plant Soil 204:57–67

Arshad M, Frankenberger WT (1991) Microbial production of plant hormones. Plant Soil 133:1–8

Arshad M, Shaharoona B, Mahmood T (2008) Inoculation with plant growth promoting rhizobacteria containing ACC-deaminase partially eliminates the effects of water stress on growth, yield and ripening of *Pisum sativum* L. Pedosphere 18:611–620

Ashraf M, Hasnain S, Berge O (2004) Bacterial exo-polysaccharides and productivity of the salt affected soils. II. Effect of exo-polysaccharides (EPS) producing bacterial inoculation on growth of roots of wheat (*Triticum aestivum* L.) plants grown in a salt-affected soil. Int J Environ Sci Tech 3:45–53

Ashrafuzzaman M, Hossen FA, Ismail MR, Hoque MA, Islam MZ, Shahidullah SM, Meon S (2009) Efficiency of plant growth promoting rhizobacteria (PGPR) for the enhancement of rice growth. Afr J Biotechnol 8:1247–1252

Babalola OO, Osir EO, Sanni AI, Odhaimbo GD, Bulimo WD (2003) Amplification of 1-aminocyclopropane-1-carboxylic (ACC) deaminase from plant growth promoting rhizobacteria in Striga-infested soils. Afr J Biotechnol 2:157–160

Banasco P, Fuente L De La, Gaultieri G, Noya F, Arias A (1998) *Fluorescent Pseudomonas* spp. as bio-control agents against forage legume root pathogenic fungi. Soil Biol Biochem 10:1317–1323

Bano A, Fatima M (2009) Salt tolerance in *Zea mays* (L). following inoculation with *Rhizobium* and *Pseudomonas*. Biol Fertil Soils 45:405–413

Barassi CA, Sueldo RJ, Creus CM, Carrozzi LE, Casanovas EM, Pereyra MA (2007) *Azospirillum* spp., a dynamic soil bacterium favourable to vegetable crop production. Dyn Soil Dyn Plant 1:68–82

Barka EA, Nowak J, Clement C (2006) Enhancement of chilling resistance of inoculated grapevine plantlets with a plant growth-promoting rhizobacterium, *Burkholderia phytoWrmans* strain PsJN. Appl Environ Microbiol 72:7246–7252

Bar-Ness E, Hadar Y, Chen Y, Romheld V, Marschner H (1992) Short-term effects of rhizosphere microorganisms on Fe uptake from microbial siderophores by maize and oat. Plant Physiol 100:451–456

Barraquio WL, Segubre EM, Gonzalez MS, Verma SC, James EK, Ladha JK, Tripathi AK (2000) Diazotrophic enterobacteria: what is their role in the rhizosphere? In: Ladha JK, Reddy PM (eds) The quest for nitrogen fixation in rice. IRRI, Manila, pp 93–118

Bashan Y, Holguin G (1998) Proposal for the division of plant growth-promoting rhizobacteria into two classifications: biocontrol-PGPB (plant growth promoting bacteria) and PGPB. Soil Biol Biochem 30:1225–1228

Basu PS, Ghosh AC (2001) Production of indole acetic acid in cultures by a *Rhizobium* species from the root nodules of a mono cotyledonous tree, *Roystonea regia*. Acta Biotechnol 1:65–72

Belimov AA, Safronova VI, Sergeyeva TA, Egorova TN, Matveyeva VA, Tsyganov VE, Borisov AY, Tikhonovich IA, Kluge C, Preisfeld A, Dietz KJ, Stepanok VV (2001) Characterization of plant growth promoting rhizobacteria isolated from polluted soils and containing 1-aminocyclopropane-1-carboxylate deaminase. Can J Microbiol 47:242–252

Bensalim S, Nowak J, Asiedu SK (1998) A plant growth promoting rhizobacterium and temperature effects on performance of 18 clones of potato. Am J Potato Res 75:145–152

Berg G (2009) Plant microbe interactions promoting plant growth and health: perspectives for controlled use of microorganisms in agriculture. Appl Microbiol Biotechnol 84:11–18

Blumwald E (2000) Sodium transport and salt tolerance in plants. Curr Opin Cell Biol 12:431–434

Bottini R, Fulchieri M, Pearce D, Pharis RP (1989) Identification of gibberellins A1, A3 and iso-A3 in culture of *Azospirillum lipoferum*. Plant Physiol 90:45–47

Bray EA (1997) Plant responses to water deficit. Trends Plant Sci 2:48–54

Cacciari I, Lippi D, Ippoliti S, Pietrosanti W, Pietrosanti W (1989) Response to oxygen of diazotrophic *Azospirillum brasiliense-Arthobacter giacomelloi* mixed batch culture. Arch Microbiol 152:111–114

Cattelana AJ, Hartela PG, Fuhrmann JJ (1999) Screening for plant growth–promoting rhizobacteria to promote early soybean growth. Soil Sci Soc Am J 63:1670–1680

Chabot R, Beauchamp CJ, Kloepper JW, Antoun H (1998) Effect of phosphorus on root colonization and growth promotion of maize by bioluminescent mutants of phosphate-solubilizing *Rhizobium leguminosarum* biovar *phaseoli*. Soil Biol Biochem 30:1615–1618

Chaiharn M, Chunhaleuchanon S, Kozo A, Lumyong S (2008) Screening of rhizobacteria for their plant growth promoting activities. KMITL Sci Technol J 8:18–23

Chakraborty U, Chakraborty B, Basnet M (2006) Plant growth promotion and induction of resistance in *Camellia sinensis* by *Bacillus megaterium*. J Basic Microbiol 46:186–195

Chen CZ, Ma S, Liu L (2008) Studies on phosphorus solubilizing activity of a strain of phosphobacteria isolated from chestnut type soil in. Bioresour Technol 99:6702–6707

Cheng Z, Park E, Glick BR (2007) 1-Aminocyclopropane-1-carboxylate (ACC) deaminase from *Pseudomonas putida* UW4 facilitates the growth of canola in the presence of salt. Can J Microbiol 53:912–918

Choi O, Kim J, Kim J, Jeong Y, Moon JS, Park CS, Hwang I (2008) Pyrroloquinoline Quinone is a plant growth promotion factor produced by *Pseudomonas fluorescens* B16. Plant Physiol 146:657–668

Cocking EC (2000) Helping plants get more nitrogen from air. Eur Rev 8:193–200

Costacurta A, Keijers V, Vanderleyden J (1994) Molecular cloning and sequence analysis of an *Azospirillum brasilense* indole-3-acetic pyruvate decarboxylase gene. Mol Gen Genet 243:463–472

Cowan AK, Cairns ALP, Bartels-Rahm B (1999) Regulation of abscisic acid metabolism: towards a metabolic basis for abscisic acid-cytokinin antagonism. J Exp Bot 50:595–603

Curl EA, Truelove B (1986) The rhizosphere. Springer, Berlin/Heidelberg, 288

de Freitas JR, Banerjee MR, Germida JJ (1997) Phosphate solubilizing rhizobacteria enhance the growth and yield but not phosphorus uptake of canola (*Brassica napus* L.). Biol Fertil Soils 24:358–364

Dileep C, Kumar BSD, Dube HC (1998) Promotion of plant growth and yield by two rhizoplane *fluorescent Pseudomonas*. Ind J Exp Biol 36:399–402

Dowling DN, O'Gara F (1994) Metabolites of *Pseudomonas* involved in the bio-control of plant disease. Trends Biotechnol 12:133–141

Duijff BJ, de Kogel WJ, Bakker PAHM, Schippers B (1994) Influence of pseudobactin 358 on the iron nutrition of barley. Soil Biol Biochem 26:1681–1688

Egamberdieva D (2008) Plant growth promoting properties of rhizobacteria isolated from wheat and pea grown in loamy sand soil. Turk J Biol 32:9–15

Egamberdiyeva D (2005) Plant-growth-promoting rhizobacteria isolated from a calcisol in a semi-arid region of Uzbekistan: biochemical characterization and effectiveness. J Plant Nutr Soil Sci 168:94–99

Etesami H, Alikhani HA, Jadidi M, Aliakbari A (2009) Effect of superior IAA producing *rhizobia* on N, P, K uptake by wheat grown under greenhouse condition. World J Appl Sci 6:1629–1633
Fallik E, Sarig S, Okon Y (1994) Morphology and physiology of plant roots associated with *Azospirillum*. In: Okon Y (ed) Azospirillum/plant associations. CRC Press, Boca Raton, pp 77–86
Figueiredo VB, Burity HA, Martinez CR, Chanway CP (2008) Alleviation of drought stress in the common bean (*Phaseolus vulgaris* L.) by co-inoculation with *Paenibacillus polymyxa* and *Rhizobium tropici*. Appl Soil Ecol 40:182–188
Galal YGM (2003) Assessment of nitrogen availability to wheat (*Triticum aestivum* L.) from inorganic and organic N sources as affected by *Azospirillum brasilense* and *Rhizobium leguminosarum* inoculation. Egypt J Microbiol 38:57–73
Galleguillos C, Aguirre C, Barea JM, Azcon R (2000) Growth promoting effect of two *Sinorhizobium meliloti* strains (a wild type and its genetically modified derivative) on a non-legume plant species in specific interaction with two arbuscular mycorrhizal fungi. Plant Sci 159:57–63
Garagulia AD, Kiprianova EA, Boiko OI (1974) Antibiotic effect of bacteria from the genus *Pseudomonas* on phytopathogenic fungi. Mikrobiol Zh (Kiev) 36:197–202
Gaur R, Shani N, Kawaljeet Johri BN, Rossi P, Aragno M (2004) Diacetyl phloroglucinol-producing *Pseudomonas* does not influence AM fungi in wheat rhizosphere. Curr Sci 86:453–457
German MA, Burdman S, Okon Y, Kigel J (2000) Effects of *Azospirillum brasilense* on root morphology of common bean (*Phaseolus vulgaris* L.) under different water regimes. Biol Fertil Soils 32:259–264
Ghosh AC, Basu PS (2002) Growth behaviour and bioproduction of indole acetic acid by *a Rhizobium* species isolated from root nodules of a leguminous tree *Dalbergia lanceolarea*. Ind Exp Biol 40:796–801
Ghosh S, Penterman JN, Little RD, Chavez R, Glick BR (2003) Three newly isolated plant growth-promoting bacilli facilitate the seedling growth of canola, *Brassica campestris*. Plant Physiol Biochem 41:277–281
Glass ADM (1989) Plant nutrition: an introduction to current concepts. Jones and Bartlett Publishers, Boston, 234
Glick BR (1995) The enhancement of plant growth by free-living bacteria. Can J Microbiol 41:109–117
Glick BR (2005) Modulation of plant ethylene levels by the bacterial enzyme ACC deaminase. FEMS Microbiol Lett 251:1–7
Grattan SR, Grieve CM (1999) Salinity mineral nutrient relations in horticultural crops. Sci Hortic 78:127–157
Gyaneshwar P, Kumar GN, Parekh LJ, Poole PS (2002) Role of soil microorganisms in improving P nutrition of plants. Plant Soil 245:83–93
Halimi MS, Raffidah Z, Ismail MR (2000) Growth responses of tomatoes (*Lycopersicon esculentum* Mill) under protected environment to rhizobacterial inoculation. Trans Malays Soc Plant Physiol 9:235–239
Han HS, Lee KD (2005) Plant growth promoting rhizobacteria effect on antioxidant status, photosynthesis, mineral uptake and growth of lettuce under soil salinity. Res J Agric Biol Sci 1:210–215
Hassanein WA, Awny NM, El-Mougith AA, Salah El-Dien SH (2009) The antagonistic activities of some metabolites produced by *Pseudomonas aeruginosa* Sha8. J Appl Sci Res 5:404–414
Hernandez M, Chailloux M (2004) Las microrizas arbusculares y las bacterias rizosfe'ricas como alternativa a la nutricion mineral del tomate. Cultivos Tropicales 25:5–16
Hiifte M, Vande Woestyne M, Verstraete W (1994) Role of siderophores in plant growth promotion and plant protection by fluorescent pseudomonads. In: Manthey JA, Crowley DE, Luster DG (eds) Biochemistry of metal micro-nutrients in the rhizosphere. Lewis Publishers, Boca Raton, pp 81–92
Holguin G, Glick BR (2001) Expression of the ACC deaminase gene from *Enterobacter cloacae* UW4 in *Azospirillum brasilense*. Microb Ecol 41:281–288
Horemans S, de Koninck K, Neuray J, Hermans R, Valassak K (1986) Production of plant growth substances by *Azospirillum* sp. and other rhizosphere bacteria. Symbiosis 2:341–346

Ingram J, Bartels D (1996) The molecular basis of dehydration tolerance in plants. Annu Rev Plant Physiol Plant Mol Biol 47:377–403

Joseph B, Patra RR, Lawrence R (2007) Characterization of plant growth promoting rhizobacteria associated with chickpea (*Cicer arietinum* L.). Int J Plant Prod 1:141–152

Jousset A, Rall B, Kalinkat G, Scheu S, Brose U (2009) Extracellular toxin production by soil bacteria cause a shift from a type III to type IV functional response by microfaunal predators [Oral presentation]. The 39th annual meeting of the ecological, Society of Germany, Austria/Switzerland

Kapsalis A, Gravanis F, Gowen S (2008) Involvement of phenazine-1-carboxylic acid, siderophores and hydrogen cyanide in suppression of *Rhizoctonia solani* and *Pythium* spp. damping-off by *Pseudomonas oryzihabitans* and *Xenorhabdus nematophila*. J Food Agric Environ 6:168–171

Khakipour N, Khavazi K, Mojallali H, Pazira E, Asadirahmani H (2008) Production of Auxin hormone by *Fluorescent Pseudomonads*. Am Eurasian J Agric Environ Sci 4:687–692

Khalid A, Arshad M, Zahir ZA (2004) Screening plant growth-promoting rhizobacteria for improving growth and yield of wheat. J Appl Microbiol 96(8):473–480

Kim KY, Jordan D, McDonald GA (1998) Effect of phosphate solubilizing bacteria and vesicular–arbuscular mycorrhizae on tomato growth and soil microbial activity. Biol Fertil Soils 26:79–87

Kim H, Sang MK, Myung I, Chun S, Kim KD (2009) Characterization of *Bacillus luciferensis* strain KJ2C12 from pepper root, a bio-control agent of *Phytophthora* blight of pepper. J Plant Pathol 25:62–69

Kloepper JW, Leong J, Teintze M, Schroth MN (1980) Enhanced plant growth by siderophores produced by plant growth promoting rhizobacteria. Nature 286:885–886

Kloepper JW, Lifshitz R, Zablotowicz RM (1989) Free-living bacterial inocula for enhancing crop productivity. Trends Biotechnol 7:39–43

Kogan FN (1997) Global drought watch from space. Bull Bridge University Press, New York

Kohler J, Caravaca F, Carrasco L, Roldan A (2006) Contribution of *Pseudomonas mendocina* and *Glomus intraradices* to aggregates stabilization and promotion of biological properties in rhizosphere soil of lettuce plants under field conditions. Soil Use Manage 22:298–304

Kohler J et al (2008) Plant-growth-promoting rhizobacteria and arbuscular mycorrhizal fungi modify alleviation biochemical mechanisms in water-stressed plants. Funct Plant Biol 35:141–151

Kraus J, Loper J (1995) Characterization of genomic region required for production of antibiotic pyoluteorin by the biological control agent *Pseudomonas fluorescens* Pf-5. Appl Environ Microbiol 61:849–854

Kremer RJ, Kennedy AC (1996) Rhizobacteria as bio-control agents of weeds. Weed Technol 10:601–609

Kumar V, Narula N (1999) Solubilization of inorganic phosphates and growth emergence of wheat as affected by *Azotobacter chroococcum* mutants. Biol Fertil Soils 28:301–305

Kumar NR, Arasu VT, Gunasekaran P (2002) Genotyping of antifungal compounds producing plant growth-promoting rhizobacteria, *Pseudomonas fluorescens*. Curr Sci 82:1465–1466

Laville J, Blumer C, Schroetter CV, Gaia V, Defago G, Keel C, Haas D (1998) Characterization of the *hcnABC* gene cluster encoding hydrogen cyanide synthase and anaerobic regulation by ANR in the strictly aerobic bio-control agent *Pseudomonas fluorescens* CHA0. J Bacteriol 180:3187–3196

Lugtenberg B, Kamilova F (2009) Plant-growth-promoting rhizobacteria. Annu Rev Microbiol 63:541–556

Malhotra M, Srivastava S (2008) Stress-responsive indole-3-acetic acid biosynthesis by *Azospirillum brasilense* SM and its ability to modulate plant growth. Eur J Soil Biol 45:73–80

Mandal SM, Mondal KC, Dey S, Pati BR (2007) Optimization of cultural and nutritional conditions for indol-3-acetic acid (IAA) production by a *Rhizobium* sp. isolated from rot nodules of *Vigna mungo* (L) hepper. Res J Microbiol 2:239–246

Mantelin S, Touraine B (2004) Plant growth-promoting bacteria and nitrate availability impacts on root development and nitrate uptake. J Exp Bot 55:27–34

Martinez-Morales LJ, Soto-Urzua L, Baca BE, Sanchez-Ahedo JA (2003) Indole-3-butyric acid (IBA) production in culture medium by wild strain *Azospirillum brasilense*. FEMS Microbiol Lett 228:167–173

Masalha J, Kosegarten H, Elmaci O, Mengel K (2000) The central role of microbial activity for iron acquisition in maize and sunflower. Biol Fertil Soils 30:433–439

Mattoo AK, Suttle JC (1991) The plant hormone ethylene. CRC Press, Boca Raton

Mauch F, Mauch-Mani B, Boller T (1988) Antifungal hydrolases in pea tissue. II. Inhibition of fungal growth by combinations of chitinase and/3-1,3-glucanase. Plant Physiol 88:936–942

Mayak S, Tirosh T, Glick BR (2004a) Plant growth-promoting bacteria that confer resistance to water stress in tomato and pepper. Plant Sci 166:525–530

Mayak S, Tirosh T, Glick BR (2004b) Plant growth-promoting bacteria confer resistance in tomato plants to salt stress. Plant Physiol Biochem 42:565–572

Molla AH, Shamsuddin ZH, Halimi MS, Morziah M, Puteh AB (2001) Potential for enhancement of root growth and nodulation of soybean co-inoculated with *Azospirillum* and *Bradyrhizobium* in laboratory systems. Soil Biol Biochem 33:457–463

Nadeem SM, Zahir ZA, Naveed M, Arshad M, Shahzad SM (2006) Variation in growth and ion uptake of maize due to inoculation with plant growth promoting rhizobacteria under salt stress. Soil Environ 25:78–84

Nagorska K, Bikowski M, Obuchowski M (2007) Multicellular behaviour and production of a wide variety of toxic substances support usage of *Bacillus subtilis* as a powerful bio-control agent. Acta Biochim Pol 54:495–508

Narula N, Deubel A, Gans W, Behl RK, Merbach W (2006) Paranodules and colonization of wheat roots by phytohormone producing bacteria in soil. Plant Soil Environ 52:119–129

Nautiyal CS, Bhadauria S, Kumar P, Lal H, Mondal R, Verma D (2000) Stress induced phosphate solubilization in bacteria isolated from alkaline soils. FEMS Microbiol Lett 182:291–296

Negi YK, Garg SK, Kumar J (2005) Cold-tolerant fluorescent *Pseudomonas* isolates from Garhwal Himalayas as potential plant growth promoting and bio-control agents in pea. Curr Sci 89:25

Noel TC, Cheng C, Yost CK, Pharis RP, Hynes FM (1996) *Rhizobium leguminosarum* as a plant-growth rhizobacterium: direct growth promotion of canola and lettuce. Can J Microbiol 42:279–283

Nowak-Thompson B, Gould SJ, Kraus J, Loper JE (1994) Production of 2, 4-diacetylphloroglucinol by the bio-control agent *Pseudomonas fluorescens* Pf-5. Can J Microbiol 40:1064–1066

O'Donnell PJ, Calvert CM, Atzorn R, Wasternack C, Leyser HMO, Bowles DJ (1996) Ethylene as a signal mediating the wound response of tomato plants. Science 274:1914–1917

Ortiz-Castro R, Valencia-Cantero E, Lopez-Bucio J (2008) Plant growth promotion by *Bacillus megaterium* involves cytokinin signalling. Plant Signal Behav 3:263–265

Pal SS (1998) Interactions of an acid tolerant strain of phosphate solubilizing bacteria with a few acid tolerant crops. Plant Soil 198:169–177

Pal KK, Tilak KVBR, Saxena AK, Dey R, Singh CS (2000) Antifungal characteristics of a *fluorescent Pseudomonas* strain involved in the biological control of *Rhizoctonia solani*. Res Microbiol 155:233–242

Pandey A, Palni LMS (1998) Isolation of *Pseudomonas corrugate* from Sikkim, Himalaya. World J Microbiol Biotechnol 14:411–413

Papavizas GC, Ayers WA (1974) Aphanomyces species and their root diseases in pea and sugarbeet. A review. US Department of Agriculture, Washington, DC

Paul D, Sarma YR (2006) Antagonistic effects of metabolites of *Pseudomonas fluorescens* strains on the different growth phases of *Phytophthora capsici*, foot rot pathogen of black pepper (*Piper nigrum* L). Arch Phytopathol Plant Prot 39:113–118

Paul D, Dineshkumar N, Nair S (2006) Proteomics of a plant growth-promoting rhizobacterium, *Pseudomonas fluorescens* MSP-393, subjected to salt shock. World J Microbiol Biotechnol 22:369–374

Rai R, Hunt PG (1993) Inoculation of maize varieties with salt tolerant mutants of *Azospirillum brasilense* and VAM fungi in saline calcareous soil. Microbiol Release 1:243–251

Rajendran G, Sing F, Desai AJ, Archana G (2008) Enhanced growth and nodulation of pigeon pea by co-inoculation of *Bacillus* strains with *Rhizobium* spp. Bioresour Technol 99:4544–4550

Rangarajan S, Loganathan P, Saleena LM, Nair S (2001) Diversity of pseudomonads isolated from three different plant rhizospheres. J Appl Microbiol 91:742–749
Reddy BP, Rao KS (2009) Biochemical and PCR- RAPD characterization of *Pseudomonas fluorescent* produced antifungal compounds inhibit the rice fungal pathogens *in vitro*. Electron J Environ Agric Food Chem 8:1062–1067
Reddy BP, Reddy KRN, Rao SM, Rao KS (2008) Efficacy of antimicrobial metabolites of *Pseudomonas fluorescens* against rice fungal pathogens. Curr Trends Biotechnol Pharm 2:178–182
Richardson AE (2001) Prospects for using soil microorganisms to improve the acquisition of phosphorus by plants. Aust J Plant Physiol 28:897–906
Rodriguez H, Fraga R (1999) Phosphate solubilizing bacteria and their role in plant growth promotion. Biotechnol Adv 17:319–339
Rokhzadi A, Asgharzadeh A, Darvish F, Nour-Mohammadi G, Majidi E (2008) Influence of plant growth promoting rhizobacteria on dry matter accumulation of chickpea (*Cicer arietinum L*) under field conditions. J Agric Environ Sci 3:253–257
Roy M, Basu PS (2004) Studies on root nodules of leguminous plants bioproduction of indole acetic acid by a *Rhizobium* sp. from a twiner *Clitorea ternatea* L. Acta Biotechnol 12:453–460
Saad MS, Shabuddin ASA, Yunus AG, Shamsuddin ZH (1999) Effects of Azospirillum inoculation on sweet potato grown on sandy tin-tailing soil. Commun Soil Sci Plant Anal 30:583–1592
Saleem M, Arshad M, Hussain S, Bhatti AS (2007) Perspective of plant growth promoting rhizobacteria (PGPR) containing ACC deaminase in stress agriculture. J Ind Microbiol Biotechnol 34:635–648
Salisbury FB, Ross CW (1992) Plant physiology. Wadsworth Pub. Co., Belmont, p 682
Saraf M, Thakker A, Patel BV (2008) Bio-control activity of different species of *Pseudomonas* against phytopathogenic fungi *In vivo* and *In vitro* conditions. Int J Biotechnol Biochem 4:223–232
Saravanakumar D, Samiyappan R (2007) ACC deaminase from *Pseudomonas Xuorescens* mediated saline resistance in groundnut (*Arachis hypogea*) plants. J Appl Microbiol 102:1283–1292
Sarig S, Blum A, Okon Y (1988) Improvement of the water status and yield of field-grown grain sorghum (*Sorghum bicolor*) by inoculation with *Azospirillum brasilense*. J Agric Sci 110:271–277
Sarwar M, Frankenberger WT (1994) Influence of L-tryptophan and auxins applied to the rhizosphere on the vegetative growth of *Zea mays* L. Plant Soil 160:97–104
Seshadri S, Ignacimuthu S, Vadivelu M, Lakshminarasimhan C (2007) Inorganic phosphate solubilization by two insect pathogenic *Bacillus* sp. Dev Plant Soil Sci 102:351–355
Shaharoona B, Naveed M, Arshad A, Zahir ZA (2008) Fertilizer-dependent efficiency of pseudomonads for improving growth, yield, and nutrient use efficiency of wheat (*Triticum aestivum* L). Appl Microbiol Biotechnol 79:147–155
Shinozaki K, Yamaguchi-Shinozaki K (1999) Molecular responses to drought stress. In: Shinozaki K, Shinozaki YK (eds) Molecular responses to cold, drought, heat and salt stress in higher plants. RG Landes, Austin, pp 11–28
Shoebitz M, Ribaudo CM, Pardo MA, Cantore ML, Ciampi L, Cura JA (2007) Plant growth promoting properties of a strain of *Enterobacter ludwigii* isolated from *Lolium perenne* rhizosphere. Biochem J 41:1768–1774
Siddiqui ZA (2006) PGPR: prospective bio-control agents of plant pathogens. In: PGPR: biocontrol and biofertilization. Springer, Dordrecht, pp 111–142
Siddiqui IA, Haas D, Heeb S (2005) Extracellular protease of *Pseudomonas fluorescens* CHA0, a bio-control factor with activity against the root-knot nematode *Meloidogyne incognita*. Appl Environ Microbiol 71:5646–5649
Singh S, Kapoor KK (1999) Inoculation with phosphate solubilizing microorganisms and a vesicular arbuscular mycorrhizal fungus improves dry matter yield and nutrient uptake by wheat grown in a sandy soil. Biol Fertil Soils 28:139–144
Singh JS, Pandey VC, Singh DP (2011) Efficient soil microorganisms: a new dimension for sustainable agriculture and environmental development. Agric Ecosyst Environ 140:339–353

Son HJ, Park GT, Cha MS, Heo MS (2006) Solubilization of insoluble inorganic phosphates by a novel salt- and pH-tolerant *Pantoea agglomerans* R-42 isolated from soybean rhizosphere. Bioresour Technol 97:204–210

Souchie EL, Azcon R, Barea JM, Saggin-Junior OJ, da Silva EMR (2007) Indolacetic acid production by P-solubilizing microorganisms and interaction with *arbuscular mycorrhizal* fungi. Acta Sci Biol Sci 29:315–320

Sridevi M, Mallaiah KV (2007) Bioproduction of indole acetic acid by *Rhizobium* strains isolated from root nodules of green manure crop, *Sesbania sesban* (L) *Merr*. Iran J Biotechnol 5:178–182

Sridevi M, Yadav NCS, Mallaiah KV (2008) Production of indol-acetic acid by *Rhizobium* isolates from *Crotalaria* species. Res J Microbiol 3:276–281

Stefan M, Mihasan M, Dunca S (2008) Plant growth promoting rhizobacteria can inhibit the *in vitro* germination of *Glycine Max* L seeds. Scientific Annals of University "Alexandru Ioan Cuza" Iasi, Section Genetics and Molecular Biology, T. IX, 3:105–110

Stevenson FJ, Cole MA (1999) Cycles of soil: carbon, nitrogen, phosphorus, sulfur, micronutrients, 2nd edn. Wiley, New York, 427

Tilak KVBR, Ranganayaki N, Pal KK, De R, Saxena AK, Nautiyal CS, Mittal S, Tripathi AK, Johri BN (2005) Diversity of plant growth and soil health supporting bacteria. Curr Sci 89:136–150

Timmusk S, Wagner GH (1999) The plant-growth-promoting rhizobacterium *Paenibacillus polymyxa* induces changes in *Arabidopsis thaliana* gene expression: a possible connection between biotic and abiotic stress responses. Mol Plant Microbe Interact 12:951–959

Toyoda H, Utsumi R (1991) Method for the prevention of *Fusarium* diseases and microorganisms used for the same. US patent no. 4, 988, p 586

Tsavkelova EA, Cherdyntseva TA, Klimova SY, Shestakov AI, Botina SG, Netrusov AI (2007) Orchid-associated bacteria produce indole-3-acetic acid, promote seed germination, and increase their microbial yield in response to exogenous auxin. Arch Microbiol 188:655–664

Upadhyay SK, Singh JS, Singh DP (2011) Exopolysaccharide-producing plant growth promoting rhizobacteria salinity condition. Pedosphere 21:214–222

Upadhyay SK, Singh JS, Saxena AK, Singh DP (2012) Impact of PGPR inoculation on growth and antioxidants status of wheat plant under saline condition. Plant Biol 14:605–611

Vazquez P, Holguin G, Puente ME, Lopez-Cortez A, Bashan Y (2000) Phosphate-solubilizing microorganisms associated with the rhizosphere of mangroves in a semiarid coastal lagoon. Biol Fertil Soils 30:460–468

Vessey JK (2003) Plant growth promoting rhizobacteria as biofertilizers. Plant Soil 255:571–586

Vessey JK, Buss TJ (2002) *Bacillus cereus* UW85 inoculation effects on growth, nodulation, and N accumulation in grain legumes. Controlled-environment studies. Can J Plant Sci 82:282–290

Voisard C, Keel C, Haas D, Defago G (1989) Cyanide production by *Pseudomonas fluorescens* helps suppress black root rot of tobacco under gnotobiotic conditions. EMBO J 8:351–358

von Wiren N, Khodr H, Hider RC (2000) Hydroxylated phytosiderophore species possess an enhanced chelate stability and affinity for iron (III). Plant Physiol 124:1149–1157

Wang Y, Brown HN, Crowley DE, Szaniszlo PJ (1993) Evidence for direct utilization of a siderophore, ferrioxamine B, in axenically grown cucumber. Plant Cell Environ 16:579–585

Weller DM (1988) Biological control of soil borne plant pathogens in the rhizosphere with bacteria. Annu Rev Phytopathol 26:379–407

Weller DM (2007) *Pseudomonas* bio-control agents of soil borne pathogens: looking back over 30 years. Annu Rev Phytopathol 40:309–348

Weller DM, Raaijmakers JM, McSpadden Gardener BB, Thomashow LS (2002) Microbial populations responsible for specific soil suppressiveness to plant pathogens. Annu Rev Phytopathol 40:309–348

Weyens N, van der Lelie D, Taghavi S, Vangronsveld J (2009) Phytoremediation: plant endophyte partnerships take the challenge. Curr Opin Biotechnol 20:248–254

Wilhite DA (2000) Drought as a natural hazard. In: Wilhite DA (ed) Drought: a global assessment. Routledge, London, pp 3–18

Zadeh HR, Khavazi K, Asgharzadeh A, Hosseinimazinani M, Mot RD (2008) Bio-control of *Pseudomonas Savastanoi*, causative agent of olive knot disease: antagonistic potential of non-pathogenic rhizosphere isolates of *Fluorescent Pseudomonas*. Commun Agric Appl Biol Sci 73:199–203

Chapter 15
Antibiotic Resistance Gene Pool and Bacterial Adaptation to Xenobiotics in the Environment

Mohd Ikram Ansari and Abdul Malik

Abstract Extreme environments are habitats that experience steady or fluctuating exposure to one or more environmental factors, such as salinity, osmolarity, desiccation, UV radiation, barometric pressure, pH, temperature, heavy metals, xenobiotics and antibiotics. The evolutionary biologist studies the steps by which the adaptations have evolved. But the general nature of such adaptive steps is still unclear. Evolution is often thought to be random and dependent on unpredictable events. This chapter focuses on antibiotic and xenobiotic stress in the environment and the microbial adaptations along with the genetic regulation of these stresses. Bacteria have come up with sophisticated modes of cooperative behaviour to cope with adverse and varying environmental conditions. They developed intricate communication capabilities, including a broad repertoire of chemical signaling mechanisms, collective activation and deactivation of genes and even exchange of genetic materials. With these tools, they can communicate and self-organize their colonies into multicellular hierarchical aggregates, out of which new abilities emerge. Many examples of bacterial mechanisms are thought to be adaptations for survival in changing environments, some of which are the mutator phenotypes, sporulation, adaptive mutation, and phase variation. The most frequently proposed hypothesis is that genes involved in antibiotic resistance originated in antibiotic-producing organisms, as part of the cluster involved in antibiotic biogenesis, to prevent self-inhibition. Eventually, these genes may have moved to the neighboring bacterial organisms, which then became resistant. Alternatively, these neighboring organisms may have introduced changes in the DNA sequence of possibly duplicate

M.I. Ansari • A. Malik (✉)
Department of Agricultural Microbiology, Faculty of Agricultural Sciences,
Aligarh Muslim University, Aligarh 202002, India
e-mail: ikram_ansari21@yahoo.com; ab_malik30@yahoo.com

A. Malik et al. (eds.), *Management of Microbial Resources in the Environment*,
DOI 10.1007/978-94-007-5931-2_15,

genes involved in functions that were thus reoriented to antibiotic elimination or detoxification. Recent studies have revealed the existence of new types of xenobiotic catabolic mobile genetic elements, such as catabolic genomic islands, which integrate into the chromosome after transfer. Molecular analysis of the catabolic pathways of xenobiotic-degrading bacteria indicated that they might have adapted to the appearance of such compounds by expressing new functions to resist the potential toxic effects of the molecules or to use their beneficiary characteristics, for example, as an alternative source of essential nutrients, such as carbon, nitrogen or energy.

Keywords Adaptation • Antibiotics • Environmental stress • Mutation • Xenobiotics

1 Introduction

The spread and evolution of resistance to antibiotic depends on the pressure exerted by the antibiotic in the microbial environment. Different effects occur in different compartments, where concentrations of particular antibiotic result in a differential growth rate of resistant bacteria. This happens even at very low antibiotic concentrations which are able to select low level resistant bacteria. When multiple antibiotics are present in the environment, then multiple and variable pressure select the bacteria which use multiple or versatile mechanisms or optimize a single mechanism of resistance to survive under the changeable environmental conditions. Host factors such as immunity add to the selective process. Antibiotics itself might support bacterial diversity, whichever mediated by the random drift effect or triggering the increase in events of mutation under bacterial stress. Analysis of selective environment related antibiotic host bacteria interactions is necessary in understanding the biology of antibiotic resistance (Baquero et al. 1998).

Xenobiotics are chemically synthesized organic compounds most of which do not occur in nature (Schlegel 1986). Xenobiotics are defined as compounds that are foreign to a living organism. Where these compounds are not easily recognized by existing degradative enzymes and they gather in soil and water (Esteve-Nenez et al. 2001). Xenobiotics include fungicides, pesticides, herbicides, insecticides, nematicides, and so on. The majority of which are substituted hydrocarbons, phenyl carbonates, and related compounds (Ojo 2007).

Among the chlorinated, polyaromatic, xenobiotics, and nitro-aromatic compounds are considered to be mutagenic, carcinogenic and toxic for living organisms. However, microbial diversity and versatility for adaptation to xenobiotics makes them the best candidates among all living organisms to convey xenobiotic compounds into natural biogeochemical cycles. Even though, more microorganisms are able to degrade these anthropogenic molecules, some xenobiotics have been shown to be uncommonly refractory (Esteve-Nenez et al. 2001; Ojo 2007).

In nature, microorganisms must survive with aggressive environmental conditions. For this purpose, they have developed complicated mutual behavior and intricate communication capabilities, such as direct cell–cell physical interactions mediated by cell surface structures, collective production of extracellular "wetting" fluid for movement on hard surfaces, long range chemical signals such as quorum sensing and chemotactic (bias of movement according to gradient of chemical agent) signaling, collective genes activation and deactivation and even exchange of genetic material. In some cases, these capabilities are exploited by bacterial colonies to develop complex spatio-temporal patterns in response to unfavorable growth conditions (Ben-Jacoba et al. 2000).

Recent studies pointed out two major breakthroughs regarding the molecular basis of bacterial adaptation to pollutants: (i) the definition of new types of xenobiotic catabolic mobile genetic elements (MGEs) and (ii) findings from microbial ecology studies that suggest that such adaptation events actually occur in a polluted environment in the field (Springael and Top 2004). Adaptation to environmental change is essential for microbial survival and proliferation. Environmental changes will shift the balance of complex microbial communities by favoring some populations, while restricting others, through mechanisms such as microbial competition (for nutrients), antibiosis and by the selection of those microorganisms best adapted to environmental stress. Thus, much research in microbial ecology focuses on the evolution of microbial populations (both taxonomic and functional groups) submitted to fluctuations in environmental conditions (Castro et al. 1997; Amoroso et al. 1998; Baath et al. 1998; Carrasco et al. 2005; Raniard et al. 2008; Baek and Kim 2009; Baek et al. 2009; Belanger et al. 2011).

Recently, several studies have suggested for a new evolutionary synthesis (Dean and Thornton 2007; Pigliucci 2007; Carroll 2008; Koonin 2009) that contain mechanisms other than mutation, natural selection and drift to elucidate evolutionary changes, like developmental constraints and epigenetic modifications among others (Boto 2010). The direct visualization of horizontal (or lateral) gene transfer, which has been achieved recently (Babic et al. 2008), is a significant method of measuring the evolution of Bacteria and Archaea, as well as that of unicellular eukaryotes, and should therefore also be considered as part of the structure of any evolutionary synthesis. This chapter highlights the mechanism of adaptations of bacterial populations to antibiotics and xenobiotics present in the environment. The chapter also insights the role of horizontal gene transfer in bacterial evolution.

2 Antimicrobial Resistance

The prokaryotic cell is proficient of adapting to the antibiotics introduced into the environment. The natural genetic variation ensures a reasonable sum of heterogeneity that ensures survivors in antibiotic charged environments. Thus surveys of bacterial isolates show the presence of resistant organisms from the pre-antibiotic days, although in small numbers (Madeinos 1997). Population dynamics would keep this proportion low enough not to influence therapeutic outcome. However, in an antibiotic charged environment a selection pressure builds up favouring the resistant organisms. This 'survival of the fittest' principle enunciated by Charles Darwin (1859) consequences in a regular rise in MICs.

The concept of the soil as a location of antibiotic resistance determinants, particularly those harboured in antibiotic producers as self-protection mechanisms has been acknowledged for decades. However, mechanistic commonalities between clinical pathogens and soil inhabiting organisms were not shown until the 1970s. In 1973, two molecular mechanisms of aminoglycoside resistance in soil-dwelling actinomycetes from the genus *Streptomyces* were determined to be the same to those in clinical pathogens (Benveniste and Davies 1973). These strains, producers of the aminoglycosides kanamycin and neomycin, were capable of drug modification by acetylation and phosphorylation, respectively as a means of self-protection (Benveniste and Davies 1973).

In recent years, approaches have been implemented to characterize the diversity and prevalence of resistance in soil bacteria, the soil antibiotic resistome as an essential reservoir of resistance (Wright 2007). Riesenfeld et al. (2004) investigated resistance in the soil, concentrating on unculturable organisms, bacteria that have yet to be characterized and thus underappreciated because of challenging culture conditions (Riesenfeld et al. 2004). By creating a functional metagenomic library (Handelsman 2004) in which cloned genomic fragments were expressed from DNA isolated directly from soil and selecting for resistance, traditional challenges associated with studying genes of unknown sequence were circumvented. Specifically, these functional analyses revealed novel antibiotic resistance proteins that were previously of unknown function and unrecognizable by sequence alone. Thus, this work not only allowed for the identification of aminoglycoside N-acetyltransferases, the O-phosphotransferases, and a putative tetracycline efflux pump but also a construct with a novel resistance determinant to the aminoglycoside butirosin (Riesenfeld et al. 2004). This work shows the power of the functional metagenomic approach when applied to a search of activity with a highly selectable phenotype such as antibiotic resistance (D'Costa et al. 2007).

Focusing on agriculturally associated resistance, Schmitt et al. characterized the diversity of tetracycline resistance determinants in soil (Schmitt et al. 2006). Using PCR-based approaches, three resistance genes were ubiquitously identified in the soil, and additional five were found in manure-supplemented soils. This work speaks to the diversity of tetracycline resistance in agricultural soils. With respect to environmental resistance to antibiotics, this ability is not simply restricted to soil-dwelling microorganisms. Both phenotypically and genetically, resistance to antibacterials has been extensively documented in genera spanning the entire bacterial domain from diverse ecosystems (Acar and Moulin 2006; Martins da Costa et al. 2006; Stepanauskas et al. 2006; De Souza et al. 2006; Schmitt et al. 2006; D'Costa et al. 2006; Ferreira da Silva et al. 2007; Jacobs and Chenia 2007; Levesque et al. 2007; Neela et al. 2007; Oliynyk et al. 2007).

3 Antibiotic Pressure and Adaptive Mutation

All previous considerations regarding the selection of antibiotic resistance were based on the classic belief that mutations occur at random during nonselective bacterial growth and that selection is essentially the "amplifying mechanism" (or amplifying field) for the organisms with mutations, resulting in an increase in fitness. Adaptive mutation can also be termed post selection mutation (Foster 1995), implying that non growing bacteria inhibited by a given antibiotic (under selection) may be able to produce increased genetic variation that eventually gives rise to mutations, thereby helping the inhibited strain survive the antibiotic challenge. It appears as if the antibiotic may "direct" resistance to itself, but in fact only the frequency of genetic variation is increased under stress (inhibition) circumstances. The study of genetic variation under stationary-phase conditions will most likely contribute to the understanding of the machinery involved in adaptive mutation (Baquero et al. 1998).

Antibiotic pressure on cells under stationary-phase conditions may contribute to the emergence of antibiotic resistance; indeed, this factor was presumed by the observation of the frequent development of resistance among biofilm-forming bacterial populations. Finally, the recent recognition of the mutator genes in promoting genetic diversity and bacterial adaptive evolution (Taddei et al. 1997) may provide key data for understanding the evolution of antibiotic resistance under antibiotic pressure (Baquero et al. 1998). The main proposed hypothesis is that genes implicated in antibiotic resistance originated in antibiotic producing organisms, as part of the cluster involved in antibiotic biogenesis, to prevent self-inhibition (Davies 1992; Baquero et al. 1998).

The advent of genome sequencing has greatly accelerated our understanding of evolution. With respect to resistance determinants, these efforts have uncovered genes responsible for resistance, cryptic genes that encode resistance but are perhaps insufficiently expressed and thus do not confer the phenotype, as well as those that serve as precursors for resistance determinants. Recent efforts have uncovered a wealth of putative resistance determinants. For example, the recently sequenced genome of the erythromycin producer *Saccharopolyspora erythraea* NRRL23338, a non-pathogenic Gram-positive bacterium resistant to a wide spectrum of antibiotics, is predicted to encode a remarkable number of putative resistance determinants representing approximately 1% of its genome (Oliynyk et al. 2007).

Horizontal gene transfer of genetic material takes the phenomenon to a different plane. Once the resistance genes are transported to plasmids, transposons or integrons then dissemination of resistance genes is rapid. In the present context transmissible resistance to fluroquinolones has already been observed in *Klebsiella pneumoniae* (Martinez-Martinez et al. 1998). If these genetic elements progress into the enteric fever causing Salmonellae, the profile of the disease would change. However, there is no details of such results having occurred. Another effective mode of production of genetic elements capable of continuous horizontal transfer is the mobilization of naturally present protective resistance determinants in antibiotic producers (Piepersberg 2001; Raghunath 2008).

4 Mechanisms of Generation of Antibiotic Resistance

Genes currently involved in antibiotic resistance have probably evolved for other purposes than antibiotic resistance. In this view, resistance can be considered as a chance product, determined by the interaction of an antibiotic and a particular genotype. This is not incompatible with the idea of a gradual modification of some genes of pre-existing cellular machinery to become resistance genes. Possession of some 'core' genes may be neutral or nearly neutral in the prevailing nonantibiotic environment but may be a latent potential for selection that can only be expressed under the appropriate conditions of antibiotic selection. The possible origin of enzymes hydrolyzing β-lactam antibiotics (β-lactamases) as an alteration of the three-dimensional structure of the target cell wall biosynthetic enzymes (transpeptidases that now increase the rates of β-lactam deacylation) may be a splendid example of evolution as tinkerer. In other cases, the mere high-level expression of genes with small activity for the purposes of resistance under normal circumstances may also result in a resistant phenotype. Antibiotic producing microorganisms under external antibiotic pressure, in the absence of effective resistance mechanisms, experience biological stress, resulting from membrane or cell wall damage, compromised protein synthesis, or altered DNA supercoiling (Baquero and Blazquez 1997).

The molecular mechanisms of resistance to antibiotics have been studied extensively and have involved investigations of the genetics and biochemistry of many different facets of bacterial cell function (Gale et al. 1981; Walsh 2003; Alekshun and Levy 2007). The study of antibiotic action and resistance has considerably contributed to our understanding of cell structure and function. Resistance processes are broadly dispersed in the microbial kingdom which have been well described for a variety of commensals (Marshall et al. 2009) and pathogens; and most of them can be spread by one or more different gene transfer mechanisms.

The principal role of human activities in the production of environmental reservoirs of antibiotic resistance cannot be neglected. Since the 1940s, ever rising amounts of antibiotics chosen for human applications have been produced, used clinically, and are released into the environment, and widely disseminated, thus provide constant selection and maintenance pressure for resistant populations in all the environments. Obtaining precise information on the quantities of antimicrobials produced by the pharmaceutical industry is not easy (it is not in the best interest of pharmaceutical companies to provide this information), but it can be predictable that many millions of metric tons of antibiotic compounds have been released into the biosphere over the last half century. Since the only existing proof indicates that little in the way of antibiotics is contributed by naturally occurring antibiotic producing strains in their native environments (Gottlieb 1976), we must assume that the commercial production gives the vast bulk of the antibiotics found in the biosphere (Davies and Davies 2010).

5 Role of Plasmids in Adaptation

Plasmid DNA may play a particularly important role in genetic adaptation in that it represents a highly mobile form of DNA that can be transferred via conjugation or transformation and can pass on novel phenotypes, including hydrocarbon-oxidizing ability, to recipient organisms. The pathways for the metabolism of salicylate, naphthalene, octane, camphor, xylene, and toluene have been found to be on plasmids in *Pseudomonas* spp. Exposure of natural microbial populations to oil or other hydrocarbons may impose a selective advantage to strains possessing plasmids encoding enzymes for hydrocarbon catabolism, resulting in an overall increase in the plasmid frequency in the community (Leahy and Colwell 1990).

Heuer and Smalla (2012) reported that the transport of conjugative plasmids across species boundaries plays a very important role in the adaptation of bacterial populations in soil. There are definite driving forces and restrictions of plasmid transfer within bacterial populations in soils. Plasmid mediated genetic variation allows bacteria to respond rapidly with adaptive responses to challenges such as irregular antibiotic or metal concentrations, or opportunities such as the utilization of xenobiotic compounds. Cultivation independent detection and capture of plasmids from soil bacteria, and whole plasmid sequences have provided new insights into the function and biology of plasmids. Broad host range plasmids such as those belonging to IncP-1 transfer accessory functions which are carried by similar plasmid backbones. Plasmids with a narrow host range can be more specifically adapted to particular species and often transfer genes which complement chromosomally encoded functions. Plasmids are ancient and successful approach to ensure survival of a soil population in spatial and temporal heterogeneous conditions with various environmental stresses or opportunities that occur unevenly or as a novel challenge in soil (Heuer and Smalla 2012).

Retrospective studies clearly indicate that mobile genetic elements (MGEs) play a major role in the *in situ* spread and even *de novo* construction of catabolic pathways in bacteria, allowing bacterial communities to rapidly adapt to new xenobiotics. The production of novel pathways seems to occur by an gathering process that involves horizontal gene transfer: different appropriate genes or gene modules that encode different parts of the novel pathway are recruited from phylogenetically related or distant hosts into one single host. Direct proof for the significance of catabolic MGEs in bacterial adaptation to xenobiotics stems from observed correlations between catabolic gene transfer and accelerated biodegradation in numerous habitats and from studies that examine catabolic MGEs in contaminated sites (Top and Springael 2003).

6 Adaptation to Xenobiotics

The spontaneous occurrence of DNA rearrangements in xenobiotic-degraders that resulted in evolution of different pathways for mineralization of synthetic compounds in natural environment is one of the principal mechanisms of adaptation to xenobiotic substrates. The evolution of catabolic pathways (that is, modified *ortho* cleavage pathway, *meta* cleavage pathway and others) in micro-organisms for xenobiotic substrates often involves different gene clusters encoding for the aromatic pathway enzymes (Ojo 2007).

The characterization of bacteria that degrade organic xenobiotics has revealed that they can adapt to these compounds by expressing 'novel' catabolic pathways. Some of them have evolved by patchwork assembly of horizontally transferred genes and following mutations and gene rearrangements. Recent studies have showed the presence of new types of xenobiotic catabolic MGEs, such as catabolic genomic islands, which incorporate into the chromosome after transfer. The significance of horizontal gene transfer and patchwork assembly for bacterial adaptation to pollutants under real environmental conditions remains uncertain, but recent publications suggest that these processes do occur in a polluted environment (Springael and Top 2004).

6.1 Recombination and Transposition

Recombination is the combining of genes (DNA) from two or more different cells. This is principally based on molecular methods involving cutting of DNA fragments from different cells harbouring desired catabolic traits. These DNA fragments are hybridized and inserted in host cells that are known as recombinants, which are seeded onto polluted environment where the expression of the catabolic trait is desired (Black 1999). This strategy is often practiced in vitro than in vivo (El-Fantroussi 2000) in the soil enrichment method for the degradation of an herbicide (Linuron) a modified strategy which can be extended for bioremediation process in soil polluted by this herbicide. Van der Meer et al. (1991) perform biodegradation experiment involving *Acinetobacter* and *Pseudomonas* spp., and a DNA rearrangement strategy was used to achieve mineralization. The sequential arrangement of the genes encoding the *ortho* cleavage pathways of *Acinetobacter calcoaceticus* and *P. putida* differs from one another and from those of other organisms, suggesting that various DNA rearrangements have occurred (Van der Meer et al. 1991). Gene rearrangements are also evident between the different operons for the modified *ortho* pathways enzymes. Rearrangement of DNA fragments is believed to be due to insertion elements which subsequently enhance gene transfer as well as activation or inactivation of silent genes (Tsuda et al. 1989; Ramezani and Hawley 2010).

Insertion elements have been shown to play an important role in rearrangement of DNA fragments, in gene transfer, and in activation or inactivation of silent genes. For catabolic pathways several examples of insertion elements are known. The TOL catabolic operons are part of a large transposable element, Tn*4651* that belongs to the family of Tn*3*-type transposons (Tsuda and Iino 1987, 1988; Tsuda et al. 1989). This transposon was later found to be part of an even larger mobile element, element, Tn*4653* (Tsuda and Iino 1988). A similar Tn*3*-type element, Tn*4655*, contains the catabolic genes on plasmid NAH7, although it is defective in its transposase function

(Tsuda and Iino 1990). *P. cepacia* 249 was shown to carry at least nine different insertion elements, which were present in 1–13 copies in its genome. These IS elements are thought to be responsible for the extraordinary adaptability and catabolic potential of this strain (Lessie et al. 1990).

6.2 Gene Duplication

This is an important mechanism for the evolution of different strains of microorganisms of the same species. Once a gene becomes duplicated, the extra gene copy thus becomes independent of selective pressures and subsequently imbibes mutations with speed. These mutations could eventually lead to full inactivation, rendering this copy silent. Reactivation of the silent gene copy could then occur through the action of insertion elements. This occurred in *Flavobacterium* sp. strain K172 that produces two isozymes of 6-aminohexanoate dimer hydrolase, one of the enzymes is involved in the degradation of nylon oligomers (Okada et al. 1983).

Studies with various TOL-type plasmids have shown that upper- and lower-pathway operons (upper-pathway" enzymes oxidize methylbenzenes to methylbenzoates while the "lower-pathway" encode enzymes for the conversion of methylbenzoates to pyruvate, acetaldehyde, and acetate via (methyl) catechols), as well as the *xylS* and *xylR* regulatory genes, sometimes switched positions, or are inverted, or increased their copy number (Chatfield and Williams 1986; Assinder and Williams 1990; Williams et al. 1990). A pair of 1.4 kb sequences present on the TOL plasmid is consider to direct several recombinational processes and the insertion of the TOL catabolic operons in the chromosome (Meulien et al. 1981; Yano et al. 2010). In other bacterial strains, genes for catechol 2, 3-dioxygenase or catechol 1, 2-dioxygenase are duplicated (Nakai et al. 1990; Schreiner et al. 1991). Plasmid pJP4, which contains the genes for the degradation of 2,4-dichlorophenoxyacetic acid, seem to undergo several gene duplications (Don et al. 1985; Perkins et al. 1990; Vedler et al. 2004; Sen et al. 2010).

6.3 Mutational Drift

Mutational drift in terms of point mutation is of much relevance in xenobiotic degradation. It is possible that a number of stress factors such as chemical pollutants induce error-prone DNA replication that subsequently accelerates DNA evolution. Point mutation involves base substitution, or nucleotide replacement, in which one base is substituted for another at a specific location in a gene (Black 1999). This kind of mutation changes a single codon in mRNA, and it may or may not change the amino acid sequence in a protein. Several examples have illustrated that single-site-mutations can alter substrate specificities of enzymes or effector specificities. Clarke (1984) isolated mutants with altered substrate specificities of the AmiE amidase of *Pseudomonas aeruginosa*, which were provoked by single-base-pair changes.

Sequential mutations in the cryptic *ebg* genes of *Escherichia coli* were shown to result in active enzymes capable of metabolizing lactose and other sugars (Ojo 2007).

Single-site-mutations are believed to arise continuously and at random as a result of errors in DNA replication or repair. Although, the important effects of single-base-pair mutation on the adaptive process have been demonstrated experimentally, the accumulation of single-base pair changes may not be the main reason for the differences in the properties of the catabolic enzymes elicited by xenobiotic-degraders. There are other factors that would precipitate changes in DNA sequences, this include gene conversion or slipped-strand mispairing (Niedle et al. 1988). The degradation mainly depends upon the adapting responses of the microbial communities which include both selective enrichment (resulting in amplification of genes) and genetic changes (mainly includes gene transfer or mutation). With the mobilization of silent sequences into the functional catabolic routes and advancement of substrate range by gradual or spontaneous mutations, the recalcitrance of several toxic synthetic pollutants would certainly decrease (Sinha et al. 2009).

6.4 Gene Transfer

Gene transfer is a process of movement of genetic information between organisms (Black 1999). The significance of gene transfer for adaptation of host cells to new compounds has been explicitly demonstrated in many studies on experimental evolution of novel metabolic activities. Such studies identified biochemical blockades in natural pathways that prevented the degradation of novel substrates and these barriers scaled by transferring appropriate genes (Reineke et al. 1982; Ojo 2007; Liu and Paroo 2010). Genetic interactions in microbial communities are effected by several mechanisms such as conjugative transfer via plasmid replicons, transduction and transformation (Saye et al. 1990). The occurrence of plasmids in bacteria in the natural environment is certainly a general phenomenon and an important pool of genetic information residing on plasmid vehicles may flow among indigenous organisms. The self transmissible plasmids that carry genes for degradation of aromatic or of other organic compounds are known and their roles in spreading these genes to other organisms is predictable (Assinder and Williams 1988; Ojo 2007). Although, the transfer of catabolic plasmids can lead to regulatory and/or metabolic problems for the cells and therefore additional mutations in the primary transconjugants are often needed to construct strains with the desired metabolic activities (Reineke et al. 1982; Davey and O'toole 2000).

Recently, approaches have been made to assemble data with reference to adaptation in bacterial populations to specific xenobiotic compounds by gene transfer and to characterize and compare the genes involved in degradation of identical or similar xenobiotic compounds in different isolated bacterial genera from different environment. The following observations have been made:

(i) Evolutionarily related catabolic genes and their clusters have been derived from very distant locations in bacterial genera;

(ii) The phylogeny of the catabolic genes is not compatible with that of the 16S rRNA genes of the related hosts;
(iii) Genes for the degradation of synthetic pollutants are often associated with plasmids and transposons and
(iv) Evolutionary related catabolic genes and entire gene modules are involved in the degradation of structurally similar but different xenobiotic compounds (Top and Springael 2003).

It, therefore, appears that such genes are mostly clustered and located either on chromosomes acting as insertion elements or on plasmids as mobile genetic elements, and they also facilitate horizontal gene transfer (Sinha et al. 2009).

7 MGEs Involved in Adaptation to Organic Xenobiotics in the Environment

From comparative studies of sequenced bacterial genomes it has become obvious that gene acquisition by HGT is one of the driving forces behind the evolution of bacterial genomes (Lawrence and Ochman 2002). During the past 15 years, numerous groups have provided strong indications that MGEs and HGT have played and still play a significant role in the distribution and construction of xenobiotic catabolic pathways (Copley 2000; Top and Springael 2003). One line of evidence for the importance of HGT is the finding that catabolic genes in organic xenobiotic-degrading bacteria are often carried by MGEs. Plasmids are the best studied catabolic MGEs, probably because they can be easily detected in a bacterial strain (Tsuda et al. 1999; Top and Springael 2003).

The role of MGEs in the distribution of catabolic pathways and in the adaptation of a microbial community to a pollutant stress has recently been shown using different microcosm studies, where donor bacteria carrying a catabolic MGE were introduced into an environment that was subsequently challenged with the xenobiotic specific for that MGE. In many cases, transfer of the MGE into members of the indigenous community was shown; in some cases, this was accompanied by community changes and enhanced degradation of the xenobiotic by the ‘adapted’ microbial community (Top et al. 2002; Sorensen et al. 2005).

Mobile genetic elements support horizontal gene transfer more accurately via conjugation or transformation that can impart novel phenotypes or may modify existing genes through mutational processes. Typical catabolic plasmids are TOL, OCT, CAM, NAH etc. (Mishra et al. 2001). Studies have confirmed that horizontal transfer of catabolic genes, mostly by means of plasmid-mediated conjugation, happens in soil microcosms, bioreactors and so on after inoculation of a donor strain containing natural catabolic genes (Fulthorpe and Wyndham 1992). The catabolism of benzoate and phthalate suggests a high degree of redundancy, as *Rhodocccus* RHA1 has three linear plasmids. The smallest one has 300 putative genes and performs one fourth of the catabolic reactions, encoding 2,3-dihydroxybiphenyl

1,2-dioxygenase. This suggests that it has potential to degrade various PCB congeners (Larkin and Kulakov La Allen 2005). Toluene degrading plasmid (TOL) is also one of the best studied catabolic plasmids, as it bears degradative genes for xylene, toluene, benzoate, salicylate, catechol, phenol and several other complex compounds (Lloyd-Jones et al. 1999).

Generally, the mechanism of action for degradation by genes varies from organism to organism, depending on the entire organization of genes as given below.

(i) In the first category, genes are organized in a single operon system e.g. phenol (Khan et al. 2001). Here the genes may be located on the plasmid genome or may be present separately on the chromosome.
(ii) In the second group, genes are organized in two operon systems e.g. polychlorinated biphenyls etc. Here the genes are located on the plasmid genome (Shimizu et al. 2001).
(iii) In the third group, genes are organized in more than two operon systems e.g. 2,4-dichlorophenoxyaceticacid degradation, etc. Here, the genes are located on the plasmid genome and the capability or incapability of bacteria to carry out the entire degradation depends on the existence of the complementary enzymes encoded by genes on the chromosome (Don and Pemberton 1981; Lykidis et al. 2010; Sinha et al. 2009).
(iv) In another group, the genes are organized on transposons, e.g. 2,4,5-trichlorophenoxyacetate (2,4,5-T is a herbicide) degradation. *Pseudomonas* AC1100 contains two insertion elements, IS*931* and IS*932*; they play a significant role in degradation of 2,4,5-T (Chaudhry and Chapalamadugu 1991).

The expression of catabolic genes is when enzymes, substrate, metabolites and structure of respective genes all are in a synchronized manner (Whyte et al. 2002; Sinha et al. 2009).

8 Genetic Composition of the Microbial Community and Adaptation

Of the three mechanisms for adaptation of microbial communities to chemical contaminants, induction and derepression of enzymes, genetic changes, and selective enrichment, only the third has been examined in detail. This has been primarily a result of limitations imposed by available methods, which have, until recently, restricted the study of adaptation of microbial communities to the phenomenon of selective enrichment, in which the numbers or proportion of microorganisms that can utilize the compound of interest increase within the community and can be enumerated by their ability to grow on a medium containing the compound as the sole carbon source.

The primary genetic mechanism for the adaptation of the microbial community is the amplification, by means of selective enrichment and gene transfer and mutation, of genes which are involved in the metabolism of the chemical contaminant

(Spain and van Veld 1983; Barkay and Pritchard 1988; Becker et al. 2006). Direct monitoring of this process with respect to adaptation to hydrocarbons has been made possible by the development of DNA probes specific for the genes encoding hydrocarbon-catabolic pathways (Trevors 1985). Sayler et al. (1985), for example, using the colony hybridization technique, showed a correlation between the enhanced rates of PAH mineralization in oil-contaminated sediments and an increase in the number of colonies containing DNA sequences which hybridized to TOL (toluate oxidation) and NAH (naphthalene oxidation) plasmid probes. The colony hybridization procedure, however, has the disadvantage of requiring the growth of organisms on laboratory media, which limits sensitivity and does not allow detection of DNA sequences in viable but nonculturable microorganisms (Roszak and Colwell 1987). Dot blot hybridization, in which DNA is extracted from environmental samples and then probed (Ogram et al. 1987; Holben et al. 1988), can be used to detect specific DNA sequences in the environment without the need for isolation and culture of microorganisms (Ansari et al. 2008; Malik et al. 2008). The use of these methods in conjunction with nucleic acid probes for genes involved in hydrocarbon metabolism will allow measurement of the frequency of those genes within the microbial community (Trevors 1985). This will permit assessment of the relative degree of adaptation of the community as well as a more detailed analysis of the dynamics of gene amplification associated with adaptation.

9 Different Strategies for Niche Adaptation

Bacteria change their physiological behavior according to signals detected in the environment. Typically, these changes are reflected in the modification of gene expression patterns. These changes can be the effect of action by a number of different types of signaling proteins, including histidine protein kinases (HPKs) and their cognate response regulators (RRs), methyl-accepting chemotaxis proteins, diguanylate and adenylate cyclases, and serine/threonine/ tyrosine protein kinases, as well as individual transcription factors or "one-component" signal transduction proteins (Ulrich et al. 2005; Galperin 2005). Out of the different protein families, HPKs are the most abundant, and historically have been considered as the main mechanism for signal transduction in bacteria (Wolanin et al. 2002). HPKs, or signal transduction proteins, are considered to play a major role in the adaptation of bacteria to new or varying environments. In agreement with this hypothesis, those bacteria that have the largest complements of signaling proteins usually tend to be bacteria with complex lifestyles such as *Myxococcus xanthus*, those that are found ubiquitously in diverse environments such as *Pseudomonas*, or bacteria with several alternative metabolic strategies such as different δ- and ε-proteobacteria (Rodrigue et al. 2000; Galperin 2005). Few HPKs have been identified in the reduced genomes of parasitic bacteria, which likely have a relatively constant external environment. While these signal transduction systems are considered to be a key part of the adaptive evolution of bacteria, few details are known about this process. Alm et al. (2006) looked specially at genes that entered into each lineage recently, making the logical supposition that current additions are more likely to provide insight into the evolutionary basis of niche adaptation. Identifying recent achievement in a environment of multiple paralogs is a complicated task. Alm et al. (2006) explained a BLAST-based procedure for classifying and establishing the age of HPK domains (Ragan and Charlebois 2002; Daubin and Ochman 2004; Price et al. 2005).

Phylogenetic analysis was based on the histidine kinase domain of each HPK only, allowing follow changes in the structure of the signaling domains (domain shuffling) that generally upstream (N-terminal) from the kinase domain. The phylogenetic procedure for inference described requires a precise species phylogeny, which they deduce from a concatenated gene profile including 15 ubiquitous bacterial genes without obvious paralogs (Alm et al. 2006).

The relative contribution of horizontal gene transfer (HGT) and gene duplication events to the evolution of new HPKs in each genome was inferred by gene histories. The result shows that some genomes acquired new HPKs primarily via HGT, while others relied mainly on lineage-specific expansion (LSE) of existing gene families (Alm et al. 2006). A careful examination of genes obtained by these two mechanisms revealed differences in the amount to which upstream signaling domains and cognate Response regulators were conserved as a result of each process, with HGT being more likely to preserve pre-existing relationships than gene duplication.

Alm et al. (2006) reported that two-component systems including histidine protein kinases represent the primary signal transduction paradigm in prokaryotic organisms. To recognize how these systems acclimatize to allow organisms to detect niche-specific signals, they analyzed the phylogenetic distribution of nearly 5,000 histidine protein kinases from 207 sequenced prokaryotic genomes. They found that many genomes carry a large repertoire of recently evolved signaling genes, which may reflect selective pressure to adapt to new environmental conditions. Both lineage-specific gene family expansion and horizontal gene transfer play important role in introducing new histidine kinases into genomes; though, there are contradiction in how these two evolutionary forces act. Genes transferredvia horizontal transfer are more expected to keep their original functionality as inferred from a comparable complement of signaling domains, while gene family extension accompanied by domain shuffling appears to be a key source of novel genetic diversity. Family extension is the main source of new histidine kinase genes in the genomes mainly enriched in signaling proteins, and detailed analysis revealed that divergence in domain organization and changes in expression patterns are hallmarks of recent expansions. Finally, while these two modes of gene transfer are widespread among bacterial taxa, there are clear species-specific preferences for which mode is used.

10 Horizontal Gene Transfer and Microbial Evolution

The effect of horizontal gene transfer in microbial evolution (Bacteria and Archaea) is dependent on a number of genes which have been transferred to and effectively maintained in microbial genomes, however is also dependent on the degree of the phenomena, in both evolutionary time framework (considering both recent and ancient events) and phylogenetic distance between organisms (Boto 2010).

Since the appearance of horizontal gene transfer as a way of explaining phylogenetic incongruence using different gene trees, a significant number of studies have been published about genes that have been acquired by horizontal gene transfer (Gogarten et al. 2002; Lerat et al. 2005), both in Bacteria (Saunders et al. 1999; Ochman et al. 2000) and Archaea (Doolittle and Logsdon 1998; Faguy and Doolittle 1999; Boto 2010), as well as in eukaryotes (Andersson 2005). These studies explain that the transfer can take place not only among but also between domains in all possible directions: from Bacteria to Archaea (Rest and Mindell 2003), from Archaea to Bacteria (Gophna et al. 2004), from Archaea to Eukarya (Andersson et al. 2003), from Bacteria to Eukarya (Watkins and Gray 2006), from Eukarya to Bacteria (Guljamow et al. 2007) and even within Eukarya (Nedelcu et al. 2008). Though, it is in bacterial and archaeal evolution that horizontal gene transfer has been more extensively documented and accepted.

10.1 Gene Transfer and the Fate of Transferred Genes

Horizontal gene transfer results both from the successful transfer of genetic material (mediated by processes such as conjugation, transduction or transfection and recombination) and from the existence of the transferred genetic material all through the generations. The presence of certain physical barriers to transfer, as well as different selective forces over the transferred genes, may explain observed differences in the type of genes involved in horizontal gene transfer. Jain et al. (1999) proposed the complexity hypothesis to elucidate the observed differences in horizontal gene transfer susceptibility between genes. This hypothesis proposes that the so-called informational genes (involved in DNA replication, transcription and translation, and whose products participate in multiple molecular interactions) are less prone to horizontal gene transfer than operational genes (involved in cell maintenance and whose products have few interactions with other molecules). This hypothesis acquired support from the characterization by Bayesian inference of recently acquired genes in prokaryotic genomes (Nakamura et al. 2004), which has shown that the portion of transferred genes is biased towards genes involved in DNA binding, pathogenicity and cell surface functions, all of them included among the functions of operational genes. This work also shows, however, that not all operational genes are participating equally in horizontal gene transfer events (Nakamura et al. 2004).

A corresponding approach to realize whether there are differences between genes, in consideration to their involvement in transfer processes, is to determine what events are involved in post-transfer gene maintenance. It is accepted that the preservation of a transferred gene is related with positive selection (Gogarten et al. 2002; Pal et al. 2005). So, genes having a useful function are conserved while useless genes are removed. Numerous recent studies shed interesting results with regard to the maintenance of transferred genes (Kuo and Ochman 2009). For example, it has been shown that the incorporation of a single transferred gene into regulatory interaction networks is very slow (Lercher and Pal 2008) in the case of genes which provides the receptor with new functions, and it is dependent on the number of partners for the gene product in the regulatory network, according to the complexity hypothesis. Furthermore, products of these genes are incorporated in the margin of the corresponding regulatory network. On the other hand, transferred genes codifying for products with few partners are more easily deleted from genomes (Rocha 2008).

Recent work also shows that the transfer of whole operons enables the incorporation and maintenance processes, allowing a quick gain of function and facilitating the coordinate regulation of the new genes in the receptor (Price et al. 2008). Considering all, these results support that the involvement of their products in multiple molecular interactions (complexity) is a more important constraint to transfer and maintenance of genes in the prokaryotic world than the functional class to which the transferred genes belong.

10.2 Incidence of Horizontal Gene Transfer

The evolutionary distance among organisms can be another significant constraint with regard to transfer because genes transferred between organisms separated a long time ago were found to contribute in very different regulatory networks. On the other hand, the ancient transfer events are difficult to detect because of the amelioration process that affects the evolution of foreign genes in the receptor genome (Almeida et al. 2008; Marri and Golding 2008; Kuo and Ochman 2009). numerous studies suggest that gene transfer could effectively be more frequent for short and intermediate evolutionary distances but infrequent between organisms that are separated by large evolutionary time frames (Ochman et al. 2000; Brugger et al. 2002; Nakamura et al. 2004; Ge et al. 2005; Choi and Kim 2007; Dagan et al. 2008). A recent study (Wagner and De la Chaux 2008) has analyzed the evolution of 2,091 insertion sequences in 438 wholly sequenced prokaryotic genomes and found only 30 cases of supposed transfer events among distantly related clades. Twenty-three of these events seemed to be ancient while only seven were recent. However, instances of gene transfer between Archaea and Bacteria have been described (Rest and Mindell 2003; Gophna et al. 2004), which shows that horizontal gene transfer could affect evolution in the prokaryotic world along the different evolutionary times.

A recent study (Kanhere and Vingron 2009) compares the distance between orthologues and the intergenomic distances to try to identify ancient transfer processes in prokaryotic genomes. The authors found that 118 of the 171 gene transfer events were between Archaea and Bacteria, and they correspond primarily to metabolic genes. Seventy four per cent of these events were transfers from Bacteria to Archaea and the remaining 26% were transfers from Archaea to Bacteria. Only 53 genes were gene transfer events between bacteria phyla and corresponded mainly to genes involved in translation. Despite the fact that this approach is restricted to protein families that evolve at a steady rate, the study shows interesting trends within the inter domain gene transfer. It seems that the majority of transfers between Bacteria and Archaea have taken place in the Bacteria to Archaea direction. On the other hand, studies by Zhaxybayeva et al. (2006, 2009) show that the intra-phylum versus inter phylum gene exchange is different between different bacteria lineages: in Cyanobacteria, intra-phylum gene change seems to be more important than inter phylum exchange (Zhaxybayeva et al. 2006). In Thermotogales, however, and in particular in Firmicutes (Zhaxybayeva et al. 2009), the inter phylum exchange is dominant over intra phylum gene transfer. In addition, the proposed multiple gene exchange between ε-proteobacteria and Aquificales seems another example of inter-phylum exchange dominance (Boussau et al. 2008).

Considering all, these studies suggest that regardless of the fact that gene transfer can be more frequent between closely associated organisms, it may also take place between distantly related organisms, contributing to evolution of Archaea and Bacteria. On the other hand, the reality that recent transfer events can be more easily detected adds a bias to the study of gene transfer in prokaryotic evolution, confounding the real impact of ancient gene transfer events (Boto 2010).

10.3 Phylogenetic Relationships, Bacterial Species Concept and Horizontal Gene Transfer

Reconstructing the phylogenetic relationships between Bacteria and Archaea, morphological characters are of limited use (Bohannon 2008) compared with metabolic and molecular markers. However, horizontal gene transfer challenges in many cases the correct reconstruction of these relationships, confounding the phylogenetic signal present in these markers. Some authors (Doolittle 1999; Martin 1999; Doolittle and Bapteste 2007) question whether it is achievable to restructure a correct phylogenetic tree for the microbial world at all, considering the existence of horizontal gene transfer events. Others support the idea that some core genes are in no way transferred (Wolf et al. 2002; Brown 2003), thus maintaining a true phylogenetic signal that enables the reconstruction of a microbial phylogenetic tree. Finally, others (Kurland et al. 2003; Kurland 2005) consider that the existence of many barriers to gene transfer between organisms lowers the impact of horizontal gene transfer in phylogenetic reconstruction.

A recent study (Sorek et al. 2007) dealing with this topic searched in all the completed bacterial and archaeal genomes for genes that cannot be cloned in *E. coli* as a proxy to the study of barriers against horizontal gene transfer. Their results propose that there are no complete barriers to gene transfer because genes in all the families considered can be cloned in *E. coli* from at least one of the genomes. In addition, the results of a network analysis of shared genes (Dagan et al. 2008) agree with the idea that horizontal gene transfer leaves no gene family untouched. Supporting the idea that horizontal gene transfer challenges the reconstruction of phylogenetic relationships among prokaryotes, another study claims that less than 0.7%

of the prokaryotic genes may be considered core genes (Bapteste et al. 2008), making the construction of a phylogenetic tree unsustainable. Recently, the pangenome concept, originally developed to establish how many genomes should be sequenced from any given bacterial species in order to get a correct representation of the whole gene repertoire (Tettelin et al. 2005, 2008; Hogg et al. 2007) and to define the complete set of genes present in a prokaryotic group has been applied to the study of the whole set of genes present in sequenced bacterial genomes (Lapierre and Gogarten 2009). In this interesting study, the authors search for the presence of homologue genes in 573 completed genomes using BLAST and conclude that only 8% of the genes in a typical bacterial genome (approx. 250 gene families in all genomes) are present in 99% of the sampled genomes and therefore can be considered to be core genes that are part of the extended core genome or set of shared genes. Lapierre and Gogarten (2009) identified two other gene categories: the so-called character genes (shared by a group of organisms) present in a subset of genomes (64% of the genes in a typical bacterial genome and 7,900 gene families in all genomes) and the so-called accessory genes present in only one or in only a few of the genomes (28% of the genes in a bacterial genome and an infinite number of gene families in all genomes). The authors also propose that character gene evolution is mostly based on mutation, gene duplication and horizontal gene transfer, while horizontal gene transfer and gene losses are involved in the evolutionary history of accessory genes. The final conclusion of this study is that the bacterial 'pan-genome' (the set of all genes present in bacteria) is of infinite size, demonstrating the plasticity of the genome evolution in prokaryotes. Disregarding the fact that the methodology used can lead to under estimations or over estimations of particular gene categories, this study strongly underlines the impact that horizontal gene transfer has had on the evolutionary history of prokaryotes and provides us with important clues towards understanding the evolution of prokaryotic genomes (Boto 2010).

11 Conclusion

Horizontal gene transfer is an important force modulating evolution in the prokaryotic world and the evolution of particular eukaryotes. Even though gene exchange is easier in closely linked organisms, horizontal gene transfer occurred between both domains in the evolution of Archaea and Bacteria. The overwhelming evidence using both functional and in silico genomic screening is that environmental organisms harbour a previously underappreciated density of antibiotic resistance genes. This unexpected conclusion should have a paradigm shifting impact on our understanding of the judicious use of antibiotics and the drug discovery process.

Further, the catabolic operons known today have evolved from MGEs, which brought diverse catabolic genes together in one host and thus allowed a wide range of organic compounds to be degraded through a few central metabolic pathways (Top and Springael 2003). The studies that have been discussed show that the genetic patrimony of catabolic MGEs reaches further than plasmids and 'conventional' transposons and strongly indicate that HGT and patchwork assembly contribute to the adaptation of bacteria to organic xenobiotics in the real environment. In conclusion, it can be expected that further study of the molecular biology of different catabolic MGEs, microbial MGEs, and the ecology of prokaryotes and their MGEs in polluted sites at the macro- and micro-scale, will provide new insights into the evolutionary processes at work during community adaptation to pollution and adaptation of microorganisms in general.

References

Acar JF, Moulin G (2006) Antimicrobial resistance at farm level. Rev Sci Technol 25:775–792

Alekshun MN, Levy SB (2007) Molecular mechanisms of antibacterial multidrug resistance. Cell 128:1037–1050

Alm E, Huang K, Arkin A (2006) The evolution of two-component systems in bacteria reveals different strategies for niche adaptation. PLoS Comput Biol 2(11):e143

Almeida FC, Leszczyniecka M, Fisher PB, DeSalle R (2008) Examining ancient inter-domain horizontal gene transfer. Evol Bioinform 4:109–119

Amoroso MJ, Castro GR, Carlino FJ, Romero NC, Hill RT, Oliver G (1998) Screening of heavy metal-tolerant actinomycetes isolated from the Sali River. J Gen Appl Microbiol 44:129–132

Andersson JO (2005) Lateral gene transfer in eukaryotes. Cell Mol Life Sci 62:1182–1197

Andersson JO, Sjogren AM, Davis LAM, Embley TM, Roger AJ (2003) Phylogenetic analysis of diplomonad genes reveals frequent lateral gene transfer affecting eukaryotes. Curr Biol 13:94–104

Ansari MI, Grohmann E, Malik A (2008) Conjugative plasmids in multi-resistant bacterial isolates from Indian soil. J Appl Microbiol 104:1774–1781

Assinder SJ, Williams PA (1988) Comparison of the *meta* pathway operons on NAH plasmids *pWW60-22* and *TOL* plasmid *pWW53-4* and its evolutionary significance. J Gen Microbiol 134:2769–2778

Assinder SJ, Williams PA (1990) The TOL plasmids: determinants of the catabolism of toluene and the xylenes. Adv Microb Physiol 31:1–69

Baath E, Diaz-Ravina M, Frostegard A, Campbell CD (1998) Effect of metal-rich sludge amendments on the soil microbial community. Appl Environ Microbiol 64:238–245

Babic A, Lindner AB, Vulic M, Stewart EJ, Radman M (2008) Direct visualization of horizontal gene transfer. Science 319:1533–1536

Baek KH, Kim HS (2009) Microbial community structure in hexadecane- and naphthalene-enriched gas station soil. J Microbiol Biotechnol 19:651–657

Baek KH, Yoon BD, Cho DH, Kim BH, Oh HM, Kim HS (2009) Monitoring bacterial population dynamics using real-time PCR during the bioremediation of crude-oil-contaminated soil. J Microbiol Biotechnol 19:339–345

Bapteste E, Susko E, Leigh J, Ruiz-Trillo I, Bucknam J, Doolittle WF (2008) Alternative methods for concatenation of core genes indicate a lack of resolution in deep nodes of the prokaryotic phylogeny. Mol Biol Evol 25:83–91

Baquero F, Blazquez J (1997) Evolution of antibiotic resistance. Tree 12(12):482–487

Baquero F, Negri MC, Morosini MI, Blazquez J (1998) Antibiotic-selective environments. Clin Infect Dis 27(1):S5–S11

Barkay T, Pritchard H (1988) Adaptation of aquatic microbial communities to pollutant stress. Microbiol Sci 5:165–169

Becker JG, Berardesco G, Rittmann BE, Stahl DA (2006) Effects of endogenous substrates on adaptation of anaerobic microbial communities to 3-chlorobenzoate. Appl Environ Microbiol 72(1):449–456

Belanger PA, Beaudin J, Roy S (2011) High-throughput screening of microbial adaptation to environmental stress. J Microbiol Methods 85:92–97
Ben-Jacoba E, Cohen I, Golding I, Gutnick DL, Tcherpakov M, Helbing D, Ron IG (2000) Bacterial cooperative organization under antibiotic stress. Phys A 282:247–282
Benveniste R, Davies J (1973) Aminoglycoside antibiotic-inactivating enzymes in actinomycetes similar to those present in clinical isolates of antibiotic-resistant bacteria. Proc Natl Acad Sci 70:2276–2280
Black JG (1999) Bioremediation. In: Microbiology: principles and explorations. Prentice Hall Publishers, Upper Saddle River, pp 751–752
Bohannon J (2008) Confusing kinships. Science 320:1031–1033
Boto L (2010) Horizontal gene transfer in evolution: facts and challenges. Proc R Soc B 277:819–827
Boussau B, Gueguen L, Gouy M (2008) Accounting for horizontal gene transfers explains conflicting hypotheses regarding the position of aquificales in the phylogeny of Bacteria. BMC Evol Biol 8:272
Brown J (2003) Ancient horizontal gene transfer. Nat Rev Genet 4:121–132
Brugger K, Redder P, She QX, Confalonieri F, Zivanovic Y, Garrett RA (2002) Mobile elements in archaeal genomes. FEMS Microbiol Lett 206:131–141
Carrasco JA, Armario P, Pajuelo E, Burgos A, Caviedes MA, Lopez R, Chamber MA, Palomares AJ (2005) Isolation and characterization of symbiotically effective Rhizobium resistant to arsenic and heavy metals after the toxic spill at the Aznalcollar pyrite mine. Soil Biol Biochem 37:1131–1140
Carroll SB (2008) Evo-devo and expanding evolutionary synthesis A genetic theory of morphological evolution. Cell 134:25–36
Castro IV, Ferreira EM, McGrath SP (1997) Effectiveness and genetic diversity of *Rhizobium leguminosarum* bv trifolii isolates in Portuguese soils polluted by industrial effluents. Soil Biol Biochem 29:1209–1213
Chatfield LK, Williams PA (1986) Naturally occurring TOL plasmids in *Pseudomonas* strains carry either two homologous or two nonhomologous catechol 23-oxygenase genes. J Bacteriol 168:878–885
Chaudhry GR, Chapalamadugu S (1991) Biodegradation of halogenated organic compounds. Microbiol Rev 55:59–79
Choi IG, Kim SH (2007) Global extent of horizontal gene transfer. Proc Natl Acad Sci 104:4489–4494
Clarke PH (1984) The evolution of degradative pathways. In: Gibson DT (ed) Microbial degradation of organic compounds. Marcel Dekker Inc, New York, pp 11–27
Copley SD (2000) Evolution of ametabolic pathway for degradation of a toxic xenobiotic: the patchwork approach. Trends Biochem Sci 25:261–265
D'Costa VM, McGrann KM, Hughes DW, Wright GD (2006) Sampling the antibiotic resistome. Science 311:374–377
D'Costa VM, Griffiths E, Wright GD (2007) Expanding the soil antibiotic resistome: exploring environmental diversity. Curr Opin Microbiol 10:481–489
Dagan T, Artzy-Randrup Y, Martin W (2008) Modular networks and cumulative impact of lateral transfer in prokaryote genome evolution. Proc Natl Acad Sci 105:10039–10044
Darwin C (1859) On the origin of species by means of natural selection (London Murray). http://wwwliteratureorg/authors/ drawin-charles/the-origin-of-species/
Daubin V, Ochman H (2004) Bacterial genomes as new gene homes: the genealogy of ORFans in *E. coli*. Genome Res 14:1036–1042
Davey ME, O'toole GA (2000) Microbial biofilms: from ecology to molecular genetics. Microbiol Mol Biol Rev 64(4):847–867
Davies J (1992) Another look at antibiotic resistance. J Gen Microbiol 138:1553–1559
Davies J, Davies D (2010) Origins and evolution of antibiotic resistance. Microbiol Mol Biol Rev 74(3):417–433
De Souza MJ, Nair S, Loka Bharathi PA, Chandramohan D (2006) Metal and antibiotic-resistance in psychrotrophic bacteria from Antarctic Marine waters. Ecotoxicology 15:379–384

Dean AM, Thornton JW (2007) Mechanistic approaches to the study of evolution: the functional synthesis. Nat Rev Genet 8:1604–1607

Don RH, Pemberton JM (1981) Properties of six pesticide degradation plasmids isolated from *Alcaligenes paradoxus* and *Alcaligenes eutrophus*. J Bacteriol 145:681–686

Don RH, Weightman AJ, Knackmuss HJ, Timmis KN (1985) Transposon mutagenesis and cloning analysis of the pathways for degradation of 24-dichlorophenoxyacetic acid and 3-chlorobenzoate in *Alcaligenes eutrophus* JMP134(pJP4). J Bacteriol 161:85–90

Doolittle WF (1999) Phylogenetic classification and the universal tree. Science 284:2124–2128

Doolittle WF, Bapteste E (2007) Pattern pluralism and the tree of life hypothesis. Proc Natl Acad Sci 104:2043–2049

Doolittle WF, Logsdon JM Jr (1998) Do archaea have a mixed heritage? Curr Biol 8:R209–R211

El-Fantroussi S (2000) Enrichment and molecular characterization of a bacterial culture that degrades methoxy-methyl urea herbicides and their aniline derivatives. Appl Environ Microbiol 66(12):5110–5115

Esteve-Nenez A, Caballero A, Ramos JL (2001) Biological degradation of 2 4 6-trinitrotoluene. Microbiol Mol Biol Rev 65:335–352

Faguy DM, Doolittle WF (1999) Lessons from the *Aeropyrum pernix* genome. Curr Biol 9:R883–R886

Ferreira da Silva M, Vaz-Moreira I, Gonzalez-Pajuelo M, Nunes OC, Manaia CM (2007) Antimicrobial resistance patterns in Enterobacteriaceae isolated from an urban wastewater treatment plant. FEMS Microbiol Ecol 60:166–176

Foster P (1995) Adaptive mutation. In: Baumberg S, Young JPW, Wellington EMH, Saunders JR (eds) Population genetics of bacteria. Cambridge University Press, Cambridge, pp 13–30

Fulthorpe RR, Wyndham RC (1992) Involvement of a chlorobenzoate catabolic transposon Tn5271 in community adaptation to chlorobiphenyl chloroaniline and 2, 4 dichlorophenoxyacetic acid in a freshwater ecosystem. Appl Environ Microbiol 58:314–325

Gale EF, Cundliffe E, Reynolds PE, Richmond M, Waring MJ (eds) (1981) The molecular basis of antibiotic action, 2nd edn. Wiley, Chichester

Galperin MY (2005) A census of membrane-bound and intracellular signal transduction proteins in bacteria: bacterial IQ extroverts and introverts. BMC Microbiol 5:35

Ge F, Wang LS, Kim J (2005) The cobweb of life revealed by genome scale estimates of horizontal gene transfer. PLoS Biol 3:e316

Gogarten JP, Doolittle WF, Lawrence JG (2002) Prokaryotic evolution in light of gene transfer. Mol Biol Evol 19:2226–2238

Gophna U, Charlebois RL, Doolittle WF (2004) Have archaeal genes contributed to bacterial virulence? Trends Microbiol 12:213–219

Gottlieb D (1976) The production and role of antibiotics in soil. J Antibiot (Tokyo) 29:987–1000

Guljamow A, Jenke-Kodama H, Saumweber H, Quillardet P, Frangeul L, Castets AM, Bouchier A, Tandeau M, de Marsac N, Dittmann E (2007) Horizontal gene transfer of two cytoskeletal elements from a eukaryote to a cyanobacterium. Curr Biol 17:R757–R759

Handelsman J (2004) Metagenomics: application of genomics to uncultured microorganisms. Microbiol Mol Biol Rev 68:669–685

Heuer H, Smalla K (2012) Plasmids foster diversification and adaptation of bacterial populations in soil. FEMS Microbiol Rev. doi:10.1111/j.1574-6976.2012.00337.x

Hogg JS, Hu FZ, Janto B, Boissy R, Hayes J, Keefe R, Post JC, Ehrlich GD (2007) Characterization and modeling of the Haemophilus influenzae core and supragenomes based on the complete genomic sequences of Rd and 12 clinical nontypeable strains. Genome Biol 8:R103

Holben WE, Jansson JK, Chelm BK, Tiedje JM (1988) DNA probe method for the detection of specific microorganisms in the soil bacterial community. Appl Environ Microbiol 54:703–711

Jacobs L, Chenia HY (2007) Characterization of integrons and tetracycline resistance determinants in *Aeromonas* spp. isolated from South African aquaculture systems. Int J Food Microbiol 114:295–306

Jain R, Rivera MC, Lake JA (1999) Horizontal gene transfer among genomes: the complexity hypothesis. Proc Natl Acad Sci 96:3801–3806

Kanhere A, Vingron M (2009) Horizontal gene transfer in prokaryotes shows differential preferences for metabolic and translational genes. BMC Evol Biol 9:9

Khan AA, Wang RF, Cao WW, Doerge DR, Wennerstrom D, Cerniglia CE (2001) Molecular cloning nucleotide sequence and expression of genes encoding a polycyclic ring dioxygenase from *Mycobacterium* sp. strain PYR-1. Appl Environ Microbiol 67:3577–3585

Koonin EV (2009) Darwinian evolution in the light of genomics. Nucleic Acids Res 37:1011–1034

Kuo CH, Ochman H (2009) The fate of new bacterial genes. FEMS Microbiol Rev 33:38–43

Kurland CG (2005) What tangled web: barriers to rampant horizontal gene transfer. Bioessays 27:741–747

Kurland CG, Canback B, Berg OB (2003) Horizontal gene transfer: a critical view. Proc Natl Acad Sci 100:9658–9662

Lapierre P, Gogarten JP (2009) Estimating the size of the bacterial pan-genome. Trends Genet 25:107–110

Larkin MJ, Kulakov La Allen CCR (2005) Biodegradation and *Rhodococcus* masters of catabolic versatility. Curr Opin Biotechnol 16:282–290

Lawrence JG, Ochman H (2002) Reconciling the many faces of lateral gene transfer. Trends Microbiol 10:1–4

Leahy JG, Colwell RR (1990) Microbial degradation of hydrocarbons in the environment. Microbiol Rev 54(3):305–315

Lerat E, Daubin V, Ochman H, Moran NA (2005) Evolutionary origins of genomic repertoires in bacteria. PLoS Biol 3:e130

Lercher MJ, Pal C (2008) Integration of horizontally transferred genes into regulatory interaction networks takes many million years. Mol Biol Evol 25:559–567

Lessie TG, Wood MS, Byrne A, Ferrante A (1990) Transposable gene-activating elements in *Pseudomonas cepacia*. In: Silver S, Chakrabarty AM, Iglewski B, Kaplan S (eds) *Pseudomonas*: biotransformations pathogenesis and evolving biotechnology. American Society for Microbiology, Washington, DC, pp 279–292

Levesque S, Frost E, Michaud S (2007) Comparison of antimicrobial resistance of *Campylobacter jejuni* isolated from humans chickens raw milk and environmental water in Quebec. J Food Prot 70:729–735

Liu Q, Paroo Z (2010) Biochemical principles of small RNA pathways. Annu Rev Biochem 79:295–319

Lloyd-Jones G, Laurie AD, Hunter DWF, Fraser R (1999) Analysis of catabolic genes for naphthalene and phenanthrene degradation in contaminated New Zealand soils. FEMS Microbiol Ecol 29:69–79

Lykidis A, Pérez-Pantoja D, Ledger T, Mavromatis K, Anderson IJ et al (2010) The complete multipartite genome sequence of *cupriavidus necator* jmp134, a versatile pollutant degrader. PLoS One 5(3):e9729

Madeinos AA (1997) Evolution and dissemination of betalactamases accelerated by generation of beta lactum antibiotics. Clin Infect Dis 29:1419–1422

Malik A, Celik E-K Bohn C, Boeckelmann U, Knobel K, Grohmann E (2008) Molecular detection of conjugative plasmids and antibiotic resistance genes in anthropogenic soils from Germany and India. FEMS Microbiol Lett 279:207–216

Marri PR, Golding GB (2008) Gene amelioration demonstrated: the journey of nascent genes in bacteria. Genome 5:164–168

Marshall BM, Ochieng DJ, Levy SB (2009) Commensals: unappreciated reservoir of antibiotic resistance. Microbe 4:231–238

Martin W (1999) Mosaic bacterial chromosomes: a challenge en route to a tree of genomes. Bioessays 21:99–104

Martinez-Martinez L, Pascnal A, Jacoby GA (1998) Quinolone resistance from a transferable plasmid. Lancet 351:797–799

Martins da Costa PM, Vaz-Pires PM, Bernardo FM (2006) Antibiotic resistance of *Enterococcus* spp. isolated from wastewater and sludge of poultry slaughterhouses. J Environ Sci Health B 41:1393–1403

Meulien P, Downing RG, Broda P (1981) Excision of the 40 kb segment of the TOL plasmid from *Pseudomonas putida* mt-2 involves direct repeats. Mol Gen Genet 184:97–101

Mishra V, Lal R, Srinivasan (2001) Enzymes and operons mediating xenobiotic degradation in bacteria. Crit Rev Microbiol 27:133–166

Nakai C, Horiike K, Kuramitsu S, Kagamilyama H, Nozaki M (1990) Three isozymes of catechol 12-dioxygenase (pyrocatechase) aa a1 and B3 from *Pseudononas arvilla* C-1. J Biol Chem 265:660–665

Nakamura Y, Itoh T, Matsuda H, Gojobori T (2004) Biased biological functions of horizontally transferred genes in prokaryotic genomes. Nat Genet 36:760–766

Nedelcu AM, Miles IH, Fagir AM, Karol K (2008) Adaptative eukaryote-to-eukaryote lateral gene transfer: stress-related genes of algal origin in the closest unicellular relatives of animals. J Evol Biol 21:1852–1860

Neela FA, Nonaka L, Suzuki S (2007) The diversity of multi-drug resistance profiles in tetracycline-resistant *Vibrio* species isolated from coastal sediments and seawater. J Microbiol 45:64–68

Niedle EL, Harnett C, Bonitz S, Ornston LN (1988) DNA sequence of *Acinetobacter calcoaceticus catechol* 12-dioxygenase-I structural gene *cat*A: evidence for evolutionary divergence of intradiol dioxygenases by acquisition of DNA sequence repetitions. J Bacteriol 170:4874–4880

Ochman H, Lawrence JG, Grolsman EA (2000) Lateral gene transfer and the nature of bacterial innovation. Nature 405:299–304

Ogram A, Sayler GS, Barkay T (1987) The extraction and purification of microbial DNA from sediments. J Microbiol Methods 7:57–66

Ojo OA (2007) Molecular strategies of microbial adaptation to xenobiotics in natural environment. Biotechnol Mol Biol Rev 2(1):001–013

Okada H, Negoro S, Kimura H, Nakamura S (1983) Evolutionary adaptation of plasmid-encoded enzymes for degrading nylon oligomers. Nature (London) 306:203–206

Oliynyk M, Samborskyy M, Lester JB, Mironenko T, Scott N, Dickens S, Haydock SF, Leadlay PF (2007) Complete genome sequence of the erythromycin-producing bacterium *Saccharopolyspora erythraea* NRRL23338. Nat Biotechnol 25:447–453

Pal C, Papp B, Lercher MJ (2005) Horizontal gene transfer depends on gene content of the host. Bioinformatics 21(S2):222–223

Perkins EJ, Gordon MP, Caceres O, Lurquin PF (1990) Organization and sequence analysis of the 24-dichlorophenol hydroxylase and dichlorocatechol oxidative operons of plasmid pJP4. J Bacteriol 172:2351–2359

Piepersberg W (2001) Endogenous antimicrobial molecules: an ecological perspective. In: Sussman M (ed) Molecular medical microbiology, vol 1. Academic, New York, pp 561–585

Pigliucci M (2007) Do we need an extended evolutionary synthesis? Evolution 61:2743–2749

Price MN et al (2005) Operon formation is driven by co-regulation and not by horizontal gene transfer. Genome Res 15:809–819

Price MN, Dehal PS, Arkin AP (2008) Horizontal gene transfer and the evolution of transcriptional regulation in *Escherichia coli*. Genome Biol 9:R4

Ragan MA, Charlebois RL (2002) Distributional profiles of homologous open reading frames among bacterial phyla: implications for vertical and lateral transmission. Int J Syst Evol Microbiol 52:777–787

Raghunath D (2008) Emerging antibiotic resistance in bacteria with special reference to India. J Biosci 33:593–603

Ramezani A, Hawley RG (2010) Strategies to insulate lentiviral vector-expressed transgenes. Methods Mol Biol 614:77–100

Raniard L, Nowak V, Echairi A, Faloya V, Chaussod R (2008) The dynamics of soil bacterial community structure in response to yearly repeated agricultural copper treatments. Res Microbiol 159:251–254

Reineke W, Jeenes DJ, Williams PA, Knackmuss HA (1982) TOL plasmid in constructed halo benzoate-degrading *Pseudomonas* strains: prevention of *meta* pathway. J Bacteriol 150:195–201

Rest JS, Mindell DP (2003) Retroids in archaea: phylogeny and lateral origin. Mol Biol Evol 20:1134–1142
Riesenfeld CS, Goodman RM, Handelsman J (2004) Uncultured soil bacteria are a reservoir of new antibiotic resistance genes. Environ Microbiol 6:981–989
Rocha EPC (2008) Evolutionary patterns in prokaryotic genomes. Curr Opin Microbiol 11:454–460
Rodrigue A, Quentin Y, Lazdunski A, Mejean V, Foglino M (2000) Two component systems in *Pseudomonas aeruginosa*: why so many? Trends Microbiol 8:498–504
Roszak DB, Colwell RR (1987) Survival strategies of bacteria in the natural environment. Microbiol Rev 51:365–379
Saunders NJ, Hood DW, Moxon ER (1999) Bacterial evolution: bacteria play pass the gene. Curr Biol 9:R180–R183
Saye DJ, Ogunseitan OA, Sayler GS, Miller RV (1990) Transduction of linked chromosomal genes between *Pseudomonas aeruginosa* strains during incubation *in situ* in a fresh water habitat. Appl Environ Microbiol 56:140–145
Sayler GS, Shields MS, Tedford ET, Breen A, Hooper SW, Sirotkin KM, Davis JW (1985) Application of DNA-DNA colony hybridization to the detection of catabolic genotypes in environmental samples. Appl Environ Microbiol 49:1295–1303
Schlegel HG (1986) Xenobiotics. In: General microbiology, 6th edn. Cambridge University Press, New York, p 433
Schmitt H, Stoob K, Hamscher G, Smit E, Seinen W (2006) Tetracyclines and tetracycline resistance in agricultural soils: microcosm and field studies. Microb Ecol 51:267–276
Schreiner A, Fuchs K, Lottspeich F, Poth H, Lingens F (1991) Degradation of 2-methylaniline in *Rhodococcus rhodochrous*: cloning and expression of two clustered catechol 2,3-dioxygenase genes from strain CTM. J Gen Microbiol 137:2041–2048
Sen D, Yano H, Suzuki H, Król JE, Rogers L, Brown CJ, Top EM (2010) Comparative genomics of pAKD4, the prototype IncP-1δplasmid with a complete backbone. Plasmid 63(2):98
Shimizu S, Kobayashi H, Masai E, Fuduka M (2001) Characterization of the 450-kb linear plasmid in a polychlorinated biphenyl degrader *Rhodococcus* sp. strain RHA1. Appl Environ Microbiol 67:2021–2022
Sinha S, Chattopadhyay P, Pan I, Chàtterjee S, Chanda P, Bandyopadhyay D, Das K, Sen SK (2009) Microbial transformation of xenobiotics for environmental bioremediation. Afr J Biotechnol 8(22):6016–6027
Sorek R, Zhu Y, Creevey CJ, Francino MP, Bork P, Rubin EM (2007) Genome-wide experimental determination of barriers to horizontal gene transfer. Science 318:1449–1452
Sorensen SJ, Bailey M, Hansen LH, Kroer N, Wuertz S (2005) Studying plasmid horizontal transfer in situ: a critical review. Nat Rev Microbiol 3:700–710
Spain JC, van Veld PA (1983) Adaptation of natural microbial communities to degradation of xenobiotic compounds: effects of concentration exposure time inoculum and chemical structure. Appl Environ Microbiol 45:428–435
Springael D, Top EM (2004) Horizontal gene transfer and microbial adaptation to xenobiotics: new types of mobile genetic elements and lessons from ecological studies. Trends Microbiol 12(2):53–58
Stepanauskas R, Glenn TC, Jagoe CH, Tuckfield RC, Lindell AH, King CJ, McArthur JV (2006) Coselection for microbial resistance to metals and antibiotics in freshwater microcosms. Environ Microbiol 8:1510–1514
Taddei F, Radman M, Maynard-Smith J, Toupance B, Gouyon PH, Godelle B (1997) Role of mutator alleles in adaptive evolution. Nature 387:700–702
Tettelin H et al (2005) Genome analysis of multiple pathogenic isolates of *Streptococus agalactiae*: implications for the microbial 'pan-genome'. Proc Natl Acad Sci 102:13950–13955
Tettelin H, Riley D, Cattuto C, Medini D (2008) Comparative genomics: the bacterial pan-genome. Curr Opin Microbiol 12:472–477
Top EM, Springael D (2003) The role of mobile genetic elements in bacterial adaptation to xenobiotic organic compounds. Curr Opin Biotechnol 14(3):262–269

Top EM et al (2002) Catabolic mobile genetic elements and their possible use in bioaugmentation of polluted soils and waters. FEMS Microbiol Ecol 42:199–208
Trevors JT (1985) DNA probes for the detection of specific genes in bacteria isolated from the environment. Trends Biotechnol 3:291–293
Tsuda M, Iino T (1987) Genetic analysis of a transposon carrying toluene degrading genes on a TOL plasmid pWWO. Mol Gen Genet 210:270–276
Tsuda M, Iino T (1988) Identification and characterization of Tn4653 a transposon covering the toluene transposon Tn4651 on TOL plasmid pWWO. Mol Gen Genet 213:72–77
Tsuda M, Iino T (1990) Naphthalene degrading genes on plasmid NAH7 are on a defective transposon. Mol Gen Genet 223:33–39
Tsuda M, Minegishi KI, Iino T (1989) Toluene transposons Tn4651 and Tn4653 are class II transposons. J Bacteriol 171:1386–1393
Tsuda M et al (1999) Mobile catabolic genes in bacteria. J Biosci Bioeng 87:401–410
Ulrich LE, Koonin EV, Zhulin IB (2005) One-component systems dominate signal transduction in prokaryotes. Trends Microbiol 13:52–56
Van der Meer JR, Eggen RIL, Zehnder AJB, De Vos WM (1991) Sequence analysis of *Pseudomonas* sp. Strain P51 *tcb* gene cluster which encodes metabolism of chlorinated catecols: evidence for specialization of catecol 1 2-dioxygenases for chlorinated substrates. J Bacteriol 173:2425–2434
Vedler E, Vahter M, Heinaru A (2004) The completely sequenced plasmid pest4011 contains a novel incp1 backbone and a catabolic transposon harboring *tfd* genes for 2, 4-dichlorophenoxyacetic acid degradation. J Bacteriol 186(21):7161–7174
Wagner A, De la Chaux N (2008) Distant horizontal gene transfer is rare for multiple families of prokaryotic insertion sequences. Mol Genet Genomics 280:397–408
Walsh C (2003) Antibiotics: actions, origins, resistance. ASM Press, Washington, DC
Watkins RF, Gray MW (2006) The frequency of eubacterium-to-eukaryote lateral gene transfer shows significant cross-taxa variation within amoebozoa. J Mol Evol 63:801–814
Whyte LG, Smits THM, Labbe D, Witholt B, Greer CW, Van Beilen JB (2002) Gene cloning and characterization of multiple alkanehydroxylase systems in *Rhodococcus* strains Q15 and NRRLB-16531. Appl Environ Microbiol 68:5933–5942
Williams PA, Assinder SJ, Shaw LE (1990) Construction of hybrid *xylE* genes between the two duplicate homologous genes from TOL plasmid pWW53: comparison of the kinetic properties of the gene products. J Gen Microbiol 136:1583–1589
Wolanin PM, Thomason PA, Stock JB (2002) Histidine protein kinases: key signal transducers outside the animal kingdom. Genome Biol 3:R3013
Wolf YI, Rogozin IB, Grishin NV, Koonin EV (2002) Genome trees and the tree of life. Trends Genet 18:472–479
Wright GD (2007) The antibiotic resistome: the nexus of chemical and genetic diversity. Nat Rev Microbiol 5:175–186
Yano H, Miyakoshi M, Ohshima K, Tabata M, Nagata Y, Hattori M, Tsuda M (2010) Complete nucleotide sequence of TOL plasmid pDK1 provides evidence for evolutionary history of IncP-7 catabolic plasmids. J Bacteriol 192(17):4337–4347
Zhaxybayeva O, Gogarten JP, Charlebois RL, Doolittle WF, Papke RT (2006) Phylogenetic analyses of cyanobacterial genomes: quantification of horizontal gene transfer events. Genome Res 16:1099–1108
Zhaxybayeva O et al (2009) On the chimeric nature thermophilic origin and phylogenetic placement of the Thermotogales. Proc Natl Acad Sci 106:5865–5870

Chapter 16
Synthetic Lethal Genetic Interaction Networks and Their Utility for Anticancer Therapy

Saman Khan, Amit Kumar Sonkar, and Shakil Ahmed

Abstract Synthetic lethality is defined as a type of genetic interaction where two simultaneous genetic defects result in cellular death. These defects could be in the form of two independent non lethal mutations, over-expression of genes or the action of a chemical compound under a certain genetic background. The synthetic lethal genetic interaction network provides information about the functional relationship between the genes in a simple and experimentally tractable model system, such as yeast and improves our understanding for the treatment of complex human diseases. The synthetic lethal interaction data can be used to identify the gene function and to elucidate the mechanism of action of drugs. A large number of yeast genes have human orthologs and the synthetic lethal interaction data obtained from the yeast can provide excellent opportunities for therapeutics in cancer cell lines that can be extended up to drug developmental stage. This chapter provides the overview of genetic diversity of yeast using a synthetic lethal interaction network in order to identify function of genes of interest, mechanism of drug action and cancer therapy in higher eukaryotes. Furthermore, yeast as a model system for viral pathogenesis studies has also been discussed.

Keywords Yeast • Synthetic lethal interaction • Genetic interaction • Cancer • Virus

A genetic interaction is defined by the emergence of a surprising phenotype when two genes are disrupted together. The genetic interaction studies have been used to understand signaling or metabolic pathways and to examine the crosstalk between different pathways in fission and budding yeast (Forsburg 2001;

S. Khan • A.K. Sonkar • S. Ahmed (✉)
Molecular and Structural Biology Division, CSIR – Central Drug Research Institute, Chattar Manzil Palace, MG Road, Lucknow 226001, India
e-mail: shakil_ahmed@cdri.res.in

A. Malik et al. (eds.), *Management of Microbial Resources in the Environment*,
DOI 10.1007/978-94-007-5931-2_16,

Avery and Wasserman 1992). Topological analysis of the yeast genetic interaction network has revealed the importance of gene interactions in phenotype modelling (Hartwell 2004). About 30 genetic interactions have been identified for non-essential genes and fivefold more for essential genes (Tong et al. 2004; Davierwala et al. 2005). Recent genetic interactions studies in the yeast model system have been able to define the general principles of genetic networks, and also pave the way for similar studies in higher eukaryotic systems. A comparative understanding of these genetic-interaction networks promises to identify the nature of quantitative traits and the basis of complex inherited diseases.

1 Genome-Wide Loss-of-Function Screening in Yeast

The availability of elegant and straightforward genetic tools makes the yeast (budding and fission yeast) the preferred organism for the development of genomic methods for systematic discovery of gene function. The availability of remarkable genetic resources, including a systematic gene deletion set has made the system more attractive. Knock out deletion strains of about 6,000 known genes in budding yeast have been constructed, resulting in 5,000 viable haploid gene deletion mutants and identifying another 1,000 essential yeast genes (Giaever et al. 2002). A genome-wide gene deletion set for the fission yeast *Schizosaccharomyces pombe* has also been constructed (Kim et al. 2010). The comparisons of orthologous gene pairs between budding and fission yeast showed that 83% of single-copy orthologs in the two yeasts had conserved dispensability. The gene dispensability between the budding and fission yeast differed for certain pathways, including mitochondrial translation and cell cycle checkpoint control. In fission yeast a relationship between gene essentiality and the presence of introns has also been identified (Kim et al. 2010) suggesting that the essential genes are less likely to be rapidly regulated (Jeffares et al. 2008). The number of essential genes in fission yeast is higher as compared to the budding yeast. About 3,492 orthologs of fission yeast genes have been identified in other eukaryotes, including humans. About 87% of these orthologs are conserved in both yeast and other eukaryotes including human. Such a high degree of conservation suggests that conclusions drawn from analyses in the two yeasts concerning molecular and cell biology will be relevant to, and improve our understanding of, metazoan cells.

The availability of deletion collections enables us to screen for complete or partial loss-of-function phenotype. To date, the collection has been surveyed for fitness defects in response to a variety of growth conditions including different drug treatments and environmental insults (Suter et al. 2006; Scherens and Goffeau 2004; Brown et al. 2006). Plating assays and parallel analysis with the deletion collection has identified the genes responsible for complex phenotypes such as chromosome cohesion (Marston et al. 2004), sporulation (Deutschbauer et al. 2002; Enyenihi and Saunders 2003) and aging defects (Powers et al. 2006).

Fig. 16.1 Model depicting synthetic lethal interactions between two genes (shown here as *gene A* and *gene B*) that can compensate for each other because of functional redundancy

Gene A	Gene B	
A	B	Viable
A	b	Viable
a	B	Viable
a	b	Lethal

2 Genetic Interactions in Yeast: Synthetic Lethality

Synthetic lethality is a type of genetic interaction in which a combination of mutations in two or more genes leads to cell death, whereas a mutation in one of these genes does not (Fig. 16.1). The classic interpretation of this interaction between two genes is that the genes (or the pathways in which they act) are at least partially functionally redundant (Hartman et al. 2001). In other words synthetic lethal relationships often occur when the products of two genes act on the same pathway and combine to control a specific essential function or when two genes function in different pathways such that one pathway functionally compensates for defects in the other.

There is another model for synthetic lethal interactions termed as 'induced essentiality'. In this model the loss of one gene results in a rearrangement of the genetic network into an alternative viable state that makes a second gene an essential gene (Fig. 16.2). In such scenario the functions of two genes or the pathways in which they act, are not necessarily redundant or related. In this scenario one gene is simply required under a condition caused by the loss of another gene. In this model, a synthetic lethal interaction is a special form of conditional lethality and may be a consequence of the evolution of adaptive responses to different environmental conditions (Yadav et al. 2011). This model is consistent with the finding that in an analysis of 276 viable deletion mutants about 95% had a perturbed expression profile relative to wild-type cells, suggesting that loss of a non-essential gene results in a network rearrangement (Hughes et al. 2000).

Despite the existence of hundreds of alleles within the human population, in most cases only rare combinations lead to a disease state. To understand the genetic interactions that compromise the function of conserved pathways, model organisms will be essential in mapping genetic interactions that are relevant to disease states. Large-scale identification of synthetic lethal interactions can provide global maps of functional relationships between genes and pathways. To facilitate such large-scale genetics projects, an automated method for the construction of double mutants called synthetic genetic array (SGA) analysis has been developed (Tong et al. 2001,

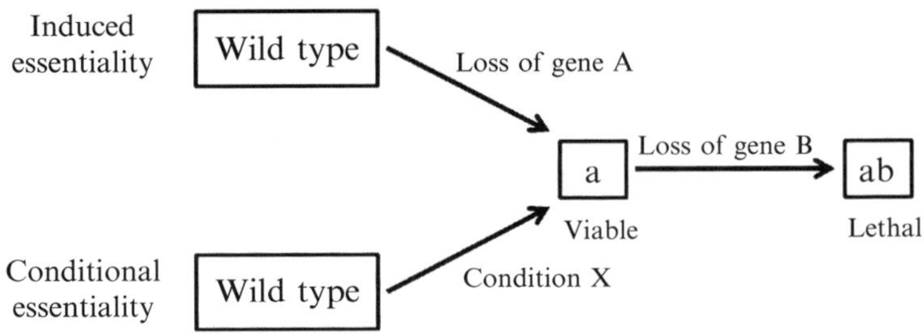

Fig. 16.2 Alternative model for synthetic lethality: the loss of one gene (*gene A*) results in a rearrangement of the genetic network in such a way that the other gene (*gene B*) becomes essential under new conditions

2004). These arrays consist of a series of sequential transfers to selective media, which enable rapid generation of haploid double mutants between a query gene and the ordered array of deletion mutants (Tong et al. 2004; Davierwala et al. 2005).

3 Synthetic Dosage Lethality (SDL) and Over-Expression

The term synthetic dosage lethality is used when a particular gene becomes essential for survival due to the over-expression of another gene. A SDL interaction occurs when overexpression of a gene is lethal in a target mutant strain but is viable in a wild type strain. At least four basic screening procedures can be applied to identify the interacting proteins using the SDL method.

1. Synthetic dosage lethality can be used as a direct test to assay for an interaction between a reference gene and a target mutation. For example, over-expression of the Cdk inhibitor protein Farlp is lethal to a strain defective in cdc34 but not in a cdcl6 mutant in *S. cerevisae* (Henchoz et al. 1997).
2. Synthetic dosage lethality can also be used as a secondary screen to uncover mutants of interest from a primary mutant collection by overexpression of a protein with a particular biological role. This screen has successfully identified a kinetochore mutant from a collection of chromosome transmission fidelity mutants by over-expression of known kinetochore proteins (Kroll et al. 1996).
3. Synthetic dosage lethality can also be adapted to a primary genetic screen to isolate mutants which interact with the reference protein. In this experiment, a wild-type strain is first mutagenized, an inducible reference gene introduced, and colonies isolated which are lethal only when the reference gene is over-expressed.
4. A variation of an SDL screen can also be performed in which a mutant strain is transformed with an over-expression library and screened for genes that when overexpressed cause cell lethality (Imiger et al. 1995).

Synthetic genetic arrays have recently been used to manipulate a genome-wide over-expression array where each strain expresses a unique yeast gene from an inducible promoter permitting systematic exploration of gain-of-function phenotypes (Sopko et al. 2006). In this manner, dosage suppressors of a conditional phenotype such as a temperature-sensitive allele of an essential gene should be readily identified (Huang et al. 2003). The combination of synthetic lethal and synthetic dosage lethality screens can also prove fruitful in the systematic dissection of specific biological processes (Measday and Hieter 2002; Measday et al. 2005).

4 Genetic Interactions and Small-Molecule Perturbation

The synthetic genetic interaction can also be used to identify the compounds that target specific pathways of interest. In this method the combination of mutations with environmental stresses and inhibition by chemical compounds mimics double deletions and hence abrogate the target pathway. Indeed, for a global understanding of complex biological processes, small molecules can serve as specific and reversible modulators of gene activity. Recently the synthetic genetics interaction network has been utilized for pathway-to-drug discovery using the DNA damage checkpoint as the target pathway (Tamble et al. 2011). Due to the availability of simple assays that can give robust fitness readouts in the yeast system, such chemical-genetic approaches on a genome-wide scale via 'chemical genomics' can be very useful for therapeutic purposes. Furthermore, many essential therapeutic targets and pathways are conserved between yeast and man.

5 The RNAi and Genome-Wide Loss-of-Function Screening

RNAi describes a process for mRNA degradation by sequence specific double-stranded RNA (dsRNA) (Moffat and Sabatini 2006). The RNA interference (RNAi) technology provides a smooth transition from sequence to function in more complex cells and metazoans. The RNAi libraries have been designed to target every predicted gene in the worm, fly, mouse and human genomes (Kamath et al. 2003; Boutros et al. 2004; Berns et al. 2004; Silva et al. 2005; Moffat et al. 2006) and offer the potential to systematically examine effects of gene perturbation on a genome-wide scale. Thus, targeted loss-of-function screens can now be used to uncover direct links between gene sequence and disease manifestation. A library of dsRNA-expressing bacterial strains targeting about 86% of the *C. elegans* genes has been used for RNAi-mediated loss-of-function screens to identify genes involved in embryonic lethality, sterility, genome instability, longevity, apoptosis, molting, transposon silencing and fat metabolism (Vastenhouw et al. 2003; Lee et al. 2003; Pothof et al. 2003; Frand et al. 2005; Ashrafi et al. 2003). Recently, adult worms were injected with a dsRNA library targeting 19,075 genes and over 650 perturbed genes were found to be required for early embryogenesis (Sonnichsen et al. 2005). The availability of different cell lines as well as dsRNA libraries and the relative ease with which cells take up exogenous dsRNA make *Drosophila* ideal for cell-based HT RNAi screens (Boutros et al. 2004; Wheeler et al. 2004; Echeverri and Perrimon 2006). High throughput *Drosophila* cell-based loss-of-function screens have identified genes involved in cell-cycle progression (Bjorklund et al. 2006), morphology (Kiger et al. 2003), cardiac and nervous system development (Kim et al. 2004; Ivanov et al. 2004) and signaling pathways (DasGupta et al. 2005; Muller et al. 2005).

6 Synthetic Lethal Interactions and Cancer Progression

Cancer arises from the consecutive acquisition of genetic alterations that produce the loss of function or transcriptional down-regulation of tumor suppressor genes and the activation or transcriptional up-regulation of oncogenes (Hernandez et al. 2007; Hanahan and Weinberg 2000). Downstream of the genetic alterations in tumor cells the expression pattern changes in many genes, which leads to abnormal regulation of biological processes such as the cell cycle and apoptosis (Green and Evan 2002). The importance of studying genetic interactions between tumor suppressor genes and oncogenes was established, that help to understand the mechanisms of tumorigenesis and metastasis (Ma et al. 2005; Xu et al. 2001; Takahashi and Ewen 2006). The genetic alterations that contribute to the tumorigenesis could be used to target cancer cells according to the concept of synthetic lethality. Under this hypothesis the genetic alteration can be identified by examining the profiles of somatic mutations, genomic alterations or gene expression changes in large series of tumors that can be utilized for the genetic interactions mapping. This genetic interaction mapping may illuminate the mechanisms of tumorigenesis with insights into prognosis.

In cases where a substantial knowledge of a particular process exists, as in DNA damage repair, checkpoint pathways, some of these synthetic lethal interactions can be predicted without the need for extensive screening (Bryant et al. 2005; Farmer et al. 2005). High-throughput screening can also be very useful in identifying more complex and unpredictable interactions. Synthetic lethal targeting of cancer cells could be therapeutically advantageous for targeting the cancer cells because only the cancer cells with a specific genetic mutation are killed. Cells without the cancer-inducing genotype are unaffected by such targeting. Exploiting synthetic lethality therefore increases selectivity towards killing only tumor cells. The tumor growth in many cases is driven by loss-of-function mutations in tumor suppressor genes. However restoring the function of tumor suppressor genes by gene therapy is not sufficient for their normal function (Harris and Hollstein 1993; Olivier et al. 2009; Liu et al. 2008). In such cases the identification of synthetic lethal drugs and pathways could enable the exploitation of these gene mutations for selective targeting. The synthetic lethality approach can also be used in combination with cytotoxic chemotherapy and/or radiotherapy, or in patients with relapsed disease (Tutt et al. 2010; Audeh et al. 2010). Moreover, the synthetic lethality could target a range of temporal mutations that occur from tumorigenesis to metastatic dissemination.

7 Screening for Mammalian Synthetic Lethality

The concept of using the synthetic lethal screen to identify new anti cancer drugs was floated by Hartwell et al. 1997. One of the limiting steps for drug discovery was the identification of tumor-selective characteristics; they suggested that loss-of-function mutations – such as those found in DNA repair genes or tumor suppressor genes- could be exploited. The idea was based on that, the first mutation could be a cancer-driving defect and highly conserved evolutionarily from model organisms to humans. The second mutation can be identified by a genome wide synthetic lethal interaction screening in yeast that can make the double mutant unable to grow. Hartwell et al. performed a small-scale screen of a panel of 70 different isogenic strains from budding yeast with deletions in DNA damage response genes against US Food and Drug Administration (FDA)-approved chemotherapies. The rationale was that the genetic instability which is a common feature of many tumors can make tumor cells more sensitive to the effects of some drugs than normal cells. This led to the identification of two putative anticancer agents: cisplatin and mitoxantrone. Cisplatin demonstrated increased specificity for yeast strains that were defective in post-replication repair, whereas mitoxantrone, a topoisomerase II poison resulted in increased sensitivity of yeast strains that were defective in double stranded DNA break repair. The applicability of synthetic lethal interaction studies in the context of mammalian cells is now more fully recognized and technological advances have been made to identify the synthetic lethal interaction genes in mammalian setting. The libraries of siRNAs and small molecules enable us to identify specific mutations on a genome-wide scale in mammalian cells. The screening of RNA interference (RNAi)-based libraries can identify genes that are important in a pathway context and thus provide a better understanding of the fundamental biology behind the interactions. In contrast the screening of small-molecule libraries can through up candidate compounds for the treatment of a given cancer genotype.

8 Conditional Synthetic Lethality Screens for Cancer Therapy

The conditional synthetic lethal screen has demonstrated great potential by using interactions based on temporary situations to further increase the therapeutic index and the selectivity for cancer cells. This kind of screening can be developed in different contexts such as, in response to ionizing radiation, cytotoxic chemotherapeutic agents, or changes in the cellular microenvironment. Recently it has been demonstrated that tumor hypoxia decreases the expression of homologous recombination proteins such as RAD51 (Bindra et al. 2004; Chan et al. 2008). Thus by suppressing the expression of DNA repair proteins, hypoxia conditionally transforms cells into a recombination-deficient state and consequently makes them sensitive to poly (ADP-ribose) polymerase (PARP) inhibitors (Chan et al. 2010). This approach represents one example of how conditional synthetic lethal interactions can potentially be manipulated and exploited for cancer therapy.

Thus the synthetic lethal interactions represent a very promising means of selectively killing tumor cells. Future synthetic lethality screens could also be used to investigate mechanisms to exploit epigenetic phenomena, the tumor microenvironment and stromal–tumour interactions. Other screens could focus on identifying genetic interactions to enhance radiotherapy or current cytotoxic chemotherapies.

9 Yeast as a Model System for Viral Pathogenic Studies

The complexity of higher eukaryotic systems makes them difficult systems for genetics and cell biology studies. Over the years scientists have utilized yeast as a simpler system for the study of various pathologies including virus proliferation and testing new drugs against these pathogenic agents. Moreover, the heterologous protein expression in yeast species has become an important tool in the production of therapeutic and commercially relevant proteins. The advantage of a yeast-based expression system is due to the fact that as a eukaryotic microorganism, it can perform complex posttranslational protein modifications combined with ease of genetic manipulation and cell growth (Buckholz and Gleeson 1991; Romanos et al. 1992; Smith et al. 1985). The simple yeast system has been utilized to elucidate the function of individual proteins from important pathogenic viruses such as *Hepatitis C* virus, and *Epstein-Barr* virus (DeMarini et al. 2003; Kapoor et al. 2005). Furthermore, studies of viruses that infect yeast have provided important information to elucidate the mechanistic aspect of the life cycle of many RNA viruses targeting eukaryotes and the host factors involved (Wickner 2008). Till now a number of viruses have been reported to undergo replication in yeast. These include RNA and DNA viruses that infect plants, insects, mammals, and humans (Angeletti et al. 2002; Panavas and Nagy 2003; Pantaleo et al. 2003; Raghavan et al. 2004; Zhao and Frazer 2002; Price et al. 2005). There are numerous ways one can utilize yeast as a model system for viral pathogenic studies.

10 AIDS Viral Protein Expression and its Effect in Yeast Cells

The yeast two-hybrid system has been extensively used to identify and characterize the potential interacting partner(s) of a given protein (Fields and Song 1989; Luban and Goff 1995; Hamdi and Colas 2012). In this method the interaction between a protein and a protein encoded by the library plasmids leads to transcriptional activation of a reporter gene. HIV-1 Integrase, which catalyzes the insertion of proviral DNA into the host cell genome, has been shown to interact with human SFN5, a transcription factor and its yeast ortholog. This interaction plays an important role during the integration step and facilitates the interaction with the host cell chromatin (Kalpana et al. 1994; Rain et al. 2009). It has been observed that the retroviral integrase is unable to perform its lethal activity in cells having disrupted the SNF5 gene, suggesting that it might be involved in the lethal effect induced by integrase in yeast cells (Parissi et al. 2000). The interaction of HIV-1 integrase with microtubules proteins has also been studied by two hybrid analysis (De Soultrait et al. 2002).

The yeast model has been extremely useful to study many aspects of the integration step in retroviral replication. The retroviral integrase gene expression in *Saccharomyces cerevisiae* yeast cells strongly increased the deleterious effects of the retroviral enzyme only in absence of the *rad52* gene, which is involved in the repair of double-strand DNA breaks. Genetic experiments suggested that the mutagenic effect and the level of recombination events were affected in cells expressing the retroviral enzyme, while expression of the mutated enzyme had no significant effect (Parissi et al. 2003). Vpr, another HIV-1 encoded protein plays several roles in the replication cycle of this retrovirus. A study on fission yeast has suggested that Vpr plays an important role during the G2 to M transition of the cell cycle (Chang et al. 2004). It has been reported that the induction of Vpr expression in the fission yeast *Schizosaccharomyces pombe* leads to several defects in the assembly and function of the mitotic spindle. It delocalizes the spindle pole body protein and affects the actin ring formation that leads to the disruption of cytokinesis (Chang et al. 2004). Similar defects in mitosis and cytokinesis have also been observed in human cells suggesting that these defects could be responsible for some of the pathological effects associated with HIV-1 infection. Vpr also triggers production of reactive oxygen species (ROS), indicating that an apoptotic-like mechanism might be mediated by these reactive species. Recently Vpr suppressor genes have been identified that are able to overcome Vpr-induced cell death in fission yeast as well as arrest Vpr-induced apoptosis in mammalian cells (Huard et al. 2008). The similarity of Vpr-induced cell death in fission yeast with the mammalian apoptotic process reinforces the idea that fission yeast may be used as a simple model system to study the apoptotic-like process induced by Vpr.

11 Interaction Studies of Plant Viruses Using Yeast as a Model Host

Using yeast as model host for some plant viruses such as bromoviruses and tombusviruses has led to the identification of replication associated factors that affect host virus interactions and viral pathology (Janda and Ahlquist 1993). The knowledge gained from such studies will lead to the development of new antiviral methods and applications in biotechnology. A number of yeast genes have been shown to affect

RNA-RNA recombination in tombusviruses. Studies with plant viruses, such as *Brome mosaic virus* (BMV), tombusviruses, and geminiviruses, as well as with animal viruses and influenza virus have shown that certain viruses can complete most of the steps required for intracellular replication in yeast cells (Naito et al. 2007; Panavas et al. 2005; Panavas and Nagy 2003; Nagy 2008).

The Y2H assay has also been used to screen cDNA libraries prepared from different plants to identify possible host factors regulating cell-to-cell movement (Desvoyes et al. 2002; Dunoyer et al. 2004; Fridborg et al. 2003), RNA encapsidation (Hofius et al. 2007; Okinaka et al. 2003), and host responses (Schaad et al. 2000; Ueda et al. 2006). The development of yeast as a model host facilitated by systems biology and proteomics approaches has led to the identification and characterization of host factors involved in RNA virus replication on a genome-wide scale. Future work based on the combined use of yeast genetics, biochemistry, and cell biology should help in further dissecting the detailed functions of the host proteins during virus replication.

12 Expression of E2 Protein in Fission Yeast

Cervical cancer is one of the leading causes of death from cancer among women. Infection with human papillomavirus (HPV) is associated with cervical cancer. Different types of HPV are associated with cervical tumor; out of these HPV16 is the most prevalent in infections (zur Hausen 1991).

Analysis of viral infected tumor cells shows disruption of the E2 gene during integration suggesting that the loss of E2p is an important step in malignant transformation. E2p regulates viral transcription via binding to palindromic DNA sequences present in the upstream regulatory region of the viral DNA (Ushikai et al. 1994; Tan et al. 1994; Steger and Corbach 1997). It is also required for the initiation of viral DNA replication (Chiang et al. 1992). The E2 gene product is also able to inhibit cell cycle progression in HPV negative cells (Frattini et al. 1997). To understand the mechanistic aspect of cell cycle progression, E2p has been expressed in fission yeast *S. pombe* and has been shown to inhibit the G2-M transition by delaying the activation of Cdc2p kinase (Fournier et al. 1999). Thus, the yeast system provides a simple method for the identification of novel E2 mutants that may give insights as to how E2p regulates cell division in higher eukaryotic cells.

13 Conclusions and Perspectives

The synthetic lethal interactions represent a very promising means of selectively killing tumor cells. The integration of the yeast synthetic-genetic interactions, physical interactions, mRNA expression profiling and functional genomics data will enable us to generate important framework for understanding human biology and disease. Future synthetic lethality screens could also be used to investigate mechanisms to exploit epigenetic phenomena, the tumor microenvironment and stromal–tumour interactions. Other screens could focus on identifying genetic interactions to enhance radiotherapy or current cytotoxic chemotherapies. Moreover, future work should lead to the discovery of new cellular factors involved in virus proliferation. Yeast can also be used to obtain recombinant viral protein to determine 3D structure as well as for high-resolution imaging. Proteomics approaches should also help us to identify various posttranslational modifications of viral and host proteins that could affect their functions during the pathogenesis. The detailed results generated using yeast model system can also be applied to dissect the interactions between plant viruses and their native plant hosts that will lead to better understanding of viral pathogenesis.

References

Angeletti P, Kim K, Fernandes F, Lambert P (2002) Stable replication of papillomavirus genomes in *Saccharomyces cerevisiae*. J Virol 76(7):3350–3358

Ashrafi K, Chang FY, Watts JL, Fraser AG, Kamath RS, Ahringer J, Ruvkun G (2003) Genome-wide RNAi analysis of *Caenorhabditis elegans* fat regulatory genes. Nature 421:268–272

Audeh MW, Carmichael J, Penson RT, Friedlander M, Powell B, Bell-McGuinn KM, Scott C, Weitzel JN, Oaknin A, Loman N, Lu K, Schmutzler RK, Matulonis U, Wickens M, Tutt A (2010) Oral poly(ADP-ribose) polymerase inhibitor olaparib in patients with BRCA1 or BRCA2 mutations and recurrent ovarian cancer: a proof-of-concept trial. Lancet 376:245–251

Avery L, Wasserman S (1992) Ordering gene function: the interpretation of epistasis in regulatory hierarchies. Trends Genet 8(9):312–316

Berns K, Hijmans EM, Mullenders J, Brummelkamp TR, Velds A, Heimerikx M, Kerkhoven RM, Madiredjo M, Nijkamp W, Weigelt B, Agami R, Ge W, Cavet G, Linsley PS, Beijersbergen RL, Bernards R (2004) A large-scale RNAi screen in human cells identifies new components of the p53 pathway. Nature 428:431–437

Bindra RS, Schaffer PJ, Meng A, Woo J, Måseide K, Roth ME, Lizardi P, Hedley DW, Bristow RG, Glazer PM (2004) Down-regulation of Rad51 and decreased homologous recombination in hypoxic cancer cells. Mol Cell Biol 24:8504–8518

Bjorklund M, Taipale M, Varjosalo M, Saharinen J, Lahdenpera J, Taipale J (2006) Identification of pathways regulating cell size and cell-cycle progression by RNAi. Nature 439:1009–1013

Boutros M, Kiger AA, Armknecht S, Kerr K, Hild M, Koch B, Haas SA, Consortium HF, Paro R, Perrimon N (2004) Genome-wide RNAi analysis of growth and viability in *Drosophila* cells. Science 303:832–835

Brown JA, Sherlock G, Myers CL, Burrows NM, Deng C, Wu HI, McCann KE, Troyanskaya OG, Brown JM (2006) Global analysis of gene function in yeast by quantitative phenotypic profiling. Mol Syst Biol 2:2006.0001

Bryant HE, Schultz N, Thomas HD, Parker KM, Flower D, Lopez E, Kyle S, Meuth M, Curtin NJ, Helleday T (2005) Specific killing of BRCA2-deficient tumors with inhibitors of poly(ADP-ribose) polymerase. Nature 434:913–917

Buckholz R, Gleeson M (1991) Yeast systems for the commercial production of heterologous proteins. Biotechnology 9:1067–1072

Chan N, Koritzinsky M, Zhao H, Bindra R, Glazer PM, Powell S, Belmaaza A, Wouters B, Bristow RG (2008) Chronic hypoxia decreases synthesis of homologous recombination proteins to offset chemoresistance and radioresistance. Cancer Res 68:605–614

Chan N, Pires IM, Bencokova Z, Coackley C, Luoto KR, Bhogal N, Lakshman M, Gottipati P, Oliver FJ, Helleday T, Hammond EM, Bristow RG (2010) Contextual synthetic lethality of cancer cell kill based on the tumor microenvironment. Cancer Res 70:8045–8054

Chang F, Sebastian F, Sazer S, Luban J (2004) HIV-1 Vpr induces defects in mitosis, cytokinesis, nuclear structure, and centrosomes. Mol Biol Cell 15(4):1793–1801

Chiang CM, Ustav M, Stenlund A, Ho T, Broker TR, Chow LT (1992) Viral E1 and E2 proteins support replication of homologous and heterologous papillomaviral origins. Proc Natl Acad Sci USA 89:5799–5803

DasGupta R, Kaykas A, Moon RT, Perrimon N (2005) Functional genomic analysis of the Wnt-wingless signaling pathway. Science 308:826–833

Davierwala AP, Haynes J, Li Z, Brost RL, Robinson MD, Yu L, Mnaimneh S, Ding H, Zhu H, Chen Y, Cheng X, Brown GW, Boone C, Andrews BJ, Hughes TR (2005) The synthetic genetic interaction spectrum of essential genes. Nat Genet 37(10):1147–1152

De Soultrait V, Caumont A, Parissi V, Morellet N, Ventura M, Lenoir C, Litvak S, Fournier M, Roques B (2002) A novel short peptide is a specific inhibitor of the human immunodeficiency virus type 1 integrase. J Mol Biol 318:45–58

DeMarini D, Johnston V, Konduri M, Gutshall L, Sarisky R (2003) Intracellular hepatitis C virus RNA-dependent RNA polymerase activity. J Virol Methods 113:65–68

Desvoyes B, Faure-Rabasse S, Chen MH, Park JW, Scholthof HB (2002) A novel plant homeodomain protein interacts in a functionally relevant manner with a virus movement protein. Plant Physiol 129:1521–1532

Deutschbauer A, Williams RM, Chu AM, Davis RW (2002) Parallel phenotypic analysis of sporulation and postgermination growth in *Saccharomyces cerevisiae*. Proc Natl Acad Sci USA 99:15530–15535

Dunoyer P, Thomas C, Harrison S, Revers F, Maule A (2004) A cysteine-rich plant protein potentiates potyvirus movement through an interaction with the virus genome-linked protein VPg. J Virol 78:2301–2309

Echeverri CJ, Perrimon N (2006) High-throughput RNAi screening in cultured cells: a user's guide. Nat Rev Genet 7:373–384

Enyenihi AH, Saunders WS (2003) Large-scale functional genomic analysis of sporulation and meiosis in *Saccharomyces cerevisiae*. Genetics 163:47–54

Farmer H, McCabe N, Lord CJ, Tutt AN, Johnson DA, Richardson TB, Santarosa M, Dillon KJ, Hickson I, Knights C, Martin NM, Jackson SP, Smith GC, Ashworth A (2005) Targeting the DNA repair defect in BRCA mutant cells as a therapeutic strategy. Nature 434:917–921

Fields S, Song O (1989) A novel genetic system to detect protein-protein interactions. Nature 340:245–246

Forsburg SL (2001) The art and design of genetic screens: yeast. Nat Rev Genet 2:659–666

Fournier N, Raj K, Saudan R, Utzig S, Sahli R, Simanis V, Beard P (1999) Expression of human papillomavirus 16 E2 protein in *Schizosaccharomyces pombe* delays the initiation of mitosis. Oncogene 18(27):4015–4021

Frand AR, Russel S, Ruvkun G (2005) Functional genomic analysis of *C. Elegans* molting. PLoS Biol 3:e312

Frattini MG, Hurst SD, Lim HB, Swaminathan S, Laimins LA (1997) Abrogation of a mitotic checkpoint by E2 proteins from oncogenic human papillomaviruses correlates with increased turnover of the p53 tumor suppressor protein. EMBO J 16:318–331

Fridborg I, Grainger J, Page A, Coleman M, Findlay K, Angell S (2003) TIP, a novel host factor linking callose degradation with the cell-to-cell movement of Potato virus X. Mol Plant Microbe Interact 16:132–140

Giaever G, Chu AM, Ni L, Connelly C, Riles L, Veronneau S, Dow S, Lucau-Danila A, Anderson K, Andre B et al (2002) Functional profiling of the *Saccharomyces cerevisiae* genome. Nature 418:387–391

Green DR, Evan GI (2002) A matter of life and death. Cancer Cell 1(1):19–30

Hamdi A, Colas P (2012) Yeast two hybrid methods and their applications in drug discovery. Trends Pharmacol Sci 33(2):109–118

Hanahan D, Weinberg RA (2000) The hallmarks of cancer. Cell 100(1):57–70
Harris CC, Hollstein M (1993) Clinical implications of the p53 tumor-suppressor gene. N Engl J Med 329:1318–1327
Hartman J, Gravik B, Hartwell L (2001) Principles for the buffering of genetic variation. Science 291:1001–1004
Hartwell L (2004) Genetics. Robust interactions. Science 303:774–775
Hartwell LH, Szankasi P, Roberts CJ, Murray AW, Friend SH (1997) Integrating genetic approaches into the discovery of anticancer drugs. Science 278:1064–1068
Henchoz S, Chi Y, Catarin B, Herskowitz I, Deshaies RJ, Peter M (1997) Phosphorylation and ubiquitin-dependent degradation of the cyclin-dependent kinase inhibitor Far1p in budding yeast. Genes Dev 11:3046–3060
Hernandez P, Huerta-Cepas J, Montaner D, Al-Shahrour F, Valls J, Gomez L, Capella G, Dopazo J, Pujana MA (2007) Evidence for systems level molecular mechanisms of tumorigenesis. BMC Genomics 8:185
Hofius D, Maier AT, Dietrich C, Jungkunz I, Bornke F, Maiss E, Sonnewald U (2007) Capsid protein-mediated recruitment of host DnaJ-like proteins is required for Potato virus Y infection in tobacco plants. J Virol 81:11870–11880
Huang ME, Rio AG, Nicolas A, Kolodner RD (2003) A genome wide screen in *Saccharomyces cerevisiae* for genes that suppress the accumulation of mutations. Proc Natl Acad Sci USA 100:11529–11534
Huard S, Chen M, Burdette K, Fenyvuesvolgyi C, Yu M, Elder RT, Zhao RY (2008) HIV-1 Vpr-induced cell death in *Schizosaccharomyces pombe* is reminiscent of apoptosis. Cell Res 18(9):961–973
Hughes TR, Marton MJ, Jones AR, Robert CJ, Stoughton R, Armour CD, Bennett HA, Coffey E, Dai H, He YD, Kidd MJ, King AM, Meyer MR, Slade D, Lum PY, Stepaniants SB, Shoemaker DD, Gachotte D, Chakraburtty K, Simon J, Bard M, Friend SH (2000) Functional discovery via a compendium of expression profiles. Cell 102:109–126
Imiger S, Piatti S, Michaelis C, Nasmyth K (1995) Genes involved in sister chromatid separation are needed for B-type cyclin proteolysis in budding yeast. Cell 81(2):269–278
Ivanov AI, Rovescalli AC, Pozzi P, Yoo S, Mozer B, Li HP, Yu SH, Higashida H, Guo V, Spencer M, Nirenberg M (2004) Genes required for *Drosophila* nervous system development identified by RNA interference. Proc Natl Acad Sci USA 101:16216–16221
Janda M, Ahlquist P (1993) RNA-dependent replication, transcription, and persistence of brome mosaic virus RNA replicons in *S. Cerevisiae*. Cell 72:961–970
Jeffares DC, Penkett CJ, Bahler J (2008) Rapidly regulated genes are intron poor. Trends Genet 24:375–378
Kalpana G, Marmon S, Wang W, Crabtree G, Goff S (1994) Binding and stimulation of HIV-1 integrase by a human homolog of yeast transcription factor SNF5. Science 266(5193):2002–2006
Kamath RS, Fraser AG, Dong Y, Poulin G, Durbin R, Gotta M, Kanapin A, Le Bot N, Moreno S, Sohrmann M, Welchman DP, Zipperlen P, Ahringer J (2003) Systematic functional analysis of the *Caenorhabditis elegans* genome using RNAi. Nature 421:231–237
Kapoor P, Lavoie B, Frappier L (2005) EBP2 plays a key role in Epstein-Barr virus mitotic segregation and is regulated by aurora family kinases. Mol Cell Biol 25(12):4934–4945
Kiger AA, Baum B, Jones S, Jones MR, Coulson A, Echeverri C, Perrimon N (2003) A functional genomic analysis of cell morphology using RNA interference. J Biol 2:27
Kim YO, Park SJ, Balaban RS, Nirenberg M, Kim Y (2004) A functional genomic screen for cardiogenic genes using RNA interference in developing *Drosophila* embryos. Proc Natl Acad Sci USA 101:159–164
Kim D, Hayles J, Kim D, Wood V, Park H, Won M, Yoo HS, Duhig T, Nam M et al (2010) Analysis of a genome-wide set of gene deletions in the fission yeast *Schizosaccharomyces pombe*. Nat Biotechnol 28(6):617–623
Kroll ES, Hyland KM, Hieter P, Li JJ (1996) Establishing genetic interactions by a synthetic dosage lethality phenotype. Genetics 143:95–102

Lee SS, Lee RY, Fraser AG, Kamath RS, Ahringer J, Ruvkun G (2003) A systematic RNAi screen identifies a critical role for mitochondria in *C. Elegans* longevity. Nat Genet 33:40–48

Liu TC, Hwang TH, Bell JC, Kirn DH (2008) Translation of targeted oncolytic virotherapeutics from the lab into the clinic, and back again: a high-value iterative loop. Mol Ther 16:1006–1008

Luban J, Goff S (1995) The yeast two-hybrid system for studying protein-protein interactions. Curr Opin Biotechnol 6(1):59–64

Ma L, Teruya-Feldstein J, Behrendt N, Chen Z, Noda T, Hino O, Cordon-Cardo C, Pandolfi PP (2005) Genetic analysis of Pten and Tsc2 functional interactions in the mouse reveals asymmetrical haploinsufficiency in tumor suppression. Genes Dev 19(15):1779–1786

Marston AL, Tham WH, Shah H, Amon A (2004) A genome-wide screen identifies genes required for centromeric cohesion. Science 303:1367–1370

Measday V, Hieter P (2002) Synthetic dosage lethality. Methods Enzymol 350:316–326

Measday V, Baetz K, Guzzo J, Yuen K, Kwok T, Sheikh B, Ding H, Ueta R, Hoac T, Cheng B, Pot I, Tong A, Yamaguchi-Iwai Y, Boone C, Hieter P, Andrews B (2005) Systematic yeast synthetic lethal and synthetic dosage lethal screens identify genes required for chromosome segregation. Proc Natl Acad Sci USA 102:13956–13961

Moffat J, Sabatini DM (2006) Building mammalian signalling pathways with RNAi screens. Nat Rev Mol Cell Biol 7:177–187

Moffat J, Grueneberg DA, Yang X, Kim SY, Kloepfer AM, Hinkle G, Piqani B, Eisenhaure TM, Luo B, Grenier JK, Carpenter AE, Foo SY, Stewart SA, Stockwell BR, Hacohen N, Hahn WC, Lander ES, Sabatini DM, Root DE (2006) A lentiviral RNAi library for human and mouse genes applied to an arrayed viral high-content screen. Cell 124:1283–1298

Muller P, Kuttenkeuler D, Gesellchen V, Zeidler MP, Boutros M (2005) Identification of JAK/STAT signalling components by genome-wide RNA interference. Nature 436:871–875

Nagy PD (2008) Yeast as a model host to explore plant virus-host interactions. Annu Rev Phytopathol 46:217–242

Naito T, Kiyasu Y, Sugiyama K, Kimura A, Nakano R, Matsukage A, Nagata K (2007) An influenza virus replicon system in yeast identified Tat-SF1 as a stimulatory host factor for viral RNA synthesis. Proc Natl Acad Sci USA 104:18235–18240

Okinaka Y, Mise K, Okuno T, Furusawa I (2003) Characterization of a novel barley protein, HCP1, that interacts with the Brome mosaic virus coat protein. Mol Plant Microbe Interact 16:352–359

Olivier M, Petitjean M, Marcel V, Petre A, Mounawar M, Plymoth A, deFromentel CC, Hainaut P (2009) Recent advances in p53 research: an interdisciplinary perspective. Cancer Gene Ther 16:1–12

Panavas T, Nagy PD (2003) Yeast as a model host to study replication and recombination of defective interfering RNA of Tomato bushy stunt virus. Virology 314:315–325

Panavas T, Hawkins CM, Panaviene Z, Nagy PD (2005) The role of the p33:p33/p92 interaction domain in RNA replication and intracellular localization of p33 and p92 proteins of Cucumber necrosis tombusvirus. Virology 338:81–95

Pantaleo V, Rubino L, Russo M (2003) Replication of Carnation Italian ringspot virus defective interfering RNA in *Saccharomyces cerevisiae*. J Virol 77(3):2116–2123

Parissi V, Caumont A, De Soultrait V, Dupont C, Pichuantes S, Litvak S (2000) Inactivation of the SNF5 transcription factor gene abolishes the lethal phenotype induced by the expression of HIV-1 integrase in yeast. Gene 247:129–136

Parissi V, Caumont A, De Soultrait VR, Desjobert C, Calmels C, Fournier M, Gourgue G, Bonneu M, Tarrago-Litvak L, Litvak S (2003) The lethal phenotype observed after HIV-1 integrase expression in yeast cells is related to DNA repair and recombination events. Gene 322:157–168

Pothof J, van Haaften G, Thijssen K, Kamath R, Fraser A, Ahringer J, Plasterk R, Tijsterman M (2003) Identification of genes that protect the *C. Elegans* genome against mutations by genome-wide RNAi. Genes Dev 17:443–448

Powers R, Kaeberlein M, Caldwell S, Kennedy B, Fields S (2006) Extension of chronological life span in yeast by decreased TOR signaling. Genes Dev 20:174–184

Price B, Eckerle L, Ball L, Johnson K (2005) Nodamura virus RNA replication in *Saccharomyces cerevisiae*: heterologous gene expression allows replication-dependent colony formation. J Virol 79(1):495–502

Raghavan V, Malik P, Choudhury N, Mukherjee S (2004) The DNA-A component of a plant geminivirus (Indian Mungbean yellow mosaic virus) replicates in budding yeast cells. J Virol 78(5):2405–2413

Rain J, Cribier A, G´erard A, Emiliani S, Benarous R (2009) Yeast two-hybrid detection of integrase-host factor interactions. Methods 47(4):291–297

Romanos M, Scorer C, Clare J (1992) Foreign protein expression in yeast: a review. Yeast 8:423–488

Schaad MC, Anderberg RJ, Carrington JC (2000) Strain-specific interaction of the tobacco etch virus NIa protein with the translation initiation factor eIF4E in the yeast two-hybrid system. Virology 273:300–306

Scherens B, Goffeau A (2004) The uses of genome-wide yeast mutant collections. Genome Biol 5:229

Silva J, Li M, Chang K, Ge W, Golding M, Rickles RJ, Siolas D, Hu G, Paddison PJ, Schlabach MR et al (2005) Second-generation shRNA libraries covering the mouse and human genomes. Nat Genet 37:1281–1288

Smith R, Duncan M, Moir D (1985) Heterologous protein secretion from yeast. Science 229:1219–1224

Sonnichsen B, Koski LB, Walsh A, Marschall P, Neumann B, Brehm M, Alleaume AM, Artelt J, Bettencourt P, Cassin E et al (2005) Full-genome RNAi profiling of early embryogenesis in *Caenorhabditis elegans*. Nature 434:462–469

Sopko R, Huang D, Preston N, Chua G, Papp B, Kafadar K, Snyder M, Oliver SG, Cyert M, Hughes TR, Boone C, Andrews B (2006) Mapping pathways and phenotypes by systematic gene overexpression. Mol Cell 21:319–330

Steger G, Corbach S (1997) Dose-dependent regulation of the early promoter of human papillomavirus type 18 by the viral E2 protein. J Virol 71:50–58

Suter B, Auerbach D, Stagljar I (2006) Yeast-based functional genomics and proteomics technologies: the first 15 years and beyond. Biotechniques 40:625–644

Takahashi C, Ewen ME (2006) Genetic interaction between Rb and Nras: differentiation control and metastasis. Cancer Res 66:9345–9348

Tamble C, St Onge R, Giaever G, Nislow C, Williams A, Stuart J, Lokey R (2011) The synthetic genetic interaction network reveals small molecules that target specific pathways in *Sacchromyces cerevisiae*. Mol Biosyst 7:2019–2030

Tan S-H, Leong LE-C, Walker PA, Bernard H-U (1994) The human papillomavirus type 16 E2 transcription factor binds with low cooperativity to two flanking sites and represses the E6 promoter through displacement of Sp1 and TFIID. J Virol 68:6411–6420

Tong AH, Evangelista M, Parsons AB, Xu H, Bader GD, Page N, Robinson M, Raghibizadeh S, Hogue CW, Bussey H, Andrews B, Tyers M, Boone C (2001) Systematic genetic analysis with ordered arrays of yeast deletion mutants. Science 294:2364–2368

Tong AH, Lesage G, Bader GD, Ding H, Xu H, Xin X, Young J, Berriz GF, Brost RL, Chang M, Chen Y, Cheng X, Chua G, Friesen H et al (2004) Global mapping of the yeast genetic interaction network. Science 303:808–813

Tutt A, Robson M, Garber JE, Domchek SM, Audeh MW, Weitzel JN, Friedlander M, Arun B, Loman N, Schmutzler RK, Wardley A, Mitchell G, Earl H, Wickens M, Carmichael J (2010) Oral poly(ADP-ribose) polymerase inhibitor olaparib in patients with BRCA1 or BRCA2 mutations and advanced breast cancer: a proof-of-concept trial. Lancet 376:235–244

Ueda H, Yamaguchi Y, Sano H (2006) Direct interaction between the Tobacco mosaic virus helicase domain and the ATP-bound resistance protein, N factor during the hypersensitive response in tobacco plants. Plant Mol Biol 61:31–45

Ushikai M, Lace MJ, Wa YY, Kono M, Anson J, Ishiji T, Parkkinen S, Wicker N, Valentine M-E, Davidson I, Turek LP, Haugen TH (1994) Transactivation by the full length E2 proteins of human papillomavirus type 16 and bovine papillomavirus type 1 in vitro and in vivo: cooperation with activation domains of cellular transcription factors. J Virol 68:6655–6666

Vastenhouw NL, Fischer SE, Robert VJ, Thijssen KL, Fraser AG, Kamath RS, Ahringer J, Plasterk RH (2003) A genome-wide screen identifies 27 genes involved in transposon silencing in *C. Elegans*. Curr Biol 13:1311–1316

Wheeler DB, Bailey SN, Guertin DA, Carpenter AE, Higgins CO, Sabatini DM (2004) RNAi living-cell microarrays for loss-of-function screens in *Drosophila melanogaster* cells. Nat Methods 1:127–132

Wickner R (2008) The yeast dsRNA virus L-A resembles mammalian dsRNA virus cores. In: Patton JT (ed) Segmented double-stranded RNA viruses: structure and molecular biology. Caister Academic Press Norfolk, UK

Xu X, Qiao W, Linke SP, Cao L, Li W, Furth P, Harris CC, Deng C (2001) Genetic interactions between tumor suppressors Brca1 and p53 in apoptosis, cell cycle and tumorigenesis. Nat Genet 28(3):266–271

Yadav S, Verma SK, Ahmed S (2011) DNA topoisomerase II mutant activates DNA damage checkpoint protein kinase Chk1 in fission yeast *S. Pombe*. Genet Res 93(4):275–283

Zhao K, Frazer I (2002) *Saccharomyces cerevisiae* is permissive for replication of bovine papillomavirus type 1. J Virol 76(23):12265–12273

zur Hausen H (1991) Human papillomaviruses in the pathogenesis of anogenital cancer. Virology 184:9–13

Chapter 17
Adaptation of *Candida albicans* for Growth Within the Host

Zuraini Zakaria, Basma Rajeh Mohammad Abu Arra, and Sumathi Ganeshan

Abstract One of the most important attributes of microorganisms is their ability to biologically adapt to every environment on earth. Microorganisms possess sophisticated signaling systems that enable them to sense and respond to environmental changes and challenges. Typically, this response results in morphological, physiological and even genetic differentiations. The genetic information associated with a microbe is capable of alterations which are sometimes reversible, and disappearing when the particular pressure is lifted. Other alterations are maintained and can even be passed on to succeeding generations of bacteria. This fact may well indicate that the structure can be modified to maintain function under environmental stress. *Candida albicans*, commonly found as a component of the normal flora of humans, residing in the gastrointestinal tract, in the genitourinary tract and on the skin, is the most common opportunistic human pathogen. The yeast is a harmless commensal in most healthy people, but it causes superficial as well as life-threatening systemic infections in immunocompromised patients. The ability of *C. albicans* to be virulent depends completely on its yeast-to-hyphae switch where the organism changes from a unicellular yeast form to a multicellular hyphal form. This switch may likely be induced by environmental conditions like temperature, pH and nutrients. This chapter presents the regulatory adaptation mechanisms that make *C. albicans* the most successful fungal pathogen of humans.

Keywords *Candida albicans* • Adaptation mechanisms • Environmental stress

---

Z. Zakaria (✉) • B.R.M.A. Arra • S. Ganeshan
Biology Programme, School of Distance Education,
Universiti Sains Malaysia, 11800 Minden, Penang, Malaysia
e-mail: zuraini@usm.my

A. Malik et al. (eds.), *Management of Microbial Resources in the Environment*,
DOI 10.1007/978-94-007-5931-2_17,

1 Introduction

The ability of microorganisms to sense and adapt to changes in the external environment is essential for their survival. This adaptation to environmental conditions is particularly important for the survival of fungal species growing with a close association with host organisms, such as pathogens, symbionts or commensals (Vylkova et al. 2011). Accordingly, the yeast *Candida albicans* is both a harmless commensal and an aggressive pathogen (Wächtler et al. 2011). Since *C. albicans* lacks any typical environmental habitat, their cells generally adapt to diverse environmental conditions, each with its own specific set of environmental pressures, with ease within very short time periods. *Candida albicans* has even adapted to grow in association with the human host whereby the human immune system has allowed it.

The fungus is found as a commensal in most healthy individuals. As commensals, *Candida* species are harmless and the population is colonized with *C. albicans* without any signs of diseases. However, if the balance of the normal flora is disrupted or the immune defenses are compromised, *C. albicans* frequently outgrows the microbial flora and causes symptoms of diseases (Mavor et al. 2005). Therefore, *C. albicans* is able to switch between being a commensal and a pathogen (Rupp 2007; Schulze and Sonnenborn 2009). However, the cause of transition from a harmless commensal to an aggressive pathogen (Wächtler et al. 2011), and the stable maintenance between commensalism and pathogenicity is still not known (Sohn et al. 2006; Rupp 2007). Nevertheless, by using the host as an environment, *C. albicans* has got to not only develop mechanisms to colonize and invade into the host, but also need to possess particular attributes to persist in the host in large numbers without causing any damages. Undoubtedly, both host and microbial factors are required to maintain the balance from commensalisms to host damage and *vice versa* (Rupp 2007).

Thus, this chapter focuses on the recently identified systems for adaptation of *C. albicans* during colonization and infection of their human host and the role of these systems in the fungal survival and virulence.

2 Epidemiology

Humans harbor *C. albicans* during or shortly after birth wherein it almost immediately colonizes the newborn and becomes a commensal (Khan and Gyanchandani 1998; Calderone and Fonzi 2001; Khan et al. 2010). The growth of the yeast is kept under control by the infant's immune system and thus produces no signs of diseases. Accordingly, it is presumed that by 6 months of age, 90% of all babies are colonized by *C. albicans*. By adulthood, almost all humans become involved in a life-long relationship and play host to *C. albicans*. Since *C. albicans* is part of the normal flora, it is normal for all humans to have controlled quantities of the organism.

Candida albicans being the most common commensal fungi in healthy individuals (Pinto et al. 2008), rarely causes diseases in immunocompetent hosts (Khan et al. 2010).

The fungi are normally found as commensals in the human oral cavity, gastrointestinal tract, genitourinary tract, diseased skin and mucosal membranes (Baron et al. 1994; Forche et al. 2011). As commensals, *C. albicans* competes with other microbes for nutrition in different body sites and adapts to different host conditions like temperature, pH and nutrients within the host (Calderone 2002). Moreover, the physical barriers such as epithelial layers and a functional immune system maintain the commensal phase of *C. albicans* in the host (Wächtler et al. 2011). Therefore, commensally growing *C. albicans* are controlled by the normal microbial flora and the immune response of the host (Mavor et al. 2005).

In order to survive in a niche, a fungus has to adapt to constantly changing parameters wherein *C. albicans* has been found to respond to these environmental changes by inducing transcriptional and translational changes for adaptation in the new environmental conditions (Khan et al. 2010). Success of survival also depends on rapid adaptations to changing micro-environments and likewise, as *C. albicans* enters the human host, their lifestyle changes from harmless commensals to opportunistic pathogens. Opportunistic pathogens, when given the opportunity, will attempt to colonize all bodily tissues. Thereby, under conditions of weakened immunity (people with Acquired Immunodeficiency Syndrome (AIDS), leukemia or cancer) or imbalance in the commensal flora, *C. albicans* becomes an opportunistic pathogen and causes diseases ranging from mild superficial infections (such as oral thrush and vaginitis) to severe, life-threatening bloodstream infections (such as disseminated candidiasis) (Forche et al. 2011). *Candida albicans* takes the advantage of certain predisposing events to cause diseases (Baron et al. 1994). These different candidal infections therefore involve adaptation of *C. albicans* to different host environmental niches and growth conditions (Forche et al. 2011).

3 Mycology

Candida albicans is an asexual oval-shaped diploid fungus of the Ascomycetes. A number of such fungi are causal agents of opportunistic infections in humans (Baron et al. 1994). Some species show phenotypic switching, a variant colony morphology (Khan et al. 2010). They grow both as yeast and filamentous cells on various mucosal surfaces of the body, including the oral cavity, gastrointestinal tract and vaginal mucosa. The predominant asexual reproductive unit of the yeast is a blastoconidium (Baron et al. 1994).

3.1 *Candida albicans* Strain Variation

Candida albicans infection in the majority of cases is a patient's own commensal flora. Evidence suggests many individuals harbor a mixture of strain types of minor variants, typically differing in levels of genetic heterozygosity (Odds 2010). As such, more than one distinctly different *C. albicans* strain type may be found in a small proportion of individuals in a single sample from a single anatomical site (Jacobsen et al. 2008), and in longitudinal samples from the same site or different anatomical sites (Odds et al. 2006a; Bougnoux et al. 2006). These interstrain differences may indicate strain replacement (Bartie et al. 2001). However, strain replacements are much less common than a single type with minor variations (Odds 2010). This single *C. albicans* strain type with minor variations may suggest microadaptations of *C. albicans* on the host.

3.2 Organism Characteristics

3.2.1 General Characteristics

An unusual feature of *C. albicans* is its ability to grow either as an unicellular budding yeast or in filamentous pseudohyphal and hyphal forms (Odds 1988; Sudbery et al. 2004). Pseudohyphae are morphologically distinguishable from the hyphae form whereby the former have constrictions at the sites of septation and are wider than the latter. By contrast, hyphae form long tube-like filaments with completely parallel sides and no constrictions at the sites of septation (Sudbery 2011).

3.2.2 Cell Wall Structure

The cell wall is crucial for colonization and infection since it defines the interface between host and pathogen. It is one of the major structures, comprising 15–25% of the dry weight of the cell of which approximately 80–90% of the cell wall of *C. albicans* consists of carbohydrates. Three basic constituents representing the major components of the cell wall are complex polymers of glucose (β-1,3- and β-1,6-glucan), chitin (N-acetylglucosamine) and polymers of mannan (mannoproteins). The glucans make up 50–60% of the total mass of the cell wall, whereas chitin constitutes 0.6–2%, and mannan moiety of mannoproteins represents 30–40% of the cell wall polysaccharides in *C. albicans*. The rigid structure of complex polymers of glucose and chitin surrounds the cell like a shield, protects it from environmental stresses like osmotic pressure and defines the shape and physical strength of the fungal cell (Sohn et al. 2006). In addition, cell walls contain 6–25% proteins and minor amounts of 1–7% lipid. However, the protein composition of the cell wall varies greatly according to the different cell morphologies like the yeast, pseudohyphal and hyphal form. Also, the expression pattern of cell wall proteins in hyphae varies with the different stimuli induced. Accordingly, it has been shown that cell surface proteins determine the adhesion (Rupp 2007), colonization and infection of host cells and any alterations in the protein composition of the cell wall may result in reduced virulence of *C. albicans* (Sohn et al. 2006). This shows that host and pathogen interact and this interaction is controlled by several signaling systems of *C. albicans*.

4 *Candida albicans* Infections

Fungal infections also commonly occur in the healthy population. Healthy persons generally encounter superficial infections but immunocompromised patients face invasive infections (Khan et al. 2010). As such, *C. albicans* has been reported to be the fourth most common cause of nosocomial infections in the USA and elsewhere in the world (Wenzel 1995). The ability of *C. albicans* to adapt and survive at different anatomic sites of the human host has also made them more harmful than other commensals of the human body (Khan et al. 2010). As commensals of the normal flora, *C. albicans* causes superficial infections and epithelial damage when it overgrows the microbial flora (Mavor et al. 2005). However, in severe cases, the fungus penetrates through epithelial layers into deeper tissues and may cause life-threatening systemic infections (Wächtler et al. 2011). Infections caused by this pathogen normally include cutaneous lesions, mucous membranes and systemic disseminated diseases.

4.1 *Cutaneous Lesions*

Superficial *Candida* infections include intertrigo (warm and moist environment like skin folds), diaper rash and nail infections. Commensal *Candida* infections rapidly colonize damaged skin or skin areas with closely opposing surfaces, such as the diaper area in infants and toddlers and abdominal fat folds and groin in older individuals. *Candida albicans* also causes candidal onchomychosis in the nails or the area around the nail and such infections produces lesions that resemble those produced by dermatophytes. Nail infection almost always includes involvement of the area around the nail leading to club-shaped fingers (Baron et al. 1994).

4.2 *Mucous Membranes*

The ability of *C. albicans* to live as a commensal on mucosal surfaces of healthy individuals often cause superficial infections of mucous membranes and may lead to a condition known as thrush (Baron et al. 1994; Mavor et al. 2005). The mucosal surface of the vagina is a frequent site of *Candida* infection (vulvovaginitis). Approximately 70% of women experience vaginal candidiasis once in a life and 20% suffer from recurrence (Baron et al. 1994; Fidel et al. 1999), however, the reasons for repeated attacks are not known.

4.3 *Systemic Disseminated Disease*

The *Candida* cells which manage to penetrate into deeper tissues cause severe systemic infections in immunocompromised patients, in children with AIDS and other immune deficiencies, as well as in very low birth weight premature infants. When a person is severely immunocompromised, *C. albicans* enters the bloodstream and causes infection of the bloodstream which leads to serious problems, especially in the kidneys, heart, lungs, eyes, brain and many other organs. Furthermore, in the severely immunocompromised host, *C. albicans* may also cause life-threatening systemic infections.

5 *Candida albicans* Pathogenesis and Virulence Factors

Although *C. albicans* is a commensal, the organism is also an important opportunistic pathogen (Rosenbach et al. 2010). Like other fungal pathogens, *C. albicans* regulates the expression of certain genes as virulence factors to produce disease. Some of virulence factors include the ability to recognize and adhere to host tissues, to respond rapidly to changes in the external environment and to secrete hydrolases (Khan et al. 2010). Some of the commonly studied virulence factors in *C. albicans* are briefly described here.

5.1 *Genetics*

Candida albicans is a diploid organism but is unable to undergo meiotic division to a haploid phase. However, it has developed unusual mechanisms for maintenance of genetic diversity in the absence of a complete sexual cycle. These include chromosomal polymorphisms, with partial ploidy changes for some chromosomes, mitotic recombination events, and gains and losses of heterozygosity (Odds 2010). Amongst these, loss of heterozygosity (LOH) has been widely observed among closely related *C. albicans* isolates undergoing microadaptation (Shepherd et al. 1985; Tavanti et al. 2004; Odds et al. 2006b). Forche et al. (2011) measured the rates of LOH and the types of LOH events that appeared in the presence and absence of physiologically relevant stresses and found that stress causes a significant increase in the rates of LOH. This increase in the rates of LOH is proportional to the degree of stress and is expected to facilitate the adaptation of *C. albicans* to changing environments within the host.

5.2 *Adherence*

The first step in infection is interaction with the host cells by adhesion. *Candida albicans* expresses various adhesins, which bind to extracellular matrix proteins of mucosal or endothelial cells. The well-known adhesins include members of the agglutin-like (Als) family of adhesins, hyphal wall protein 1 (Hwp1), which forms covalent bonds with the host cell through the action of host cell transglutaminases (Staab et al. 1999) and elF4E-associated protein 1 (Eap1), which is a hypha-specific protein that confers adhesive properties to *Saccharomyces cerevisiae* cells when heterologously expressed (Li and Palecek 2008). Most *Candida albicans* adhesins are glycoproteins (Mavor et al. 2005). Adhesins promote the binding of the organism to host cells via hydrophobic interactions and the variability of adhesins gives diversity to host cells invasion. The adhesin genes are also differentially expressed according to the environmental conditions.

5.3 Filament Formation

Morphogenesis in *C. albicans* is defined as a switch from unicellular yeast form to filamentous form (pseudohyphae or hyphae) (Khan et al. 2010). As such, many human fungal pathogens are morphogenetic. Accordingly, Calderone and Fonzi (2001) found filamentation to increase the ability of *C. albicans* to cause infection and associated the yeast form with asymptomatic carriage and the hyphae form with active infection. Similarly, Mavor et al. (2005) suggested that yeast cells are better suited for dissemination while hyphae are important for tissue and organ invasion. Furthermore, strains of *C. albicans* that are unable to switch between yeast and hyphal growth forms were avirulent (Sohn et al. 2006). All these suggested that the yeast and hyphal growth forms play different roles in causing infections (Mavor et al. 2005) and therefore, each morphological form has a varying virulence factor which differs with particular environments.

5.4 Proteases

Secreted aspartly proteinases (Saps) from *Candida* have been reported to hydrolyze many proteins. To date, ten proteins have been recognized as Sap family (Saps 1–10) and found to be responsible for tissue invasion (Khan et al. 2010). Phospholipases are enzymes that hydrolyze ester linkages of glycophospholipids and in *C. albicans*, four types of phospholipases have been found. Phospholipase activity was observed in 99.4% of the strains of *C. albicans* tested (Pinto et al. 2008) and that results were similar to those obtained by Kantarcioğlu and Yücel (2002) who described 94% of the strains of *C. albicans* as phospholipase producers. Accordingly, phospholipases and Saps secretion in *Candida albicans* has been considered as relevant virulence factors (Pinto et al. 2008).

5.5 Phenotypic Switching

Phenotypic switching also plays a role in altering the yeast's adherence properties, antigen expression and tissue affinity (Mavor et al. 2005).This phenotypic switching provides cells with the flexibility for adaptation of the organism to the hostile conditions imposed by the host and aids survival in different microenvironments. This switching is reversible, occurs spontaneously in stress, and results in changes in cell surface behaviour, colony appearance, and metabolic, biochemical and molecular attributes to become more virulent and effective during infection (Odds et al. 2006b). Strains isolated from vaginitis or systemically infected patients showed higher frequencies of switching, indicating a strong role for the switching phenomenon in establishing diseases (Kvaal et al. 1999).

6 Adaptation of *Candida albicans* to the Host

For being a successful human commensal and pathogen, *C. albicans* has developed host adaptation mechanisms on various levels. The regulated expression of virulence and other genes in response to environmental signals allows effective adaptation to new host niches during the course of an infection. When fungi invade a mammalian host their lifestyle converts from saprophytic to parasitic. As saprophytes, fungi survive in an environment with a moderate ambient temperature and pH, defined sources for crucial nutrients such as carbon and metal ions, and atmospheric concentrations of gases like carbon dioxide and oxygen. By invading a human host, these environmental factors undergo a sudden and drastic change where ambient temperature is suddenly replaced with the restrictively high temperature of the human body. Ambient pH is replaced with acidic mucosal surfaces or neutral blood and tissues. Known sources of carbon and metal ions are missing in an environment where essential nutrients are sequestered from microbes to support host survival. Carbon dioxide and oxygen concentrations are reversed in host tissues, leaving *C. albicans* to adapt to hypoxia and high levels of carbon dioxide. Herein, we concentrate on recently studied adaptation mechanisms to the abiotic stresses that *C. albicans* encounter during colonization and infection of their human hosts and review the functions of these mechanisms in *C. albicans* survival and virulence.

6.1 Thermal Adaptation

Fungal survival at the high temperature of a human host is essential for virulence. Fungi often develop morphogenetic virulence mechanisms, e.g., formation of yeasts, hyphae, and spherules that facilitate their multiplication within the host at higher temperature. Yeast cells of many *Candida* species produce filamentous pseudohyphae and hyphae in tissues (Khan et al. 2010). *C. albicans* alters from commensal yeast to invasive hyphae at this elevated temperature.

The exact mechanisms by which thermal adaptation is regulated in eukaryotic cells have been widely studied, but are still not yet completely understood. However, it was found that when yeast cells are exposed to an acute thermal stress, proteins unfold, the heat shock transcription factor becomes activated by phosphorylation and this promotes the expression of heat shock genes (Leach et al. 2012). Recently, researchers identified Ryp1, a homolog of the *C. albicans* transcriptional regulator Wor1, as essential regulator for thermal dimorphism in response to elevated temperature. Ryp1 binds its own promoter and might act as an autoregulatory transcriptional regulator in *C. albicans*. Ryp1 mutants grow as hyphae and are unable to induce expression of most yeast phase-specific genes at 37°C (Cooney and Klein 2008).

The heat shock response in *C. albicans* has been of interest for few reasons. First, temperature up-shifts induce morphological changes from the yeast to hyphae and this cellular morphogenesis is a major virulence factor in *C. albicans*. Second, mutations that block heat shock transcription factor (Hsf1) activation in *C. albicans* prevent thermal adaptation and significantly decrease the virulence of *C. albicans*. Third, antifungal drug resistance is abrogated by Hsp90 inhibitors and by elevated temperatures equivalent to those in febrile patients. Fourth, *C. albicans* heat shock proteins are immunogenic, so, directly affecting host-pathogen interactions during infection. Last, autoantibodies against Hsp90 are immunoprotective against *C. albicans* infections (Leach et al. 2012).

To summarize, the thermal adaptation of fungal pathogens is of high importance because it is crucial for virulence, and because heat shock proteins represent targets for novel therapeutic strategies.

6.2 pH Adaptation

The capability of fungal pathogens to cause disease depends on their ability to survive within the human host environment. Generally, ambient pH is replaced with acidic conditions of mucosal surfaces or neutral to slightly alkaline pH of blood and tissues, and the ability of fungi to grow at this pH is crucial for pathogenesis. The Rim101 signal transduction pathway is the primary pH sensing pathway described in the pathogenic fungi, and in *C. albicans*, it is essential for a variety of diseases (Davis 2009). Rim101 is essential for *C. albicans* virulence in models of mucosal invasion and systemic candidiasis (Davis et al. 2000). Rim101 is activated downstream of a signaling cascade involving a plasma membrane complex, an endosomal membrane complex, and the proteasomes

The proteins of the plasma membrane signaling complex undergo pH-dependent activation. PalI/Rim9 is responsible for localization of the pH-sensor PalH/Rim21 to the plasma membrane (Calcagno-Pizarelli et al. 2007). This facilitates PalH-dependent phosphorylation and ubiquitylation of PalF/Rim8. PalF is an arrestin-like protein thought to trigger endocytosis of the plasma membrane complex upon PalH-dependent modification, mediating transduction of the pH signal from the plasma membrane to the endosomal membrane (Herranz et al. 2005). PalH is also required for localization of the endosomal complex protein PalC to the endosomal membrane (Galindo et al. 2007). This occurs through interaction of PalC with the endosomal sorting complex protein Vps32, which is essential for both the pH response and the virulence of

C. albicans (Cornet et al. 2005). Similarly, expression of ferric reductase genes occurs downstream of Rim101 activation in *C. albicans* (Baek et al. 2008). This connects the pH response to known virulence factors and is probably one reason why PacC/Rim101 mutants show decreased virulence. Mucosal invasion by *C. albicans* requires degradation of epithelial cell junctions by the protease Sap5p, which is upregulated downstream of Rim101. Expression of Sap5p in a Rim101 mutant strain restored the ability of *C. albicans* to invade an epithelial barrier in an *in vitro* model (Villar et al. 2007). Changes in cell wall composition also occur downstream of Rim101 activation in *C. albicans*, with over-expression of several cell wall-modulating proteins partially restoring virulence to a Rim101 mutant strain (Nobile et al. 2008). These results indicate the importance of pH sensing in fungal adaptation to mammalian hosts and highlight components of the pH signaling cascade as potential drug targets.

On the other hand, calcineurin, Crz1, and Crz2 are required for the growth of *C. albicans* at acidic pH. In fact, Crz1 and Crz2 act together to enhance the growth at acidic pH (Davis 2009).

6.3 Gas Tension

6.3.1 Carbon Dioxide

During human infection, fungi are exposed to carbon dioxide (CO_2) concentrations from low atmospheric CO_2 on epithelial surfaces to higher physiological CO_2 within host tissues. Work in *C. albicans* identified both carbonic anhydrase (CA) and adenylyl cyclase (AC) as CO_2 sensors. Conversion of atmospheric CO_2 to bicarbonate by CA is crucial for fatty acid biosynthesis and growth. Physiological CO_2 levels spontaneously generate enough bicarbonate for both growth and stimulation of AC. This activates the cAMP pathway and begins filamentation and expression of virulence traits (Bahn and Muhlschlegel 2006). Simultaneously, CA is required for virulence in models with ambient CO_2 levels, such as *C. albicans* epithelial invasion (Klengel et al. 2005). In contrast, CA is dispensible and AC is important in models with increased CO_2, like systemic infections by *C. albicans* (Bahn et al. 2005; Klengel et al. 2005; Mogensen et al. 2006).

6.3.2 Oxygen

In order to maintain metabolic and biosynthetic functions in the host, fungal pathogens must also be able to adapt to hypoxia within host tissues. Oxygen levels in mammalian tissues are below atmospheric levels. Furthermore, inflammation, thrombosis, and necrosis accompanied with infection are thought to rise degrees of hypoxia. In *C. albicans*, the response to hypoxia depends on the coordination of specific transcriptional regulators. In hypoxic conditions, the transcription factor Ace2 represses oxidative metabolic processes and promotes filamentation (Mulhern et al. 2006). The transcriptional regulator Egf1p, however, antagonizes Ace2 by repressing filamentation during hypoxia (Setiadi et al. 2006). The hypoxic environment of the vaginal mucosa has been shown to induce iron uptake protein expression, probably linking the hypoxic response to both iron acquisition and virulence in *C. albicans* (Sosinska et al. 2008).

6.4 Nutrients

6.4.1 Carbon Metabolism

Like most fungi, *C. albicans* uses sugars, especially glucose, as the preferred carbon source. While, a wide variety of nonfermentable carbon sources may also satisfy cellular requirements, including but not restricted to ethanol, acetate, glycerol, amino acids, and fatty acids. Collectively, these compounds are sometimes known as nonpreferred or alternative carbon sources since fungi only use them in the absence of sugars. These alternative sources are metabolized by three main pathways: β-oxidation of fatty acids, the glyoxylate cycle, and gluconeogenesis (Ramirez and Lorenz 2007). The main goals of these three interconnected pathways are to provide energy, replenish tricarboxylic acid (TCA) cycle intermediates and acetyl-coenzyme A (CoA), and ultimately convert lipids to acetate to glucose (Lorenz and Fink 2001; Lorenz et al. 2004).

Due to the fact that alternative carbon metabolism in *C. albicans* are important during systemic infections, deletions of genes encoding key enzymes in each pathway, the -oxidation multifunctional protein (FOX2), isocitrate lyase (ICL1), and fructose-1, 6-bisphosphatase (FBP1) in the pathways of β-oxidation of fatty acids (FOX2), the glyoxylate cycle (ICL1), and gluconeogenesis (FBP1) allow virulence defects from moderate to severe (Ramirez and Lorenz 2007). Therefore, alternative carbon metabolism in *C. albicans* plays an important role in survival within the host. On the other hand, some aspects of alternative carbon metabolism are unique to microorganisms, the identification of relevant carbon sources *in vivo* may highlight enzymes or pathways as attractive candidates for antifungal drug discovery.

6.4.2 Iron Acquisition

Generally, Iron is required for the survival of most organisms, primarily due to its role as a cofactor in essential metabolic functions (Lan et al. 2004). However, within the mammalian host environment, iron is sequestered away from microbes by iron carrier proteins, being stored in intracellular ferritin complexes; the trace amounts of extracellular iron bound by transferrin in the tissues and lactoferrin on mucosal surfaces and body secretions (Davis 2009), creating an iron-limited environment in which fungal pathogens must encode mechanisms for iron acquisition in order to survive (Haas et al. 2008).

Many fungi possess high affinity iron chelators, called siderophores, which efficiently bind host iron in the extracellular space and store it within the fungal cytoplasm. Fungal pathogens must also possess mechanisms for controlling and coordinating the utilization of acquired iron. For instance, *C. albicans* has multiple mechanisms for utilizing iron sources from the environment, including a reductive pathway and transport of heterologous siderophores. Moreover, Als3, an adhesin, binds to ferritin, enabling its use as a source of iron (Ding et al. 2011). The recent identification of the transcriptional regulator Hap43 in *C. albicans* has provided view for a mechanism by which *C. albicans* can adapt to iron limitation by reducing iron utilization. Hap43 was found to act as a transcriptional activator for the ferric reductases, which are crucial for the removal and utilization of iron from chelators including both siderophores and host carrier proteins (Baek et al. 2008). A novel mechanism by which *C. albicans* scavenges iron from host hemoglobin was also recently described. Receptor mediated endocytosis of hemoglobin facilitates extraction of iron, probably by a heme oxygenase in the vacuole (Weissman et al. 2008). Some Bcr1 targets in *C. albicans* play a role in acquiring iron from host proteins. These include two CFEM proteins, Rbt5 and Pga10, which act as receptors for hemoglobin, allowing endocytosis of the host iron complex (Ding et al. 2011). This hemoglobin utilization system considered an additional iron acquisition system that will probably be linked to survival of *C. albicans* within the host.

In addition to the specific sensory and regulatory adaptation mechanisms briefed above, *C. albicans* pathogens must also adapt to changes in nitrogen, calcium, magnesium, and copper sources, pressure, and fluid flow rates.

7 Prevention of Candidiasis

In general, like other fungal infections, most *Candida* infections can be prevented by keeping the skin clean, dry, and free from abrasions or cuts. Moreover, by using antibiotics according to doctor's directions since our bodies contain Acidophilus bacteria as a normal flora in the gut and this friendly bacteria helps keep our body in balance and able to fight pathogenic bacteria and fungus, like *Candida*. Uncontrolled intake of antibiotics may kill *Acidophilus* bacteria and enhance *Candida* growth. On the other hand, resistance of *Candida* to the used antifungal drugs can be decreased by following the doctor's directions and taking the described antibiotic completely. Furthermore, following a healthy lifestyle, including proper nutrition ensures good immunity. People with diabetes should try to keep their blood sugar under tight control since sugars are food for *Candida* and promote its growth.

8 Treatment of Candidiasis

Now a day, several antifungal agents are available, these include: Polyene derivatives such as Amphotericin B, Lipid based amphotericin, Nystatin, Azoles, and Griseofulvin. Amphotericin B is a polyene antimycotic, has been the drug of choice for most systemic fungal infections. It has a greater affinity for ergosterol in the cell membranes of fungi than for the cholesterol in the host's cells; once bound to ergosterol it causes disruption of the cell membrane and death of the fungal cell. Amphotericin B is usually administered intravenously, often for 2–3 months. The drug has side effects of being toxic; thrombo-phlebitis, nephrotoxicity, fever, chills and anemia frequently occur during administration. Although newer drugs have been shown to be as efficacious and less toxic, amphotericin B is still the gold standard for comparison as well as the therapy of last resort for severe infections. However, Lipid based amphotericin is effective, less toxic, and more expensive (Khan and Jain 2000; Di salvo 2007). Next, nystatin is considered the drug of choice for vaginitis and cutaneous infections (Hector 1993). On the other hand, the azoles (imidazoles and triazoles), including ketoconazole, fluconazole, itraconozole, voriconazole and posaconazole are being used for muco-cutaneous candidiasis, dermatophytosis, and for some systemic fungal infections. The general mechanism of action of the azoles is the inhibition of ergosterol synthesis which affects cell wall synthesis. Oral administration and reduced toxicity are distinct advantages. Griseofulvin is a very slow-acting drug which is used for severe skin and nail infections. Its effect depends on its accumulation in the *stratum corneum* where it is incorporated into the tissue and forms a barrier, which stops further fungal penetration and growth. It is administered orally (Khan and Jain 2000; Di salvo 2007). New agents with different mechanisms of action are under development (Hector 1993). Echinocandins (caspofungin), a new antifungal agent recently approved by the FDA (Di salvo 2007). Importantly, it is active against *Candida* isolates that are refractory to azole treatment. It kills fungi by inhibiting the synthesis of ß-1, 3-glucan, a major component of the fungal cell wall. Thus, it has potent *in vitro* activity against *C. albicans* biofilms (Bachmann 2002).

Recently, thousands of products have been screened *in vitro* for antimicrobial activity and promising molecules have been evaluated in various animal models for new drug development. Much research focuses on plant sources.

9 Conclusions and Future Recommendations

During growth within the intestinal tract, *C. albicans* senses pH, oxygen, carbon sources, and the presence of surfaces allowing it to optimize gene expression for a particular environment. With these mechanisms for sensing, *C. albicans* is able to efficiently survive in humans starting from infancy, establishing itself in its most important natural niche.

References

Bachmann SP, Walle KV, Ramage G, Patterson TF, Wickes BL, Graybill JR, López-Ribot JL (2002) In vitro activity of caspofungin against Candida albicans biofilms. Antimicrobial Agents Chemother 46(11):3591–3596

Baek YU, Li M, Davis D (2008) *Candida albicans* ferric reductases are differentially regulated in response to distinct forms of iron limitation by the Rim101 and CBF transcription factors. Eukaryot Cell 7:1168–1179

Bahn YS, Muhlschlegel F (2006) CO_2 sensing in fungi and beyond. Curr Opin Microbiol 9:572–578

Bahn YS, Cox GM, Perfect JR, Heitman J (2005) Carbonic anhydrase and CO_2 sensing during *Cryptococcus neoformans* growth, differentiation, and virulence. Curr Biol 15:2013–2020

Baron EJ, Chang RS, Howard DH, Miller JN, Turner JA (1994) Medical microbiology: a short course. Wiley, New York, pp 537–543

Bartie KL, Williams DW, Wilson MJ, Potts AJC, Lewis MAO (2001) PCR fingerprinting of *Candida albicans* associated with chronic hyperplastic candidosis and other oral conditions. J Clin Microbiol 39(11):4066–4075

Bougnoux M-E, Diogo D, Francois N, Sendid B, Veirmere S, Colombel JF, Van Kruiningen H, d'Enfert C, Poulain D (2006) Multilocus sequence typing reveals intrafamilial transmission and microevolutions of *Candida albicans* isolates from the human digestive tract. J Clin Microbiol 44(5):1810–1820

Calcagno-Pizarelli AM, Negrete-Urtasun S, Denison SH, Rudnicka JD, Bussink HJ, Munera-Huertas T, Stanton L, Hervas-Aguilar A, Espeso EA, Tilburn J, Arst HN, Penalv MA (2007) Establishment of the ambient pH signaling complex in *Aspergillus nidulans*: PalI assists plasma membrane localization of PalH. Eukaryot Cell 6:2365–2375

Calderone R (2002) *Candida* and candidiasis. American Society for Microbiology, Washington, DC

Calderone RA, Fonzi WA (2001) Virulence factors of *Candida albicans*. Trends Microbiol 9(7):327–335

Cooney NM, Klein BS (2008) Fungal adaptation to the mammalian host: it is a new world, after all. Curr Opin Microbiol 11:511–516

Cornet M, Bidard F, Schwarz P, Da Costa G, Blanchin-Roland S, Dromer F, Gaillardin C (2005) Deletions of endocytic components VPS28 and VPS32 affect growth at alkaline pH and virulence through both RIM101-dependent and RIM101-independent pathways in *Candida albicans*. Infect Immun 73:7977–7987

Davis DA (2009) How human pathogenic fungi sense and adapt to pH: the link to virulence. [Review]. Curr Opin Microbiol 12:365–370

Davis D, Edwards JE Jr, Mitchell AP, Ibrahim A (2000) *Candida albicans* RIM101 pH response pathway is required for hostpathogen interactions. Infect Immun 68:5953–5959

Di Salvo A (2007) Medical microbiology (MBIM 650/720) Course. Medical mycology. L:82–86

Ding C, Vidanes GM, Maguire SL, Guida A, Synnott JM, Andes DR et al (2011) Conserved and divergent roles of Bcr1 and CFEM proteins in *Candida parapsilosis* and *Candida albicans*. PLoS One 6(12):e28151

Fidel PL Jr, Vanquez JA, Sobel JD (1999) *Candida glabrata*: a review of epidemiology, pathogenesis and clinical disease with comparison to *Candida albicans*. Clin Microbiol Rev 12:80–96

Forche A, Abbey D, Pisithkul T, Weinzierl MA, Ringstrom T, Bruck D, Petersen K, Berman J (2011) Stress alters rates and types of loss of heterozygosity in *Candida albicans*. mbio 2(4):1–8, asm.org

Galindo A, Hervas-Aguilar A, Rodriguez-Galan O, Vincent O, Arst HN Jr, Tilburn J, Penalva MA (2007) PalC, one of two Bro1 domain proteins in the fungal pH signalling pathway, localizes to cortical structures and binds Vps32. Traffic 8:1346–1364

Haas H, Eisendle M, Turgeon B (2008) Siderophores in fungal physiology and virulence. Annu Rev Phytopathol 46:149–187

Hector RF (1993) Compounds active against cell walls of medically important fungi. Clin Microbiol Rev 6:1–21

Herranz S, Rodriguez JM, Bussink HJ, Sanchez-Ferrero JC, Arst HN Jr, Penalva MA, Vincent O (2005) Arrestin-related proteins mediate pH signaling in fungi. Proc Natl Acad Sci USA 102:12141–12146

Jacobsen MD, Duncan AD, Bain J, Johnson EM, Naglik JR, Shaw DJ, Gow NAR, Odss FC (2008) Mixed *Candida albicans* strain populations in colonized and infected mucosal tissues. FEMS Yeast Res 8(8):1334–1338

Kantarcioğlu AS, Yücel A (2002) Phospholipase and protease activities in clinical *Candida* isolates with reference to the sources of strains. Mycoses 45:160–165

Khan ZK, Gyanchandani A (1998) Candidiasis: a review. PINSA 64:1–34

Khan ZK, Jain P (2000) Antifungal agents and immunomodulators in systemic mycoses. Indian J Chest Dis Allied Sci 42:345–3551

Khan MSA, Ahmad I, Aqil F, Owais M, Shahid M, Musarrat J (2010) Virulence and pathogenicity of fungal pathogens with special reference to *Candida albicans*. In: Ahmad I, Owais M, Shahid M, Aqil F (eds) Combating fungal infections, problems and remedy. Springer, Berlin/Heidelberg, pp 21–45

Klengel T, Liang WJ, Chaloupka J, Ruoff C, Schroppel K, Naglik JR, Eckert SE, Moqensen EG, Haynes K, Tuite MF, Levin LR, Buck J, Muhlschlegel FA (2005) Fungal adenylyl cyclase integrates CO_2 sensing with cAMP signaling and virulence. Curr Biol 15:2021–2026

Kvaal C, Lachke SA, Srikantha T, Daniels K, McCoy J, Soll DR (1999) Misexpression of the opaque phase specific gene PEP1 (SAP1) in the white phase of *Candida albicans* confers increased virulence in a mouse model of cutaneous infection. Infect Immun 67:6652–6662

Lan C-Y, Rodarte G, Murillo LA, Jones T, Davis RW, Dungan J, Newport G, Agabian N (2004) Regulatory networks affected by iron availability in *Candida albicans*. Mol Microbiol 5:1451–1469

Leach MD, Tyc KM, Brown AJP, Klipp E (2012) Modelling the regulation of thermal adaptation in *Candida albicans*, a major fungal pathogen of humans. PLoS One 7(3):e32467

Li F, Palecek SP (2008) Distinct domains of the *Candida albicans* adhesin Eap 1 p mediate cell-cell and cell-substrate interactions. Microbiology 154:1193–1203

Lorenz MC, Fink G (2001) The glyoxylate cycle is required for fungal virulence. Nature 412:83–86

Lorenz M, Bender JA, Fink GR (2004) Transcriptional response of *Candida albicans* upon internalization by macrophages. Eukaryot Cell 3:1076–1087

Mavor AL, Thewes S, Hube B (2005) Systemic fungal infections caused by *Candida* species: epidemiology, infection process and virulence attributes. Curr Drug Targets 6:863–874

Mogensen EG, Janbon G, Chaloupka J, Steegborn C, Fu MS, Moyrand F, Klengel T, Pearson DS, Geeves MA, Buck J, Levin LR, Muhlschlegel FA (2006) *Cryptococcus neoformans* senses CO_2 through the carbonic anhydrase Can2 and the adenylyl cyclase Cac1. Eukaryot Cell 5:103–111

Mulhern SM, Logue ME, Butler G (2006) Candida albicans transcription factor Ace2 regulates metabolism and is required for filamentation in hypoxic conditions. Eukaryot Cell 5:2001–2013

Nobile CJ, Solis N, Myers CL, Fay AJ, Deneault JS, Nantel A, Mitchell AP, Filler SG (2008) *Candida albicans* transcription factor Rim101 mediates pathogenic interactions through cell wall functions. Cell Microbiol 10:2180–2196

Odds FC (1988) Candida and candidosis, 2nd edn. Baillière Tivdall, London, pp 42–59

Odds FC (2010) Molecular phylogenetics and epidemiology of *Candida albicans*. Future Microbiol 5(1):67–79

Odds FC, Davidson AD, Jacobsen MD, Tavanti A, Whyte JA, Kibbler CC, Ellis DH, Maiden MCJ, Shaw DJ, Gow NAR (2006a) *Candida albicans* strain maintenance, replacement, and microvariation demonstrated by multilocus sequence typing. J Clin Microbiol 44(10):3647–3658

Odds FC, Gow NAR, Brown AJP (2006b) Toward a molecular understanding of *Candida albicans* virulence. In: Heitman J, Filler SG, Edwards JE Jr, Mitchell AP (eds) Molecular principles of fungal pathogenesis. ASM, Washington, DC, pp 305–319

Pinto E, Ribeiro IC, Ferreira NJ, Fortes CE, Fonseca PA, Figueiral MH (2008) Correlation between enzyme production, germ tube formation and susceptibility to fluconazole in *Candida* species isolated from patients with denture-related stomatitis and control individuals. J Oral Pathol Med 37:587–592

Ramirez MA, Lorenz M (2007) Mutations in alternative carbon utilization pathways in *Candida albicans* attenuate virulence and confer pleiotropic phenotypes. Eukaryot Cell 6:280–290

Rosenbach A, Dignard D, Pierce JV, Whiteway M, Kumamoto CA (2010) Adaptations of *Candida albicans* for growth in the mammalian intestinal tract. Eukaryot Cell 9(7):1075–1086

Rupp S (2007) Review: interactions of the fungal pathogen *Candida albicans* with the host. Future Microbiol 2(2):141–151

Schulze J, Sonnenborn U (2009) Yeast in the gut: from commensals to infectious agents. Dtsch Arztebl Int 106(51–52):837–842

Setiadi ER, Doedt T, Cottier F, Noffz C, Ernst J (2006) Transcriptional response of *Candida albicans* to hypoxia: linkage of oxygen sensing and Efg1p-regulatory networks. J Mol Biol 361:399–411

Shepherd MG, Poulter RTM, Sullivan PA (1985) *Candida albicans*: biology, genetics and pathogenicity. Annu Rev Microbiol 39:579–614

Sohn K, Schwenk J, Urban C, Lechner J, Schweikert M, Rupp S (2006) Getting in touch with *Candida albicans*: the cell wall of a fungal pathogen. Curr Drug Targets 7(4):505–512

Sosinska GJ, de Groot PW, Teixeira de Mattos MJ, Dekker HL, de Koster CG, Hellingwerf KJ, Klis FM (2008) Hypoxic conditions and iron restriction affect the cell-wall proteome of *Candida albicans* grown under vagina-simulative conditions. Microbiology 154:510–520

Staab JF, Bradway SD, Fidel PL, Sundstrom P (1999) Adhesive and mammalian transglutaminase substrate properties of *Candida albicans* Hwp1. Science 283:1535–1538

Sudbery PE (2011) Growth of *Candida albicans* hyphae. Nat Rev Microbiol 9:737–748. www.nature.com/reviews/micro

Sudbery PE, Gow NAR, Berman J (2004) The distinct morphogenic states of *Candida albicans*. Trends Microbiol 12:317–324

Tavanti A, Gow NAR, Maiden MCJ, Odds FC, Shaw DJ (2004) Genetic evidence for recombination in *Candida albicans* based on haplotype analysis. Fungal Genet Biol 41:553–562

Villar CC, Kashleva H, Nobile CJ, Mitchell AP, Dongari-Bagtzoglou A (2007) Mucosal tissue invasion by *Candida albicans* is associated with E-cadherin degradation, mediated by transcription factor Rim101p and protease Sap5p. Infect Immun 75:2126–2135

Vylkova S, Carman AJ, Danhof HA, Collette JR, Zhou H, Lorenz MC (2011) The fungal pathogen *Candida albicans* autoinduces hyphal morphogenesis by raising extracellular pH. mBio.asm.org 2(3):1–12. http://mbio.asm.org/content/2/3/e00055-11.full.html

Wächtler B, Wilson D, Haedicke K, Dalle F, Hube B (2011) From attachment to damage: defined genes of *Candida albicans* mediate adhesion, invasion and damage during interaction with oral epithelial cells. PLoS One 6(2):1–14. www.plosone.org

Weissman Z, Shemer R, Conibear E, Kornitzer D (2008) An endocytic mechanism for haemoglobin-iron acquisition in *Candida albicans*. Mol Microbiol 69:201–217

Wenzel RP (1995) Nosocomial candidiasis: risk factors and attributable mortality. Clin Infect Dis 20:1531–1534

Chapter 18
The Role of Marine Anaerobic Bacteria and Archaea in Bioenergy Production

A.J. Cavaleiro, A.A. Abreu, D.Z. Sousa, M.A. Pereira, and M.M. Alves

Abstract The development of products from marine bioresources is gaining importance in the biotechnology sector. The global market for Marine Biotechnology products and processes was, in 2010, estimated at € 2.8 billion with a cumulative annual growth rate of 5–10% (Børresen et al., Marine biotechnology: a new vision and strategy for Europe. Marine Board Position Paper 15. Beernem: Marine Board-ESF, 2010). Marine Biotechnology has the potential to make significant contributions towards the sustainable supply of food and energy, the solution of climate change and environmental degradation issues, and the human health. Besides the creation of jobs and wealth, it will contribute to the development of a greener economy. Thus, huge expectations anticipate the global development of marine biotechnology. The marine environment represents more than 70% of the Earth's surface and includes the largest ranges of temperature, light and pressure encountered by life. These diverse marine environments still remain largely unexplored, in comparison with terrestrial habitats. Notwithstanding, efforts are being done by the scientific community to widespread the knowledge on ocean's microbial life. For example, the J. Craig Venter Institute, in collaboration with the University of California, San Diego (UCSD), and Scripps Institution of Oceanography have built a state-of-the-art computational resource along with software tools to catalogue and interpret microbial life in the world's oceans. The potential application of the marine biotechnology in the bioenergy sector is wide and, certainly, will evolve far beyond the current interest in marine algae. This chapter revises the current knowledge on marine anaerobic bacteria and archaea with a role in bio-hydrogen production, syngas fermentation and bio-electrochemical processes, three examples of bioenergy production routes.

A.J. Cavaleiro and A.A. Abreu contributed equally to this chapter.

A.J. Cavaleiro • A.A. Abreu • D.Z. Sousa • M.A. Pereira • M.M. Alves (✉)
IBB – Institute for Biotechnology and Bioengineering,
Centre of Biological Engineering, University of Minho,
Campus de Gualtar, 4710-057 Braga, Portugal
e-mail: acavaleiro@deb.uminho.pt; angela_abreu@deb.uminho.pt;
dianasousa@deb.uminho.pt; alcina@deb.uminho.pt; madalena.alves@deb.uminho.pt

A. Malik et al. (eds.), *Management of Microbial Resources in the Environment*,
DOI 10.1007/978-94-007-5931-2_18,

1 Introduction

Energy is essential for life and has been a critical and decisive factor for the development of human civilizations. The ever-increasing use of energy, and particularly the overexploitation of fossil fuels and the depletion of easily accessible oil resources, have prompted several environmental problems and pressed the search for alternative renewable energy sources.

Renewable and sustainable bioenergy can be produced from any organic material that stored sunlight in the form of chemical energy. This includes lignocellulosic biomass, crops, agricultural and animal wastes, organic fraction of municipal solid wastes and some industrial organic wastes. Heat and electricity can be generated from these materials, as well as several liquid and gaseous biofuels, namely bioethanol, biobutanol, biodiesel, biogas and biohydrogen.

Chemical, thermo-chemical and biological technologies are currently used for biofuel production. Among these, processes performed by microorganisms are the most cost effective (Barnard et al. 2010). Nevertheless, a more comprehensive understanding of the microbial processes involved and the identification of novel microorganisms capable of producing biofuels is needed for process development and optimization.

Marine habitats are particularly attractive for bioprospection, due to the vast microbial abundance and diversity. Microorganisms that grow in deep oceanic environments have unique characteristics, necessary for thriving under extreme conditions of light, temperature and pressure. Therefore, the oceans contain an immense pool of genetic information with high biotechnology potential (Børresen et al. 2010).

Studies in deep oceans are technically challenging and expensive, but recent advances in genome sequencing techniques, metagenomics, remote sensing of microorganisms and bioinformatics have contributed to the intensification of marine biotechnology. This has resulted in different applications with societal importance, such as new therapeutics, chemical products and enzymes (Glöckner et al. 2012). Several genes of marine organisms have been patented and are mainly applied in pharmacology and human health (55%), agriculture or aquaculture (26%) and food industry (17%) (Arrieta et al. 2010). Applications in ecotoxicology, bioremediation and biofuel production are now emerging.

Marine non-phototrophic anaerobic microorganisms present an extensive and almost unexplored potential for biofuels production. The microbial communities present in hydrothermal fields use the energy of organic compounds (*e.g.* decaying biomass) but also reduced inorganic compounds (*e.g.* carbon monoxide) for growth, producing hydrogen, methane and alcohols (Sokolova et al. 2001). In the last decades, several hydrogen-producing bacteria have been screened from different marine environments, although few could be cultivated in the laboratory (Jayasinghearachchi et al. 2010). The use of extremophiles as catalysts may improve hydrogen yields in dark fermentation processes or the energy yield in microbial electro-chemical

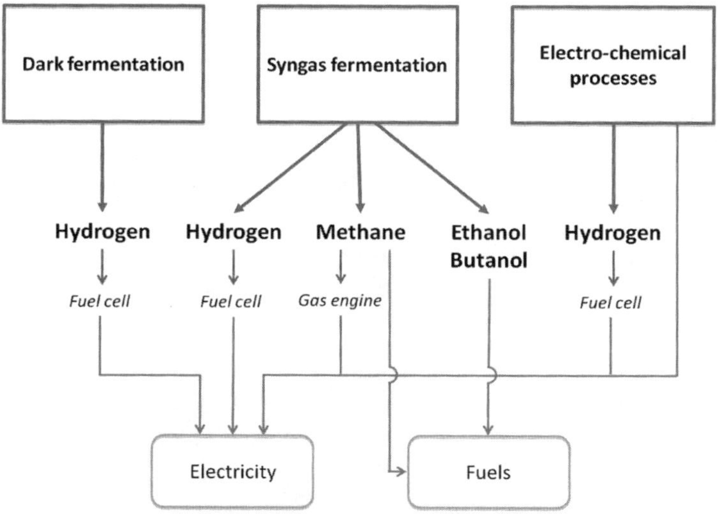

Fig. 18.1 Bioenergy production through dark fermentation, syngas fermentations and electrochemical processes catalyzed by anaerobic marine microorganisms

processes, since these microorganisms possess unique metabolic and physiological characteristics (Mathis et al. 2008).

In this chapter, the current knowledge on anaerobic microorganisms, or microbial cultures, retrieved from marine habitats with potential application in biotechnology systems for bioenergy production are reviewed, specially focusing on (1) dark fermentation for hydrogen production, (2) syngas (CO, CO_2, H_2) fermentation, and (3) electro-chemical processes (Fig. 18.1).

1.1 The Marine Environment

The marine environment covers more than 70% of the Earth's surface and comprises 97% of all the water on the planet, much of which (75% of the ocean's volume) at depths higher than 1,000 m. The ocean's average depth is 4,000 m, but it may reach 11,000 m deep. The pelagic zone (from the Greek "open sea") can be divided into several ecological sections based on depth (Fig. 18.2). The photic zone of the ocean, i.e. the volume actually penetrated by sunlight, is located in the upper layer, 200–300 m deep, and accounts only for approximately 2% of the total water volume. Therefore, the majority of the ocean (approx. 1.3×10^{18} m^3) is deprived of light (Orcutt et al. 2011).

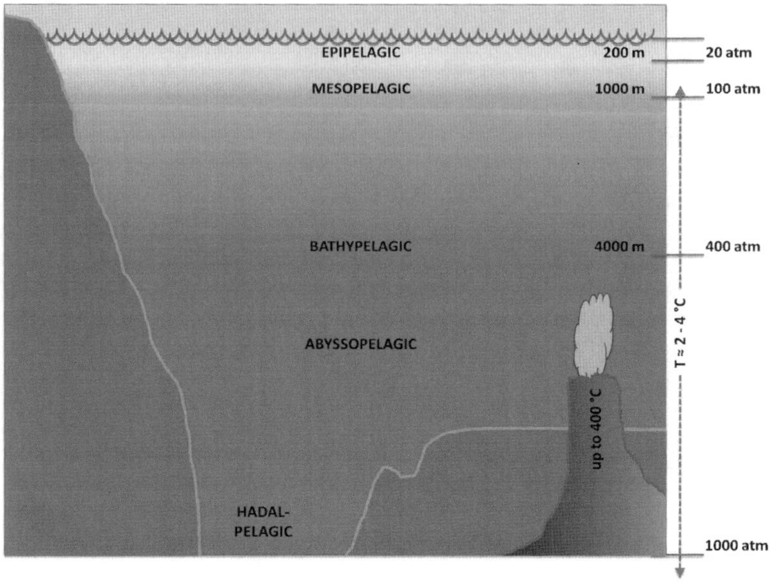

Fig. 18.2 The marine environment and its principal characteristics

Besides the ocean water column, the marine environment includes other microbial habitats, namely marine sediments, the oceanic crust and hydrothermal vents. These environments are characterized by extreme pressure and temperature conditions (Fig. 18.2).

Pressure increases approximately 0.1 atm m^{-1} of water column, and more than 62% of the ocean is exposed to pressure values higher than 100 atm (Orcutt et al. 2011). Barophilic or piezophilic microorganisms exhibit an optimal growth at pressure higher than 400 atm, while barotolerant microorganisms have their optimal growth at pressures below 400 atm. Inhibitory effects of increased pressure on biochemical processes have been reported, and deep-sea microorganisms present specific features that allow them to live and grow under these high-pressure conditions (Horikoshi 1998; Kato and Bartlett 1997).

Temperature in the water column decreases from the surface until 100 m deep, from which it remains more or less constant around 2–4°C (Fig. 18.2). In hydrothermal vents, however, thermophilic and hyperthermophilic conditions are present and temperatures up to 400°C may occur (Orcutt et al. 2011). In these geological places, chemically reduced compounds are released at high temperatures from the earth subsurface. Mineral deposition occurs in these zones and may cause the formation of chimney structures surrounding the advecting fluids. Higher biological activity is present in these areas, relatively to the majority of the deep sea, with chemosynthetic archaea forming the base of a diverse food chain. Hydrogenotrophic methanogenesis and sulfate reduction are the dominant anaerobic processes in high temperature vent fluids (Orcutt et al. 2011).

Table 18.1 Standard reduction potentials at 25°C and pH 7 for some possible electron acceptors in marine microbial metabolism

Electron acceptor (ox/red)	Half-reaction	$E^{0\prime}$ (V)
O_2/H_2O	$O_2 (g) + 4H^+ + 4e^- \rightarrow 2H_2O$	+0.815
NO_3^-/N_2	$2NO_3^- + 12H^+ + 10e^- \rightarrow N_2 (g) + 6H_2O$	+0.747
NO_2^-/N_2	$2NO_2^- + 8H^+ + 6e^- \rightarrow N_2 + 4H_2O$	+0.958
NO_3^-/NO_2^-	$NO_3^- + 2H^+ + 2e^- \rightarrow NO_2^- + H_2O$	+0.432
SO_4^{2-}/HS^-	$SO_4^{2-} + 9H^+ + 8e^- \rightarrow HS^- + 4H_2O$	−0.217
HCO_3^-/CH_4	$HCO_3^- + 9H^+ + 8e^- \rightarrow CH_4 (g) + 3H_2O$	−0.238

Reduction potentials calculated from Thauer et al. (1977) at standard conditions (1 M concentration for each solute, partial pressure of 1 atm for each gas, 25 °C and pH 7)

The movement of chemical compounds (*e.g.* oxygen, nutrients, waste products) through the water column is another important factor that influences marine microbial communities. Caused by differences in temperature and salinity, vertical gradients occur in the water masses, creating zones with diverse physical and chemical characteristics. The metabolic activity of marine microorganisms is also greatly dependent on the availability and speciation of electron donors and acceptors. Organic matter and reduced inorganic compounds, such as hydrogen, methane, reduced sulfur compounds, reduced iron and manganese, and ammonium, are the main sources of electrons for the microbial metabolic reactions. Molecular hydrogen oxidation is energetically favorable, and thus marine microorganisms intensely compete for this compound (Orcutt et al. 2011).

Inorganic compounds such as oxygen, nitrate, nitrite, manganese and iron oxides, oxidized sulfur compounds and carbon dioxide may serve as electron sinks in marine microbial metabolisms (Table 18.1). The availability of terminal electron acceptors influences significantly the dominant microbial metabolic pathways. Microorganisms preferable use the electron acceptors that provide higher thermodynamic energy yields, generally following a redox cascade.

Microorganisms living in deep-sea habitats have different requirements in terms of salt concentration for growth. Halotolerant microbes prefer low salt concentration, but are able to survive and grow in the presence of relatively high Na^+ concentrations, while halophilic microorganisms require salt concentrations from 0.5 to 2.5 M or even higher (saturated solutions) (Kivistö and Karp 2011).

1.2 Microbial Abundance and Diversity in the Marine Environment

The discovery of high microbial activity in deep ocean and sediments, previously considered devoid of life, significantly changed the understanding of marine ecosystems. In the last decades, genomic and metagenomic approaches and the use of high throughput sequencing techniques have revealed the remarkable diversity of marine microbes, and contributed to the identification of novel ecological processes and functions in the marine environments. Dedicated research programs were launched, specifically focusing on the assessment of genetic diversity and function in marine microbial communities. For example, the Global Ocean Sampling expedition (GOS) by Craig Venter started in 2004 with the aim of improving the number of whole genome sequences of ecologically relevant marine microorganisms (Sun et al. 2011).

Total number of bacteria and archaea in the ocean is estimated to be around 10^{29} cells. Prokaryotic biomass concentration in the pelagic zone is approximately 10^{3}–10^{5} cells mL^{-1} and tends to decrease with depth, contrasting with diversity that typically is higher at higher profundities. In deep sediments microbial abundance exceeds 10^{5} cells mL^{-1}. In these habitats, higher abundance and diversity occur at the surface and then decrease sharply, reflecting changes in the geochemical characteristics of the sediments (Nagata et al. 2010; Orcutt et al. 2011; Zhang et al. 2012).

A relatively high abundance of Archaea (10–50% of total prokaryote cell abundance), has been reported in the bathypelagic communities (Nagata et al. 2010; Orcutt et al. 2011). The structure of these communities appears to be more closely related with depth than with the local of origin, since microbial communities collected in different places at the same depth present higher similarity than the microbial assemblages inhabiting a specific ocean at different depths. This fact suggests the existence of vertical stratification patterns at global scale (Orcutt et al. 2011).

Similar marine habitats appear to be dominated by similar microbial groups (at the phylum level) (Inagaki et al. 2006; Orcutt et al. 2011). *Alpha-*, *Delta-* and *Gamma-Proteobacteria* prevail in typical deep-ocean bacterial communities, whereas the *Crenarchaeota* marine group I dominates the archaeal communities (Eloe et al. 2011; Fuhrman and Davis 1997; Orcutt et al. 2011). In deep sediments, predominant phylotypes of the Archaea domain include the *Crenarchaeota* clades Marine Benthic Group B and Miscellaneous *Crenarchaeota* Group, for which there are no cultivated members. These habitats also include as dominant bacteria, members of the candidate OP9/JS1 phylum or of the *Chloroflexi* phylum (Inagaki et al. 2006; Orcutt et al. 2011; Schippers et al. 2012; Zhang et al. 2012). *Gamma-* and *Epsilonproteobacteria* dominate the hydrothermal habitats, as well as members of the *Euryarchaeota* phylum, from genera *Archaeoglobus*, *Thermococcales* and *Methanococcales* (Orcutt et al. 2011; Sokolova et al. 2001).

Despite the recent advances achieved with metagenomic approaches, an enormous pool of previously unknown, uncultured microorganisms and genes has been revealed. As an example, Venter et al. (2004) identified over 1.2 million previously unknown genes through genome shotgun sequencing of samples from the Sargasso Sea. The function of many of these genes was not identified thus far (Glöckner et al. 2012; Pedrós-Alió 2006). In order to fully exploit the genomic potential of marine microbes, with possible applications in biotechnology, the development of novel research tools for the characterization of these previously unknown marine microorganisms is absolutely necessary (Glöckner et al. 2012).

In line with this need, the J. Craig Venter Institute (JCVI) launched the project CAMERA, the Community Cyber Infrastructure for Advanced Marine Microbial Ecology Research and Analysis Database, which is a continually evolving open access tool for general access to raw environmental sequence data, associated metadata, pre-computed search results, and high-performance computational resources (http://camera.calit2.net/). The aim of this project is to serve the needs of the microbial ecology research community by creating a rich, distinctive data repository and a bioinformatics tools resource that will address many of the unique challenges of metagenomics analysis. Initially, CAMERA is making available all the metagenomic data being collected by the J. Craig Venter Institute's Sorcerer II Global Ocean Sampling (GOS) expeditions, which have sampled microbial communities around the globe, plus 150 new full genome maps of ocean microbes. It also includes other data sets: a large-scale metagenomic survey of marine viral organisms collected from sites around the North American continent by Forest Rohwer and his research team at San Diego State University, and a vertical profile of marine microbial communities collected at the Hawaii Ocean Time-Series (HOTS) station ALOHA by Ed DeLong and his research team at Massachusetts Institute of Technology (MIT) (http://www.jcvi.org/cms/research/projects/camera/overview/).

2 Dark Fermentation Process

Hydrogen, a high-energy, non-polluting and environmental friendly compound can be biologically produced through biophotolysis, indirect biophotolysis, photofermentation and dark fermentation of organic matter. In dark fermentation, carbohydrate-rich substrates are converted by anaerobic microorganisms into organic acids and alcohols, releasing hydrogen and CO_2 in the process. A variety of microorganisms can be involved in this process, either as pure or in mixed cultures.

This process has several advantages relatively to other biological hydrogen production methods, namely the use of a wide range of organic substrates, including organic wastes, and higher hydrogen production rates (Hallenbeck et al. 2012). However, the economical feasibility of dark fermentation still needs to be improved, for example through technological developments and increased knowledge on microorganisms capable of efficient hydrogen production.

In dark fermentation, strictly anaerobic or facultative microorganisms break down complex organic matter under anaerobic conditions, producing organic acids and alcohols, and releasing H_2 and CO_2. Carbohydrates are hydrolyzed to monomeric simple sugars, that are further converted to acetyl-coenzyme A (acetyl-CoA) mainly through pentose phosphate and glycolysis pathways (Fig. 18.3).

The maximum theoretical yield of fermentative hydrogen production is 4 mol H_2 mol^{-1} hexose, estimated from the reaction of glucose conversion to acetate. The production of pyruvate from glucose yields 2 moles of reduced nicotinamide adenine dinucleotide (NADH), which can be regenerated with the formation of 2 mol of hydrogen (Fig. 18.3). Further breakdown of pyruvate generates acetyl-CoA and

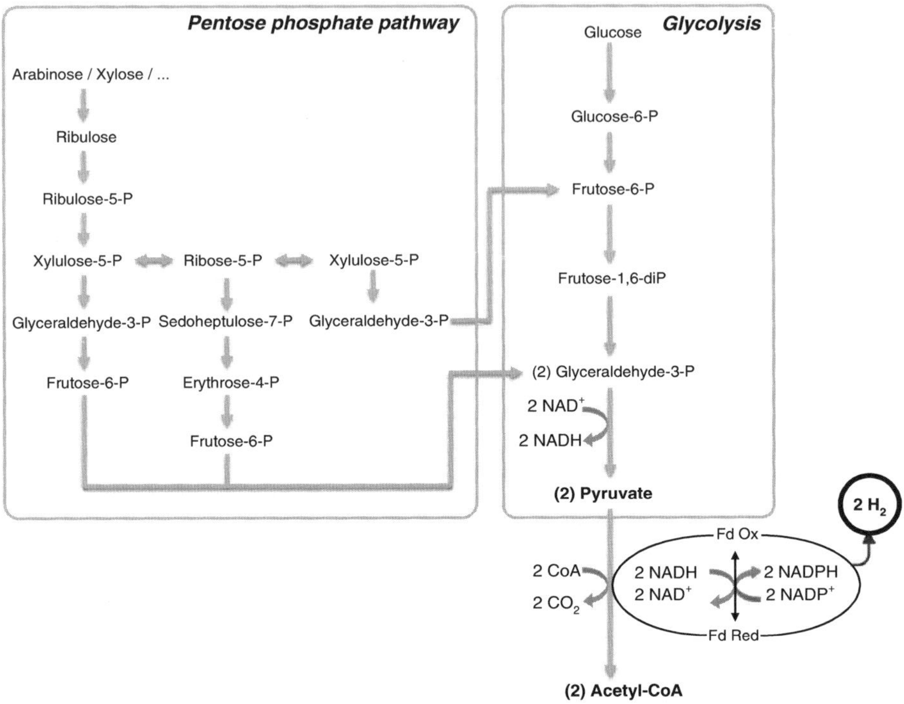

Fig. 18.3 Major catabolic pathways involved in the fermentation of hexoses and pentoses

reduced ferredoxin (Fd_{red}). Re-oxidation of ferredoxin yields hydrogen (1 mol H_2 mol^{-1} Fd), making a total hydrogen yield of 4 moles per mol of glucose consumed.

Alternative pathways for hydrogen production have also been reported in microorganisms isolated from marine habitats, such as *Vibrio aerogenes* and *Pantoea agglomerans* (Shieh et al. 2000; Zhu et al. 2008), although yielding lower amounts of hydrogen (around 2 mol H_2 mol^{-1} hexose).

Fermentative hydrogen production can be carried out by a wide range of marine microorganisms, with diverse requirements in terms of substrate preference, pH and temperature (Wang and Wan 2009). Those parameters do not only determine the growth of the microorganisms, but also have a crucial role on the metabolic pathway that will prevail, severely affecting the final hydrogen yield (Table 18.2). Another important parameter, with significant influence on hydrogen production, is hydrogen partial pressure (PH_2). High PH_2 tend to divert the metabolic pathway towards more reduced end products (e.g. lactate and ethanol), with consequently lower H_2 yields. However, at higher temperatures this effect is not so severe. For example, at 25°C, PH_2 needs to be lower than 0.022 kPa to allow for the conversion of glucose to acetate (reaction 2 in Table 18.2), while at 100°C this reaction becomes exergonic at PH_2 lower than 2.2 kPa (Verhaart et al. 2010).

Several archaea and bacteria capable of efficient hydrogen production were isolated from marine environments, mostly from hydrothermal vents and deep-sea

Table 18.2 Hydrogen-producing reactions in anaerobic processes, highlighting the theoretical hydrogen yields

	Substrate	Reaction	$\Delta G^{0'}$ (kJ reaction^{-1})	Reference
(1)	Glucose	1 glucose + 12 $H_2O \rightarrow$ 6 HCO_3^- + 6 H^+ + 12 H_2	+3	Thauer et al. (1977)
(2)	Glucose	1 glucose + 2 $H_2O \rightarrow$ 2 acetate$^-$ + 2 CO_2 + 2 H^+ + 4 H_2	−206	Thauer et al. (1977)
(3)	Glucose	1 glucose $\rightarrow$ 1 butyrate$^-$ + 2 CO_2 + 2 H^+ + 2 H_2	−254	Thauer et al. (1977)
(4)	Arabinose	1 arabinose + 1.67 $H_2O \rightarrow$ 1.67 acetate$^-$ + 1.67 CO_2 + 1.67 H^+ + 3.33 H_2	−192	Abreu et al. (2012)
(5)	Arabinose	1 arabinose $\rightarrow$ 0.83 butyrate$^-$ + 1.66 CO_2 + 0.83 H^+ + 1.66 H_2	−228	Abreu et al. (2012)

Gibbs free energies (at 25 °C) calculated at standard conditions (solute concentrations of 1 M and gas partial pressure of 10^5 Pa).

geothermal heated sediments (Table 18.3). These microorganisms grow at very high temperatures (extreme thermophilic and hyperthermophilic), with enhanced hydrolysis and thermodynamically more favorable metabolic reactions (Verhaart et al. 2010). This high potential for biotechnology applications has been explored at lab-scale, mainly in batch or fed-batch conditions, as shown in Table 18.3.

Clostridium amygdalinum grows at mesophilic temperatures between 20 and 60°C, with an optimum around 45°C. These spore-forming anaerobic aerotolerant bacteria are able to tolerate and grow in the presence of up to 50% air in the gas phase (Parshina et al. 2003). *Thermotoga* species are marine microorganisms that are able to produce hydrogen and grow at temperatures up to 80°C (Frock et al. 2010).

The anaerobic archaea reported in Table 18.3 belong to the order *Thermococcales* that include two major genera, *Thermococcus* and *Pyrococcus*. *Thermococcus kodakaraensis* and *Thermococcus onnurineus* grow between 60 and 100°C. *Pyrococcus furiosus* grows at 70–103°C with an optimum around 100°C. A fast doubling time of 37 min was reported for this microorganism. This fermentative anaerobe is capable of utilizing maltose, starch, glycogen or cellobiose (Fiala and Stetter 1986).

As shown in Table 18.3, hydrogen yields around 3–4 mol H_2 mol^{-1} hexose were obtained for the different microorganisms and substrates tested. Maximum hydrogen yields were achieved in pure cultures of *Thermotoga* species (Table 18.3), probably related to a bifurcating hydrogenase. This hydrogenase, recently characterized in *T. maritima* (Schut and Adams 2009), uses the reducing equivalents from both NADH and reduced ferredoxin in a 1:1 ratio to produce hydrogen. Ferredoxin oxidation is thermodynamically unfavorable at standard temperature and pressure conditions ($\Delta G^{0'}$ = +3 kJ reaction^{-1}), but becomes exergonic for low PH_2 (e.g. $\Delta G'$ = −25 kJ reaction^{-1} for PH_2 = 1 Pa) (Abreu et al. 2012). Apparently, the exergonic oxidation of ferredoxin is used to drive the unfavorable oxidation of NADH. To maintain the 1:1 ratio of reducing equivalents from Fd and NADH, *T. maritima* seems to use a Fd:NADH oxireductase to supply the bifurcating hydrogenase.

Table 18.3 Hydrogen-producing microorganisms isolated from marine habitats

Organism	Domain	Isolation source	Growth temp. (°C)	Culturing type	Substrate	Yield (mol H_2 mol^{-1} hexose)	Genome size (Mb)	Ref.
Clostridium amygdalinum strain C9	Bacteria	Sea buried crude oil pipelines	37	Batch	Xylose	3.0	([a])	[1]
Thermotoga maritima DSM 3109	Bacteria	Geothermal heated sea sediments	80	Batch	Glucose	4.0	1.86	[2,3]
Thermotoga neapolitana DSM 4359	Bacteria	Shallow submarine hot springs	80–85	Batch	Glucose	3.8	~1.8	[4]
				Batch	Glucose and Xylose	3.3		[5]
			80	Batch	Potato steam peels	3.8		[6]
			77	Batch	Glucose	4.1		[3]
Pyrococcus furiosus DSM 3638	Archaea	Shallow marine geothermal beach	90	Batch	Cellobiose	2.8	1.9	[7,8]
				Batch	Maltose	3.5		[7]
				Chemostat	Cellobiose	3.8		[9]
				Chemostat	Maltose	2.9		[10]
			100	Batch	Glucose	3.5		[11]
Thermococcus kodakaraensis KOD1	Archaea	Geothermal spring in a coastal area of Japan	85	Chemostat	Starch	3.3	~2.1	[12,13]
Thermococcus onnurineus NA1	Archaea	Deep-sea hydrothermal vent around Papua New Guinea	80	Batch	Starch	3.1	1.85	[14,15]

[a]Genome being sequenced by DOE Joint Genome Institute

[1] (Jayasinghearachchi et al. 2010). [2] (Nelson et al. 1999). [3] (Schroder et al. 1994). [4] (Munro et al. 2009). [5] (de Vrije et al. 2009). [6] (Mars et al. 2010). [7] (Kengen and Stams 1994a). [8] (Robb et al. 2001). [9] (Chou et al. 2007). [10] (Schicho et al. 1993). [11] (Kengen and Stams 1994b). [12] (Fukui et al. 2005). [13] (Kanai et al. 2005). [14] (Bae et al. 2012). [15] (Lee et al. 2008)

Recently, the genome of several of these microorganisms has been sequenced (Fukui et al. 2005), providing new information and allowing the discovery of novel functions. For example, the genome sequencing of *Thermococcus onnurineus* allowed the identification of genes coding for multiple hydrogenases (Lee et al. 2008). The genome sequencing data of *T. maritima* DSM 3109 showed the presence of numerous metabolic pathways involved in the degradation of many simple and complex carbohydrates, namely glucose, xylose, mannose, starch, carboxymethyl-cellulose (CMC), xylan, and pectin (Huber et al. 1986). Moreover, 8–11% of *T. maritima* genes were found to be most similar to Archaea, whereas 42–48% of genes were most similar to *Firmicutes*, suggesting the occurrence of lateral gene transference between these microbial groups (Nelson et al. 1999). This mechanism may contribute to the development of new capacities by marine bacteria, improving the ability to live in extreme habitats and cope with environmental changes.

With genomic information, chromosomal gene disruption or replacement technology can be applied to expand the range of substrates utilized, as well as to knock-out competitive metabolic pathways or increase species resilience. For instance, *T. kodakaraensis* KOD1 cannot utilize maltose as a carbon source because of the lack of maltose transporter (Fukui et al. 2005). Introduction of the genes encoding the respective transporter of other hyperthermophiles (such as *P. furiosus* and *T. litoralis* (DiRuggiero et al. 1999)) could develop a new strain with the ability to produce H_2 from maltose.

3 Syngas Fermentation

In the last decades, the biological process of synthesis gas fermentation has gained interest as a sustainable technology for the production of valuable compounds and biofuels, namely methane, ethanol and butanol. Low biodegradable materials, such as lignocellulosic components of biomass, and other recalcitrant wastes (*e.g.* naphta, residual oil and petroleum coke) are used as substrate for syngas production through gasification (Ragauskas et al. 2006; Lawson et al. 2011). This process is performed at temperatures higher than 700°C with controlled supply of oxygen and/or steam. Formed syngas is mainly composed of CO, CO_2 and H_2 (McKendry 2002). Syngas can be further converted to fuels, either through catalytic processes (*e.g.* Fischer-Tropsch for olefins and gasoline synthesis) (Spath and Dayton 2003) or microbial processes.

Biological processes, although generally slower than chemical reactions, have several advantages over chemically catalyzed processes, such as higher specificity, higher yields, and generally greater resistance to poisoning (Klasson et al. 1992). These microbiological reactions occur at moderately high temperature and pressure conditions, resulting in minimum energy requirements and comparative lower cost.

Acetogenic anaerobes oxidize CO to CO_2 and H_2 via CO dehydrogenase (CODH) through a water-gas shift reaction (reaction 6 in Table 18.4). CODH is linked to the Wood-Ljungdahl pathway (or reductive acetyl-CoA pathway), which plays a central role in acetogenic CO metabolism (Fig. 18.4). In this pathway, H_2 (or CO) is used as

Table 18.4 Stoichiometry and Gibbs free energy changes of some possible reactions for the conversion of syngas components to biofuels (Henstra 2006; Sipma 2006)

	Reaction	$\Delta G^{0'}$ (kJ reaction^{-1})
(6)	$CO + H_2O \rightarrow H_2 + CO_2$	−20
(7)	$6CO + 3H_2O \rightarrow$ ethanol $+ 4CO_2$	−222
(8)	$2CO + 4H_2 \rightarrow$ ethanol $+ H_2O$	−288
(9)	$4CO + 8H_2 \rightarrow$ n-butanol $+ 3H_2O$	−324
(10)	$12CO + 5H_2O \rightarrow$ n-butanol $+ 8CO_2$	−480
(11)	$4CO + 2H_2O \rightarrow CH_4 + 3CO_2$	−211
(12)	$CO + 3H_2 \rightarrow CH_4 + H_2O$	−151
(13)	$HCO_3^- + 4H_2 + H^+ \rightarrow CH_4 + 3H_2O$	−135

Gibbs free energies (at 25°C) calculated at standard conditions (solute concentrations of 1 M and gas partial pressure of 10^5 Pa)

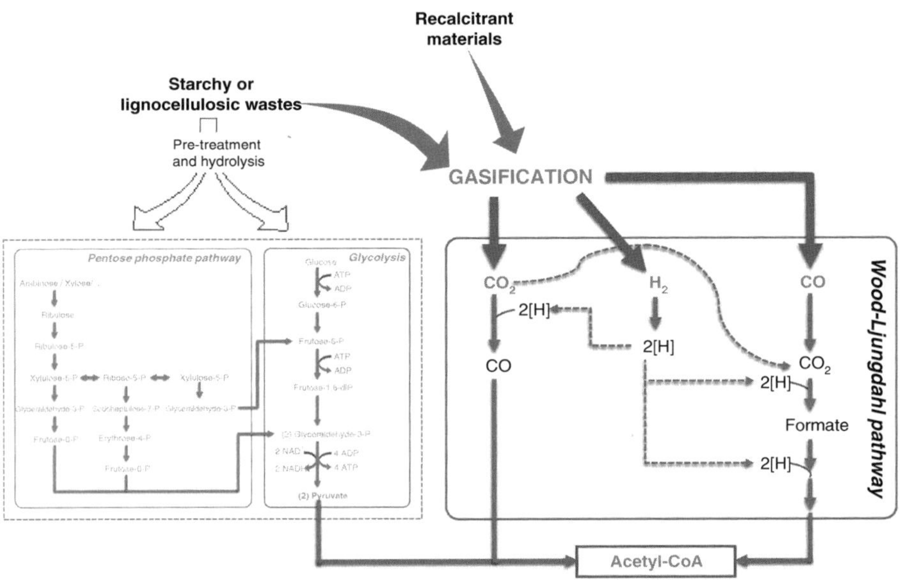

Fig. 18.4 Pathways for the metabolic conversion of biomass and recalcitrant materials into acetyl coenzyme A and further to biofuels

an electron donor and CO_2 as an electron acceptor, with the formation of acetyl-Coenzyme A (acetyl-CoA) (Fischer et al. 2008). Alternatively, hydrogen and CO_2 can also react to form CO, which is further converted to acetyl-CoA via acetyl-CoA synthase system, by the activity of a CO dehydrogenase (Kopke et al. 2010). Acetyl-CoA is the precursor of volatile fatty acids (*e.g.* acetate, butyrate) and alcohols, namely ethanol and butanol. Therefore, Wood-Ljungdahl pathway represents a possible alternative to glycolysis in the formation of this precursor, and gasification of biomass constitutes a more direct pathway towards the production of acetyl-CoA (Fig. 18.4).

Additionally, the different syngas components (CO, H_2 and CO_2) can be biologically converted to methane (CH_4), as show in reactions 11, 12 and 13 in Table 18.4. Methanogenesis is a biological process performed by prokaryotic microorganisms from the Archaea domain, observed in a wide range of environments, such as oceans, lakes, sediments, hydrothermal vents, anaerobic bioreactors and human and animal gut (Sowers and Ferry 2003).

CO utilization seems to be a quite widespread feature distributed among different phyla of bacteria and archaea (Henstra et al. 2007a). The presence of CO-consuming anaerobic microorganisms was detected in sediments and hydrothermal vents, where CO is one of the main components (*e.g.* Svetlichny et al. 1991). The number of carboxydotrophic isolates retrieved from marine environments is relatively low (Table 18.5), but considering the microbial diversity in CO-containing environments this number is probably underestimated.

One of the reported CO-consuming microorganisms isolated from submarine hot vent is the *Caldanaerobacter subterraneus* (Table 18.5). This Gram-positive, non-motile bacterium grows chemolitotrophically on CO with the production of equimolar quantities of CO_2 and H_2. *C. subterraneus* is an extreme-thermophile with optimum growth temperature at 70°C, belonging to *Clostridia* class. Another bacterium, *Acetobacterium woodii*, is capable of utilizing CO or H_2+CO_2 at 30°C with the formation of acetate. *A. woodii* is a Gram-positive bacterium with oval-shaped cells, and was isolated from marine sediments (Balch et al. 1977).

The first carboxydotrophic archaea described was isolated from hydrothermal vents and belongs to the *Thermococcus* genera (Sokolova et al. 2004). *Thermococcus* strain AM4 presents coccoid cells and grows at hyperthermophilic conditions (80°C), producing H_2 and CO_2 from CO (Table 18.5). *Thermococcus onnurineus* is also able to grow in media supplemented with carbon monoxide, producing H_2 (Lee et al. 2008). A maximum hydrogen production rate of 1.55 mmol H_2 L^{-1} h^{-1} and a yield of 0.98 mol H_2 mol^{-1} substrate were obtained from the degradation of CO in batch bioreactors inoculated with pure culture of *T. onnurineus* (Bae et al. 2012). Complete genome sequencing of this microorganism allowed the identification of genes coding for a CO dehydrogenase (Codh) (TON_1018), besides multiple hydrogenases (Lee et al. 2008).

Carbon monoxide is also utilized by *Archaeoglobus fulgidus*, a hypherthermophilic archaea (optimal growth at 83°C) with irregular coccoid cells, isolated from submarine hot spring. *A. fulgidus* can grow chemolitoautotrophically with CO forming acetate (Henstra et al. 2007b).

The mesophilic *Methanosarcina acetivorans*, isolated from marine sediments, and the hyper-thermophilic *Methanocaldococcus jannaschii*, isolated from hydrothermal vent, can utilize CO for the formation of methane. Additionally, these archaea can also convert H_2+CO_2 to methane.

Other marine archaea are able to produce methane from the gaseous mixture of CO_2 and H_2 (Sowers and Ferry 2003). Mesophilic and extreme-thermophilic microorganisms were isolated from sediments and hydrothermal vents, respectively. These microorganisms mainly belong to the genera *Methanococcus, Methanocaldococcus, Methanoculleus, Methanolacinia, Methanogenium* and *Methanopyrus* (Sowers and Ferry 2003).

Table 18.5 Marine anaerobic microorganisms able to utilize carbon monoxide for growth

Microorganism	Domain	Isolation source	Growth temp. (°C)	Main Product	Genome Size (Mb)	References
Acetobacterium woodii	Bacteria	black sediment, Oyster Pond (marine estuary), Woods Hole, Mass	30	acetate	4,04	[1,2,3]
Caldanaerobacter subterraneus subsp. *pacificus*	Bacteria	submarine hot vent, Okinawa Trough	70	hydrogen	–	[4,5]
Archaeoglobus fulgidus	Archaea	submarine hot spring, Vulcano Island, Italy	83	acetate	2,18	[6,7,8]
Methanosarcina acetivorans	Archaea	Marine sediment, Scripps Canyon, La Jolla, Calif	37	methane	5,75	[9,10,11]
Thermococcus strain AM4	Archaea	East Pacific Rise	82	hydrogen	–	[12]
Thermococcus onnurineus NA1	Archaea	Deep-sea hydrothermal vent around Papua New Guinea	80	hydrogen	1.85	[13]

[1] (Balch et al. 1977). [2] (Genthner and Bryant 1987). [3] (Poehlein et al. 2012). [4] (Sokolova et al. 2001). [5] (Fardeau et al. 2004). [6] (Stetter 1988). [7] (Klenk et al. 1997). [8] (Henstra 2006). [9] (Sowers et al. 1984). [10] (Galagan et al. 2002). [11] (Rother and Metcalf 2004). [12] (Sokolova et al. 2004). [13] (Lee et al. 2008)

Despite the high potential of marine microorganisms for bioenergy production from syngas through biological fermentation, this process is still not sufficiently studied. Practical applications in biotechnological process have just emerged, requiring further research for full exploitation of its potential.

4 Electro-Chemical Processes

Harvesting electrons from bacterial metabolism as a potential sustainable energy source has been for long subject of research interest, but only in the last few years the prospects for practical applications improved considerably due to the development of microbial fuel cells (MFC) with enhanced power output (Liu et al. 2004; Liu and Logan 2004; Rabaey et al. 2003, 2004).

In a MFC, biological and electrochemical processes are combined to convert dissolved organic matter directly into electrical current. This is achieved by diverting the electrons produced by electrochemically active bacteria, during oxidation of the organic matter, towards an insoluble acceptor, i.e. the anode electrode. The produced protons diffuse through a membrane into a cathode compartment, where they react with oxygen generating water and an electrical current from the anode towards the cathode. In general, a MFC is a two-chamber structure, one containing the anode and electrochemically active bacteria growing under anaerobic conditions, and another containing the cathode (Fig. 18.5). The cathode chamber is kept aerobic by sparging air in the water. Simplified designs were developed with single-chamber configurations where the cathode is fused to the proton exchange membrane and directly exposed to air (Liu et al. 2004; Liu and Logan 2004; Park and Zeikus 2003). Simple systems without proton exchange membranes have been developed as well (Liu et al. 2005; Liu and Logan 2004). The performance of the MFC depends on several parameters,

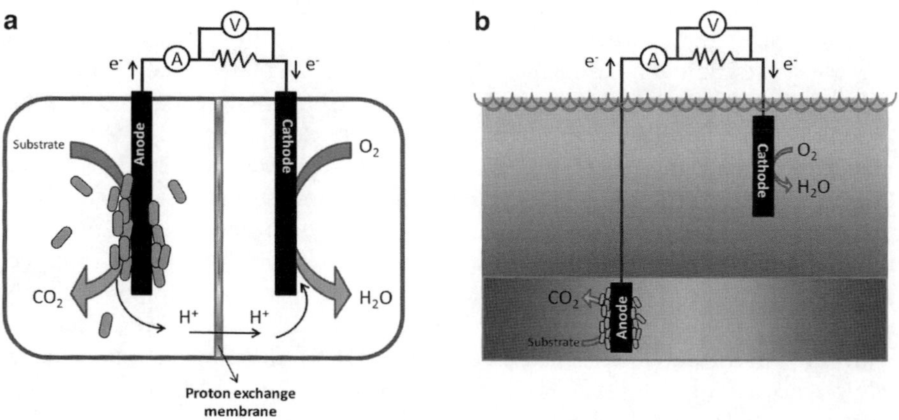

Fig. 18.5 Schematic representation of a microbial fuel cell (**a**) and a marine microbial fuel cell (**b**)

such as substrate conversion rate, performance of the proton exchange membrane and internal resistance of the MCF (Rabaey and Verstraete 2005).

A diverse range of microorganisms has been found capable of interacting with electrodes, usually referred as anodophiles (Park and Zeikus 2003), exoelectrogens (Logan and Regan 2006), electrogenic (Debabov 2008), anode-respiring (Torres et al. 2007) or electrochemically active microorganisms (Chang et al. 2006). The term electricigens was also proposed specifically for microorganisms that completely oxidize organic compounds to carbon dioxide with an electrode serving as the sole electron acceptor (Lovley 2006). Several mechanisms by which microorganisms may transfer electrons to the anode of microbial fuel cells have been proposed, including direct electron transfer through outer-surface c-type cytochromes (Busalmen et al. 2008), conductive biofilm matrix containing cytochromes (Marcus et al. 2007) or microbial nanowires (Reguera et al. 2005), and indirect electron transfer through soluble electron shuttles (Marsili et al. 2008).

Microbial fuel cells were first developed to produce power from the electrical current generated by bacteria, but there has been an evolution of the system for other applications. Additional voltage added to the potential generated by the bacteria allow for various products to be generated at the cathode, such as hydrogen (Rozendal et al. 2006), methane (Cheng et al. 2009) and hydrogen peroxide (Rozendal et al. 2009). The terms bio-electrochemical systems (BES), microbial electrolysis cell (MEC) and microbial electrochemical systems (MxC) have been used to describe those technologies. Other applications include sensoring, remediation and wastewater treatment (Clauwaert et al. 2007; Li et al. 2008; Liu et al. 2004; Liu and Logan 2004; Tront et al. 2008; Zhang et al. 2010).

The first practical application of marine microbial fuels cells was to power low-energy consuming marine instrumentation, *e.g.* meteorological buoys capable of measuring air temperature, pressure, relative humidity and water temperature (Tender et al. 2008). Anaerobic marine MFC are generally composed of graphite electrodes that are placed *in situ*: the anode is introduced in anaerobic marine sediments and the cathode is positioned in the oxygen-rich seawater (Fig. 18.5). The naturally occurring microorganisms colonize the anode and oxidize the organic substrates present in the sediments, while in the cathode seawater constituents are reduced. These systems are based on the natural redox gradient that occurs at the water-sediment interface due to the microbial metabolic activity, and do not require proton exchange membrane (Dumas et al. 2007; Tender et al. 2002). Addition to the sediments of insoluble slowly degrading organic substrates, such as chitin or cellulose, or the use of anodes modified with charge transfer mediators, has resulted in power output increase (Rezaei et al. 2007, 2008, 2009). For example, Lowy and Tender (2008) reported a maximum power density of approximately 98 mW m^{-2} (of anode area) at a cell voltage of 0.24 V in a marine MFC operating with an anthraquinone-1,6-disulfonic acid (AQDS)-modified graphite anode, while a maximum value around 20 mW m^{-2} at 0.30 V was attained in a similar system assembled with a plain graphite anode.

Anoxic marine sediments have been frequently used as source of electrogenic microorganisms (Table 18.6). Sequences from a denaturing gradient gel electrophoresis

Table 18.6 Electricigenic anaerobic microorganisms isolated from marine sediments or MFC

Organism	Class	Isolated from	Growth temp. (°C)	Main substrate	Reference
Desulfuromonas acetoxidans	*Deltaproteobacteria*	Sediments from Antarctic ocean	30	Acetate, ethanol, propanol	Pfennig and Biebl (1976)
Geopsychrobacter electrodiphilus	*Deltaproteobacteria*	Anode surface of marine sediment MFC	22	Acetate, several OA, AA, LCFA and Arom	Holmes et al. (2004b)
Prolixibacter bellariivorans	*Bacteroidia*	Marine sediment MFC	22	Several sugars	Holmes et al. (2007)
Rhodoferax ferrireducens	*Betaproteobacteria*	Isolated from coastal aquifer sediment	25	Glucose	Finneran et al. (2003)

OA organic acids, *AA* amino acids, *LCFA* long-chain fatty acids, *Arom* aromatic compounds

(DGGE)-screened 16S rDNA clone library showed that a marine sediment used to inoculate an MFC fed with cysteine resulted in a bacterial community in which 97% of the sequences detected belong to the *Gammaproteobacteria* and were similar to *Shewanella affinis* KMM 3686 (40% of clones), with *Vibrio* spp. and *Pseudoalteromonas* spp. being the next most frequently detected (Logan et al. 2005).

The *Shewanella* genus has a wide environmental distribution and several species have been retrieved from marine and freshwater habitats. These mesophilic facultative anaerobic bacteria are capable of dissimilatory metal reduction (DMR). Member of the genus *Shewanella* grow forming a biofilm on insoluble metal oxides, which facilitates the contact between bacteria and the metal allowing direct electron transfer. Alternative mechanisms of electron transfer by *Shewanella* (nanowires or soluble electron shuttles) have also been proposed (Flynn et al. 2012; Venkateswaran et al. 1999). Pure cultures of *Shewanella oneidensis* DSP10 and *Shewanella oneidensis* MR-1 were used as inocula for power generation from lactate in marine sediment MFC (Kim et al. 2002; Ringeisen et al. 2006). High output power per device cross-section and volume (3 W m^{-2}, 500 W m^{-3}, respectively) was achieved in a miniature microbial fuel cell (mini-MFC) inoculated with *Shewanella oneidensis* DSP10. Current and power was enhanced 30–100% by the addition of electron mediators (Ringeisen et al. 2006). The genome of this bacterium was completely sequenced in 2002, contributing for the understanding of the mechanisms involved in *Shewanella* metabolism (Heidelberg et al. 2002).

Other studies report a clear dominance of *Deltaproteobacteria* in sediment MFC. Bond et al. (2002) verified that 71% of the sequences obtained in a 16S rDNA clone library from an anode electrode were *Deltaproteobacteria*, and 70% of these belonged to the family *Geobacteraceae*. In a similar system, *Deltaproteobacteria* accounted for 76% of the sequences retrieved, from which 59% were from the family *Geobacteraceae* and presented more than 95% similarity to *Desulfuromonas acetoxidans* (Tender et al. 2002). Another group of *Deltaproteobacteria* sequences, most closely related to sulfate-reducing bacteria from the family *Desulfobulbaceae*, was also consistently enriched on the anodes of marine sediment fuel cells (Holmes et al. 2004a). In fact, in one field experiment, organisms in this cluster accounted for all of the deltaproteobacterial sequences and for approximately 62% of the bacterial 16S rRNA gene sequences recovered from the current-harvesting anode (Holmes et al. 2004a).

Desulfuromonas acetoxidans, from the *Desulfuromonadacea* family of *Deltaproteobacteria*, was isolated from Antarctic Ocean marine sediments and is a strictly anaerobic, rod-shaped, Gram-negative bacterium (Pfennig and Biebl 1976). In this bacterium, the complete oxidation of organic compounds, such as acetate, ethanol or propanol, is coupled with the reduction of a wide range of soluble and insoluble electron acceptors (e.g. sulfur, fumarate, ferric iron or manganese) (Table 18.6). Electron transfer to the anode appears to be related with a complex network of multiheme cytochromes, from which only few have been characterized (Alves et al. 2011). Despite the reported abundance of *D. acetoxidans* in microbial communities colonizing the anode of sediment MFC (*e.g.* Tender et al. 2002), inoculation of MFC with pure cultures of these microorganisms has not been tested.

Several mesophilic psychrotolerant bacteria, capable of DMR metabolism, have been isolated from marine sediment MFC, namely *Geopsychrobacter electrodiphilus*,

Prolixibacter bellariivorans, *Rhodoferax ferrireducens* (Finneran et al. 2003; Holmes et al. 2004b, 2007). These mesophilic microorganisms can grow and reduce metals, e.g. iron(III), manganese (IV), at low temperatures (*ca.* 4°C), although exhibiting optimum growth around 22–25°C. Lab-scale experiments were performed with MFC inoculated with a pure culture of *Rhodoferax ferrireducens*, achieving electric current intensity of 31 mA m^{-2} (Chaudhuri and Lovley 2003). Higher values of electric current (209–254 mA m^{-2}) were obtained at 60°C with a MFC prepared with a thermophilic electrogenic microbial community recovered from marine sediments (South Carolina, USA) (Mathis et al. 2008). The main ribotype retrieved (approximately 68% of the clones) was found to be closely related to *Thermincola carboxydophila* (99% similarity), with uncultured microorganisms belonging to the *Firmicutes* and *Deferribacteres* phyla as the remaining 16S rRNA genes.

In the last years MFC systems have developed considerably toward a simple and robust technology. The main research efforts have been focused on the materials and configuration of electrodes and proton exchange membranes, as well as the design of the MFC system. More inputs on the selection of electricigenic microbial communities are required for optimal electricity production in MFC. Marine environments constitute a powerful source of potential electrochemically active microorganisms.

5 Conclusions and Future Prospects

Marine habitats can offer unique microbial and metabolic features with huge potential for application in bioenergy production processes. Dark fermentation, syngas fermentation and bioelectrochemical processes were selected as demonstration examples of the potential application of marine anaerobic bacteria and archaea in the bioenergy field.

The advances in "omics" in line with newly designed and optimized biotechnology processes, for example operating at extreme conditions of temperature and pressure or electrically assisted, will turn marine biotechnology a field of growing interest and increasing application in the sector of bioenergy.

References

Abreu A, Karakashev D, Angelidaki I, Sousa D, Alves M (2012) Biohydrogen production from arabinose and glucose using extreme thermophilic anaerobic mixed cultures. Biotechnol Biofuels 5:6. doi:10.1186/1754-6834-5-6

Alves AS, Paquete CM, Fonseca BM, Louro RO (2011) Exploration of the 'cytochromome' of *Desulfuromonas acetoxidans*, a marine bacterium capable of powering microbial fuel cells. Metallomics 3:349–353. doi:10.1039/C0MT00084A

Arrieta JM, Arnaud-Haond S, Duarte CM (2010) What lies underneath: conserving the oceans' genetic resources. Proc Natl Acad Sci USA 107:18318–18324. doi:10.1073/pnas.0911897107

Bae SS, Kim TW, Lee HS, Kwon KK, Kim YJ, Kim M-K, Lee J-H, Kang SG (2012) H_2 production from CO, formate or starch using the hyperthermophilic archaeon, *Thermococcus onnurineus*. Biotechnol Lett 34:75–79. doi:10.1007/s10529-011-0732-3

Balch WE, Schoberth S, Tanner RS, Wolfe RS (1977) *Acetobacterium*, a new genus of hydrogen-oxidizing, carbon dioxide-reducing, anaerobic bacteria. Int J Syst Bacteriol 27:355–361. doi:10.1099/00207713-27-4-355

Barnard D, Casanueva A, Tuffin M, Cowan D (2010) Extremophiles in biofuel synthesis. Environ Technol 31:871–888. doi:10.1080/09593331003710236

Bond DR, Holmes DE, Tender LM, Lovley DR (2002) Electrode-reducing microorganisms that harvest energy from marine sediments. Science 295:483–485. doi:10.1126/science.1066771

Børresen T, Boyen C, Dobson A, Höfle M, Ianora A, Jaspars M, Kijjoa A, Olafsen J, Querellou J, Rigos G, Wijffels R (2010) In: McDonough N (ed) Marine biotechnology: a new vision and strategy for Europe. Marine Board Position Paper 15. Marine Board-ESF, Beernem

Busalmen JP, Esteve-Núñez A, Berná A, Feliu JM (2008) C-type cytochromes wire electricity-producing bacteria to electrodes. Angew Chem Int Ed 47:4874–4877. doi:10.1002/anie.200801310

Chang IS, Moon H, Bretschger O, Jang JK, Park HI, Nealson KH, Kim BH (2006) Electrochemically Active Bacteria (EAB) and mediator-less microbial fuel cells. J Microbiol Biotechnol 16:163–177

Chaudhuri SK, Lovley DR (2003) Electricity generation by direct oxidation of glucose in mediatorless microbial fuel cells. Nat Biotechnol 21:1229–1232. doi:10.1038/nbt867

Cheng S, Xing D, Call DF, Logan BE (2009) Direct biological conversion of electrical current into methane by electromethanogenesis. Environ Sci Technol 43:3953–3958. doi:10.1021/es803531g

Chou CJ, Shockley KR, Conners SB, Lewis DL, Comfort DA, Adams MWW, Kelly RM (2007) Impact of substrate glycoside linkage and elemental sulfur on bioenergetics of and hydrogen production by the hyperthermophilic archaeon *Pyrococcus furiosus*. Appl Environ Microbiol 73:6842–6853. doi:10.1128/AEM.00597-07

Clauwaert P, Rabaey K, Aelterman P, De Schamphelaire L, Pham TH, Boeckx P, Boon N, Verstraete W (2007) Biological denitrification in microbial fuel cells. Environ Sci Technol 41:3354–3360. doi:10.1021/es062580r

de Vrije T, Bakker RR, Budde MAW, Lai MH, Mars AE, Claassen PAM (2009) Efficient hydrogen production from the lignocellulosic energy crop *Miscanthus* by the extreme thermophilic bacteria *Caldicellulosiruptor saccharolyticus* and *Thermotoga neapolitana*. Biotechnol Biofuels 2:12. doi:10.1186/1754-6834-2-12

Debabov V (2008) Electricity from microorganisms. Microbiology 77:123–131. doi:10.1134/S002626170802001X

DiRuggiero J, Brown JR, Bogert AP, Robb FT (1999) DNA repair systems in archaea: mementos from the last universal common ancestor? J Mol Evol 49:474–484. doi:10.1007/PL00006570

Dumas C, Mollica A, Féron D, Basséguy R, Etcheverry L, Bergel A (2007) Marine microbial fuel cell: use of stainless steel electrodes as anode and cathode materials. Electrochim Acta 53:468–473. doi:10.1016/j.electacta.2007.06.069

Eloe EA, Fadrosh DW, Novotny M, Zeigler Allen L, Kim M, Lombardo MJ, Yee-Greenbaum J, Yooseph S, Allen EE, Lasken R, Williamson SJ, Bartlett DH (2011) Going deeper: metagenome of a hadopelagic microbial community. PLoS ONE 6:e20388. doi:10.1371/journal.pone.0020388

Fardeau ML, Salinas MB, L'Haridon S, Jeanthon C, Verhe F, Cayol JL, Patel BKC, Garcia JL, Ollivier B (2004) Isolation from oil reservoirs of novel thermophilic anaerobes phylogenetically related to *Thermoanaerobacter subterraneus*: reassignment of *T. subterraneus*, *Thermoanaerobacter yonseiensis*, *Thermoanaerobacter tengcongensis* and *Carboxydibrachium pacificum* to *Caldanaerobacter subterraneus* gen. nov., sp nov., comb. nov as four novel subspecies. Int J Syst Evol Microbiol 54:467–474. doi:10.1099/ijs.0.02711-0

Fiala G, Stetter KO (1986) *Pyrococcus furiosus* sp. nov. represents a novel genus of marine heterotrophic archaebacteria growing optimally at 100°C. Arch Microbiol 145:56–61. doi:10.1007/BF00413027

Finneran KT, Johnsen CV, Lovley DR (2003) *Rhodoferax ferrireducens* sp. nov., a psychrotolerant, facultatively anaerobic bacterium that oxidizes acetate with the reduction of Fe(III). Int J Syst Evol Microbiol 53:669–673. doi:10.1099/ijs.0.02298-0

Fischer CR, Klein-Marcuschamer D, Stephanopoulos G (2008) Selection and optimization of microbial hosts for biofuels production. Metab Eng 10:295–304. doi:10.1016/j.ymben.2008.06.009

Flynn CM, Hunt KA, Gralnick JA, Srienc F (2012) Construction and elementary mode analysis of a metabolic model for *Shewanella oneidensis* MR-1. Biosystems 107:120–128. doi:10.1016/j.biosystems.2011.10.003

Frock AD, Notey JS, Kelly RM (2010) The genus *Thermotoga*: recent developments. Environ Technol 31:1169–1181. doi:10.1080/09593330.2010.484076

Fuhrman JA, Davis AA (1997) Widespread archaea and novel bacteria from the deep sea as shown by 16S rRNA gene sequences. Mar Ecol Prog Ser 150:275–285. doi:10.3354/meps150275

Fukui T, Atomi H, Kanai T, Matsumi R, Fujiwara S, Imanaka T (2005) Complete genome sequence of the hyperthermophilic archaeon *Thermococcus kodakaraensis* KOD1 and comparison with *Pyrococcus* genomes. Genome Res 15:352–363. doi:10.1101/gr.3003105

Galagan JE, Nusbaum C, Roy A, Endrizzi MG, Macdonald P, FitzHugh W, Calvo S, Engels R, Smirnov S, Atnoor D, Brown A, Allen N, Naylor J, Stange-Thomann N, DeArellano K, Johnson R, Linton L, McEwan P, McKernan K, Talamas J, Tirrell A, Ye WJ, Zimmer A, Barber RD, Cann I, Graham DE, Grahame DA, Guss AM, Hedderich R, Ingram-Smith C, Kuettner HC, Krzycki JA, Leigh JA, Li WX, Liu JF, Mukhopadhyay B, Reeve JN, Smith K, Springer TA, Umayam LA, White O, White RH, de Macario EC, Ferry JG, Jarrell KF, Jing H, Macario AJL, Paulsen I, Pritchett M, Sowers KR, Swanson RV, Zinder SH, Lander E, Metcalf WW, Birren B (2002) The genome of *M. acetivorans* reveals extensive metabolic and physiological diversity. Genome Res 12:532–542. doi:10.1101/gr.223902

Genthner BRS, Bryant MP (1987) Additional characteristics of one-carbon-compound utilization by *Eubacterium limosum* and *Acetobacterium woodii*. Appl Environ Microbiol 53:471–476

Glöckner FO, Stal LJ, Sandraa R-A, Gasol JM, O'Gara F, Hernandez F, Labrenz M, Stoica E, Varela MM, Bordalo A, Pitta P (2012) In: Calewaert JB, McDonough N (eds) Marine microbial diversity and its role in ecosystem functioning and environmental change. Marine Board Position Paper 17. Marine Board-ESF, Ostend

Hallenbeck PC, Abo-Hashesh M, Ghosh D (2012) Strategies for improving biological hydrogen production. Bioresour Technol 110:1–9. doi:10.1016/j.biortech.2012.01.103

Heidelberg JF, Paulsen IT, Nelson KE, Gaidos EJ, Nelson WC, Read TD, Eisen JA, Seshadri R, Ward N, Methe B, Clayton RA, Meyer T, Tsapin A, Scott J, Beanan M, Brinkac L, Daugherty S, DeBoy RT, Dodson RJ, Durkin AS, Haft DH, Kolonay JF, Madupu R, Peterson JD, Umayam LA, White O, Wolf AM, Vamathevan J, Weidman J, Impraim M, Lee K, Berry K, Lee C, Mueller J, Khouri H, Gill J, Utterback TR, McDonald LA, Feldblyum TV, Smith HO, Venter JC, Nealson KH, Fraser CM (2002) Genome sequence of the dissimilatory metal ion-reducing bacterium *Shewanella oneidensis*. Nat Biotechnol 20:1118–1123. doi:10.1038/nbt749

Henstra AM (2006) CO metabolisms of *Carboxydothermus hydrogenoformans and Archaeoglobus fulgidus*. PhD thesis, Wageningen University

Henstra AM, Sipma J, Rinzema A, Stams AJM (2007a) Microbiology of synthesis gas fermentation for biofuel production. Curr Opin Biotechnol 18:200–206. doi:10.1016/j.copbio.2007.03.008

Henstra AM, Dijkema C, Stams AJ (2007b) *Archaeoglobus fulgidus* couples CO oxidation to sulfate reduction and acetogenesis with transient formate accumulation. Environ Microbiol 9:1836–1841. doi:10.1111/j.1462-2920.2007.01306.x

Holmes DE, Bond DR, O'Neil RA, Reimers CE, Tender LR, Lovley DR (2004a) Microbial communities associated with electrodes harvesting electricity from a variety of aquatic sediments. Microb Ecol 48:178–190. doi:10.1007/s00248-003-0004-4

Holmes DE, Nicoll JS, Bond DR, Lovley DR (2004b) Potential role of a novel psychrotolerant member of the family *Geobacteraceae*, *Geopsychrobacter electrodiphilus* gen. nov., sp. nov., in electricity production by a marine sediment fuel cell. Appl Environ Microbiol 70:6023–6030. doi:10.1128/AEM.70.10.6023-6030.2004

Holmes DE, Nevin KP, Woodard TL, Peacock AD, Lovley DR (2007) *Prolixibacter bellariivorans* gen. nov., sp. nov., a sugar-fermenting, psychrotolerant anaerobe of the phylum *Bacteroidetes*, isolated from a marine-sediment fuel cell. Int J Syst Evol Microbiol 57:701–707. doi:10.1099/ijs.0.64296-0

Horikoshi K (1998) Barophiles: deep-sea microorganisms adapted to an extreme environment. Curr Opin Microbiol 1:291–295. doi:10.1016/S1369-5274(98)80032-5

Huber R, Langworthy TA, Kunig H, Thomm M, Woese CR, Sleytr UB, Stetter KO (1986) *Thermotoga maritima* sp. nov. represents a new genus of unique extremely thermophilic eubacteria growing up to 90°C. Arch Microbiol 144:324–333. doi:10.1007/BF00409880

Inagaki F, Nunoura T, Nakagawa S, Teske A, Lever M, Lauer A, Suzuki M, Takai K, Delwiche M, Colwell FS, Nealson KH, Horikoshi K, D'Hondt S, JØrgensen BB (2006) Biogeographical distribution and diversity of microbes in methane hydrate-bearing deep marine sediments on the Pacific Ocean Margin. Proc Natl Acad Sci USA 103:2815–2820. doi:10.1073/pnas.0511033103

Jayasinghearachchi HS, Singh S, Sarma PM, Aginihotri A, Lal B (2010) Fermentative hydrogen production by new marine *Clostridium amygdalinum* strain C9 isolated from offshore crude oil pipeline. Int J Hydrogen Energy 35:6665–6673. doi:10.1016/j.ijhydene.2010.04.034

Kanai T, Imanaka H, Nakajima A, Uwamori K, Omori Y, Fukui T, Atomi H, Imanaka T (2005) Continuous hydrogen production by the hyperthermophilic archaeon, *Thermococcus kodakaraensis* KOD1. J Biotechnol 116:271–282. doi:10.1016/j.jbiotec.2004.11.002

Kato C, Bartlett DH (1997) The molecular biology of barophilic bacteria. Extremophiles 1:111–116. doi:10.1007/s007920050023

Kengen SWM, Stams AJM (1994a) Growth and energy conservation in batch cultures of *Pyrococcus furiosus*. FEMS Microbiol Lett 117:305–309. doi:10.1111/j.1574-6968.1994.tb06784.x

Kengen SWM, Stams AJM (1994b) Formation of l-alanine as a reduced end-product in carbohydrate fermentation by the hyperthermophilic archaeon *Pyrococcus furiosus*. Arch Microbiol 161:168–175. doi:10.1007/BF00276479

Kim HJ, Park HS, Hyun MS, Chang IS, Kim M, Kim BH (2002) A mediator-less microbial fuel cell using a metal reducing bacterium, *Shewanella putrefaciens*. Enzyme Microb Technol 30:145–152. doi:10.1016/S0141-0229(01)00478-1

Kivistö AT, Karp MT (2011) Halophilic anaerobic fermentative bacteria. J Biotechnol 152:114–124. doi:10.1016/j.jbiotec.2010.08.014

Klasson KT, Ackerson MD, Clausen EC, Gaddy JL (1992) Bioconversion of synthesis gas into liquid or gaseous fuels. Enzyme Microb Technol 14:602–608. doi:10.1016/0141-0229(92)90033-K

Klenk HP, Clayton RA, Tomb JF, White O, Nelson KE, Ketchum KA, Dodson RJ, Gwinn M, Hickey EK, Peterson JD, Richardson DL, Kerlavage AR, Graham DE, Kyrpides NC, Fleischmann RD, Quackenbush J, Lee NH, Sutton GG, Gill S, Kirkness EF, Dougherty BA, McKenney K, Adams MD, Loftus B, Peterson S, Reich CI, Mcneil LK, Badger JH, Glodek A, Zhou LX, Overbeek R, Gocayne JD, Weidman JF, McDonald L, Utterback T, Cotton MD, Spriggs T, Artiach P, Kaine BP, Sykes SM, Sadow PW, DAndrea KP, Bowman C, Fujii C, Garland SA, Mason TM, Olsen GJ, Fraser CM, Smith HO, Woese CR, Venter JC (1997) The complete genome sequence of the hyperthermophilic, sulphate-reducing archaeon *Archaeoglobus fulgidus*. Nature 390(10.1038/37052):364–370. doi:10.1038/37052

Kopke M, Held C, Hujer S, Liesegang H, Wiezer A, Wollherr A, Ehrenreich A, Liebl W, Gottschalk G, Durre P (2010) *Clostridium ljungdahlii* represents a microbial production platform based on syngas. Proc Natl Acad Sci USA 107:13087–13092. doi:10.1073/pnas.1004716107

Lawson PA, Allen TD, Caldwell ME, Tanner RS (2011) Anaerobes: a piece in the puzzle for alternative biofuels. Anaerobe 17:206–210. doi:10.1016/j.anaerobe.2011.05.011

Lee HS, Kang SG, Bae SS, Lim JK, Cho Y et al (2008) The complete genome sequence of *Thermococcus onnurineus* NA1 reveals a mixed heterotrophic and carboxydotrophic metabolism. J Bacteriol 190:7491–7499. doi:10.1128/JB.00746-08

Li Z, Zhang X, Lei L (2008) Electricity production during the treatment of real electroplating wastewater containing Cr^{6+} using microbial fuel cell. Process Biochem 43:1352–1358. doi:10.1016/j.procbio.2008.08.005

Liu H, Logan BE (2004) Electricity generation using an air-cathode single chamber microbial fuel cell in the presence and absence of a proton exchange membrane. Environ Sci Technol 38:4040–4046. doi:10.1021/es0499344

Liu H, Ramnarayanan R, Logan BE (2004) Production of electricity during wastewater treatment using a single chamber microbial fuel cell. Environ Sci Technol 38:2281–2285. doi:10.1021/es034923g

Liu H, Cheng S, Logan BE (2005) Production of electricity from acetate or butyrate using a single-chamber microbial fuel cell. Environ Sci Technol 39:658–662. doi:10.1021/es048927c

Logan BE, Regan JM (2006) Electricity-producing bacterial communities in microbial fuel cells. Trends Microbiol 14:512–518

Logan BE, Murano C, Scott K, Gray ND, Head IM (2005) Electricity generation from cysteine in a microbial fuel cell. Water Res 39:942–952. doi:10.1016/j.watres.2004.11.019

Lovley DR (2006) Bug juice: harvesting electricity with microorganisms. Nat Rev Microbiol 4:497–508. doi:10.1038/nrmicro1442

Lowy DA, Tender LM (2008) Harvesting energy from the marine sediment-water interface: III. Kinetic activity of quinone- and antimony-based anode materials. J Power Sources 185:70–75. doi:10.1016/j.jpowsour.2008.06.079

Marcus AK, Torres CI, Rittmann BE (2007) Conduction-based modeling of the biofilm anode of a microbial fuel cell. Biotechnol Bioeng 98:1171–1182. doi:10.1002/bit.21533

Mars AE, Veuskens T, Budde MAW, van Doeveren PFNM, Lips SJ et al (2010) Biohydrogen production from untreated and hydrolyzed potato steam peels by the extreme thermophiles *Caldicellulosiruptor saccharolyticus* and *Thermotoga neapolitana*. Int J Hydrogen Energy 35:7730–7737. doi:10.1016/j.ijhydene.2010.05.063

Marsili E, Baron DB, Shikhare ID, Coursolle D, Gralnick JA, Bond DR (2008) *Shewanella* secretes flavins that mediate extracellular electron transfer. Proc Natl Acad Sci USA 105:3968–3973. doi:10.1073/pnas.0710525105

Mathis B, Marshall C, Milliken C, Makkar R, Creager S, May H (2008) Electricity generation by thermophilic microorganisms from marine sediment. Appl Microbiol Biotechnol 78:147–155. doi:/10.1007/s00253-007-1266-4

McKendry P (2002) Energy production from biomass (part 3): gasification technologies. Bioresour Technol 83:55–63. doi:10.1016/S0960-8524(01)00120-1

Munro SA, Zinder SH, Walker LP (2009) The fermentation stoichiometry of *Thermotoga neapolitana* and influence of temperature, oxygen, and pH on hydrogen production. Biotechnol Prog 25:1035–1042. doi:10.1002/btpr.201

Nagata T, Tamburini C, Arístegui J, Baltar F, Bochdansky AB, Fonda-Umani S, Fukuda H, Gogou A, Hansell DA, Hansman RL, Herndl GJ, Panagiotopoulos C, Reinthaler T, Sohrin R, Verdugo P, Yamada N, Yamashita Y, Yokokawa T, Bartlett DH (2010) Emerging concepts on microbial processes in the bathypelagic ocean – ecology, biogeochemistry, and genomics. Deep Sea Res Pt II 57:1519–1536. doi:10.1016/j.dsr2.2010.02.019

Nelson KE, Clayton RA, Gill SR, Gwinn ML, Dodson RJ et al (1999) Evidence for lateral gene transfer between archaea and bacteria from genome sequence of *Thermotoga maritima*. Nature 399:323–329. doi:10.1038/20601

Orcutt BN, Sylvan JB, Knab NJ, Edwards KJ (2011) Microbial ecology of the dark ocean above, at, and below the seafloor. Microbiol Mol Biol Rev 75:361–422. doi:10.1128/MMBR.00039-10

Park DH, Zeikus JG (2003) Improved fuel cell and electrode designs for producing electricity from microbial degradation. Biotechnol Bioeng 81:348–355. doi:10.1002/bit.10501

Parshina SN, Kleerebezem R, Sanz JLS, Lettinga G, Nozhevnikova AN et al (2003) *Soehngenia saccharolytica* gen. nov., sp. nov. and *Clostridium amygdalinum* sp. nov., two novel anaerobic, benzaldehyde-converting bacteria. Int J Syst Evol Microbiol 53:1791–1799. doi:10.1099/ijs.0.02668-0

Pedrós-Alió C (2006) Genomics and marine microbial ecology. Int Microbiol 9:191–197

Pfennig N, Biebl H (1976) *Desulfuromonas acetoxidans* gen. nov. and sp. nov., a new anaerobic, sulfur-reducing, acetate-oxidizing bacterium. Arch Microbiol 110:3–12. doi:10.1007/BF00416962

Poehlein A, Schmidt S, Kaster AK, Goenrich M, Vollmers J, Thurmer A, Bertsch J, Schuchmann K, Voigt B, Hecker M, Daniel R, Thauer RK, Gottschalk G, Muller V (2012) An ancient pathway combining carbon dioxide fixation with the generation and utilization of a sodium ion gradient for ATP synthesis. PLoS One 7. doi:10.1371/journal.pone.0033439

Rabaey K, Verstraete W (2005) Microbial fuel cells: novel biotechnology for energy generation. Trends Biotechnol 23:291–298

Rabaey K, Lissens G, Siciliano SD, Verstraete W (2003) A microbial fuel cell capable of converting glucose to electricity at high rate and efficiency. Biotechnol Lett 25:1531–1535. doi:10.1023/A:1025484009367

Rabaey K, Boon N, Siciliano SD, Verhaege M, Verstraete W (2004) Biofuel cells select for microbial consortia that self-mediate electron transfer. Appl Environ Microbiol 70:5373–5382. doi:10.1128/AEM.70.9.5373-5382.2004

Ragauskas AJ, Williams CK, Davison BH, Britovsek G, Cairney J, Eckert CA, Frederick WJ, Hallett JP, Leak DJ, Liotta CL, Mielenz JR, Murphy R, Templer R, Tschaplinski T (2006) The path forward for biofuels and biomaterials. Science 311:484–489. doi:10.1126/science.1114736

Reguera G, McCarthy KD, Mehta T, Nicoll JS, Tuominen MT, Lovley DR (2005) Extracellular electron transfer via microbial nanowires. Nature 435:1098–1101. doi:10.1038/nature03661

Rezaei F, Richard TL, Brennan RA, Logan BE (2007) Substrate-enhanced microbial fuel cells for improved remote power generation from sediment-based systems. Environ Sci Technol 41:4053–4058. doi:10.1021/es070426e

Rezaei F, Richard TL, Logan BE (2008) Enzymatic hydrolysis of cellulose coupled with electricity generation in a microbial fuel cell. Biotechnol Bioeng 101:1163–1169. doi:10.1002/bit.22015

Rezaei F, Richard TL, Logan BE (2009) Analysis of chitin particle size on maximum power generation, power longevity, and Coulombic efficiency in solid–substrate microbial fuel cells. J Power Sources 192:304–309. doi:10.1016/j.jpowsour.2009.03.023

Ringeisen BR, Henderson E, Wu PK, Pietron J, Ray R, Little B, Biffinger JC, Jones-Meehan JM (2006) High power density from a miniature microbial fuel cell using *Shewanella oneidensis* DSP10. Environ Sci Technol 40:2629–2634. doi:10.1021/es052254w

Robb FT, Maeder DL, Brown JR, DiRuggiero J, Stump MD et al (2001) Genomic sequence of hyperthermophile, *Pyrococcus furiosus*: implications for physiology and enzymology. Hyperthermophilic Enzymes Pt A 330:134–157. doi:10.1016/S0076-6879(01)30372-5

Rother M, Metcalf WW (2004) Anaerobic growth of *Methanosarcina acetivorans* C2A on carbon monoxide: an unusual way of life for a methanogenic archaeon. Proc Natl Acad Sci USA 101:16929–16934. doi:10.1073/pnas.0407486101

Rozendal RA, Hamelers HVM, Euverink GJW, Metz SJ, Buisman CJN (2006) Principle and perspectives of hydrogen production through biocatalyzed electrolysis. Int J Hydrogen Energy 31:1632–1640. doi:10.1016/j.ijhydene.2005.12.006

Rozendal RA, Leone E, Keller J, Rabaey K (2009) Efficient hydrogen peroxide generation from organic matter in a bioelectrochemical system. Electrochem Commun 11:1752–1755. doi:10.1016/j.elecom.2009.07.008

Schicho RN, Ma K, Adams MWW, Kelly RM (1993) Bioenergetics of sulfur reduction in the hyperthermophilic archaeon *Pyrococcus furiosus*. J Bacteriol 175:1823–1830

Schippers A, Kock D, Höft C, Köweker G, Siegert M (2012) Quantification of microbial communities in subsurface marine sediments of the Black Sea and off Namibia. Front Microbiol 3:1–11. doi:10.3389/fmicb.2012.00016

Schroder C, Selig M, Schonheit P (1994) Glucose fermentation to acetate, CO_2 and H_2 in the anaerobic hyperthermophilic eubacterium *Thermotoga maritima*: involvement of the Embden-Meyerhof pathway. Arch Microbiol 161:460–470. doi:10.1007/BF00307766

Schut GJ, Adams MWW (2009) The iron-hydrogenase of *Thermotoga maritima* utilizes ferredoxin and NADH synergistically: a new perspective on anaerobic hydrogen production. J Bacteriol 191:4451–4457. doi:10.1128/JB.01582-08

Shieh WY, Chen A-L, Chiu H-H (2000) *Vibrio aerogenes* sp. nov., a facultatively anaerobic marine bacterium that ferments glucose with gas production. Int J Syst Evol Microbiol 50:321–329. doi:10.1099/00207713-50-1-321

Sipma J (2006) Microbial hydrogenogenic CO conversions: applications in synthesis gas purification and biodesulfurization. PhD thesis, Wageningen University.

Sokolova TG, González JM, Kostrikina NA, Chernyh NA, Tourova TP, Kato C, Bonch-Osmolovskaya EA, Robb FT (2001) *Carboxydobrachium pacificum* gen. nov., sp. nov., a new anaerobic, thermophilic, CO-utilizing marine bacterium from Okinawa Trough. Int J Syst Evol Microbiol 51:141–149. doi:10.1099/00207713-51-1-141

Sokolova TG, Jeanthon C, Kostrikina NA, Chernyh NA, Lebedinsky AV, Stackebrandt E, Bonch-Osmolovskaya EA (2004) The first evidence of anaerobic CO oxidation coupled with H_2

production by a hyperthermophilic archaeon isolated from a deep-sea hydrothermal vent. Extremophiles 8:317–323. doi:10.1007/s00792-004-0389-0

Sowers KR, Ferry JG (2003) Methanogenesis in the marine environment. Encyclopedia of environmental microbiology. Wiley, New York

Sowers KR, Baron SF, Ferry JG (1984) *Methanosarcina acetivorans* sp. nov., an acetotrophic methane-producing bacterium isolated from marine sediments. Appl Environ Microbiol 47:971–978

Spath PL, Dayton DC (2003) Preliminary screening – technical and economic assessment of synthesis gas to fuels and chemicals with emphasis on the potential for biomass-derived syngas. National Renewable Energy Laboratory, U.S. Department of Energy

Stetter KO (1988) *Archaeoglobus fulgidus* gen. nov., sp. nov. a new taxon of extremely thermophilic archaebacteria. Syst Appl Microbiol 10:172–173

Sun S, Chen J, Li W, Altintas I, Lin A, Peltier S, Stocks K, Allen EE, Ellisman M, Grethe J, Wooley J (2011) Community cyberinfrastructure for advanced microbial ecology research and analysis: the CAMERA resource. Nucleic Acids Res 39:D546–D551. doi:10.1093/nar/gkq1102

Svetlichny VA, Sokolova TG, Gerhardt M, Kostrikina NA, Zavarzin GA (1991) Anaerobic extremely thermophilic carboxydotrophic bacteria in hydrotherms of Kuril Islands. Microbiol Ecol 21:1–10

Tender LM, Reimers CE, Stecher HA, Holmes DE, Bond DR, Lowy DA, Pilobello K, Fertig SJ, Lovley DR (2002) Harnessing microbially generated power on the seafloor. Nat Biotechnol 20:821–825. doi:10.1038/nbt716

Tender LM, Gray SA, Groveman E, Lowy DA, Kauffman P, Melhado J, Tyce RC, Flynn D, Petrecca R, Dobarro J (2008) The first demonstration of a microbial fuel cell as a viable power supply: powering a meteorological buoy. J Power Sources 179:571–575. doi:10.1016/j.jpowsour.2007.12.123

Thauer RK, Jungermann K, Decker K (1977) Energy conservation in chemotrophic anaerobic bacteria. Bacteriol Rev 41:100–180

Torres CI, Marcus AK, Rittmann BE (2007) Kinetics of consumption of fermentation products by anode-respiring bacteria. Appl Microbiol Biotechnol 77:689–697. doi:10.1007/s00253-007-1198-z

Tront JM, Fortner JD, Plötze M, Hughes JB, Puzrin AM (2008) Microbial fuel cell biosensor for *in situ* assessment of microbial activity. Biosens Bioelectron 24:586–590. doi:10.1016/j.bios.2008.06.006

Venkateswaran K, Moser DP, Dollhopf ME, Lies DP, Saffarini DA, MacGregor BJ, Ringelberg DB, White DC, Nishijima M, Sano H, Burghardt J, Stackebrandt E, Nealson KH (1999) Polyphasic taxonomy of the genus *Shewanella* and description of *Shewanella oneidensis* sp. nov. Int J Syst Bacteriol 49:705–724. doi:10.1099/00207713-49-2-705

Venter JC, Remington K, Heidelberg JF, Halpern AL, Rusch D, Eisen JA, Wu D, Paulsen I, Nelson KE, Nelson W, Fouts DE, Levy S, Knap AH, Lomas MW, Nealson K, White O, Peterson J, Hoffman J, Parsons R, Baden-Tillson H, Pfannkoch C, Rogers YH, Smith HO (2004) Environmental genome shotgun sequencing of the Sargasso Sea. Science 304:66–74. doi:10.1126/science.1093857

Verhaart MRA, Bielen AAM, van der Oost J, Stams AJM, Kengen SWM (2010) Hydrogen production by hyperthermophilic and extremely thermophilic bacteria and archaea: mechanisms for reductant disposal. Environ Technol 31:993–1003. doi:10.1080/09593331003710244

Wang JL, Wan W (2009) Factors influencing fermentative hydrogen production: a review. Int J Hydrogen Energy 34:799–811. doi:10.1016/j.ijhydene.2008.11.015

Zhang T, Gannon SM, Nevin KP, Franks AE, Lovley DR (2010) Stimulating the anaerobic degradation of aromatic hydrocarbons in contaminated sediments by providing an electrode as the electron acceptor. Environ Microbiol 12:1011–1020. doi:10.1111/j.1462-2920.2009.02145.x

Zhang Y, Su X, Chen F, Wang Y, Jiao L, Dong H, Huang Y, Jiang H (2012) Microbial diversity in cold seep sediments from the northern South China Sea. Geosci Front 3:301–316. doi:10.1016/j.gsf.2011.11.014

Zhu D, Wang G, Qiao H, Cai J (2008) Fermentative hydrogen production by the new marine *Pantoea agglomerans* isolated from the mangrove sludge. Int J Hydrogen Energy 33:6116–6123. doi:10.1016/j.ijhydene.2008.07.008

Chapter 19
Bacteriocins: Natural Weapons for Control of Food Pathogens

Nabil Ben Omar, Hikmate Abriouel, Ismail Fliss, Miguel Ángel Ferandez-Fuentes, Antonio Galvez, and Djamel Drider

Abstract Research on antimicrobial peptides is continuously growing because of the possibilities of applications they offer in different domains including food safety, human medicine, and plant biocontrol (phytosanitary).

The present chapter is aiming to shed lights on diversity, function and structure of ribosomally synthesized antimicrobial peptides from Gram positive bacteria usually referred to as bacteriocins. In bacterial systems, competition is often driven by the production of bacteriocins; narrow spectrum proteinaceous toxins that serve to kill closely related species providing the producer better access to limited resources. Despite high levels of bacteriocin diversity, these proteins share many general characteristics. They are generally high molecular weight protein antibiotics that kill

N.B. Omar • H. Abriouel • A. Galvez
Área de Microbiología, Departamento de Ciencias de la Salud, Facultad de Ciencias Experimentales, Edif. B3, Universidad de Jaén,
Campus Las Lagunillas s/n. 23071, Jaén, Spain

I. Fliss
STELA Dairy Research Center, Nutraceuticals and Functional Foods Institute,
Université Laval, Québec, QC, Canada

M.Á. Ferandez-Fuentes
Área de Microbiología, Departamento de Ciencias de la Salud, Facultad de Ciencias Experimentales, Edif. B3, Universidad de Jaén,
Campus Las Lagunillas s/n. 23071, Jaén, Spain

Laboratoire des Procédés Biologiques, Génie Enzymatique et Microbien (ProBioGEM),
UPRES-EA 1026, Polytech'Lille/IUTA, Université Lille Nord de France,
Avenue Paul Langevin, 59655, Villeneuve d'Ascq Cedex, France

D. Drider (✉)
Laboratoire des Procédés Biologiques, Génie Enzymatique et Microbien (ProBioGEM),
UPRES-EA 1026, Polytech'Lille/IUTA, Université Lille Nord de France,
Avenue Paul Langevin, 59655 Villeneuve d'Ascq Cedex, France
e-mail: djamel.drider@univ-lille1.fr

A. Malik et al. (eds.), *Management of Microbial Resources in the Environment*,
DOI 10.1007/978-94-007-5931-2_19, © Springer Science+Business Media Dordrecht 2013

closely related strains or species. The bacteriocin gains entry into the target cell by recognizing specific cell surface receptors and then kills the cell by forming ion-permeable channels in the cytoplasmic membrane, by nonspecific degradation of cellular DNA, by inhibition of protein synthesis through the specific cleavage of 16s rRNA, or by cell lysis. In this chapter, the limits and performances of production will be presented. Further clear evidences on their aptitudes to master growth of microbes will be discussed as well as the main achievements and perspectives of their applications in food, environment and medical domains.

1 Lactic Acid Bacteria Bacteriocins: Brief Overview

This chapter is focused on the role of lactic acid bacteria (LAB) bacteriocins in the control of food pathogens, which constitute a serious public concern worldwide. LAB-bacteriocins are produced by strains belonging to the genera of *Lactococcus, Lactobacillus, Leuconostoc, Enterococcus, Pediococcus, Carnobacterium* and *Streptococcus*. In 1976, Tagg and coworkers have defined bacteriocins as proteinaceous compounds with inhibition activity against related bacteria. The continuing research on LAB-bacteriocins has shed light on their capacities of antagonism. Remarkably, LAB-bacteriocins were reported to be active against Gram negative bacteria such as *Campylobacter jejuni* (Cole et al. 2006; Stern et al. 2006; Nazef et al. 2008; Messaoudi et al. 2011). LAB bacteriocins should be defined as naturally and ribosomally-synthesized antimicrobial peptides displaying a large antagonism spectrum. The LAB bacteriocins are documented in an online database, named BACTIBASE (Hammami et al. 2010) that is available at http://bactibase.pfba-lab-tun.org. Overall, BACTIBASE provides physicochemical, structural, microbiological, and taxonomic informations about bacteriocins produced by both Gram-positive and Gram-negative bacteria.

LAB bacteriocins have been subjected to different classification schemes due to their biochemical and genetic diversities and their bioactivities. The first classification scheme was provided by Klaenhammer (1993). As new bacteriocin members were identified and being characterized, this classification was amended at different instances (Tagg et al. 1976; Cotter et al. 2005). In 2001, Cintas and collaborators have proposed a classification scheme including four main classes, among which: Class I lantibiotics are post-translationally modified, heat-stable, low molecular mass peptides (<5 kDa) characterized by the presence of unusual amino acids, such as lanthionine or β-methyllanthionine. Class II bacteriocins are small heat-stable, unmodified peptides (<10 kDa) and are subdivided into three subclasses, namely, class IIa (pediocin-like), class IIb (two-peptide), and IIc (other [i.e., non-pediocin-like], one-peptide bacteriocins). Class III bacteriocins are large (>30 kDa) and heat-labile proteins. Lastly, class IV bacteriocins include cyclic peptides with covalently linked N- and C-termini. Recently, an updated classification was reported by Rea et al. (2011). This novel classification is briefly described below.

1.1 Class Ia (Lantibiotics)

Bacteriocins of this class are <5 kDa and 28 amino acids in length. Lantibiotics undergo post-translational modifications leading to unusual aminoacids such as lanthionine (Lan), and/or B-methyllanthioonine (meLan) and dehydroalanine (Dha). The linear or type A lantibiotics comprise bacteriocins such as nisin, subtilin or epidermin. The type A lantibiotics are known to be elongated, cationic and amphiphilic and could contain until 34 amino acids in length. They act by pore forming leading to the death of the target cell upon a cascade of damages like the dissipation of membrane potential and efflux of small molecules. The globular lantibiotics or type B lantibiotics such as merscaidin, mutacin and lacticin 481 are structurally more compact, they are small peptides less than 19 aminoacids. Their mode of action is based on the inhibition of lipid II, which is the key precursor of peptidoglycan in the cell wall. Remarkably, nisin, which is a linear lantibiotic (type A lantibiotic) could act by pore forming or lipid II inhibition (Breukink et al. 1999). As this classification has become misleading, the novel classification of lantibiotics contains four subclasses (subclass I, subclass II, subclass III and subclass IV).

1.2 Class Ib (Labyrinthopeptins)

Recently identified (Meindl et al. 2010), these peptide are characterized by their "labyrinthine" structure and the presence of "labionin", which is a carbocyclic, post-translationally modified amino acid.

1.3 Class Ic (Sactibiotics)

Bacteriocins of this class are subtilosin A and Thuricin CD produced by *Bacillus subtilis* and *Bacillus thuringiensis* 6431, respectively. Subtilosin A is a circular peptide, post-translationally modified with cross-linkages between the sulphur and cystein residues. Thuracidin CD is a dipeptide (Trnα, Trnβ) with intramolecular crosslinkages between three cysteine residues in each peptide and the α-carbons.

1.4 Class II: Unmodified Bacteriocins

We find in this class peptides less than 10 kDa with linear or cyclic structures. This class has been committed to intensive investigation and despite its heterogeneous traits, different classifications were suggested. Currently, four subclasses (Class IIa, Class IIb, Class IIc and Class IVd) are proposed and they are briefly described below.

1.4.1 Class IIa (Pediocin-Like Bacteriocins)

Bacteriocins of this subclass are certainly of major interest because of their potential as food preservatives but also as alternatives to current antibiotics. Class IIa bacteriocins display activity against *Listeria* strains and even against other pathogens such as *Clostridium* spp. There are more than 30 bacteriocins in this repertoire with specificities. Class IIa bacteriocins are ranking from 37 to 57 amino acids in length and all of them share a YGNGV box at the N-terminal moitiey. Class IIa bacteriocins are also named cystibiotics because they contain at least two cysteines in the C-terminal part leading to disulphide bond formation. It happens that models of class IIa bacteriocins such as divercin V41 could harbor four cystein residues, in which two residues are located in the C-terminal part and the other two in the N-terminal parts, respectively.

1.4.2 Class IIb (Two Peptide Bacteriocins)

Class IIb bacteriocins consist of two distinct peptides, which are necessary to obtain high antimicrobial activity. The antimicrobial activity requires the presence of both peptides at equal amounts. There are at least 16 bacteriocins nowadays known as class IIb bacteriocins. Studies of the mode of action of class IIb bacteriocins have revealed a leakage in the membrane of the sensitive target bacteria. However, specificities in the mode of action, mainly in the movement of the ions across the membrane, have been reported for plantaricin E/F and plantaricin J/K. All class IIb bacteriocins comprise between 30 and 50 aminoacids, they are cationic, amphiphatic, membrane active and synthesized as pre-peptide cut of and vehiculed outcell by the dedicated ABC transporter. It was also demonstrated that the synthesis of class IIb bacteriocins is regulated in some bacteria by three component regulatory systems.

1.4.3 Class IIc (Circular Bacteriocins)

Structurally, the circular bacteriocins are characterized by the head –to-tail cyclization of their backbone. They are produced by LAB as well as by non-LAB strains. The best characterized are Enterocin AS-48, grassericin, carnocyclin A and lactocyclicin A. They are known to have potent antimicrobial activity, which is thought to be attributed to their circular structure. In this sense, it has been established that the enzymatic hydrolysis of Enterocin AS-48 by thermolysin releases a linear component lacking bioactivity despite its helical structure. Overall, circular bacteriocins (class IIc bacteriocins) display broad spectra including activity against food spoilage pathogens. Enterocin AS-48 and lactocyclicin Q are also active against Gram negative bacteria.

1.4.4 Class IId (Linear and Non-pediocin Like Bacteriocins)

Class IId bacteriocins have no significant similarities to the other class II bacteriocins. Class IId bacteriocins are synthesized by the *sec*-independent double glycine motif and then are transported by ABC transporters. Some of the class IId bacteriocins are synthesized without an N-terminal leader sequence or signal peptide. For this reason, they are named “leaderless bacteriocins”.

Enterocin L50, produced by *Enterococcus faecium* L50, consists of two peptides named Enterocin L50A (EntL50A) and Enterocin L50B (EntL50B), which are highly similar (70%).

1.5 Bacteriolysins

They were formerly denominated class III bacteriocins. Overall, they are large, heat labile proteins. In this class, we can find helveticin J from *Lactobacillus helveticus* J and enterolysin from *Enterococcus faecalis*. It should be noted that other class IId bacteriocins from non LAB exist, such as zoocin A, linocin and millericin.

2 Bacteriocins and Control of Food Pathogens

In the last years, bacteriocins have been considered very promising agents for fighting foodborne pathogens (García et al. 2010; Mills et al. 2011). Even though bacteriocins are produced by either Gram-positive or Gram-negative bacteria, the most accepted peptides for food preservation or even clinical applications are those produced by Gram-positive bacteria, especially lactic acid bacteria (LAB). Those bacteriocins are produced by a bacterial group which is generally present in different foodstuffs and even intensively used in foods, moreover, LAB are generally recognized as safe (GRAS status) (Carr et al. 2002; Pedersen et al. 2005) and their Qualified Presumption of Safety (QPS) has been proposed by the EFSA (2007). These peptides were described in detail and many applications have been proposed (Acuña et al. 2010; De Vuyst and Leroy 2007). Although a number of peptides were described in the literature so far, nisin and pediocin PA-1 are better positioned than the other peptides as food preservatives.

This book chapter focuses on the role ascribed to the LAB bacteriocins in the control of foodborne pathogens able to grow in different food matrices. The main bacterial pathogens of concern in food industry are those able to survive and multiply in the raw materials, such as *Listeria monocytogenes*, *Staphylococcus aureus*, *Escherichia coli, Salmonella* spp., *Bacilli* and *Clostridia*.

2.1 *Listeria monocytogenes*

Through the last two decades, many studies have been carried out to search for bacteriocin producing strains able to inhibit *Listeria monocytogenes*. The application of bacteriocins or bacteriocin producing starter cultures in food can provide an additional hurdle to *L. monocytogenes* and can possibly ensure the safety of food products in the future. Among all bacteriocins tested, nisin (in the commercial form Nisaplin) has been tested extensively in foods. In the dairy industry, nisin has found many applications, especially in processed cheeses and cheese products (e.g., hard cheese, soft white cheeses, slices, spreads, sauces, dips) to prevent proliferation of *L. monocytogenes* (Davies and Delves-Broughton 1999; Thomas and Delves-Broughton 2001). In skim milk, whey, or simulated milk ultrafiltrate media, the use of a combination of nisin and pulsed electric fields (PEFs) resulted in a significant inhibition of *L. monocytogenes* (Calderón-Miranda et al. 1999).

Addition of nisin to pasteurized liquid whole egg reduced the viable counts of *L. monocytogenes* and increased the shelf-life of the refrigerated product (Delves-Broughton et al. 1992; Knight et al. 1999; Schuman and Sheldon 2003). Both nisin and pediocin PA-1/Ach acted synergistically with heat treatments against *L. monocytogenes* (Knight et al. 1999; Muriana 1996) in liquid whole egg and in egg white during pasteurization (Boziaris et al. 1998).

Although the application of nisin in meats is limited due to several factors such as its poor solubility, interaction with phospholipids, and inactivation by glutathione (Montville et al. 1995; Rayman et al. 1983; Rose et al. 1999; Stergiou et al. 2006), addition of nisin has shown to extend the lag phase of *L. monocytogenes* inoculated into minced buffalo meat (Pawar et al. 2000). In meat marketing, prime cuts are often vacuum-packaged in order to extend their shelf-life during distribution before preparation of retail cuts. As a result of these practices, meat can become contaminated with *L. monocytogenes* and spoilage bacteria that may shorten the shelf-life of retail meats. Under modified atmosphere packaging, nisin was able to completely inhibit the growth of *L. monocytogenes* in pork (Fang and Lin 1994a, b). In sausages and other fermented meat products, addition of nisin induced a significant inhibition of *L. monocytogenes*. The lower pH of sausages compared to fresh meats may increase the solubility of nisin, and probably the antimicrobial activity as well. Addition of nisin alone was effective in inhibiting *L. monocytogenes* in sucuk, a Turkish fermented sausage (Hampikyan and Ugur 2007).

The effectiveness of nisin in sausages increases in combination with other antimicrobials. The combination of nisin and a grape seed extract showed an enhanced antibacterial activity in refrigerated turkey frankfurters with reduction of *L. monocytogenes* populations to undetectable levels (Sivarooban et al. 2007). In ham and/or bologna sausages, a mixture of nisin-lysozyme-EDTA inhibited the growth of *Leuconostoc mesenteroides* and *L. monocytogenes* (Gill and Holley 2000a, b). Activity of nisin against *L. monocytogenes* in minced beef was potentiated with thyme essential oil, decreasing the impact of the oil on the meat organoleptic properties (Solomakos et al. 2008).

Although effective, the addition of bacteriocin as food ingredients may be limited by the degradation of the active compound by different bacterial proteases especially in fermented foods or by the interaction of the bacteriocin with several food compounds such as fat. The use of bacteriocin-producing starter or adjunct cultures in foods represents therefore an efficient alternative to overcome these problems. It may also significantly reduce costs associated with nisin production, purification and processing. Nisin-producing strains have been reported to inhibit *L. monocytogenes* in several types of cheeses, such as cheddar cheese (Benech et al. 2003), cottage cheese (Benkerroum and Sandine 1988) or Camembert (Maisnier-Patin et al. 1992; Sulzer and Busse 1991). In Manchego cheese made from raw ewe's milk, *Lactococcus lactis* subsp. *lactis* ESI 515 reduced viable counts of *L. innocua* by 4.08 log units after 60 days of ripening, and the produced nisin was detected in cheese through the ripening period.

In vegetables, *L. monocytogenes* can proliferate rapidly and several listeriosis outbreaks have been associated with fresh produce, such as raw celery, tomatoes, and lettuce (Beuchat 1996). Leverentz et al. (2003) showed that nisin reduced *L. monocytogenes* populations on honeydew melon slices and apple slices. The listericidal effect was enhanced by application of nisin in combination with a phage mixture (Leverentz et al. 2003). Exposure of *L. monocytogenes* Scott A to nisin in tofu resulted in an initial reduction of viable counts followed by regrowth of survivors to nisin during further incubation (Schillinger et al. 2001). Nisin was tested alone or in combination with sodium lactate, potassium sorbate, phytic acid, and citric acid as possible sanitizer treatments for reducing the population of *L. monocytogenes* on cabbage, broccoli, and mung bean sprouts (Bari et al. 2005). After a 1-min wash, a significant reduction of *L. monocytogenes* was observed on cabbage and broccoli, with nisin-phytic acid combination (Bari et al. 2005).

Lacticin 3147 produced by *L. lactis* subsp. *lactis* DPC3147 is another bacteriocin with a high potential for application in the preservation of foods (Ross et al. 1999). Lacticin 3147 powder was shown to rapidly inactivate *L. monocytogenes* Scott A in an infant milk formulation (Morgan et al. 1999). In natural yogurt and in cottage cheese supplemented with lacticin 3147 powder, viable cell numbers of *L. monocytogenes* were reduced by 99 and by 85%, respectively, within 2 h (Morgan et al. 2001). An increased bactericidal effect was reported for the combined treatment of lacticin 3147 concentrates and HHP against *L. monocytogenes* in milk and whey (Morgan et al. 2000). The lactocin 705 is another bacteriocin produced by *Lactobacillus casei* CRL 705 (Vignolo et al. 1996) which was highly effective against *L. monocytogenes* in beef slurry (Vignolo et al. 1998) and also in a meat system when used in combination with enterocin CRL35 produced by *E. faecium* CRL35 (Farías et al. 1994) and nisin (Vignolo et al. 2000).

Application of enterococcal bacteriocins on foods to inhibit the growth of *L. monocytogenes* has been the focus of many investigations (reviewed by Foulquié Moreno et al. 2006; Giraffa 1995). An early report indicated that bacteriocin from *E. faecium* DPC1146 had a rapid bactericidal effect on *L. monocytogenes* in milk (Parente and Hill 1992). A decrease of viable *L. monocytogenes* was also reported in enterocin-added yogurt and in Saint-Paulin cheese (Lauková et al. 2001).

However, after 6 weeks and at the end of the experiment, the difference in surviving listeria was only 1 or 0.7 log units compared to the control cheese (Lauková et al. 2001). In "bryndza" (a traditional Slovak dairy product from sheep milk), the addition of enterocin CCM 4231 has reduced the levels of *L. monocytogenes* Li1 during a 7-day ripening period (Lauková and Czikková 2001). In contrast, when concentrated enterocin CRL35 was added to goat cheese, the population of *L. monocytogenes* diminished by 9 log units by the end of the ripening period without affecting the cheese quality (Farías et al. 1999). Similarly, cultured broths obtained from raw ewe's milk containing enterocin 4 (enterocin AS-48) significantly reduced viable counts of *L. monocytogenes* (Rodríguez et al. 1997), whilst in soy milk the enterocin CCM 4231 has completely eliminated *L. monocytogenes*. The enterococcal *faecal* CCM4231 was able to grow and produce enterocin in soy milk (Lauková and Czikková 1999). In fermented meat, enterocins can inhibit *Listeria*, as shown for enterocin CCM 4231 when incorporated in dry fermented Hornád salami (Lauková et al. 1999c), and enterocins A and B in espetect (traditional Spanish sausages; Aymerich et al. 2000). In a meat sausage model system, added enterocin AS-48 inhibited the growth of *L. monocytogenes* (Ananou et al. 2005a, b).

Bacteriocinogenic enterococci could be used as cocultures for preservation of meat products (e.g., fermented sausages and sliced vacuum-packed cooked meat products) and for the control of emergent pathogenic and spoilage bacteria (Foulquié Moreno et al. 2006; Hugas et al. 2003). When used as starter cultures in sausage fermentation, the bacteriocinogenic strains *E. faecium* CCM 4231 and *E. faecium* RZS C13 were partially competitive and strongly inhibited the growth of *Listeria* spp. (Callewaert et al. 2000). *E. faecium* CTC492 (producer of enterocins A and B) partially prevented ropiness due to *Lactobacillus sakei* CTC746 in sliced vacuum-packaged cooked ham (Aymerich et al. 2002). The strain *E. casseliflavus* IM 416K1 (producer of enterocin 416 K1) was able to eliminate *L. monocytogenes* in artificially inoculated "cacciatore" Italian sausages (Sabia et al. 2003). The cyclic bacteriocin enterocin AS-48 produced *in situ* by an *E. faecalis* strain or a food-grade *E. faecium* transconjugant controlled the growth of *L. monocytogenes* in a meat model system (Ananou et al. 2005a, b).

On the other hand, a limited number of studies have focused on the application of pediocin-producing strains in dairy foods, given the poor adaptation of pediococci to dairy substrates. Early experiments indicated that inhibition of *L. monocytogenes* in milk required a high cell concentration of pediococci (Raccach and Geshell 1993). For this reason, genetically engineered pediocin-producing LAB were developed, such as *L. lactis* subsp. *lactis* or the yogurt starter culture *Streptococcus thermophilus* (Coderre and Somkuti 1999; Somkuti and Steinberg 2003). In cheddar cheese, the pediocin PA-1 producer *L. lactis* subsp. *lactis* MM217 reduced the counts of inoculated *L. monocytogenes* from 10^6 to 10^2 CFU/g within 1 week of ripening, and then to about 10 CFU/g within 3 months (Buyong et al. 1998). It was concluded that pediocin-producing starter cultures have significant potential for protecting cheese against *L. monocytogenes* (Buyong et al. 1998). In a more recent study, the pediocin PA-1-producing derivatives *L. lactis* CL1 and *L. lactis* CL2 also reduced the counts of *L. monocytogenes* during cheese ripening (Rodríguez et al. 2005).

Similarly, heterologous production of pediocin PA-1/AcH in nisin-producing and non nisin-producing *L. lactis* strains previously selected as starters because of their technological properties for cheese making reduced viable counts of *L. monocytogenes* in cheese below 50 or 25 CFU/g at the end of the ripening period (Reviriego et al. 2007).

Pediocin production has also been reported in non-pediococcal LAB from dairy environments. Spraying with a cell suspension of the pediocin AcH producer strain *Lactobacillus plantarum* WHE92 on the surface of Muenster cheese was reported to prevent the growth of *L. monocytogenes* (Ennahar et al. 1998). Because *L. plantarum* WHE 92 exists naturally in Muenster cheese, it did not adversely affect the ripening process (Ennahar et al. 1998). On red smear cheese, an almost complete inhibition of *L. monocytogenes* by pediocin-producing *L. plantarum* was also reported (Loessner et al. 2003). However, pediocin-resistant listeria were readily detected, they were able to proliferate in the cheese, regardless of the produced bacteriocin. It was concluded that the continuous use of pediocin AcH does not appear to be suitable as a primary means of food preservation (Loessner et al. 2003).

Besides the previously mentioned bacteriocins, there are other bacteriocins which have been shown to be very effective against listeriae. Among those bacteriocins, we can mention propionicin PLG-1 (produced by *Propionibacterium thoenii* P127) (Lyon and Glatz 1993) which was shown to kill or inhibit several psychrotrophic spoilage or pathogenic bacteria including *L. monocytogenes, Pseudomonas fluorescens, Vibrio parahaemolyticus, Yersinia enterocolitica*, and *Corynebacterium* sp., suggesting its potential use as an antibacterial food preservative (Lyon et al. 1993). The bacteriocin-producer strain *Streptococcus salivarius* subsp. *thermophilus* B was tested as a thermophilic starter in yogurt to control *L. monocytogenes.* Use of the Bac+starter was reported to extend the product shelf-life by 5 days (Benkerroum et al. 2002).

2.2 *Escherichia coli and Salmonella* spp.

Escherichia coli and *Salmonella enterica* are of major concern in a wide variety of foods that have not undergone a germ reducing process. Enteric bacteria are especially tolerant towards adverse environmental conditions such as low pH, high salt concentrations (Small et al. 1994; Cheville et al. 1996; Brown et al. 1997) and have been shown to survive during storage in acidic (or low pH) foods or products with high concentrations of salt or organic acids (Presser et al. 1998; Glass et al. 1992; Leyer et al. 1995; Reitsma and Henning 1996).

Several reports suggest that bacteriocins of LAB may contribute to the inactivation of Gram-negative microorganisms in foods if these are applied in combination with chelating agents (Shefet et al. 1995; Scannell et al. 1997). The architecture of the outer membrane (OM) of Gram-negative organisms prevents penetration of the bacteriocins to their target cells, the cytoplasmic membrane, and therefore confers a high degree of resistance (Stevens et al. 1991; Schved et al. 1994). Chelating agents such as EDTA as well as the application of sublethal stress such as heating or freezing were shown to disrupt the permeability barrier of the LPS leading to an increased sensitivity of Gram-negative bacteria towards LAB bacteriocins (Stevens et al. 1991; Kalchayanand et al. 1992; Cutter and Siragusa 1995; Murdock et al. 2007).

Nisin was proposed as a hurdle in association with chelating agents for controlling *Salmonella* and *E. coli*. The first report using this approach conclusively showed that at least 20 *Salmonella* serovars were inhibited with simultaneous treatment of 50 μg/ml nisin and 20 mM EDTA. In fact, the population was reduced by up to 5.3 log CFU units after an hour of treatment. Neither EDTA nor nisin alone were able to inhibit the growth of *Salmonella* (Stevens et al. 1991). Later, Cutter and Siragusa (1995) showed a significant inhibition effect by combining 50 μg/ml nisin with different chelators such as 500 mM lactate, 100 mM citrate, 50 mM EDTA or 1% (w/v) sodium hexametaphosphate in buffer. *Salmonella typhimurium* population was reduced up to 5.5 log CFU units. On the other hand, the combination of 1 μM nisin (~3.35 μg/ml) with 0.5–5 mM trisodium phosphate was successfully used in controlling *S. enteritidis*. Cell counts were reduced by 6 log CFU units after only 30 min of treatment (Carneiro de Melo et al. 1998). The main flaw of these reports though, is that experiments were performed under cell starvation conditions, which seem to constitute a different model from food products and hence the conclusions may not be accurate. Cutter and Siragusa (1995) only got 0.4-log units reduction of *Salmonella* upon nisin-lactate treatment when experiments were carried out in beef instead of buffer. In the same trend, Carneiro de Melo et al. (1998) only found over 1 log unit reduction when nisin–trisodium phosphate was applied in chicken skins instead of buffer. Furthermore, although Branen and Davidson (2004) working with trypticase soy broth instead of buffer they still observed a synergistic effect with nisin–EDTA combination on two pathogenic *E. coli* strains, the effect on *S. enteritidis* was not synergistic at all. However, they did observe some bactericidal activity with a minimal bactericidal concentration of 46.9 μg/ml nisin and 1.25 mg/ml EDTA (~3.4 mM). It is important to note that Branen and Davidson (2004) used a low concentration of the chelator, which does not allow direct comparisons between them. Besides, it was shown that *S. typhimurium* OM was stabilized by nisin pre-treatment when cells were suspended in 0.1 mM EDTA. Therefore, nisin should not be used in combination with chelating agents at low concentrations. On the other hand, high concentrations of these agents would be able to completely disrupt the OM and therefore, nisin would reach the inner membrane (IM) and exert its bactericidal effect (Helander and Mattila-Sandholm 2000).

Nisin also showed a positive effect in association with other chemicals. For example in ham and/or bologna sausages, a mixture of nisin-lysozyme-EDTA inhibited the growth of *E. coli* O157:H7 (Gill and Holley 2000a). In fresh pork sausages, a combination of nisin and organic acids reduced the viable counts of *Salmonella* Kentucky and *S. aureus* (Scannell et al. 1997). In apple juice, a combination of nisin and cinnamon accelerated death of *Salmonella Typhimurium* and *E. coli* O157:H7, enhancing the safety of the product (Yuste and Fung 2004). When nisin and lysozyme were tested for inactivation of *Salmonella typhimurium* in orange juice in combination with PEFs, the combination of the two antimicrobials had a more pronounced bactericidal effect than either nisin or lysozyme alone (Liang et al. 2002).

The combined effect of bacteriocins and other hurdles had been shown to be highly effective against Gram-negative bacteria, such as nisin and curvacin A in combination with low pH, 5% NaCl, or propylparabene. This combination also leads to an increased sensitivity of *E. coli* and *S. enterica* towards nisin and curvacin A (Gänzle et al. 1999). These results suggested that bacteriocins may be active against *E. coli* at environmental conditions near the growth limiting factor levels even if a functional outer membrane is present. In another study, an inhibitory synergetic effect was obtained against *Salmonella enteritidis* PT4 in liquid whole egg and in egg white when both nisin or pediocin Pa1/Ach were applied in combination with pasteurization (Boziaris et al. 1998). Sakacin P produced by *L. sakei* strains Lb674 and LHT673 (Holck et al. 1994; Tichaczek et al. 1994) also acted synergistically against *E. coli* when tested in combination with the fish antimicrobial peptide pleurocidin (Lüders et al. 2003).

Recently *in vitro* experiments (in growth media) performed by Smaoui et al. (2010) showed that BacTN635, a peptide produced by *Lactobacillus plantarum* TN635 was able to kill *Salmonella.* The plantaricin-producing strain *L. plantarum* 2.9 (isolated from ben saalga, a traditional pearl millet fermented food from Burkina Faso) produced a strong inhibitory activity in malted millet flour, decreasing the survival of *E. coli* O157:H7, and *S. enterica* (Valenzuela et al. 2008). This strain could be used as a starter culture to improve the safety of ben saalga (Ben Omar et al. 2006). Paracin 1.7, a bacteriocin produced by *Lactobacillus paracasei* HD1.7, was also reported to be effective in inhibiting *Salmonella* (Ge et al. 2009).

Regarding enterocins, enterocin AS-48 has been shown to be active against some Gram-negative bacteria especially when combined with outer-membrane permeabilizing agents (Abriouel et al. 1998). This bacteriocin was evaluated on *S. choleraesuis* LT2 in combination with EDTA and Tris. The cell survival was reduced proportionally to the enterocin concentration. This positive effect could be enhanced either by using acidic (pH 4) or alkaline conditions (pH 9) or mild heat treatment (Abriouel et al. 1998). Both nisin and enterocin AS-48 were successfully used for surface decontamination of fruits and vegetables. In this regard, a positive effect of nisin-chelating agent treatments in foods was reported by Ukuku and Fett (2004). They combined 50 μg/ml nisin with 20 mM EDTA, 3% sodium lactate or 2% potassium sorbate as sanitizer treatments on whole and fresh-cut cantaloupe. All the combinations reduced *Salmonella* by 3 log units/cm^2 at day 0 of treatment and by 2 log units/cm^2 after 3–7 days of treatment. Combined treatment of enterocin AS-48 with each of the following preservatives (lactic, polyphosphoric and peracetic acids, sodium hypochlorite, hexadecylpyridinium chloride and hydrocinnamic acid) significantly reduced (P 0.05) the *S. enterica* counts during storage at 15°C for at least 48 h. This synergistic effect was further explored for other Gram-negative bacteria (*E. coli* O157:H7, *S. sonnei, Shigella flexneri, E. aerogenes, Y. enterocolitica, A. hydrophila* and *P. fluorescens*), using a combination treatment of enterocin AS-48 and polyphosphoric acid. The enterocin AS-48 (25 μg/ml) alone did not significantly (P 0.05) reduce the growth of these bacteria during storage. However, the combination of enterocin AS-48 (25 μg/ml) and polyphosphoric acid (0.1–0.2% range) significantly reduced the viable counts of all of the above mentioned Gram-negative bacteria during storage at 6 and 15°C, as compared to treatment with either AS-48 or polyphosphoric acid alone. These results indicate the potential of combination treatments of AS-48 along with other preservatives to effectively control Gram-negative bacteria in vegetable foods (Molinos et al. 2008).

Another enterocin which has shown best results against Gram-negative bacteria was enterocin E 50–52, a pediocin-like bacteriocin produced by *E. faecium* (NRRL B-30746) that has been shown to be very effective in controlling *S. enteritidis* as well as *E. coli* O157:H7 among others. It should be noted that the anti-*Salmonella* activity was not only demonstrated *in vitro* but also in therapeutic tests in chickens (Svetoch et al. 2008). Kang and Lee (2005) reported that an enterocin P-like bacteriocin produced by *E. faecium* GM-1 had a broad antimicrobial spectrum including *S. typhimurium*. This finding is in sharp contrast to enterocin P itself, which is unable to kill Gram-negative bacteria (Cintas et al. 1997). Another recently reported bacteriocin with broad inhibitory spectrum is the enterocin produced by an *E. faecium* strain isolated from mangrove environment. In particular, it was shown that this enterocin was active against *S. paratyphi* (Annamalai et al. 2009). Ferreira et al. (2007) screened 70 strains of *Enterococcus mundtii* and found that only four of them produced bacteriocins active against *Salmonella Enteritidis*. These bacteriocins were only partially purified and characterized so far (Ferreira et al. 2007). One of the newest bacteriocins from LAB reported to be active against *Salmonella* spp. is lactococcin BZ, which was produced by *L. lactis* subsp. *lactis* BZ. This peptide is relatively heat labile because its activity was abolished after 15 min at 110°C. It is also sensitive to beta mercaptoethanol. The anti-*Salmonella* activity did not seem to be that high, at least as compared to other food pathogens tested (Şahingil et al. 2011). Enterocin 012 and acidophilin 801 are among the few long known bacteriocins able to kill *Salmonella*. On one hand, acidophilin 801, a bacteriocin produced by *L. acidophilus* IBB 801 with a very narrow inhibitory spectrum, surprisingly inhibits *Salmonella Panama* 1467 and *E. coli* Row (Zamfir et al. 1999; Jennes et al. 2000). On the other hand, enterocin 012 is a bacteriocin produced by *Enterococcus gallinarum* 012, a strain isolated from the duodenum of an ostrich. Remarkably, enterocin O12 has a lytic activity to the bactericidal or bacteriostatical activities usually reported for the LAB-bacteriocins (Jennes et al. 2000).

2.3 Endospore-Forming Bacteria

Endospore-forming bacteria represent an important threat to the safety and shelf-life of many foods and foodstuffs, because endospores may survive heat treatments applied to foods. After germination under suitable storage conditions, the resulting vegetative cells may propagate and produce food-poisoning toxins or cause food spoilage. Frequently, the cooking process does not inactivate heat-resistant bacterial spores and, consequently, endospore-forming *Bacillus* spp. and *Clostridum* spp. can be found in the final products, even during storage at low temperature (5°C) (Gould 1995). *Clostridium botulinum* and *Bacillus cereus* belong to these genera that are reported as the cause of several food-borne outbreaks (Angulo et al. 1998; Doan and Davidson 2000), thus, their inhibition by bacteriocins has a high significance.

In dairy food, nisin has been tested extensively. One of the earliest applications was to prevent gas blowing in cheese caused by *C. tyrobutyricum* (De Vuyst and Vandamme 1994; Hirsch et al. 1951). More recently, a strain of *L. lactis* ssp. *lactis* IPLA 729 isolated from raw milk cheese producing the natural variant nisin Z was reported to reduce the levels of the spoilage strain *C. tyrobutyricum* CECT 4011 in Vidiago cheese (a semihard farmhouse variety manufactured in Asturias, Northern Spain) during ripening. The produced nisin Z activity was stable in the cheese at least until 15 days of ripening. The nisin-producing strain was used in combination with a suitable starter to achieve desired acidification (Rilla et al. 2003).

Nisin has been tested for its useful contribution to control *Bacillus* and *Clostridium* growth in potato-based products by Thomas et al. (2002). Addition of 6.25 μg of nisin per gram of cooked mashed potatoes retarded the growth of *B. cereus* and *B. subtilis*, previously inoculated in the product not vacuum packaged, for at least 27 days at 8°C and the growth of *C. sporogenes* and *Clostridium tyrobutiricum*, added as spores in the product and then vacuum packaged, for at least 58 days at 25°C. Nisin remained at active levels after pasteurization, but the authors highlighted that, in order to be effective against temperature abuse and in extending shelf-life of final products, nisin must be well mixed to the various ingredients.

Incorporation of nisin in canned vegetables can prevent spoilage caused by nonaciduric (*Bacillus stearothermophilus* and *Clostridium thermosaccharolyticum*) and aciduric (*Clostridium pasteurianum, Bacillus macerans* and *Bacillus coagulans*) spore formers (Thomas et al. 2000). Nisin was also an effective preservative in fresh pasteurized "home-made"–type soups (Thomas et al. 2000) and in the control of *Bacillus* and *Clostridium* in cooked potato products (Thomas et al. 2002). In one example, in nisin-added, pasteurized, vacuum-packaged mashed potatoes inoculated with a cocktail of *Clostridium sporogenes* and *C. tyrobutyricum* spores, no bacterial growth was observed and the shelf-life of the mashed potatoes was extended by at least 30 days (Thomas et al. 2002). Similar results were reported in trials involving a cocktail of *B. cereus* and *B. subtilis* strains (Thomas et al. 2002). In heat-treated cream, growth of *Bacillus cereus* during storage was completely inhibited by low concentrations of nisin (Nissen et al. 2001; Pol et al. 2001).

Concerning the effect of nisin on Gram-positive spores like *Bacillus* and *Clostridium* spp., several reports showed that spores were particularly susceptible to nisin, being more sensitive than vegetative cells (Delves-Broughton et al. 1996). Nisin action against spores was caused by binding to sulfhydryl groups of protein residues (Morris et al. 1984). It was observed that spores became more sensitive to nisin the more heat damaged they are, and it is an important factor in the use of nisin as a food preservative in heat processed foods. For example, spores of *Clostridium anaerobe* PA3679 which have survived heat treatment of 3 min at 121.1°C were 10 times more sensitive to nisin than those which had not been heat damaged (Delves-Broughton et al. 1996). Sensitivity of spores to nisin varied, those of species like *Bacillus stearothermophilus* and *Clostridium thermosaccharolyticum* being particularly susceptible, as were all spores which open their coats by mechanical rupture.

Enterocin AS-48 added to a rice-based infant formula dissolved in whole milk completely inactivated *B. cereus* and prevented its growth for at least 15 days at 37°C (Grande et al. 2006). Enterocin AS-48 was also able to suppress *B. coagulans* vegetative cells in tomato paste, syrup from canned peaches, and juice from canned pineapple for at least 15 days of storage at 37°C (Lucas et al. 2006). In a nonfat hard cheese, the strain *E. faecalis* A-48-32 produced enough enterocin AS-48 to inhibit *B. cereus* and reduce the cell counts of bacilli by 5.6 log units after 30 days of ripening (Muñoz et al. 2004). Growth of starter cultures used in cheese making was not affected by the bacteriocin-producing strain. Similarly, the same strain A-48-32 successfully inhibited *B. cereus* in skim milk (Muñoz et al. 2004, 2007).

The enterocin EJ97 produced by *E. faecalis* EJ97 (Gálvez et al. 1998) had a bactericidal effect on *Bacillus macroides*/*Bacillus maroccanus* after several incubation conditions (4 h at 37°C, 24 h at 15°C and 48 h at 4°C); its activity was reduced at pH 5.0 and 9.0 and enhanced by sodium nitrite, sodium benzoate, sodium lactate and sodium tripolyphosphate. The *in situ* efficacy of pure enterocin EJ97 was obtained with a 10-fold higher concentration, whereas no inhibition was detected with the application of *E. faecalis* EJ97 as a developing bacterium in purée, although it was able to produce the bacteriocin *in situ*. Thus, the enterocin EJ97 has a potential to preserve food spoiled by *B. macroides*/*B. maroccanus* if used in concentrated pure form.

Thermophilin from *Streptococcus thermophilus* ST580 is active against *C. tyrobutyricum* (Mathot et al. 2003). Strain ST580 could be used as thermophilic starter for hard cheese making because the bacteriocin is not active against thermophilic lactobacilli. Furthermore, curds made with strain ST580 and inoculated with *C. tyrobutyricum* endospores showed no gas production for up to 20 days (Mathot et al. 2003). The strain *S. macedonicus* ACA-DC 198 isolated from Greek Kasseri cheese produced the food-grade lantibiotic macedocin in skim milk supplemented with nitrogen sources (Georgalaki et al. 2002; Tsakalidou et al. 1998) as well as in cheese (Anastasiou et al. 2007; Van den Berghe et al. 2006). Since macedocin showed inhibitory activity toward *C. tyrobutyricum*, it could be used as a nisin substitute to inhibit gas formation in cheese (Georgalaki et al. 2002). O'Mahony et al. (2001) showed that added variacin, a bacteriocin produced by *Kocuria varians* (in the form of a milk-based ingredient) inhibited the proliferation of *B. cereus* in chilled dairy products, vanilla, and chocolate desserts in a concentration-dependent way.

Alicyclobacillus acidoterrestris is a spore-forming bacterium known to cause problems in fruit juices and fruit juice-based drinks either not heat-treated or pasteurized (Pettipher et al. 1997). Komitopoulou et al. (1999) studied the growth of *A. acidoterrestris* in fruit juice and its sensitivity to heat treatment and nisin. The spores were confirmed to be heat-resistant after 10 min at 80°C, 2 min at 90°C and 1 min at 95°C in orange, grapefruit and apple juice. The resistance was reduced with decrement of pH of juices, although the effect waless marked at higher temperatures. Nisin addition (100 AU/ml) completely prevented *A. acidoterrestris* under all temperature and time of storage conditions. In particular, the presence of nisin during heating decreased the decimal reduction time up to 40% and its minimal inhibition concentration against *A. acidoterrestris* spores was only 5 AU/ml at 25°C.

Control of *A. acidoterrestris* in fruit juice was also approached with enterocin AS-48 (Grande et al. 2005). Vegetative cells of *A. acidoterrestris* DSMZ 2498 were inactivated by 2.5 μg/ml of this bacteriocin in natural orange and apple juices incubated at 37°C. No growth was detected in both juices until the 15th day of observation. Commercial orange, apple, pineapple, peach and grapefruit juices were then added with the same concentration of enterocin AS-48 and inoculated with vegetative cells or endospores of strain DSMZ 2498 and maintained at different incubation temperatures (4, 15 and 37°C) for 3 months. In those cases, no viable cells were observed during the whole incubation, except for apple, peach and grapefruit juices at 37°C containing vegetative cells which, however, were stable for up to 60 days. Treatment with enterocin AS-48, as revealed by electron microscopy, determined cell damage and bacterial lysis and disorganization of endospore structure in all fruit juices object of the study. These findings showed that enterocin AS-48 can be a valid substitute of the intense heat treatments necessary for inactivation of *A. acidoterrestris* endospores without altering the chemical composition of fruit juices.

2.4 *Staphylococcus aureus*

Staphylococcal food poisoning is among the most common causes of reported food borne diseases (Tirado and Schimdt 2001; WHO 2002; Le Loir et al. 2003; EFSA 2010), requiring hospital attention by up to 19.5% of the affected individuals (EFSA 2010). In many countries, *S. aureus* is the second or third most common pathogen responsible for outbreaks of food poisoning (Veras et al. 2008). *S. aureus* is found in the nostrils as well as on the skin and hair of warm-blooded animals, and up to 30 e 50% of human population are carriers (Le Loir et al. 2003). *S. aureus* has been isolated from several foods including meat and meat products, chicken, milk and dairy products, fermented food items, salads, vegetables, fish products, etc. (Tamarapu et al. 2001; Jørgensen et al. 2005; Seo and Bohach 2007). Most strains were capable of producing one or more heat stable enterotoxins (Balaban and Rasooly 2000; Ortega et al. 2010) which were the cause of the gastrointestinal symptoms observed during intoxications (Tamarapu et al. 2001). One of the approaches proposed for the control of *S. aureus* in foods was the application of bacteriocins either singly or in combination with other antimicrobials (Gálvez et al. 2008).

To control the development of *S. aureus* in foods, in addition to traditional chemical and physical preservatives, several bacteriocins of LAB, either alone or combined with other hurdles, have been used with varying degrees of success. In this sense, many recent studies have shown the effect of nisin against *S. aureus*. In sliced cheese, immobilized nisin in a polyethylene/polyamide packaging was shown to reduce the population of *S. aureus* (Scannell et al. 2000b). In fresh pork sausages, a combination of nisin and organic acids reduced the viable counts of *S. aureus* (Scannell et al. 1997). The combination of sodium citrate or sodium lactate with lacticin 3147 was also reported to be an effective biopreservative (Scannell et al. 2000a).

In skim milk, whey or simulated milk ultrafiltrate media, increased nisin activity in combination with pulsed electric fields (PEFs) has been reported against *S. aureus* (Sobrino-López and Martín Belloso 2006).

Compared to nisin, pediocin has been shown to be more effective against *S. aureus* (Cintas et al. 1998; Eijsink et al. 1998). Moreover, the evaluation of antibacterial efficacy of the bacteriocins, nisin and pediocin AcH revealed that they had better antibacterial property in combination due to synergistic effect than when used singly (Hanlin et al. 1993; Mulet-Powell et al. 1998).

Lacticin 3147 powder was shown to rapidly reduce *S. aureus* viable cell counts in an infant milk formulation (Morgan et al. 1999). Similarly, as was reported with nisin an increased bactericidal effect was shown for the combined treatment of lacticin 3147 concentrates and HHP against *S. aureus* in milk and whey (Morgan et al. 2000).

Regarding enterocins, enterocin CCM 4231 reduced the viable counts of *S. aureus* SA1 in skim milk, Sunar (milk nourishment for suckling babies), and yogurt (Lauková et al. 1999a, b). Enterocin AS-48 may also be an interesting bacteriocin to inhibit the growth of *S. aureus*. Several reports demonstrated the susceptibility of *S. aureus* to AS-48 in BHI broth, a sausage model system, milk and cheese and in vegetable sauces (Ananou et al. 2004; Grande et al. 2007; Muñoz et al. 2007). Muñoz et al. (2007) indicated that bacteriocin AS-48 was effective at controlling *S. aureus* in milk whether added exogenously or produced by a bacteriocinogenic strain. The efficacy of AS-48 was greatly enhanced by combination with a moderate heat treatment, which is of great technological relevance. In unripened cheese, AS-48 was also effective in controlling staphylococci when added as an adjunct culture during the manufacture of cheese.

In vegetable sauces anti-staphylococcal activity of AS-48 was significantly improved when the enterocin was used in combination with different phenolic compounds and even some of the combinations of enterocin AS-48 and phenolic compounds served to completely inactivate *S. aureus* in sauces. Nevertheless the effect depended largely on the type of food, which in turn had a great influence on the activity of AS-48 as well as the phenolic compounds tested individually (Grande et al. 2007). The storage temperature was also an important factor in the inhibition of *S. aureus* by AS-48 in sauces being more effective at high (22°C) than at low (10°C) storage temperatures.

References

Abriouel H, Valdivia E, Gálvez A, Maqueda M (1998) Response of *Salmonella choleraesuis* LT2 spheroplasts and permeabilized cells to the bacteriocin AS-48. Appl Environ Microbiol 64:4623–4626

Acuña L, Morero RD, Bellomio A (2010) Development of wide spectrum hybrid bacteriocins for food biopreservation. Food Bioprocess Technol 4:1029–1049

Ananou S, Valdivia E, Martínez-Bueno M, Gálvez A, Maqueda M (2004) Effect of combined physico-chemical preservatives on enterocin AS-48 activity against the enterotoxigenic *Staphylococcus aureus* CECT 976 strain. J Appl Microbiol 97:48–56

Ananou S, Garriga M, Hugas M, Maqueda M, Martínez-Bueno M, Gálvez A, Valdivia E (2005a) Control of *Listeria monocytogenes* in model sausages by enterocin AS-48. Int J Food Microbiol 103:179–190
Ananou S, Maqueda M, Martínez-Bueno M, Gálvez A, Valdivia E (2005b) Control of *Staphylococcus aureus* in sausages by enterocin AS-48. Meat Sci 71:549–576
Anastasiou R, Georgalaki M, Manolopoulou E, Kandarakis I, De Vuyst L, Tsakalidou E (2007) The performance of *Streptococcus macedonicus* ACA-DC 198 as starter culture in Kasseri cheese production. Int Dairy J 17:208–217
Angulo FJ, Getz J, Taylor JP, Hendricks KA, Hatheway CL, Barth SS, Solomon HM, Larson AE, Johnson EA, Nickey LN, Reis AA (1998) A large outbreak of botulism: the hazardous baked potato. J Infect Dis 178:172–177
Annamalai N, Manivasagan P, Balasubramanian T, Vijayalakshmi S (2009) Enterocin from *Enterococcus faecium* isolated from mangrove environment. Afr J Biotechnol 8:6311–6316
Aymerich MT, Garriga M, Ylla J, Vallier J, Monfort JM, Hugas M (2000) Application of enterocins as biopreservatives against *Listeria innocua* in meat products. J Food Prot 63:721–726
Aymerich MT, Garriga M, Costa S, Monfort JM, Hugas M (2002) Prevention of ropiness in cooked pork by bacteriocinogenic cultures. Int Dairy J 12:239–246
Balaban N, Rasooly A (2000) Staphylococcal enterotoxins. Int J Food Microbiol 61:1–10
Bari ML, Ukuku DO, Kawasaki T, Inatsu Y, Isshiki K, Kawamoto S (2005) Combined efficacy of nisin and pediocin with sodium lactate, citric acid, phytic acid, and potassium sorbate and EDTA in reducing the *Listeria monocytogenes* population of inoculated fresh-cut produce. J Food Prot 68:1381–1387
Ben Omar N, Abriouel H, Lucas R, Martínez-Cañamero M, Guyot JP, Gálvez A (2006) Isolation of bacteriocinogenic *Lactobacillus plantarum* strains from ben saalga, a traditional fermented gruel from Burkina Faso. Int J Food Microbiol 112:44–50
Benech RO, Kheadr EE, Lacroix C, Fliss I (2003) Impact of nisin producing culture and liposome-encapsulated nisin on ripening of Lactobacillus added-Cheddar cheese. J Dairy Sci 86:1895–1909
Benkerroum N, Sandine WE (1988) Inhibitory action of nisin against *Listeria monocytogenes*. J Dairy Sci 71:3237–3245
Benkerroum N, Oubel H, Mimoun LB (2002) Behavior of *Listeria monocytogenes* and *Staphylococcus aureus* in yogurt fermented with a bacteriocin-producing thermophilic starter. J Food Prot 65:799–805
Beuchat LR (1996) Pathogenic microorganisms associated with fresh produce. J Food Prot 59:204–216
Boziaris IS, Humpheson L, Adams MR (1998) Effect of nisin on heat injury and inactivation of *Salmonella enteritidis* PT4. Int J Food Microbiol 43:7–13
Branen JK, Davidson PM (2004) Enhancement of nisin, lysozyme, and monolaurin antimicrobial activities by ethylenediaminetetraacetic acid and lactoferrin. Int J Food Microbiol 90:63–74
Breukink E, Wiedemann I, van Kraaij C, Kuipers OP, Sahl H, de Kruijff B (1999) Use of the cell wall precursor lipid II by a pore-forming peptide antibiotic. Science 5448:2361–2364
Brown JL, Ross T, McMeekin TA, Nichols PD (1997) Acid habituation of *Escherichia coli* and the potential role of cyclopropane fatty acids in low pH tolerance. Int J Food Microbiol 37:163–173
Buyong N, Kok J, Luchansky JB (1998) Use of a genetically enhanced, pediocin-producing starter culture, *Lactococcus lactis* subsp. *lactis* MM217, to control *Listeria monocytogenes* in cheddar cheese. Appl Environ Microbiol 64:4842–4845
Calderón-Miranda ML, Barbosa-Cánovas GV, Swanson BG (1999) Inactivation of Listeria innocua in skim milk by pulsed electric fields and nisin. Int J Food Microbiol 51:19–30
Callewaert R, Hugas M, De Vuyst L (2000) Competitiveness and bacteriocin production of enterococci in the production of Spanish-style dry fermented sausages. Int J Food Microbiol 57:33–42
Carneiro de Melo AM, Cassar CA, Miles RJ (1998) Trisodium phosphate increases sensitivity of gram-negative bacteria to lysozyme and nisin. J Food Prot 61:839–843
Carr FJ, Chill D, Maida N (2002) The lactic acid bacteria: a literature survey. Crit Rev Microbiol 28:281–370

Cheville AM, Arnold KW, Buchrieser C, Cheng CM, Kaspar CW (1996) rpoS regulation of acid, heat, and salt tolerance in *Escherichia coli* O157:H7. Appl Environ Microbiol 62:1822–1824
Cintas LM, Casaus P, Håvarstein LS, Hernández PE, Nes IF (1997) Biochemical and genetic characterization of enterocin P, a novel sec-dependent bacteriocin from *Enterococcus faecium* P13 with a broad antimicrobial spectrum. Appl Environ Microbiol 63:4321–4330
Cintas LM, Casaus P, Fernandez MF, Hernandez PE (1998) Comparative antimicrobial activity of enterocin L50, pediocin PA-1, nisin A and lactocin S against spoilage and food borne pathogenic bacteria. Food Microbiol 15:289–298
Coderre PE, Somkuti GA (1999) Cloning and expression of the pediocin operon in *Streptococcus thermophilus* and other lactic fermentation bacteria. Curr Microbiol 39:295–301
Cole K, Farnell MB, Donoghue AM, Stern NJ, Svetoch EA, Eruslanov BN, Volodina LI, Kovalev YN, Perelygin VV, Mitsevich EV, Mitsevich IP, Levchuk VP, Pokhilenko VD, Borzenkov VN, Svetoch OE, Kudryavtseva TY, Reyes-Herrera I, Blore PJ, Solis de los Santos F, Donoghue DJ (2006) Bacteriocins reduce *Campylobacter* colonization and alter gut morphology in turkey poults. Poult Sci 85:1570–1575
Cotter PD, Hill C, Ross RP (2005) Bacteriocins: developing innate immunity for food. Nat Rev Microbiol 3:777–788
Cutter CN, Siragusa GR (1995) Treatments with nisin and chelators to reduce *Salmonella* and *Escherichia coli* on beef. J Food Prot 58:1028–1030
Davies EA, Delves-Broughton J (1999) Nisin. In: Robinson R, Batt C, Patel P (eds) Encyclopedia of food microbiology. Academic, London, pp 191–198
De Vuyst L, Leroy F (2007) Bacteriocins from lactic acid bacteria: production, purification, and food applications. J Mol Microbiol Biotechnol 13:194–199
De Vuyst L, Vandamme EJ (1994) Bacteriocins of lactic acid bacteria: microbiology, genetics and applications. Blackie Academic & Professional, London
Delves-Broughton J, Williams GC, Wilkinson S (1992) The use of bacteriocin, nisin, as a preservative in pasteurized liquid whole egg. Lett Appl Microbiol 15:133–136
Delves-Broughton J, Blackburn P, Evans RJ, Hugenholtz J (1996) Applications of the bacteriocin, nisin. Antonie Van Leeuwenhoek 69:193–202
Doan CH, Davidson PM (2000) Microbiology of potatoes and potato products: a review. J Food Prot 63:668–683
EFSA (2007) Introduction of a Qualified Presumption of Safety (QPS) approach for assessment of selected microorganisms referred to EFSA1. The EFSA J 587:1–16
EFSA (2010) The community summary report on trends and sources of zoonoses and zoonotic agents and food-borne outbreaks in the European Union in 2008. J EFSA 8(1):1496 [pp. 410]. doi:10. 2903/j.efsa.2010.1496
Eijsink VG, Skeie M, Middelhoven PH, Brurberg MB, Nes IF (1998) Comparative studies of Class IIa bacteriocins of lactic acid bacteria. Appl Environ Microbiol 64:3275–3281
Ennahar S, Assobhel O, Hasselmann C (1998) Inhibition of *Listeria monocytogenes* in a smear-surface soft cheese by *Lactobacillus plantarum* WHE 92, a pediocin AcH producer. J Food Prot 61:186–191
Fang TJ, Lin LW (1994a) Growth of *Listeria monocytogenes* and *Pseudomonas fragi* on cooked pork in a modified atmosphere packaging/nisin combination system. J Food Prot 57:479–485
Fang TJ, Lin LW (1994b) Inactivation of *Listeria monocytogenes* on raw pork treated with modified atmosphere packaging and nisin. J Food Drug Anal 2:189–200
Farías ME, de Ruiz Holgado AP, Sesma F (1994) Bacteriocin production by lactic acid bacteria isolated from regional cheeses: inhibition of foodborne pathogens. J Food Prot 57:1013–1015
Farías ME, Nuñez de Kairuz M, Sesma F, Palacios J, de Ruiz Holgado AP, Oliver G (1999) Inhibition of *Listeria monocytogenes* by the bacteriocin enterocin CRL35 during goat cheese making. Milchwissenschaft 54:30–32
Ferreira A, Canal N, Morales D, Fuentefria D, Corção G (2007) Characterization of enterocins produced by *Enterococcus mundtii* isolated from humans feces. Braz Arch Biol Technol 50:249–258

Foulquié Moreno MR, Sarantinopoulos P, Tsakalidou E, De Vuyst L (2006) The role and application of enterococci in food and health. Int J Food Microbiol 106:1–24

Gálvez A, Lucas-López R, Abriouel H, Valdivia E, Ben Omar N (2008) Application of bacteriocins in the control of foodborne pathogenic and spoilage bacteria. Crit Rev Biotechnol 28:125–152

Gálvez A, Valdivia E, Abriouel H, Camafeita E, Mendez E, Martinez-Bueno M, Maqueda M (1998) Isolation and characterization of enterocin EJ97, a bacteriocin produced by Enterococcus faecalis EJ97. Arch Microbiol 171(1):59–65

Gänzle MG, Weber S, Hammes WP (1999) Effect of ecological factors on the inhibitory spectrum and activity of bacteriocins. Int J Food Microbiol 46:207–217

García P, Rodríguez L, Rodríguez A, Martínez B (2010) Food biopreservation: promising strategies using bacteriocins, bacteriophages and endolysins. Trends Food Sci Technol 21:373–382

Ge J, Ping W, Song G, Du C, Ling H, Sun X (2009) Paracin 1.7, a bacteriocin produced by *Lactobacillus paracasei* HD1.7 isolated from Chinese cabbage sauerkraut, a traditional Chinese fermented vegetable food. Wei sheng wu xue bao. Acta Microbiol Sin 49:609–616

Georgalaki MD, Van den Berghe E, Kritikos D, Devreese B, Van Beeumen J, Kalantzopoulos G, De Vuyst L, Tsakalidou E (2002) Macedocin, a food-grade lantibiotic produced by *Streptococcus macedonicus* ACA-DC 198. Appl Environ Microbiol 68:5891–5903

Gill AO, Holley RA (2000a) Inhibition of bacterial growth on ham and bologna by lysozyme, nisin and EDTA. Food Res Int 33:83–90

Gill AO, Holley RA (2000b) Surface application of lysozyme, nisin, and EDTA to inhibit spoilage and pathogenic bacteria on ham and bologna. J Food Prot 63:1338–1346

Giraffa G (1995) Enterococcal bacteriocins: their potential as anti- *Listeria* factors in dairy technology. Food Microbiol 12:291–299

Glass KA, Loeffelholz JM, Ford JP, Doyle MP (1992) Fate of *Escherichia coli* O157:H7 as affected by pH or sodium chloride and in fermented, dry sausage. Appl Environ Microbiol 58:2513–2516

Gould GW (1995) Homeostatic mechanisms during food preservation by combined methods. In: Welti-Chanes J, Barbosa-Canovas G (eds) Food preservation by moisture control – fundamentals and applications. Technomic Publishing Co., Lancaster, pp 397–410

Grande MJ, Lucas R, Valdivia E, Abriouel H, Maqueda M, Martínez-Cañamero M, Ben Omar N, Gálvez A (2005) Stability of enterocin AS-48 in fruit and vegetable juices. J Food Prot 68:2085–2094

Grande MJ, Lucas R, Abriouel H, Valdivia E, Ben Omar N, Maqueda M, Martínez-Bueno M, Martínez-Cañamero M, Gálvez A (2006) Inhibition of toxicogenic *Bacillus cereus* in rice-based foods by enterocin AS-48. Int J Food Microbiol 106:185–194

Grande MJ, Lucas R, Abriouel H, Valdivia E, Ben Omar N, Maqueda M (2007) Treatment of vegetable sauces with enterocin AS-48 alone or in combination with phenolic compounds to inhibit proliferation of *Staphylococcus aureus*. J Food Prot 70:405–411

Hammami R, Zouhir A, Le Lay C, Ben Hamida J, Fliss I (2010) BACTIBASE second release: a database and tool platform for bacteriocin characterization. BMC Microbiol 10:22

Hampikyan H, Ugur M (2007) The effect of nisin on *L. monocytogenes* in Turkish fermented sausages (sucuks). Meat Sci 76:327–332

Hanlin MB, Kalchayanand N, Ray P, Ray B (1993) Bacteriocins of lactic acid bacteria in combination have a greater antibacterial activity. J Food Prot 56:252–255

Helander IM, Mattila-Sandholm T (2000) Permeability barrier of the gram-negative bacterial outer membrane with special reference to nisin. Int J Food Microbiol 60:153–161

Hirsch A, Grinsted E, Chapman HR, Mattick ATR (1951) A note on the inhibition of an anaerobic spore former in Swiss-type cheese by a nisin-producing *Streptococcus*. J Dairy Res 18:205–206

Holck AL, Axelsson L, Huhme K, Krockel L (1994) Purification and cloning of sakacin 674, a bacteriocin from *Lactobacillus sake* Lb674. FEMS Microbiol Lett 115:143–150

Hugas M, Garriga M, Aymerich T (2003) Functionality of enterococci in meat products. Int J Food Microbiol 88:223–233

Jennes W, Dicks LM, Verwoerd DJ (2000) Enterocin 012, a bacteriocin produced by *Enterococcus gallinarum* isolated from the intestinal tract of ostrich. J Appl Microbiol 88:349–357

Jørgensen HJ, Maticen T, Løvseth A, Oboe K, Qvale KS, Loncarevic S (2005) An outbreak of staphylococcal food poisoning caused by enterotoxin H in mashed potato made with raw milk. FEMS Microbiol Lett 252:267–272

Kalchayanand N, Hanlin MB, Ray B (1992) Sublethal injury makes Gram-negative and resistant Gram-positive bacteria sensitive to the bacteriocins, pediocin AcH and nisin. Lett Appl Microbiol 15:239–243

Kang JH, Lee MS (2005) Characterization of a bacteriocin produced by *Enterococcus faecium* GM-1 isolated from an infant. J Appl Microbiol 98:1169–1176

Klaenhammer TR (1993) Genetics of bacteriocins produced by lactic acid bacteria. FEMS Microbiol Rev 12:39–85

Knight KP, Bartlett FM, McKellar RC, Harris LJ (1999) Nisin reduces the thermal resistance of *Listeria monocytogenes* Scott A in liquid whole egg. J Food Prot 62:999–1003

Komitopoulou E, Boziaris IS, Davies EA, Delves-Broughton J, Adams MR (1999) *Alicyclobacillus acidoterrestris* in fruit juices and its control by nisin. Int J Food Sci Technol 34:81–85

Lauková A, Czikková S (1999) The use of enterocin CCM 4231 in soy milk to control the growth of *Listeria monocytogenes* and *Staphylococcus aureus*. J Appl Microbiol 87:182–186

Lauková A, Czikková S (2001) Antagonistic effect of enterocin CCM 4231 from *Enterococcus faecium* on "bryndza", a traditional Slovak dairy product from sheep milk. Microbiol Res 156:31–34

Lauková A, Czikková S, Burdova O (1999a) Anti-staphylococcal effect of enterocin in Sunar and yogurt. Folia Microbiol (Praha) 44:707–711

Lauková A, Czikková S, Dóbransky T, Burdova O (1999b) Inhibition of *Listeria monocytogenes* and *Staphylococcus aureus* by enterocin CCM4231 in milk products. Food Microbiol 16:93–99

Lauková A, Czikková S, Laczková S, Turek P (1999c) Use of enterocin CCM 4231 to control *Listeria monocytogenes* in experimentally contaminated dry fermented Horn'ad salami. Int J Food Microbiol 52:115–119

Lauková A, Vlaemynick G, Czikková S (2001) Effect of enterocin CCM 4231 on *Listeria monocytogenes* in Saint-Paulin cheese. Folia Microbiol 46:157–160

Le Loir Y, Baron F, Gautier M (2003) *Staphylococcus aureus* and food poisoning. Genet Mol Res 2:63–76

Leverentz B, Conway WS, Camp MJ, Janisiewicz WJ, Abuladze T, Yang M, Saftner R, Sulakvelidze A (2003) Biocontrol of *Listeria monocytogenes* on fresh-cut produce by treatment with lytic bacteriophages and a bacteriocin. Appl Environ Microbiol 69:4519–4526

Leyer GJ, Wang LL, Johnson EA (1995) Acid adaptation of *Escherichia coli* O157:H7 increases survival in acidic foods. Appl Environ Microbiol 61:3752–3755

Liang Z, Mittal GS, Griffiths MW (2002) Inactivation of *Salmonella Typhimurium* in orange juice containing antimicrobial agents by pulsed electric field. J Food Prot 65:1081–1087

Loessner M, Guenther S, Steffan S, Scherer S (2003) A pediocin-producing *Lactobacillus plantarum* strain inhibits *Listeria monocytogenes* in a multispecies cheese surface microbial ripening consortium. Appl Environ Microbiol 69:1854–1857

Lucas R, Grande MJ, Abriouel H, Maqueda M, Ben Omar N, Valdivia E, Martínez-Cañamero M, Gálvez A (2006) Application of the broad-spectrum bacteriocin enterocin AS-48 to inhibit *Bacillus coagulans* in low-pH canned fruit and vegetable foods. Food Chem Toxicol 44:1774–1781

Lüders T, Birkemo GA, Fimland G, Nissen-Meyer J, Nes IF (2003) Strong synergy between a eukaryotic antimicrobial peptide and bacteriocins from lactic acid bacteria. Appl Environ Microbiol 69:1797–1799

Lyon WJ, Glatz BA (1993) Isolation and purification of propionicin PLG-1, a bacteriocin produced by a strain of *Propionibacterium thoenii*. Appl Environ Microbiol 59:83–88

Lyon WJ, Sethi JK, Glatz BA (1993) Inhibition of psychrotrophic organisms by propionicin PLG-1, a bacteriocin produced by *Propionibacterium thoenii*. J Dairy Sci 76:1506–1513

Maisnier-Patin S, Deschamps N, Tatini SR, Richard J (1992) Inhibition of *Listeria monocytogenes* in Camembert cheese made with a nisin-producing starter. Lait 72:249–263

Mathot AG, Beliard E, Thuault D (2003) *Streptococcus thermophilus* 580 produces a bacteriocin potentially suitable for inhibition of *Clostridium tyrobutyricum* in hard cheese. J Dairy Sci 86:3068–3074
Meindl K, Schmiederer T, Schneider K, Reicke A, Butz D, Keller S, Gühring H, Vértesy L, Wink J, Hoffmann H, Brönstrup M, Sheldrick GM, Süssmuth RD (2010) Labyrinthopeptins: a new class of carbacyclic lantibiotics. Angew Chem Int Ed Engl 49:1151–1154
Messaoudi S, Kergourlay G, Rossero A, Ferchichi M, Prévost H, Drider D, Manai M, Dousset X (2011) Identification of lactobacilli residing in chicken ceca with antagonism against *Campylobacter*. Int Microbiol 14:103–110
Mills S, Stanton C, Hill C, Ross RP (2011) New developments and applications of bacteriocins and peptides in foods. Annu Rev Food Sci Technol 2:299–329
Molinos AC, Abriouel H, López RL, Valdivia E, Ben Omar N, Gálvez A (2008) Combined physico-chemical treatments based on enterocin AS-48 for inactivation of Gram-negative bacteria in soybean sprouts. Food Chem Toxicol 46:2912–2921
Montville TJ, Winkowski K, Ludescherer RD (1995) Models and mechanisms for bacteriocin action and application. Int Dairy J 5:797–814
Morgan SM, Garvin M, Kelly J, Ross RP, Hill C (1999) Development of a lacticin 3147-enriched whey powder with inhibitory activity against foodborne pathogens. J Food Prot 62:1011–1016
Morgan SM, Ross RP, Beresford T, Hill C (2000) Combination of hydrostatic pressure and lacticin 3147 causes increased killing of *Staphylococcus* and *Listeria*. J Appl Microbiol 88:414–420
Morgan SM, Garvin M, Ross RP, Hill C (2001) Evaluation of a spray-dried lacticin 3147 powder for the control of *Listeria monocytogenes* and *Bacillus cereus* in a range of food systems. Lett Appl Microbiol 33:387–391
Morris SL, Walsh RC, Hansen JN (1984) Identification and characterization of some bacterial membrane sulfhydryl groups which are targets of bacteriostatic and antibiotic action. J Biol Chem 259:13590–13594
Mulet-Powell N, Lactoste-Armynot AM, Vinas M, Simeon De Buochberg M (1998) Interactions between pairs of bacteriocins from lactic acid bacteria. J Food Prot 61:1210–1212
Muñoz A, Maqueda M, Gálvez A, Martínez-Bueno M, Rodríguez A, Valdivia E (2004) Biocontrol of psychrotrophic enterotoxigenic *Bacillus cereus* in a non fat hard type cheese by an enterococcal strain-producing enterocin AS-48. J Food Prot 67:1517–1521
Muñoz A, Ananou S, Gálvez A, Martínez-Bueno M, Rodríguez A, Maqueda M, Valdivia E (2007) Inhibition of *Staphylococcus aureus* in dairy products by enterocin AS-48 produced in situ and ex situ: bactericidal synergism through heat and AS-48. Int Dairy J 17:760–769
Murdock CA, Cleveland J, Matthews KR, Chikindas ML (2007) The synergistic effect of nisin and lactoferrin on the inhibition of *Listeria monocytogenes* and *Escherichia coli* O157:H7. Lett Appl Microbiol 44:255–261
Muriana PM (1996) Bacteriocins for control of*Listeria* spp. in food. J Food Prot 59:S54–S63
Nazef L, Belguesmia Y, Tani A, Prévost H, Drider D (2008) Identification of lactic acid bacteria from poultry feces: evidence on anti-*campylobacter* and anti-*listeria* activities. Poult Sci 87:329–334
Nissen H, Holo H, Axelsson L, Blom H (2001) Characterization and growth of *Bacillus* spp. in heat-treated cream with and without nisin. J Appl Microbiol 90:530–534
O'Mahony T, Rekhif N, Cavadini C, Fitzgerald GF (2001) The application of a fermented food ingredient containing 'variacin', a novel antimicrobial produced by *Kocuria varians*, to control the growth of *Bacillus cereus* in chilled dairy products. J Appl Microbiol 90:106–114
Ortega E, Abriouel H, Lucas R, Gálvez A (2010) Multiple roles of *Staphylococcus aureus* enterotoxins: pathogenicity, superantigenic activity, and correlation to antibiotic resistance. Toxins 2:2117–2131
Parente E, Hill C (1992) Inhibition of *Listeria* in buffer, broth, and milk by enterocin 1146, a bacteriocin produced by *Enterococcus faecium*. J Food Prot 55:503–508
Pawar DD, Malik SVS, Bhilegaonkar KN, Barbuddhe SB (2000) Effect of nisin and its combination with sodium chloride on the survival of *Listeria monocytogenes* added to raw buffalo meat mince. Meat Sci 56:215–219

Pedersen MB, Iversen SL, Sørensen KI, Johansen E (2005) The long and winding road from the research laboratory to industrial applications of lactic acid bacteria. FEMS Microbiol Rev 29:611–624

Pettipher GL, Osmundson ME, Murphy JM (1997) Methods for the detection and enumeration of *Alicyclobacillus acidoterrestris* and investigations of growth and taint in fruit juice and fruit-containing drinks. Lett Appl Microbiol 24:185–189

Pol IE, Mastwijk HC, Slump RA, Popa ME, Smith EJ (2001) Influence of food matrix on inactivation of *Bacillus cereus* by combinations of nisin, pulsed electric field treatment, and carvacrol. J Food Prot 64:1012–1018

Presser KA, Ross T, Ratkowsky DA (1998) Modelling the Angrowth limits (growth/no growth interface) of *Escherichia coli* as a function of temperature, pH, lactic acid concentration, and water activity. Appl Environ Microbiol 64:1773–1779

Raccach M, Geshell DJ (1993) The inhibition of *Listeria monocytogenes* in milk by pediococci. Food Microbiol 10:181–186

Rayman K, Malik N, Hurst A (1983) Failure of nisin to inhibit outgrowth of *Clostridium botulinum* in a model cured meat system. Appl Environ Microbiol 46:1450–1452

Rea MC, Ross P, Cotter PD, Hill C (2011) Classifications of bacteriocins from Gram positive bacteria. In: Drider D, Rebuffat S (eds) Prokaryotic antimicrobial peptides: from Geenes to applications. Springer, New York, 29–53. doi:10. 1007/978-1-4419-7692-5_3

Reitsma CJ, Henning DR (1996) Survival of enterohemor- rhagic *Escherichia coli* O157:H7 during the manufacture and curing of cheddar cheese. J Food Prot 59:460–464

Reviriego C, Fernández L, Rodríguez JM (2007) A food-grade system for production of pediocin PA-1 in nisin-producing and nonnisin- producing *Lactococcus lactis* strains: application to inhibit *Listeria* growth in a cheese model system. J Food Prot 70:2512–2517

Rilla N, Martinez B, Delgado T, Rodriguez A (2003) Inhibition of *Clostridium tyrobutyricum* in Vidiago cheese by *Lactococcus lactis* ssp. Lactis IPLA 729, a nisin Z producer. Int J Food Microbiol 85(1–2):23–33

Rodríguez JL, Gaya P, Medina M, Núñez M (1997) Bactericidal effect of enterocin 4 on *Listeria monocytogenes* in a model dairy system. J Food Prot 60:28–32

Rodríguez E, Calzada J, Arqués JL, Rodríguez JM, Núñez M, Medina M (2005) Antimicrobial activity of pediocin-producing, *Lactococcus lactis* on *Listeria monocytogenes*, *Staphylococcus aureus* and *Escherichia coli* O157:H7 in cheese. Int Dairy J 15:51–57

Rose NL, Sporns P, Stiles ME, McMullen LM (1999) Inactivation of nisin by glutathione in fresh meat. J Food Sci 64:759–762

Ross RP, Galvin M, McAuliffe O, Morgan S, Ryan M, Twomey DP, Meaney WJ, Hill C (1999) Developing applications for lactococcal bacteriocins. Antonie Leeuwenhoek 76:337–346

Sabia C, de Niederhäusern S, Messi P, Manicardi G, Bondi M (2003) Bacteriocin-producing *Enterococcus casseliflavus* IM 416K1, a natural antagonist for control of *Listeria monocytogenes* in Italian sausages ("cacciatore"). Int J Food Microbiol 87:173–179

Şahingil D, Işlero lu H, Yildirim Z, Akçelik M, Yildirim M (2011) Characterization of lactococcin BZ produced by *Lactococcus lactis* subsp. *lactis* BZ isolated from boza. Turk J Biol 35:21–33

Scannell AGM, Hill C, Buckley DJ, Arendt EK (1997) Determination of the influence of organic acids and nisin on shelflife and microbiological safety aspects of fresh pork sausage. J Appl Microbiol 83:407–412

Scannell AG, Ross RP, Hill C, Arendt EK (2000a) An effective lacticin biopreservative in fresh pork sausage. J Food Prot 63:370–375

Scannell AGM, Hill C, Ross RP, Marx S, Hartmeier W, Arendt EK (2000b) Development of bioactive food packaging materials using immobilised bacteriocins lacticin 3147 and Nisaplin. Int J Food Microbiol 60:241–249

Schillinger U, Becker B, Vignolo G, Holzapfel WH (2001) Efficacy of nisin in combination with protective cultures against *Listeria monocytogenes* Scott A in tofu. Int J Food Microbiol 71:159–168

Schuman JD, Sheldon BW (2003) Inhibition of *Listeria monocytogenes* in pH-adjusted pasteurized liquid whole egg. J Food Prot 66:999–1006

Schved F, Henis Y, Juven BJ (1994) Response of spheroplasts and chelator-permeabilized cells of Gram-negative bacteria to the action of the bacteriocins pediocin SJ-1 and nisin. Int J Food Microbiol 21:305–314

Seo KS, Bohach GA (2007) Staphylococcus aureus. In: Doyle MP, Beuchat LR (eds) Food microbiology: fundamentals and frontiers, 3rd edn. ASM Press, Washington, DC, pp 493–518

Shefet SM, Sheldon BW, Klaenhammer TR (1995) Efficacy of optimized nisin-based treatments to inhibit *Salmonella typhimurium* and extend shelf life of broiler carcasses. J Food Prot 58:1077–1082

Sivarooban T, Hettiarachchy NS, Johnson MG (2007) Inhibition of *Listeria monocytogenes* using nisin with grape seed extract on turkey frankfurters stored at 4 and 10°C. J Food Prot 70:1017–1020

Small P, Blankenhorn D, Welty D, Zinser E, Slonczewski JL (1994) Acid and base resistance in *Escherichia coli* and *Shigella flexneri*: role of *rpoS* and growth pH. J Bacteriol 176:1729–1737

Smaoui S, Elleuch L, Bejar W, Karray-Rebai I, Ayadi I, Jaouadi B (2010) Inhibition of fungi and gram-negative bacteria by bacteriocin BacTN635 produced by *Lactobacillus plantarum* sp. TN635. Appl Biochem Biotechnol 162:1132–1146

Sobrino-López A, Martín Belloso O (2006) Enhancing inactivation of *Staphylococcus aureus* in skim milk by combining highintensity pulsed electric fields and nisin. J Food Prot 69:345–353

Solomakos N, Govaris A, Koidis P, Botsoglou N (2008) The antimicrobial effect of thyme essential oil, nisin, and their combination against *Listeria monocytogenes* in minced beef during refrigerated storage. Food Microbiol 25:120–127

Somkuti GA, Steinberg DH (2003) Pediocin production by recombinant lactic acid bacteria. Biotechnol Lett 25:473–477

Stergiou VA, Thomas LV, Adams MR (2006) Interactions of nisin with glutathione in a model protein system and meat. J Food Prot 69:951–956

Stern NJ, Svetoch EA, Eruslanov BV, Perelygin VV, Mitsevich EV, Mitsevich IP, Pokhilenko VD, Levchuk VP, Svetoch OE, Seal BS (2006) Isolation of a *Lactobacillus salivarius* strain and purification of its bacteriocin, which is inhibitory to *Campylobacter jejuni* in the chicken gastrointestinal system. Antimicrob Agents Chemother 50:3111–3116

Stevens KA, Sheldon BW, Klapes NA, Klaenhammer TR (1991) Nisin treatment for inactivation of *Salmonella* species and other Gram-negative bacteria. Appl Environ Microbiol 57:3612–3615

Sulzer G, Busse M (1991) Growth inhibition of *Listeria* spp. On Camembert cheese by bacteria producing inhibitory substances. Int J Food Microbiol 14:287–296

Svetoch EA, Eruslanov BV, Perelygin VV, Mitsevich EV, Mitsevich IP, Borzenkov VN (2008) Diverse antimicrobial killing by *Enterococcus faecium* E 50–52 bacteriocin. J Agric Food Chem 56:1942–1948

Tagg JR, Dajani AS, Wannamaker LW (1976) Bacteriocins of gram-positive bacteria. Bacteriol Rev 40:722–7256

Tamarapu S, McKillip JL, Drake M (2001) Development of a multiplex polymerase chain reaction assay for detection and differentiation of *Staphylococcus aureus* in dairy products. J Food Prot 64:664–668

Thomas LV, Delves-Broughton J (2001) New advances in the application of the food preservative nisin. Adv Food Sci 2:11–22

Thomas LV, Clarkson MR, Delves-Broughton J (2000) Nisin. In: Naidu AS (ed) Natural food antimicrobial systems. CRC Press, Boca Raton, pp 463–524

Thomas LV, Ingram RE, Bevis HE, Davies EA, Milne CF, Delves-Broughton J (2002) Effective use of nisin to control *Bacillus* and *Clostridium* spoilage of a pasteurized mashed potato product. J Food Prot 65:1580–1585

Tichaczek PS, Vogel RF, Hammes WP (1994) Cloning and sequencing of sakP encoding sakacin P, the bacteriocin produced by *Lactobacillus sake* LHT673. Microbiol 140:361–367

Tirado C, Schimdt K (2001) WHO surveillance programme for control of foodborne infections and intoxications: preliminary results and trends across greater Europe. J Infect 43:80–84

Tsakalidou E, Zoidou E, Pot B, Wassill L, Ludwig W, Devriese LA, Kalantzopoulos G, Schleifer KH, Kersters K (1998) Identification of streptococci from Greek Kasseri cheese and description of *Streptococcus macedonicus* sp. nov. Int J Syst Bacteriol 48:519–527

Ukuku DO, Fett WF (2004) Effect of nisin in combination with EDTA, sodium lactate, and potassium sorbate for reducing *Salmonella* on whole and fresh-cut cantaloupet. J Food Prot 67:2143–2150

Valenzuela AS, Ben Omar N, Abriouel H, Lucas Lopez R, Ortega E, Martinez Canamero M, Gálvez A (2008) Risk factors in enterococci isolated from foods in Morocco: determination of antimicrobial resistance and incidence of virulence traits. Food Chem Toxicol 46:2648–2652

Van den Berghe E, Skourtas G, Tsakalidou E, DeVuyst L (2006) *Streptococcus macedonicus* ACA-DC 198 produces the lantibiotic, macedocin, at temperature and pH conditions that prevail during cheese manufacture. Int J Food Microbiol 107:138–147

Veras JF, Carmo LS, Tong LC, Shupp JW, Cummings C, Santos DA (2008) A study of the enterotoxigenicity of coagulase negative and coagulase positive staphylococcal isolates from food poisoning outbreaks in Minas Gerais, Brazil. Int J Infect Dis 12:410–415

Vignolo G, Fadda S, de Kairuz MN, de Ruiz Holgado AA, Oliver G (1996) Control of *Listeria monocytogenes* in ground beef by 'Lactocin 705', a bacteriocin produced by *Lactobacillus casei* CRL 705. Int J Food Microbiol 29:397–402

Vignolo G, Fadda S, de Kairuz MN, Holgado APR, Oliver O (1998) Effects of curing additives on the control of *Listeria monocytogenes* by lactocin 705 in meat slurry. Food Microbiol 15:259–264

Vignolo G, Palacios J, Farias ME, Sesma F, Schillinger U, Holzapfel W, Oliver G (2000) Combined effect of bacteriocins on the survival of various *Listeria* species in broth and meat system. Curr Microbiol 41:410–416

WHO (2002) Statistical information on food-borne disease in Europe. Microbiological and chemical hazards. In: FAO/WHO Pan-European conference on food quality and safety, Budapest. Springer, pp 29–53. ISBN 978-1-4419-7691-8

Yuste J, Fung DY (2004) Inactivation of *Salmonella Typhimurium* and *Escherichia coli* O157:H7 in apple juice by a combination of nisin and cinnamon. J Food Prot 67:371–377

Zamfir M, Callewaert R, Cornea PC, Savu L, Vatafu I, De Vuyst L (1999) Purification and characterization of a bacteriocin produced by *Lactobacillus acidophilus* IBB 801. J Appl Microbiol 87:923–931

Chapter 20
Anaerobic Degradation of Lindane and Other HCH Isomers

Farrakh Mehboob, Alette A.M. Langenhoff, Gosse Schraa, and Alfons J.M. Stams

Abstract Lindane (γ-HCH) is a pesticide that has mainly been used in agriculture. Lindane and the other HCH isomers are highly chlorinated hydrocarbons. The presence of a large number of electron withdrawing chlorine groups makes some of the HCH isomers rather recalcitrant in oxic environments. Especially β-HCH is poorly degraded by aerobic bacteria. The chlorine groups make HCH isomers more accessible for an initial reductive attack, a common mechanism in anoxic environments. Among the HCH isomers, γ-HCH is degraded most easily while β-HCH is most persistent. Little is known about the diversity of the microorganisms involved in anaerobic HCH degradation. Thus far, species within the genera *Clostridium* and *Bacillus*, two *Desulfovibrio* species, and one species each of *Desulfococcus*, *Desulfobacter*, *Citrobacter* and *Dehalobacter* have been found to metabolize lindane

F. Mehboob (✉)
Laboratory of Microbiology, Wageningen University,
Dreijenplein 10, 6703 HB Wageningen, The Netherlands

Ecotoxicology Research Programme, National Institute of Bio-remediation,
National Agriculture Research Centre, Islamabad, Pakistan
e-mail: farrakhmehboob@gmail.com

A.A.M. Langenhoff
Sub-department of Environmental Technology, Wageningen University,
Bornse Weilanden 9, 6708 WG Wageningen, The Netherlands

G. Schraa
Laboratory of Microbiology, Wageningen University,
Dreijenplein 10, 6703 HB Wageningen, The Netherlands

A.J.M. Stams
Laboratory of Microbiology, Wageningen University,
Dreijenplein 10, 6703 HB Wageningen, The Netherlands

Centre of Biological Engineering, University of Minho,
Campus de Gualtar, 4710-057 Braga, Portugal

A. Malik et al. (eds.), *Management of Microbial Resources in the Environment*,
DOI 10.1007/978-94-007-5931-2_20, © Springer Science+Business Media Dordrecht 2013

and other HCH isomers. Benzene and monochlorobenzene are the end products of anaerobic degradation, while in some studies pentachlorocyclohexane, tetrachlorocyclohexene, chlorobenzenes and chlorophenols have been detected as intermediates. Enzymes and coding genes involved in the reductive dechlorination of HCH isomers are largely unknown. Recently, a metagenomic analysis has indicated the presence of numerous putative reductive dehalogenase genes in the genome of β-HCH degrading *Dehalobacter* sp. High-throughput omics techniques can help to explore the key players and enzymes involved in the reductive dehalogenation of lindane and other HCH isomers.

Keywords Anaerobic • Reductive dechlorination • Halorespiration • Hexchlorocyclohexane (HCH) • Lindane • *Dehalobacter*

1 Introduction

Lindane is the gamma isomer of the 1,2,3,4,5,6-hexachlorocyclohexane. Hexachlorocyclohexane (HCH) or benzene hexachloride (BHC) has eight stereoisomers i.e. α, β, γ, δ, ε, ζ, η and θ. α-HCH is the only chiral isomer of HCH and has two enantiomeric forms (Buser and Müller 1995). Only γ-HCH has strong insecticidal properties (Deo et al. 1994). It is a broad spectrum pesticide used against mosquitoes, aphids, grasshoppers, flies, boil weevils, mites, termites, ants, leafminers, thrips, armyworms, wireworms, stem borers (Phillips et al. 2005) and even in the form of lotion and shampoo for topical treatment for head and body lice and scabies (US-FDA 2009).

Michael Faraday synthesized HCH for the very first time in 1825 (Li 1999) and the insecticidal properties of the gamma isomer were discovered in 1942 (Metcalf 1955; Li 1999). HCH is commercially produced by photochemical chlorination of benzene (Walker et al. 1999). Technical mixtures contain 55–70% α-HCH, 5–14% β-HCH, 10–15% γ-HCH, 2–10% δ-HCH and 1–4% ε-HCH (van Eekert et al. 1998). Though HCH is now banned in many countries it has been applied in huge quantities in the past. The total global technical HCH usage between 1948 and 1997 has been estimated to be 10 million tons (Li 1999). Approximately 382,000 tons of technical HCH and 81,000 tons of lindane were used in Europe from 1970 to 1996 (Breivik et al. 1999). Initially, HCH was applied as a mixture, but later on during the late 70's only γ-HCH was applied and the rest of the isomers were separated and dumped.

There are many health related issues associated with HCH. Alpha, beta and gamma isomers of HCH mainly act as depressant of the nervous system (Nagata et al. 1996). The isomers of HCH can cause cancer in mice (Sagelsdorff et al. 1983). The β-HCH may be the toxicologically most significant due to the high persistence in mammalian tissues and its estrogenic effects in mammalian cells and fish (Willett et al. 1998). β-HCH is thought to be the representative of a new class of xenobiotics that produce estrogen like effects through non-classical mechanisms and, therefore, may be of concern with regard to breast and uterine cancer risk (Steinmentz et al. 1996). In human fat the bioconcentration factor of β-HCH is nearly 30 times higher than that of γ-HCH (Geyer et al. 1987). The median half-life of β-HCH in the blood is 7.2 years (Jung et al. 1997) in comparison to only 1 day for γ-HCH (Feldmann and Maibach 1974).

HCH is volatile and its residues have been found in glaciers (Berg et al. 2004). It is fat soluble, which may result in bio-accumulation in fat tissue. The UNEP Stockholm Convention (2008) has listed the alfa, beta and gamma isomers of HCH as new Persistent Organic Pollutants (POPs).

2 Persistence

Lindane and the other HCH-isomers are recalcitrant and can persist for more than 11 years (Lichtenstein and Polivka 1959). Mineralization of some HCH isomers has been reported to occur in the presence of oxygen (Senoo and Wada 1989; Sahu et al. 1990, 1995; Nagasawa et al. 1993). Generally, oxygenases are involved in aerobic degradation of organic chlorine compounds. However, molecules that are substituted with numerous electron withdrawing groups such as chlorine-, nitro- and azo- groups are relatively resistant to aerobic biodegradation (Okey and Bogan 1965; Field et al. 1995). Such molecules are more susceptible to an initial reductive attack, which is common in anoxic environments (Field et al. 1995).

The alfa and gamma isomers can be degraded by aerobic organisms but the beta and delta isomers are thought to be more recalcitrant. In a long term field plot study, the alpha isomer was the least persistent with only 4% remaining in the soil, while the most persistent beta isomer still remained 44% in the soil after 15 years (Stewart and Chisholm 1971). The beta isomer is the most persistent of all the HCH isomers (Bachmann et al. 1988a, b; Doelman et al. 1988; Beurskens et al. 1991). Generally, the rate of dechlorination follows the order γ-HCH > α-HCH > δ-HCH > β-HCH (Jagnow et al. 1977; Buser and Müller 1995; Quintero et al. 2005). The persistence of HCH isomers is due to their chemical structure and isomerization.

2.1 *Chemical Structure*

The structures of the most common HCH isomers are shown in Fig. 20.1. The relative persistence of each isomer is generally attributed to the orientation of the chlorine atoms on the molecule (Beurskens et al. 1991; Phillips et al. 2005). A lower number of equatorial chlorine atoms or a higher number of axial chlorine atoms renders persistence to the compound. The axial chlorine atoms are thought to provide the specific sites for

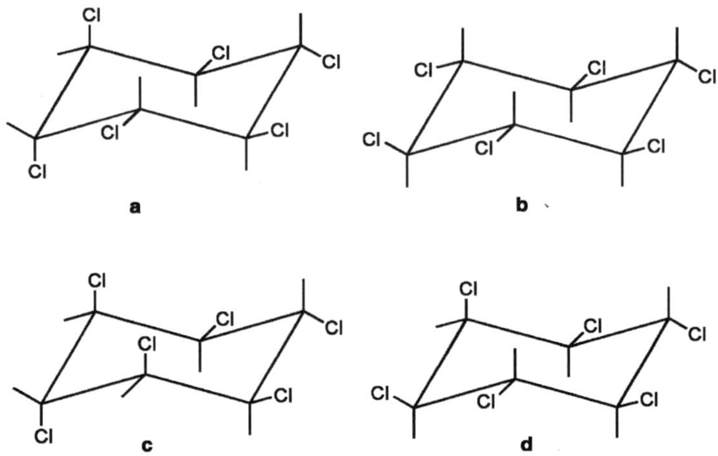

Fig. 20.1 The structures of the most common HCH isomers showing the position of axial and equatorial chlorine atoms. (**a**) Represents the α-HCH with 2 axial and 4 equatorial chlorine atoms; (**b**) represents the β-HCH with all equatorial chlorine atoms; (**c**) represents the γ-HCH with the 3 axial and 3 equatorial chlorine atoms and (**d**) represents δ-HCH with one axial and 5 equatorial chlorine atoms

enzymatic attack. So γ- and α-HCH, having three and two axial chlorine atom, are degraded faster than the one axial chlorine containing δ-HCH and β-HCH (Buser and Müller 1995). β-HCH, which has six equatorially oriented chlorine atoms is the most stable compound (Jagnow et al. 1977; Buser and Müller 1995; Phillips et al. 2005).

2.2 *Isomerization*

HCH isomers are interconvertible to each other. Photo-isomerization and bio-isomerization can change their relative proportions, and this may explain the higher concentrations of α-HCH relative to γ-HCH in specific environments. γ-HCH isomerizes into α-HCH in aqueous solutions placed in UV light (Malaiyandi et al. 1982). Similarly, γ-HCH bio-isomerizes to α-HCH by *Pseudomonas putida* under anoxic conditions (Benezet and Matsumura 1973) and by *P. aeruginosa* in the presence of oxygen (Lodha et al. 2007). However, during anoxic conditions *Escherichia coli* has been found to catalyze the isomerization of α-HCH to γ-HCH (Vonk and Quirijns 1979). Based on thermodynamic stabilities, a complete transformation of γ-HCH first to α-HCH, then to δ-HCH and finally to β-HCH could be expected. This isomerization may stop at the alpha isomer in case of volatilization to the atmosphere. However, under anoxic conditions where isomerization of γ-HCH was occurring rapidly, the accumulation of δ-HCH was observed. It was suggested that the isomerization results in a continuous detoxification of a compound (Newland et al. 1969). A very small percentage of γ-HCH was converted into α- and δ-HCH in the activated sludge. The isomerization of γ-HCH into α-HCH was nearly 240 times slower than the overall degradation rate (Buser and Müller 1995). Such studies indicate the insignificant role of isomerization in the overall persistence of HCH. Yet, in an extensive study Wu et al. (1997) demonstrated that α-HCH isomerizes predominantly into the β-HCH along with small quantities of gamma and delta isomers in both oxic and anoxic environments. The authors suggested that this bulk isomerization, besides its chemical stability, is a reason for β-HCH persistence.

3 Aerobic Biodegradation of Lindane and Other HCH Isomers

Organic molecules with a large number of electron withdrawing groups are relatively resistant to oxidative attack by oxygenases and thus resistant to aerobic degradation (Field et al. 1995). For a long time it was assumed that HCH is best degraded under anoxic conditions (MacRae et al. 1967; Bachmann et al. 1988b). However, there are a number of reports on the aerobic degradation of lindane and other HCH isomers. More than 30 aerobic HCH degrading *Sphingomonads* have been described (Lal et al. 2010). Apart from *Sphingomonas* spp., *Sphingobium* and *Pseudomonas* are other two dominant genera involved in the aerobic degradation HCH (Lal et al. 2010). Mainly due to ease in handling and faster growth of aerobic microorganisms, a lot of work on the biochemistry and genetics of the aerobic HCH degradation has been done. The genes and enzymes involved in the aerobic HCH degradation pathway have been characterized. The enzymes involved in the initial aerobic degradation of HCH include LinA i.e. a 16.5 kDa homotetrameric protein and dehydrochlorinase in function (Nagata et al. 1993; Suar et al. 2005; Wu et al. 2007), LinB i.e. a 32 kDa monomeric enzyme having an haloalkane dehalogenase function (Nagata et al. 2005; Sharma et al. 2006; Wu et al. 2007) and LinC that has a dehydrogenase activity (Nagata et al. 1994; Dogra et al. 2004). Other enzymes that catalyze the downstream process of the aerobic HCH degradation include LinD, LinE, LinF, LinGH and LinJ, which have reductive dechlorinase, ring cleaving dioxygenase, maleyl acetate reductase, acyl-CoA transferase and thiolase activity, respectively (Miyauchi et al. 1999; Nagata et al. 1999; Dogra et al. 2004). The initial enzymes together with the downstream enzymes lead to the aerobic degradation of HCH (reviewed by Lal et al. 2010; Camacho-Pérez et al. 2012).

4 Anaerobic Biodegradation of Lindane and Other HCH Isomers

Anaerobic degradation of lindane and other HCH isomers is known for about 50 years. Table 20.1 lists various studies in which biodegradation of lindane and other HCH isomers have been reported. Lindane was rapidly degraded in non-sterile

Table 20.1 Summary of HCH degradation studies, with known microorganisms

Microorganism involved	Compound	External source of C or energy or supplement	Metabolite formed	End product	Reference
Clostridium sphenoides UQM 780 and its washed cell suspensions	γ-HCH α-HCH PCCHa γ-PeCCH PCCOL β-TeCCH δ-TeCCH γ-TeCCH	0.5% YE	α-HCH via δ-3,4,5,6-tetrachloro-1-cyclohexene and γ-HCH via γ-3,4,5,6-tetrachloro-1-cyclohexene	Approx. 75% of the theoretical amount of the chlorine of lindane was released as chloride	MacRae et al. (1969), Heritage and MacRae (1977a), Heritage and MacRae (1979)
Clostridium butyricum *C. pasteurianum,* *Citrobacter freundii,* *B. polymyxa,* *E. coli,* *Enterobacter aerogenes,* *Enterobacter cloacae,* *B. macerans,* *B. laterosporus,* *B. alvei,* *B. circulans,* *Serratia marcescens,* *Proteus mirabilis,* *P. vulgaris*	γ-HCH α-HCH β-HCH δ-HCH γ-HCH	Glucose, pyruvate, formate, soluble starch, yeast extract	γ-tetrachlorocyclohexene (γ-TeCCH) as main intermediate & small amount of tri-chlorobenzene and tetra-chlorobenzene found too but not confirmed	Chloride	Haider and Jagnow 1975, Jagnow et al. (1977), Haider (1979)

Clostridium rectum strain S-17 and its washed cell suspensions	γ-HCH α-HCH γ-1,2,3,4,5,6-hexachlorocyclohexene and γ-l,3,4,5,6-pentachlorocyclohexene	1% peptone and 0.1% yeast extract. leucine, alanine, pyruvate, leucine, proline, hydrogen	γ-3,4,5,6-tetrachlorocyclohexene (TeCCH) γ-1,2,3,4,5,6-hexachlorocyclohexene is converted into 1,2,4-trichlorobenzene and γ-l,3,4,5,6-pentachlorocyclohexene is converted to and 1,4-dichlorobenzene	Chlorobenzene	Ohisa and Yamaguchi (1978b), Ohisa et al. (1980) Ohisa et al. (1982)
Various *Clostridia* designated as O-1, O-2, H-1, H-2, H-6, H-7, H-8, H-9, H-13, S-1, S-5, S-6, S-7, S-8, S-13	γ-HCH	0.2% peptone	–	–	Ohisa and Yamaguchi (1978a)
O-4, H-3, H-4, H-10, S-3, S-10, S-11, S-12, S-14, S-15, S-17	γ-HCH and α-HCH				
*Clostridium rectum*strain S-17, *C. bifermentans,* C. *sporogenes,* C. *putrefaciens,* *C. lentoputrescens,* *C. butyricum,* *C. acetobutylicum,* C. *histolyticum*	γ-HCH α-HCH γ-HCH	1% peptone, 0.1% YE, whenever needed glucose added	–	–	Ohisa and Yamaguchi (1979)
Mixed anaerobic microbial populations containing *Clostridium* bifermentans, *Clostridium* glycolium, *Clostridium* sp.	γ-HCH and dieldrin	Acetate, formate, yeast extract and peptone	Tetrachlocyclohexene (TeCCH)	–	Maule et al. (1987)

(continued)

Table 20.1 (continued)

Microorganism involved	Compound	External source of C or energy or supplement	Metabolite formed	End product	Reference
Desulfovibrio gigas ATCC 19364 *Desulfovibrio africanus* ATCC 19997 *Desulfobacter curvatus* ATCC 43919 *Desulfococcus multivorans* ATCC 33890	γ-HCH	Acetate, citrate, lactate, yeast extract and sulfate Also benzoate for *D. multivorans*	γ-3,4,5,6-tetrachlorocyclohexene (γ-TeCCH)	Chlorobenzene and benzene	Boyle et al. (1999), Badea et al. (2009)
Gram positive organisms (*Desufomicrobium sp.* & several uncultured bacteria)	γ-HCH	Lactate, methanol & H_2	–	Benzene and chlorobenzene	Elango et al. (2011)
Dehalobacter species	β-HCH α-HCH γ-HCH	H_2, lactate, pyruvate, acetate, (0.005%) peptone	–	Benzene, chlo-robenzene and chloride	Van Doesburg et al. (2005)

–: not described/not detected; YE: yeast extract; tetrachlorocyclohexene (TeCCH); pentachlorocyclohexane (PCCHa); γ-2,3,4,5,6-pentachloro-1-cyclohexene (γ-PeCCH); 2,3,4,5,6-pentachloro-2-cyclohexane-l-ol (PCCOL) soils but not in sterile soils (autoclaved or azide treated) showing the biological nature of the degradation (Lichtenstein et al. 1966; Raghu and MacRae 1966). The half life of lindane in an active culture was in the order of 1 day, while in a sterile culture it was nearly 170 days (Hill and McCarty 1967). Rapid degradation after a second addition of lindane was observed, suggesting that lindane degrading anaerobic microorganisms had been enriched (Raghu and MacRae 1966). All four isomers (alpha, beta, gamma and delta) of HCH were degraded faster in flooded soil than in sterilized soil and C^{14} labeled CO_2 was evolved from non-sterile flooded soils treated with C^{14} labeled γ-HCH (MacRae et al. 1967). Similarly, it has been observed that degradation of lindane occurs faster in flooded soil than in unflooded soils (Yoshida and Castro 1970), because of anoxic conditions (Panda et al. 1988). A slightly higher anaerobic degradation rate of lindane was observed than the aerobic degradation rate (Hill and McCarty 1967) and it was suggested that anaerobic microorganisms were more active in the degradation of lindane than their aerobic counterparts (MacRae et al. 1967).

In a waste dump site polluted with HCH isomers, 35% of α-HCH was degraded in 20 weeks (Doelman et al. 1985). The (+)-enantiomer of α-HCH was degraded faster than the (–)-enantiomer when incubated with the sludge indicating the high enantioselectivity of degradation. This shows the biotic nature of the process. Nearly 80–95% of the degradation was attributed to microbes present inside the sludge. The calculated half lives were 35, 99, 20, 178, 126 h for (+)-α-HCH, (–)-α-HCH, γ-HCH, β-HCH and δ-HCH showing that γ-HCH is degraded the fastest and β-HCH the slowest. The degradation rates were in the following order γ-HCH > α-HCH > δ-HCH > β-HCH (Buser and Müller 1995). In another study it was found that in 34 groundwater samples, collected from beneath a pesticide reformulating and packaging facility, the (–)-enantiomer of α-HCH was degraded faster than the (+)-enantiomer of α-HCH (Law et al. 2004). Under methanogenic conditions, after a lag phase of about 30 days, α-HCH was (bio)converted at a rate of 13 mg kg^{-1} of soil per day. Denitrifying and sulfate-reducing conditions did not result in any biotransformation (Bachmann et al. 1988b).

Significant reductive dechlorination of β-HCH was demonstrated by Middeldorp et al. (1996) under methanogenic condition. The enrichments also dechlorinated the other three isomers of HCH. A long lag phase (60 days) was observed for δ- HCH. γ-HCH was degraded without any lag phase but at a slower rate. α-HCH was dechlorinated at a rate comparable to that of β-HCH. In a flowthrough column experiment an increasing breakthrough of β-HCH was observed in the first 20 days. When 0.1 mM lactate was added as electron donor on day 21, a complete disappearance of β-HCH was observed. Since no dechlorination occurred in pasteurized batches and dechlorination was not inhibited after addition of 2-bromoethansulfonic acid (BrES), it was concluded that neither spore-forming bacteria nor methanogens were involved in the dechlorination of HCH (Middeldorp et al. 1996). Benzene and chlorobenzene were detected as end product, which are quite recalcitrant in anoxic environments. There is not a single bacterial isolate that can degrade chlorobenzene anaerobically and there are only five bacterial isolates that can degrade benzene anaerobically (Mehboob et al. 2010). So a sequential anaerobic-aerobic strategy has been suggested for the mineralization of HCH in polluted soils and aquifers (Middeldorp et al. 1996).

Bacteria that have the ability to reductively dechlorinate the β-HCH are present in different anoxic environments. Anaerobic degradation of β-HCH was observed by methanogenic granular sludge grown on methanol, volatile fatty acids or sucrose. Contaminated soil samples were also capable of reductive dechlorination of β- and α-HCH (van Eekert et al. 1998). Anaerobic reduction was also demonstrated in batches with material from an HCH-contaminated site, where 8 uM of HCH was degraded within 4 months, without the addition of an external carbon source to prove the intrinsic bioremediation occurring at the site (Langenhoff et al. 2002).

Soil from a heavy polluted site rapidly dechlorinated β-HCH upon addition of lactate as electron donor while slightly polluted soil did not show degradation within 4 months. All the four isomers, α-, β-, γ- and δ-HCH added separately or together, were reductively dechlorinated by the soil sample. Acetate, propionate, lactate, methanol, H_2, yeast extract and landfill leachate were used as electron donors. The lag phase for β-HCH degradation was 20 days and less than 10 days for the degradation of the others HCH isomers. However, the degradation rate for β-HCH was higher than rest of the isomers (Middeldorp et al. 2005).

The four isomers of HCH were found to be degraded in the liquid and soil slurry system. The degradation rates were found in the following order γ-HCH > α-HCH > β-HCH > δ-HCH. α- and γ-HCH were dechlorinated in 20–40 days while it took 102 days for the dechlorination of β- and δ-HCH. In liquid medium the degradation rates were found to be higher (Quintero et al. 2005) due to mass transfer limitations associated to the soils (Rijnaarts et al. 1990). In soil assays only partial degradation of HCH isomers was observed. Quintero et al. (2005) suggested that slurry systems with anaerobic sludge could be an alternative in stimulating the bioremediation of HCH polluted soils.

Almost 78% of the γ-HCH was found to be co-metabolically degraded by a bacterial culture enriched from a pesticide contaminated soil. A comparable lindane degradation occurred with cultures enriched on lindane or on endosulfan or on mixed pesticides. However, cultures enriched on methyl parathion and carbofuran were not effective in degrading lindane. It was suggested that lindane and endosulfan might share a common dechlorinating enzyme for their degradation (Krishna and Philip 2008).

4.1 Reactor Studies

There are only a few studies where the dechlorination of lindane or other HCH isomers was studied in the reactors. A complete removal of α- and γ-HCH within 10 days and 90% removal of β- and δ-HCH was observed in anaerobic slurry reactor when it was operated under optimal conditions with a sludge concentration of 8 g VSS l^{-1}, a starch concentration of 2 g COD l^{-1} and soil replacements of 10–20% (Quintero et al. 2006).

Most of the studies on lindane and other HCH isomers were conducted at low concentrations of HCH. Bhat et al. (2006) demonstrated the reductive dechlorination of HCH at a much higher concentration of 100–200 mg l^{-1}. A 85% removal of technical HCH (175 mg l^{-1}) and a complete removal of β-HCH was observed in an up-flow anaerobic sludge blanket reactor (UASB) under continuous mode of operation at a constant HRT of 48 h. Methanol was used as co-substrate and electron donor. The degradation rate was found to be 6.66 mg g^{-1} day^{-1} (Bhat et al. 2006).

Sequential methanogenic-sulfate reducing slurry bioreactors showed 98% lindane removal efficiency against 41 and 82% lindane removal efficiency of only methanogenic and sulphate reducing slurry bioreactors, respectively. The sulfate-reducing stage of operation contributed the most to lindane reduction (Camacho-Pérez et al. 2010).

4.2 *Field Study*

Langenhoff et al. (2002) conducted a field study for *in situ* intrinsic and stimulated biodegradation of HCH. At an industrial site, intermediates of HCH degradation, i.e. benzene, chlorobenzene and chlorophenol were detected in the groundwater, indicating that intrinsic bioremediation was occurring. The aquifer showed an average dissolved organic carbon concentration of 27 mg l^{-1}, demonstrating that electron donor capacity for reductive dechlorination was present. In laboratory experiments, these intermediates were shown to be degraded by adding a small amount of oxygen or nitrate. The addition of either compost percolate or landfill leachate led to a rapid degradation of HCH. Hence, the anaerobic degradation of HCH can be stimulated via the addition of an electron donor, followed by aerobic degradation of the formed intermediates (Langenhoff et al. 2002). This was tested at the site, to increase the intrinsic degradation capacity. Infiltration was performed with methanol, and as a result, the amount of HCH in the groundwater decreased to zero. The produced monochlorobenzene and benzene could be degraded in an on-site reactor (Langenhoff 2009).

Many studies have shown that 80–90% degradation of lindane and other HCH isomers can be attributed to the microbiota present in diverse type of inocula ranging from contaminated soils to anaerobic sludges. A more rapid degradation was found for the flooded soils than the non-flooded soils due to anoxic conditions. A longer lag phase and slower dechlorination of β- and δ-HCH was observed compared to the smaller or no lag phase and rapid dechlorination of γ- and α-HCH. Generally, the rate of dechlorination follows the following order γ-HCH>α-HCH>δ-HCH>β-HCH (Jagnow et al. 1977; Buser and Müller 1995; Quintero et al. 2005).

5 Diversity of Microorganisms Involved in the Anaerobic Dechlorination of Lindane and Other HCH Isomers

Microorganisms involved in the anaerobic dechlorination of lindane and other HCH isomers are quite diverse and are present under diverse anoxic conditions. Many species of *Clostridium, Bacillus, Enterobacter, Desulfovibrio, Desulfococcus, Desulfobacter, Citrobacter*, *Escherichia coli and Dehalobacter* have been found to metabolize lindane and other HCH isomers. Initially, only co-metabolic degradation of lindane and other HCH isomers was observed (MacRae et al. 1969; Jagnow et al. 1977; Boyle et al. 1999). Later on dehalorespiration by specialized genera like *Dehalobacter* was demonstrated (van Doesburg et al. 2005).

Clostridium sphenoides UQM 780 is a gram negative, spore-forming, non-motile organism that was isolated from a flooded soil in medium containing 3–4 ppm of lindane and 0.5% yeast extract as carbon source (MacRae et al. 1969). *C. sphenoides* reduced lindane in concentrations from 3.70 to 0.03 ppm in 11 days. A washed cell suspension degraded lindane within 27 h. Nearly 75% of the theoretical amount of lindane was recovered as chloride (MacRae et al. 1969). This bacterium was also capable of degradation of α-HCH via the formation of δ-3,4,5,6-tetrachloro-1-cyclohexene (δ-TeCCH). Lindane was degraded via γ-3,4,5,6-tetrachloro-1-cyclohexene (γ-TeCCH) (Heritage and MacRae 1977b). Optimal conditions for metabolism were 40°C and a pH of 8. Lindane adapted cells were also capable of metabolizing the beta, gamma and delta isomers of 3,4,5,6-tetrachloro-l-cyclohexene (β-, γ-, δ-TeCCH), 2,3,4,5,6-pentachloro-2-cyclohexen-l-ol (PCCOL), γ-l,2,3,4,5,6-pentachlorocyclohex-l-ene (PeCCH) and 1,2,3,4,5-pentachlorocyclohexane (PCCHa) (Heritage and MacRae 1979). Cell free extracts (CFE) of *C. sphenoides* were also capable of converting lindane into γ-TeCCH. The CFE activity was associated with membranes and glutathionine was required suggesting the involvement of glutathionine conjugation in the process. Membrane fractions showed a faster conversion of lindane into γ-TeCCH and subsequently γ-TeCCH into another unknown product than cell free extract (Heritage and MacRae 1977a).

An anaerobic mixed bacterial culture consisting of *Bacilli* and *Clostridia* degraded 90% of the applied lindane within 4–5 days. α-HCH was degraded slower than the γ-HCH (Haider and Jagnow 1975). In this study many strict and facultative anaerobes were screened and it was found that many *Clostridia, Bacilli, Enterobacter* species are involved in anaerobic degradation of lindane. *C. butyricum, C. pasteurianum* and *Citrobacter freundii* showed good degradation of lindane. *B. polymyxa, E. coli, Enterobacter aerogenes, Enterobacter cloacae, B. macerans* moderately degraded lindane. However, poor degradation of lindane was observed with *B. laterosporus, B. alvei, B. circulans, Serratia marcescens, Proteus mirabilis,* and *P. vulgaris* (Jagnow et al. 1977). Aerobically grown cultures that were subsequently incubated anaerobically with different electron donors also degraded lindane. Mixed anaerobic cultures, *C. butyricum, C. pasteurianum* and *Citrobacter freundii* were also shown to degrade the alpha, beta and delta isomers of HCH. Degradation rates of different isomers were in the following order γ-HCH > α-HCH > β-HCH = δ-HCH (Jagnow et al. 1977). It was suggested that organisms which evolve H_2 during fermentation are involved in the dechlorination process. Similarly, *Clostridia* are known to produce vitamin B_{12} (Neujahr and Rossi-Ricci 1960; Sugita et al. 1991) and reductive dechlorination of lindane and other HCH isomers can also occur with the corrins, porphrins (Marks et al. 1989) and hydroxocobalamin (vitamin B_{12a}) (Rodriguez-Garrido et al. 2010). Reductive dehalogenation of lindane and other HCH isomers could be due to the vitamin B_{12} produced by the *Clostridia* (Jagnow et al. 1977).

Clostridium rectum strain S-17 was isolated from a paddy field soil on medium containing 1 ppm of lindane, 1% peptone and 0.1% yeast extract. Rapid degradation was observed during the growth phase as compared with resting states (Ohisa and Yamaguchi 1978b). Degradation of lindane requires the addition of electron donors like leucine, alanine, pyruvate, a leucine-proline mixture, and molecular hydrogen. Since almost all the *Clostridium* sp. which carry out a Stickland reaction were able to degrade lindane and the process was accompanied with concomitant formation of isovaleric acid so the authors suggested a close relation between the Stickland reaction and lindane degradation (Ohisa et al. 1980; Ohisa and Yamaguchi 1979). Though at first the process was thought to be co-metabolic, later on ATP synthesis in *C. rectum* was found to be associated with the metabolism of lindane, showing the metabolic nature of the process. Hence it was concluded that HCH can act as electron acceptor for the Stickland reaction (Ohisa et al. 1982). However, this was demonstrated with cells pre-grown on a rich medium without γ-HCH and use of γ-HCH as the sole terminal electron acceptor in successive transfers of the culture was not evaluated (Elango et al. 2011). CFE of *C. rectum* S-17 dechlorinated lindane, hexachlorocyclohexenes (HeCCH), pentachlorocyclohexenes (PeCCH), and tetrachlorocyclohexenes (TeCCH) in the presence of dithiothreitol (DTT) (Kurihara et al. 1981; Ohisa et al. 1982).

Sixteen strains capable of γ-HCH degradation and 11 strains capable of both γ- and α-HCH degradation were enriched in peptone containing medium. Due to common properties of being rod shaped, spore forming, strict anaerobes which are catalase negative, isolates were tentatively identified as belonging to the genus *Clostridium*. The optimum temperature for their metabolism was found to be 40°C. Since addition of lindane did not increase their number, it was thought that these strains did not use the lindane as a source of carbon and energy and instead fortuitously degraded lindane. Since many species of *Clostridium* use amino acids in their metabolism via the Stickland reaction, the peptone might have been a good source for their growth (Ohisa and Yamaguchi 1978a).

C. rectum S-17, *C. sporogenes*, C. *putrefaciens, C. bifermentans, C. lentoputrescens, C. butyricum, C. acetobutylicum and* C. *histolyticum* were all capable of γ-HCH degradation, but none of them could decompose β- and δ-HCH. *C. rectum* S-17 and *C. bifermentans* were also able to degrade α-HCH. It seems that the activity of degradation of lindane and other HCH isomers is widespread among *Clostridia* and they are the most dominant HCH co-metabolizing organisms in soils (Ohisa and Yamaguchi 1979). It can also be concluded that the large numbers of *Clostridia* could be the reason for the rapid decomposition of lindane in submerged soils (Ohisa and Yamaguchi 1978a).

Mixed anaerobic microbial populations were enriched from soil, freshwater mud, sheep rumen, and chicken litter in medium reduced with cysteine and sodium sulfide. The bacteria were originally enriched in a medium containing 10 ug ml^{-1} of dieldrin which were later screened for lindane degradation. 100% removal of lindane was observed from the enrichment within 6 h. The mixed anaerobic population contained ten different obligate anaerobic bacteria including *C. bifermentans, C. glycolium, and Clostridium sp.* The parent culture was more efficient in lindane degradation than the isolates (Maule et al. 1987).

Marine sediments around the dwellings of a tribromopyrrole-producing marine hemichordate, were used to enrich cultures capable of lindane degradation co-metabolically. Citrate, lactate, yeast extract and sulfate containing enrichments dehalogenated lindane. Lindane degradation was inhibited with molybdate, which is the specific inhibitor for sulfate reduction indicating the involvement of sulfate reducers. Elimination of sulfate from the medium reduced lindane degradation by only 5% showing the involvement of non-sulfate reducers in the process. More than 90% of the added lindane was dechlorinated within 20 days by the enrichments. In subsequent experiments, pure cultures of *Desulfovibrio gigas* ATCC 19364, *Desulfovibrio africanus* ATCC 19997, *Desulfococcus multivorans* ATCC 33890 and *Desulfobacter curvatus* ATCC 43919 were found to dechlorinate more than 50% of lindane within 4 h of incubation. Autoclaved cultures of *Desulfovibrio gigas* were also able to dechlorinate lindane indicating that it is a co-factor rather than an enzyme mediated process (Boyle et al. 1999).

The first defined anaerobic bacterial culture that is able to dehalorespire with β-HCH was reported by van Doesburg et al. (2005). The culture was enriched from HCH polluted soil using lactate as electron donor, which was later replaced by the H_2, and with β-HCH as electron acceptor. The co-culture was maintained for over 3 years with hydrogen as the electron donor and β-HCH as the only terminal electron acceptor through successive transfers. The co-culture consisted of a *Dehalobacter* sp. and a *Sedimentibacter* sp. The *Dehalobacter* sp. is the dehalorespiring bacterium which grows with β-HCH and a suitable electron donor, but only in the presence of the *Sedimentibacter* sp. (van Doesburg et al. 2005). Formerly the genus *Sedimentibacter* was classified inside the genus *Clostridium* (Breitenstein et al. 2002). We have observed that the *Sedimentibacter* sp., which we found in co-culture with *Dehalobacter*, can also carry out the Stickland reaction. This reaction was originally thought to be the electron donating process for the γ-HCH dechlorination in various *Clostridium spp.*, but our *Sedimentibacter* sp. was unable to grow with β-HCH. This may suggest that Stickland reactions are not necessary for electron donation and instead it can be that only the presence of peptone and/or yeast extract in the medium is important. It has been suggested that *Sedimentibacter* stimulates the dechlorination of β-HCH by providing some kind of growth factors like vitamins, amino acids or other compounds. H_2, lactate, pyruvate and acetate were the electron donors required for the dechlorination (van Doesburg et al. 2005). When the bicarbonate buffer was replaced by phosphate and the CO_2 from headspace was replaced by H_2 only, the co-culture still grew and dechlorinated the β-HCH suggesting that instead of bicarbonate, either HCH itself or peptone is used as carbon source. α-HCH was dechlorinated at the rates comparable with β-HCH degradation, while it took 130 days for dechlorination of the γ-HCH. It was also shown that the benzene: chlorobenzene ratio decreased with increasing number of axial chlorine atoms in the HCH molecule. β-HCH, which does not contain an axial chlorine has a ratio of 0.7, while α-HCH (containing 2 axial Cl atoms) has 0.3 and γ-HCH (containing 3 axial Cl atoms) has a ratio of 0.2 (van Doesburg et al. 2005).

Based upon the fraction of electron equivalents from the electron donor to be used for reductive dechlorination, Elango et al. (2011) demonstrated that γ-HCH acts as a terminal electron acceptor for an enrichment culture in HEPES-buffered anaerobic mineral salt medium. 79–90% of the electron equivalents from hydrogen (electron donor) were used by the enrichment culture for reductive dechlorination of the γ-HCH. The rate of dechlorination of lindane was found to be 1.0–1.5 mg l^{-1} day^{-1}. An enrichment culture with hydrogen as electron donor was successfully maintained for 1 year through three transfers (1% v/v), suggesting that γ-HCH dechlorination is linked to the growth of microbes performing this reaction. Benzene and chlorobenzene were the only volatile transformation products detected, accounting for 25 and 75% of the γ-HCH degradation. Addition of vancomycin significantly decreased the rate of γ-HCH dechlorination, showing that Gram-positive organisms are responsible. A denaturing gradient gel electrophoresis (DGGE) of the enrichment culture did not show any known chlororespiring genera, including *Dehlaobacter*. However, a *Desulfomicrobium* sp. and several uncultured bacteria were found to be present (Elango et al. 2011).

6 Detection of Metabolites and Mechanism of Anaerobic Dechlorination

The mechanism of anaerobic degradation of lindane and other HCH isomers is explained on the basis of detection of intermediates of the presumed pathway. However, expected intermediates of HCH degradation such as tetrachlorocyclohexenes (TeCCHs), pentachlorocyclohexenes (PeCCHs) and pentachlorocyclohexanes (PCCHa) have not always been detected presumably because of different conversion rates for the individual transformation steps (Buser and Müller 1995). van Eekert et al. (1998) suggested that since the first step in the transformation of HCH in their study was the rate-limiting step, so intermediates were not observed. Tables 20.1 and 20.2 list a number of intermediates formed during HCH degradation by various pure cultures and enrichments.

Intermediates were used as indicator for the occurrence of lindane biodegradation in many studies. The formation of TeCCH was observed in active flooded soil treated with lindane while TeCCH was not found in soils without lindane treatment or soils treated with sodium azide (Tsukano and Kobayashi 1972).

Rapid formation of γ-TeCCH was observed as the main intermediate (Jagnow et al. 1977) and was suggested as the reason for the faster degradation under anoxic conditions, compared with the slower aerobic degradation where PeCCH was detected as intermediate. Small amounts of tri- and tetra-chlorobenzene have also been found, but they were not confirmed as intermediates. A dichloro elimination reaction has been proposed as a first step in the degradation of lindane (Jagnow et al. 1977).

Formation of a pentachloro compound (PeCCHs and PCCHa) was ruled out during HCH dechlorination with *C. sphenoides*. Neither these pentachloro compounds were detected during lindane degradation nor TeCCH formation was observed during the dechlorination of these two pentachloro compounds, while TeCCH is already a known

Table 20.2 Summary of biodegradation studies of the HCH isomers, by unknown microorganisms

Compound	External source of C or energy or supplement	Metabolite formed	End product	Reference
γ-HCH	Organic matter	–	–	Raghu and MacRae (1966)
	Organic matter	–	–	Lichtenstein et al. (1966)
	Organic matter	–	–	Hill and McCarty (1967)
	Organic matter	–	–	Yoshida and Castro (1970)
	1.95% total carbon content	–	–	Tsukano and Kobayashi (1972)
	Organic matter	γ-TeCCH (>5%), γ-PeCCH (<2%) and small amounts of 1,2,4-TCB, 1,2,3,5 and/or 1,2,4,5 tetrachlorobenzene and 1,2,3,4-tetrachlorobenzene	–	Mathur and Saha (1975)
	Citrate, lactate, yeast extract and sulfate containing media	–	Benzene and CB	Boyle et al. (1999)
	Dextrose	unknown	Chloride released	Krishna and Philip (2008)
	Lactate	–	–	Baczynski et al. (2010)
	Organic matter, silicone oil added in selected units	PeCCH and 1,2,4-TCB	–	Comacho-Perez et al. (2010)
β-HCH α-HCH	Methanol, VFAs (acetate, propionate, Butyrate) or sucrose	–	Benzene and CB	Van Eekert et al. (1998)
α-HCH	Organic matter, glucose, glutamic acid, peptone	–	–	Doelman et al. (1985)
	Glucose, acetate	3,5-dichlorophenol, and a trichlorophenol (possibly 2,4,5-trichlorophenol)	CB	Bachmann et al. (1988b)
HCH	Lactate, landfill leachate, compost leachate	–	Benzene and CB and chlorophenol	Langenhoff et al. (2002)

Tech. HCH	Methanol	–	Methane and CO_2	Bhat et al. (2006)
	Acetate, butyrate, formate, ethanol and methanol	–	–	Bhatt et al. (2008)
γ-HCH	Organic matter	–	$C^{14}O_2$	MacRae et al. (1967)
δ-HCH β-HCH α-HCH	Soil with 3.98% carbon contents and amended with urea, inositol, YE, a mixture of glutamine, serine, proline, leucine	–	–	MacRae et al. (1984)
	Soluble starch and bakers' yeast as additional nutrients	–	–	Buser and Müller (1995)
	Lactate	TeCCH	Benzene and CB	Middeldorp et al. (1996)
	Acetate, propionate, lactate, methanol, H2, yeast extract and landfill leachate[a]	–	Benzene and CB	Middeldrop et al. (2005)
	The columns were fed with lactate			
	Potato starch in liquid medium	α- and γ-HCH via PCCHa, from which the 1,2-DCB and 1,3-DCB isomers and finally CB are formed	CB	Quintero et al. (2005)
	In slurry acetic-propionic-butyric acid (4:1:1)	In β- and δ-HCH via PCCHa, TeCCH, 1,2,3-TCB, 1,4-DCB and finally CB		
	Glucose + methanol, glucose only, methanol only, acetate, and ethanol	–	CB	Cui et al. 2012

– not described/not detected, *1,2,3-TCB* 1,2,3-trichlorobenzene, *1,2,4-TCB* 1,2,4-trichlorobenzene, *1,2-DCB* 1,2-dichlorobenzene, *1,4-DCB* 1,4-dichlorobenzene, *CB* chlorobenzene, *PCCHa* pentachlorocyclohexane, *TeCCH* tetrachlorocyclohexene, *γ-PeCCH* γ-2,3,-4,5,6-pentachloro-1-cyclohexene

[a]Mainly a mix of different fatty acids

intermediate in the lindane degradation (Heritage and MacRae 1977b, 1979). Due to the formation of TeCCH (Heritage and MacRae 1977b) and ruling out a pentachloro intermediate, it was indicated that two chlorine atoms are removed in the first step of degradation of lindane (Heritage and MacRae 1979) as was also proposed by Jagnow et al. (1977).

Clostridium rectum S-17 metabolizes lindane to chlorobenzene via the formation of γ-TeCCH (Ohisa and Yamaguchi 1978b; Ohisa et al. 1980). This strain can also metabolize the γ-1,2,3,4,5,6-hexachlorocyclohexene and γ-1,3,4,5,6-pentachlorocyclohexene via 1,2,4-trichlorobenzene and 1,4-dichlorobenzene, respectively (Ohisa et al. 1982). TeCCH was also detected in mixed anaerobic microbial populations metabolizing lindane (Maule et al. 1987).

A complete mechanism of β-HCH degradation was proposed by Middeldorp et al. (1996). 3,4,5,6-tetrachlorocyclohexane (TeCCH) was detected as intermediate and benzene and chlorobenzene as the end products of the degradation. It was proposed that β-HCH was converted into TeCCH via the dichloro-elimination. Another dichloro-elimination reaction yielded a dichlorocyclohexadiene, which was either converted to chlorobenzene via an abiotic dehydrohalogenation reaction or biotically to benzene via another dichloro-elimination reaction (Baker et al. 1985; Middeldorp et al. 1996). After 49 days the chlorobenzene formed corresponded to about 67% and benzene corresponds to 19% of initial amount of β-HCH added. The remaining fraction (15%) was attributed to loss of volatile products in sampling and to analysis variance (Middeldorp et al. 1996).

Mass balance calculations indicated that nearly 85% of the initial α-HCH was metabolized to chlorobenzene (CB), 3,5-dichlorophenol, and a trichlorophenol isomer, possibly 2,4,5-trichlorophenol (Bachmann et al. 1988b). Equal amounts of benzene and chlorobenzene were formed by the soil microbiota during the reductive dechlorination of β-HCH (van Eekert et al. 1998). However, a higher ratio of benzene to chlorobenzene was observed in granular sludge as compared with the crushed granules. This was attributed to the syntrophic interaction between the microorganisms (van Eekert et al. 1998). During the lindane degradation by sulfate reducing enrichments a 3:1 ratio of chlorobenzene and benzene was formed and in total it represented nearly 60% of the added lindane. The incomplete mass balance was attributed to the loss of benzene and chlorobenzene to the headspace (Boyle et al. 1999).

Based upon the intermediates detection Quintero et al. (2005) suggested two different pathways for anaerobic degradation of HCH isomers. The degradation of α- and γ-HCH is initiated with a dechlorination to form PCCHa, from which the 1,2-DCB (1,2-dichlorobenzene) and 1,3-DCB (1,2-dichlorobenzene) and finally CB is formed. In contrast, for β- and δ-HCH the biodegradation pathway is initiated with PCCHa, followed by one sequential chlorine removal at each step forming TeCCH, 1,2,3-TCB (1,2,3-trichlorobenzene), 1,2-DCB, 1,4-dichlorobenzene (1,4-DCB) and finally CB (Quintero et al. 2005).

Based upon the intermediates formed by the pure cultures (Table 20.1) and the conclusions of Jagnow et al. (1977) and Heritage and MacRae (1979), a pathway for the reductive dehalogenation of HCH is presented in Fig. 20.2.

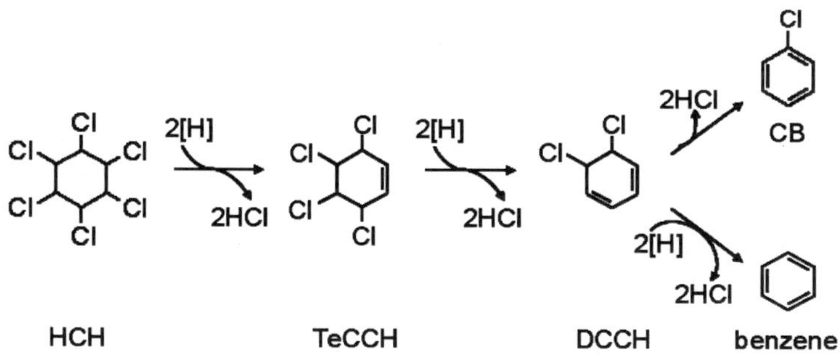

Fig. 20.2 Proposed pathway for the reductive dehalogenation of HCH (re-drawn from van Doesburg et al. 2005). Biotic transformation takes place via three successive dihaloelimination reactions forming tetrachlorocyclohexene (*TeCCH*), dichlorocyclohexadiene (*DCCH*) and finally benzene. Chlorobenzene (*CB*) is formed abiotically via a dehydrohalogenation reaction

7 Factors Affecting the Microbial Dechlorination of Lindane and Other HCH Isomers

Factors such as type of electron donor or substrate, electron acceptor, redox condition and temperature etc. can affect the dechlorination of lindane and other HCH isomers.

7.1 *Effect of Electron Donor or Substrate*

The type of electron donor or substrate affected the dechlorination of the lindane and other HCH isomers in different studies. Rice straw was found to be the best followed by cellulose and then glucose during degradation of the four isomers of HCH in submerged soils. The positive effect of rice straw may be due to the high content of available carbon and nitrogen (Castro and Yoshida 1974). During the dechlorination of lindane by *Citrobacter freundii*, glucose was found to be the best substrate followed by pyruvate, formate and, to a little extent, succinate (Jagnow et al. 1977). Similarly, addition of glucose also slightly stimulated the lindane degradation in flooded soils (Ohisa and Yamaguchi 1978a). However, in contrast to this, glucose inhibited the dechlorination of lindane in an anaerobic mixed population (Maule et al. 1987). Here, formate was found to be the best substrate for dechlorination followed by Stickland type amino acids and then acetate, pyruvate, and mannitol (Maule et al. 1987).

Peptone and yeast extract had a positive effect on HCH degradation. Peptone might also have helped to lower the redox potential. A mixture of amino acids, yeast extract and inositol were found to stimulate the degradation of α-, β-, γ- and δ-HCH. Amino acids and inositol can be fermented by *Clostridia* and yeast extract has a growth stimulating effect on microbes thus increasing the degradation rate of the four isomers of HCH (MacRae et al. 1984).

A shorter lag phase of 20 days was observed with lactate, yeast extract, and fermented landfill leachate for dechlorination of β-HCH while a longer lag phase of >20 days was found with methanol, hydrogen, acetate and propionate. The ratio of benzene and chlorobenzene produced with different electron donors was stable during the dechlorination and was between 0.2 and 0.5 for all electron donors tested, except in fermented landfill leachate (0.8) and acetate (1.3) (Middeldorp et al. 2005).

In an experiment with a β-HCH dechlorinating enrichment, a shorter lag phase of 11 days was observed with acetate and lactate while with propionate and methanol it was found to be 19 and 28 days, respectively. Only the lactate enrichment continued reductive dechlorination on further transfer. It took 2 weeks for complete dechlorination of β-HCH with lactate or acetate as the electron donor, while it took 3–4 weeks when propionate and methanol were used as electron donor (van Doesburg et al. 2005).

The order of technical HCH degradation in a UASB reactor for different electron donors was found to be methanol > ethanol > acetate > butyrate > formate (Bhatt et al. 2008). The addition of electron donors significantly enhanced the dechlorination of α-, β-, γ- and δ-HCH by an enrichment culture. α- and γ-HCH in methanol and ethanol fed cultures were completely dechlorinated in 8 and 12 days while δ- and β-HCH with the same electron donors were dechlorinated in 28 days. Dechlorination with glucose and acetate was found to be slower (Cui et al. 2012).

The organic matter level is directly related to the rate of decomposition and a more pronounced effect of organic matter addition was observed with soils having a lower level of organic matter (Castro and Yoshida 1974). Organic matter helps in lowering the redox potential of the soil (Yoshida and Castro 1970) and may provide the electrons for lindane degradation.

Siddaramappa and Sethunathan (1975) found that the transformation of β-HCH in soil required a redox potential lower than –40 mV. The low redox potential caused by methanogenic conditions, and/or the presence of sulfide in the incubations was helpful for a complete biotransformation of β-HCH to benzene and chlorobenzene (van Eekert et al. 1998).

7.2 *Effect of Temperature*

Anaerobic lindane degradation increases with an increase in temperature (Hill and McCarty 1967; Yoshida and Castro 1970). The rate of lindane degradation was 0.3 ug ml^{-1} day^{-1} at 35°C and 0.1 ug ml^{-1} day^{-1} at 20°C (Hill and McCarty 1967). In an anaerobic β-HCH dechlorinating mixed culture, no measurable β-HCH dechlorination occurred at 4 and 10°C over a period of 100 days. After 248 days β-HCH was completely dechlorinated in the incubations at 10°C but not at 4°C. An optimum temperature of 30°C was found for this mixed culture (Middeldorp et al. 2005). A 95% loss of lindane was observed at 30°C within 2 weeks as compared to only 63% at 12°C (Baczynski et al. 2010).

7.3 *Effect of Electron Acceptor*

Molecular oxygen, nitrate, and manganese oxide retarded the rate of γ-HCH degradation in rice soils. The probable reasons are that these compounds inhibit the lowering of the redox potential of the soil and that they can act as electron acceptor (Yoshida and Castro 1970).

A decrease in degradation of lindane was observed in an anaerobic mixed bacterial culture when oxygen contents were increased (Haider and Jagnow 1975). Anoxic conditions were found to be the most suitable for the dechlorination of lindane among the oxic (shake cultures under air), semi-oxic (stationary cultures under air) and anoxic (stationary cultures under N_2) conditions (Jagnow et al. 1977). Lindane degradation by *C. rectum* S-17 was immediately inhibited when anaerobically growing cultures were exposed to oxygen (Ohisa and Yamaguchi 1979).

The addition of sulfate had no effect on β-HCH dechlorination and led to a simultaneous reduction of β-HCH and sulfate, while the addition of nitrate stopped the β-HCH degradation. When nitrate was omitted, it took 1 month to restore the dechlorination activity in a soil percolation column (Middeldorp et al. 2005). By contrast, nitrate was used to stimulate the degradation of intermediates of HCH degradation (Langenhoff et al. 2002). So nitrate inhibits the HCH degradation in the reaction mixture as the competing electron acceptor, but when used sequentially it can stimulate the degradation by metabolizing the intermediates in the HCH degradation pathway.

8 Metagenomics of Anaerobic HCH Dechlorination

Molecular studies for the anaerobic dechlorination of lindane and other HCH isomers are totally absent. It is pertinent to mention here that even most of the isolates capable of reductive dechlorination of HCH have not been identified by the most widely used 16S rRNA gene marker. No information is available on the genes and enzymes responsible for reductive dechlorination of HCH. The reason could be that since the bacteria which can dehalorespire with chlorinated compounds are difficult to grow and are very sensitive to oxygen. Also due to their small genome size, they are mostly dependent on another syntrophic partner and are difficult to isolate. Only very recently, Maphosa (2010) adopted a metagenomic approach to gain insight in the genomes of *Dehalobacter* sp. E1 and *Sedimentibacter* sp. B4 in a co-culture capable of β-HCH dechlorination, which was originally isolated by van Doesburg et al. (2005). This is the only co-culture available which can dehalorespire with β-HCH. The draft genome of *Dehalobacter* is approximately 2.6 Mbp in size and contains ten putative reductive dehalogenase-encoding gene clusters. No reductive dehalogenase (Rdh) genes were found on the *Sedimentibacter* genome confirming the previous results that *Dehalobacter* is responsible for reductive dehlaogenation of β-HCH. Phylogenetic analysis revealed that seven out of ten putative reductive dehalogenases (Rdh) are closely related to chlorophenol Rdh genes and three to tetrachloroethene Rdh genes. This indicates that *Dehalobacter* sp. E1 has a higher dehalogenation potential than previously thought. This has been confirmed in degradation experiments where the dechlorination of trichloroethene could be coupled to growth by this co-culture (Maphosa 2010). Previously, this culture was unable to dechlorinate tetrachloroethene (PCE) within 5 months (van Doesburg et al. 2005). A transcription analysis revealed that after 11 days of growth on β-HCH, eight out of ten Rdh genes were induced. However, after 30 days of incubation, three of the Rdhs having the highest expression during the first 11 days were not transcribed leading to the suggestion that those three Rdhs are only involved in the initial dechlorination of β-HCH. However, four other Rdhs, which are supposed to be involved in the subsequent chloro-elimination reactions, were up-regulated after 30 days. The 4.2 Mbp genome of *Sedimentibacter* contains genes for carbohydrates catabolism, for biosynthesis of amino acids, and for the production of several vitamins. Genes for the fermentative conversion of acetyl-CoA to butyrate and lactate and genes for cobalamin (vitamin B12) biosynthesis on the *Sedimentibacter* sp. genome could be the reason for dependence of *Dehalobacter* on *Sedimentibacter* (Maphosa 2010).

9 Conclusions and Future Recommendations

Since lindane and other HCH isomers are persistent, toxic, bio-accumulating, carcinogenic substances and are included in the new POPs list, studies for the enrichment and isolation of HCH dechlorinating microorganisms are essential. Studying the physiology of HCH-degrading microorganisms will help to have a better understanding of the HCH dehalogenation mechanism. This will help to develop an effective strategy for biological removal of lindane and other HCH isomers from the environment. Most of the studies have focused on the dechlorination of HCH via co-metabolic reactions that are often slow and may result in incomplete degradation of the HCH isomers. The need is to focus more on the enrichment and isolation of more specialized, dehalorespiring microorganisms like *Dehalobacter*. One such enrichment was obtained and studied by Elango et al. (2011), but the responsible microorganisms are not yet known. Not one single isolate is known, which can couple the dechlorination of all isomers of HCH to growth. The lack of such an organism is a bottleneck in studying the genetics and biochemistry of the anoxic microbial HCH dechlorination process. An increase in the number of these HCH dechlorinating enrichments and ultimately pure bacterial cultures will help to understand the process and will open more options for bioremediation

Metagenomic and metatranscriptomic approaches such as adopted by Maphosa (2010) could in the meantime help to better understand the mechanism of anaerobic dechlorination of HCH. Additional metaproteomic and metabolomic studies with this co-culture might help better understand the mechanism. More field studies like the one conducted by Langenhoff et al. (2002) are needed to monitor and if required stimulate the *in situ* bioremediation of lindane and other HCH isomers. A sequential anaerobic-aerobic treatment strategy is suggested for the complete mineralization of lindane and other HCH isomers (Langenhoff 2009).

Acknowledgements We would like to thank Wim van Doesburg (WUR) and Dr. Sher (AIOU) for help in making figures. We also thank the Dutch Center for Soil Quality Management and Knowledge Transfer, (SKB; www.skbodem.nl) The Netherlands, and the Wageningen Institute for Environment and Climate Research (WIMEK), The Netherlands for providing funds for the present study.

References

Bachmann A, de Bruin W, Jumelet JC, Rijnaarts HHM, Zehnder AJB (1988a) Aerobic biomineralization of alphahexachlorocyclohexane in contaminated soil. Appl Environ Microbiol 54:548–554

Bachmann A, Wijnen P, de Bruin W, Huntjens JLM, Roelofsen W, Zehnder AJB (1988b) Biodegradation of alpha- and beta-hexachlorocyclohexane in a soil slurry under different redox conditions. Appl Environ Microbiol 54:143–149

Baczynski TP, Pleissner D, Grotenhuis T (2010) Anaerobic biodegradation of organochlorine pesticides in contaminated soil-significance of temperature and availability. Chemosphere 78:22–28

Badea S-L, Vogt C, Weber S, Danet A-F, Richnow H-H (2009) Stable isotope fractionation of γ-hexachlorocyclohexane (lindane) during reductive dechlorination by two strains of sulfate-reducing bacteria. Environ Sci Technol 43:3155–3161

Baker MT, Nelson RM, van Dijke RA (1985) The formation of chlorobenzene and benzene by the reductive metabolism of lindane in rat liver microsomes. Arch Biochem Biophys 236:506–514

Benezet HJ, Matsumura F (1973) Isomerization of γ-BHC to α -BHC in the environment. Nature 243:480–481

Berg T, Kallenborn R, Manø S (2004) Challenges in arctic-alpine environmental research. Arct Antarct Alp Res 36:284–291

Beurskens JEM, Stams AJM, Zehnder AJB, Bachmann A (1991) Relative biochemical reactivity of three hexachlorocyclohexane isomers. Ecotoxicol Environ Safety 21:128–136

Bhat P, Kumar MS, Mudliar SN, Chakrabarti T (2006) Biodegradation of tech-hexachlorocyclohexane in a upflow anaerobic sludge blanket (UASB) reactor. Bioresour Technol 97:824–830

Bhatt P, Kumar MS, Mudliar SN, Chakrabarti T (2008) Enhanced biodegradation of hexachlorocyclohexane in upflow anaerobic sludge blanket reactor using methanol as an electron donor. Bioresour Technol 99:2594–2602

Boyle AW, Haggblom MM, Young LY (1999) Dehalogenation of lindane (γ-hexachlorocyclohexane) by anaerobic bacteria from marine sediments and by sulfate-reducing bacteria. FEMS Microbiol Ecol 29:379–387

Breitenstein A, Wiegel J, Haertig C, Weiss N, Andreesen JR, Lechner U (2002) Reclassification of *Clostridium hydroxybenzoicum* as *Sedimentibacter hydroxybenzoicus* gen. nov., comb. nov., and description of *Sedimentibacter saalensis* sp. nov. Int J Sys Evol Microbiol 52:801–807

Breivik K, Pacyna JM, Munch J (1999) Use of α, β and γ-hexachlorocyclohexane in Europe, 1970–1996. Sci Total Environ 239:151–163

Buser H-R, Müller MD (1995) Isomer and enantioselective degradation of hexachlorocyclohexane isomers in sewage sludge under anaerobic conditions. Environ Sci Technol 29:664–672

Camacho-Pérez B, Ríos-Leal E, Esparza-García F, Cortes JB, Fava F, Poggi-Varaldo HM (2010) Bioremediation of an agricultural soil polluted with lindane in triphasic, sequential methanogenic-sulfate reducing slurry bioreactors. J Biotechnol 150:S561–S562

Camacho-Pérez B, Ríos-Leal E, Rinderknecht-Seijas N, Poggi-Varaldo HM (2012) Enzymes involved in the biodegradation of hexachlorocyclohexane: a mini review. J Environ Manag 95:S306–S318

Castro TF, Yoshida T (1974) Effect of organic matter on the biodegradation of some organochlorine insecticides in submerged soils. Soil Sci Plant Nutr 20:363–370

Cui Z, Meng F, Hong J, Li X, Ren X (2012) Effects of electron donors on the microbial reductive dechlorination of hexachlorocyclohexane and on the environment. J Biosci Bioeng 113:765–770

Deo PG, Karanth NG, Karanth NGK (1994) Biodegradation of hexachlorocyclohexane isomers in soil and food environment. CRC Crit Rev Microbiol 20:57–78

Doelman P, Haanstra L, deRuiter E, Slange J (1985) Rate of microbial degradation of high concentrations of a-hexachlorocyclohexane in soil under aerobic and anaerobic conditions. Chemosphere 14:565–570

Doelman P, Haanstra L, Vos A (1988) Microbial degradation by the autochthonous soil population of alpha and beta HCH under anaerobic field conditions in temperate regions. Chemosphere 17:481–487

Dogra C, Raina V, Pal R, Suar M, Lal S, Gartemann KH, van der Meer JR, Holliger C, Lal R (2004) Organization of lin genes and IS6100 among different strains of hexachlorocyclohexane degrading *Sphingomonas paucimobilis*: evidence for natural horizontal transfer. J Bacteriol 186:2225–2235

Elango V, Kurtz HD, Anderson C, Freedman DL (2011) Use of γ-hexachlorocyclohexane as a terminal electron acceptor by an anaerobic enrichment culture. J Hazard Mater 197:204–210

Feldmann RJ, Maibach HI (1974) Percutaneous penetration of some pesticides and herbicides in man. Toxicol Appl Pharmacol 28:126–132

Field JA, Stams AJM, Kato M, Schraa G (1995) Enhanced biodegradation of aromatic pollutants in cocultures of anaerobic and aerobic bacterial consortia. Antonie Van Leeuwenhoek 67:47–77

Geyer HJ, Scheunert I, Korte F (1987) Correlation between the bioconcentration potential of organic environmental chemicals in humans and their n-octanol/water partition coefficients. Chemosphere 16:239–252

Haider K (1979) Degradation and metabolization of lindane and other hexachlorocyclohexane isomers by anaerobic and aerobic soil microorganisms. Z Naturforsch 34c:1066–1069

Haider K, Jagnow G (1975) Abbau von ^{14}C-, ^{3}H- und ^{36}Cl-markierten γ-Hexachlorcyclohexan durch anaerobe Bodenmikroorganismen. Arch Microbiol 104:113–121

Heritage AD, MacRae IC (1977a) Degradation of lindane by cell-free preparations of *Clostridium sphenoides*. Appl Environ Microbiol 34:222–224

Heritage AD, MacRae IC (1977b) Identification of intermediates formed during the degradation of hexachlorocyclohexanes by *Clostridium sphenoides*. Appl Environ Microbiol 33:1295–1297

Heritage AD, MacRae IC (1979) Degradation of hexachlorocyclohexanes and structurally related substance by *Clostridium sphenoides*. Appl Environ Microbiol 33:1295–1297

Hill DW, McCarty PL (1967) Anaerobic degradation of selected chlorinated hydrocarbon pesticides. JWPCF 39:1259–1277

Jagnow G, Haider K, Ellwardt P (1977) Anaerobic dechlorination and degradation of hexachlorocyclohexane isomers by anaerobic and faculative anaerobic bacteria. Arch Microbiol 115:285–292

Jung D, Becher H, Edler L, Flesch-Janys D, Gurn P, Konietzko J, Manz A, Papke O (1997) Elimination of beta-hexachlorocyclohexane in occupationally exposed persons. J Toxicol Environ Health 51:23–34

Krishna KR, Philip L (2008) Biodegradation of lindane, methyl parathion and carbofuran by various enriched bacterial isolates. J Environ Sci Health B 43:157–171

Kurihara N, Ohisa N, Nakajima M, Kakutani T, Senda M (1981) Relation between microbial degradation and polarographic half-wave potential of polychlorocyclohexenes and BHC isomers. Agric Biol Chem 45:1229–1235

Lal R, Pandey G, Sharma P et al (2010) Biochemistry of microbial degradation of hexachlorocyclohexane and prospects for bioremediation. Microbiol Mol Bio Rev 74:58–80

Langenhoff AAM (2009) Bioremediation of areas polluted with chlorinated and non-chlorinated hydrocarbons. Land Contam Recl 17:295–302

Langenhoff AAM, Staps JJM, Pijls C, Alphenaar A, Zwiep G, Rijnaarts HHM (2002) Intrinsic and stimulated *in situ* biodegradation of hexachlorocyclohexane (HCH). Water, Air, Soil Poll Focus 2:171–181

Law SA, Bidleman TF, Martin MJ, Ruby MV (2004) Evidence of enantioselective degradation of α-hexachlorocyclohexane in groundwater. Environ Sci Technol 38:1633–1638

Li YF (1999) Global technical hexachlorocyclohexane usage and its contamination consequences in the environment: from 1948 to 1997. Sci Total Environ 232:121–158

Lichtenstein EP, Polivka JB (1959) Persistence of some chlorinated hydrocarbon insecticides in turf soils. J Econ Ent 52:289–293

Lichtenstein EP, Schulz KR, Skrentny RF, Tsukano Y (1966) Toxicity and fate of insecticide residues in water. Arch Environ Health 12:199–212

Lodha B, Bhat P, Kumar SM, Vaidya AN, Mudliar S, Killedar DJ, Chakrabarti T (2007) Bioisomerization kinetics of γ-HCH and biokinetics of *Pseudomonas aeruginosa* degrading technical HCH. Biochem Eng J 35:12–19

MacRae IC, Raghu K, Castro TF (1967) Persistence and biodegradation of four common isomers of benzene hexachloride in submerged soils. J Agric Food Chem 15:911–914

MacRae IC, Raghu K, Bautista EM (1969) Anaerobic degradation of the insecticide lindane by *Clostridium* sp. Nature 221:859–860

MacRae IC, Yamaya Y, Yoshida T (1984) Persistence of hexachlorocyclohexane isomers in soil suspensions. Soil Biol Biochem 16:285–286

Malaiyandi M, Shah SM, Lee P (1982) Fate of alpha- and gamma-hexachlorocyclohexane isomers under simulated environmental conditions. J Environ Sci Health A A17:283–297

Maphosa F (2010) Chasing organohalide respirers: ecogenomics approaches to assess the bioremediation capacity of soils. PhD dissertation, Wageningen University, The Netherlands

Marks TS, Allpress JD, Maule A (1989) Dehalogenation of lindane by a variety of porphyrins and corrins. Appl Environ Microbiol 55:1258–1261

Mathur SP, Saha JG (1975) Microbial degradation of lindane-C^{14} in a flooded sandy loam soil. Soil Sci 120:301–307

Maule A, Plyte S, Quirk AV (1987) Dehalogenation of organochlorine insecticides by mixed anaerobic microbial populations. Pestic Biochem Physiol 27:229–236

Mehboob F, Weelink S, Saia FT, Junca H, Stams AJM, Schraa G (2010) Microbial degradation of aliphatic and aromatic hydrocarbons with (per)chlorate as electron acceptor. In: Timmis KN (ed) Handbook of hydrocarbon and lipid microbiology. Springer, Berlin/Heidelberg

Metcalf RL (1955) Organic insecticides, their chemistry and mode of action. Interscience, New York

Middeldorp PJ, Jaspers M, Zehnder AJB, Schraa G (1996) Biotransformation of α-, β-, γ-, δ-hexachlorocyclohexane under methaonogenic conditions. Environ Sci Technol 30:2345–2349

Middeldorp PJ, van Doesburg W, Schraa G, Stams AJ (2005) Reductive dechlorination of hexachlorocyclohexane (HCH) isomers in soil under anaerobic conditions. Biodegradation 16:283–290

Miyauchi K, Adachi Y, Nagata Y, Takagi M (1999) Cloning and sequencing of a novel *meta*-cleavage dioxygenase gene whose product is involved in degradation of γ-hexachlorocyclohexane in *Sphingomonas paucimobilis*. J Bacteriol 181:6712–6719

Nagasawa S, Kikuchi R, Nagata Y, Takagi M, Matsuo M (1993) Aerobic mineralization of gamma-HCH by *Pseudomonas paucimobilis* UT26. Chemosphere 26:1719–1728

Nagata YT, Hatta IR, Kimbara K, Fukuda M, Yano K, Takagi M (1993) Purification and characterization of γ-hexachlorocyclohexane (γ-HCH) dehydrochlorinase (LinA) from *Pseudomonas paucimobilis*. Biosci Biotechnol Biochem 59:1582–1583

Nagata YT, Ohtomo R, Miyauchi K, Fukuda M, Yano K, Takagi M (1994) Cloning and sequencing of a 2,5-dichloro-2,5-cyclohexadiene-1,4-diol dehydrogenase gene involved in the degradation of γ-hexachlorocyclohexane in *Pseudomonas paucimobilis*. J Bacteriol 176:3117–3125

Nagata K, Huang CS, Hamilton BJ, Carter DB, Narahashi T (1996) Differential effects of hexachlorocyclohexane on the GABA receptor subunits expressed in human embryonic kidney-cell line. Brain Res 738:131–137

Nagata YT, Miyauchi K, Takagi M (1999) Complete analysis of genes and enzymes for γ-hexachlorocyclohexane degradation in *Sphingomonas paucimobilis* UT26. J Ind Microbiol Biotechnol 23:380–390

Nagata YT, Prokop Z, Sato Y, Jerabek P, Kumar A, Ohtsubo Y, Tsuda M, Damborsky J (2005) Degradation of γ-hexachlorocyclohexane by haloalkane dehalogenase LinB from *Sphingomonas paucimobilis* UT26. Appl Environ Microbiol 71:2183–2185

Neujahr HY, Rossi-Ricci G (1960) On vitamins in sewage sludge XII. Production of vitamin B_{12} by certain *Clostridia*. Acta Chem Scand 14:43–47

Newland LW, Chesters G, Lee GB (1969) Degradation of γ-BHC in simulated lake impoundments as affected by aeration. JWPCF 41:R174–R188

Ohisa N, Yamaguchi M (1978a) Degradation of gamma-BHC in flooded soils enriched with peptone. Agric Biol Chem 42:1983–1987

Ohisa N, Yamaguchi M (1978b) Gamma-BHC degradation accompanied by the growth of *Clostridium rectum* isolated from paddy field soil. Agric Biol Chem 42:1819–1823

Ohisa N, Yamaguchi M (1979) *Clostridium* species and γ-BHC degradation in paddy soil. Soil Biol Biochem 11:645–649

Ohisa N, Yamaguchi M, Kurihara N (1980) Lindane degradation by cell-free extracts of *Clostridium rectum*. Arch Microbiol 125:221–225

Ohisa N, Kurihara N, Yamaguchi M (1982) ATP synthesis associated with the conversion of hexachlorocyclohexane related compounds. Arch Microbiol 131:330–333

Okey RW, Bogan RH (1965) Apparent involvement of electron mechanism in limiting metabolism of pesticides. JWPCF 37:692–712

Panda S, Sharmila M, Ramanand K, Panda D, Sethunathan N (1988) Persistence of hexachlorocyclohexane isomers and carbofuran applied to surface and sub-surface layers of a flooded soil. Pestic Sci 23:199–207

Phillips TM, Seech AG, Lee H, Trevors JT (2005) Biodegradation of hexachlorocyclohexane (HCH) by microorganisms. Biodegradation 16:363–392

Quintero JC, Moreira MT, Feijoo G, Lema JM (2005) Anaerobic degradation of hexachlorocyclohexane isomers in liquid and soil slurry systems. Chemosphere 61:528–536

Quintero JC, Moreira MT, Lema JM, Feijoo G (2006) An anaerobic bioreactor allows the efficient degradation of HCH isomers in soil slurry. Chemosphere 63:1005–1013

Raghu K, MacRae IC (1966) Biodegradation of the gamma isomer of benzene hexachloride in submerged soils. Science 154:263–264

Rijnaarts HHM, Bachmann A, Jumelet JC, Zehnder AJB (1990) Effect of desorption and intraparticle mass transfer on the aerobic biomineralization of α-hexachlorocyclohexane in a contaminated calcareous soil. Environ Sci Technol 24:1349–1354

Rodriguez-Garrido B, Lu-Chau TA, Feijoo G, Macias F, Monterrroso MC (2010) Reductive dechlorination of α, β-, γ-, δ-hexachlorocyclohexane isomers with hydroxocobalamin, in soil slurry systems. Environ Sci Technol 44:7063–7069

Sagelsdorff P, Lutz WK, Schlatter S (1983) The relevance of covalent binding to mouse liver DNA to the carcinogenic action of hexachlorocyclohexane isomers. Carcinogenesis 4:1267–1273

Sahu SK, Patnaik KK, Sharmila M, Sethunathan N (1990) Degradation of alpha-, beta-, and gamma-hexachlorocyclohexane by a soil bacterium under aerobic conditions. Appl Environ Microbiol 56:3620–3622

Sahu SK, Patnaik KK, Bhuyan S, Sreedharan B, Kurihara N, Adhya TK, Sethunathan N (1995) Mineralization of α-, γ-, and β-isomers of hexachlorocyclohexane by a soil bacterium under aerobic conditions. J Agric Food Chem 43:833–837

Senoo K, Wada H (1989) Isolation and identification of an aerobic gamma-HCH-decomposing bacterium from soil. Soil Sci Plant Nutr 35:79–87

Sharma P, Raina V, Kumari R, Malhotra S, Dogra C, Kumari H, Kohler H-PE, Buser H-R, Holliger C, Lal R (2006) Haloalkane dehalogenase LinB is responsible for beta- and delta-hexachlorocyclohexane transformation in Sphingobium indicum B90A. Appl Environ Microbiol 72:5720–5727

Siddaramappa R, Sethunathan N (1975) Persistence of gamma-BHC and beta-BHC in Indian rice soils under flooded conditions. Pestic Sci 6:395–403
Steinmentz R, Young PCM, Caperellgrant A, Gize EA, Madhukar BV, Benjonathan N, Bigsby RM (1996) Novel estrogenic action of the pesticide residue beta-hexachlorocyclohexane in human breast cancer cells. Cancer Res 56:5403–5409
Stewart DKR, Chisholm D (1971) Long-term persistence of BHC, DDT and Chlordane in a sandy loam soil. Can J Soil Sci 51:379–383
UNEP Stockholm Convention (2008) The new POPs under the Stockholm Convention. http://chm.pops.int. Cited 18 May 2012
Suar M, Hauser A, Poiger T, Buser H-R, Muller MD, Dogra C, Raina V, Holliger C, van der Meer JR, Lal R, Kohler H-PE (2005) Enantioselective transformation of chiral alpha HCH by HCH dehydrochlorinases (LinA1 and LinA2) from *Sphingomonas paucimobilis* B90A. Appl Environ Microbiol 71:8514–8518
Sugita H, Miyajima C, Deguchi Y (1991) The vitamin B_{12}-producing ability of the intestinal microflora of freshwater fish. Aquaculture 92:267–276
Tsukano Y, Kobayashi A (1972) Formation of γ-BTC in flooded rice field soils treated with γ-BHC. Agric Biol Chem 36:166–167
United States Food and Drug Administration (2009) Lindane shampoo and lindane lotion. http://www.fda.gov. Cited 18 May 2012
van Doesburg W, van Eekert MH, Middeldorp PJ, Balk M, Schraa G, Stams AJ (2005) Reductive dechlorination of β-hexachlorocyclohexane (β-HCH) by a *Dehalobacter* species in coculture with a *Sedimentibacter* sp. FEMS Microbiol Ecol 54:87–95
van Eekert MHA, van Ras NJP, Mentink GH, Rijnaarts HHM, Stams AJM, Field JA, Schraa G (1998) Anaerobic transformation of β-hexachlorocyclohexane by methanogenic granular sludge and soil microflora. Environ Sci Technol 32:3299–3304
Vonk JW, Quirijns JK (1979) Anaerobic formation of α-hexachlorocyclohexane from γ- hexachlorocyclohexane in soil and by *Escherichia coli*. Pest Biochem Physiol 12:68–74
Walker K, Vallero DA, Lewis RG (1999) Factors influencing the distribution of lindane and other hexachlorocyclohexanes in the environment. Environ Sci Technol 33:4373–4378
Willett KL, Ulrich EM, Hites RA (1998) Differential toxicity and environmental fates of hexachlorocyclohexane isomers. Environ Sci Technol 32:2197–2207
Wu WZ, Xu Y, Schramm KW, Kettrup A (1997) Study of sorption, biodegradation and isomerization of HCH in stimulated sediment/water system. Chemosphere 35:1887–1894
Wu J, Hong Q, Sun Y, Hong Y, Yan Q, Li S (2007) Analysis of the role of LinA and LinB in biodegradation of δ-hexachlorocyclohexane. Environ Microbiol 9:2331–2340
Yoshida T, Castro TE (1970) Degradation of γ-BHC in rice soils. Soil Sci Soc Am Proc 34:440–448

Index

A. Malik et al. (eds.), *Management of Microbial Resources in the Environment*,
DOI 10.1007/978-94-007-5931-2, © Springer Science+Business Media Dordrecht 2013

B

C

Printed by Printforce, the Netherlands